VOLUME FIVE HUNDRED AND EIGHTY SEVEN

METHODS IN ENZYMOLOGY

Molecular Characterization of Autophagic Responses, Part A

METHODS IN ENZYMOLOGY

VOLUME FIVE HUNDRED AND EIGHTY SEVEN

METHODS IN ENZYMOLOGY

Molecular Characterization of Autophagic Responses, Part A

Edited by

LORENZO GALLUZZI

Department of Radiation Oncology, Weill Cornell Medical College, New York, NY, United States; Université Paris Descartes/Paris V, Paris, France

JOSÉ MANUEL BRAVO-SAN PEDRO

Université Paris Descartes/Paris V, Paris; Gustave Roussy Cancer Campus, Villejuif, France

GUIDO KROEMER

Université Paris Descartes/Paris V, Paris; Gustave Roussy Cancer Campus, Villejuif, France

ELSEVIER

ACADEMIC PRESS
An imprint of Elsevier
elsevier.com

Academic Press is an imprint of Elsevier
50 Hampshire Street, 5th Floor, Cambridge, MA 02139, United States
525 B Street, Suite 1800, San Diego, CA 92101–4495, United States
The Boulevard, Langford Lane, Kidlington, Oxford OX5 1GB, United Kingdom
125 London Wall, London, EC2Y 5AS, United Kingdom

First edition 2017

ISBN: 978-0-12-809675-8
ISSN: 0076-6879

For information on all Academic Press publications
visit our website at https://www.elsevier.com/books-and-journals

Publisher: Zoe Kruze
Acquisition Editor: Zoe Kruze
Editorial Project Manager: Helene Kabes
Production Project Manager: Magesh Kumar Mahalingam
Cover Designer: Mark Rogers

CONTENTS

Contributors xv
Preface xxiii

1. Correlative Live Cell and Super Resolution Imaging of Autophagosome Formation **1**

S.A. Walker, E. Karanasios, and N.T. Ktistakis

1. Introduction 2
2. Live Cell Imaging 3
3. Correlative Super Resolution Imaging of Autophagosome Formation 6

Acknowledgments 19
References 19

2. Quantifying Autophagic Structures in Mammalian Cells Using Confocal Microscopy **21**

C.A. Lamb, J. Joachim, and S.A. Tooze

1. Introduction 22
2. Detection and Quantification of Autophagic Puncta in Fixed Mammalian Cells 24
3. Quantifying Starvation-Induced ATG9 Redistribution by Indirect Immunofluorescence and Confocal Microscopy 31
4. Quantification of ATG9 Compartment/Autophagosome Contact in Live Cells 36

Acknowledgments 41
References 41

3. The Use of DQ-BSA to Monitor the Turnover of Autophagy-Associated Cargo **43**

L.S. Frost, A. Dhingra, J. Reyes-Reveles, and K. Boesze-Battaglia

1. Introduction 44
2. Materials 45
3. Establishment of Polarized Epithelial Cell Cultures 47
4. Incorporation of DQ™-BSA Conjugates 47
5. Monitoring Autolysosome Formation 47
6. Monitoring LC3-Associated Phagolysosome Formation 50
7. Immunofluorescence Analysis 51

8. Summary 52
Acknowledgments 53
References 53

4. Turnover of Lipidated LC3 and Autophagic Cargoes in Mammalian Cells **55**

M. Rodríguez-Arribas, S.M.S. Yakhine-Diop, R.A. González-Polo, M. Niso-Santano, and J.M. Fuentes

1. Introduction 56
2. Materials 58
3. Cell Culture, Treatments, and Sample Collection 62
4. Electrophoresis, Western Blot, and Data Analysis 64
5. Notes 67

Acknowledgments 69
References 69

5. High-Throughput Quantification of GFP-LC3^{+} Dots by Automated Fluorescence Microscopy **71**

J.M. Bravo-San Pedro, F. Pietrocola, V. Sica, V. Izzo, A. Sauvat, O. Kepp, M.C. Maiuri, G. Kroemer, and L. Galluzzi

1. Introduction 72
2. Cell Culture 75
3. Generation of Stable GFP-LC3-Expressing Cells 75
4. Treatments 76
5. Image Analysis 77
6. Concluding Remarks 77
7. Notes 80

Acknowledgments 84
References 85

6. Use of pHlurorin-mKate2-human LC3 to Monitor Autophagic Responses **87**

I. Tanida, T. Ueno, and Y. Uchiyama

1. Introduction 88
2. Estimation of Autophagic Process Using pHlurorin-mKate2-human LC3 90
3. Concluding Remarks 95

Acknowledgments 95
References 95

7. Production of Human ATG Proteins for Lipidation Assays 97

Y. Zheng, Y. Qiu, J.E.C. Gunderson, and B.A. Schulman

1. Introduction 98
2. Expression of Human LC3B, ATG7, ATG3, ATG12~ATG5, and ATG16L1 99
3. Protein Purification 105
4. Storing Purified Proteins 111
5. Concluding Remarks 111

Acknowledgments 111
References 111

8. Investigating Structure and Dynamics of Atg8 Family Proteins 115

O.H. Weiergräber, M. Schwarten, B. Strodel, and D. Willbold

1. Introduction 116
2. X-Ray Crystallography 117
3. NMR Spectroscopy 122
4. MD Simulations 126
5. Applications 130
6. Future Prospects 138

References 139

9. Methods for Studying Interactions Between Atg8/LC3/GABARAP and LIR-Containing Proteins 143

T. Johansen, Å.B. Birgisdottir, J. Huber, A. Kniss, V. Dötsch, V. Kirkin, and V.V. Rogov

1. Introduction 144
2. Discovering LIR-Containing Proteins and Defining LIRs 146
3. Characterization of Interactions Between Atg8/LC3/GABARAP and LIR-Containing Peptides 158
4. Summary and Future Outlook 165

Acknowledgments 166
References 166

10. Assessment of Posttranslational Modifications of ATG proteins 171

Y. Xie, R. Kang, and D. Tang

1. Introduction 172
2. Monitoring PTMs of ATG Proteins Using Western Blotting 173
3. Monitoring PTMs of ATG Proteins Using Simple Western™ Assays 175
4. Monitoring PTMs of ATG Proteins Using Immunoprecipitation Assays 181

5. Monitoring PTMs of ATG Proteins Using Immunofluorescence Assays 183
6. Concluding Remarks 186
Acknowledgments 187
References 187

11. Tagged ATG8-Coding Constructs for the In Vitro and In Vivo Assessment of ATG4 Activity 189

C. López-Otín and G. Mariño

1. Introduction 190
2. The Atg4–Atg8 System and Its Unexpected Evolutionarily Acquired Complexity 191
3. Monitoring Atg8-Like Proteins Processing in the Context of Autophagic Flux Analysis 194
4. Construction of Mammalian Expression Vectors Including Tagged Atg8 Proteins for Cleavage Assays 195
5. Transfection of Cultured Cells for the Analysis of Atg4-Like Activity 197
6. Hydrodynamic Delivery-Based Expression of Atg8-Like Tagged Constructs for the Analysis of Atg4-Like Activity in Live Animals 200
7. Concluding Remarks 203
Acknowledgments 204
References 204

12. Measurement of the Activity of the Atg4 Cysteine Proteases 207

M. Li, Y. Fu, Z. Yang, and X.-M. Yin

1. Introduction 208
2. The Structure and Regulatory Machinery of Atg4 210
3. Overview of the Methods to Detect the Atg4 Activity In Vitro and Ex Vivo 211
4. Expression and Purification of Atg4 and Atg8 Proteins for In Vitro Assays 213
5. Measurement of Atg4 Activity In Vitro and Ex Vivo 215
6. Kinetics Analysis of Atg4 Enzymes 219
7. Summary 222
Acknowledgments 222
References 223

13. Crystallographic Characterization of ATG Proteins and Their Interacting Partners 227

Y. Qiu, Y. Zheng, A.M. Taherbhoy, S.E. Kaiser, and B.A. Schulman

1. Introduction 228
2. Domain Structures of Atg Proteins in the UBL Conjugation Pathway 230

3. Proteins That Have Been Crystallized in the UBL Conjugation System in Autophagy 231
4. Crystallization of *Saccharomyces cerevisiae* Atg7 Alone or in Complex With Atg3 or Atg10 232
5. Crystallization of Human ATG5–ATG16L1 (1–69) E122D Disease-Associated Mutant 239
Acknowledgment 244
References 244

14. Dynamics of Atg5–Atg12–Atg16L1 Aggregation and Deaggregation **247**

P. Kharaziha and T. Panaretakis

1. Introduction 248
2. Experimental Conditions 250
3. *Atg5–Atg12–Atg16L1* Reporter Systems 251
4. Conclusion 254
References 255

15. Fluorescent FYVE Chimeras to Quantify PtdIns3P Synthesis During Autophagy **257**

S.M.S. Yakhine-Diop, G. Martínez-Chacón, R.A. González-Polo, J.M. Fuentes, and M. Niso-Santano

1. Introduction 258
2. Cell Culture Conditions 261
3. Transient Transfection of Plasmid 262
4. Autophagy Assay Conditions 264
5. Detection by Fluorescence Microscopy 264
6. Quantification of PtdIns3P Puncta Formation 265
7. Concluding Remarks 265
8. Notes 266
Acknowledgments 267
References 267

16. Quantification of Phosphatidylinositol Phosphate Species in Purified Membranes **271**

A. Jeschke, N. Zehethofer, D. Schwudke, and A. Haas

1. Introduction 272
2. Detection and Quantification of PI(3)P and PI(4)P in Lysosomes or Autophagosomes by PIP-Binding Proteins 274

3. Quantification of PIPs in Lysosome Preparations by RP-HPLC-MS 282
4. Results and Discussion 285
Acknowledgments 289
References 289

17. Mass Assays to Quantify Bioactive PtdIns3P and PtdIns5P During Autophagic Responses 293

J. Viaud, G. Chicanne, R. Solinhac, K. Hnia, F. Gaits-Iacovoni, and B. Payrastre

1. Introduction 293
2. Preparation of Recombinant Proteins 298
3. Lipid Extraction From Biological Samples 301
4. Purification of Phosphatidylinositol Monophosphates 304
5. Quantification of PtdIns3P by Mass Assay 306
6. Quantification of PtdIns5P by Mass Assay 308
Acknowledgments 309
References 309

18. Fluorescence-Based Assays to Analyse Phosphatidylinositol 5-Phosphate in Autophagy 311

M. Vicinanza, M.J. Gratian, M. Bowen, and D.C. Rubinsztein

1. PI(5)P Role in Autophagy 312
2. Microscopy-Based Detection of PI(5)P 315
3. Manipulations of PI(5)P Levels to Visualize PI(5)P During Autophagy 318
4. Super-Resolution Structured Illumination Microscopy (SR-SIM) to Visualize PI(5)P During Autophagy 321
5. Concluding Remarks 328
Acknowledgments 328
References 328

19. Ultrastructural Characterization of Phagophores Using Electron Tomography on Cryoimmobilized and Freeze Substituted Samples 331

J. Biazik, H. Vihinen, E. Jokitalo, and E.-L. Eskelinen

1. Introduction 332
2. Cryoimmobilization 334
3. Materials and Methods 335
4. Results 337
5. Discussion 342

Acknowledgments 346
References 346

20. A Simple Cargo Sequestration Assay for Quantitative Measurement of Nonselective Autophagy in Cultured Cells 351

M. Luhr, P. Szalai, F. Sætre, L. Gerner, P.O. Seglen, and N. Engedal

1. Introduction 352
2. Measuring Nonselective Autophagic Sequestration of Cytosol in Cultured Cells 353
3. Concluding Remarks 362

Acknowledgments 363
References 363

21. In Vitro Reconstitution of Autophagosome–Lysosome Fusion 365

J. Diao, L. Li, Y. Lai, and Q. Zhong

1. Introduction 366
2. SNARE Protein Purification 367
3. Protein Reconstitution 370
4. Fluorescent Measurement 372
5. Single-Vesicle Assay 373
6. Summary 374

Acknowledgments 375
References 375

22. In Vitro Reconstitution of Atg8 Conjugation and Deconjugation 377

D. Fracchiolla, B. Zens, and S. Martens

1. Introduction 378
2. Methods for In Vitro Atg8 Lipidation 379

References 388

23. Study of ULK1 Catalytic Activity and Its Regulation 391

B. Stork and J. Dengjel

1. Introduction 392
2. Detection of Phospho-ULK1 Variants by Immunoblotting 393
3. Analysis of ULK1 Phosphorylation by Mass Spectrometry 397
4. Analysis of ULK1 Inhibitors by In Vitro Kinase Assays 400
5. Summary 402

Acknowledgments 403
References 403

24. Evaluating the mTOR Pathway in Physiological and Pharmacological Settings **405**

S. Hong and K. Inoki

1. Introduction 406
2. Functional Readouts and Inhibitors for the mTOR Pathway 409
3. Methods 410
4. Concluding Remarks 424

Acknowledgments 425
References 425

25. Methods to Study the BECN1 Interactome in the Course of Autophagic Responses **429**

M. Antonioli, F. Ciccosanti, J. Dengjel, and G.M. Fimia

1. Introduction 430
2. Materials 431
3. Methods 433

Acknowledgments 443
References 443

26. In Vitro Characterization of VPS34 Lipid Kinase Inhibition by Small Molecules **447**

F. Fassy, C. Dureuil, A. Lamberton, M. Mathieu, N. Michot, B. Ronan, and B. Pasquier

1. Introduction 448
2. Purification of Recombinant VPS34 Proteins 450
3. Catalytic Assay 451
4. Binding Assay 454
5. Crystallization 457
6. Cell Assay 458
7. Conclusion 461

Acknowledgments 462
References 462

27. Methods to Study Lysosomal AMPK Activation **465**

C.-S. Zhang, M. Li, and S.-C. Lin

1. Introduction 466
2. Analysis of Lysosomal Localization of AXIN/LKB1 467
3. In Vitro Reconstitution of Lysosomal AMPK Activation 473

Acknowledgment 477
References 477

28. Allosteric Modulation of AMPK Enzymatic Activity: In Vitro Characterization 481

J. Ward, A.R. Reyes, and R.G. Kurumbail

1. Introduction 482
2. Notes About Key Reagents 489
3. Assays for Measuring Allosteric Activation of AMPK 491
4. Steady-State Kinetic Analysis of AMPK Activators 496
5. Assays for Monitoring Phosphorylation at Thr172 of the α-Subunit 498
6. Activation–Protection Assay 501
7. Summary 503

References 503

29. Assessing the Catalytic Activity of Transglutaminases in the Context of Autophagic Responses 511

M. D'Eletto, M.G. Farrace, M. Piacentini, and F. Rossin

1. Introduction 512
2. TG Transamidating Activity Assay 513
3. Analysis of TG2 Degradation During Autophagy 516
4. Analysis of TG2 Interaction With p62 During Autophagy 517

References 519

Author Index *521*
Subject Index *549*

CONTRIBUTORS

M. Antonioli
Freiburg Institute for Advanced Studies (FRIAS), University of Freiburg, Freiburg, Germany; National Institute for Infectious Diseases I.R.C.C.S. 'Lazzaro Spallanzani', Rome, Italy

J. Biazik
University of Helsinki, Helsinki, Finland

Å.B. Birgisdottir
Molecular Cancer Research Group, Institute of Medical Biology, University of Tromsø—The Arctic University of Norway, Tromsø, Norway

K. Boesze-Battaglia
SDM, University of Pennsylvania, Philadelphia, PA, United States

M. Bowen
Cambridge Institute for Medical Research, Wellcome/MRC Building, Cambridge Biomedical Campus, Cambridge, United Kingdom

J.M. Bravo-San Pedro
Gustave Roussy Cancer Campus, Villejuif; INSERM, U1138; Equipe 11 labellisée par la Ligue Nationale contre le Cancer, Centre de Recherche des Cordeliers; Université Paris Descartes/Paris V, Sorbonne Paris Cité; Université Pierre et Marie Curie/Paris VI, Paris, France

G. Chicanne
INSERM, U1048 and Université Toulouse 3, I2MC, Toulouse, France

F. Ciccosanti
National Institute for Infectious Diseases I.R.C.C.S. 'Lazzaro Spallanzani', Rome, Italy

M. D'Eletto
University of Rome "Tor Vergata", Rome, Italy

J. Dengjel
Freiburg Institute for Advanced Studies (FRIAS), University of Freiburg, Freiburg, Germany

A. Dhingra
SDM, University of Pennsylvania, Philadelphia, PA, United States

J. Diao
University of Cincinnati College of Medicine, Cincinnati, OH, United States

V. Dötsch
Institute of Biophysical Chemistry and Center for Biomolecular Magnetic Resonance, Goethe University, Frankfurt am Main, Germany

C. Dureuil
Sanofi, Quai Jules Guesde, Vitry Sur Seine, Paris, France

N. Engedal
Centre for Molecular Medicine Norway, Nordic EMBL Partnership, University of Oslo, Oslo, Norway

E.-L. Eskelinen
University of Helsinki, Helsinki, Finland

M.G. Farrace
University of Rome "Tor Vergata", Rome, Italy

F. Fassy
Sanofi, Quai Jules Guesde, Vitry Sur Seine, Paris, France

G.M. Fimia
National Institute for Infectious Diseases I.R.C.C.S. 'Lazzaro Spallanzani', Rome; University of Salento, Lecce, Italy

D. Fracchiolla
Max F. Perutz Laboratories, University of Vienna, Vienna Biocenter, Vienna, Austria

L.S. Frost
SDM, University of Pennsylvania, Philadelphia, PA, United States

J.M. Fuentes
Centro de Investigación Biomédica en Red en Enfermedades Neurodegenerativas (CIBERNED), Universidad de Extremadura, Cáceres; Facultad de Enfermería y Terapia Ocupacional, Universidad de Extremadura, Cáceres, Spain

Y. Fu
School of Pharmaceutical Sciences, Sun Yat-Sen University, Guangzhou, Guangdong, China

F. Gaits-Iacovoni
INSERM, U1048 and Université Toulouse 3, I2MC, Toulouse, France

L. Galluzzi
Gustave Roussy Cancer Campus, Villejuif; INSERM, U1138; Equipe 11 labellisée par la Ligue Nationale contre le Cancer, Centre de Recherche des Cordeliers; Université Paris Descartes/Paris V, Sorbonne Paris Cité; Université Pierre et Marie Curie/Paris VI, Paris, France; Weill Cornell Medical College, New York, NY, United States

L. Gerner
Centre for Molecular Medicine Norway, Nordic EMBL Partnership, University of Oslo, Oslo, Norway

R.A. González-Polo
Centro de Investigación Biomédica en Red en Enfermedades Neurodegenerativas (CIBERNED), Universidad de Extremadura, Cáceres; Facultad de Enfermería y Terapia Ocupacional, Universidad de Extremadura, Cáceres, Spain

M.J. Gratian
Cambridge Institute for Medical Research, Wellcome/MRC Building, Cambridge Biomedical Campus, Cambridge, United Kingdom

J.E.C. Gunderson
St. Jude Children's Research Hospital, Memphis, TN; Hendrix College, Conway, AR, United States

A. Haas
Cell Biology Institute, University of Bonn, Bonn, Germany

K. Hnia
INSERM, U1048 and Université Toulouse 3, I2MC, Toulouse, France

S. Hong
Life Sciences Institute, University of Michigan, Ann Arbor, MI, United States

J. Huber
Institute of Biophysical Chemistry and Center for Biomolecular Magnetic Resonance, Goethe University, Frankfurt am Main, Germany

K. Inoki
Life Sciences Institute, University of Michigan; University of Michigan Medical School, Ann Arbor, MI, United States

V. Izzo
Gustave Roussy Cancer Campus, Villejuif; INSERM, U1138; Equipe 11 labellisée par la Ligue Nationale contre le Cancer, Centre de Recherche des Cordeliers; Université Paris Descartes/Paris V, Sorbonne Paris Cité; Université Pierre et Marie Curie/Paris VI, Paris, France

A. Jeschke
Cell Biology Institute, University of Bonn, Bonn, Germany

J. Joachim
Molecular Cell Biology of Autophagy Group, The Francis Crick Institute, London, United Kingdom

T. Johansen
Molecular Cancer Research Group, Institute of Medical Biology, University of Tromsø—The Arctic University of Norway, Tromsø, Norway

E. Jokitalo
Electron Microscopy Unit, Institute of Biotechnology, University of Helsinki, Helsinki, Finland

S.E. Kaiser
St. Jude Children's Research Hospital, Memphis, TN, United States

R. Kang
University of Pittsburgh, Pittsburgh, PA, United States

E. Karanasios
Signalling Programme, The Babraham Institute, Cambridge, United Kingdom

O. Kepp
INSERM, U1138; Equipe 11 labellisée par la Ligue Nationale contre le Cancer, Centre de Recherche des Cordeliers; Université Paris Descartes/Paris V, Sorbonne Paris Cité;

Université Pierre et Marie Curie/Paris VI, Paris; Metabolomics and Cell Biology Platforms, Gustave Roussy Cancer Campus, Villejuif, France

P. Kharaziha
Cancer Centrum Karolinska, Karolinska Institutet and University Hospital, Stockholm, Sweden

V. Kirkin
Merck KGaA, Darmstadt, Germany

A. Kniss
Institute of Biophysical Chemistry and Center for Biomolecular Magnetic Resonance, Goethe University, Frankfurt am Main, Germany

G. Kroemer
INSERM, U1138; Equipe 11 labellisée par la Ligue Nationale contre le Cancer, Centre de Recherche des Cordeliers; Université Paris Descartes/Paris V, Sorbonne Paris Cité; Université Pierre et Marie Curie/Paris VI; Pôle de Biologie, Hôpital Européen Georges Pompidou, AP-HP, Paris; Metabolomics and Cell Biology Platforms, Gustave Roussy Cancer Campus, Villejuif, France; Karolinska University Hospital, Stockholm, Sweden

N.T. Ktistakis
Signalling Programme, The Babraham Institute, Cambridge, United Kingdom

R.G. Kurumbail
Worldwide Medicinal Chemistry, Pfizer Worldwide Research and Development, Groton, CT, United States

Y. Lai
Stanford University, Stanford, CA, United States

C.A. Lamb
Molecular Cell Biology of Autophagy Group, The Francis Crick Institute, London, United Kingdom

A. Lamberton
Sanofi, Quai Jules Guesde, Vitry Sur Seine, Paris, France

L. Li
Center for Autophagy Research, University of Texas Southwestern Medical Center; University of Texas Southwestern Medical Center, Dallas, TX, United States

M. Li
School of Pharmaceutical Sciences, Sun Yat-Sen University, Guangzhou, Guangdong; State Key Laboratory of Cellular Stress Biology, Innovation Center for Cell Signaling Network, School of Life Sciences, Xiamen University, Xiamen, China

S.-C. Lin
State Key Laboratory of Cellular Stress Biology, Innovation Center for Cell Signaling Network, School of Life Sciences, Xiamen University, Xiamen, China

C. López-Otín
Instituto Universitario de Oncología (IUOPA), de Bioquímica y Biología Molecular, Universidad de Oviedo, Oviedo, Spain

M. Luhr
Centre for Molecular Medicine Norway, Nordic EMBL Partnership, University of Oslo, Oslo, Norway

M.C. Maiuri
Gustave Roussy Cancer Campus, Villejuif; INSERM, U1138; Equipe 11 labellisée par la Ligue Nationale contre le Cancer, Centre de Recherche des Cordeliers; Université Paris Descartes/Paris V, Sorbonne Paris Cité; Université Pierre et Marie Curie/Paris VI, Paris, France

G. Mariño
Instituto para la investigación sanitaria del Principado de Asturias, Universidad de Oviedo, Oviedo, Spain

S. Martens
Max F. Perutz Laboratories, University of Vienna, Vienna Biocenter, Vienna, Austria

G. Martínez-Chacón
Centro de Investigación Biomédica en Red en Enfermedades Neurodegenerativas (CIBERNED); Facultad de Enfermería y Terapia Ocupacional, Universidad de Extremadura, Cáceres, Spain

M. Mathieu
Sanofi, Quai Jules Guesde, Vitry Sur Seine, Paris, France

N. Michot
Sanofi, Quai Jules Guesde, Vitry Sur Seine, Paris, France

M. Niso-Santano
Centro de Investigación Biomédica en Red en Enfermedades Neurodegenerativas (CIBERNED), Universidad de Extremadura, Cáceres; Facultad de Enfermería y Terapia Ocupacional, Universidad de Extremadura, Cáceres, Spain

T. Panaretakis
Cancer Centrum Karolinska, Karolinska Institutet and University Hospital, Stockholm, Sweden

B. Pasquier
Sanofi, Quai Jules Guesde, Vitry Sur Seine, Paris, France

B. Payrastre
INSERM, U1048 and Université Toulouse 3, I2MC; CHU (Centre Hospitalier Universitaire) de Toulouse, Laboratoire d'Hématologie, Toulouse, France

M. Piacentini
University of Rome "Tor Vergata"; National Institute for Infectious Diseases, IRCCS "Lazzaro Spallanzani", Rome, Italy

F. Pietrocola
Gustave Roussy Cancer Campus, Villejuif; INSERM, U1138; Equipe 11 labellisée par la Ligue Nationale contre le Cancer, Centre de Recherche des Cordeliers; Université Paris Descartes/Paris V, Sorbonne Paris Cité; Université Pierre et Marie Curie/Paris VI, Paris, France

Y. Qiu
St. Jude Children's Research Hospital, Memphis, TN, United States

J. Reyes-Reveles
SDM, University of Pennsylvania, Philadelphia, PA, United States

A.R. Reyes
Cardiovascular and Metabolic Diseases Research Unit, Pfizer Worldwide Research and Development, Cambridge, MA, United States

M. Rodríguez-Arribas
Centro de Investigación Biomédica en Red en Enfermedades Neurodegenerativas (CIBERNED); Facultad de Enfermería y Terapia Ocupacional, Universidad de Extremadura, Cáceres, Spain

V.V. Rogov
Institute of Biophysical Chemistry and Center for Biomolecular Magnetic Resonance, Goethe University, Frankfurt am Main, Germany

B. Ronan
Sanofi, Quai Jules Guesde, Vitry Sur Seine, Paris, France

F. Rossin
University of Rome "Tor Vergata", Rome, Italy

D.C. Rubinsztein
Cambridge Institute for Medical Research, Wellcome/MRC Building, Cambridge Biomedical Campus, Cambridge, United Kingdom

A. Sauvat
INSERM, U1138; Equipe 11 labellisée par la Ligue Nationale contre le Cancer, Centre de Recherche des Cordeliers; Université Paris Descartes/Paris V, Sorbonne Paris Cité; Université Pierre et Marie Curie/Paris VI, Paris; Metabolomics and Cell Biology Platforms, Gustave Roussy Cancer Campus, Villejuif, France

B.A. Schulman
St. Jude Children's Research Hospital; University of Tennessee Health Sciences Center; Howard Hughes Medical Institute, St. Jude Children's Research Hospital, Memphis, TN, United States

M. Schwarten
Institute of Complex Systems, ICS-6 (Structural Biochemistry), Forschungszentrum Jülich, Jülich, Germany

D. Schwudke
Priority Research Area Infections, Research Center Borstel, Leibniz-Center for Medicine and Biosciences, Borstel, Germany

P.O. Seglen
Centre for Molecular Medicine Norway, Nordic EMBL Partnership, University of Oslo, Oslo, Norway

V. Sica
Gustave Roussy Cancer Campus, Villejuif; INSERM, U1138; Equipe 11 labellisée par la Ligue Nationale contre le Cancer, Centre de Recherche des Cordeliers; Université Paris Descartes/Paris V, Sorbonne Paris Cité; Université Pierre et Marie Curie/Paris VI, Paris; Faculté de Medicine, Université Paris Saclay/Paris XI, Le Kremlin-Bicêtre, France

R. Solinhac
INSERM, U1048 and Université Toulouse 3, I2MC, Toulouse, France

B. Stork
Institute of Molecular Medicine I, Medical Faculty, Heinrich-Heine-University, Düsseldorf, Germany

F. Sætre
Centre for Molecular Medicine Norway, Nordic EMBL Partnership, University of Oslo, Oslo, Norway

B. Strodel
Institute of Complex Systems, ICS-6 (Structural Biochemistry), Forschungszentrum Jülich, Jülich, Germany

P. Szalai
Centre for Molecular Medicine Norway, Nordic EMBL Partnership, University of Oslo, Oslo, Norway

A.M. Taherbhoy
St. Jude Children's Research Hospital, Memphis, TN, United States

D. Tang
University of Pittsburgh, Pittsburgh, PA, United States

I. Tanida
Juntendo University School of Medicine, Bunkyo-ku, Tokyo, Japan

S.A. Tooze
Molecular Cell Biology of Autophagy Group, The Francis Crick Institute, London, United Kingdom

Y. Uchiyama
Juntendo University School of Medicine, Bunkyo-ku, Tokyo, Japan

T. Ueno
Juntendo University Graduate School of Medicine, Bunkyo-ku, Tokyo, Japan

J. Viaud
INSERM, U1048 and Université Toulouse 3, I2MC, Toulouse, France

M. Vicinanza
Cambridge Institute for Medical Research, Wellcome/MRC Building, Cambridge Biomedical Campus, Cambridge, United Kingdom

H. Vihinen
Electron Microscopy Unit, Institute of Biotechnology, University of Helsinki, Helsinki, Finland

S.A. Walker
Signalling Programme, The Babraham Institute, Cambridge, United Kingdom

J. Ward
Cardiovascular and Metabolic Diseases Research Unit, Pfizer Worldwide Research and Development, Cambridge, MA, United States

O.H. Weiergräber
Institute of Complex Systems, ICS-6 (Structural Biochemistry), Forschungszentrum Jülich, Jülich, Germany

D. Willbold
Institute of Complex Systems, ICS-6 (Structural Biochemistry), Forschungszentrum Jülich, Jülich; Institut für Physikalische Biologie und BMFZ, Heinrich-Heine-Universität Düsseldorf, Düsseldorf, Germany

Y. Xie
University of Pittsburgh, Pittsburgh, PA, United States; Xiangya Hospital, Central South University, Changsha, China

S.M.S. Yakhine-Diop
Centro de Investigación Biomédica en Red en Enfermedades Neurodegenerativas (CIBERNED), Universidad de Extremadura, Cáceres; Facultad de Enfermería y Terapia Ocupacional, Universidad de Extremadura, Cáceres, Spain

Z. Yang
School of Pharmaceutical Sciences, Sun Yat-Sen University, Guangzhou, Guangdong, China

X.-M. Yin
Indiana University School of Medicine, Indianapolis, IN, United States

N. Zehethofer
Priority Research Area Infections, Research Center Borstel, Leibniz-Center for Medicine and Biosciences, Borstel, Germany

B. Zens
Max F. Perutz Laboratories, University of Vienna, Vienna Biocenter, Vienna, Austria

C.-S. Zhang
State Key Laboratory of Cellular Stress Biology, Innovation Center for Cell Signaling Network, School of Life Sciences, Xiamen University, Xiamen, China

Y. Zheng
St. Jude Children's Research Hospital; University of Tennessee Health Sciences Center, Memphis, TN, United States

Q. Zhong
Center for Autophagy Research, University of Texas Southwestern Medical Center; University of Texas Southwestern Medical Center, Dallas, TX, United States

PREFACE

L. Galluzzi*[†,‡,§,¶,‖,1], J.M. Bravo-San Pedro[†,‡,§,¶,‖,1], G. Kroemer[†,‡,§,¶,#,,††,1]**

*Department of Radiation Oncology, Weill Cornell Medical College, New York, NY, United States
[†]Equipe 11 labellisée Ligue contre le Cancer, Centre de Recherche des Cordeliers, Paris, France
[‡]INSERM, U1138, Paris, France
[§]Université Paris Descartes/Paris V, Sorbonne Paris Cité, Paris, France
[¶]Université Pierre et Marie Curie/Paris VI, Paris, France
[‖]Gustave Roussy Comprehensive Cancer Institute, Villejuif, France
[#]Department of Women's and Children's Health, Karolinska Institute, Karolinska University Hospital Q2:07, Stockholm, Sweden
**Metabolomics and Cell Biology Platforms, Gustave Roussy Comprehensive Cancer Institute, Villejuif, France
[††]Pôle de Biologie, Hôpital Européen George Pompidou, AP-HP, Paris, France
[1]Corresponding authors: e-mail address: deadoc@vodafone.it; chemabsp@gmail.com; kroemer@orange.fr

1. INTRODUCTION

Autophagy is an evolutionary ancient biological process by which eukaryotic cells dispose of unwanted cytosolic entities through lysosomal degradation (Kaur & Debnath, 2015). Three distinct forms of autophagy have been characterized so far, differing from each other in the mechanisms through which autophagic substrates are delivered to lysosomes. In the course of *microautophagy*, cytosolic entities destined to disposal are taken up by lysosomes directly, upon invagination of the lysosomal membrane (Li, Li, & Bao, 2012). *Chaperone-mediated autophagy* relies on the recognition of cytosolic proteins bearing a KFERQ motif by members of the heat-shock protein (HSP) family, coupled to the lysosome-associated protein 2 (LAMP2)-dependent translocation of such cargo:chaperone complexes across the lysosomal membrane (Cuervo & Wong, 2014). *Macroautophagy* involves the progressive sequestration of autophagic cargoes by a double-membraned organelle (commonly known as autophagosome) that—upon sealing—becomes able to fuse with lysosomes (Lamb, Yoshimori, & Tooze, 2013). Although the pathophysiological relevance of microautophagy and chaperone-mediated autophagy is being increasingly recognized (Schneider & Cuervo, 2014; Uytterhoeven et al., 2015), macroautophagy arguably remains the best-characterized form of autophagy (and will be referred to as autophagy here below) (Galluzzi, Pietrocola, et al., 2015).

Autophagy continuously operates at low rates in virtually all eukaryotic cells, ensuring the degradation of normal by-products of life that may become dangerous upon accumulation, like damaged mitochondria and redox-active protein aggregates (Green, Galluzzi, & Kroemer, 2011). Such a baseline autophagic activity is critical for the maintenance of normal cellular functions, as demonstrated by the fact that defects in essential components of the autophagic machinery are associated with clinically relevant disorders including neurodegeneration, aging, and cancer (Choi, Ryter, & Levine, 2013; Galluzzi, Pietrocola, et al., 2015; Menzies, Fleming, & Rubinsztein, 2015). Moreover the autophagic flux, i.e., the actual degradation of autophagic substrates within lysosomes, increases in response to potentially lethal perturbations of intracellular and extracellular homeostasis, including nutritional, metabolic, physical, and chemical cues (Galluzzi, Pietrocola, Levine, & Kroemer, 2014; Green, Galluzzi, & Kroemer, 2014). In this setting, autophagy plays a key adaptive role as it supports the recovery of homeostasis and the survival of stressed cells. Accordingly, the pharmacological or genetic inhibition of essential constituents of the molecular machinery for autophagy generally (but not always) accelerates the demise of cells exposed to adverse microenvironmental conditions (Galluzzi, Bravo-San Pedro, et al., 2015). This said, autophagy can also contribute to cellular demise in a causal manner, both in developmental scenarios (e.g., in the maturation of salivary glands in *Drosophila melanogaster* larvae) (Berry & Baehrecke, 2007) and in cells responding to specific perturbations of homeostasis (e.g., neonatal neurons succumbing to hypoxia/ischemia) (Galluzzi, Bravo-San Pedro, Blomgren, & Kroemer, 2016). In both these cases, the pharmacological or genetic inhibition of autophagy de facto retards the cellular demise, warranting the use of the term "autophagic cell death" (Galluzzi, Bravo-San Pedro, et al., 2015). One particular instance of autophagic cell death that also impinges on the plasma membrane Na^{+}/K^{+} ATPase is commonly known as autosis (Liu et al., 2013). In summary, autophagy mediates robust cytoprotective effects in multiple pathophysiologically relevant settings, but can also exert cytotoxic activity in some scenarios (Galluzzi, Bravo-San Pedro, et al., 2015).

Of note, autophagy can be relatively unselective or exquisitely specific (Sica et al., 2015). Rather unselective forms of autophagy dispose of various, nonessential cytoplasmic constituents and are generally activated by an increased demand of autophagic products (e.g., macromolecules for bioenergetic or anabolic purposes). Conversely, highly selective forms of autophagy target to lysosomal degradation-specific cytoplasmic entities and are often triggered by an increased availability in autophagic substrates (e.g., mitochondria or bacteria decorated with ubiquitin moieties) and/or functions (e.g., the removal of damaged mitochondria or pathogens) (Okamoto, 2014). In particular, "mitophagy" specifically degrades damaged mitochondria, "reticulophagy" nonfunctional portions of the endoplasmic reticulum, "nucleophagy" nuclear buds containing damaged chromatin, "ribophagy" supernumerary ribosomes, "aggrephagy" redox-active protein aggregates, "xenophagy" invading bacteria that escape endocytic vacuoles, "virophagy" constituents of the viral capsid, and "lipophagy" lipid droplets destined to lipolysis (Sica et al., 2015). The existence of all these autophagic programs further underscores the importance of autophagy in the maintenance of cellular homeostasis (Galluzzi, Pietrocola, et al., 2015). In addition, autophagy is critically required for the optimal secretion of ATP by malignant cells undergoing so-called immunogenic cell death (Kroemer, Galluzzi, Kepp, & Zitvogel, 2013), hence playing a key role in the elicitation of anticancer immune responses triggered by some forms of chemotherapy (Galluzzi, Buque, Kepp, Zitvogel, & Kroemer, 2015). Thus, autophagy is also involved in cell-extrinsic mechanisms that ensure the preservation of organismal homeostasis.

Autophagy is regulated by a complex network of interacting signal transduction cascades (Galluzzi et al., 2014; Sica et al., 2015). One of the major regulators of autophagic responses is mechanistic target of rapamycin (MTOR) complex 1 (MTORC1), a supramolecular complex with protein kinase activity that tonically suppresses autophagy in physiologically conditions (Laplante & Sabatini, 2012). Various inducers of autophagy, including nutrient deprivation, operate indeed by shutting down MTORC1 signaling (Mihaylova & Shaw, 2011). Thus, when intracellular ATP levels drop, ADP and AMP accumulate and become able to activated AMP-dependent protein kinase (AMPK), which catalyzes the inactivating phosphorylation of MTORC1 and the activating phosphorylation of unc51-like kinase 1 (ULK1) (Laplante & Sabatini, 2012; Mihaylova & Shaw, 2011). ULK1 initiates a beclin 1 (BECN1)-dependent signal transduction cascade that ultimately results in the formation of autophagosomes (Egan et al., 2011). This process is generally accompanied by the formation of phosphatidylinositol-3-phosphate by phosphatidylinositol 3-kinase catalytic subunit type 3 (PIK3C3, best known as VPS34), as well as by the lipidation of microtubule-associated protein 1 light chain 3 (MAP1LC3, best known as LC3), which accumulates in the autophagosomal membrane to operate as an autophagic adaptor (Lamb et al., 2013). Additional proteins that are required for canonical autophagic responses are autophagy-related 4 (ATG3), ATG4, ATG5, ATG7, ATG9, ATG10, ATG12, and ATG16L1

(Lamb et al., 2013). ATG3, ATG4, and ATG7 cooperate to catalyze the lipidation of LC3 and other members of the Atg8 protein family. ATG7 also cooperates with ATG10 to generate an ATG5–ATG12–ATG16L1 complex that contributes to the elongation of forming autophagosomes. Finally, ATG9 appears to play a critical role in the first steps of autophagosome formation (Lamb et al., 2013). Of note, noncanonical BECN1- and VPS34-independent variants of autophagy have been reported (Codogno, Mehrpour, & Proikas-Cezanne, 2012; Niso-Santano et al., 2015), suggesting the existence of at least some degree of redundancy in the molecular mechanisms that initiate autophagic responses.

Not surprisingly, autophagy has been the subject of an ever more intense wave of investigation throughout the past 2 decades, a progression that has been accompanied by a significant methodological evolution (Ohsumi, 2014). Indeed, dozens of techniques are nowadays available for the characterization of virtually all aspects of autophagic responses, in vitro and in vivo (Klionsky et al., 2016). However, various conceptual/methodological problems that have permeated autophagy research in the past persist, and should be taken into attentive consideration when experimental determinations are designed as well as when results are interpreted. Consensus guidelines recently published in *Autophagy* provide a very comprehensive and detailed analysis of most (if not all) of these problems (Klionsky et al., 2016). Here, we would like to briefly recall three of them, which we feel still have a major negative impact on current studies on autophagy. First, investigators have tended (and still tend) to misemploy end-point biomarkers of autophagy, like the accumulation of autophagosomes in the cytoplasm or the lipidation of LC3, as indicators of an ongoing autophagic response (Klionsky et al., 2016). This is particularly problematic as many of these biomarkers can also accumulate when the last steps of the autophagic program are inhibited. Several approaches are currently available to properly measure autophagic flux in vitro, whereas it remains complicated to perform such measurements in fixed samples (Klionsky et al., 2016). Second, many of the chemicals that have been (and still are) employed to modulate autophagy in experimental settings, including the MTORC1 inhibitor rapamycin, the PI3K inhibitors 3-methyladenine and wortmannin, as well as the lysosomal inhibitor chloroquine, are relatively unspecific and affect various cellular processes other than autophagy, in an on-target or off-target manner (Klionsky et al., 2016). Genetic tools (e.g., homologous recombination, RNA interference, the CRISP/Cas9 system) provide a comparatively more robust approach to ascribe a particular observation or phenotype to autophagy (Dominguez, Lim, & Qi, 2016; Ghildiyal & Zamore, 2009). However, (1) several components of the autophagic machinery exert autophagy-unrelated functions; and (2) the stable absence of a specific protein (as in knockout animals) may cause a considerable rewiring of signaling and metabolism, and/or the activation of compensatory processes. Genetically targeting at least two, if not three, distinct constituents of the autophagic machinery with RNA interference may be sufficient to circumvent the first of these issues (Ghildiyal & Zamore, 2009). Conversely, more refined strategies including the use of activatable Cre-coding constructs and the timely administration of (tissue-specific) Cre-encoding viral vectors are required to avoid cellular and organismal adaptation to the lifelong absence of one or more proteins (Masiero et al., 2009; Rao et al., 2014). We surmise that the development of novel, highly specific chemical inhibitors of autophagy and ever more refined strategies for the timed, tissue-specific genetic blockage of autophagic responses will shed new light on the implication of autophagy in several human disorders.

In *Molecular Characterization of Autophagic Responses* (an issue of the successful *Methods in Enzymology* series) world-leading experts in the field offer a comprehensive panel of techniques that can be used to measure virtually all aspects of autophagy, in vitro and in vivo. Thus, each of the 55 chapters of *Molecular Characterization of Autophagic Responses* provides a detailed description of one or a few protocols that are suitable for the characterization of autophagy in models as diverse as cultured human cancer cells, mice, fish, insects, worms, plants, and yeasts. The book (which consists of two parts, part A and part B) is equally addressed to expert investigators who may wish to expand their methodological competences, and to beginners in this exciting and rapidly expanding area of research.

ACKNOWLEDGMENTS

Authors are supported by the Ligue contre le Cancer (équipe labellisée); Agence National de la Recherche (ANR)—Projets blancs; ANR under the frame of E-Rare-2, the ERA-Net for Research on Rare Diseases; Association pour la recherche sur le cancer (ARC); Cancéropôle Ile-de-France; Institut National du Cancer (INCa); Fondation Bettencourt-Schueller; Fondation de France; Fondation pour la Recherche Médicale (FRM); the European Commission (ArtForce); the European Research Council (ERC); the LabEx Immuno-Oncology; the SIRIC Stratified Oncology Cell DNA Repair and Tumor Immune Elimination (SOCRATE); the SIRIC Cancer Research and Personalized Medicine (CARPEM); and the Paris Alliance of Cancer Research Institutes (PACRI).

Author disclosure: The authors have no conflicts of interest to disclose.

REFERENCES

Berry, D. L., & Baehrecke, E. H. (2007). Growth arrest and autophagy are required for salivary gland cell degradation in Drosophila. *Cell*, *131*, 1137–1148.

Choi, A. M., Ryter, S. W., & Levine, B. (2013). Autophagy in human health and disease. *The New England Journal of Medicine*, *368*, 651–662.

Codogno, P., Mehrpour, M., & Proikas-Cezanne, T. (2012). Canonical and non-canonical autophagy: Variations on a common theme of self-eating? *Nature Reviews. Molecular Cell Biology*, *13*, 7–12.

Cuervo, A. M., & Wong, E. (2014). Chaperone-mediated autophagy: Roles in disease and aging. *Cell Research*, *24*, 92–104.

Dominguez, A. A., Lim, W. A., & Qi, L. S. (2016). Beyond editing: Repurposing CRISPR-Cas9 for precision genome regulation and interrogation. *Nature Reviews. Molecular Cell Biology*, *17*, 5–15.

Egan, D. F., Shackelford, D. B., Mihaylova, M. M., Gelino, S., Kohnz, R. A., Mair, W., et al. (2011). Phosphorylation of ULK1 (hATG1) by AMP-activated protein kinase connects energy sensing to mitophagy. *Science*, *331*, 456–461.

Galluzzi, L., Bravo-San Pedro, J. M., Blomgren, K., & Kroemer, G. (2016). Autophagy in acute brain injury. *Nature Reviews. Neuroscience*, *17*, 467–484.

Galluzzi, L., Bravo-San Pedro, J. M., Vitale, I., Aaronson, S. A., Abrams, J. M., Adam, D., et al. (2015). Essential versus accessory aspects of cell death: Recommendations of the NCCD 2015. *Cell Death and Differentiation*, *22*, 58–73.

Galluzzi, L., Buque, A., Kepp, O., Zitvogel, L., & Kroemer, G. (2015). Immunological effects of conventional chemotherapy and targeted anticancer agents. *Cancer Cell*, *28*, 690–714.

Galluzzi, L., Pietrocola, F., Bravo-San Pedro, J. M., Amaravadi, R. K., Baehrecke, E. H., Cecconi, F., et al. (2015). Autophagy in malignant transformation and cancer progression. *The EMBO Journal*, *34*, 856–880.

Galluzzi, L., Pietrocola, F., Levine, B., & Kroemer, G. (2014). Metabolic control of autophagy. *Cell*, *159*, 1263–1276.

Ghildiyal, M., & Zamore, P. D. (2009). Small silencing RNAs: An expanding universe. *Nature Reviews. Genetics*, *10*, 94–108.

Green, D. R., Galluzzi, L., & Kroemer, G. (2011). Mitochondria and the autophagy-inflammation-cell death axis in organismal aging. *Science*, *333*, 1109–1112.

Green, D. R., Galluzzi, L., & Kroemer, G. (2014). Cell biology. Metabolic control of cell death. *Science*, *345*, 1250256.

Kaur, J., & Debnath, J. (2015). Autophagy at the crossroads of catabolism and anabolism. *Nature Reviews. Molecular Cell Biology*, *16*, 461–472.

Klionsky, D. J., Abdelmohsen, K., Abe, A., Abedin, M. J., Abeliovich, H., Acevedo Arozena, A., et al. (2016). Guidelines for the use and interpretation of assays for monitoring autophagy (3rd edition). *Autophagy*, *12*, 1–222.

Kroemer, G., Galluzzi, L., Kepp, O., & Zitvogel, L. (2013). Immunogenic cell death in cancer therapy. *Annual Review of Immunology*, *31*, 51–72.

Lamb, C. A., Yoshimori, T., & Tooze, S. A. (2013). The autophagosome: Origins unknown, biogenesis complex. *Nature Reviews. Molecular Cell Biology*, *14*, 759–774.

Laplante, M., & Sabatini, D. M. (2012). mTOR signaling in growth control and disease. *Cell*, *149*, 274–293.

Li, W. W., Li, J., & Bao, J. K. (2012). Microautophagy: Lesser-known self-eating. *Cellular and Molecular Life Sciences*, *69*, 1125–1136.

Liu, Y., Shoji-Kawata, S., Sumpter, R. M., Jr., Wei, Y., Ginet, V., Zhang, L., et al. (2013). Autosis is a Na+, K+-ATPase-regulated form of cell death triggered by autophagy-inducing peptides, starvation, and hypoxia-ischemia. *Proceedings of the National Academy of Sciences of the United States of America*, *110*, 20364–20371.

Masiero, E., Agatea, L., Mammucari, C., Blaauw, B., Loro, E., Komatsu, M., et al. (2009). Autophagy is required to maintain muscle mass. *Cell Metabolism*, *10*, 507–515.

Menzies, F. M., Fleming, A., & Rubinsztein, D. C. (2015). Compromised autophagy and neurodegenerative diseases. *Nature Reviews. Neuroscience*, *16*, 345–357.

Mihaylova, M. M., & Shaw, R. J. (2011). The AMPK signalling pathway coordinates cell growth, autophagy and metabolism. *Nature Cell Biology*, *13*, 1016–1023.

Niso-Santano, M., Malik, S. A., Pietrocola, F., Bravo-San Pedro, J. M., Marino, G., Cianfanelli, V., et al. (2015). Unsaturated fatty acids induce non-canonical autophagy. *The EMBO Journal*, *34*, 1025–1041.

Ohsumi, Y. (2014). Historical landmarks of autophagy research. *Cell Research*, *24*, 9–23.

Okamoto, K. (2014). Organellophagy: Eliminating cellular building blocks via selective autophagy. *The Journal of Cell Biology*, *205*, 435–445.

Rao, S., Tortola, L., Perlot, T., Wirnsberger, G., Novatchkova, M., Nitsch, R., et al. (2014). A dual role for autophagy in a murine model of lung cancer. *Nature Communications*, *5*, 3056.

Schneider, J. L., & Cuervo, A. M. (2014). Liver autophagy: Much more than just taking out the trash. *Nature Reviews. Gastroenterology & Hepatology*, *11*, 187–200.

Sica, V., Galluzzi, L., Bravo-San Pedro, J. M., Izzo, V., Maiuri, M. C., & Kroemer, G. (2015). Organelle-specific initiation of autophagy. *Molecular Cell*, *59*, 522–539.

Uytterhoeven, V., Lauwers, E., Maes, I., Miskiewicz, K., Melo, M. N., Swerts, J., et al. (2015). Hsc70-4 deforms membranes to promote synaptic protein turnover by endosomal microautophagy. *Neuron*, *88*, 735–748.

CHAPTER ONE

Correlative Live Cell and Super Resolution Imaging of Autophagosome Formation

S.A. Walker[1], E. Karanasios, N.T. Ktistakis[1]
Signalling Programme, The Babraham Institute, Cambridge, United Kingdom
[1]Corresponding authors: e-mail address: simon.walker@babraham.ac.uk; nicholas.ktistakis@babraham.ac.uk

Contents

1. Introduction 2
2. Live Cell Imaging 3
   2.1 Brief Overview of Protocol 3
3. Correlative Super Resolution Imaging of Autophagosome Formation 6
   3.1 Fixation of Samples on the Microscope Stage 7
   3.2 Labeling of Samples 10
   3.3 Prepare the Microscope for Structured Illumination Imaging 10
   3.4 Relocate Cell(s) of Interest 12
   3.5 Acquire 3D Structured Illumination Raw Image Data 13
   3.6 Reconstruct 3D Structured Illumination Images and Assess for Artifacts 14
   3.7 Acquire Raw dSTORM Image Data 14
Acknowledgments 19
References 19

Abstract

Autophagy is a highly dynamic intracellular process involving interactions between protein complexes and membranes. Direct observation of these components in living cells provides information on how they interact and when and where they are involved in the autophagy pathway. This chapter provides an overview of methods used to acquire images of fluorescently labeled components of the autophagy pathway in living cells using wide-field microscopy. Due to the diffraction-limited nature of this technique further details are provided on how to acquire postfixation correlative super resolution images from the same cells that have previously been imaged live. Combining these techniques offers an opportunity to follow the processes of autophagy in living cells with unprecedented detail.

Methods in Enzymology, Volume 587
ISSN 0076-6879
http://dx.doi.org/10.1016/bs.mie.2016.09.050

1. INTRODUCTION

Membrane rearrangements underpin many important cellular functions including autophagy. In this context autophagy may be unique in that it can be induced very rapidly (within minutes) by a simple switch in growth medium, it produces a structure visible by light microscopy (the autophagosome), and it is near synchronous, yielding tens of visible autophagosomal structures within a few minutes of starvation. The membrane movements leading to the formation of an autophagosome are highly complex and to some extent still mysterious (Ktistakis et al., 2014); moreover, the exact dynamics of each autophagosomal birthing event exhibit cell to cell and even intracellular variability. For these reasons we think that autophagy is an excellent pathway to be studied by live-imaging approaches (reviewed in Karanasios & Ktistakis, 2015). It is only by studying specific components of the autophagy pathway in a living system that we can simultaneously measure both their temporal and spatial organization. However, this diffraction-limited imaging technique can only resolve structure down to ~300 nm, meaning that investigating the biology occurring at smaller scales (i.e., the processes where autophagy is initiated) requires alternative techniques (Mizushima, Yoshimori, & Ohsumi, 2011; Yang & Klionsky, 2010). Traditionally this has meant electron microscopy; however, the molecule-specific contrast provided by fluorescence microscopy is lost with this method; consequently, identifying the key players in the melee of the cell ultrastructure is highly problematic. An alternative is to use one of the increasingly available super resolution fluorescence imaging techniques. Although some of these methods have been applied to live cells, commercial systems do not yet have the speed and sensitivity to image weakly expressed fluorescent proteins over the time course useful to elucidate the autophagy pathway. Therefore, it is necessary to adopt a correlative approach, studying the live cell dynamics of autophagy using wide-field fluorescence and then postfixation imaging structures of known provenance using super resolution microscopy.

In this chapter, we will provide up-to-date information on various protocols that have been used in our lab for the study of autophagy dynamics by live imaging. Since we have described extensive protocols for conventional time-lapse microscopy in recent years (Karanasios & Ktistakis, 2015; Karanasios, Stapleton, Manifava, & Ktistakis, 2014; Karanasios, Stapleton, Walker, Manifava, & Ktistakis, 2013), we will only outline a short summary here. Instead, we will provide more detailed protocols for combining live imaging with super resolution microscopy in order to identify the morphology of early autophagy structures in the context of their lifetime and provenance. This level of analysis is especially necessary for characterization of the early structures (before the emergence of the double-membrane crescent/phagophore) where it can help identify protein or membrane assemblies that make a dynamic contribution to autophagosome formation but fall below the resolution limit of conventional light microscopy.

2. LIVE CELL IMAGING

The two most important considerations when setting up live-imaging experiments are the choice of microscope and the choice of reporter system (Walker, Chandra, Manifava, Axe, & Ktistakis, 2008). *Microscopes*: We have examined autophagy dynamics in living cells using many different light microscopy techniques including wide-field, point-scanning confocal, spinning disk confocal, and total internal reflection fluorescence (TIRF). All produce useful data, but for studies using weakly expressed fluorescent proteins in flat-cultured cells, we believe that wide-field systems offer the best compromise between resolution, speed, and sensitivity. Using this method we typically capture high-resolution images every 5 s for up to 90 min and see minimum photo toxicity or bleaching. *Reporter system*: We have examined autophagy dynamics of the following autophagy-related proteins tagged with fluorescent reporters: ULK1, FIP200, ATG13, ATG101, ATG9, ATG14, VPS34, BECLIN1, VMP1, DFCP1, WIPI2, ATG5, ATG16, SQSTM1, LC3. With the exception of LC3, VMP1, and possibly SQSTM1 (see later), we believe reliable information can be obtained from the remainder, but only when they are expressed at the lowest possible level, in stable cell lines (Fig. 1). VMP1 cannot be expressed stably because it is toxic (Ropolo et al., 2007). LC3 can be expressed transiently but care must be taken with red-tagged versions (mRFP or mCherry), because we have observed continuous accumulation in lysosomes, causing a high basal level of punctae even before starvation is induced. In contrast, we have found GFP- and CFP-tagged LC3 to work really well for transient transfection where basal punctae are very low. Similar considerations hold for SQSTM1.

2.1 Brief Overview of Protocol

a. *Transfection*: Avoid reagents that, on their own, produce fluorescent particles (especially liposome-based ones). We use X-tremeGENE 9

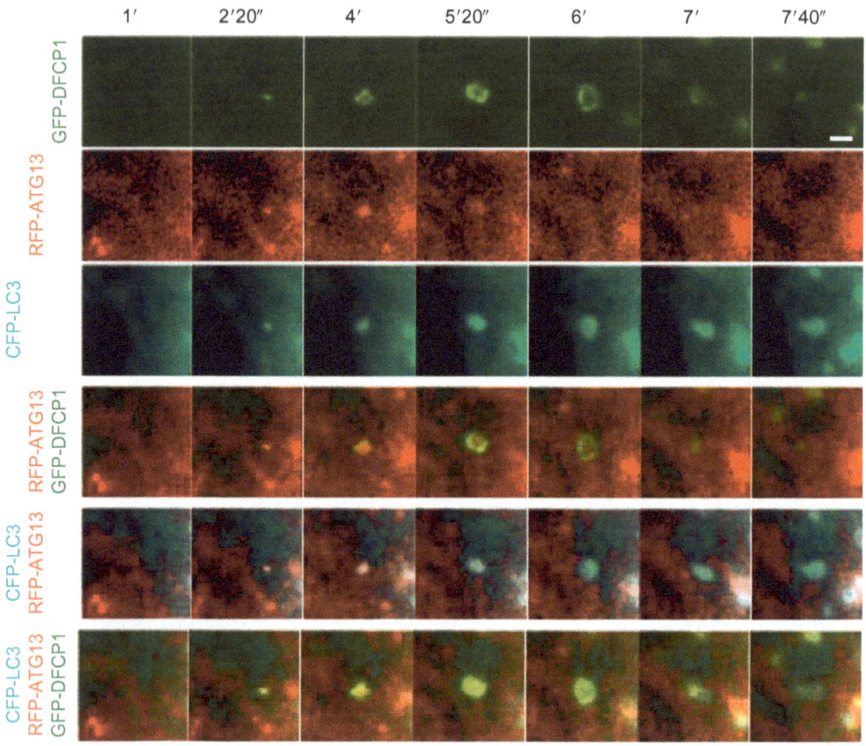

Fig. 1 Montage of images captured during a live cell imaging experiment showing a small region from a HEK cell where an autophagosome is developing. Three components involved in the autophagy pathway (DFCP1, ATG13, and LC3) have been tagged with different fluorescent proteins (GFP, mCherry, and CFP, respectively) and expressed at low level to ensure a physiological response. Scare bar = 1 μm.

(Roche, St. Louis, MO, Sigma Cat# XTG9-RO) which is a polymer-based reagent using exactly the manufacturer's recommendation. It is best to transfect on plastic on day 0 and replate after trypsinization on coverslips on day 1, ready for imaging on day 2. With this method one can better estimate optimal confluency for the imaging step and have a clean starting coverslip devoid of transfection-related debris. For live imaging followed by SIM and STORM approaches (see Section 3), cells are plated onto 35 mm dishes with a gridded glass coverslip base (MatTek, Cat# P35G-1.5-14-CGRD-D.S).

b. *Imaging medium*: To produce movies where cells are first grown in normal conditions and are then induced for autophagy during image capture, we have used two variations. For amino acid starvation, we first grow cells in starvation medium: Na-HEPES 20 m*M*, pH 7.4 (Sigma Cat# H3784), 140 m*M* NaCl (Sigma Cat# S7653), 1 m*M* CaCl (Sigma Cat# C1016)$_2$, 1 m*M* $MgCl_2$ (Sigma Cat# M2393), 5 m*M* glucose (Sigma Cat# G7528), and 1% BSA (Sigma Cat# A9418) supplemented with essential amino acids (1.75 × from a 50 × stock, ThermoFisher Cat# 11130036) and nonessential amino acids (1 × from a 100 × stock, ThermoFisher, Madison, WI Cat# 11140035). In this medium autophagy is minimal and any change observed in cell morphology will be independent of starvation. Cells in this medium are then starved during imaging by a simple wash on the microscope stage with starvation medium devoid of amino acids, applying using a pipette and aspirating off the excess. For autophagy induction using mTOR chemical inhibition, we grow the cells in Ham's F12 Nutrient Mixture (ThermoFisher Cat# 21765029) supplemented with 10% dialyzed serum (ThermoFisher Cat# 26400044). We avoid DMEM because it contains high amounts of riboflavin and pyridoxal which can cause a problematic autofluorescent background and have previously been shown to decrease GFP photostability (Bogdanov, Kudryavtseva, & Lukyanov, 2012). Cells in this medium are imaged for various lengths of time (they show minimal autophagy under these conditions), and then they are washed on stage with fresh medium containing 0.9–1 μ*M* PP242 (Sigma Cat# P0037), an mTOR inhibitor. A few minutes after changing into the two media (starvation or PP242 containing), the cells start to show autophagy-related structures.

c. *Microscopy*: All major microscope manufacturers offer complete solutions for live cell imaging. We routinely use either an Olympus IX81 Cell R imaging system (now replaced by IX83 CellSens) or a Nikon Ti-E-based system. The Cell R system has a Solent Scientific full-enclosure chamber, whereas the Nikon uses one manufactured by OKO labs. The Cell R imaging system is equipped with a 100 × 1.4 N.A. objective (Olympus, Tokyo), Polychrome V monochromator (TILL Photonics), and a Hamamatsu Orca ER CCD camera. CFP, GFP, and mRFP are excited at 425, 488, and 585 nm, respectively, typically using 15 nm bandwidth and 20% transmission. Bandpass filters at 467–500, 505–545, and 605–680 nm are used to collect fluorescent light from CFP, GFP, and mRFP, respectively. The Nikon Ti-E-based system comprises a Nikon Ti-E microscope, 100 × 1.4 N.A. objective (Nikon, Japan), SpecraX LED illuminator (Lumencor, Beaverton, OR), Hamamatsu Flash 4.0 sCMOS camera, emission filter wheel (Sutter Instruments, Novato, CA) and is controlled using Nikon

Elements software. Excitation and emission filters (all from Semrock) are as follows: CFP 434/17 (ex) 480/17 (em), GFP 488/10 (ex) 525/30 (em), mRFP 560/25 (ex) 607/36 (em).

d. *Imaging parameters*: We routinely set the camera exposure time between 100 and 300 ms with an acquisition rate of one frame every 5–10 s. We use 2×2 binning on the camera, which combined with a $100\times$ lens gives pixels of ~ 130 μm in x/y. This meets the Nyquist sampling criterion (https://en.wikipedia.org/wiki/Nyquist%E2%80%93Shannon_sampling_theorem) to obtain the maximum resolution of the objective. Higher binning will allow shorter exposure times, faster image acquisition, and lower levels of illumination (less phototoxicity and bleaching) but will result in insufficient pixel size to achieve the maximum resolution of the objective, necessary to see early autophagy events. We image flat regions of cells that provide the best chance to obtain good quality movies and avoid the necessity for optical sectioning. For multichannel imaging it is desirable to configure the filter system in order to minimize delay between channels. Even with this caveat, it is important to remember that delay between channels may produce artifacts that could be wrongly interpreted as important dynamic changes. This can be avoided by the use of multicamera adapters, although care needs to be taken to choose the correct combinations of fluorescent proteins and filter sets to minimize spectral bleed through.

3. CORRELATIVE SUPER RESOLUTION IMAGING OF AUTOPHAGOSOME FORMATION

The choice of super resolution technique is likely to be limited by local availability, and readers are advised to search out some of the excellent reviews on the different methods to find out how well they can work for their particular application. We have a Nikon combined Structured Illumination Microscopy (N-SIM) and Stochastic Optical Reconstruction Microscopy (N-STORM) system permitting either or both of these techniques on the same sample. Similar systems are available from GE (DeltaVision OMX) and Carl Zeiss (Elyra); stimulated emission depletion systems may also be appropriate. SIM has the advantages of relatively quick image acquisition and compatibility with common fluorophores in multiple combinations. Reconstructed images have high contrast and low noise, but SIM images can "only" offer up to twice the resolution of the optical system and are prone to artifact. STORM has the potential to resolve structure down to tens of nanometers, but requires use of specific dyes in a special buffer and typically involves the acquisition of thousands of raw images to reconstruct the super resolution image. STORM images are a map of localizations and can be difficult to interpret in an analogous way to standard fluorescence images.

The following guidelines describe the use of our Nikon N-SIM/N-STORM system (Nikon Ti-E microscope, 100 × 1.49 N.A. lens, Andor iXon 897 EM-CCD camera, Nikon laser combiner, Nikon SIM illuminator, Nikon STORM TIRF condenser) to obtain both SIM and STORM images from the same sample (Fig. 2). However, it is perfectly feasible to use either techniques in isolation and much of the strategies and advice can be applied to the systems from different manufacturers.

3.1 Fixation of Samples on the Microscope Stage

The main challenge with fixation of samples that are being imaged live is to minimize the time it takes for all intracellular movement to stop as the fixation (in reality a cross-linking process) takes hold. In our experience, fixation with formaldehyde works very well and is compatible with all downstream methods such as SIM and STORM labeling and imaging. The following protocol assumes cells are being imaged in a 35-mm dish with a gridded coverslip base (see Section 2.1a) to enable relocation on a different microscope. Cells of interest are more easily relocated if they are in regions of lower confluency, with neighboring cells forming distinct shapes and patterns.

1. Prepare a 3.7% solution of formaldehyde (Sigma Cat# P6148) in 200 m*M* Na-HEPES, pH 7.4, buffer and maintain at room temperature covered in foil during live imaging.
2. Be sure that the live-imaging medium in the 35-mm dish is exactly 2 mL and that the dish is uncovered during the imaging step. Monitor as many channels as possible during live imaging (e.g., protein of interest + LC3 + endoplasmic reticulum marker) to make sure that the cells are responding well and that appropriate membranous structures are evident. At the appropriate time during live imaging, add exactly 2 mL of the formaldehyde solution to the dish, while continuously imaging. Care must be taken to add the formaldehyde solution as unobtrusively as possible, without interfering with the optical path and without moving the dish containing the cells that are being imaged. As the formaldehyde hits the dish, the intensity of the acquired images changes but not

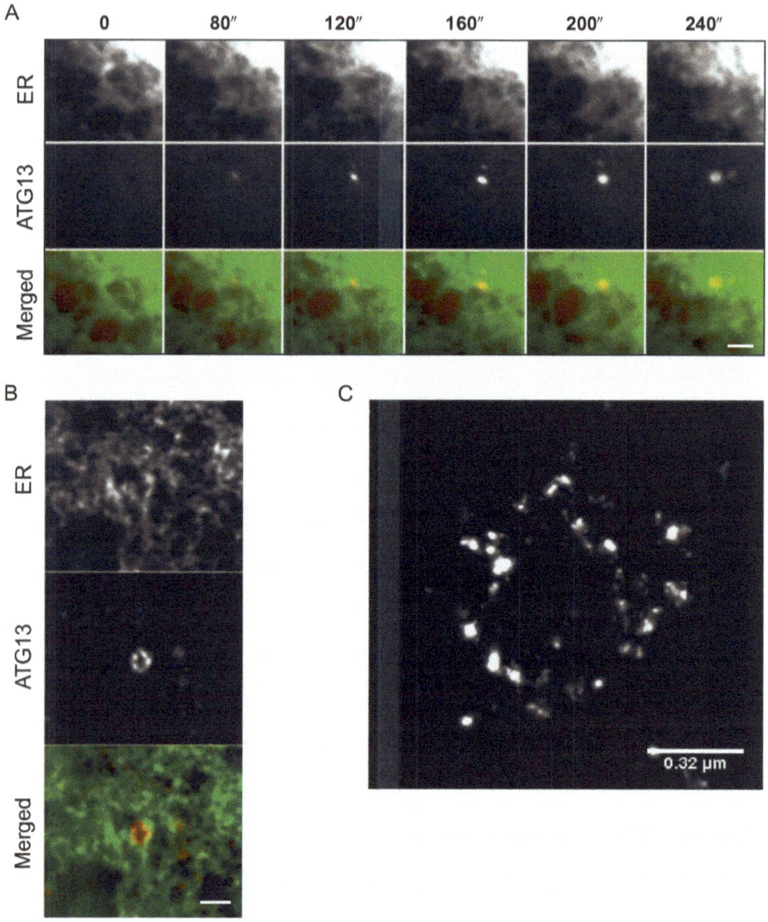

Fig. 2 Correlative live cell and super resolution imaging can provide information on both the temporal evolution of an autophagosome, its detailed structure and interactions with other intracellular components. (A) Frames taken from a live cell imaging experiment of GFP-ATG13 and mCherry-ER expressed in a HEK-293 cell. (B) SIM reconstruction of the same autophagosome postfixation. (C) STORM imaging of the same autophagosome. Scale bars in (A) and (B) are 2 and 1 μm, respectively.

enough to prevent following the various events for the next few frames. In our experience intracellular movement stops within a few seconds.

3. After a few frames, stop image capture and determine the position of the imaged cells on the gridded coverslip. Use brightfield illumination to capture images of the etched grid, adjusting the field diaphragm, condenser height, and condenser aperture diaphragm to maximize contrast of the etched markings (Fig. 3). DIC optics can help enhance the contrast

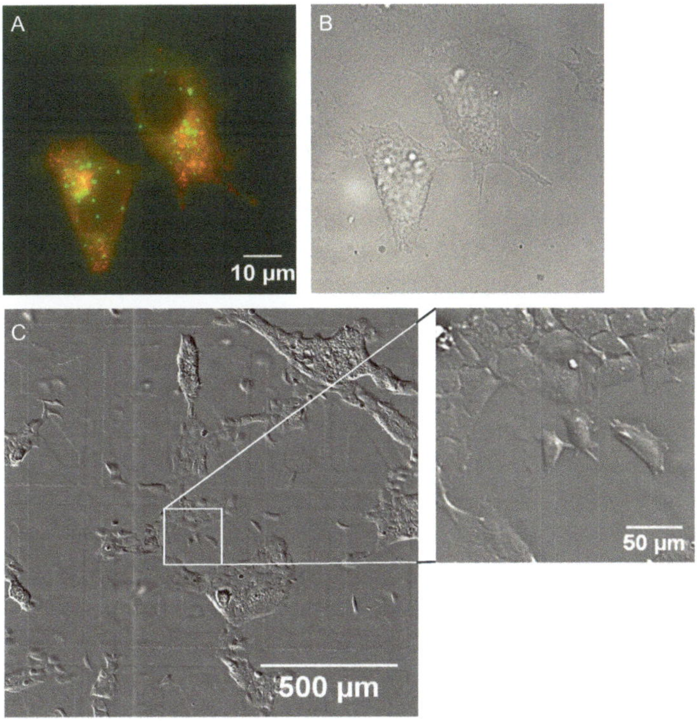

Fig. 3 Typical reference images used to aid relocation of cells that have been imaged using wide-field fluorescence microscopy. Cells are grown on a 35-mm dish containing a gridded cover glass base. (A) Live cell fluorescence image of HEK cells expressing GFP-DFCP1 and mCherry-ATG9. (B) Transmission image of the cells shown in (A) using 100 × objective. (C) Postfixation montage of DIC images giving an overview of the region surrounding cells imaged in (A), acquired using a 10 × objective. Distinguishing the grid pattern can be difficult, but optical contrast enhancement techniques such as DIC and digital processing of the captured images should help. Raw images acquired by Sigurdur Gudmundsson.

as can defocusing the objective and/or the condenser. Note that in our experience the gridded coverslip is sometimes mounted onto the 35-mm dish upside down, so the letters and numbers will be back to front. Switch to a low-magnification lens, e.g., 10 × and capture brightfield images of a larger field of view to aid relocation. It is important to make sure that the dish is left completely undisturbed during this step. Using a marker pen put a small mark on the outside of the 35-mm dish in, e.g., the 12 o'clock position as a rotation guide.

With practice (best to be done a few times without formaldehyde) the whole procedure from adding the fixative to acquiring the brightfield reference images can be done in less than 3 min. The dish is then covered and removed from the microscope stage for the next steps.

3.2 Labeling of Samples

SIM imaging does not necessarily require antibody labeling as the technique is compatible with standard fluorescent proteins. However, amplification of weak fluorescence with antibody labeling may be needed and combined SIM/STORM WILL require antibody labeling.

1. Complete incubation with formaldehyde for 20 min, then rinse cells three times with 20 m*M* HEPES-buffered DMEM, pH 7.4 (ThermoFisher Cat# 41965039), rocking cells gently for 10 min with each wash.
2. Wash cells twice in NETgel [150 m*M* NaCl, 5 m*M* EDTA (Sigma Cat# E6758), 50 m*M* Tris–Cl, pH 7.4 (Sigma Cat# 93352), 0.05% Nonidet P-40 (ThermoFisher Cat# 28324), 0.25% gelatin (Sigma Cat# G1393), and 0.02% NaAzide (Sigma Cat# 08591)] and permeabilize for 5 min in NETgel supplemented with 0.25% NP-40.
3. Wash cells 2 × with NETgel and keep in this solution for all subsequent staining steps.
4. Samples for SIM/STORM and STORM are labeled with appropriate primary antibodies for 45 min at room temperature followed by secondary antibodies conjugated to Alexa Fluor 647 (ThermoFisher Scientific) or CF 568 (Biotium), also for 45 min at room temperature. A good control for STORM imaging is to produce two sets of samples that have been labeled with the two secondary antibodies switched. This will help ensure that no artifacts are introduced with double labeling as a result of the different properties of the two fluorophores.
5. Wash cells extensively in seven changes of PBS and keep in PBS stored at 4°C covered with aluminum foil until ready for imaging.

3.3 Prepare the Microscope for Structured Illumination Imaging

1. SIM imaging is an exercise in obtaining the best possible optical performance from the microscope; therefore, it is critical to ensure the objective lens is performing optimally. Check the top lens of the objective is free from damage and perfectly clean; unscrew the objective from the turret and make sure the bottom lens is free from dust and grease. Nikon recommend cleaning using petroleum ether, but we only use this if there

is stubborn dirt/grease (removing the lens from the microscope and cleaning in a fume hood); routinely we use 100% isopropyl alcohol or 70% ethanol.

2. Turn on the Nikon N-SIM microscope several hours before use to ensure temperature stabilization. This minimizes the risk of thermal expansion during data acquisition and the resultant loss of grid focus (see step 4 below).
3. Once the temperature of the system has stabilized (usually around 29°C inside the isolation chamber) the objective correction collar position should be optimized to minimize spherical aberration. This can be done by measuring the point spread function (PSF) using subresolution fluorescent beads, e.g., TetraSpeck 0.1 μm (ThermoFisher Scientific, Cat# T7279), ideally mounted in an identical fashion to the cells to be imaged (i.e., in a 35-mm dish with a gridded coverslip base and overlaid with appropriate buffer/mounting medium), or added at low concentration into the dish containing the cells. PSF measurements are made by acquiring a 3-μm deep *z*-stack of isolated beads adhered to the cover glass, with images acquired at 0.2 μm steps using the wide-field illumination setting. Nikon advise a visual interpretation of the symmetry of the PSF to determine whether the correction collar needs adjustment; however, we have found it difficult to accurately assess the PSF in this manner and instead use PSFj (Theer, Mongis, & Knop, 2014) to make a quantitative measure of the PSF and record the optimized values for future reference. Using PSFj, it is possible to incrementally adjust the correction collar until the smallest lateral and axial PSF measurements are achieved. If it is not possible to achieve PSF measurements close to the theoretical values then ensure, there are no unnecessary optical elements in the light path, e.g., DIC prism, clean the objective and the sample holder and try again. If it is still not possible, there may be a defect with the objective lens in which case it is advisable to test an alternative.
4. For 3D-SIM imaging the grid focusing needs to be calibrated. This is done using the same subresolution fluorescent beads used to measure the PSF. The Nikon software guides the user through the grid focusing procedure; however, we would additionally note that (a) it is important to use isolated beads (no other signal within at least a 3-μm radius), (b) the exposure time/laser power/EM gain/preamplifier gain need to be set to give sufficient signal for an accurate focus calibration; however, too much signal and the software cannot locate the peak intensity in *z*, (c) the profile of the *z* intensity plot should be symmetrical around the

peak, if it is not then the correction collar needs to be adjusted (see step 3 above), and (d) the peak of the z intensity profile should be in the center of the x-axis on the plot, if it is not then the sample is drifting in z during the calibration data acquisition. This can usually be minimized by leaving the sample on the stage for a few minutes before going through the grid focus calibration procedure. If this does not rectify the problem, then it is likely that thermal drift is occurring (see step 2 above).

3.4 Relocate Cell(s) of Interest

1. Remove PBS from the 35 mm sample dish and ensure the cover glass base is perfectly clean, e.g., by wiping using 70% ethanol. Add 200 μL antifade mounting medium onto the cells, e.g., VectaShield (Vector Labs, Cat# H-1000). Using an antifade mounting medium will reduce spherical aberration and preserve the fluorescence signal during imaging; however, we have successfully used aqueous buffer, e.g., STORM imaging buffer, when imaging structures close to the coverslip surface.
2. Apply immersion oil to the high-magnification objective used for SIM. We use Nikon Type NF due to the high viscosity, which limits dripping when the lens is rotated out of position.
3. Move the objective turret to a low power lens, e.g., 10×. Place the 35-mm dish onto the stage ensuring the rotation guide mark (see Section 3.1.3) on the dish is in the correct position.
4. Use brightfield illumination to visualize the etched grid (Fig. 3), adjusting the field diaphragm, condenser height, and condenser aperture diaphragm to maximize contrast of the etched markings. If DIC optics are not available, then setting up the light path for Köhler illumination is not necessarily the best strategy to achieve the best contrast on the grid; defocusing the condenser and minimizing the condenser aperture diaphragm can help maximize contrast. Additionally faint markings can sometimes be seen better using the microscope camera rather than looking directly by eye, although bear in mind the image may be rotated 90 degree or inverted compared to the reference.
5. Move the microscope stage so that the reference grid coordinates are in the field of view and then use reference images acquired in Section 3.1.3 (above) to identify the cell(s) of interest and position into the center of the field of view.
6. Rotate the microscope objective turret back to the high-magnification objective used for SIM. Go live on the microscope camera and use

appropriate epi-illumination (or live SIM illumination) to focus on the sample fluorescence, noting that there may be a shift in the focus plane when moving between lenses. Due to the significantly smaller field of view, the cell(s) of interest may now be located outside of the region displayed by the camera. Finding the relevant cell(s) may require moving the stage, in which case it is a good idea to mark the current stage position using the stage control software before moving so that it is always possible to relocate back to the original position. Locating the cell(s) of interest on the Nikon system can be aided by rotating the relay lens in front of the camera to the 1 × position to increase the field of view and signal, but remember to rotate back to the SIM position before acquiring SIM data! Minimizing the light exposure used to locate the cell(s) of interest will help preserve the signal for super resolution imaging.

3.5 Acquire 3D Structured Illumination Raw Image Data

1. Use the moving grid 3D-SIM illumination setting to optimize the acquisition settings. The moving grid setting is unique to the Nikon platform and avoids the stripy illumination pattern being burnt into the sample. The acquisition settings should be adjusted such that no pixel records a value higher than ~4000 gray levels, ensuring the camera is operating over its linear range. Achieving the desired signal requires adjusting the camera exposure time, camera EM gain, camera preamplifier gain, and laser illumination intensity. We have found the best SIM reconstructions (i.e., those with the best resolution and lowest artifact) come from images acquired using short exposure times (<100 ms), with EM gain 150 or less. Surprisingly the preamplifier gain setting seems to make little difference to the final reconstructed image, with the reconstruction highly resilient to noise in the raw image data, therefore it is preferable to operate the system with reduced laser power, short exposure time, and high preamplifier gain (i.e., 5.1), rather than high laser power, long exposure time, and low preamplifier gain (i.e., 1.0). A more quantitative assessment of the raw image data quality can be made using tools such as SIMcheck (Ball et al., 2015).
2. Capture the 3D-SIM images. 3D-SIM super resolution images are reconstructed from 15 raw SIM images, representing five phase shifts and three rotations of the illumination pattern. On the Nikon platform SIM images can be acquired from a single focal plane, or as a *z*-stack to

image a volume. In this latter case the step interval of the images has to be set to 120 μm to be compatible with the Nikon volumetric reconstruction algorithm. If imaging a volume in more than one channel, it is possible to set the acquisition to capture multiple channels at each focal plane, or the complete volume of one channel and then repeat for the next. In the former it can take a long time to image the volume, particularly if the grid focus calibration is different for the different channels, in the latter there is a risk that the sample will have drifted between the capture of the channels so they will not be correctly aligned.

3.6 Reconstruct 3D Structured Illumination Images and Assess for Artifacts

1. Using appropriate software, e.g., Nikon Elements, reconstructs the raw SIM image data to give the super resolution images. The Nikon software has two reconstruction algorithms, one for single focal plane 3D-SIM data (Reconstruct Slice) and one optimized for z-stack data (Reconstruct Stack). The former has settings for Illumination Modulation Contrast, Out-of-Focus Blur Suppression, and High-Resolution Noise Suppression, whereas the latter does not have the option to suppress out-of-focus blur. For most reconstructions, we leave the settings at their default values, except for Illumination Modulation Contrast which we manually set at 1.0 to avoid an "insufficient contrast" error. The Out-of-Focus Blur Suppression setting can be increased to reduce artifact from out-of-focus light at the expense of resolution.
2. Assess the reconstructed super resolution images for artifact. SIM super resolution images are produced from the processing of multiple raw structured illumination images. The SIM reconstruction algorithm assumes a certain optical performance of the imaging system, if this is not met, e.g., due to spherical aberration, then reconstruction artifacts will appear in the reconstructed SIM image (Fig. 4A and B). Visual assessment of the reconstructed images for the presence of these artifacts is an important quality control step; however, SIMcheck (Ball et al., 2015) will give further confidence that the reconstructions are truly representative of the structures present in the sample and not containing misleading information.

3.7 Acquire Raw dSTORM Image Data

If compatible labeling has been used (and a dual-mode imaging system is available), then the SIM imaging steps described earlier can be done in

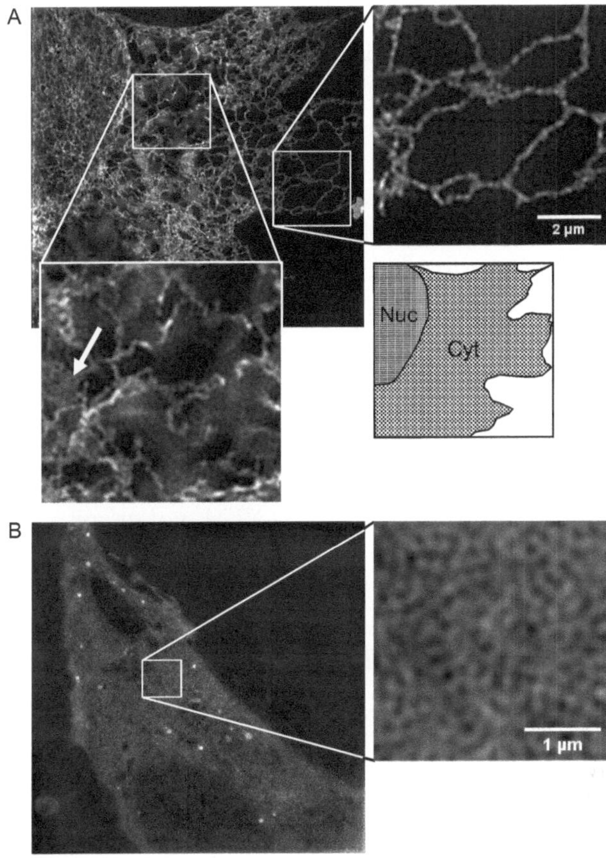

Fig. 4 Reconstructed SIM images are prone to artifact and care has to be taken when interpreting the data. (A) Slice reconstructed SIM image of mCherry-ER expressed in a HEK-293 cell. *Insets* show that signal from the flat periphery of the cell reconstructs very well; in areas closer to the nucleus, where there is a greater bulk of cytoplasm and more out-of-focus signal, a characteristic "honeycomb" pattern appears in the slice reconstruction (the *arrow points* to one of many such regions in the *inset*). The *line drawing* shows the location of the nucleus (Nuc) and the cytoplasm (Cyt) in the wider reconstruction. (B) Stack reconstructed SIM image of mCherry-ATG13 in a HEK-293 cell. *Inset* shows an area of the cytoplasm where there is uniform distribution of the ATG13 signal, but the reconstruction shows heavy patternation.

advance of dSTORM imaging, using STORM imaging buffer as the mounting medium (see Section 3.4, step 1) and ignoring steps 3 and 4 below. The Nikon N-STORM microscope is designed for use with the dye-pair labeling method first described by Rust, Bates, and Zhuang (2006); however, a much more convenient method is direct STORM (dSTORM), first described by Heilemann et al. (2008) which uses conventional single dye labels. Unfortunately dSTORM only works well with a limited number of dyes, so the possibilities of multicolour labeling are limited. Moreover, dSTORM requires high levels of laser illumination and suitable lasers may not be present on a given system. On our version 1.0 N-STORM system the 300 mW 647 nm laser has sufficient power for dSTORM with many far-red dyes (e.g., Alexa Fluor 647, Cy5), the 150 mW 561 nm laser has just enough power for dSTORM with some red dyes (e.g., CF555, CF568), but the 50 mW 488 argon laser does not have sufficient power for dSTORM. The latest iterations of the commercial STORM imaging systems are configured with higher-power lasers to permit dSTORM with a wider variety of dyes.

1. Prepare the microscope for dSTORM imaging. On the Nikon Ti-E microscope set the camera relay lens to 1 × (not the STORM position which is an astigmatic lens for 3D STORM, not discussed here). Ensure the objective lens is clean and that the correction collar is set correctly (see Section 3.3, step 3). The 2 × magnifying lens should be inserted into the illumination path in the TIRF condenser, and the laser beam exiting the objective should be checked for focusing and centering. We do this by placing a piece of card ~2 cm above the objective lens, setting the TIRF angle to 0 degree, and turning on one of the STORM lasers at low power. The spot of laser light should be a tight dot and positioned centrally; if not make suitable adjustments on the TIRF condenser.
2. Prepare the STORM imaging buffer required to create the environment in which the dye(s) exhibit on/off blinking behavior (see Rust et al., 2006). The optimum buffer composition will vary with the fluorophores used to label the sample, but for dSTORM using Alexa Fluor 647 and/or CF568, we have found the following composition to work well: catalase 1 μg/mL (Sigma Cat# C100), TCEP 0.8 μ*M* (Sigma Cat# C4706), glucose 40 mg/mL (Sigma Cat# G7528), glucose oxidase 50 μg/mL (Sigma Cat# G2133), glycerine 12.5% (Sigma Cat# G2289), KCl 1.25 m*M* (Sigma Cat# P9541), Tris–HCl 1 m*M* (Sigma Cat# 93352), and MEA-HCl 100 m*M* (Sigma Cat# M6500). For convenience, the catalase, TCEP, KCl, Tris–HCl, and glucose oxidase can be made up as a 20 × stock solution and the glucose and glycerine as a separate 2.5 × stock solution, both stored at −20°C in aliquots and made up in PBS as required. Stored like this we have found these buffer components to be stable for >12 months. The MEA-HCl should be made fresh

on the day of the experiment. Once made, the STORM buffer should be kept in a sealed tube on ice and used within a few hours.

3. Clean the gridded coverslip base of the 35 mm sample dish using 70% ethanol and place on the microscope stage. Aspirate off any existing buffer and overlay the cells with 200 μL STORM imaging buffer. The glass-bottomed 35-mm dishes have a hole in the plastic base which makes a convenient well into which the STORM buffer can be placed. Overlay the STORM buffer with ~2 mL of mineral oil (Sigma Cat# M5904) to reduce gas exchange which preserves the lifetime of the buffer.
4. Follow steps described in Section 3.4 to relocate cell(s) of interest, noting that the default live view in the Nikon software is a 2× zoom, representing the area from which dSTORM data will be collected. To remove this zoom, when locating cells click on the ROI button at the top of the live view window.
5. Fine-tune the focusing of the sample and if possible apply a dynamic focus lock. The Nikon Ti-E microscope has a perfect focus system, using infrared light to continuously monitor the distance between the objective and the sample, correcting for any drift as necessary. This is a very useful feature as image acquisition can take several minutes and any axial drift will compromise the final STORM image. If this feature is not available, then maintaining the imaging system in a room with minimal air temperature fluctuations should minimize drift.
6. While viewing the sample live on the camera and using low laser power (i.e., 1% transmission), adjust the illumination angle on the TIRF condenser so that the maximum level of signal is seen from the sample, i.e., when approaching the angle used for true TIRF where there is a dramatic increase in the intensity of illumination.
7. Set the image capture regime. The Nikon N-STORM system offers three modes of data acquisition: Normal Mode (i.e., dye-pair STORM), Single Color Continuous (i.e., dSTORM with one dye), Multicolor Continuous (i.e., dSTORM with multiple dyes). If using this latter option, then it is necessary to have an appropriate multibandpass/multiedge filter set/dichroic fitted in the filter cube carousel which can reflect all the required laser lines and pass the fluorescence; we use a Semrock LF405/488/561/635-A-000. The iXon 897 camera is set to acquire images at 256 × 256 pixels (0.16 μm/px) at a frame rate of 55 fps with the preamplifier gain set to 2.4 and the EM gain typically set at 100. We usually acquire a total of 15,000–20,000 images.

8. Start the image capture. Increase primary laser transmission to 100%. The sample as seen in the live view window should show a dramatic but brief increase in brightness, which rapidly fades to predominantly black with frequent blinks/flashes seen from individual fluorophores as they transiently come out of the dark state, fluoresce, and then return to the dark state. This should continue for the duration of the image acquisition, although the frequency of blinks will diminish over time as the proportion of dye molecules that have bleached increases and the buffer acidifies. If the sample does not rapidly fade to black when initially exposed to maximum laser transmission, then there is insufficient light reaching the sample. This may be due to many reasons, but the first things to check are the TIRF angle, the condenser focusing/centring, the presence of 2 × lens in TIRF condenser, and additional power settings for individual lasers. If the sample rapidly fades to black but does not exhibit blinking behavior, or shows limited blinking, then the most likely reason is poor buffer quality. One option to increase blinking frequency is to illuminate the sample with 405 nm light in addition to the primary laser. This can boost the number of dye molecules exiting the dark state, but if taken too high will increase the background signal and compromise localization precision.
9. Process the image stack to localize blinks. The Nikon system has built in software to process the raw dSTORM data with various blink identification parameters (we tend to use the default settings) and automatic correction for lateral drift. There is also a convenient tool for correcting registration between different color channels (using a calibration made with fluorescent beads), which is an important consideration for multicolour dSTORM. Alternatively the raw image data can be exported and processed in one of many freely available single-molecule localization microscopy analysis packages (see http://bigwww.epfl.ch/smlm/software/directory of resources). Processed raw STORM data will yield a coordinate list of blink localizations and associated information such as the precision of the localization. The blinks can be displayed to represent an image of the sample; we typically use the Gaussian display setting where the spot intensity and diameter are representative of the precision of the localization (better localized blinks are brighter and smaller) which gives an image similar to that obtained from a standard light microscope, although it is important to understand the differences between the two types of image. Interpreting the reconstructed STORM image can be a challenge. Insufficient blinks from, e.g., poor labeling density, will lead

to undersampling, making it hard to visualize the underlying structure. Target proteins will typically be labeled using primary antibodies, which themselves will be labeled with secondary antibodies. Each antibody adds ~10 nm uncertainty onto the exact location of the target, moreover multiple secondary antibodies may be bound to the target and multiple dyes conjugated to the secondary antibody. This can create a cloud effect surrounding a single target molecule. Labeling primary antibodies, using nanobodies as a replacement for traditional secondary antibodies, understanding the photochemistry of the dye used, and knowing the exact stoichiometry of the dye:antibody can help reduce these problems.

Using the methods described earlier it should be possible to acquire correlative super resolution images of proteins involved in autophagy that have been earlier studied in a living cell. The super resolution data will reveal structure and associations that are obscured by the diffraction limit of wide-field microscopy. SIM images can provide up to twice the resolution of conventional fluorescence microscopy and can be used to acquire 3D volumetric data. STORM images can reveal structure in the highest detail. Combined, SIM/STORM data complement each other both in terms of offering different capabilities, but also by minimizing the risk of misinterpreting artifact, as features seen with both techniques are highly likely to be genuine.

ACKNOWLEDGMENTS

Our work is supported by the Biotechnology and Biological Sciences Research Council.

REFERENCES

Ball, G., Demmerle, J., Kaufmann, R., Davis, I., Dobbie, I. M., & Schermelleh, L. (2015). SIMcheck: A toolbox for successful super-resolution structured illumination microscopy. *Scientific Reports*, *5*. 15915.

Bogdanov, A. M., Kudryavtseva, E. I., & Lukyanov, K. A. (2012). Anti-fading media for live cell GFP imaging. *PloS One*, *7*(12), e53004.

Heilemann, M., van de Linde, S., Schüttpelz, M., Kasper, R., Seefeldt, B., Mukherjee, A., et al. (2008). Subdiffraction-resolution fluorescence imaging with conventional fluorescent probes. *Angewandte Chemie (International Ed. in English)*, *47*(33), 6172–6176.

Karanasios, E., & Ktistakis, N. T. (2015). Live-cell imaging for the assessment of the dynamics of autophagosome formation: Focus on early steps. *Methods*, *75*, 54–60.

Karanasios, E., Stapleton, E., Manifava, M., & Ktistakis, N. T. (2014). Imaging autophagy. *Current Protocols in Cytometry*, *69*, 12.34.1–12.34.16.

Karanasios, E., Stapleton, E., Walker, S. A., Manifava, M., & Ktistakis, N. T. (2013). Live cell imaging of early autophagy events: Omegasomes and beyond. *Journal of Visualized Experiments*, *27*(77), e50484.

Ktistakis, N. T., Karanasios, E., & Manifava, M. (2014). Dynamics of autophagosome formation: A pulse and a sequence of waves. *Biochemical Society Transactions*, *42*(5), 1389–1395.

Mizushima, N., Yoshimori, T., & Ohsumi, Y. (2011). The role of Atg proteins in autophagosome formation. *Annual Review of Cell and Developmental Biology*, *27*, 107–132.

Ropolo, A., Grasso, D., Pardo, R., Saccheti, M. L., Archange, C., Lo Re, A., et al. (2007). The pancreatitis-induced vacuole membrane protein 1 triggers autophagy in mammalian cells. *The Journal of Biological Chemistry*, *282*, 37124–37133.

Rust, M. J., Bates, M., & Zhuang, X. (2006). Sub-diffraction-limit imaging by stochastic optical reconstruction microscopy (STORM). *Nature Methods*, *3*, 793–796.

Theer, P., Mongis, C., & Knop, M. (2014). PSFj: Know your fluorescence microscope. *Nature Methods*, *11*, 981–982.

Walker, S., Chandra, P., Manifava, M., Axe, E., & Ktistakis, N. T. (2008). Making autophagosomes: Localized synthesis of phosphatidylinositol 3-phosphate holds the clue. *Autophagy*, *4*, 1093–1096.

Yang, Z., & Klionsky, D. J. (2010). Eaten alive: A history of macroautophagy. *Nature Cell Biology*, *12*, 814–822.

Note added in proof

For a detailed application of this methodology to the problem of autophagy induction see Karanasios, E., Walker, S. A., Okkenhaug, H., Manifava, M., Hummel, E., Zimmermann, H., Domart, M. C., Collinson, L., & Ktistakis, N. T. (2016). Autophagy initiation requires ER to Golgi traffic and ULK complex assembly on distinct tubulovesicular ER-ATG9 regions. *Nature comm.*, Aug 11;7:12420.

CHAPTER TWO

Quantifying Autophagic Structures in Mammalian Cells Using Confocal Microscopy

C.A. Lamb[1], J. Joachim, S.A. Tooze[1]
Molecular Cell Biology of Autophagy Group, The Francis Crick Institute, London, United Kingdom
[1]Corresponding authors: e-mail address: christopheralamb101@gmail.com; sharon.tooze@crick.ac.uk

Contents

1. Introduction 22
2. Detection and Quantification of Autophagic Puncta in Fixed Mammalian Cells 24
   2.1 Preparation of Coverslips for Confocal Immunofluorescence Microscopy 24
   2.2 Identifying Autophagic Puncta Using Imaris 27
3. Quantifying Starvation-Induced ATG9 Redistribution by Indirect Immunofluorescence and Confocal Microscopy 31
   3.1 Background 31
   3.2 Cell Culture and Indirect Immunofluorescence Labeling 32
   3.3 Image Analysis 33
4. Quantification of ATG9 Compartment/Autophagosome Contact in Live Cells 36
   4.1 Preparation of Cell Cultures for Live Cell Confocal Microscopy 37
   4.2 Analysis of Proximity Between mRFP-ATG9 and GFP-LC3B Structures Using Imaris 38
Acknowledgments 41
References 41

Abstract

Autophagy relies on the sequential, hierarchical association of proteins with phagophores, and forming autophagosomes to allow completion of the process. Additionally, the trafficking of the unique transmembrane autophagy-related protein ATG9 is vital for autophagy progression. In this chapter, we discuss methods to monitor autophagosome number using confocal microscopy, by following the association of different autophagosomal markers with the phagophore and completed autophagosome. We also discuss methods to monitor the trafficking of ATG9 in mammalian cells under starvation conditions.

Methods in Enzymology, Volume 587
ISSN 0076-6879
http://dx.doi.org/10.1016/bs.mie.2016.09.051

1. INTRODUCTION

Autophagosome formation, cargo recruitment, and degradation require sequential association of protein factors with the phagophore and autophagosomes. For example, initiation of autophagy induces formation of phosphatidylinositol 3-phosphate (PtdIns(3)P)-enriched structures on the endoplasmic reticulum called omegasomes, labeled with the lipid-binding protein DFCP1 (Axe et al., 2008).

This pool of PtdIns(3)P in turn recruits an effector protein, WIPI2b, which acts as a platform to recruit the ATG12–5-16L1 complex (Dooley et al., 2014; Polson et al., 2010). This E3-ligase-like protein complex catalyzes the final step in the covalent linking of the ATG8 family (in mammals, called LC3 and GABARAP proteins) to the lipid phosphatidylethanolamine (PE). LC3/GABARAP-PE conjugates associate with the phagophore membrane and remain associated with it as it closes to form an autophagosome (Kabeya et al., 2004). The inducible association of these proteins with the phagophore and autophagosome can be monitored in cells stably expressing GFP-tagged LC3/GABARAP (Fig. 1).

An important function of LC3/GABARAP is to recruit autophagic cargos into the growing phagophore. This is achieved in part through recruitment of autophagy receptors including p62, which bind ubiquitinated proteins and link them to the phagophore membrane via LIR (LC3-interacting region) motifs (Birgisdottir, Lamark, & Johansen, 2013). LC3/GABARAP proteins and autophagy receptors bound to the

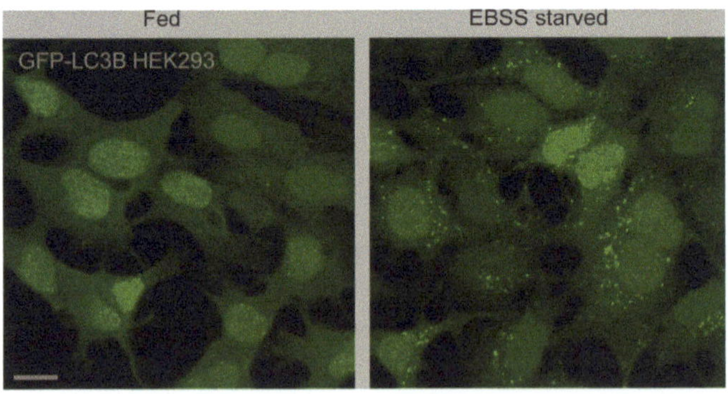

Fig. 1 HEK293A cells stably expressing GFP-LC3B were incubated in full medium (Fed) or EBSS starvation medium for 2 h prior to fixation and confocal microscopy. Scale bar 20 μm. Note the formation of cytoplasmic autophagosomes upon starvation.

inner membrane of the autophagosome are degraded once autophagosomes fuse with lysosomes, along with the autophagosomal cargo.

WIPI2b, the lipidation machinery, and the LC3/GABARAP proteins are peripherally membrane associated, or covalently attached to lipids, while only integral membrane protein required for mammalian autophagy is ATG9 (Young et al., 2006). Under basal (nutrient replete) conditions ATG9 largely resides in the juxtanuclear Golgi region of the cell. However, during amino acid starvation ATG9 translocates to peripheral endosomal compartments, making transient contacts with autophagosomes, without becoming incorporated into the autophagosome membrane (Orsi et al., 2012). ATG9 has also been localized to the plasma membrane (Puri, Renna, Bento, Moreau, & Rubinsztein, 2013), recycling endosomes (Knævelsrud et al., 2013; Longatti et al., 2012), and to a unique tubulovesicular membrane compartment, the "ATG9 compartment" (Orsi et al., 2012), which is conserved in yeast (Mari et al., 2010). ATG9-positive structures may function to deliver lipids or other factors to growing autophagosomes. Starvation-induced redistribution of ATG9 is intimately linked with canonical autophagy signaling, as this trafficking event requires ULK1 (Young et al., 2006).

The sequential association of these proteins with the forming autophagosomes allows them to be used as tools to determine at what point a particular treatment affects the autophagic pathway. Indeed, these proteins—both tagged and endogenous—are regularly used to monitor the progression of autophagy, as these proteins show a marked increase in membrane association upon induction of autophagy, and thus the number of marker-positive spots or puncta is a robust and reliable readout (Klionsky et al., 2016). In this chapter, we describe protocols for the preparation of cells labeled for both tagged and endogenous autophagosomal proteins, and the quantification of confocal images captured from these cells using the Imaris software package (Bitplane). It is important to note that using automated image analysis software, such as Imaris, offers several advantages over manual quantification methods: it reduces unconscious bias, allows storage and reuse of analysis parameters, and greatly increases the speed with which an investigator can process large data sets. In particular, Imaris provides many analysis tools in one package, allowing numerous parameters to be extracted from a single image.

These methods not only provide a basis for studying biochemically intractable trafficking steps such as those mediated by ATG9, but they also may reveal subtle autophagy phenotypes which may not be readily evident through biochemical methods, such as the typical LC3-I conversion to LC3-II assays.

2. DETECTION AND QUANTIFICATION OF AUTOPHAGIC PUNCTA IN FIXED MAMMALIAN CELLS

In this section, we describe protocols for preparing cells for confocal microscopy, inducing autophagy and analyzing the number of autophagic puncta present in cells using the Imaris 7.5 software package (Bitplane). These protocols have been optimized for HEK293A cells, however, they are widely applicable to numerous cell lines. Amino acid starvation HEK293A cells provide a robust system for detecting autophagy, having a low basal level in fed conditions, and an easily detectable starvation response.

2.1 Preparation of Coverslips for Confocal Immunofluorescence Microscopy

2.1.1 Cell Culture and Autophagy Induction

1. Place sterile, clean coverslips (13 mm round) into a 12-well tissue culture cluster dish. If using weakly adhering cells, coat wells and coverslips with 1 mL 0.1 mg/mL poly-D-lysine in sterile dH_2O for 10 min and wash twice with sterile dH_2O.
2. Plate cells in appropriate complete media (for example, DMEM supplemented with 10% fetal bovine serum and 4 m*M* glutamine) such that on the following day, there are approximately 2×10^5 cells per well (80% confluent). The cells may be incubated in the plates overnight at 37°C in a humidified incubator at 10% CO_2.
3. Induce autophagy in the cells. For example, we induce amino acid deprivation by washing the cells three times with Earle's balanced salt solution (EBSS; 5.56 m*M* D-glucose, 123.08 m*M* NaCl, 5.37 m*M* KCl, 1.82 m*M* CaCl2, 0.81 m*M* $MgSO_4$, 0.99 m*M* Na_2HPO_4, 13.1 m*M* $NaHCO_3$) equilibrated to 37°C, followed by further 2 h incubation in EBSS in a humidified incubator at 10% CO_2. Note that EBSS starvation deprives the cells of amino acids and growth factors, but not glucose. Canonical autophagy may also be induced by pharmacological treatment with drugs that inhibit the activity of the mTORC1 complex, including rapamycin and Torin1, for 2 h in complete media.
4. If autophagosome maturation is being investigated, addition of lysosomal inhibitors, such as Bafilomycin A1 (BafA1) at 100 n*M*, to the EBSS will block autophagosomal maturation during starvation. This treatment

shows whether an observed increase in autophagosomal number is due to promotion of autophagosome formation, or inhibition of autophagosome clearance.

5. 2 h after starvation, remove the media without washing and fix all coverslips in 3% paraformaldehyde, 0.1 m*M* $CaCl_2$, 0.1 m*M* $MgCl_2$ in PBS for 20 min at room temperature. The following steps are performed at room temperature unless otherwise stated. Remove the fix solution and wash twice with PBS. Volumes should be adjusted for the cell culture dish. For one well in a 12-well dish we use 1 mL.

2.1.2 Permeablization and Antigen Staining

To allow access of antibodies to internal antigens, cells must be permeabilized. This is usually achieved with mild detergents; however, staining of the LC3/GABARAP family of proteins is particularly sensitive to these treatments. Here, we describe two techniques for permeabilization of PFA fixed cells.

For general immunofluorescence:

1. Quench remaining PFA in the wells using 50 m*M* NH_4Cl for 10 min at room temperature. Wash twice with PBS.

For immunofluorescence of LC3/GABARAP family proteins:

1. Incubate coverslips in 1 mL room temperature methanol for 5 min to permeabilize. Wash three to four times in PBS to remove excess methanol.

Then for either fixation protocol:

2. Incubate the cells for 3 min with 1 mL 50 μg/mL digitonin or 0.2% Triton X-100 in PBS, then wash three times in PBS to remove excess detergent.

3. To block nonspecific antibody binding, add 1 mL of 5% BSA (bovine serum albumin fraction V) in PBS to the wells and incubate for at least 20 min.

4. During this incubation, prepare dilutions of appropriate primary antibodies in 1% BSA. Set up a parafilm lined humidified chamber. Aliquot the antibody dilution in 60–100 μL drops (one per coverslip to be stained) on the parafilm.

5. Transfer coverslips from cluster plates to the humidified chamber using watchmaker's forceps. Remove excess liquid using a tissue and place the coverslip cell-side down on the drop. Incubate in the chamber for 45–60 min.

6. During this incubation, dilute the secondary antibody in 1% BSA. We use Alexa-dye conjugated secondary antibodies from Life Technologies (Paisley, UK) at 1:1000. To stain nuclear DNA, add Hoechst DNA dye to a final concentration of 1 μg/mL. Aliquot the diluted secondary antibody/Hoechst mixture in 60-100 μl drops (one per coverslip) on the parafilm.
7. Pick up the coverslips using watchmaker's forceps and wash by dipping into a beaker of PBS three to four times. Remove excess liquid using a tissue and transfer the coverslip cell-side down onto a drop of secondary antibody solution. Incubate for 45–60 min in the chamber in the dark.
8. Wash coverslips in PBS three to four times as in step 7 and then three to four times in a beaker of distilled water to remove salt from the PBS washes.
9. Remove excess liquid, place cell-side down on a microscope slide on a 10 μL drop of mounting medium (for example, Mowiol), and allow to dry in a dark place at room temperature for several hours.

2.1.3 Direct Fluorescence

For cell lines expressing a fluorescently tagged autophagy protein, when no endogenous proteins are being stained:

1. After treatment and fixation quench the remaining PFA for 10 min using 50 m*M* NH_4Cl, and wash twice with PBS.
2. Incubate the coverslips in 1-mL PBS with 1 μg/mL Hoechst DNA dye for 5 min.
3. Wash coverslips in PBS three to four times and then three to four times in a beaker of distilled water to remove PBS salts from the coverslips. Remove excess liquid from the coverslip and place cell-side down on a 10 μL drop of mounting medium on a microscope slide and allow it to dry in a dark place at room temperature for several hours.

2.1.3 Confocal Imaging

Imaging conditions may vary between microscopes. However, as a guideline, to image cells for quantification of autophagic puncta we use a 40 × objective oil immersion lens, 1.3 numerical aperture, and a confocal slice of 0.8 μm. To ensure robust quantification, imaging at least 10 fields of view per condition is recommended. Hoechst DNA dye can be used to locate the cells, as this removes bias in selecting fields of view. It is important to use identical microscope settings for all images in an experiment; otherwise, quantification of the number of autophagic puncta will be unreliable.

2.2 Identifying Autophagic Puncta Using Imaris

The Imaris software package "Spots" function is used to enumerate punctate structures within cells. This function identifies foci-like structures in the cells, based on a diameter entered by the user, and extrapolates a margin of error based on a normal distribution. This margin of error can be adjusted by the user to remove false-positive puncta in control images. Depending on the conditions used in the experiment (for example, if the cells of interest are expressing a particular protein and the nonexpressing cells need to be excluded, or if a protein has irrelevant staining in the nucleus), it can be helpful to use the "Cell" function to identify the cytosol and nuclei (see later).

To identify the puncta within an image using the Imaris "Spots" function:

1. Open Imaris 7.5 and open an image from control cells where autophagosomes are expected to form. For example, in an experiment where a comparison is made between fed and starved cells transfected with a nontargeting siRNA and an siRNA targeting a gene of interest, choose an image from the nontargeting siRNA + starved set.
2. Using the Imaris "Slice" function, measure the diameter (μm) of several visible puncta by clicking the mouse to draw a line transecting the spots. Note the size of the measured spots and calculate the mean size. Thus, you will now have an approximate punctum size.
3. Using the "Surpass" function, choose "Spots" from the left hand Surpass Scene menu.
4. In the panel below the Surpass Scene menu, a dialog box for setting up a spot recognition algorithm will appear. Leave the settings as Default and leave the check-boxes empty. Click "next" (designated by a blue single right "play" symbol) at the bottom of window to proceed (Fig. 2A).
5. In the following dialog box, select which channel is being imaged, and in the "Estimated XY diameter" box, enter the approximate spot size as determined in point 2. Check the box marked "Background subtraction." Click "next" to proceed (Fig. 2B). At this stage, spots recognized by the program should appear on the image on the cells.
6. The following dialog box is a quality filter, and the histogram shows the number of spots recognized by the automatically detected settings. By dragging left or right on the histogram, the spot selection criteria become more or less stringent. Thus, the algorithm can be fine-tuned to minimize the number of false-positive spots. Once this has been achieved, the green double arrow should be clicked to complete the setup (Fig. 2C).

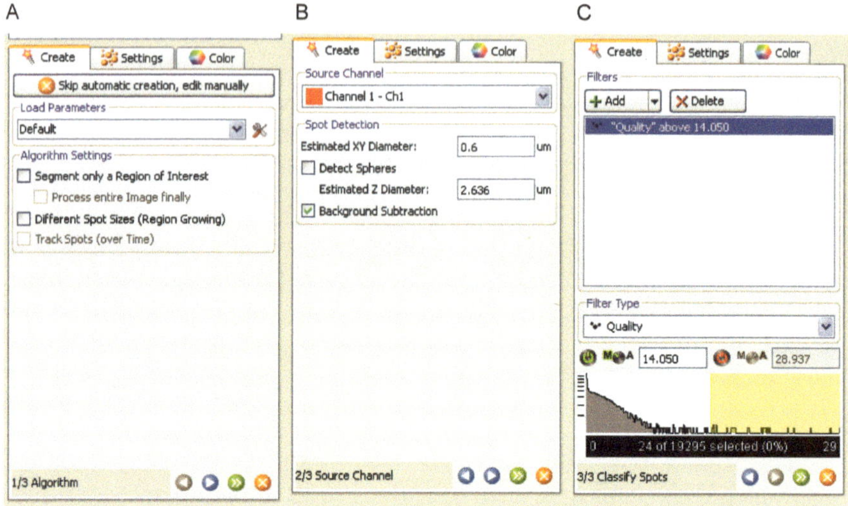

Fig. 2 (A–C) Screenshots of Imaris 7.5 showing spot segmentation.

7. Select the "Statistics" tab. Various dialog boxes appear with the statistics associated with the Spots algorithm. The total number of spots in the image is available in the "Overall" tab.
8. Count the number of cells in the field by eye, using the Hoechst stain as a landmark. The average number of spots per cell is given by dividing the total spots by the total number of cells. All these values should be noted in a spreadsheet.
9. The analysis can be repeated for other images in the dataset. It is important that the settings are not changed between analyses of different images. Simply open a new image and click the "Rebuild" button under the "Create" tab of the Spots algorithm (Fig. 6E). The "Create" tab is designated by a magic wand symbol. If all the images cannot be analyzed in one sitting, save the original control image with the Spots algorithm set up to start the process again.

2.2.1 Segmenting Cells, Cytosol, and Nuclei Using Imaris

The earlier protocol works well for quantification of endogenous autophagy proteins (e.g., WIPI2, LC3B, or p62). However, cell lines expressing fluorescently tagged LC3/GABARAP proteins, particularly GFP-LC3B, are commonly used as model systems. One problem with this is that GFP-LC3B accumulates in the nucleus, leading to recognition of irrelevant nuclear puncta. To avoid this, we use the Imaris Cell function to define areas of cytosol and nucleus. This function permits recognition of large areas of fluorescence as cell bodies or nuclei, and thus exclusion of puncta in either region.

To use the Imaris Cell function after generating a population of spots (after step 6 earlier):

1. In Slice mode, use the line tool to measure the diameter of several cells and nuclei in the field of view to obtain an estimate of their size. Make a note of these measurements and calculate the mean to give you an approximate size of cells and nuclei.
2. In Surpass mode, select the Cell function. A dialog box will appear below the Surpass Scene menu. With parameters again set as Default, leave "Segment only a region of interest" unchecked and click "next" (Fig. 3A).
3. In the following dialog box, there are options for "Cell body detection" (Fig. 3B). Select the color of the "Source Channel" to be used to identify the cell—this is usually the channel with which the spots have been identified (see step 5).
4. Leave "Smooth" unchecked and check the "Background subtraction" box. In the "Estimated sphere Diameter" box, enter the value for the size of the cell (Fig. 3B).
5. Check the box marked "Detect Nuclei" to allow nucleus detection. Repeat steps 3 and 4 to define initial parameters to identify nuclei in the image. Click "next" (Fig. 3B).
6. In the next dialog box, there are two histograms with movable indicators, allowing more accurate identification of nuclei and cells. The nuclei and cells are highlighted based on the original parameters. The indicators should be moved up or down to provide the best coverage as possible of the cell bodies and nuclei in the image. Once this has been achieved, click "next" to proceed (Fig. 3C).
7. Select "Expand Cell on Nucleus" and ensure "Fill holes in cell and nucleus" is unchecked. Click "next" to proceed (Fig. 3D).
8. The following step allows further filtering of the cells and nuclei, again displayed as histograms which can be adjusted to exclude unwanted cells, for example, mitotic or apoptotic cells which could bias results. Click "next" when this is complete (Fig. 3E).
9. The final option is to identify vesicles—this is not required as the spots we have previously identified will be imported as vesicles. Uncheck "detect vesicles" and click the green arrow to finish the analysis (Fig. 3F).

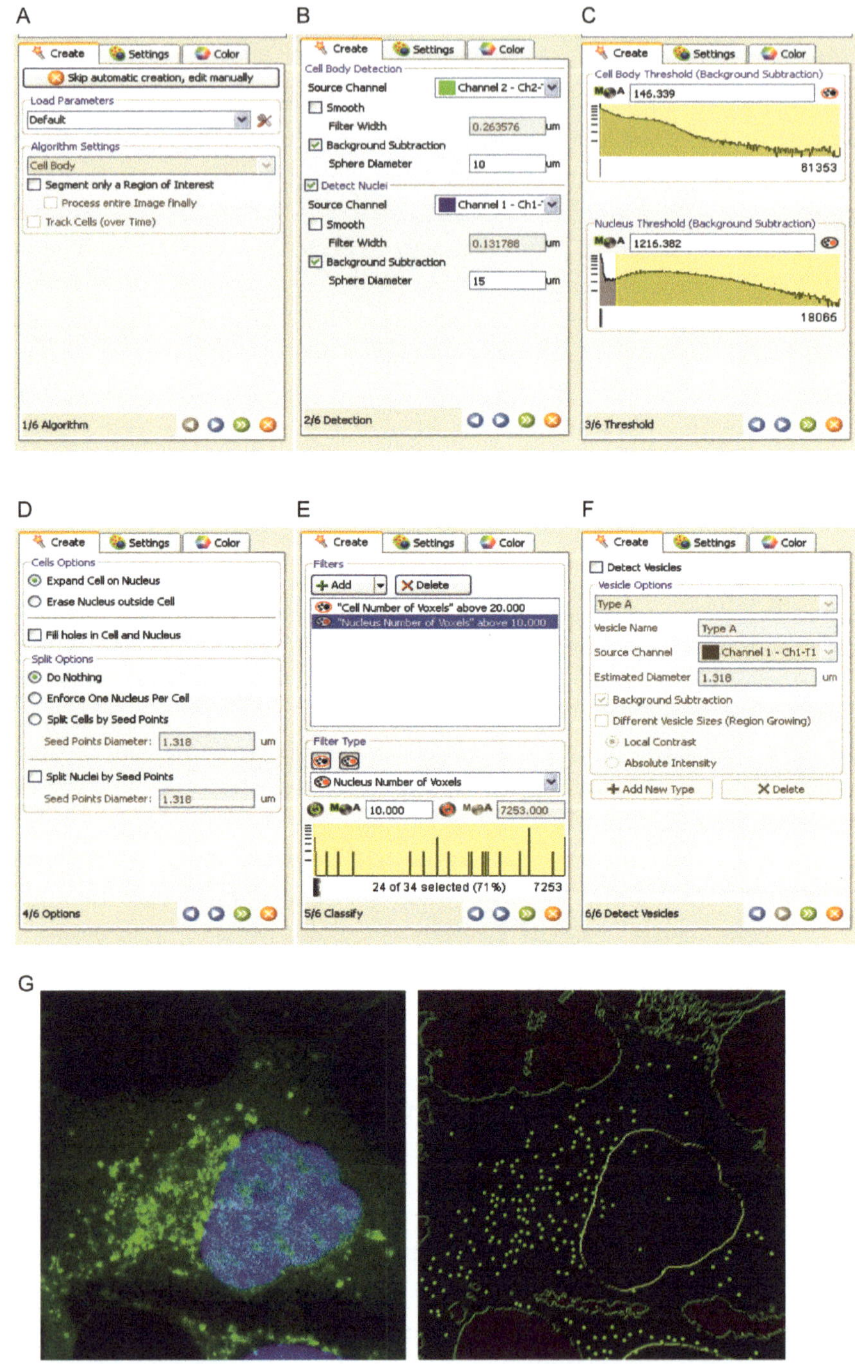

Fig. 3 (A–F) Screenshots of Imaris 7.5 showing generation of a cell segmentation. (G) HEK293A cells stably expressing EGFP-LC3B were starved for 2 h in EBSS medium prior to analysis by confocal microscopy. *Left*: *Green* channel, GFP; *blue* channel, Hoechst. Cytoplasmic EGFP-LC3B spots can be seen. *Right*: Imaris mask around cell boundary (using EGFP-LC3B channel) and mask around cell nucleus (using Hoechst channel). EGFP-LC3B puncta have been identified in the cytoplasm and the nucleus have been categorized as such by the software quantitation.

At this point, we have cells, nuclei, and spots identified in our image (see Fig. 3G). For Imaris to locate the spots in the cytoplasm or nucleus, the Spots must be converted to Vesicles in the analysis:

1. To convert the Spots to Vesicles, click the Edit (pencil) tab after the cells have been defined.
2. Select the option "Import spots to vesicles." This opens a dialog box showing the types of spots (depending on number of antigens probed) that have been created during the analysis. Select the spots you wish to turn into vesicles and choose to import them as "New Vesicle Type."
3. The number of spots located in the nucleus or cytosol may be accessed by exporting all the analysis parameters using the "Export Statistics" tab, saving all the data as a series of Excel files. Open the file named "Vesicle Export Location in Cytosol."
4. This lists every vesicle (spot or autophagic punctum) in the cytosol region defined by Imaris. The vesicles are classified as 1 (cytosol) or 0 (nucleus). These values can be added to give a total number of spots in the image and divided by the total number of cells in the image to give spots per cell.
5. The analysis should be saved now, preserving the Cell settings for analysis of the rest of the experiment.

3. QUANTIFYING STARVATION-INDUCED ATG9 REDISTRIBUTION BY INDIRECT IMMUNOFLUORESCENCE AND CONFOCAL MICROSCOPY

3.1 Background

During amino acid starvation, or rapamycin treatment, the redistribution of the transmembrane protein ATG9 is required for the early phagophore formation stage in autophagy (Orsi et al., 2012). The relocalization of ATG9 is visually most striking as a decrease in the signal intensity of the juxtanuclear pool upon starvation (Fig. 4). This pool would ordinarily show overlap with Golgi or TGN markers such as GM130 or TGN46, respectively, during basal conditions.

The increase of ATG9 in the peripheral regions of the cell upon starvation is less striking. This is probably because the same amount of ATG9 protein occupies a larger endosomally-distributed volume upon starvation, as ATG9 protein levels are unaffected by 2-h starvation. The fluorescence signal is, therefore, less concentrated in the *XY*-axes and, crucially for confocal

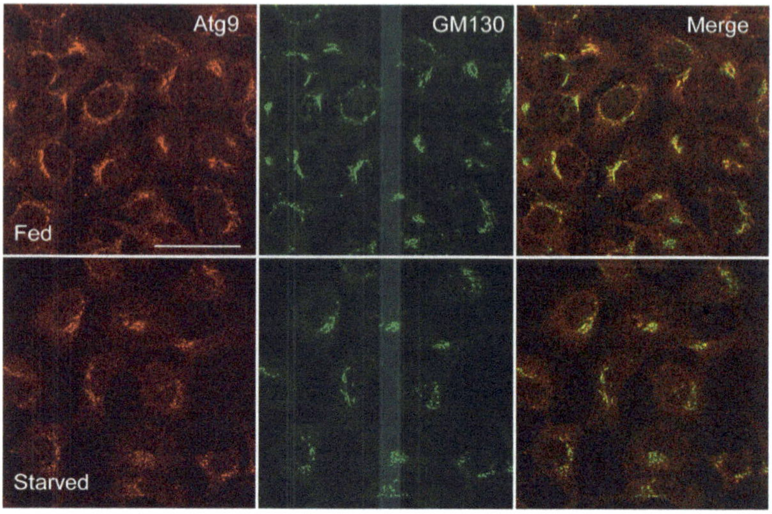

Fig. 4 HEK293A cells were incubated in full medium (Fed) or EBSS starvation medium for 2 h prior to immunofluorescence staining cells for ATG9 and GM130. Note the reduction of juxtanuclear ATG9 staining upon starvation.

microscopy, in the *Z*-axis. Here, we demonstrate a method for quantifying juxtanuclear concentrated ATG9 in HEK293A cells, and hence for showing ATG9 redistribution upon starvation.

3.2 Cell Culture and Indirect Immunofluorescence Labeling

3.2.1 Cell Culture and Induction of Autophagy

Cell culture should be carried out as in Section 2.1.1.

3.2.2 Indirect Immunofluorescence Labeling

Indirect immunofluorescence should be carried out as in Section 2.1.2. We have found the best permeabilization method for detecting ATG9 is to use 50 μg/mL digitonin in PBS for 3 to 5 min. We have also found that the best primary antibody is Armenian hamster monoclonal IgM ATG9A antibody clone 14F2 8B1 (from ThermoFisher, Paisley, UK or Abcam, Cambridge, UK) at a final concentration of 0.9 μg/mL diluted in 1% BSA (bovine serum albumin fraction V, Roche), although this may require optimization for different systems. As a secondary antibody, use Cy™3-conjugated AffiniPure Goat Anti-Armenian Hamster IgG (heavy and light) from Jackson ImmunoResearch (West Grove, PA, USA), code number 127-165-160. As this is a red fluorophore, it is important to select appropriate secondary antibodies for other proteins being stained.

3.2.3 Confocal Imaging

Acquire confocal images of ATG9 and nuclei (Hoechst) and ensure there are no saturating brightness values in your images. We use a 63 × magnification, 1.4 numerical aperture oil immersion lens. Set the gain, offset and laser power values according to the fed control coverslip, where the juxtanuclear ATG9 signal will be more pronounced compared to the starved coverslips. Use the same settings for acquisition of all your images. For statistical power, we usually image a minimum of 100 cells per condition per experiment, or around 10 images per coverslip. For confocal imaging, note that we sometimes open to the pinhole to obtain sections >1 μm *z*-thickness and substantially increase the laser power. This is useful to obtain a robust ATG9 signal; however, one must be careful about making assumptions about colocalization with other organelle markers under these conditions.

3.3 Image Analysis

1. Open Imaris 7.5.0 software and click the open button. Open a fed control image of ATG9 in Imaris and select Surpass mode by clicking the "Surpass" button (Fig. 5A).
2. Create a new Surface by selecting Surpass > Surfaces in the top dropdown menu (Fig. 5B). This will create a new algorithm called "Surfaces 1" (Fig. 5C).
3. Now you have the option to define an algorithm that will select juxtanuclear ATG9 in the whole image and create a mask on ATG9 (Fig. 7). Click on "Surfaces 1" to select it, and make sure "Surfaces 1" is also checked (Fig. 5C). All other options under "Surpass Scene" should be checked but not selected (not highlighted) (Fig. 5C). Click "next" (designated by a blue single right arrow) on the Create tab (Fig. 6A). The create tab can be found in the bottom left corner of the Imaris window. If you want to analyze each cell individually, check the "Segment only a region of interest" box, before clicking "next." This will then prompt you to select your region of interest. However, in this method we will be analyzing the whole field of view and will leave this option unchecked.
4. Next select the source channel you want to analyze (Fig. 6B). In this case the red (ATG9) staining has been selected. Check the "Smooth" box and set "Surfaces Area Detail Level" to 0.3 μm. Under "Thresholding" check "Absolute Intensity." Click the "next" button.

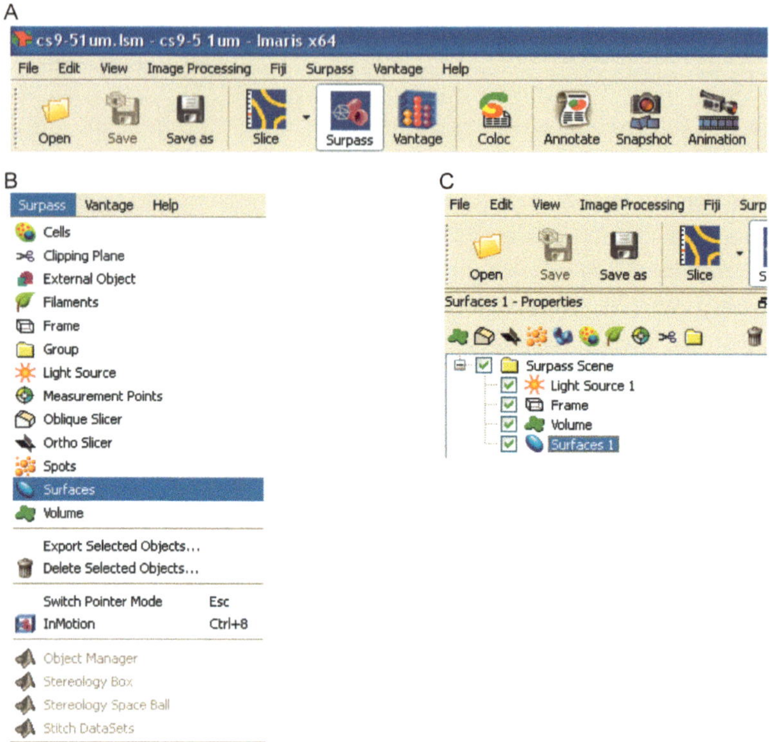

Fig. 5 Screenshots from Imaris 7.5 showing how to create a new surfaces algorithm.

5. In this section of the algorithm settings (Fig. 6C) you have the option to adjust the threshold, which determines what is recognized as juxtanuclear ATG9 staining. The threshold can be adjusted by dragging the vertical line on the graph to the left or right, or by entering numerical values in the boxes. In gray you can see what has been called as juxtanuclear ATG9. Leave "Split touching Objects (Region Growing)" unchecked. Click the "next" button twice to finish the algorithm setup. Note that after setting the threshold, you will be prompted to adjust the minimum number of voxels (3D or volumetric pixels) that are recognized as juxtanuclear ATG9. In this case, we have not adjusted this. However, this option can be used to exclude small or large structures if necessary; for example, small cytoplasmic ATG9 puncta. Now you will be able to see the surfaces mask over the ATG9 signal in the surpass image window (Fig. 7). This mask can be toggled on or off by checking or unchecking the "Surfaces 1" option, respectively (Fig. 5C).

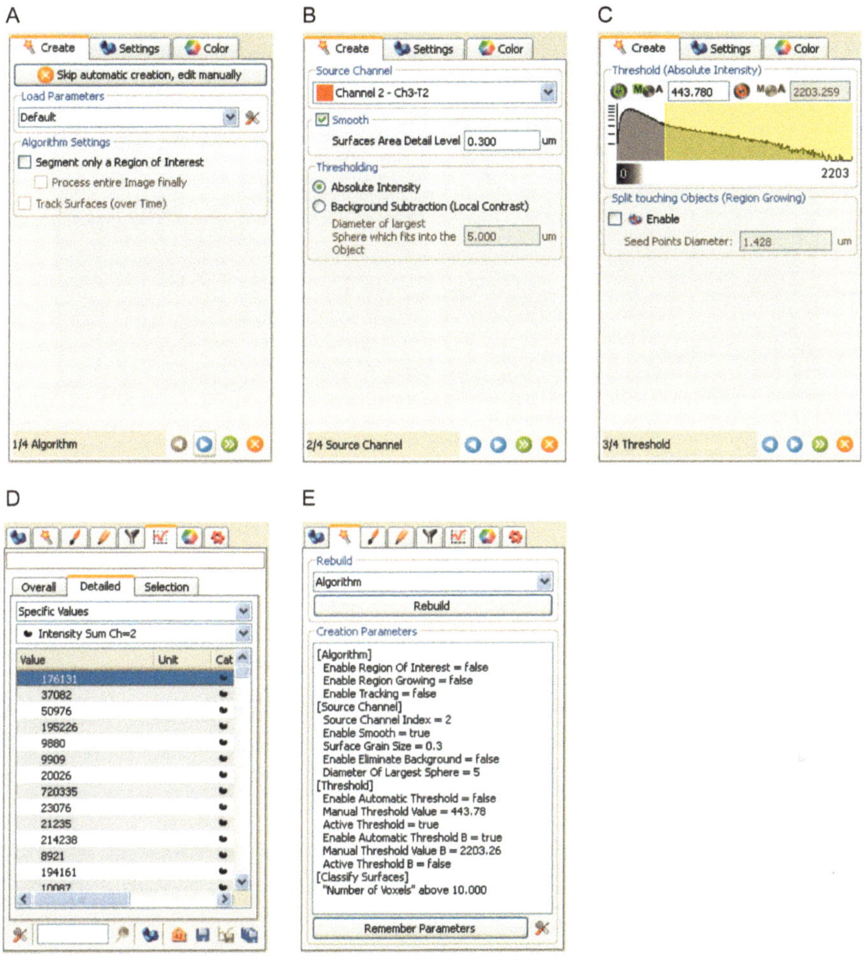

Fig. 6 (A–E) Screenshots from Imaris 7.5 showing how to setup a surface mask around the ATG9 stained region.

6. Select statistics tab > Detailed tab (Fig. 6D). From the dropdown boxes select "Specific values" and "Intensity Sum Ch = 2" (or whichever channel is ATG9). Here, you can see a list of intensity values for every contiguous surface. Click the export statistics button (designated by a single floppy disk symbol) to save the numerical data. Note that many other values can be obtained from this analysis from the dropdown menu, for example, one could investigate the amount of signal of another protein localized at the ATG9 compartment.

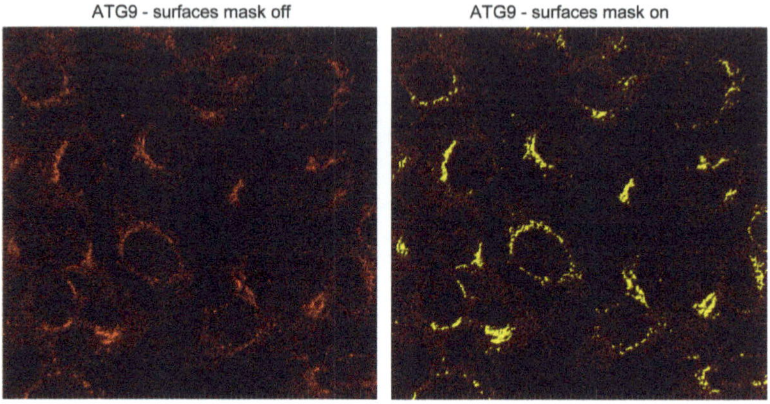

Fig. 7 Screenshots from Imaris 7.5 showing original ATG9 staining image and the finished surfaces mask applied.

7. Click "Save as" to save the settings of the algorithm (Fig. 5A). Use the same settings to analyze images from the same experiment. Image analysis can be continued during another session by opening the saved file in Imaris and then opening a new image to analyze. You will then have the option to select "Surfaces 1" (Fig. 5C) and under the "magic wand" tab you can now select the "Rebuild" button (Fig. 6E). Simply hit the fast forward button (green button with double arrow) to analyze subsequent images with the same algorithm settings and export data as in step 6 above.
8. Exported data can be imported and tabulated in spreadsheet software such as Microsoft Excel® and the ATG9 signal intensity values for each field of view (image) can be summed. These summed values can be divided by the number of cells per field (determined using the Hoecsht stain) to give the juxtanuclear ATG9 signal per cell, for each field of view. More resolution of the data can be obtained if individual cells have been analyzed by selecting a region of interest.

4. QUANTIFICATION OF ATG9 COMPARTMENT/ AUTOPHAGOSOME CONTACT IN LIVE CELLS

As discussed in the previous section, the redistribution of ATG9 from a juxtanuclear, Golgi associated compartment to peripheral endosomal structures is one of the earliest responses during autophagy induction. Work from our group has shown in live cells, mRFP-ATG9 containing vesicles (referred to as the ATG9 compartment) transiently contact

GFP-DFCP1-positive omegasomes and GFP-LC3-positive phagophores and autophagosomes (Orsi et al., 2012). More recently, we developed a method to monitor this process via live cell confocal microscopy with good spatiotemporal resolution (submicron and subminute) (Lamb et al., 2016). Here, we describe this protocol to generate movies of live cells and subsequently count the contact events between the ATG9 compartment and forming autophagosomes.

The cell line used should stably express a fluorescent ATG9 and a fluorescent marker of the phagophore/autophagosome. For convenience, in this section, we shall refer to mRFP-ATG9 and GFP-LC3B as the proteins used, although the experimenter may wish to use other autophagosomal markers.

4.1 Preparation of Cell Cultures for Live Cell Confocal Microscopy

1. Coat a 35-mm glass-bottomed tissue culture dish (14-mm glass microwell, MatTek Corporation, Ashland, MD, USA) with 0.1 mg/mL poly-D-lysine solution for 10 min, wash three times with sterile distilled water and add 3 mL culture medium. Plate approximately 2×10^5 cells in each dish and leave overnight in a humidified incubator at 37°C with 10% CO_2. It is advisable to leave the cells in an incubator close to the confocal system being used, as transporting the cells at room temperature may induce stress responses and unanticipated autophagosome formation.
2. The next day, prepare the microscope for imaging. We use the inverted Zeiss LSM 880 with 63 × objective oil immersion lens, 1.4 numerical aperture. The microscope should have a temperature controlled chamber around the stage, and this should be equilibrated to 37°C for >2 h before the start of the experiment to avoid expansion of components and loss of focus. While microscope systems vary, we have found that taking confocal sections of 0.6 μm at 10 s intervals generates good quality movies with HEK293A-derived cell lines.
3. Equilibrate EBSS supplemented with 30 m*M* HEPES, pH 7.4 (EBSS + HEPES) to 37°C. The addition of HEPES will buffer the pH of the EBSS in the absence of CO_2.
4. Wash one dish of cells once with 3 mL EBSS and replace with 3 mL EBSS + HEPES. Place the dish on the microscope stage and locate the cells using one of the fluorescent markers. Now is a good time to set the parameters of the microscope such that the images are not overexposed.

5. Leave the dish in the microscope chamber for a further 10–15 min, then change the focus to relocate the cells. This step is important as, upon starvation, cells often round up slightly and this can move them out of focus.
6. Start filming the cells. By 10–15 min from start of starvation, mRFP-ATG9 should have redistributed from the juxtanuclear store and the first GFP-LC3B positive structures will have formed. We have previously taken images for 1 h, generating a movie comprising 360 still images.

4.2 Analysis of Proximity Between mRFP-ATG9 and GFP-LC3B Structures Using Imaris

By using the "Distance transformation" plugin in Imaris 8.2, it is possible to analyze the proximity of defined structures within cells to another structure. This function labels all vesicles in one population (GFP-LC3B) with the distance to the nearest vesicle of another population (in this case, mRFP-ATG9) and lists the values in microns associated with each vesicle. This protocol explains how to manually generate cell surfaces in confocal movies, and subsequently analyze the proximity of the two populations of vesicles within a cell during a timecourse using the "Distance transformation" plugin.

4.2.1 Manually Creating Cell Surfaces in Imaris

1. Open the movie to be analyzed in the Surpass view and select "Add new surfaces" algorithm. In the dialog box that opens, choose "Skip automatic creation, edit manually" (Fig. 8A).
2. Ensure the cursor is in "select" rather than "navigate" mode on the "Pointer" tab. In the surfaces dialog, select "Draw." Go to the mode tab and select the "distance" option. Click the "Draw" button (Fig. 8B).
3. Click where you wish to start drawing on the still from the movie. This will switch the cursor to drawing mode and generate a polygon with evenly spaced vertices when dragged. Drag the cursor around the cell of interest to close the polygon (Fig. 8B).
4. Once the area is enclosed, click "Create surface" (Fig. 8B). To extrapolate this surface to the whole timecourse, go to the "Edit" tab and choose "duplicate to all time points" (Fig. 8C). This will create a second Surface object, "surfaces selection."
5. To transfer this surface to a cell, click the "Add new cells" tab. Instead of automatically creating cells, choose "Skip automatic creation, edit

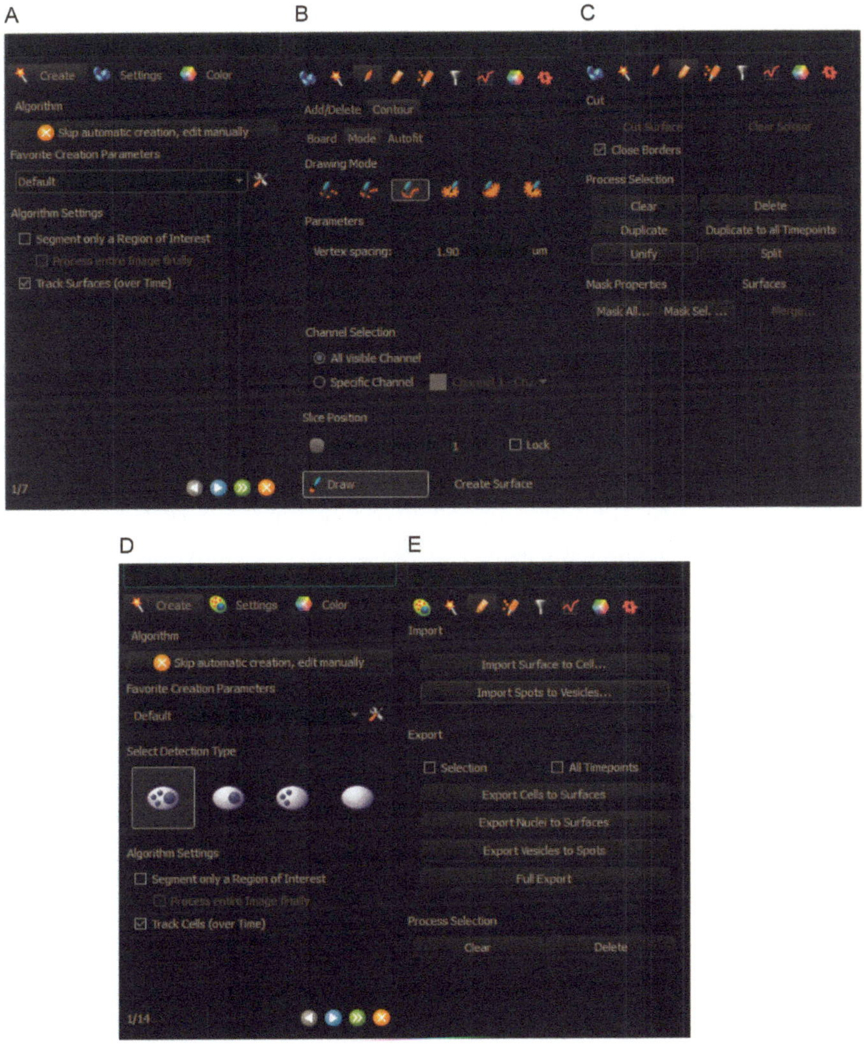

Fig. 8 (A–E) Generating a manually drawn surface and importing as a cell using Imaris 8.0.

manually" (Fig. 8D). Instead of basing the cell body on a density of fluorescent labeling (as in Section 2.2.1) this process generates a cell area based on the manually created surface.

6. Go to the Edit (pencil) tab. Select "Import surface to cell" (Fig. 8E). Choose the newly generated surface present in all time points, "Surfaces selection." This will generate a cell surface throughout all time points in the movie and allow analysis of vesicular content.

4.2.2 Measuring Distance Between Two Vesicle Populations Using the Distance Transformation Function

1. Generate spots for both fluorescent proteins in the timecourse (mRFP-ATG9 and GFP-LC3) as in Section 2.2. To avoid using too much processing power and speed up the analysis, it is advisable to select "Segment only a region of interest" and "process entire image finally." This allows the experimenter to select a few frames of the movie to prepare the analysis, and finally extend this to the whole movie. Once the spots are generated it is advisable to rename them to match the appropriate fluorescent protein.
2. Import both sets of spots to vesicles as in Section 2.2.1. This will generate two populations of mRFP-ATG9 and GFP-LC3 vesicles associated with the cell of interest.
3. In the cells tab, click "Export vesicles to spots" and ensure this is set to "all time points." This will generate two new types of spots (one for each vesicle type), specifically associated with our cell of interest. These can then be analyzed using the "Distance transformation" plugin.
4. Before using Distance transformation, select the "Edit" menu on the toolbar at the top of the Imaris window, choose "Change data type" and switch the data to "32 bit float."
5. Select the mRFP-ATG9 spots generated in step 3. In the dialog box that opens, select the Tools menu and click on "Distance transformation." This will take a few moments to calculate the values. Once the calculation is complete, a new channel will appear as a blue shading across the image. This shading represents the distance of each pixel from the nearest mRFP-ATG9 spot.
6. Select the GFP-LC3 spots. Under the "Statistics" tab, select "Intensity Centre Channel 3"—the number may vary depending on the number of channels originally in the movie, but it will be the highest number.
7. This will give a list of values for each frame. These represent the distance of each GFP-LC3 spot to the nearest mRFP-ATG9 spot. The total list of values can be saved by clicking the "Export all statistics to file" tab.
8. Open the file named "Intensity Centre Channel 3" using Microsoft Excel®. This is a list of all the GFP-LC3 vesicles in each frame and the value column lists their distance from the nearest mRFP-ATG9 vesicle. These can be analyzed to find the total number of instances when GFP-LC3 and mRFP-ATG9 vesicles have come into contact—this can be achieved by setting a numerical threshold below which the vesicles

are assumed to be in contact, usually less than 1 μm. The number of "contact events" can then be normalized against the maximum number of GFP-LC3 spots in the cell during the timecourse.

ACKNOWLEDGMENTS

We would like to thank Matt Renshaw and Peter Jordan (Francis Crick Institute Light Microscopy Facility) and Delisa Garcia (Bitplane) for assistance developing these protocols. This work was supported by Cancer Research UK and the Francis Crick Institute which receives its core funding from Cancer Research UK, the UK Medical Research Council, and the Wellcome Trust.

REFERENCES

Axe, E. L., Walker, S. A., Manifava, M., Chandra, P., Roderick, H. L., Habermann, A., … Ktistakis, N. T. (2008). Autophagosome formation from membrane compartments enriched in phosphatidylinositol 3-phosphate and dynamically connected to the endoplasmic reticulum. *The Journal of Cell Biology*, *182*(4), 685–701. http://dx.doi.org/10.1083/jcb.200803137.

Birgisdottir, Å., Lamark, T., & Johansen, T. (2013). The LIR motif—Crucial for selective autophagy. *Journal of Cell Science*, *126*(Pt. 15), 3237–3247. http://dx.doi.org/10.1242/jcs.126128.

Dooley, H. C., Razi, M., Polson, H. E., Girardin, S. E., Wilson, M. I., & Tooze, S. A. (2014). WIPI2 links LC3 conjugation with PI3P, autophagosome formation, and pathogen clearance by recruiting Atg12-5-16L1. *Molecular Cell*, *55*(2), 238–252. http://dx.doi.org/10.1016/j.molcel.2014.05.021.

Kabeya, Y., Mizushima, N., Yamamoto, A., Oshitani-Okamoto, S., Ohsumi, Y., & Yoshimori, T. (2004). LC3, GABARAP and GATE16 localize to autophagosomal membrane depending on form-II formation. *Journal of Cell Science*, *117*(Pt. 13), 2805–2812. http://dx.doi.org/10.1242/jcs.01131.

Klionsky, D. J., Abdelmohsen, K., Abe, A., Abedin, M. J., Abeliovich, H., Acevedo Arozena, A., … Zughaier, S. M. (2016). Guidelines for the use and interpretation of assays for monitoring autophagy (3rd edition). *Autophagy*, *12*(1), 1–222. http://dx.doi.org/10.1080/15548627.2015.1100356.

Knævelsrud, H., Søreng, K., Raiborg, C., Håberg, K., Rasmuson, F., Brech, A., … Simonsen, A. (2013). Membrane remodeling by the PX-BAR protein SNX18 promotes autophagosome formation. *The Journal of Cell Biology*, *202*(2), 331–349. http://dx.doi.org/10.1083/jcb.201205129.

Lamb, C. A., Nühlen, S., Judith, D., Frith, D., Snijders, A. P., Behrends, C., & Tooze, S. A. (2016). TBC1D14 regulates autophagy via the TRAPP complex and ATG9 traffic. *The EMBO Journal*, *35*(3), 281–301. http://dx.doi.org/10.15252/embj.201592695.

Longatti, A., Lamb, C. A., Razi, M., Yoshimura, S., Barr, F. A., & Tooze, S. A. (2012). TBC1D14 regulates autophagosome formation via Rab11- and ULK1-positive recycling endosomes. *The Journal of Cell Biology*, *197*(5), 659–675. http://dx.doi.org/10.1083/jcb.201111079.

Mari, M., Griffith, J., Rieter, E., Krishnappa, L., Klionsky, D. J., & Reggiori, F. (2010). An Atg9-containing compartment that functions in the early steps of autophagosome biogenesis. *The Journal of Cell Biology*, *190*(6), 1005–1022. http://dx.doi.org/10.1083/jcb.200912089.

Orsi, A., Razi, M., Dooley, H. C., Robinson, D., Weston, A. E., Collinson, L. M., & Tooze, S. A. (2012). Dynamic and transient interactions of Atg9 with autophagosomes, but not membrane integration, are required for autophagy. *Molecular Biology of the Cell*, *23*(10), 1860–1873. http://dx.doi.org/10.1091/mbc.E11-09-0746.

Polson, H. E., de Lartigue, J., Rigden, D. J., Reedijk, M., Urbé, S., Clague, M. J., & Tooze, S. A. (2010). Mammalian Atg18 (WIPI2) localizes to omegasome-anchored phagophores and positively regulates LC3 lipidation. *Autophagy*, *6*(4), 506–522. http://dx.doi.org/10.4161/auto.6.4.11863.

Puri, C., Renna, M., Bento, C. F., Moreau, K., & Rubinsztein, D. C. (2013). Diverse autophagosome membrane sources coalesce in recycling endosomes. *Cell*, *154*(6), 1285–1299. http://dx.doi.org/10.1016/j.cell.2013.08.044.

Young, A. R., Chan, E. Y., Hu, X. W., Köchl, R., Crawshaw, S. G., High, S., & Tooze, S. A. (2006). Starvation and ULK1-dependent cycling of mammalian Atg9 between the TGN and endosomes. *Journal of Cell Science*, *119*(Pt. 18), 3888–3900. http://dx.doi.org/10.1242/jcs.03172.

CHAPTER THREE

The Use of DQ-BSA to Monitor the Turnover of Autophagy-Associated Cargo

L.S. Frost, A. Dhingra, J. Reyes-Reveles, K. Boesze-Battaglia[1]
SDM, University of Pennsylvania, Philadelphia, PA, United States
[1]Corresponding author: e-mail address: battagli@upenn.edu

Contents

1. Introduction 44
2. Materials 45
3. Establishment of Polarized Epithelial Cell Cultures 47
4. Incorporation of DQ™-BSA Conjugates 47
5. Monitoring Autolysosome Formation 47
   5.1 Rapamycin Stimulation 48
   5.2 Serum Starvation 48
   5.3 Evaluation of Autolysosome Formation 48
6. Monitoring LC3-Associated Phagolysosome Formation 50
7. Immunofluorescence Analysis 51
   7.1 Labeling Endogenous LC3 51
   7.2 Confocal Imaging 51
   7.3 Important Considerations in Experimental Design 51
8. Summary 52
Acknowledgments 53
References 53

Abstract

There is increasing evidence documenting the critical role played by autophagic and autophagy-associated processes in maintaining cell homeostasis and overall systemic health. Autophagy is considered a degradative as well as a recycling pathway that relies on encapsulated intracellular components trafficking to and fusing with degradative compartments, including lysosomes. In this chapter, we describe the use of DQ™-BSA to study autophagosome–lysosome fusion as well as a means by which to analyze hybrid autophagic pathways. Such noncanonical pathways include LC3-associated phagocytosis, better known as LAP. Both autophagosomes and LAPosomes (LC3-associated phagosomes) deliver cargo for degradation. The use of fluorescent DQ™-BSA in conjugation with autophagic makers and biomarkers of hybrid autophagy offers a reliable technique to monitor the formation of autolysosomes and LAPo-lysosomes in both fixed- and

Methods in Enzymology, Volume 587
ISSN 0076-6879
http://dx.doi.org/10.1016/bs.mie.2016.09.052

live-cell studies. This technique relies on cleavage of the self-quenched DQ™ Green- or DQ™ Red BSA protease substrates in an acidic compartment to generate a highly fluorescent product.

1. INTRODUCTION

The cell turns over its own constituents in a regulated manner to maintain homeostasis, in a process called autophagy, derived from the Greek word for "self-eating." Numerous components of the autophagic pathway are multifunctional, forming a nexus of cross talk with other cellular processes to help define cell type-specific functions of autophagy and processes associated with autophagy-related proteins. Such processes include metabolism, hybrid phagocytosis, membrane transport, and host defense strategies.

Degradation of both intracellular and extracellular material is a crucial function of phagocytic cells. Two seemingly independent engulfment pathways, phagocytosis and autophagy, deliver material to lysosomes for degradation. Macroautophagy, hereafter referred to as autophagy, is the major catabolic pathway required for the lysosome/vacuolar degradation of cytoplasmic proteins and organelles. Through canonical autophagy, intracellular substrates are enwrapped as cargo by double membrane structures known as autophagosomes; thus allowing for bulk turnover of cytoplasmic components, enabling among other functions, the survival of nutrient-deprived cells. Upon nutrient deprivation, a serine-threonine kinase, UNC-51-like kinase (ULK), is released from its mammalian target of rapamycin (mTOR)-mediated inhibition (Jung et al., 2009; Kim, Kundu, Viollet, & Guan, 2011); in concert with the class III phosphatidylinositol-3 kinase, Vps-34 complexed with Beclin1, it recruits and activates components of the Atg5/12/16L conjugation system (Funderburk, Wang, & Yue, 2010). The Atg5/12/16L multimeric complex regulates LC3 lipidation by phosphatidylethanolamine to form lipidated LC3, called LC3II. LC3II is necessary for autophagosome elongation and closure in cargo engulfment. LC3-containing autophagosomes subsequently fuse with lysosomes in an apparent LC3II-dependent manner to facilitate degradation of intracellular components (Esclatine, Chaumorcel, & Codogno, 2009; Jahreiss, Menzies, & Rubinsztein, 2008). Details of autophagy-associated proteins and their specific functions are reviewed in Klionsky et al. (2016).

Not all LC3-containing intracellular vesicles are autophagosomes; phagosomes can be targeted by autophagy proteins in an autophagosome-independent manner. In some epithelial cells and macrophages, phagocytosis activates the Vps34/beclin1 and Atg5/12/16L conjugation systems resulting in lipidation of LC3 directly onto the single membrane (nascent) phagosome (Florey, Kim, Sandoval, Haynes, & Overholtzer, 2011; Florey & Overholtzer, 2012). In this hybrid pathway, commonly known as LC3-associated phagocytosis (LAP), the LC3II-decorated phagosome fuses with lysosomes for degradation. Autophagosome-independent, LC3-associated degradative events exhibit common themes that define the process of LAP: LC3 recruitment to phagosomes occurs under nutrient replete conditions in which mTOR is active with canonical autophagy inhibited. LAP is however dependent on Vps34/beclin1 and Atg5/12/16L. It is becoming increasingly evident that LAP processes are a means by which phagocytes monitor their contents to ensure complete degradation of ingested materials. LAP requiring processes include phagocytosis of dead cells (Martinez et al., 2011), the degradation of photoreceptors, and recycling of visual pigments (Frost et al., 2015; Frost, Mitchell, & Boesze-Battaglia, 2014; Kim et al., 2013) and pathogen degradation (Sanjuan et al., 2007). Most recently, LAP has been shown to inhibit the autoimmune response (Martinez et al., 2016).

Independent of whether LC3 decorates the inner and outer bilayers of double membrane autophagosomes or just the outside of ingested phagosomes, the ultimate fate of the contents of these structures is degradation upon fusion with lysosomes. Historically, the function of degradative compartments was often measured as proteolytic activity; highly substituted fluorescein conjugated derivatives of proteins served as biomarkers of a cell's degradative efficiency and capacity. A limitation of such fluorescein derivatives is their pH sensitivity over physiological lysosomal pH ranges; they provide limited detection below pH 8.0. Newer fluorescent compounds have emerged including the BODIPY© family of probes which are insensitive to pH changes from pH 3 to 11. Here we provide a method by which the bovine serum albumin derivative DQ™-BSA is utilized to detect the formation of the autolysosomes and the quantitation of degradative cargo using confocal imaging.

2. MATERIALS

The DQ™ BSA conjugate is a derivative of BSA that is labeled to such a high degree with either the green-fluorescent BODIPY® FL or the red-fluorescent BODIPY® TR-X dye that the fluorescence is self-quenched. Spectral properties of DQ™ Green and DQ™ Red BSA allow for extensive utility in a variety of applications. They both provide low background fluorescence and provide high signal-to-noise ratios upon digestion. The cleavage of DQ™ Green BSA results in the release of fragments which have $\lambda_{ex} = 505$ nm and $\lambda_{em} = 515$ nm. In the case of DQ™ Red BSA the proteolytic fragments have $\lambda_{ex} = 590$ nm and $\lambda_{em} = 620$ nm. As the DQ™-BSA enters an acidic environment proteases cleave the previously quenched polypeptide to generate fluorescent peptide fragments as illustrated schematically in Fig. 1.

DQ-BSA probes are available as a lyophilized powder that if protected from light are stable for up to 6 months at −20°C. After reconstitution, the solution is stable for 3–4 weeks at 4°C (protected from light; wrap vial in foil). We routinely reconstitute with sterile PBS.

The methods below describe the use of DQ™ Green BSA in the visualization and semiquantitation of autolysosome formation as well as the delivery of LC3-associated cargo to lysosomes (Frost et al., 2015). Ha et al. (2010), utilized DQ™ Red BSA to measure the role of autophagic flux in Anthrax Toxin lethal factor delivery using confocal imaging and FACS. Both approaches effectively measure the convergence between an autophagic or autophagocytic compartment and a functional lysosome; in the case of ingested phosphatidylserine positive photoreceptor outer segments (POS), the autophagy-associated process is LAP (LC3-phagolysosomes) while for anthrax toxin it is autolysosome formation. Here we describe DQ™-BSA's use for autolysosomes as well as a modification in which we measure LAP.

The individual steps for monitoring autolysosome and LC3-phagolysosome formation are described later. This protocol can be simplified for nonpolarized cells by ignoring steps in Section 3.

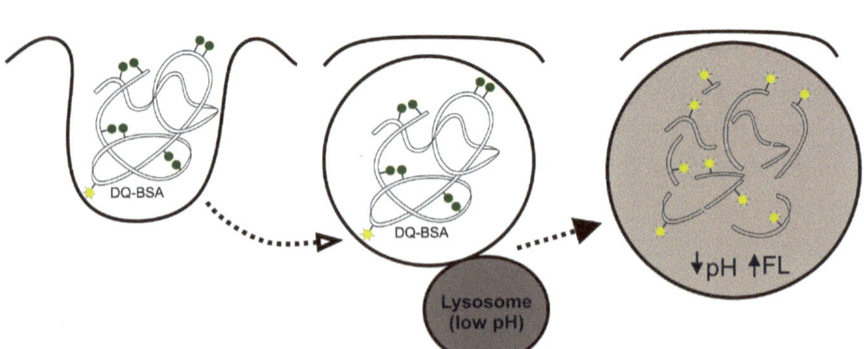

Fig. 1 Schematic representation of self-quenched DQ™ Green BSA upon cleavage by proteases in acidic compartments. Proteolysis results in fragment formation with fluorescence dequenching that is observed as an increase in fluorescence intensity.

3. ESTABLISHMENT OF POLARIZED EPITHELIAL CELL CULTURES

1. ARPE19 cells (ATCC CRL-2302) or human fetal RPE (hfRPE) cells are seeded on 12-well Transwell filters with 0.4 μ*M* pore size (Corning 3460, Oneonta, NY) at an initial seeding density of 1.6×10^5 cells/well in the apical chamber. Routinely, 0.5 and 1.5 mL of DMEM/F12 (1:1) (Gibco 11330032) with 10% FBS is added to the apical and basal chambers, for ARPE19 cells, while hfRPE cells utilized Advanced MEM (Gibco 12492013) with 1% pen/strep (Gibco 15140122, Grand Island, NY), 1% glutamine (Corning 250054), 125 mg taurine (Sigma T0625), hydrocortisone (Tocris Bioscience 4093), 0.0065 μg triiodo-thyronin (Sigma T5516) with 5% FBS. Cells are subsequently incubated at 37°C and 5% CO_2. Frost et al. (2015) and Adijanto et al. (2014) provide detailed protocols for ARPE19 and hfRPE cultures, respectively.
2. After initial plating, the formulation of the media is changed to 1% FBS, and cells are fed twice a week.
3. To confirm barrier function, the transepithelial resistance (TER) of mature cells is measured using an Epithelial Volt–Ohm Meter (EVOM2). Upon cell maturity, usually within 4 weeks, the TER is routinely $\sim 50\ \Omega\ cm^2$ for ARPE19 cells and $\leq 500\ \Omega\ cm^2$ for hfRPE.

4. INCORPORATION OF DQ™-BSA CONJUGATES

1. Cells at 65–85% confluence are incubated with 10 μg/mL DQ™-BSA in complete media at 37°C for 30–60 min. When using polarized cells, DQ™-BSA is added to the apical chamber and incubated at 37°C for 1 h. DQ™ Green BSA (D12050) and DQ™ Red BSA (12051) are both obtained from Molecular Probes.
2. Excess probe is removed by washing cells three times with PBS containing 0.5 m*M* $MgCl_2$ and 0.9 m*M* $CaCl_2$ (PBS-CM, Thermo-Fisher, Cat# 14040133).
3. Autolysosome formation can be monitored by confocal imaging as indicated later.

5. MONITORING AUTOLYSOSOME FORMATION

Monitoring of autolysosome formation requires that the DQ™-BSA traffic to lysosomes where acidic proteases generate a highly fluorescent product. Autolysosomes are defined herein as intracellular structures in which there is colocalization of cleaved DQ™-BSA fragments (a marker of lysosomes) with fluorescently tagged-LC3 or endogenous LC3 (a marker of autophagic structures).

1. To identify autophagosomes or LAPosomes, fluorescently labeled LC3 can be expressed using transient transfection techniques. In addition, numerous cells lines have been developed that stably express GFP-LC3 or mCHERRY-LC3. In addition, several mouse models express eGFP-LC3 or mCHERRY-LC3 (Iwai-Kanai et al., 2008). GFP-LC3 mice are available from RIKEN BioResource Center (RBRC No. 00806, Japan), and mCHERRY-LC3 mice are available from Jackson Laboratories (Stock No. 021853). In addition, a tandem tagged LC3 mouse is available from Jackson Laboratories (Stock No. 027139) (Li, Wang, Hill, & Lin, 2014). Endogenous LC3 is labeled as described in more detail later (Section 7.1).
2. When monitoring autolysosomes using transiently transfected, fluorescently tagged LC3, transfections should be performed 24–36 h prior to DQ™-BSA addition.
3. Autophagosome formation can be stimulated using several different experimental protocols depending on the nature of the studies proposed; we will describe pharmacologic stimulation with rapamycin and serum starvation.

5.1 Rapamycin Stimulation

Rapamycin acts to remove the mTOR-mediated inhibition of the Ulk1 complex thereby resulting in an increase in autophagy. Cells are incubated with 100 n*M* rapamycin (Enzo Life Sciences, BML-A275-0005) for 4 or 24 h and subsequently incubated with DQ™-BSA, follow steps 1–4 in Section 5.3.

1. Rapamycin activity is confirmed as loss of phospho-S6 by immunoblot using anti-phosphoS6 (Cell Signaling, 2211s).

5.2 Serum Starvation

To remove media wash cells three times in PBS. To serum starve change to 0.25% FBS. To control cells, add complete culture media DMEM/F12 with 1% fetal bovine serum (Sigma, 12003C) and 5% Penicillin–Streptomycin (Sigma, P4333). Incubate cells at 37°C for 4 h.

5.3 Evaluation of Autolysosome Formation

1. Incubate control, rapamycin treated or serum-starved cells with DQ™-BSA as described earlier (Section 4) and wash three times with PBS.

2. For confocal image analysis, plate cells on coverslips prior to treatment and fix in 4% paraformaldehyde (PFA) after treatment.
3. The fluorescence degradation products of DQ™-BSA are imaged using standard techniques. A representative example of a multifluor confocal image is shown in Fig. 2A. In this example, imaging was performed

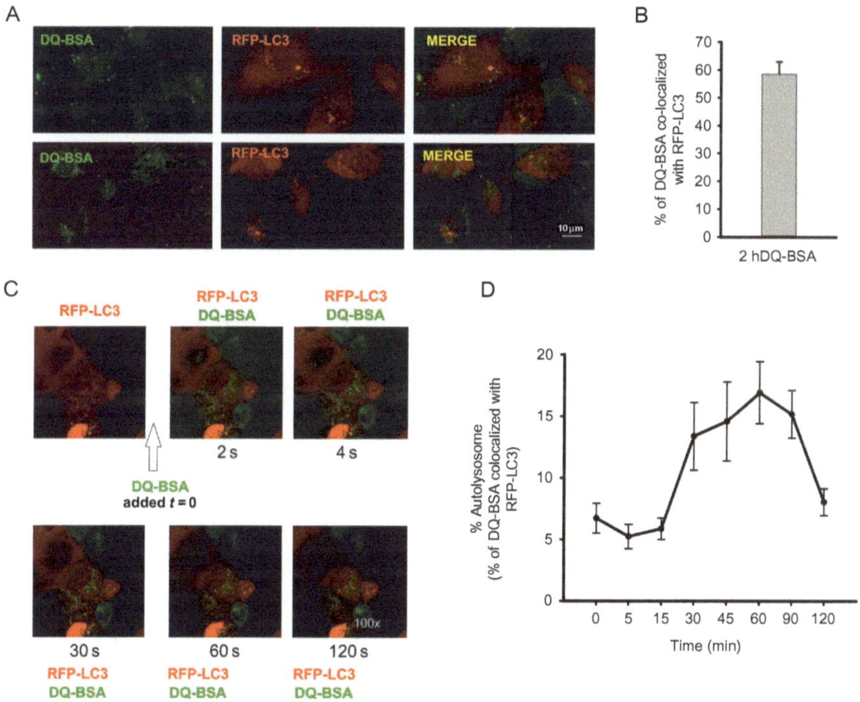

Fig. 2 Serum-starved ARPE19 cells accumulate autolysosomes. (A) RFP-LC3 transfected cells were serum starved for 4 h in 0.25% FBS. Starved cells were incubated with DQ™ Green BSA. Autolysosomes were identified as LC3-positive (*red*)—green DQ-containing structures, with a Pearson's coefficient cut-off of 0.55. (B) Percentage of total lysosomes representing autolysosomes is shown at 2 h after rapamycin treatment (100 n*M*). (C) Live-cell confocal imaging, serum-starved (for 4 h) RFP-LC3-transfected ARPE19 cells grown on coverslips were placed in a temperature-controlled stage and loaded with DQ™ Green BSA. The uptake of DQ™ Green BSA and subsequent fusion with RFP-LC3 autophagosome was recorded using time-lapse microscopy. Images were captured every 2 s, images represent DQ-BSA and RFP-LC3, at the indicated time points. (D) In another series of experiments, the extent of colocalization between DQ-BSA and RFP-LC3 was determined exactly as described in (C) except cells were serum starved for 1 h. The results are mean ± SEM of five cells in each of three individual live-cell assays, with a Pearson's coefficient cut-off of 0.55.

on a Nikon A1R confocal microscope and images analyzed using Nikon Elements software 4.1. In codistribution analyses with LC3, cells are imaged and codistribution analyzed using a binary submask Pearson's coefficient cut-off of at least >0.55. The extent of autolysosome formation is indicated by DQ™-BSA colocalized with RFP-LC3 as shown for a 2-h incubation in Fig. 2B.

4. Alternatively, serum-starved cells are analyzed using live-cell confocal imaging. In this way, the uptake and localization of DQ™-BSA is measured in real-time in combination with fluorescent-LC3. The extent of RFP-LC3-DQ™-BSA uptake observed by live-cell confocal imaging, beginning at 2 min is depicted in Fig. 2C.

6. MONITORING LC3-ASSOCIATED PHAGOLYSOSOME FORMATION

The generation of highly fluorescent proteolytic products in lysosomes allows us to monitor the delivery of autophagy-dependent cargo to these organelles. In the example below, we provide a semiquantitative method by which to analyze a hybrid phagocytosis–autophagy-dependent process; LC3-associated phagocytosis known as LAP. Here LC3-phagolysosomes are defined as intracellular structures in which there is a triple colocalization between cleaved DQ™-BSA fragments, fluorescently tagged-LC3 or endogenous LC3 and fluorescently labeled phagosomes (Frost et al., 2015).

1. Incubate polarized RPE cells with 10 μg/mL DQ™-BSA in the apical chamber for 1 h at 37°C.
2. Wash three times in media prior to the addition of fluorescently labeled outer segments or other phagocytosed material, i.e., latex beads, zymogens, etc. In this example, we add purified POS, the endogenous particles phagocytosed by the RPE.
3. Add Alexa Fluor 647-labeled POS (AF647-POS) at a density of ~10 particles per cell directly to the media on the apical side for 2 h at 37°C.
4. After 2 h quench the unincorporated AF647-POS with 0.2% Trypan Blue (Corning, 25-900-cl) for 10 min at 37°C.
5. Wash three times in PBS-CM.
6. Fix inserts in 4% PFA, wash in PBS, and mount or process for immunofluorescence staining.

7. IMMUNOFLUORESCENCE ANALYSIS

7.1 Labeling Endogenous LC3

1. Permeabealize inserts with ice-cold methanol at −20°C for 10 min.
2. Wash with PBS and incubate in PBS for 30 min to rehydrate.
3. Incubate inserts in 4% BSA (Sigma A7284) in PBS containing 0.05% Triton TX-100 (PBST) for 60 min at 37°C to block prior to labeling with appropriate antibodies.
4. After blocking, incubate inserts at 37°C for 60 min with anti-LC3 rabbit polyclonal antibody 1:150 (Cell Signaling, Catalog no. 2775s).
5. Wash inserts three times in PBST.
6. Incubate with secondary antibody, Alexa Fluor 594 donkey anti-rabbit IgG (Invitrogen R37119) at 1:1000 and Hoechst 33258 (Thermo-Fisher, H3569) at 1:10,000 (37°C for 60 min) and wash three times in PBS.

7.2 Confocal Imaging

1. In preparation for microscopy, cut the filters from the inserts and mount in Cytoseal (Electron Microscopy Sciences, 18006).
2. For polarized cell analysis, capture *Z*-stack images of 1 μm apical to basolateral sections on a Nikon A1R laser scanning confocal or comparable microscope with a 60 × water objective (NA 1.2).
3. A representative example of the triple colocalization profiles of Texas-Red (TR)-OS, LC3, and DQ™ Green BSA is shown in Fig. 3A, with subsequent quantitation of OS association with DQ™ Green BSA-LC3 positive structures in Fig. 3B.

7.3 Important Considerations in Experimental Design

1. The internalization and transport of DQ™-BSA to the lysosomes is compromised in the presence of PI3K inhibitors, such as wortmanin and upon the overexpression of motor proteins and Rabs (dominant negative mutants) as well as upon alterations in phospholipid phosphatases, including SH-2-containing inositol 5′-polyphosphatase 1 (SHIP1) and phosphatase and tensin homolog (PTEN). To avoid inefficient localization of DQ™-BSA, cells should be incubated overnight with DQ™-BSA prior to pharmacologic manipulation. However, if DQ-BSA is used as an indicator of endolysosomal trafficking the levels of protein(s) or toxin of interest should be altered prior to DQ™-BSA. A detailed description of such as approach is provided (Corrotte, Fernandes, Tam, & Andrews, 2012).

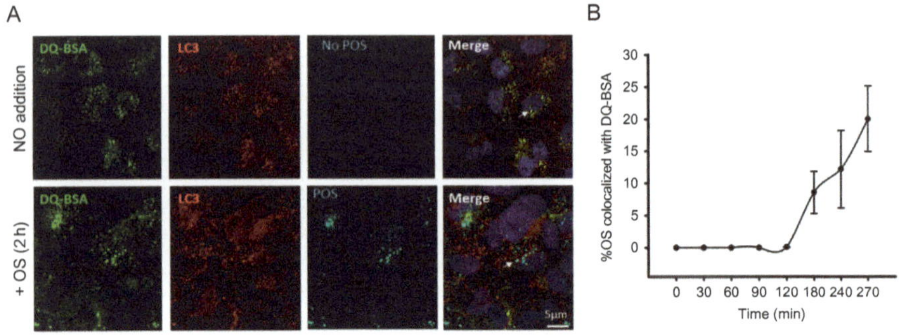

Fig. 3 LC3-OS-lysosome association. (A) DQ™ Green BSA containing ARPE19 cells were fed Alexa Fluor 647-labeled POS at a density of 10 particles per cell directly to the media in the apical chamber for 2 h at 37°C. Phagocytosis was terminated and extracellular fluorescence quenched. Cells were fixed and stained for endogenous LC3. A representative image of each channel is shown and triple colocalization indicated with *arrows*. The average size of the particles quantified as LAPo-lysosomes was 0.65–0.89 μm. Scale bar is 5 μm. (B) Live-cell confocal imaging, ARPE19 cells grown on coverslips were loaded with DQ™ Green BSA. The coverslips, placed in a temperature-controlled stage, were fed Alexa Fluor 647-labeled POS and images recorded every 2 s for 4.5 h. The extent of colocalization was determined at each indicated time point 422 with Pearson's coefficient >0.55.

2. Treatment with inhibitors, including chloroquine or bafilomycin A, a v-ATPase inhibitor, that block endosomal acidification or autophagosome–lysosome fusion led to a strong reduction in DQ-BSA fluorescence confirming that cleavage of this dye requires residence in acidic compartments (Corrotte et al., 2012; Goeritzer et al., 2015).
3. A caveat to interpretation of results with DQ™-BSA is that this probe may be dequenched in late endosomes (Authier, Posner, & Bergeron, 1996; Goebeler, Poeter, Zeuschner, Gerke, & Rescher, 2008). Given the dynamic nature of the late-endosome to lysosome conversion, we recommend several complementary approaches that utilize both functional convergence and organelle biomarker identification, including cathepsin D, Lysotracker, or LAMP1 (Frost et al., 2015). Complementary LC3-DQ™-BSA codistribution analysis can be performed using FACS (Ha et al., 2010).

8. SUMMARY

We describe basic protocols to assay the last step of autophagy, one shared with LAP, the fusion of an LC3 positive compartment with a degradative lysosome. Incorporation of the fluorogenic DQ™ Green or Red BSA probe is an easy and fast method by which to monitor starvation or pharmacologically induced autophagy in fixed or live cells. Although the colocalization of phagocytosed particles with LC3 assesses the formation of a LAPosome, it does not address whether that structure then appropriately traffics to a lysosome for degradation. Therefore, detecting the colocalization of LC3-associated cargo with DQ™-BSA is the one of the most reliable methods for monitoring fusion between a LAPosome and an active proteolytic compartment, either late-endosome or lysosome. This method is in many ways superior to colocalization studies that rely only on the presence of lysosomal enzyme. Typically, such immunofluorescence studies do not assess the compartment's function, as the presence of a protease, for example, cathepsin D does not necessarily confirm that the compartment is functional. The use of DQ™-BSA circumvents this problem since fluorescence is only observed in a proteolytically active compartment.

ACKNOWLEDGMENTS

This work was supported in whole or in part by the National Institutes of Health Grants EY10420, EY026525, and DE022465 (to K.B.B.) and P30EY00331 (Penn-Vision Core). The authors would like to acknowledge the use of the PDM live-cell imaging core.

REFERENCES

Adijanto, J., Du, J., Moffat, C., Seifert, E. L., Hurle, J. B., & Philp, N. J. (2014). The retinal pigment epithelium utilizes fatty acids for ketogenesis. *Journal of Biological Chemistry*, *289*(30), 20570–20582.

Authier, F., Posner, B. I., & Bergeron, J. J. (1996). Endosomal proteolysis of internalized proteins. *FEBS Letters*, *389*(1), 55–60.

Corrotte, M., Fernandes, M. C., Tam, C., & Andrews, N. W. (2012). Toxin pores endocytosed during plasma membrane repair traffic into the lumen of MVBs for degradation. *Traffic*, *13*(3), 483–494.

Esclatine, A., Chaumorcel, M., & Codogno, P. (2009). Macroautophagy signaling and regulation. *Current Topics in Microbiology and Immunology*, *335*, 33–70.

Florey, O., Kim, S. E., Sandoval, C. P., Haynes, C. M., & Overholtzer, M. (2011). Autophagy machinery mediates macroendocytic processing and entotic cell death by targeting single membranes. *Nature Cell Biology*, *13*(11), 1335–1343.

Florey, O., & Overholtzer, M. (2012). Autophagy proteins in macroendocytic engulfment. *Trends in Cell Biology*, *22*(7), 374–380.

Frost, L. S., Lopes, V. S., Bragin, A., Reyes-Reveles, J., Brancato, J., Cohen, A., … Boesze-Battaglia, K. (2015). The contribution of melanoregulin to microtubule-associated protein 1 light chain 3 (LC3) associated phagocytosis in retinal pigment epithelium. *Molecular Neurobiology*, *52*(3), 1135–1151.

Frost, L. S., Mitchell, C. H., & Boesze-Battaglia, K. (2014). Autophagy in the eye: Implications for ocular cell health. *Experimental Eye Research*, *124*, 56–66.

Funderburk, S. F., Wang, Q. J., & Yue, Z. (2010). The Beclin 1-VPS34 complex—At the crossroads of autophagy and beyond. *Trends in Cell Biology*, *20*(6), 355–362.
Goebeler, V., Poeter, M., Zeuschner, D., Gerke, V., & Rescher, U. (2008). Annexin A8 regulates late endosome organization and function. *Molecular Biology of the Cell*, *19*(12), 5267–5278.
Goeritzer, M., Vujic, N., Schlager, S., Chandak, P. G., Korbelius, M., Gottschalk, B., … Kratky, D. (2015). Active autophagy but not lipophagy in macrophages with defective lipolysis. *Biochimica et Biophysica Acta*, *1851*(10), 1304–1316.
Ha, S. D., Ham, B., Mogridge, J., Saftig, P., Lin, S., & Kim, S. O. (2010). Cathepsin B-mediated autophagy flux facilitates the anthrax toxin receptor 2-mediated delivery of anthrax lethal factor into the cytoplasm. *Journal of Biological Chemistry*, *285*(3), 2120–2129.
Iwai-Kanai, E., Yuan, H., Huang, C., Sayen, M. R., Perry-Garza, C. N., Kim, L., & Gottlieb, R. A. (2008). A method to measure cardiac autophagic flux in vivo. *Autophagy*, *4*(3), 322–329.
Jahreiss, L., Menzies, F. M., & Rubinsztein, D. C. (2008). The itinerary of autophagosomes: From peripheral formation to kiss-and-run fusion with lysosomes. *Traffic*, *9*(4), 574–587.
Jung, C. H., Jun, C. B., Ro, S. H., Kim, Y. M., Otto, N. M., Cao, J., … Kim, D. H. (2009). ULK-Atg13-FIP200 complexes mediate mTOR signaling to the autophagy machinery. *Molecular Biology of the Cell*, *20*(7), 1992–2003.
Kim, J., Kundu, M., Viollet, B., & Guan, K. L. (2011). AMPK and mTOR regulate autophagy through direct phosphorylation of Ulk1. *Nature Cell Biology*, *13*(2), 132–141.
Kim, J. Y., Zhao, H., Martinez, J., Doggett, T. A., Kolesnikov, A. V., Tang, P. H., … Ferguson, T. A. (2013). Noncanonical autophagy promotes the visual cycle. *Cell*, *154*(2), 365–376.
Klionsky, D. J., Abdelmohsen, K., Abe, A., Abedin, M. J., Abeliovich, H., Acevedo Arozena, A., … Zughaier, S. M. (2016). Guidelines for the use and interpretation of assays for monitoring autophagy (3rd edition). *Autophagy*, *12*(1), 1–222.
Li, L., Wang, Z. V., Hill, J. A., & Lin, F. (2014). New autophagy reporter mice reveal dynamics of proximal tubular autophagy. *Journal of the American Society of Nephrology*, *25*(2), 305–315.
Martinez, J., Almendinger, J., Oberst, A., Ness, R., Dillon, C. P., Fitzgerald, P., … Green, D. R. (2011). Microtubule-associated protein 1 light chain 3 alpha (LC3)-associated phagocytosis is required for the efficient clearance of dead cells. *Proceedings of the National Academy of Sciences of the United States of America*, *108*(42), 17396–17401.
Martinez, J., Cunha, L. D., Park, S., Yang, M., Lu, Q., Orchard, R., … Green, D. R. (2016). Noncanonical autophagy inhibits the autoinflammatory, lupus-like response to dying cells. *Nature*, *533*(7601), 115–119.
Sanjuan, M. A., Dillon, C. P., Tait, S. W., Moshiach, S., Dorsey, F., Connell, S., … Green, D. R. (2007). Toll-like receptor signalling in macrophages links the autophagy pathway to phagocytosis. *Nature*, *450*(7173), 1253–1257.

CHAPTER FOUR

Turnover of Lipidated LC3 and Autophagic Cargoes in Mammalian Cells

M. Rodríguez-Arribas*,†, S.M.S. Yakhine-Diop*,†, R.A. González-Polo*,†, M. Niso-Santano*,†,1, J.M. Fuentes*,†,1

*Centro de Investigación Biomédica en Red en Enfermedades Neurodegenerativas (CIBERNED), Universidad de Extremadura, Cáceres, Spain

†Facultad de Enfermería y Terapia Ocupacional, Universidad de Extremadura, Cáceres, Spain

[1]Corresponding authors: e-mail address: mireianiso@hotmail.com; jfuentes@unex.es

Contents

1. Introduction 56
2. Materials 58
   2.1 Cell Lines 58
   2.2 Reagents 59
   2.3 Antibodies 60
   2.4 Solutions 60
   2.5 Equipment 61
   2.6 Labware 62
3. Cell Culture, Treatments, and Sample Collection 62
   3.1 Cell Culture 62
   3.2 Treatments 63
   3.3 Sample Preparation 63
4. Electrophoresis, Western Blot, and Data Analysis 64
   4.1 Electrophoresis 64
   4.2 Western Blotting 64
   4.3 Interpretation 66
5. Notes 67
Acknowledgments 69
References 69

Abstract

Macroautophagy (usually referred to as *autophagy*) is the most important degradation system in mammalian cells. It is responsible for the elimination of protein aggregates, organelles, and other cellular content. During autophagy, these materials (i.e., cargo) must be engulfed by a double-membrane structure called an *autophagosome*, which delivers the cargo to the lysosome to complete its degradation. Autophagy is a very dynamic pathway called *autophagic flux*. The process involves all the steps that are

Methods in Enzymology, Volume 587
ISSN 0076-6879
http://dx.doi.org/10.1016/bs.mie.2016.09.053

implicated in cargo degradation from autophagosome formation. There are several techniques to monitor autophagic flux. Among them, the method most used experimentally to assess autophagy is the detection of LC3 protein processing and p62 degradation by Western blotting. In this chapter, we provide a detailed and straightforward protocol for this purpose in cultured mammalian cells, including a brief set of notes concerning problems associated with the Western-blotting detection of LC3 and p62.

1. INTRODUCTION

Autophagy, for the Greek term meaning "self-eating," is a highly dynamic, multistep process (Klionsky et al., 2016) that is conserved in eukaryotes. During autophagy, the intracellular contents within lysosomes are degraded to obtain energy or essential nutrients for cell homeostasis (Jiang & Mizushima, 2015; Yoshii & Mizushima, 2015). There are three main types of autophagy, classified according to the manner in which the cargo is internalized into lysosomes: macroautophagy, microautophagy, and chaperone-mediated autophagy (Cuervo & Wong, 2014). In macroautophagy (hereafter referred to as simply *autophagy*), the material to be degraded is engulfed by a double-membrane structure called an *autophagosome* that will fuse with lysosomes to degrade their cargo (fully reviewed in Feng, He, Yao, & Klionsky, 2014). Because growing evidence suggests that deregulation of autophagy may be implicated in several illnesses such as cancer (White, Mehnert, & Chan, 2015) and neurodegenerative (Wong & Cuervo, 2010) and inflammatory diseases (Sridhar, Botbol, Macian, & Cuervo, 2012), an accurate and comprehensive analysis of the results obtained must be performed to characterize the autophagic flux properly and avoid misinterpretation (Klionsky et al., 2016).

Accordingly, one of the most common analysis methods to assess autophagic behavior in mammalian models is to quantify the LC3 levels by immunoblotting. The microtubule-associated protein MAP1LC3 (hereafter referred to as *LC3*), is probably the most studied mammalian ortholog of the yeast *atg8* (Wesselborg & Stork, 2015). This protein has two isoforms: LC3-I, which is primarily cytosolic, and LC3-II, which is conjugated with phosphatidylethanolamine (PE). *Atg4* cleaves pro-LC3 to obtain LC3-I, which is conjugated with PE through the involvement of the *Atg3–Atg7* and *Atg5–Atg12* complexes (Coutts & La Thangue, 2016). This conjugated isoform is present in autophagosomes and autolysosome membranes and is degraded during the autophagy process. Thus, analyzing the LC3-I levels is a very meaningful method of monitoring the autophagic flux.

Accordingly, as a general pattern, the LC3-I level diminishes and the LC3-II level increases when the cells are in a short starvation period, which corresponds to an increase in autophagosome formation (Fig. 1) (Mizushima & Yoshimori, 2007). If the starvation period persists, both LC3 isoforms tend to disappear (Mizushima & Yoshimori, 2007). However, we have to be cautious when assessing mammalian cells because we can observe rapid degradation of LC3-II compared to the other isoform, depending on the model and tissue considered (Klionsky et al., 2016). Lysosomal degradation of the LC3-II isoform is avoided if lysosomal protease inhibitors such as *E64d* + pepstatin are used or if fusion is blocked between the autophagosomes and lysosomes using compounds such as chloroquine or bafilomycin A1 (Baf.A1). In this situation, their levels will increase, as was observed in several cell lines (Gomez-Sanchez et al., 2015; Jiang & Mizushima, 2015; Mizushima & Yoshimori, 2007). Because there are two different biological issues that provide an increase in LC3-II (i.e., enhanced autophagic flux and blockade

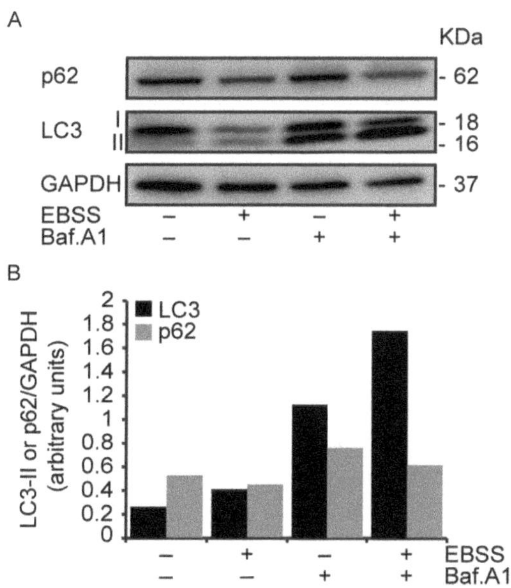

Fig. 1 Autophagy flux on MEFs WT (*ATG5*$^{+/+}$). Panel (A) represents the LC3-II, LC3-I, and p62 protein blots. The cells were treated with or without EBSS (autophagy inducer), Baf. A1 (autophagy blocker), or both for 4 h. Panel (B) shows the densitometry of a representative LC3-II and p62 blot. GAPDH is used as a loading control. Data are represented in arbitrary units. The molecular mass is indicated in kDa.

of autophagy), it is worthwhile to perform more analyses to obtain a better understanding of the autophagic flux of the sample. One good approach consists of performing the assay in the presence of lysosomal inhibitors. If the autophagic flux is enhanced in the presence of these compounds, we will observe further accumulation of LC3-II; if an increase in accumulation is not observed, the accumulation of LC3-II under basal conditions may originate from a blockade of lysosomal fusion rather than an increased autophagic flux (Mizushima & Yoshimori, 2007). In addition, this assay would determine the amount of LC3-II delivered to the lysosomes (Yoshii & Mizushima, 2015). The knock-down of several autophagic regulator proteins increased the LC3-II levels, suggesting the possible existence of an autophagy independent LC3-II synthesis mechanism (Jiang & Mizushima, 2015).

Another approach consists of monitoring a protein that is normally degraded during the autophagy process, thereby gaining a strong idea of whether complete autophagic flux occurs or whether there is a blockade during the steps. SQSTM1 (hereafter referred to as *p62*) is one of the most well known autophagy substrates. The substrate interacts directly with LC3 through an LC3-interacting region and with polyubiquitinated substrates through its UBA domain; thus, p62 acts as a bridge between LC3 and cargo targeted for degradation (Pankiv et al., 2007). Because p62 is primarily degraded during the autophagy process (Lippai & Low, 2014), its measurement can provide a better idea of the autophagic flux than analyzing LC3 alone. Accordingly, p62 is not degraded when lysosomal inhibitors are added to mouse embryonic fibroblasts (MEFs), N27 rat dopaminergic cells, or human fibroblasts (HFs) (Gomez-Sanchez et al., 2015, 2016; Jiang & Mizushima, 2015; Sahani, Itakura, & Mizushima, 2014) and even slightly accumulates in SH-SY5Y cells (Gomez-Sanchez et al., 2016). It is found that p62 tends to be degraded during short starvation periods (Gomez-Sanchez et al., 2015, 2016; Jiang & Mizushima, 2015; Sahani et al., 2014); however, its levels are restored through the upregulation of its expression after a starvation period of more than 4 h, at least in MEFs (Sahani et al., 2014).

2. MATERIALS

2.1 Cell Lines

1. MEFs wt ($ATG5^{+/+}$) and $ATG5$ deficient ($ATG5^{-/-}$).

2.2 Reagents

1. Antibiotic solution (10^4 U/mL of penicillin and 10^4 μg/mL of streptomycin) (Hyclone, GH Healthcare Live Sciences, Buckinghamshire, UK, SV30010).
2. Bafilomycin A1 (LC Laboratories, Woburn, Canada, B-1080).
3. Bicinchoninic acid (BCA) assay (Sigma-Aldrich, St. Louis, MO, USA, B9643) and copper (II) sulfate solution (Sigma-Aldrich, St. Louis, MO, USA, C2284).
4. Bovine serum albumin (BSA) (Sigma-Aldrich, St. Louis, MO, USA, A7906).
5. Bromophenol blue (Sigma-Aldrich, St. Louis, MO, USA, B126).
6. Chemiluminescent reagent: Pierce ECL WB substrate (Thermo Scientific, Waltham, MA, USA, 32106).
7. 3-(Cyclohexylamino)-1-propanesulfonic acid (CAPS) (Sigma-Aldrich, St. Louis, MO, USA, C2632).
8. Dulbecco's modified Eagle's medium (DMEM)-high glucose (Sigma-Aldrich, St. Louis, MO, USA, D6546).
9. Earle's balanced salt solution (EBSS) (Sigma-Aldrich, St. Louis, MO, USA, E2888).
10. Fetal bovine serum (FBS) (Sigma-Aldrich, St. Louis, MO, USA, F7524).
11. L-Glutamine (200 m*M*) (Sigma-Aldrich, St. Louis, MO, USA, G7513).
12. Glycerol (Panreac, Castellar del Vallès, Barcelona, Spain, 211339).
13. KCl (Panreac, Castellar del Vallès, Barcelona, Spain, 131494).
14. KH_2PO_4 (Panreac, Castellar del Vallès, Barcelona, Spain, 141509).
15. β-Mercaptoethanol (Sigma-Aldrich, St. Louis, MO, USA, M7522).
16. Methanol (Panreac, Castellar del Vallès, Barcelona, Spain, 131091).
17. Mini-PROTEAN TGX Precast Protein Gels (12%) (Bio-Rad, Hercules, CA, USA, 456-1043).
18. NaCl (Panreac, Castellar del Vallès, Barcelona, Spain, 141659).
19. Na_2PO_4 (Panreac, Castellar del Vallès, Barcelona, Spain, 141679).
20. Nonidet P40 (NP-40) (Roche, Basilea, Switzerland, 11332473001).
21. Phosphatase inhibitor: PhosSTOP inhibitor cocktail tablets (Roche, Basilea, Switzerland, 04906837001).
22. Polyvinylidene difluoride (PVDF) membranes (Bio-Rad, Hercules, CA, USA, 162-0177).
23. Ponceau S solution (0.1% (w/v) in 5% acetic acid) (Sigma-Aldrich, St. Louis, MO, USA, P7170).

24. Precision Plus Protein Dual Color Standard (Bio-Rad, Hercules, CA, USA, 161-0374).
25. Protease inhibitor: cOmplete, Mini, ethylenediaminetetraacetic acid (EDTA)–free tablets (Roche, Basilea, Switzerland, 11836170001).
26. Sodium dodecyl sulfate (SDS) (Bio-Rad, Hercules, CA, USA, 161-0311).
27. Tris base [Tris(hydroxymethyl)aminomethane] (Sigma-Aldrich, St. Louis, MO, USA, T6066).
28. Tris/glycine/SDS (TGS) 10 × (250 m*M* Tris, 1.92 *M* glycine, and 1% SDS, pH 8.3) (Bio-Rad, Hercules, CA, USA, 161-0772).
29. Tris/glycine 10 × (250 m*M* Tris and1.92 *M* glycine, pH 8.3) (Bio-Rad, Hercules, CA, USA, 161-0771).
30. Trypsin–EDTA (Sigma-Aldrich, St. Louis, MO, USA, T4049).
31. Tween 20 (polyethylene glycol sorbitan monolaurate) (Sigma-Aldrich, St. Louis, MO, USA, P1379).

2.3 Antibodies

1. HRP-conjugated goat antirabbit IgG (Bio-Rad, Hercules, CA, USA, 170-6515).
2. HRP-conjugated goat antimouse IgG (Bio-Rad, Hercules, CA, USA, 170-5047).
3. Monoclonal anti-p62 (SQSTM1 monoclonal antibody (M01), clone 2C11, (Abnova, Taipei, Taiwan, H00008878-M01).
4. Monoclonal antiglyceraldehyde 3-phosphate dehydrogenase (GAPDH) (Merck Millipore, Darmstadt, Germany, MAB374).
5. Polyclonal anti-LC3B (Sigma-Aldrich, St. Louis, MO, USA, L7543).

2.4 Solutions

1. Blocking/dilution antibodies solution.

   Dissolve 10 g of nonfat dried milk (10% w/v) in 100 mL of TBST solution (see below).
2. Cell culture media.

   The cell culture media for the MEFs is as follows: To obtain 500 mL of DMEM, remove 57 mL from the bottle and add the other constituent supplements with 50 mL (10% v/v) of FBS, 5 mL of L-glutamine (2 m*M* working concentration), and 2 mL of antibiotic solution (working solution 40 U/mL of penicillin and 40 μg/mL of streptomycin).

   The starvation medium is EBSS.

3. Electrophoresis buffer.
   1 × TGS diluted 10-fold from the commercial 10 × TGS with ddH_2O.
4. NP-40 lysis buffer.
   To make 10 mL, combine the following reagents: 0.05 mL of NP-40 (0.5% v/v), 1 mL of Tris–HCl (0.5 *M*, pH 8; 50 m*M*), and 0.3 mL of NaCl (5 *M*; 150 m*M*). Complete with 8.65 mL of double-distilled water (ddH_2O).
5. Phosphate-buffered saline (PBS) solution.
   Dissolve 8 g of NaCl, 0.2 g of KCl, 1.44 g of Na_2HPO_4, and 0.24 g of KH_2PO_4 in 800 mL of ddH_2O (working concentrations of 137, 2.7, 10, and 1.8 m*M*, respectively). Adjust the pH to 7.4 with HCl and then add H_2O to 1 L. Sterilize the solution by autoclaving for 20 min at 121°C.
6. Sample loading buffer (5 ×).
   To make 10 mL, combine 0.025 g of bromophenol blue (0.025% w/v), 1 g of SDS (10% w/v), 0.5 mL of β-mercaptoethanol (5% v/v), 2.5 mL of 1 *M* Tris–HCl (pH 6.8) (final concentration of 0.25 *M*), and 5 mL of glycerol (50% v/v). Complete to 10 mL with ddH_2O.
7. Transfer buffer.
   1 × CAPS/methanol, composed (per liter) of 2.21 g of CAPS in 700 mL ddH_2O (10 m*M*). Adjust the pH to 11 with concentrated NaOH, add 100 mL of methanol (10% v/v), and complete to 1 L with ddH_2O.
8. Tris buffer, 1 *M*, pH 6.8.
   Dissolve 121.14 g of Tris in ddH_2O. Adjust the pH to 6.8 with concentrated HCl and complete to 1 L with ddH_2O.
9. Tris-buffered saline.
   Dissolve 2.42 g of Tris and 8 g of NaCl in ddH_2O, adjust the pH to 7.4 with concentrated HCl, and complete to 1 L with ddH_2O.
10. Tris-buffered saline with Tween 20 (TBST).
    Dilute 2 mL of Tween 20 (0.2% v/v) in 0.98 L of Tris-buffered saline, pH 7.4.

2.5 Equipment

1. Autoclave (Raypa, Terrassa, Spain, Stericlav-S 150, 13330015).
2. DirectQ3UV (Merck-Millipore, Darmstadt, Germany) for ddH_2O production.

3. CO_2 incubator (Thermo Fisher Scientific, Waltham, MA, USA, Series 8000 water-jacketed).
4. Imaging system (Amersham Imager 600QC, GE Healthcare Live Sciences, Waltham, MA, USA, 29-0834-64).
5. Microplate reader TECAN SUNRISE (Tecan, Männedorf, Zürich, Switzerland, F039300).
6. Power supply (Power Pac HC, Bio-Rad, Waltham, MA, USA, 164-5052).
7. Protein electrophoresis apparatus (Bio-Rad, Hercules, CA, USA, MiniProtean Tetra Cell, 165-8004).
8. Refrigerated MicroCentrifuge (Sorvall ST8R, Fisher Scientific, Waltham, MA, USA, 75007204).
9. WB transfer apparatus (Bio-Rad, Hercules, CA, USA, Mini Trans-Blot Electrophoretic Transfer Cell, 170-3930).

2.6 Labware

1. Blotting pads (Bio-Rad, Hercules, CA, USA, 170-3933).
2. Cell culture flask (Thermo Scientific Biolite, Waltham, MA, USA, 11884235).
3. Six-well cell culture dishes (Thermo Scientific Biolite, Waltham, MA, USA, 11825275).
4. Centrifuge tubes (15-mL) (Fisherbrand, Waltham, MA, USA, 11819650).
5. Centrifuge tubes (50-mL) (Fisherbrand, Waltham, MA, USA, 11849650).
6. Extra Thick Blot Filter Paper (Bio-Rad, Hercules, CA, USA, 1703965).
7. Microcentrifuge tubes (1.5-mL) (Daslab, Waltham, MA, USA, 1755089).
8. Pipettes (5-mL) (Costar Stripette, St. Louis, MO, USA, 4487).
9. Pipettes (10-mL) (Costar Stripette, St. Louis, MO, USA, 4488).

3. CELL CULTURE, TREATMENTS, AND SAMPLE COLLECTION

3.1 Cell Culture

1. MEFs are incubated at 37°C under saturating humidity in 5% CO_2/95% air in 75-cm^2 cell culture flasks with DMEM supplemented, as described in Section 2.4 (Notes 1 and 2).
2. The MEFs are grown to a density of $3 \times x10^5$ cells/mL (Note 3).
3. When the cells are confluent (~80%; see Notes 4 and 5), the spent culture medium is discarded. The cells are washed carefully once with prewarmed PBS (Note 6) and trypsinized with 0.25% trypsin/EDTA

(Note 7). The flask is rotated to cover the entire surface with trypsin. The trypsinization is stopped by the addition of two volumes of complete DMEM (Note 8).

4. The cells are transferred to 15-mL centrifuge tubes and centrifuged at 400 × *g* for 10 min.
5. The pelleted cells are diluted and subcultured 2 mL in a six-well plate at a concentration of 10^5 cells/mL (Note 9).
6. The six-well plates are incubated overnight under the abovementioned conditions to allow the cells to adhere to the plate.

3.2 Treatments

1. After 24 h, the confluence (about 80%) is checked by light microscopy (Note 10).
2. The culture medium is removed and the cells are washed with prewarmed PBS (1 mL/well).
3. A freshly made medium is added (2 mL/well) containing different treatments (Control, EBSS, Baf.A1, and Baf.A1 + EBSS; see steps 4–7).
4. Fusion between the autophagosomes and lysosomes is blocked with 100 n*M* of Baf.A1 diluted in a prewarmed culture medium.
5. To establish starvation conditions, the cells are washed and the culture medium replaced with EBSS.
6. For a combined treatment, the cells are washed three times with 1 mL of PBS and finally treated with 100 n*M* of Baf.A1 dissolved in EBSS.
7. All treatments have a duration of 4 h prior to sample preparation.

3.3 Sample Preparation

1. Once treatment is complete, the culture medium is carefully aspirated.
2. A six-well plate is placed in an ice bucket and washed three times with 1 mL of ice-cold PBS.
3. Add 250 μL of trypsin to each well to detach cells. Stop trypsin by adding two volumes of cell culture media. Centrifuge at 600 × *g* 7 min at 4°C. Then remove the supernatant and add rinse pellets with 1 mL cold PBS 1 ×. Centrifuge at 3000 × *g* for 4 min. Remove the supernatant.
4. After the third wash, 1 mL of NP-40 lysis buffer is added and the cells are incubated for 5 min. The samples are mechanically disrupted by pipetting until homogenized.

5. The lysed cells are transferred to cold microcentrifuge tubes and centrifuged at 1314 × *g* for 13 min at 4°C in a precooled centrifuge.
6. The resulting supernatants are transferred to a new microcentrifuge tube (kept on ice) and quantified by the BCA assay. The pellets are discarded.
7. For BCA protein quantification, a calibration curve (0–10 μg) is prepared using a 2-mg/mL BSA stock solution. The BCA reagent is composed of 50 parts reagent A and 1 part reagent B (i.e., a 50:1 ratio). A small amount (up to 25 μL) of lysate is mixed with 200 μL of BCA solution to be used for the determination of protein concentration in cell lysate measuring absorbance at 570 nm in a microplate reader after 30 min of incubation.
8. An appropriate volume of each sample (containing 20 μg of protein) is added to 5 × sample buffer (Note 11) and heated at 95°C for 10 min in microcentrifuge tubes (Note 12).

4. ELECTROPHORESIS, WESTERN BLOT, AND DATA ANALYSIS

4.1 Electrophoresis

1. Use Mini-PROTEAN TGX Precast Protein Gel containing a linear 12% acrylamide gradient to separate both LC3 isoforms and p62 (Note 13).
2. Remove the gels from the storage pouch and prepare them for assembly. For this step, carefully add the green comb (upper) and tape (lower).
3. Exhaustively wash each cell with electrophoresis buffer and place the gel cassettes into the electrode assembly.
4. Place the electrophoresis module into the tank and fill the buffer chambers with electrophoresis buffer (approximately 200 mL for the inner buffer reservoir and 500 mL for the outer buffer reservoir).
5. Carefully load the samples into the cells of the gel, including one cell with a molecular weight marker.
6. Resolve the electrophoresis for 90 min at 100 V (constant) (Note 14).
7. After electrophoresis is complete, remove the gels from the cell and submerge it for 15 min in a transfer buffer at room temperature.

4.2 Western Blotting

1. Use transfer PVDF membranes (Note 15) previously wetted for 1 min in methanol, and subsequently rinse it with ddH_2O.

2. Soak the PVDF membrane in a transfer buffer for 15 min at room temperature.
3. Make a sandwich containing (in this order) blotting foam, Extra Thick Blot Filter Paper, PVDF membrane, gel, Extra Thick Blot Filter Paper, and blotting foam (Notes 16 and 17).
4. Carefully use a roller to remove air trapped between the layers and introduce them into the assembled cassette.
5. Place the assembled cassette into the transfer module and tank and add a transfer buffer to cover the transfer unit completely (approximately 700 mL).
6. Transfer for 90 min at a constant amperage of 300 mA (Notes 18 and 19).
7. Disassemble the sandwich. Immediately place the membrane in an adequate plastic container and wash two or three times with a TBST buffer.
8. Incubate the membrane with the appropriate volume (usually 5–10 mL) of blocking solution for 60 min at room temperature with rolling (Notes 20 and 21).
9. Rinse the membrane three times for 5 min with the abovementioned volume of TTBS buffer.
10. After the last wash, incubate the membrane with the primary antibody diluted 1:5000 in blocking solution (for all antibodies, LC3, p62, and GAPDH), overnight at 4°C (except for GAPDH for 1 h at room temperature) with rolling (Notes 22 and 23).
11. Rinse the membrane three times for 5 min with the same volume of TBST buffer with rolling.
12. Incubate the membrane with the respective HRP-conjugated secondary antibody diluted 1:10,000 in blocking solution for 1 h at room temperature with rolling.
13. Remove the secondary antibody and exhaustively wash the membrane three or four times for 10 min each time with a TTBS buffer with rolling.
14. Place the membrane on a clean surface and add ECL solution, making sure to cover the membrane completely (approximately 2 mL). Incubate the membrane for 5 min at room temperature.
15. Prior to imaging, drain the excess substrate and place the membrane in a protective sleeve (such as plastic wrap) to prevent drying.
16. Finally, the capture of chemoluminescence is performed in an Amersham Imager 600 system. The quantification analysis is completed with the ImageQuant TL software using the GAPDH level as a loading control.

4.3 Interpretation

The autophagy machinery is inhibited through inactivation of the *Atg5* gene (Fig. 2) because the Atg5 protein is part of the complex involved in the conversion and lipidation of LC3-II (Coutts & La Thangue, 2016; Klionsky et al., 2016). Consequently, we cannot observe an accumulation of the LC3-II isoform in any treatment being considered (Jiang & Mizushima, 2015) or the degradation of the p62 protein (Jiang & Mizushima, 2015; Klionsky et al., 2016). These results were previously reported after 2 h of starvation treatment in MEFs (Sahani et al., 2014) and 6 h of treatment in *Atg5* knockout MEFs (Ni et al., 2011).

Fig. 1 represents a common experiment of autophagy flux in which we analyze the LC3 and p62 levels through Western blotting under the following situations: basal condition and induction of the mechanism through starvation. We used WT MEFs (*ATG5*$^{+/+}$). When the WT MEFs are starved, we observe an increase in the LC3-II isoform and a reduction of LC3-I. This increase is also apparent when we block lysosome fusion with

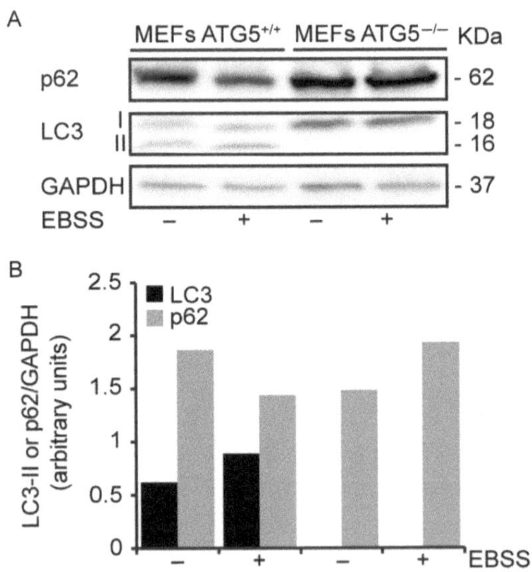

Fig. 2 Determination of the LC3 and p62 protein levels in WT and *ATG5*$^{-/-}$ MEFs. The cells were with or without EBSS (the starvation medium) for 4 h. Panel (A) represents the LC3 (isoforms I and II) and p62 protein blots. The *ATG5*$^{-/-}$ MEFs did not display LC3 isoform II. (B) Densitometry of the representative LC3-II and p62 blots. GAPDH is used as a loading control. Data are represented in arbitrary units. The molecular mass is indicated in kDa.

autophagosomes using Baf.A1, which is a v-ATPase proton pump inhibitor and is even higher when we combine these treatments, resulting in an increase in autophagosome formation and the inability to degrade them. These results have been confirmed in MEFs and HeLa cells after 2 h of treatment (Jiang & Mizushima, 2015) and in SH-SY5Y neuroblastoma cells, MEFs and HFs after 4 h of treatment (Gomez-Sanchez et al., 2015, 2016). The p62 levels are not degraded in the presence of Baf.A1, and they are reduced when the cells are starved with EBSS (also observed in Fig. 2), which correspond to a blockade in its degradation and a higher autophagic flux, respectively. Interestingly, when we do the dual treatment, we found less p62 protein than those found in Baf.A1 treatment alone. This finding has been previously reported in MEFs and HeLa cells after 2 h of treatment; the authors attributed differences between this amount of p62 and those found in starvation treatment as the quantity of protein that is not degraded because of the blockade of autophagy (Jiang & Mizushima, 2015). In addition, this hallmark has been reported in MEFs, SH-SY5Y, N27 rat dopaminergic cells, and HFs after 4 h of treatment (Gomez-Sanchez et al., 2015, 2016). As mentioned previously, p62 may restore the amount of protein during long starvation periods as a consequence of the upregulation of its messenger RNA (mRNA) expression, at least in MEFs and HepG2 cells (Sahani et al., 2014). This upregulation has also been observed under 4 h starvation conditions in HFs and to a lesser extent in MEFs and SH-SY5Y cells (Gomez-Sanchez et al., 2016). Thus, because the mammalian machinery tries to counteract p62 degradation, we recommend performing p62 mRNA measurement in addition to the protein quantification to understand the autophagic flux properly (Sahani et al., 2014).

Although these are general considerations when interpreting LC3 and p62 immunoblots, there are some limitations when analyzing these proteins by immunoblotting, and these parameters may vary depending on the cell line and state considered (fully discussed in Klionsky et al., 2016; Mizushima & Yoshimori, 2007). Adequate experiments and interpretation of the results will shed some light on how to address the autophagic flux in specific cell lines or tissues.

5. NOTES

1. MEFs deficient in *Atg5* were kindly provided by Dr. Noboru Mizushima (Department of Biochemistry and Molecular Biology,

Graduate School and Faculty of Medicine, University of Tokyo, Tokyo, Japan).

2. The passage number is the number of subcultures that the cells have undergone. The passage number should be recorded and should not be too high.
3. These cellular density can be optimized depending on cell lines.
4. Discard the cells if they are detaching in large numbers (attached lines), look shriveled, or both.
5. Do not let the cells become overconfluent because they will start to exhibit changes in their metabolic or oxidative status or die off.
6. Use approximately half the culture medium volume. Repeat this wash step if necessary. All serum must be removed prior to the addition of trypsin.
7. Add 2.5 mL of trypsin to each flask.
8. Return the flask to the incubator and incubate for 3–5 min or until the cells are detached. Prolonged exposure of the cells to trypsin may result in damage to cell surface receptors.
9. Warm fresh culture media to 37°C in a water bath or incubator for at least 30 min.
10. Although the time varies, this state is normally reached 24 h after seeding.
11. This buffer can be saved in 0.5-mL aliquots and maintained at −20°C.
12. Samples can be conserved for several months at −20°C.
13. A 12% acrylamide gel is sufficient to resolve the LC3 isoforms, and an 8–10% gel is sufficient for p62. However, use a 12% gel to ensure a proper resolution of both proteins in the same gel.
14. The resolution improves significantly if the uncharged gel is run for 15 previously loaded samples.
15. Although nitrocellulose can be used, in our experience, the best results are achieved with PVDF.
16. The blotting foam and Extra Thick Blot Filter Paper must be presoaked in transfer buffer.
17. Place the sandwich in the assembled cassette with the gel closest to the black plate and the PVDF membrane closest to the red plate of the cassette.
18. These conditions are sufficient for LC3 and p62.
19. Add the cooling unit and the stir bar and fill the tank with transfer buffer and run at 4°C. Place the tank on a stir plate and begin stirring to maintain an even buffer temperature and ion concentration during the transfer.

20. Staining the membrane with Ponceau S reagent for 5 min prior to blocking is highly recommended to verify a correct transfer. Ponceau S is easily removed by washing with ddH_2O.
21. The blocking solution must be freshly prepared.
22. Primary antibodies can be saved, maintained at 4°C, and reused several times by the addition of 0.05% (w/v) sodium azide. Repeated freeze/thaw cycles must be avoided because they can denature the antibody and cause it to form aggregates that reduce its binding capacity.
23. Correct agitation must be ensured to obtain a homogeneous distribution of the antibody in the membrane.

ACKNOWLEDGMENTS

M.R.-A. was supported by an FPU predoctoral fellowship (FPU13/01237; Ministerio de Educación, Cultura y Deporte, Spain); R.A.G.-P. was supported by a "Contrato para la Retención y Atracción de Talento Investigador del Gobierno de Extremadura, TA13009" (Junta de Extremadura, Spain) and received research support from the ISCIII, Ministerio de Economía y Competitividad, Spain (PI14/00170). Dr. J.M.F. received research support from the ISCIII, Ministerio de Economía y Competitividad, Spain, CIBERNED (CB06/05/004 and PI15/00034), and the Consejería de Economía e Infraestructuras, Junta de Extremadura, Spain (GR15045). M.N.-S. was supported by a "Contrato Juan de la Cierva" (JCI-2012-14383) from Ministerio de Economia y Competitividad, Spain. This work is also supported by "Fondo Europeo de Desarrollo Regional" (FEDER) from the European Union. The authors thank P. Delgado and FUNDESALUD for invaluable technical assistance.

REFERENCES

Coutts, A. S., & La Thangue, N. B. (2016). Regulation of actin nucleation and autophagosome formation. *Cellular and Molecular Life Sciences*, *73*(17), 3249–3263.

Cuervo, A. M., & Wong, E. (2014). Chaperone-mediated autophagy: Roles in disease and aging. *Cell Research*, *24*(1), 92–104.

Feng, Y., He, D., Yao, Z., & Klionsky, D. J. (2014). The machinery of macroautophagy. *Cell Research*, *24*(1), 24–41.

Gomez-Sanchez, R., Pizarro-Estrella, E., Yakhine-Diop, S. M., Rodriguez-Arribas, M., Bravo-San Pedro, J. M., Fuentes, J. M., & Gonzalez-Polo, R. A. (2015). Routine Western blot to check autophagic flux: Cautions and recommendations. *Analytical Biochemistry*, *477*, 13–20.

Gomez-Sanchez, R., Yakhine-Diop, S. M., Rodriguez-Arribas, M., Bravo-San Pedro, J. M., Martinez-Chacon, G., Uribe-Carretero, E., … Gonzalez-Polo, R. A. (2016). mRNA and protein dataset of autophagy markers (LC3 and p62) in several cell lines. *Data in Brief*, *7*, 641–647.

Jiang, P., & Mizushima, N. (2015). LC3- and p62-based biochemical methods for the analysis of autophagy progression in mammalian cells. *Methods*, *75*, 13–18.

Klionsky, D. J., Abdelmohsen, K., Abe, A., Abedin, M. J., Abeliovich, H., Acevedo Arozena, A., … Zughaier, S. M. (2016). Guidelines for the use and interpretation of assays for monitoring autophagy (3rd edition). *Autophagy*, *12*(1), 1–222.

Lippai, M., & Low, P. (2014). The role of the selective adaptor p62 and ubiquitin-like proteins in autophagy. *BioMed Research International*, *2014*, 832704.
Mizushima, N., & Yoshimori, T. (2007). How to interpret LC3 immunoblotting. *Autophagy*, *3*(6), 542–545.
Ni, H. M., Bockus, A., Wozniak, A. L., Jones, K., Weinman, S., Yin, X. M., & Ding, W. X. (2011). Dissecting the dynamic turnover of GFP-LC3 in the autolysosome. *Autophagy*, 7(2), 188–204.
Pankiv, S., Clausen, T. H., Lamark, T., Brech, A., Bruun, J. A., Outzen, H., ... Johansen, T. (2007). p62/SQSTM1 binds directly to Atg8/LC3 to facilitate degradation of ubiquitinated protein aggregates by autophagy. *The Journal of Biological Chemistry*, *282*(33), 24131–24145.
Sahani, M. H., Itakura, E., & Mizushima, N. (2014). Expression of the autophagy substrate SQSTM1/p62 is restored during prolonged starvation depending on transcriptional upregulation and autophagy-derived amino acids. *Autophagy*, *10*(3), 431–441.
Sridhar, S., Botbol, Y., Macian, F., & Cuervo, A. M. (2012). Autophagy and disease: Always two sides to a problem. *The Journal of Pathology*, *226*(2), 255–273. http://dx.doi.org/10.1002/path.3025.
Wesselborg, S., & Stork, B. (2015). Autophagy signal transduction by ATG proteins: From hierarchies to networks. *Cellular and Molecular Life Sciences*, 72(24), 4721–4757.
White, E., Mehnert, J. M., & Chan, C. S. (2015). Autophagy, metabolism, and cancer. *Clinical Cancer Research*, *21*(22), 5037–5046.
Wong, E., & Cuervo, A. M. (2010). Autophagy gone awry in neurodegenerative diseases. *Nature Neuroscience*, *13*(7), 805–811.
Yoshii, S. R., & Mizushima, N. (2015). Autophagy machinery in the context of mammalian mitophagy. *Biochimica et Biophysica Acta*, *1853*(10 Pt. B), 2797–2801.

CHAPTER FIVE

High-Throughput Quantification of GFP-LC3$^+$ Dots by Automated Fluorescence Microscopy

J.M. Bravo-San Pedro[*,†,‡,§,¶,1], **F. Pietrocola**[*,†,‡,§,¶], **V. Sica**[*,†,‡,§,¶,||], **V. Izzo**[*,†,‡,§,¶], **A. Sauvat**[†,‡,§,¶,#], **O. Kepp**[†,‡,§,¶,#], **M.C. Maiuri**[*,†,‡,§,¶], **G. Kroemer**[†,‡,§,¶,#,**,††], **L. Galluzzi**[*,†,‡,§,¶,‡‡,1]

*Gustave Roussy Cancer Campus, Villejuif, France
†INSERM, U1138, Paris, France
‡Equipe 11 labellisée par la Ligue Nationale contre le Cancer, Centre de Recherche des Cordeliers, Paris, France
§Université Paris Descartes/Paris V, Sorbonne Paris Cité, Paris, France
¶Université Pierre et Marie Curie/Paris VI, Paris, France
||Faculté de Medicine, Université Paris Saclay/Paris XI, Le Kremlin-Bicêtre, France
#Metabolomics and Cell Biology Platforms, Gustave Roussy Cancer Campus, Villejuif, France
**Pôle de Biologie, Hôpital Européen Georges Pompidou, AP-HP, Paris, France
††Karolinska University Hospital, Stockholm, Sweden
‡‡Weill Cornell Medical College, New York, NY, United States
[1]Corresponding authors: e-mail address: chemabsp@gmail.com (J.M.B.-S.P.); deadoc@vodafone.it (L.G.)

Contents

1. Introduction 72
2. Cell Culture 75
3. Generation of Stable GFP-LC3-Expressing Cells 75
4. Treatments 76
5. Image Analysis 77
6. Concluding Remarks 77
7. Notes 80
Acknowledgments 84
References 85

Abstract

Macroautophagy is a specific variant of autophagy that involves a dedicated double-membraned organelle commonly known as autophagosome. Various methods have been developed to quantify the size of the autophagosomal compartment, which is an indirect indicator of macroautophagic responses, based on the peculiar ability of microtubule-associated protein 1 light chain 3 beta (MAP1LC3B; best known as LC3) to accumulate in forming autophagosomes upon maturation. One particularly convenient method to monitor the accumulation of mature LC3 within autophagosomes relies on a green fluorescent protein (GFP)-tagged variant of this protein and fluorescence microscopy. In physiological conditions, cells transfected temporarily or

Methods in Enzymology, Volume 587
ISSN 0076-6879
http://dx.doi.org/10.1016/bs.mie.2016.10.022

stably with a GFP-LC3-encoding construct exhibit a diffuse green fluorescence over the cytoplasm and nucleus. Conversely, in response to macroautophagy-promoting stimuli, the GFP-LC3 signal becomes punctate and often (but not always) predominantly cytoplasmic. The accumulation of GFP-LC3 in cytoplasmic dots, however, also ensues the blockage of any of the steps that ensure the degradation of mature autophagosomes, calling for the implementation of strategies that accurately discriminate between an increase in autophagic flux and an arrest in autophagic degradation. Various cell lines have been engineered to stably express GFP-LC3, which—combined with the appropriate controls of flux, high-throughput imaging stations, and automated image analysis—offer a relatively straightforward tool to screen large chemical or biological libraries for inducers or inhibitors of autophagy. Here, we describe a simple and robust method for the high-throughput quantification of GFP-LC3^{+} dots by automated fluorescence microscopy.

1. INTRODUCTION

Macroautophagy is a peculiar type of autophagy that relies on dedicated, double-membraned organelles commonly known as autophagosomes (Codogno, Mehrpour, & Proikas-Cezanne, 2012; Galluzzi, Pietrocola, et al., 2015; Kaur & Debnath, 2015). Autophagosomes derive from so-called phagophores (also known as isolation membranes), which in turn appear to bud from specialized portions of the endoplasmic reticulum in the context of complex supramolecular structures that have been dubbed omegasomes (Lamb, Yoshimori, & Tooze, 2013; Noda & Inagaki, 2015). Irrespective of their source (which remains a matter of debate), forming autophagosomes engulf cytoplasmic material destined to degradation (e.g., damaged organelles, redox-active protein aggregates, invading pathogens) (Deretic, Saitoh, & Akira, 2013; Green, Galluzzi, & Kroemer, 2011; Levine, Mizushima, & Virgin, 2011; Menzies, Fleming, & Rubinsztein, 2015), close to generate perfectly sealed double-membraned vesicles, and deliver their cargo to lysosomes upon fusion (Codogno et al., 2012; Galluzzi, Pietrocola, et al., 2015). The autophagic cargo is eventually degraded by hydrolases upon acidification of the lysosomal lumen, and the breakdown products are recycled back into the cytoplasm by transporters of the lysosomal membrane (Lamb et al., 2013; Sica et al., 2015). Although additional types of autophagy exist (e.g., microautophagy, chaperone-mediated autophagy) (Choi, Ryter, & Levine, 2013; Cuervo & Wong, 2014), from here onward, we will refer to macroautophagy as to autophagy, for the sake of simplicity.

Multiple assays have been developed to detect autophagy based on the increase in number of autophagosomes that generally accompanies (at least the first phases of) a proficient autophagic response (Klionsky et al., 2016; Tooze et al., 2015). For instances, when biochemical assays were not yet available, autophagic cells were identified based on the accumulation of vacuoles in their cytoplasm, a process that could readily be monitored on light microscopy (Kroemer et al., 2009). More recently, as biochemical and functional approaches took over morphological determinations (Fuchs & Steller, 2015; Galluzzi, Buque, Kepp, Zitvogel, & Kroemer, 2015; Galluzzi et al., 2012), immunoblotting and immunofluorescence microscopy have been preferentially harnessed to monitor the size of the autophagosomal compartment, mostly based on the peculiar ability of microtubule-associated protein 1 light chain 3 beta (MAP1LC3B; best known as LC3) to accumulate in forming autophagosomes upon cleavage and condensation with phosphatidylethanolamine (PE) (Kabeya et al., 2004; Klionsky et al., 2016; Mizushima et al., 2001). For instance, the maturation of endogenous LC3 can be readily assessed by immunoblotting. Indeed, while immature LC3 (which is commonly referred to LC3-I) migrates on SDS-PAGE with an apparent molecular weight of approx. 18 kDa, its mature counterpart (which is commonly known as LC3-II) displays an increased electrophoretic mobility and an apparent molecular weight of approx. 15 kDa (Niso-Santano et al., 2015). Although this is a reliable approach to determine LC3 maturation in vitro and ex vivo, immunoblotting is time consuming, and incompatible with high-throughput applications (Klionsky et al., 2016).

To circumvent (at least partially) these issues, Kabeya and colleagues cloned the cDNA coding for rat LC3 into the *Bgl*II and *Eco*RI sites of pEGFP-C1, a commercial vector to generate fluorescent proteins under the control of the immediate-early cytomegalovirus promoter (Kabeya et al., 2000). The resulting plasmid de facto encoded an enhanced variant of *Aequorea victoria* green fluorescence protein (EGFP) fused to the N-terminus of rat LC3 and could be used to conveniently monitor the subcellular localization of LC3 in human cervical carcinoma HeLa cells (Kabeya et al., 2000). Since then, several other constructs encoding GFP-LC3 have been generated and employed to create stable transfectants for investigation (Klionsky et al., 2016). In 2004, Mizushima and collaborators went one step further and generated a mouse strain expressing GFP-LC3 at the whole-body level (Mizushima, Yamamoto, Matsui, Yoshimori, & Ohsumi, 2004).

Irrespective of model, GFP-LC3 is distributed rather homogenously throughout the cytoplasm and nucleus in physiological conditions. Conversely, GFP-LC3 relocalizes to forming autophagosomes in the course of autophagic responses, resulting in the formation of bright GFP-$LC3^+$ cytoplasmic dots and often (but not always) exclusion of the GFP signal from the nuclear compartment (Klionsky et al., 2016). Both the number of GFP-$LC3^+$ dots per cell and the percentage of cells containing more GFP-$LC3^+$ dots than cells maintained in control conditions have been extensively employed as indicators of an autophagic response (Klionsky et al., 2016). Coupled to modern imaging stations and software for automated image analysis, cells stably expressing a GFP-LC3 chimera provide a convenient system to query libraries of chemicals or biologicals for agents that induce or inhibit autophagy in the context of high-throughput screening campaigns (Kepp et al., 2014; Niso-Santano et al., 2015; Shen et al., 2012).

Importantly, high-throughput screening approaches based on GFP-LC3-expressing cancer cells must take into attentive consideration (and rule out) an aspect of autophagic responses that may compromise the interpretation of results, which is: autophagosomes bearing lipidated LC3 (and hence GFP-$LC3^+$ dots) accumulate not only when the initiation of autophagy is increased (increased on-rate), but also when lysosomal degradation is inhibited (decreased off-rate) (Galluzzi, Bravo-San Pedro, Blomgren, & Kroemer, 2016; Klionsky et al., 2016). To discriminate between these diametrically opposed situations (and hence investigate autophagic flux), GFP-LC3 aggregation (as well as LC3 lipidation and other autophagosomal markers) must be quantified in cells responding to a chemical or biological stimulus both in the absence and in the presence of lysosomal inhibitors (e.g., bafilomycin A1, pepstatin + E64d) (Klionsky et al., 2016). Implementing such an experimental setup on primary screens may increase considerably costs and runtimes without obvious advantages. However, it is fundamental to evaluate the ability of lysosomal inhibitors to increase (bona fide autophagic response) or not (autophagic blockage), the accumulation of GFP-$LC3^+$ dots in cells responding to a putative inducer of autophagy as soon as primary hits have been confirmed. Of note, this stays true for all other assays that monitor autophagic responses based on parameters that reflect (directly or indirectly) the size of the autophagosomal compartment (Galluzzi et al., 2016; Klionsky et al., 2016).

Here, we describe in detail a method for quantifying GFP-LC3 aggregation in human cancer cells stably expressing GFP-LC3, including the steps of image acquisition and analysis. With minor variations, this protocol can be adapted to several cell lines that have been engineered for the stable expression of GFP-LC3 and can be readily scaled up to high-throughput workflows.

2. CELL CULTURE

1. Human brain neuroglioma H4 cells are routinely maintained at 37°C (5% CO_2), in Dulbecco's Modified Eagle Medium (DMEM) supplemented with 100 m*M* 2-[4-(2-hydroxyethyl)piperazin-1-yl] ethanesulfonic acid (HEPES) buffer and 10% fetal calf serum (FCS) (see Notes 1–3).
2. Once maintenance cultures reach 75–80% confluence, exhausted culture medium is discarded, and cells are gently washed with prewarmed PBS (see Notes 4 and 5). Thereafter, the cell monolayer is detached with 0.25% trypsin/EDTA, diluted, and reseeded (see Notes 6–10). This provides cells for maintenance culture as well as for experimental procedures.
3. For stable transfection, 300×10^3 wild-type H4 cells are seeded in standard 6-well plates for cell culture, allowed to adapt, and recover standard proliferation for 12–15 h (see Note 11), and processed as detailed in Section 3.
4. For the high-throughput assessment of GFP-LC3 aggregation, 12×10^3 GFP-LC3-expressing H4 cells are seeded in Falcon 96-Well Black with Clear Flat Bottom TC-Treated Imaging Microplate with Lid (Corning Inc., Corning, NY, USA), allowed to adapt, and recover normal proliferation rate for 12 h (see Notes 11–13), and processed as detailed in Sections 4 and 5.

3. GENERATION OF STABLE GFP-LC3-EXPRESSING CELLS

1. Appropriate confluence is rapidly inspected on light microscopy (see Note 14).
2. One microgram of GFP-LC3-coding plasmid compatible with stable transfection (see Note 15) is diluted in 100 μL FCS-free culture medium (Solution A) (see Note 16).
3. Five microliter of Attractene Transfection Reagent (Qiagen, Hilden, Germany) (see Note 17) are carefully mixed with Solution A, and

transfection complexes are allowed to form for 20 min at room temperature (see Notes 18 and 19).
4. Meanwhile, exhausted culture medium is substituted with fresh, complete culture medium (2.0 mL/well) (see Note 20).
5. Transfection complexes are added dropwise over the entire surface of the culture medium (see Note 21), and plates are returned to standard culture conditions (37°C, 5% CO_2).
6. Five to eight hours later (see Note 22), transfection medium is replaced with 2 mL DMEM supplemented with 100 m*M* HEPES buffer and 10% FCS (2.0 mL/well).
7. Twenty-four hours later, culture medium is replaced with 2 mL DMEM supplemented with 100 m*M* HEPES buffer, 10% FCS, and 0.5–1.0 mg/mL G418 (2.0 mL/well) (see Note 23).
8. Four to five days later, live cells are detached with 0.25% trypsin/EDTA and used to generate clones stably expressing GFP-LC3 by the limiting dilution technique or cytofluorometry-assisted sorting based on the GFP signal (see Note 24).

4. TREATMENTS

1. Appropriate confluence is rapidly inspected on light microscopy (see Note 25).
2. Exhausted culture medium is removed and cells are rapidly washed once with prewarmed PBS (200 μL/well) (see Note 26).
3. The following control culture conditions are established (see Note 27): 200 μL fresh, complete culture medium (see Note 28), 200 μL fresh, complete culture medium supplemented with 10 μ*M* rapamycin (see Note 29), 200 μL Hanks' balanced salt solution (HBSS) (see Note 29), alone or in the presence of 50 n*M* bafilomycin A1 (see Note 30).
4. Experimental conditions that are expected to promote or inhibit autophagy, alone or in combination with 50 n*M* bafilomycin A1 (see Note 30), are established in the rest of the 96-well plate (see Note 27).
5. Once the stimulation time has elapsed (see Note 31), supernatants are discarded, cells are washed once with cold PBS (200 μL/well), and incubated in 4% paraformaldehyde supplemented with 5–10 μg/mL Hoechst 33342 (w/v in PBS; 200 μL/well) (see Note 32) overnight at 4°C (see Note 33).

6. The reagent for fixation and nuclear counterstaining is discarded and replaced with PBS (200 μL/well). In these conditions, plates can be stored at 4°C (see Note 34) or processed for imaging, as described in Section 5.

5. IMAGE ANALYSIS

1. Plates are imaged on an ImageXpress Micro XLS Widefield High-Content Analysis System operated by the MetaXpress® Image Acquisition and Analysis Software (Molecular Devices, Sunnyvale, CA, US). Imaging is performed at 20 × magnification to obtain 9 fields per well in the GFP (407 nm) and Hoechst 33342 (461 nm) channels (see Notes 35 and 36).
2. MetaXpress® is employed to process each image as follows:
   a. Nuclear segmentation based on the Hoechst 33342 signal (Fig. 1A).
   b. Cytoplasmic segmentation, upon subtraction of the Hoechst 33342 signal from the GFP-LC3 signal, which generates cytoplasmic regions of interest (ROIs) (Fig. 1B).
   c. Count of GFP-LC3$^+$ dots within each cytoplasmic ROI (see Note 37), based on differential pixel intensity of the GFP-LC3 signal (Fig. 1C).
3. Autophagic responses are evaluated based on (1) the number of GFP-LC3$^+$ dots/cell and/or (2) the percentage of cells containing more GFP-LC3$^+$ dots than a predetermined threshold number (normally 5–10) (see Notes 38 and 39).
4. When a specific treatment stimulates autophagic flux, the addition of bafilomycin A1 provokes a statistically significant increase in either or both these parameters (Fig. 2).

6. CONCLUDING REMARKS

The method presented here provides a rapid and reliable approach for quantifying the size of the autophagosomal compartment (through GFP-LC3 aggregation) in cultured cells responding to chemical or biological cues, which is compatible with high-throughput workflows and automation (Klionsky et al., 2016). Both the number of GFP-LC3$^+$ dots per cells and the amount of cells exhibiting more GFP-LC3$^+$ dots than cells maintained in physiological conditions potentially constitute indicators of an ongoing autophagic response. However, both these parameters are also

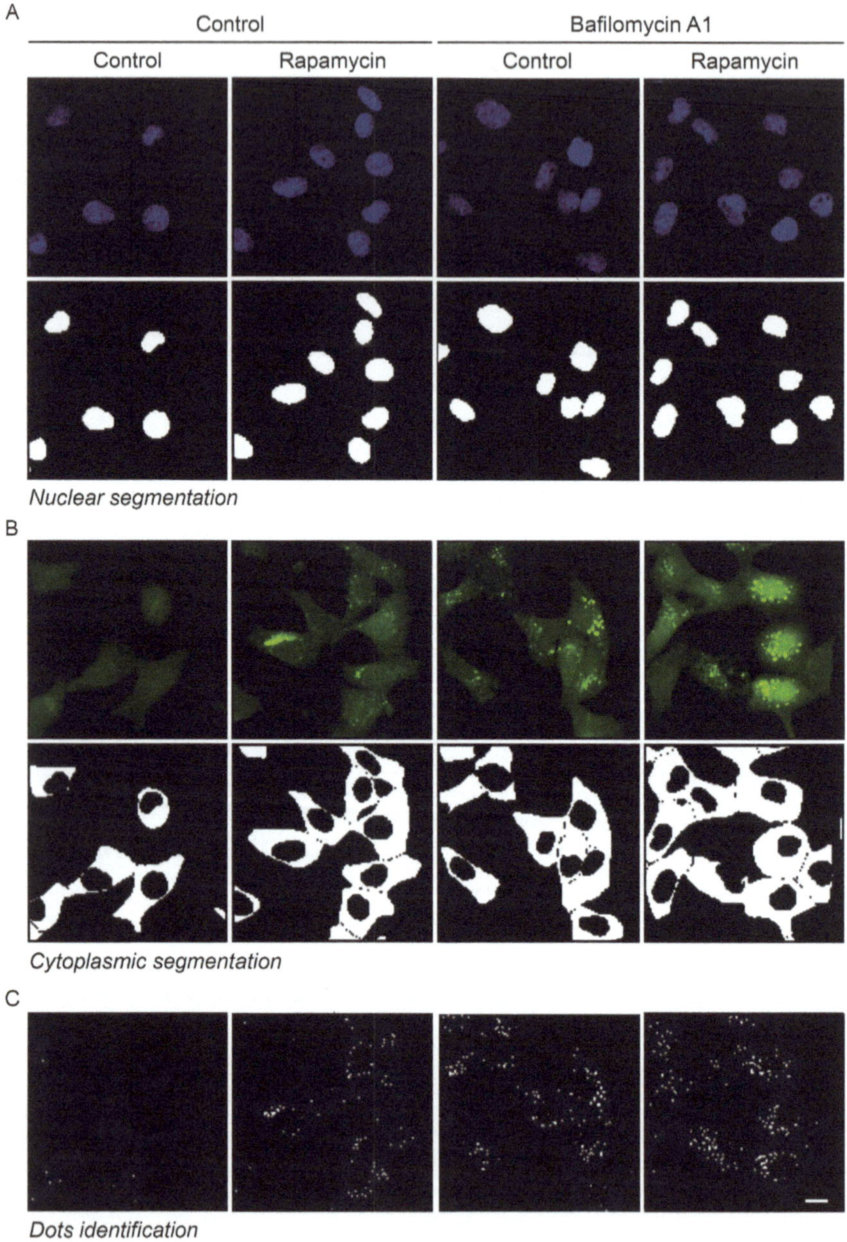

Fig. 1 Image segmentation. Human brain neuroglioma H4 cells engineered to stably express GFP-LC3 as described in Sections 2 and 3 were seeded in 96-well plates, and allowed to recover normal growth for 12 h. Thereafter, cells were maintained in control conditions, treated with 10 μ*M* rapamycin or 50 n*M* bafilomycin A1, alone or in the indicated combinations, for additional 6 h, as described in Sections 2 and 4. Finally, plates were processed for imaging on an ImageXpress Micro XLS Widefield High-Content Analysis System as described in Sections 4 and 5. Images representative of nuclear segmentation (A), cytoplasmic segmentation (B), and dots identification (C) are reported (scale bars = 10 μm).

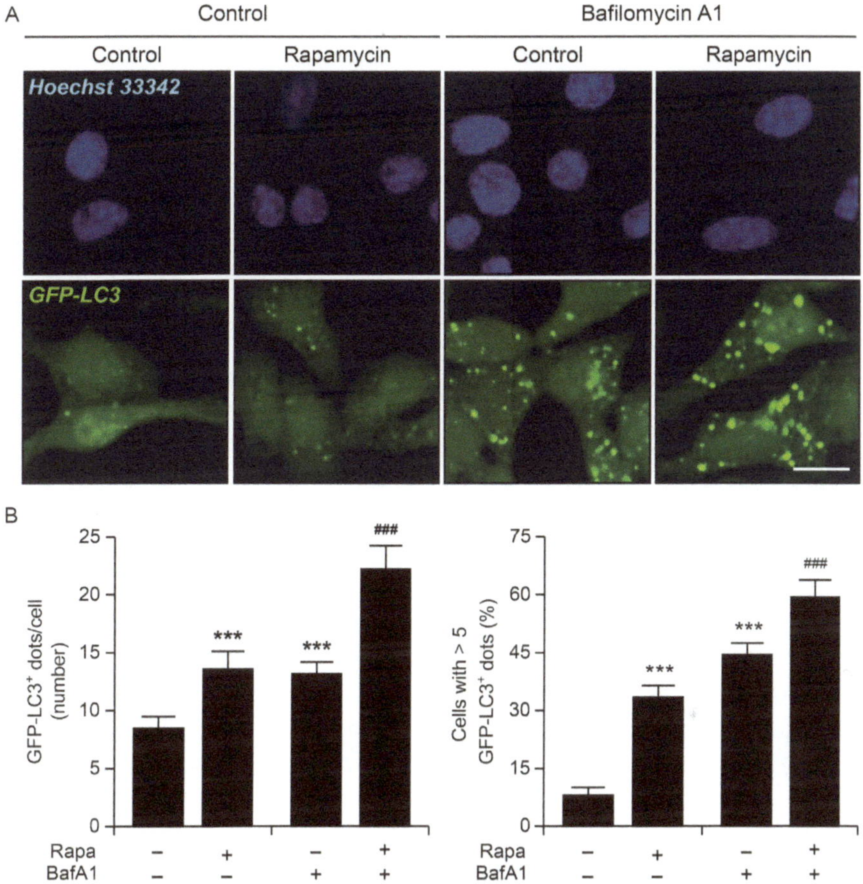

Fig. 2 High-throughput quantification of GFP-LC3 dots. GFP-LC3-expressing human brain neuroglioma H4 cells were seeded in 96-well plates, allowed to adapt for 12 h, and maintained in control conditions or exposed to 10 μ*M* rapamycin (Rapa), alone or in the presence of 50 n*M* bafilomycin A1 (BafA1), for additional 6 h, as described in Sections 2 and 4. Thereafter, plates were processed and imaged on an ImageXpress Micro XLS Widefield High-Content Analysis System as detailed in Sections 4 and 5. Representative images (scale bars = 10 μm) and quantitative data are reported (mean ± SD; $n = 10$ parallel assessments). $^{***}p < 0.001$ as compared to cells maintained in control conditions (paired Student's *t* test). ###$p < 0.001$ as compared to cells treated with Rapa in the absence of BafA1 (paired Student's *t* test).

sensitive to the blockage of autophagic degradation, calling for the assessment of stimulus-induced GFP-LC3 aggregation in a specific set of control conditions (i.e., in the absence and in the presence of lysosomal inhibitors) (Klionsky et al., 2016). Only such an experimental setup allows for the discrimination between the accrual of GFP-LC3-containing autophagosomes as a consequence of increased on-rate vs decreased off-rate, and hence for the proper interpretations of results (Klionsky et al., 2016). This said, additional approaches should be undertaken to confirm the bona fide activation or inhibition of autophagic flux in low-throughput experimental settings, including assessing the levels of specific autophagy substrates by immunoblotting (see chapter "Turnover of lipidated LC3 and autophagic cargoes in mammalian cells" by Rodríguez-Arribas et al.) and monitoring long-term protein degradation (see chapter "Long-lived protein degradation during autophagy" by Dupont et al. in MiE Volume 588). Irrespectively, monitoring GFP-LC3 aggregation by high-throughput fluorescence microscopy and automated image analysis is suitable to identify chemicals or biologicals with autophagy-modulatory potential within large libraries.

7. NOTES

1. As per official recommendations of the American Type Culture Collection (ATCC, Manassas, VA, US) (see also http://www.lgcstandards-atcc.org/products/all/HTB-148.aspx?geo_country=fr#culturemethod).
2. H4 cells can be conveniently maintained in 25-, 75-, or 175-cm^2 flasks, as well as in 6- or 10-cm Ø Petri dishes for cell culture. As an indication, 25-, 75-, and 175-cm^2 flasks exhibiting 80% confluence contain around 2–2.5 × 10^6, 6–7 × 10^6, and 14–15 × 10^6 H4 cells, respectively.
3. Optimal culture conditions may differ for other cell lines.
4. Cells should not be detached for passaging at excessively low (<30%) confluence, as this may favor the emergence of subclones and hence result in the genetic drift of the original cell population. Similarly, cells should not be allowed to grow >80% confluence, because excessively dense cultures may suffer from metabolic, nutritional, and/or oxidative stress.
5. Washing the cell layer with PBS minimizes FCS leftover, which strongly inhibit the catalytic activity of trypsin. With H4 cells, which are not excessively sensitive to detachment, this step can also be performed with Trypsin/EDTA (see also Note 8).
6. Other reagents are commercially available to perform this step, including TrypLE™ Express (from Gibco®-Thermo Fisher Scientific, Carlsbad, CA, USA). TrypLE™ Express is generally more stable and more efficient than Trypsin/EDTA and—at odds with Trypsin/EDTA—does not require postdetachment inactivation (see also Note 7).
7. Detachment time varies quite considerably with multiple factors, including (but not limited to): cell type, confluence, presence of

experimental agents in the culture medium. Untreated H4 cells at 70–80% confluence, as most other malignant cells of human origin in the same conditions, are efficiently detached upon 1–3 min incubation in Trypsin/EDTA. According to our experience, the use of TrypLE™ Express reduces the detachment time of a H4 cell monolayer to <1 min (see also Note 6).

8. It is good practice to ensure detachment by visually inspecting cell culture supports on light microscopy. FCS leftover may indeed inhibit Trypsin/EDTA, hence greatly reducing the yield of detachment (see also Note 5).
9. Incubation in Trypsin/EDTA should not be prolonged beyond the time required for the detachment of the cell monolayer. Indeed, while so-called "overtrypsinization" normally does not have a negative impact on viability, it may alter the properties of the cell surface and postpone complete adaptation to the new cell culture support (see also Note 11).
10. For H4 cells, the ATCC recommends a subcultivation ratio of 1:10–1:15 (see also http://www.lgcstandards-atcc.org/products/all/HTB-148.aspx?geo_country=fr#culturemethod). In our personal experience, a H4 cell culture at 80% confluence diluted 1:10 reaches 80% confluence approximately 2–3 days later.
11. The time required for detached cells to adhere and resume normal proliferation depends on cell type. HeLa cells display a relatively short adaptation time of 12–16 h. Conversely, other malignant cells like human colorectal HCT 116 require 24–30 h to completely adhere to the cell culture substrate and restart proliferation. Importantly, before complete adaptation cells may display increased sensitivity to various experimental interventions (see also Note 9).
12. Since some stable transfectants tend to lose transgene expression over time, it is recommended to intermittently expose GFP-LC3-expressing H4 cell cultures to 0.5–1 mg/mL G418, or to routinely maintain these cells in complete culture medium supplemented with 400 μg/mL G418 (see also Notes 15 and 23).
13. To limit possible biases linked to the preferential evaporation of culture medium from wells located at the outer rim of the plate, we recommend to fill wells from line A, line H, column 1, and column 12 with 200 μL culture medium only.
14. In our hands, liposome-based plasmid transfection attains maximal efficiency when cells exhibit 80–90% confluence. Transfection

efficiency drops when confluence is <60% and may be associated with significant toxicity when cells display <40% confluence.

15. Multiple GFP-LC3-encoding constructs are available from commercial providers, including Addgene (Cambridge, MA, US), InvivoGen (San Diego, CA, US), and others. These constructs differ not only on source of the *MAP1LC3B* sequence (human and mouse variants are available), but also on promoter and selection cassette. We normally employ #11546 from Addgene (https://www.addgene.org/11546/), which contains a G418 resistance cassette (see also Notes 12 and 23).
16. Quantities are suitable for the transfection of a single well of a 6-well plate, but can be readily scaled up for the transfection of a higher number of cells if needed.
17. Several other liposomal transfection reagents are commercially available and can be used instead of Attractene. Using alternative liposomal formulations, though, may require some modification of the transfection protocol.
18. Transfection efficiency may drop considerably if transfection complexes are allowed insufficient time to form.
19. The resulting solution may be cloudy, but this will not affect negatively transfection efficiency.
20. While serum-free medium may be required for the efficient transfection of small-interfering RNAs (siRNAs), plasmids are efficiently transfected in complete culture medium.
21. Transfection complexes have a very high affinity for the plasma membrane and may be toxic to cells if present at locally increased concentrations. It is therefore important to add transfection complexes as slowly as possible and to attempt covering the entire surface of the well to avoid local toxicity.
22. Transfection time can be safely prolonged to 12–15 h.
23. G418 is added as a selective pressure on nontransfected cells (see also Notes 12 and 15).
24. In both cases, clones expressing GFP-LC3 at medium-to-high levels should be selected and validated for their ability to undergo GFP-LC3 aggregation in bright cytoplasmic dots in response to autophagy-inducing conditions prior to use in high-throughput assessments.
25. Optimal confluence levels for the assessment of GFP-LC3 aggregation depend on the duration of the assay, which generally does not exceed 12 h (see also Note 31). As a note, underconfluence (<40%) should

be avoided as excessively subconfluent cells generally display increased sensitivity to various forms of stress as compared to cells at optimal confluence levels (Green, Galluzzi, & Kroemer, 2014). Moreover, excessive underconfluence may have a negative impact on the number of objects/well at imaging. Overconfluence (>80%) should also be avoided as excessively dense cells may have already initiated an autophagic response to cope with nutritional stress. Ideally cells at 50–70% confluence should be employed for experimental determinations.

26. This step is important because serum leftover may inhibit autophagic responses by virtue of the ability of multiple growth factors to stimulate mechanistic target of rapamycin 1 (MTORC1) signaling (Galluzzi, Pietrocola, Levine, & Kroemer, 2014; Harrison et al., 2009).

27. To compensate for interwell variability and obtain statistically robust results, we recommend to establish control conditions as well as experimental conditions in at least 6 wells/condition. Ideally control conditions should be scattered throughout the plate in a random manner, to avoid positional bias.

28. This is an appropriate negative control condition as the abundant glucose, amino acid, and growth factors contained in fresh medium potently suppress autophagic responses to baseline levels (Efeyan, Comb, & Sabatini, 2015; Galluzzi et al., 2014).

29. Both rapamycin and HBSS are appropriate to constitute positive control conditions, as they stimulate potent autophagic responses by inhibiting MTORC1 (Galluzzi et al., 2014). Earle's balanced salt solution (EBSS) can be used instead of HBSS, at the investigator's discretion.

30. It is important to monitor GFP-LC3 aggregation in control and experimental conditions combined alone AND combined with bafilomycin A1 to discriminate between increased autophagic flux and inhibited lysosomal degradation (see also Sections 2 and 5). Bafilomycin A1 can be replaced with 10 μg/mL pepstatin (a potent inhibitor of aspartyl proteases) + 10 μg/mL E64d (an irreversible, membrane-permeable inhibitor of lysosomal, and cytosolic cysteine proteases).

31. The time required for the activation of an autophagic response depends on multiple parameters, including cell type and intensity of the initiating stimulus. As an indication, rapamycin and HBSS trigger robust autophagic responses in H4 cells in 6–12 h. Importantly, prolonging stimulation time may cause cells maintained in negative control conditions to mount an autophagic response to dropping levels of glucose, amino acids, or growth factors. This (which may constitute an issue at data analysis) can be limited (at least to some extent) by using an increased amount of culture medium (up to 320 μL per well) (see also Note 25).

32. Hoechst 33342 is a cell-permeant dye that stains double-strand nucleic acids.
33. As an alternative, fixation and nuclear counterstaining can be performed by incubating cells in 4% paraformaldehyde supplemented with 5–10 μg/mL Hoechst 33342 (v/v in PBS; 200 μL/well) for 20 min at room temperature.
34. Once sealed with adhesive aluminum foils (which simultaneously limit PBS evaporation, possible bacterial contamination, and photobleaching), plates can be stored at 4°C for up to 18 months and imaged thereafter without significant loss in GFP and Hoechst 33342 fluorescence.
35. Common GFP variants have excitation/emission peaks at 488/507 nm, respectively. Hoechst 33342 has excitation/emission peaks at 350/461 nm, respectively.
36. As an indication, imaging an entire 96-well plate with these settings requires approximately 10–20 min. Runtimes can vary to significant degrees with other instruments and/or if other filters are employed.
37. The nuclear compartment contains a considerable amount of LC3, whose specific contribution to autophagic responses remains to be elucidated. For this reason, we exclude nuclear ROIs from the analysis of GFP-LC3 aggregation.
38. The number of GFP-$LC3^+$ dots/cell that discriminates between baseline autophagy levels and the activation of an adaptive autophagic response varies in different cell types and/or in the presence of specific culture conditions. Generally, a healthy mammalian cell growing in fresh, complete medium displays 3–5 cytoplasmic GFP-$LC3^+$ dots. However, automatically counting GFP-$LC3^+$ dots/cell in cells exposed to negative control conditions (as detailed in Sections 4 and 5) can assist in the precise identification of such threshold.
39. In some circumstances, the area of GFP-$LC3^+$ dots can provide additional insights into the magnitude of autophagic responses.

ACKNOWLEDGMENTS

G.K. is supported by the Ligue contre le Cancer (équipe labellisée); Agence National de la Recherche (ANR)—Projets blancs; ANR under the frame of E-Rare-2, the ERA-Net for Research on Rare Diseases; Association pour la recherche sur le cancer (ARC); Cancéropôle

Ile-de-France; Institut National du Cancer (INCa); Fondation Bettencourt-Schueller; Fondation de France; Fondation pour la Recherche Médicale (FRM); the European Commission (ArtForce); the European Research Council (ERC); the LabEx Immuno-Oncology; the SIRIC Stratified Oncology Cell DNA Repair and Tumor Immune Elimination (SOCRATE); the SIRIC Cancer Research and Personalized Medicine (CARPEM); the Swiss Bridge Foundation, ISREC, and the Paris Alliance of Cancer Research Institutes (PACRI).

REFERENCES

Choi, A. M., Ryter, S. W., & Levine, B. (2013). Autophagy in human health and disease. *The New England Journal of Medicine*, *368*, 651–662.

Codogno, P., Mehrpour, M., & Proikas-Cezanne, T. (2012). Canonical and non-canonical autophagy: Variations on a common theme of self-eating? *Nature Reviews. Molecular Cell Biology*, *13*, 7–12.

Cuervo, A. M., & Wong, E. (2014). Chaperone-mediated autophagy: Roles in disease and aging. *Cell Research*, *24*, 92–104.

Deretic, V., Saitoh, T., & Akira, S. (2013). Autophagy in infection, inflammation and immunity. *Nature Reviews. Immunology*, *13*, 722–737.

Efeyan, A., Comb, W. C., & Sabatini, D. M. (2015). Nutrient-sensing mechanisms and pathways. *Nature*, *517*, 302–310.

Fuchs, Y., & Steller, H. (2015). Live to die another way: Modes of programmed cell death and the signals emanating from dying cells. *Nature Reviews. Molecular Cell Biology*, *16*, 329–344.

Galluzzi, L., Bravo-San Pedro, J. M., Blomgren, K., & Kroemer, G. (2016). Autophagy in acute brain injury. *Nature Reviews. Neuroscience*, *17*, 467–484.

Galluzzi, L., Buque, A., Kepp, O., Zitvogel, L., & Kroemer, G. (2015). Immunological effects of conventional chemotherapy and targeted anticancer agents. *Cancer Cell*, *28*, 690–714.

Galluzzi, L., Pietrocola, F., Bravo-San Pedro, J. M., Amaravadi, R. K., Baehrecke, E. H., Cecconi, F., et al. (2015). Autophagy in malignant transformation and cancer progression. *The EMBO Journal*, *34*, 856–880.

Galluzzi, L., Pietrocola, F., Levine, B., & Kroemer, G. (2014). Metabolic control of autophagy. *Cell*, *159*, 1263–1276.

Galluzzi, L., Vitale, I., Abrams, J. M., Alnemri, E. S., Baehrecke, E. H., Blagosklonny, M. V., et al. (2012). Molecular definitions of cell death subroutines: Recommendations of the Nomenclature Committee on Cell Death 2012. *Cell Death and Differentiation*, *19*, 107–120.

Green, D. R., Galluzzi, L., & Kroemer, G. (2011). Mitochondria and the autophagy-inflammation-cell death axis in organismal aging. *Science*, *333*, 1109–1112.

Green, D. R., Galluzzi, L., & Kroemer, G. (2014). Cell biology. Metabolic control of cell death. *Science*, *345*, 1250256.

Harrison, D. E., Strong, R., Sharp, Z. D., Nelson, J. F., Astle, C. M., Flurkey, K., et al. (2009). Rapamycin fed late in life extends lifespan in genetically heterogeneous mice. *Nature*, *460*, 392–395.

Kabeya, Y., Mizushima, N., Ueno, T., Yamamoto, A., Kirisako, T., Noda, T., et al. (2000). LC3, a mammalian homologue of yeast Apg8p, is localized in autophagosome membranes after processing. *The EMBO Journal*, *19*, 5720–5728.

Kabeya, Y., Mizushima, N., Yamamoto, A., Oshitani-Okamoto, S., Ohsumi, Y., & Yoshimori, T. (2004). LC3, GABARAP and GATE16 localize to autophagosomal membrane depending on form-II formation. *Journal of Cell Science*, *117*, 2805–2812.

Kaur, J., & Debnath, J. (2015). Autophagy at the crossroads of catabolism and anabolism. *Nature Reviews. Molecular Cell Biology*, *16*, 461–472.

Kepp, O., Senovilla, L., Vitale, I., Vacchelli, E., Adjemian, S., Agostinis, P., et al. (2014). Consensus guidelines for the detection of immunogenic cell death. *Oncoimmunology*, *3*, e955691.

Klionsky, D. J., Abdelmohsen, K., Abe, A., Abedin, M. J., Abeliovich, H., Acevedo Arozena, A., et al. (2016). Guidelines for the use and interpretation of assays for monitoring autophagy (3rd edition). *Autophagy*, *12*, 1–222.

Kroemer, G., Galluzzi, L., Vandenabeele, P., Abrams, J., Alnemri, E. S., Baehrecke, E. H., et al. (2009). Classification of cell death: Recommendations of the Nomenclature Committee on Cell Death 2009. *Cell Death and Differentiation*, *16*, 3–11.

Lamb, C. A., Yoshimori, T., & Tooze, S. A. (2013). The autophagosome: Origins unknown, biogenesis complex. *Nature Reviews. Molecular Cell Biology*, *14*, 759–774.

Levine, B., Mizushima, N., & Virgin, H. W. (2011). Autophagy in immunity and inflammation. *Nature*, *469*, 323–335.

Menzies, F. M., Fleming, A., & Rubinsztein, D. C. (2015). Compromised autophagy and neurodegenerative diseases. *Nature Reviews. Neuroscience*, *16*, 345–357.

Mizushima, N., Yamamoto, A., Hatano, M., Kobayashi, Y., Kabeya, Y., Suzuki, K., et al. (2001). Dissection of autophagosome formation using Apg5-deficient mouse embryonic stem cells. *The Journal of Cell Biology*, *152*, 657–668.

Mizushima, N., Yamamoto, A., Matsui, M., Yoshimori, T., & Ohsumi, Y. (2004). In vivo analysis of autophagy in response to nutrient starvation using transgenic mice expressing a fluorescent autophagosome marker. *Molecular Biology of the Cell*, *15*, 1101–1111.

Niso-Santano, M., Malik, S. A., Pietrocola, F., Bravo-San Pedro, J. M., Marino, G., Cianfanelli, V., et al. (2015). Unsaturated fatty acids induce non-canonical autophagy. *The EMBO Journal*, *34*, 1025–1041.

Noda, N. N., & Inagaki, F. (2015). Mechanisms of autophagy. *Annual Review of Biophysics*, *44*, 101–122.

Shen, S., Niso-Santano, M., Adjemian, S., Takehara, T., Malik, S. A., Minoux, H., et al. (2012). Cytoplasmic STAT3 represses autophagy by inhibiting PKR activity. *Molecular Cell*, *48*, 667–680.

Sica, V., Galluzzi, L., Bravo-San Pedro, J. M., Izzo, V., Maiuri, M. C., & Kroemer, G. (2015). Organelle-specific initiation of autophagy. *Molecular Cell*, *59*, 522–539.

Tooze, S. A., Dooley, H. C., Jefferies, H. B., Joachim, J., Judith, D., Lamb, C. A., et al. (2015). Assessing mammalian autophagy. *Methods in Molecular Biology*, *1270*, 155–165.

CHAPTER SIX

Use of pHlurorin-mKate2-human LC3 to Monitor Autophagic Responses

I. Tanida*,[1], T. Ueno†, Y. Uchiyama*
*Juntendo University School of Medicine, Bunkyo-ku, Tokyo, Japan
†Juntendo University Graduate School of Medicine, Bunkyo-ku, Tokyo, Japan
[1]Corresponding author: e-mail address: tanida@juntendo.ac.jp

Contents

1. Introduction 88
2. Estimation of Autophagic Process Using pHlurorin-mKate2-human LC3 90
   2.1 Plasmids for an Expression of pHlurorin-mKate2-hLC3 and Its Negative Control 90
   2.2 Monitoring Autophagic Responses Using pHluorin-mKate2-hLC3 in Mammalian Cells 92
   2.3 Interpretation of the Results 93
3. Concluding Remarks 95
Acknowledgments 95
References 95

Abstract

In this chapter, we introduce the usage of pHluorin-mKate2-human LC3 for monitoring autophagy. Using EGFP and RFP, tandem fluorescent protein-tagged LC3 has been generated for monitoring autophagic structures. A critical point for this purpose is the sensitivity of the green fluorescent protein to acidic pH. A super-ecliptic pHluorin is most sensitive to acidic pH among EGFP, mWasabi, and pHluorin, indicating pHluorin is most suitable for monitoring autophagic structures. During autophagy, green-positive and red-positive fluorescent puncta of pHluorin-mKate2-human LC3 indicate signals of preautophagosomes and autophagosomes. After fusion of autophagosomes with lysosomes to form autolysosomes, green fluorescence of this intraautophagosomal protein is abolished according to acidification of autolysosomes. Therefore, these green-negative and red-positive fluorescent puncta reflect autolysosomes, in which intraluminal proteins are finally degraded by lysosomal proteases. To monitor autophagic flux, the accumulation of its green-negative and red-positive fluorescent puncta is monitored by inhibiting major lysosomal proteases, cathepsins. In addition, a mutant pHluorin-mKate2-human LC3ΔG is also introduced as a negative control probe.

Methods in Enzymology, Volume 587
ISSN 0076-6879
http://dx.doi.org/10.1016/bs.mie.2016.09.054

1. INTRODUCTION

Autophagy is a bulk of degradation for intracellular components including organelles. When autophagy is induced, phagophores appear in the cytosol (Fig. 1) (Mizushima & Komatsu, 2011). Phagophores elongate to form cup-shaped structures (preautophagosomes/the isolation membrane) for engulfing the cytoplasmic components. After its elongation, preautophagosomes are enclosed to form autophagosome. Autophagosomes fuse with lysosomes to form autolysosomes. The intraluminal pH is acidified to about pH 5.0 via this autophagosome–lysosome fusion to activate lysosomal hydrolases. Thereafter, the engulfed components are degraded by these hydrolases.

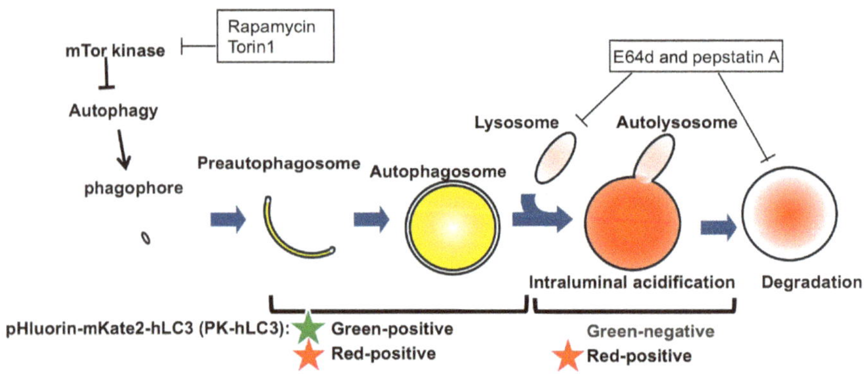

Fig. 1 Schematic representation of fluorescence of pHluorin-mKate2-hLC3 on autophagic structures during autophagy. When autophagy is induced, phagophore appears in the cytosol. Phagophore is elongated to form a cup-shaped structure (preautophagosome/the isolation membrane) for engulfing intracellular compartments. Finally, preautophagosomes are enclosed to form autophagosomes. LC3-II (LC3-phosphatidylethanolamine conjugate/a membrane-bound form of LC3) is localized on preautophagosomes and autophagosomes. Autophagosomes fuse with lysosomes (autolysosomes). LC3-II on a luminal face is localized on autolysosomes, while it is on a cytoplasmic face is delipidated by a cysteine protease, hATG4B. Intraluminal pH is acidified via V-type H^+-ATPase after autophagosome–lysosome fusion, and intraluminal components are degraded by lysosomal hydrolases according to the acidification. Therefore, pHluorin-mKate2-hLC3 on preautophagosomes and autophagosomes shows green-positive and red-positive fluorescent puncta, while it is on autolysosomes shows green-negative and red-positive puncta. mTor kinase negatively regulates autophagy. Rapamycin and torin1 activate autophagy via inhibition of mTor-signaling pathway. A cocktail of E64d and pepstatin A inhibits degradation in lysosomes, resulting in accumulation of intraautolysosomal LC3-II.

LC3 is a definitive marker of autophagic structures in mammalian cells (Kabeya et al., 2000). When preautophagosomes form during autophagy, LC3-I (a cytosolic form of LC3) is conjugated to phosphatidylethanolamine to form LC3-II (the LC3-phosphatidylethanolamine conjugate) (a membrane-bound form of LC3) (Tanida, Ueno, & Kominami, 2004). The LC3-II localizes to preautophagosomes and autophagosomes. LC3-II on the cytoplasmic face is delipidated by Atg4B to form LC3-I during the fusion of autophagosomes with lysosomes, while intraluminal LC3-II is localized on the luminal face (Tanida, 2011). The intraluminal LC3-II is finally degraded by lysosomal hydrolases after the formation of autolysosomes. Therefore, lysosomal turnover of LC3-II is an index for monitoring autophagic activity (Tanida, Minematsu-Ikeguchi, Ueno, & Kominami, 2005).

Intraluminal LC3-II on autolysosomes is under the acidic pH conditions before its degradation, while LC3-II on autophagosomes is under the neutral pH conditions. The tandem fluorescent EGFP-RFP-tagged LC3 (tfLC3) has been developed to monitor the maturation step of mammalian autophagic structures in mammalian cells, based on the change in intraluminal pH of autolysosomes and the sensitivity of EGFP to acidic pH (Kimura, Noda, & Yoshimori, 2007). EGFP shows an acidic pH-sensitive green fluorescence, while RFP shows a pH-resistant fluorescence. Intracellular puncta of tfLC3 on autophagosomes (nonacidic compartments) shows green-positive and red-positive fluorescence. After the fusion of autophagosomes with lysosomes, tfLC3 on autolysosomes decreases green fluorescence. Therefore, the green-negative and red-positive puncta of tfLC3 reflect the signals of autolysosomes (Kimura et al., 2007). However, EGFP still has a weak fluorescence in autolysosomes and lysosomes until its degradation (Katayama, Yamamoto, Mizushima, Yoshimori, & Miyawaki, 2008; Patterson, Knobel, Sharif, Kain, & Piston, 1997; Tanida, Ueno, & Uchiyama, 2014), since the pK_a of EGFP is 5.9. As an improved version of tfLC3, mTagRFP-mWasabi-LC3 was generated, since the pK_a of mWasabi (pK_a = 6.5) was higher than that of EGFP (Ai, Olenych, Wong, Davidson, & Campbell, 2008; Zhou et al., 2012). However, mWasabi continues to emit a weak, but significant, fluorescence at a pH of approximately 6 (Fig. 2) (Tanida et al., 2014). Therefore, these probes potentially tend to overestimate the amount of autophagosomes as green-positive and red-positive puncta.

A critical point for these tandem fluorescent protein-tagged LC3s to monitor autophagy was the sensitivity of green fluorescent protein at an acidic pH. We focused on a green fluorescent protein, super-ecliptic pHluorin (pK_a = 7.6) (Miesenbock, De Angelis, & Rothman, 1998), since its fluorescence

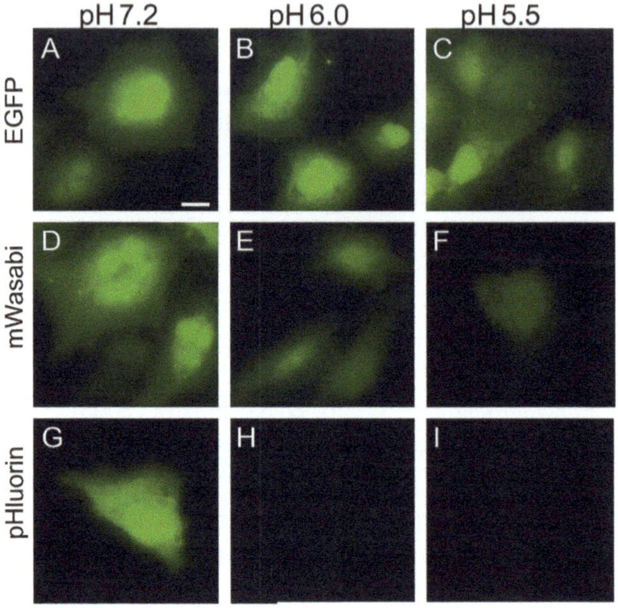

Fig. 2 Comparison of the pH sensitivity of the *green fluorescence* among EGFP, mWasabi, and super-ecliptic pHluorin. The *green fluorescent* proteins, EGFP (A–C), mWasabi (D–F), and pHluorin (G–I), were expressed in the Huh7.5.1 cells. After fixation and permeabilization of the cells, the cells were buffered at pH 7.2 (A, D, G), 6.0 (B, E, H), and 5.5 (C, F, I).

is significantly quenched at pH 6.5, and is almost completely abolished at pH 5.5–6.0 (Tanida et al., 2014). pHluorin-mKate2 fusion protein shows pH-sensitive green fluorescence and pH-resistant far-red fluorescence (Fig. 3). Therefore, we generated expression plasmids for pHluorin-mKate2-human LC3 (PK-hLC3) and its negative control to overcome the problem of tfLC3 and mTagRFP-mWasabi-LC3. In this chapter, we describe the basic protocol for monitoring autophagy with pHluorin-mKate2-hLC3.

2. ESTIMATION OF AUTOPHAGIC PROCESS USING pHlurorin-mKate2-HUMAN LC3

2.1 Plasmids for an Expression of pHlurorin-mKate2-hLC3 and Its Negative Control

We have generated expression vectors for wild-type pHluorin-mKate2-hLC3 (PK-hLC3), pEX-PK-hLC3WT (Addgene, Cambridge, MA, Plasmid # 61458) and FUGW-PK-hLC3WT (Addgene, Plasmid # 61460)

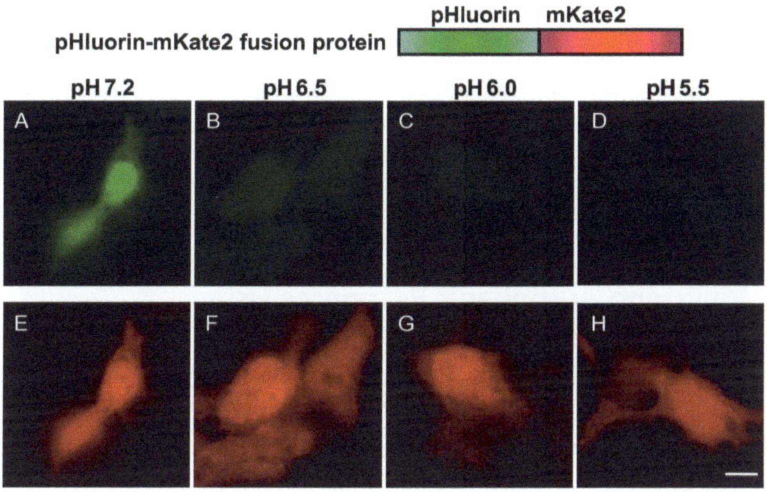

Fig. 3 The pH sensitivity of the pHluorin-mKate2 fusion protein. The pHluorin-mKate2 fusion protein was expressed in the Huh7.5.1 cells. After fixation and permeabilization, cells were buffered at pH 5.5 (D, H), 6.0 (C, G), 6.5 (B, F), and 7.2 (A, E). The green and far-red fluorescence in the cells were obtained with Biozero BZ-8000 using the filter sets for GFP (A–D) and Texas Red (E–H). *Bar* indicates 10 μm. *This figure is modified from Tanida, I., Ueno, T., & Uchiyama, Y. (2014). A super-ecliptic, pHluorin-mKate2, tandem fluorescent protein-tagged human LC3 for the monitoring of mammalian autophagy.* PLoS One, 9, *e110600.*

(Tanida et al., 2014). The former is an expression plasmid of PK-hLC3 under the control of a CAG promoter (a hybrid promoter of the cytomegalovirus enhancer fused to the chicken β-actin promoter), and is suitable for the transient expression. The latter is a third-generation lentiviral plasmid for pHluorin-mKate2-hLC3 under the control of human ubiquitin c promoter. This is suitable for production of a lentivirus for the expression of pHluorin-mKate2-hLC3 in addition to its transient expression. As respective negative controls, pEX-PK-hLC3ΔG (Addgene, Plasmid # 61459) and FUGW-PK-hLC3ΔG (Addgene, Plasmid # 61461) were generated. These plasmids are expression plasmids for mutant pHluorin-mKate2-hLC3ΔG (Fig. 4) that lacks carboxyl-terminal Gly essential for the formation of LC3-II. The use of negative control is important, since it has been reported that GFP-LC3 shows autophagy-independent puncta (Tanida et al., 2008).

Regarding lentiviral production, Dr. Torono's website (http://tcf.epfl.ch/page-6764-en.html) (Dull et al., 1998) and Addgene websites (https://www.addgene.org/viral-vectors/lentivirus/lenti-guide/#production) are available on line.

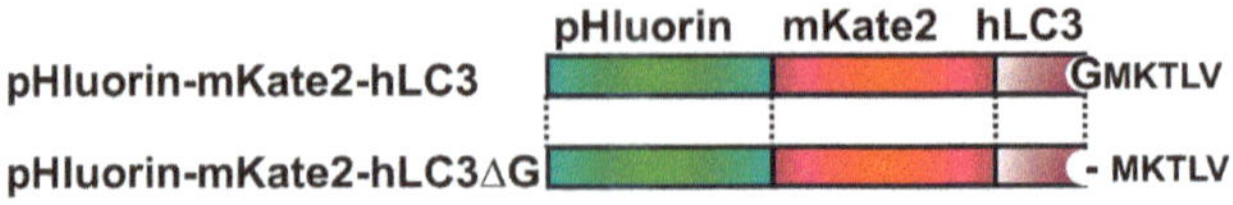

Fig. 4 Schematic representation of pHluorin-mKate2-hLC3. Wild type pHluorin-mKate2-hLC3 and mutant pHluorin-mKate2-hLC3ΔG are shown. Mutant pHluorin-mKate2-hLC3ΔG lacks the carboxyl terminal glycine (G), which is essential for the conjugation of LC3 with phosphatidylethanolamine.

2.2 Monitoring Autophagic Responses Using pHluorin-mKate2-hLC3 in Mammalian Cells

To monitor autophagic activity, the amount of LC3 degradation in lysosomes should be considered (Tanida et al., 2005). To investigate the amount, the degradation in lysosomes is inhibited by an inhibitor cocktail for cathepsins (E64d and pepstatin A). This cocktail inhibits activities of cathepsins B, D, E, H, and L (major lysosomal proteases) without loss of lysosomal acidification.

In some cases, the lysosomotropic agents and bafilomycin A1 are employed to inhibit degradation in lysosomes. However, these reagents are not suitable for monitoring autophagic structures when pHlurorin-mKate2-hLC3 is used to distinguish autolysosomes from autophagosomes and preautophagosomes via its green fluorescence. The lysosomotropic agents (ammonium chloride and chloroquine) inhibit acidification of the intracellular compartments, leading to a defect in the fusion of autophagosomes with lysosomes during autophagy. Bafilomycin A1 is an inhibitor for vacuolar H^+-ATPase, and also inhibits acidification of the intracellular compartments. In theses cases, lysosomal acidification is impaired, resulting in an increase in green-positive and red-positive fluorescence of pHluorin-mKate2-hLC3 on autolysosomes.

1. On the day before the experiment, inoculate cells in appropriate culture medium (e.g., Dulbecco's modified Eagle's medium (Sigma-Aldrich, St. Louis, MO, D6546) containing 10% fetal bovine serum) on a glass-bottom dish (AGC Techno Glass IWAKI, Tokyo, Japan, 3961-035). Incubate the cells at 37°C, 5% CO_2 for 24 h.
2. Transfect pEX-PK-hLC3WT (or pEX-PK-hLC3ΔG as a negative control) into the cells using a lipofection reagent such as FugeneHD (Promega, Madison, WI, E2311) according to the manufacturer's protocol. Incubate the cells at 37°C, 5% CO_2 for 24–48 h.
3. For the starvation-induced autophagy, cells are incubated in Krebs-Ringer bicarbonate buffer (118.5 m*M* NaCl, 4.74 m*M* KCl, 1.18 m*M* KH_2PO_4, 23.4 m*M* $NaHCO_3$, 6 m*M* glucose, 2.5 m*M* $CaCl_2$, and

1.18 m*M* $MgSO_4$, pH 7.6) (Ueno, Muno, & Kominami, 1991) or Earle's balanced salt solution (Thermo Fisher Scientific, Waltham, MA, 24010043). If Hanks' balanced salt solution (Thermo Fisher Scientific, 24020117) is employed, its buffering and incubation conditions should be cared. Hanks' balanced salt solution should be buffered by adding 10 m*M* hydroxyethyl piperazine ethane sulfonate (HEPES), and cells are incubated without control of CO_2. In addition, the effect of endocytosed HEPES on acidification in intracellular components should be considered.

When an inhibitor for mTORC1 such as rapamycin (or an inhibitor for mTor kinase such as torin1) is used to induce autophagy, cells are incubated in the culture medium, and a reagent is added to the medium. 100 n*M* of rapamycin (or 100 n*M* torin1) is sufficient for this purpose in general (Klionsky et al., 2016; Tanida & Waguri, 2010; Thoreen et al., 2009).

To inhibit degradation in lysosomes, E64d (inhibitor for cathepsin B, H, and L) (Peptide Institute, Ibaraki, Osaka, Japan, 4321-v) and pepstatin A (inhibitor for cathepsin D and E) (Peptide Institute, 4397-v) are added to the medium at a final concentration of 10 μg/mL.

4. Incubate the cells at 37°C, 5% CO_2 for 2–6 h. The fluorescence of the fluorescent proteins in the living cells expressing wild-type pHluorin-mKate2-hLC3 is monitored using a confocal laser-scanning microscope such as Olympus FluoView FV1000 (Shinjuku, Tokyo, Japan).
5. To estimate the colocalization of pHluorin-signals (green fluorescence) with mKate2-signals (red-fluorescence), the obtained images are analyzed with Pearson's correlation coefficient (PCC) analysis with Costes' method (Costes et al., 2004). For this purpose, we used an ImageJ software (http://imagej.nih.gov/ij/) (Schneider, Rasband, & Eliceiri, 2012) with a JACoP (just another colocalization) plugin (http://www.blackwell-synergy.com/doi/pdf/10.1111/j.1365-2818.2006.01706.x) (Bolte & Cordelieres, 2006).

2.3 Interpretation of the Results

When autophagy is activated, intracellular puncta of pHluorin-mKate2-hLC3 (PK-hLC3) is increased (Fig. 5). At the same time, degradation of pHluorin-mKate2-hLC3 in lysosomes is also activated. Therefore, a temporally increased number of puncta of pHluorin-mKate2-hLC3 under starvation conditions is insufficient to estimate autophagic activity (Tanida et al.,

2005). To estimate autophagic flux using pHluorin-mKate2-hLC3, the number of its green-negative and red-positive puncta in the presence of E64d and pepstatin A should be investigated (Fig. 5). When autophagy is activated, the number of green-negative and red-positive puncta (i.e., accumulated autolysosomes) is increased in the presence of these inhibitors. In this case, the PCC value between green-positive image and red-positive image is decreased, indicating an increase in green-negative and red-positive puncta.

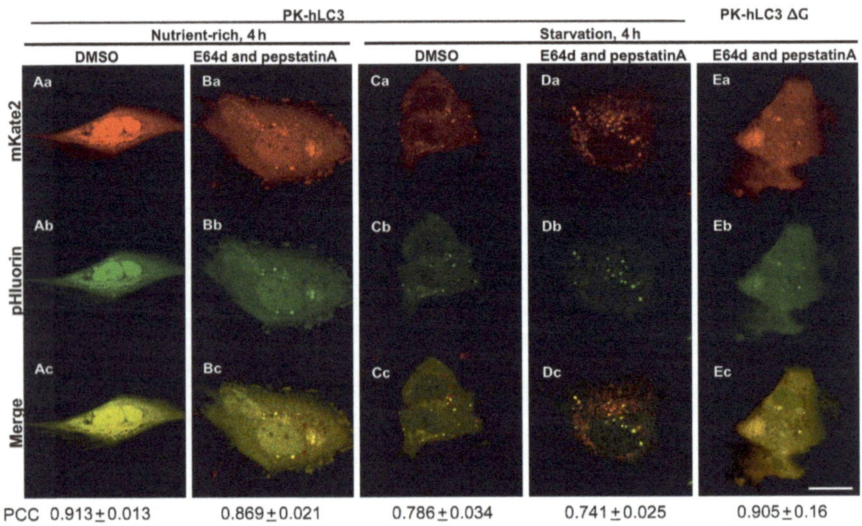

Fig. 5 Formation of the puncta of pHluorin-mKate2-hLC3 during autophagy. The pHluorin-mKate2-hLC3 (PK-hLC3) was expressed in Huh7.5.1 cells. The cells were incubated in the Krebs-Ringer buffer for 4 h as starvation conditions in the presence (E64d and pepstatin A) (Da–Dc) or absence (DMSO) (Ca-Cc) of 10 μg/mL E64d and 10 μg/mL pepstatin A (Starvation, 4 h). As nutrient-rich conditions, cells were incubated in the cultured medium (Nutrient-rich, 4 h) (Aa–Ac, Ba–Bc). The pHluorin-mKate2-hLC3△G (PK-hLC3△G) (Ea–Ec) was expressed in the cells instead of the pHluorin-mKate2-hLC3 under the same conditions (Da–Dc). The *far-red* (mKate2) and *green* (pHluorin) fluorescence in the cells were monitored using an Olympus FluoView FV1000 confocal laser scanning microscope. "Merge" indicates the merging of the *green* (pHluorin) and *far-red* images (mKate2). Pearson's correlation coefficient (PCC) analysis with Costes' method was used as a measure of colocalization of mKate2 signals with pHluorin signals. The mean PCC value ± S.E. of at least 20 cells is shown on the bottom. *This figure is modified from Tanida, I., Ueno, T., & Uchiyama, Y. (2014). A super-ecliptic, pHluorin-mKate2, tandem fluorescent protein-tagged human LC3 for the monitoring of mammalian autophagy.* PLoS One, 9, *e110600.*

Of course, when a mutant PK-hLC3ΔG as well as wild-type PK-hLC3 shows punctate patterns under your experimental conditions, the formation of puncta is independent of autophagy.

3. CONCLUDING REMARKS

Using a pH-sensitive green fluorescent protein, pHluorin, we generated a third-generation of tandem fluorescent protein-tagged LC3, pHluorin-mKate2-hLC3, for monitoring autophagic structures. The fluorescence of pHluorin is significantly quenched at pH 6.5, and is almost completely abolished at pH 5.5–6.0. This pH-sensitivity is important and useful to monitor autophagosome–lysosome fusion and autolysosomes, since intraluminal pH of autophagosomes is acidified after its fusion with lysosomes. In addition to "conventional" autophagy, organelle-specific autophagy (mitophagy/mitochondria-specific autophagy, pexophagy/peroxisome-specific autophagy, reticulophagy/ER-specific autophagy, etc.) has important roles against mammalian diseases. Organelle-specific autophagy contributes the quality control of organelle, resulting in the effective organelle-specific functions. Defects in organelle-specific autophagy lead to cell death and diseases including neurodegenerative diseases, peroxisomal disease, and lysosomal diseases. A super-ecliptic pHluorin will be applied for monitoring the organelle-specific autophagy in the future.

ACKNOWLEDGMENTS

This work was supported by a grant JSPS KAKENHI 15K12749 (IT), a grant from Institute for Environmental & Gender-specific Medicine, Juntendo University (IT), and from Research Institute for Diseases of Old Age, Juntendo University School of Medicine (IT).

REFERENCES

Ai, H. W., Olenych, S. G., Wong, P., Davidson, M. W., & Campbell, R. E. (2008). Hue-shifted monomeric variants of Clavularia cyan fluorescent protein: Identification of the molecular determinants of color and applications in fluorescence imaging. *BMC Biology*, *6*, 13.

Bolte, S., & Cordelieres, F. P. (2006). A guided tour into subcellular colocalization analysis in light microscopy. *Journal of Microscopy*, *224*, 213–232.

Costes, S. V., Daelemans, D., Cho, E. H., Dobbin, Z., Pavlakis, G., & Lockett, S. (2004). Automatic and quantitative measurement of protein-protein colocalization in live cells. *Biophysical Journal*, *86*, 3993–4003.

Dull, T., Zufferey, R., Kelly, M., Mandel, R. J., Nguyen, M., Trono, D., & Naldini, L. (1998). A third-generation lentivirus vector with a conditional packaging system. *Journal of Virology*, *72*, 8463–8471.

Kabeya, Y., Mizushima, N., Ueno, T., Yamamoto, A., Kirisako, T., Noda, T., … Yoshimori, T. (2000). LC3, a mammalian homologue of yeast Apg8p, is localized in autophagosome membranes after processing. *The EMBO Journal*, *19*, 5720–5728.

Katayama, H., Yamamoto, A., Mizushima, N., Yoshimori, T., & Miyawaki, A. (2008). GFP-like proteins stably accumulate in lysosomes. *Cell Structure and Function*, *33*, 1–12.

Kimura, S., Noda, T., & Yoshimori, T. (2007). Dissection of the autophagosome maturation process by a novel reporter protein, tandem fluorescent-tagged LC3. *Autophagy*, *3*, 452–460.

Klionsky, D. J., Abdelmohsen, K., Abe, A., Abedin, M. J., Abeliovich, H., Acevedo Arozena, A., … Zughaier, S. M. (2016). Guidelines for the use and interpretation of assays for monitoring autophagy (3rd edition). *Autophagy*, *12*, 1–222.

Miesenbock, G., De Angelis, D. A., & Rothman, J. E. (1998). Visualizing secretion and synaptic transmission with pH-sensitive green fluorescent proteins. *Nature*, *394*, 192–195.

Mizushima, N., & Komatsu, M. (2011). Autophagy: Renovation of cells and tissues. *Cell*, *147*, 728–741.

Patterson, G. H., Knobel, S. M., Sharif, W. D., Kain, S. R., & Piston, D. W. (1997). Use of the green fluorescent protein and its mutants in quantitative fluorescence microscopy. *Biophysical Journal*, *73*, 2782–2790.

Schneider, C. A., Rasband, W. S., & Eliceiri, K. W. (2012). NIH Image to ImageJ: 25 years of image analysis. *Nature Methods*, *9*, 671–675.

Tanida, I. (2011). Autophagosome formation and molecular mechanism of autophagy. *Antioxidants & Redox Signaling*, *14*, 2201–2214.

Tanida, I., Minematsu-Ikeguchi, N., Ueno, T., & Kominami, E. (2005). Lysosomal turnover, but not a cellular level, of endogenous LC3 is a marker for autophagy. *Autophagy*, *1*, 84–91.

Tanida, I., Ueno, T., & Kominami, E. (2004). LC3 conjugation system in mammalian autophagy. *The International Journal of Biochemistry & Cell Biology*, *36*, 2503–2518.

Tanida, I., Ueno, T., & Uchiyama, Y. (2014). A super-ecliptic, pHluorin-mKate2, tandem fluorescent protein-tagged human LC3 for the monitoring of mammalian autophagy. *PLoS One*, *9*, e110600.

Tanida, I., & Waguri, S. (2010). Measurement of autophagy in cells and tissues. *Methods in Molecular Biology*, *648*, 193–214.

Tanida, I., Yamaji, T., Ueno, T., Ishiura, S., Kominami, E., & Hanada, K. (2008). Consideration about negative controls for LC3 and expression vectors for four colored fluorescent protein-LC3 negative controls. *Autophagy*, *4*, 131–134.

Thoreen, C. C., Kang, S. A., Chang, J. W., Liu, Q., Zhang, J., Gao, Y., … Gray, N. S. (2009). An ATP-competitive mammalian target of rapamycin inhibitor reveals rapamycin-resistant functions of mTORC1. *The Journal of Biological Chemistry*, *284*, 8023–8032.

Ueno, T., Muno, D., & Kominami, E. (1991). Membrane markers of endoplasmic reticulum preserved in autophagic vacuolar membranes isolated from leupeptin-administered rat liver. *The Journal of Biological Chemistry*, *266*, 18995–18999.

Zhou, C., Zhong, W., Zhou, J., Sheng, F., Fang, Z., Wei, Y., … Lin, J. (2012). Monitoring autophagic flux by an improved tandem fluorescent-tagged LC3 (mTagRFP-mWasabi-LC3) reveals that high-dose rapamycin impairs autophagic flux in cancer cells. *Autophagy*, *8*, 1215–1226.

CHAPTER SEVEN

Production of Human ATG Proteins for Lipidation Assays

Y. Zheng[*,†,1], **Y. Qiu**[*,1], **J.E.C. Gunderson**[*,‡], **B.A. Schulman**[*,†,§,2]

[*]St. Jude Children's Research Hospital, Memphis, TN, United States
[†]University of Tennessee Health Science Center, Memphis, TN, United States
[‡]Hendrix College, Conway, AR, United States
[§]Howard Hughes Medical Institute, St. Jude Children's Research Hospital, Memphis, TN United States
[2]Corresponding author: e-mail address: brenda.schulman@stjude.org

Contents

1. Introduction 98
2. Expression of Human LC3B, ATG7, ATG3, ATG12~ATG5, and ATG16L1 99
   2.1 General Overview of Protein Expression and Purification 99
   2.2 Expression Constructs 100
   2.3 Protein Expression 103
3. Protein Purification 105
   3.1 Preparation of Cell Lysate 105
   3.2 Glutathione-Affinity Chromatography for LC3B, ATG7, ATG3, and ATG12~ATG5-ATG16L1 105
   3.3 Protease-Mediated Elution From Beads for Purification of LC3B, ATG3, and ATG7 106
   3.4 Glutathione-Mediated ATG12~ATG5-ATG16L1 Elution From GS4B 106
   3.5 Nickel-Affinity Chromatography, Imidazole-Mediated Elution From Nickel Beads, and TEV Cleavage to Obtain ATG12 (52–140)~ATG5-ATG16L1 (1–69) Complex 108
   3.6 Ion Exchange Chromatography on ATG3 and ATG12 (52–140)~ATG5-ATG16L1 (1–69) 109
   3.7 Gel Filtration Chromatography for LC3B, ATG7, ATG3, and ATG12~ATG5-ATG16L1 110
4. Storing Purified Proteins 111
5. Concluding Remarks 111
Acknowledgments 111
References 111

Abstract

Humans express several orthologs of yeast Atg8, in the LC3 and GABARAP families, which play crucial roles in autophagy through their covalent ligation to lipids, typically

[1] These authors contributed equally.

Methods in Enzymology, Volume 587
ISSN 0076-6879
http://dx.doi.org/10.1016/bs.mie.2016.09.055

phosphatidylethanolamine (PE), in a process known as lipidation. Lipidation of LC3 and GABARAP regulates numerous facets of the autophagy process, including regulating expansion of the phagophore membrane, recruiting selected cargoes for degradation, and providing an autophagosome membrane-bound platform mediating dynamic interactions with other regulatory proteins. LC3 and GABARAP are families of related ubiquitin-like proteins (UBLs) (referred to here collectively as LC3/GABARAP), and their lipidation involves a divergent UBL conjugation cascade including ATG7, ATG3, and ATG12~ATG5-ATG16L1 acting as E1, E2, and E3 enzymes, respectively. ATG7 initiates LC3/GABARAP conjugation by catalyzing their C-terminal adenylation and conjugation to the catalytic cysteine of ATG3. Ultimately, the ATG12~ATG5-ATG16L1 complex catalyzes LC3/GABARAP ligation to a primary amino group on PE or other acceptor lipids. This chapter describes methods for expressing and purifying human LC3 or GABARAP, ATG7, ATG3, and the ATG12~ATG5-ATG16L1 complex for in vitro studies of LC3/GABARAP lipidation.

1. INTRODUCTION

In eukaryotic cells, autophagy is a tightly regulated process responsible for degrading and recycling cellular components, and eliminating toxic assemblies and pathogens by sequestering cargo within an autophagosome, which fuses with a lysosome (or vacuole in yeast) harboring hydrolytic enzymes (Klionsky & Emr, 2000; Kobayashi, 2015; Wileman, 2013). Several key aspects of this regulation involve ubiquitin-like proteins (UBLs) in the Atg8 family, which form LC3 and GABARAP subfamilies in higher eukaryotes. LC3/GABARAP proteins have molecular weights in solution of approximately 14–17 kDa, but their electrophoretic migration is relatively accelerated upon ligation to lipids such as in autophagic membranes. LC3/GABARAP ligation to lipids (lipidation) plays numerous roles in regulation, including influencing the formation, maturation, and size of autophagosomes, as well as recruiting selected cargoes and regulatory proteins (Ichimura et al., 2000; Ohsumi & Mizushima, 2004; Nakatogawa, Ichimura, & Ohsumi, 2007; Tanida, Ueno, & Kominami, 2008.

LC3 and GABARAP lipidation is orchestrated by a UBL ligation system, specifically conjugating the carboxyl group of the C-terminal glycine of LC3/GABARAP to the amine group of phosphatidylethanolamine (PE) (or potentially other lipids such as phosphatidyl serine) in autophagic membranes (Hanada et al., 2007; Ichimura et al., 2000; Klionsky & Emr, 2000; Sou, Tanida, Komatsu, Ueno, & Kominami, 2006; Tanida, Ueno, & Kominami, 2004a, 2004b). This involves a specialized enzyme cascade, where ATG7 acts as the E1 and ATG3 acting as the E2 to produce covalent, reactive thioester-linked ATG3~LC3/GABARAP ("~" denotes a covalent protein interaction and "-" denotes a noncovalent complex) intermediates. Efficient lipidation requires an ATG12~ATG5-ATG16L1 multiprotein complex acting as the E3. Notably, formation of this E3-like assembly first requires a related conjugation cascade, wherein the UBL ATG12 is covalently conjugated to ATG5 by ATG7 acting as the E1 and ATG10 acting as the E2 (Hanada et al., 2007; Noda, Fujioka, Hanada, Ohsumi, & Inagaki, 2013), and then the ATG12~ATG5 binds ATG16L1 to achieve full E3-like activity toward ATG3~LC3/GABARAP intermediates. Mutations or dysfunction of proteins in the LC3/GABARAP lipidation cascade results in severe diseases and poor prognosis in various eukaryotic organisms including humans (Cadwell, Patel, Komatsu, & Virgin, 2009; Dezelak, Repnik, Koder, Ferkolj, & Potocnik, 2016; Kim et al., 2016; Lenz, Vierstra, Nurnberger, & Gust, 2011; Moloughney et al., 2011; Reed, Morris, Owczarczyk, & Lukacs, 2015; Tindwa et al., 2015; Wu et al., 2014; Xi et al., 2016; Xue, Chiu, & Oleinick, 2010).

Given the importance of LC3 and GABARAP lipidation, and potential interest in targeting their conjugation for therapeutic purposes (Cheng, Ren, Hait, & Yang, 2013; Puri & Chandra, 2014), purified proteins for assaying LC3/GABARAP lipidation and performing structural analyses would be useful (Bae & Park, 2010; Kaiser et al., 2013; Metlagel, Otomo, Ohashi, Takaesu, & Otomo, 2014; Metlagel, Otomo, Takaesu, & Otomo, 2013; Nath et al., 2014; Otomo, Metlagel, Takaesu, & Otomo, 2013; Sakoh-Nakatogawa et al., 2013; Taherbhoy et al., 2011). However, our optimal expression and purification procedures differ between the human LC3/GABARAP lipidation components and their yeast counterparts (Qiu, Hofmann, Coats, Schulman, & Kaiser, 2013). Thus, in this chapter, we describe detailed methods for expression and purification of human ATG proteins involved in LC3/GABARAP lipidation, including ATG7, ATG3, ATG12~ATG5-ATG16L1, and a method for expressing LC3B as exemplary for the human Atg8 orthologs.

2. EXPRESSION OF HUMAN LC3B, ATG7, ATG3, ATG12~ATG5, AND ATG16L1

2.1 General Overview of Protein Expression and Purification

LC3B (emblematic of LC3/GABARAP family members), ATG3, and ATG16L1 were expressed as GST-fusion proteins in *Escherichia coli*, whereas ATG7 and the ATG12~ATG5 conjugate were expressed through distinct

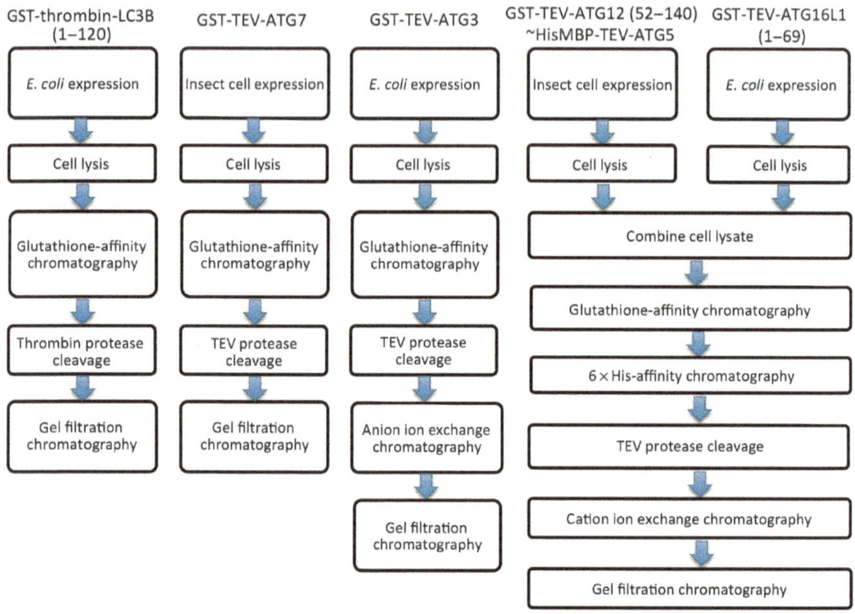

Fig. 1 Flow-chart indicating the expression system and purification strategies for production of human ATG proteins for lipidation assays.

methods in insect cells (Berger, Fitzgerald, & Richmond, 2004). An overview of the subsequent purification strategies is shown in Fig. 1.

2.2 Expression Constructs

1. A truncated version of LC3B, corresponding to residues 1–120 that form the activated form with a C-terminal glycine that is the site of chemical reactions (Tanida et al., 2004a, 2004b), is expressed from pGEX-4T-1 (Table 1, Fig. 2A) in *E. coli*.
2. Full-length ATG7 was expressed from a modified version of pFastBac, with TEV-cleavable N-terminal GST tag (pFastBac-GST) (Table 1, Fig. 2B) for expression in insect cells.
3. Full-length ATG3 was expressed from a modified version of pGEX-4T-1 harboring a cleavage site for TEV protease instead of thrombin between GST and ATG3 (pGEX-TEV) (Table 1, Fig. 2C), for expression in *E. coli*.
4. A truncated version of ATG16L1 corresponding to residues 1–69 was expressed from pGEX-TEV (Table 1, Fig. 2D) in *E. coli*.

Table 1 Plasmids Used in This Chapter

Proteins Expressed	Plasmids Generated	Functions
LC3B (1–120)	pGEX-4T-1-LC3B	Protein expression in *E. coli*
ATG7	pFastBac-GST-ATG7	To generate bacmid for protein expression in insect cells
ATG3	pGEX-TEV-ATG3	Protein expression in *E. coli*
ATG12 (52–140)~ATG5	pLIB-GST-TEV-ATG12	To generate pBIG1a plasmid
	pLIB-His-TEV-ATG5	To generate pBIG1a plasmid
	pLIB-ATG10	To generate pBIG1a plasmid
	pLIB-ATG7	To generate pBIG1a plasmid
	pBIG1a-ATG12-ATG5-ATG10-ATG7	To generate bacmid for protein expression in insect cells
ATG16L1 (1–69)	pGEX-TEV-ATG16L1	Protein expression in *E. coli*

5. The ATG12~ATG5 subcomplex corresponds to a truncated version of ATG12 (residues 52–140) that comprises the ubiquitin-like domain (Hanada & Ohsumi, 2005) linked via an isopeptide bond between its C-terminus to Lys130 on full-length ATG5 (Hanada et al., 2007; Otomo et al., 2013). The linkage between ATG12 and ATG5 is achieved by their coexpression with ATG7 and ATG10 in insect cells using the biGBac system (Weissmann et al., 2016). First, four separate plasmids were generated. The coding sequences were inserted into pLIB vectors, with ATG7, and ATG10 untagged, ATG5 harboring a N-terminal TEV cleavable His-tag, and ATG12 (52–140) with a N-terminal TEV cleavable GST-tag (Table 1, Fig. 2E). Next, GST-TEV-ATG12 (52–140), His-TEV-ATG5, ATG10, and ATG7 along with the promoter and terminator from the pLIB vectors were sequentially amplified using primers described in Weissmann et al. (2016), and coassembled into the pBIG1a vector (Table 1, Fig. 2F) using the method of Gibson Assembly (Gibson, 2011; Gibson et al., 2009; Weissmann et al., 2016).
6. All constructs were generated through transformation into DH5alpha (NEB), bacterial growth, minipreps, and verified by automated sequencing. The recommended antibiotic concentrations are: 100 μg/mL for ampicillin, 50 μg/mL for kanamycin and spectinomycin, 25 μg/mL for chloramphenicol, 10 μg/mL for tetracyclin, and 7 μg/mL for gentamycin.

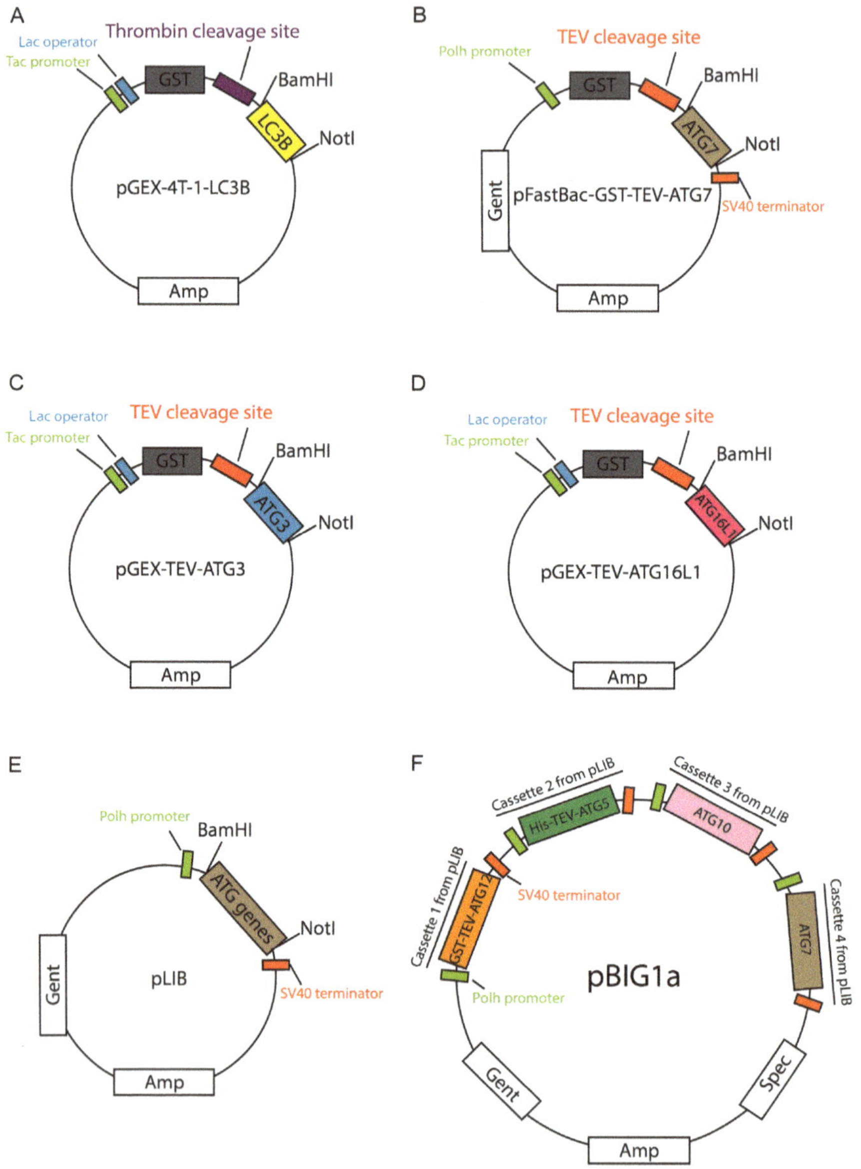

Fig. 2 Graphical representation of plasmids used in this chapter. (A) pGEX-4T-1-LC3B for LC3B (1–120) expression in *E. coli*. (B) pFastBac-GST-ATG7 to construct bacmid for ATG7 expression in insect cells. (C) pGEX-TEV-ATG3 for ATG3 expression in *E. coli*. (D) pGEX-TEV-ATG16L1 for ATG16L1 (1–69) expression in *E. coli*. (E) pLIB plasmid containing a GST-TEV-ATG12 (52–140), His-TEV-ATG5, ATG10, or ATG7 expression cassette to generate pBIG1a plasmid. (F) pBIG1a-ATG12-ATG5-ATG10-ATG7 plasmid to construct bacmid for ATG12 (52–140)∼ATG5 expression in insect cells.

2.3 Protein Expression

2.3.1 *Expression in* E. coli

1. Transform plasmid into BL21-CodonPlus (DE3)-RIL *E. coli* competent cells (Agilent Technologies), plate on LB-agar with appropriate antibiotics. Incubate at 37°C for 12–18 h.
2. Prepare a starter culture from a single colony in 5 mL LB medium with antibiotics. Shake culture at 37°C, 200 rpm for 10–16 h.
3. Dilute the 5 mL starter culture into 1 L LB medium with appropriate antibiotics. Grow cells at 37°C, 200 rpm, and monitor closely by measuring the absorbance at 600 nm. When OD_{600} reaches 0.8–1.0, add IPTG at a final concentration of 0.6 m*M*, change shaker temperature to 18°C, and grow with shaking at 200 rpm for 16–20 h.

2.3.2 *Expression in Insect Cells*

1. Transform pFastBac or pBIG1a vector containing expression cassettes into DH10EMBacY *E. coli* competent cells (Geneva Biotech). This enables transposase-mediated recombination resulting in a bacmid containing expression cassettes (Bac-to-Bac Baculovirus Expression System, Thermo Fisher Scientific).
2. Screen for colonies harboring bacmids that have undergone recombination using the blue/white identification method described in the Bac-to-Bac Baculovirus Expression System manual. Briefly, a LB plate containing ampicillin, kanamycin, tetracycline, and gentamycin, is spread with 40 μL 40 mg/mL X-Gal (dissolved in dimethylformamide) and 40 μL 0.1 *M* IPTG is incubated at 37°C for 4 h to let the dimethylformamide evaporate before plating the transformation product.
3. Incubate at 37°C for at least 24 h until colonies grow and background colonies with a clear blue color appear. White colonies are candidates for harboring a correctly recombined bacmid. PCR screen for candidates that contains all the target genes. At the same time, restreak white colonies on a fresh plate with X-gal, IPTG, and the four antibiotics. Include a blue colony as a control for color development. Perform bacmid extraction from culture of a single white colony with positive PCR screening result.
4. All insect cell medium used in this chapter must be prewarmed at 27°C. Transfect to obtain baculovirus in five steps. (a) Coat 6 well flat bottom culture plate (CoStar) with 5×10^6 Sf9 insect cells (Thermo Fisher

Scientific) per well and incubate at 27°C for 30 min for cell attachment. (b) Dilute 200 ng bacmid with 100 μL SFX serum-free insect cell medium (GE Healthcare Life Sciences) in a 15 mL polystyrene tube (Falcon) (do not use polypropylene tube). Mix 10 μL FuGENE (Promega) with diluted bacmid. After incubation at 27°C for 30 min, add 900 μL medium to achieve a 1 mL volume. (c) Aspirate medium from Sf9 cells. Evenly apply 1 mL bacmid-FuGENE mixture onto cells and incubate for 4 h. (d) Add another 1 mL medium and incubate for 4 days. (e) Collect supernantant that contains P1 virus by syringe and pass through a 0.2 μm PES syringe filter (Thermo Fisher Scientific). Virus can be stored at 4°C.

5. Amplify virus to generate a P2 stock: Plate 3×10^7 Sf9 cells on a 150 mm × 25 mm dish (Corning) and incubate 30 min at 27°C to allow cell adherence (add more SFX serum-free insect cell medium if liquid volume is below 12 mL or fails to cover the whole dish surface). Apply 0.5 mL P1 virus dropwise to adhered cells, gently rock to spread virus evenly, and incubate plate at 27°C for 2 h. Then add more medium to a total volume of 25 mL per dish and incubate at 27°C for another 3 days. Collect supernatant that contains virus into a 50 mL conical tube followed by centrifugation at 500 × *g* for 10 min to remove floating cells. Save the clarified supernatant as P2 amplified virus.
6. Amplify virus to generate a P3 stock. Use 1 mL P2 virus for infection using the same procedure for P2 amplification, except may scale up depending on the ultimate culture size for protein production. The volume of P3 amplified virus produced by each 150 mm cell culture dish supports approximately 2 L Hi5 cells during infection to produce protein. Repeat if further virus amplification is needed.
7. Infection for protein production. Pellet Hi5 insect cell (Thermo Fisher Scientific) culture at log phase by centrifugation at 500 × *g* for 10 min. Then resuspend the cell pellet by medium to reach a cell density of 1.25×10^7/mL in PC Erlenmeyer flasks (VWR). Add 5 mL P3 amplified virus per 100 mL resuspended cells, then incubate cells in shaker at 27°C, 75 rpm for 2 h with ambient CO_2.
8. Diluted cell culture by adding medium to a cell density of 2.5×10^6/mL (make sure the culture volume is less than 50% of the flask volume). Let cells grow and express target protein in shaker at 27°C, 155 rpm for 3 more days with ambient CO_2.

3. PROTEIN PURIFICATION

3.1 Preparation of Cell Lysate

1. Harvest bacteria cells by centrifugation at 5900 × *g* for 12 min at 4°C, or insect cells at 500 × *g* for 15 min at 4°C.
2. Resuspend cell pellet in 15 mL lysis buffer per liter bacteria culture, or 50 mL per liter insect cell culture. Lysis buffer for all proteins except the ATG12~ATG5 subcomplex is: 25 m*M* Tris–HCl pH 7.6, 200 m*M* NaCl, 2.5 m*M* phenylmethanesulfonyl fluoride (PMSF), 5 m*M* 2-Mercaptoethanol (BME). Lysis buffer for ATG12~ATG5 complex is: 1 × PBS, 400 m*M* NaCl, 2.5 m*M* PMSF, 5 m*M* BME. Also add 20 mg/L aprotinin, 10 mg/L leupeptin, or a protease inhibitor cocktail of choice to insect cell lysis buffer.
3. Lyse bacteria cells by sonicating 7 times, or insect cells 3 times, 10 s each time with intermittent cooling on ice.
4. Clear lysate by centrifugation at 32,000 × *g* for 1 h at 4°C. Collect the supernatant and keep it on ice.

3.2 Glutathione-Affinity Chromatography for LC3B, ATG7, ATG3, and ATG12~ATG5-ATG16L1

For GST-thrombin-LC3B, GST-TEV-ATG7, and GST-TEV-ATG3, glutathione-affinity chromatography is directly applied to their corresponding cell lysates. To obtain GST-TEV-ATG12 (52–140)~His-TEV-ATG5-GST-TEV-ATG16L1 (1–69) complex, cell lysates from insect cells and bacteria are mixed at 1:1 volume ratio and incubated at 4°C for 10 min before processing.

1. Wash Glutathione Sepharose 4B (GS4B, GE Healthcare Life Sciences) 3 times with 10 column volumes of wash buffer in a chromatography column at 4°C. Alternatively, wash the beads with 3 volumes of wash buffer by centrifugation at 500 × *g* for 5 min at 4°C, decanting the wash buffer, and repeating twice. Suggested bead volumes: 0.75 mL per 1 L bacterial culture, or 2 mL per 1 L Hi5 insect cell culture, which is appropriate for both GST and nickel beads and all species of ATG proteins described in this chapter. Wash buffer is the same as lysis buffer but without PMSF, aprotinin, leupeptin, or any other proteasome inhibitor.

2. Incubate lysate with washed beads. Gently rock the mixture at 4°C for 2 h for GS4B beads or 1 h for nickel beads to allow full binding.
3. Transfer lysate with beads into a Poly-Prep chromatography column (Bio-Rad) at 4°C. Allow lysate to flow through by gravity to generate an even resin surface in column. Although the flow-through is unneeded in principle, it is a good practice to save the flow-through for trouble-shooting in the event of problems.
4. If the lysate-bead solution is too thick to pass through the column, centrifugate the solution in 50 mL tubes at 500 × *g* for 5 min at 4°C. Collect the supernatant as flow through. Gently resuspend pelleted beads in 10 column volumes of wash buffer and follow step 3.
5. Wash resin in column with 10 column volumes of wash buffer without disturbing the flat resin surface. Repeat 3 times.

3.3 Protease-Mediated Elution From Beads for Purification of LC3B, ATG3, and ATG7

1. Measure rough protein concentration of bead turbid solution using Bio-Rad protein assay with a standard curve. Bovine serum albumin is a suitable standard.
2. Store 10 μL of uncleaved sample as a negative control for cleavage efficiency check before adding protease. Add thrombin protease to LC3B (1 mg thrombin per 100 mg of total protein), or TEV protease to ATG7 and ATG3 (2 mg TEV per 100 mg total protein). Keep gently rotating on a rocking shaker overnight to allow full cleavage.
3. To be certain that all desired protein is cleaved from the GST-tag, examine the bead slurry by SDS-PAGE, using a 15% acrylamide gel and Coomassie-staining, comparing cleaved samples to uncleaved samples (LC3B: Fig. 3, Lanes a and b; ATG7: Fig. 4, Lanes a and b; ATG3: Fig. 5, Lanes a and b).
4. Transfer slurry into an empty Poly-Prep chromatography column prerinsed with wash buffer, and collect flow through. When liquid completely passes through, apply another 1 column volume of wash buffer on the top of resin, and collect this second fraction containing residual protein.

3.4 Glutathione-Mediated ATG12~ATG5-ATG16L1 Elution From GS4B

1. Gently apply 0.7 resin volume of GS4B elution buffer on the top of resin surface while keeping stopcock closed. After a 10-min incubation, open

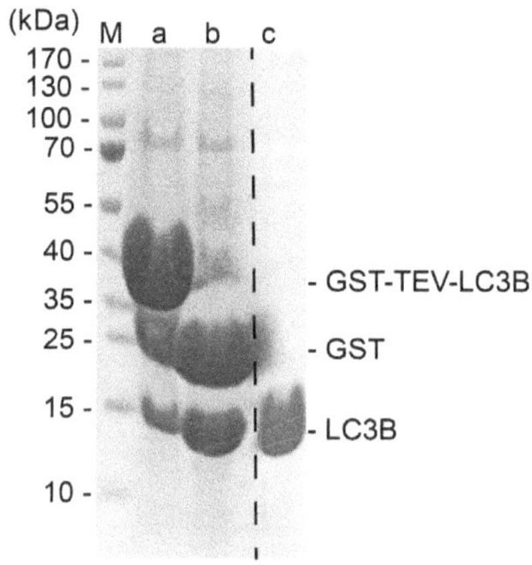

Fig. 3 15% SDS-PAGE showing LC3B (1–120) at different steps of purification. Lane M shows the molecular weight markers. Lane a shows constituents of bead slurry after binding GS4B resin. Lane b shows constituents of bead slurry after thrombin treatment. Lane c shows final purified LC3B (1–120) after gel filtration chromatography.

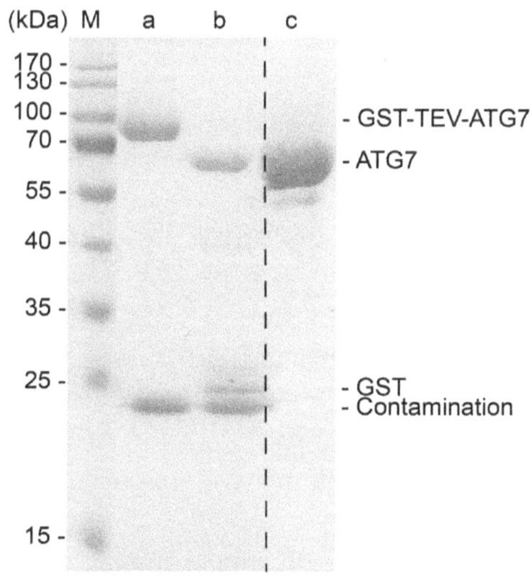

Fig. 4 15% SDS-PAGE showing ATG7 at different steps of purification. Lane M shows the molecular weight markers. Lane a shows constituents of bead slurry after binding GS4B resin. Lane b shows constituents of bead slurry after TEV treatment. Lane c shows final purified ATG7 after gel filtration chromatography.

the stopcock and collect the eluate in a tube. Repeat 6 times and keep elution fractions in separate tubes. GS4B elution buffer: wash buffer supplemented with 20 m*M* glutathione (reduced, free acid) (EMD Millipore), adjust pH by NaOH solution to the same as wash buffer.

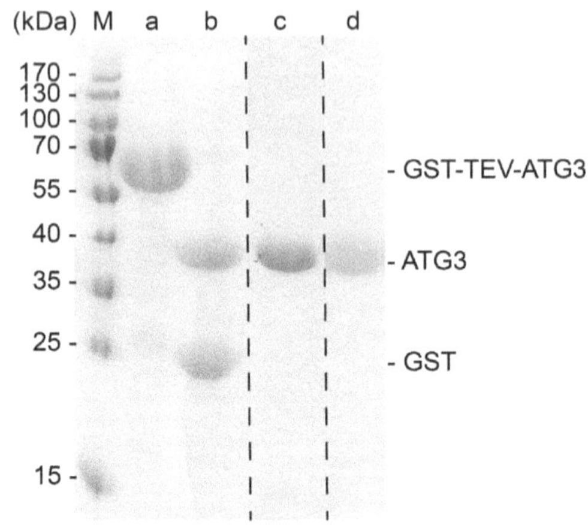

Fig. 5 15% SDS-PAGE showing ATG3 at different steps of purification. Lane M shows the molecular weight markers. Lane a shows constituents of bead slurry after binding GS4B resin. Lane b shows constituents of bead slurry after TEV treatment. Lane c shows purified ATG3 after anion ion exchange chromatography. Lane d shows final purified ATG3 after gel filtration chromatography.

2. Run each eluted fraction on a 15% SDS-PAGE gel to determine which fractions contain the majority of target protein. Pool desired fractions.

3.5 Nickel-Affinity Chromatography, Imidazole-Mediated Elution From Nickel Beads, and TEV Cleavage to Obtain ATG12 (52–140)~ATG5-ATG16L1 (1–69) Complex

After glutathione chromatography and glutathione mediated elution, the excessive GST-TEV-ATG16L1 (1–69) is further removed by nickel affinity chromatography.

1. Wash HIS-Select Nickel Affinity Gel (Sigma-Aldrich) 3 times with 10 column volumes of wash buffer in a chromatography column at 4°C. Then perform nickel-affinity chromatography to enrich the ATG5-bound complex. Apply the glutathione elution of GST-TEV-ATG12 (52–140)~His-TEV-ATG5-GST-TEV-ATG16L1 (1–69) to the HIS-Select Nickel Affinity Gel beads and incubate following the same procedure described above for glutathione-affinity chromatography.
2. Perform an imidazole elution following steps described in glutathione elution using nickel bead elution buffer (wash buffer supplemented with 100 m*M* imidazole pH 7.5).

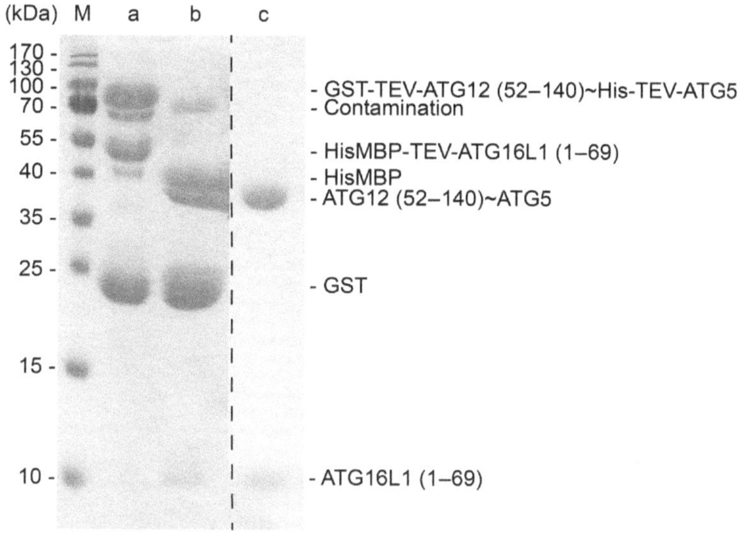

Fig. 6 15% SDS-PAGE showing ATG12 (52–140)~ATG5-ATG16L1 (1–69) complex at different stages of purification. Lane M shows the molecular weight markers. Lane a shows elution from nickel-affinity chromatography, capturing His- and His-MBP tags on ATG5 and ATG16L1 (1–69), respectively. This is the second step in the purification, following an initial purification by glutathione affinity chromatography to capture GST-tag on ATG12. Lane b shows product of TEV cleavage reaction preformed during dialysis. Lane c shows final purified ATG12 (52–140)~ATG5-ATG16L1 (1–69) complex after subsequent cation ion exchange and gel filtration chromatography.

3. Store 10 μL uncleaved sample as a negative control to check cleavage efficiency. Add TEV protease (2 mg TEV per 100 mg total protein) to imidazole eluate, then dialyze in wash buffer while performing cleavage overnight at 4°C.
4. Examine cleavage completeness by SDS-PAGE, using a 15% acrylamide gel and Coomassie-staining (Fig. 6, Lanes a and b).
5. The cleaved GST and HisMBP tags are removed from ATG12 (52–140)~ATG5-ATG16L1 (1–69) complex by further cation ion exchange purification described below.

3.6 Ion Exchange Chromatography on ATG3 and ATG12 (52–140)~ATG5-ATG16L1 (1–69)

Perform anion ion exchange to tag-free ATG3 and cation ion exchange to tag-free ATG12 (52–140)~ATG5-ATG16L1 (1–69) accordingly to their isoelectric points, before ultimately purifying by gel filtration chromatography. This step is optional for LC3B and ATG7.

1. Equilibrate a 5 mL HiTrap SP (cation exchange) or a Q (anion exchange) HP column (GE Healthcare Life Sciences) on an AKTA FPLC (GE Healthcare Life Sciences) in 95% buffer A with 5% buffer B. Cation exchange buffer A: 25 m*M* HEPES pH 7.0, 5 m*M* DTT. Anion exchange buffer A: 25 m*M* Tris–HCl pH 8.0, 5 m*M* DTT. Buffer B corresponds to buffer A with 1 *M* NaCl.
2. Dilute protein samples with buffer A to a final NaCl concentration of 50 m*M*.
3. Clarify samples by centrifugation at 4000 × *g* for 5 min at 4°C before loading onto a FPLC system. If loading is through a sample pump, equilibrate the sample pump in buffer A. The ATG proteins described in this chapter can be concentrated to 10 mg/mL if desired to reduce volume prior to loading onto FPLC.
4. Load sample onto column with a suggested loading rate of 2 mL/min. After loading, wash column by 5 column volumes of 95% buffer A with 5% buffer B.
5. After all unbound sample is washed out, initiate gradient, raising buffer B proportion from 5% to 50% over 25 column volumes at 2 mL/min flow rate, and collecting fractions with a volume of 2 mL.
6. Examine peak fractions by SDS-PAGE with Coomassie-stained 15% acrylamide gels to determine which fractions contain the majority of the ATG proteins (the combined fractions of ATG3 is shown in Fig. 5, Lane c; ATG12~ATG5-ATG16L1 is not shown).

3.7 Gel Filtration Chromatography for LC3B, ATG7, ATG3, and ATG12~ATG5-ATG16L1

Superdex 200 10/300 GL columns (GE Healthcare Life Sciences) are used for all the ATG proteins described here, except for LC3B using a Superdex 75 10/300 GL column (GE Healthcare Life Sciences). Protein samples of LC3B and ATG7 used in this step are directly collected from protease cleavage, while ATG3 and ATG12~ATG5-ATG16L1 are prepurified by ion exchange.

1. Equilibrate a Superdex 75/200 column on an AKTA FPLC system with storage buffer (25 m*M* Tris–HCl pH 7.6, 200 m*M* NaCl, 5 m*M* DTT). Increase NaCl concentration to 500 m*M* in the buffer used for LC3B, or to 300 m*M* in the buffer used for ATG12 (52–140)~ATG5-ATG16L1 (1–69).
2. Concentrate protein samples by Amicon Ultra Centrifugal Filter Units (EMD Millipore) with appropriate molecular weight cutoff. Transfer

concentrated protein samples to one or multiple 1.7 mL microcentrifuge tubes. Perform centrifugation at 16,000 × *g* for 10 min at 4°C to clarify samples before loading onto a FPLC system.

3. Load 1–2 mL clarified protein sample onto a Superdex 200 column, run gel filtration in storage buffer, and collect fractions with a fraction volume of 0.5 mL.
4. Examine peak fractions by SDS-PAGE with Coomassie-stained 15% acrylamide gels to determine which fractions contain the majority of the ATG proteins. Pool desired fractions (the pooled fractions of LC3B, ATG7, ATG3, and ATG12~ATG5-ATG16L1 are shown in Fig. 3, Lane c; Fig. 4, Lane c; Fig. 5, Lane d; Fig. 6, Lane c, respectively).

4. STORING PURIFIED PROTEINS

Purified proteins can be concentrated to approximately 10 mg/mL and stored in 20 μL aliquots at −80°C after flash-freezing in liquid nitrogen. Thaw on ice for use and refreeze for storage up to twice to ensure protein activity.

5. CONCLUDING REMARKS

Proteins prepared as described earlier are eligible for multiple biochemical assays, including LC3B lipidation. In our experience, aliquots of purified ATG proteins stored at −80°C are stable for up to a year, giving consistent results in assays for LC3B lipidation, formation of thioester-bonded ATG7~LC3B and ATG3~LC3B intermediates, binding using BioLayer Interferometry (Octet Red, ForteBio) or isothermal titration calorimetry (ITC), and for structural studies using NMR and crystallography.

ACKNOWLEDGMENTS

This work was supported by ALSAC, the NIH R01GM077053, the Howard Hughes Medical Institute, and AHA 14POST19890021. B.A.S. is an Investigator of the Howard Hughes Medical Institute.

REFERENCES

Bae, J. Y., & Park, H. H. (2010). Purification and characterization of a ubiquitin-like system for autophagosome formation. *Journal of Microbiology and Biotechnology*, *20*(12), 1647–1652.

Berger, I., Fitzgerald, D. J., & Richmond, T. J. (2004). Baculovirus expression system for heterologous multiprotein complexes. *Nature Biotechnology*, *22*(12), 1583–1587.

Cadwell, K., Patel, K. K., Komatsu, M., Virgin, H. W. T., & Stappenbeck, T. S. (2009). A common role for Atg16L1, Atg5 and Atg7 in small intestinal Paneth cells and Crohn disease. *Autophagy*, *5*(2), 250–252.

Cheng, Y., Ren, X., Hait, W. N., & Yang, J. M. (2013). Therapeutic targeting of autophagy in disease: Biology and pharmacology. *Pharmacological Reviews*, *65*(4), 1162–1197.

Dezelak, M., Repnik, K., Koder, S., Ferkolj, I., & Potocnik, U. (2016). A prospective pharmacogenomic study of Crohn's disease patients during routine therapy with anti-TNF-alpha drug adalimumab: Contribution of ATG5, NFKB1, and CRP genes to pharmacodynamic variability. *Omics: A Journal of Integrative Biology*, *20*(5), 296–309.

Gibson, D. G. (2011). Enzymatic assembly of overlapping DNA fragments. *Methods in Enzymology*, *498*, 349–361.

Gibson, D. G., Young, L., Chuang, R. Y., Venter, J. C., Hutchison, C. A., 3rd, & Smith, H. O. (2009). Enzymatic assembly of DNA molecules up to several hundred kilobases. *Nature Methods*, *6*(5), 343–345.

Hanada, T., Noda, N. N., Satomi, Y., Ichimura, Y., Fujioka, Y., Takao, T., et al. (2007). The Atg12–Atg5 conjugate has a novel E3-like activity for protein lipidation in autophagy. *The Journal of Biological Chemistry*, *282*(52), 37298–37302.

Hanada, T., & Ohsumi, Y. (2005). Structure–function relationship of Atg12, a ubiquitin-like modifier essential for autophagy. *Autophagy*, *1*(2), 110–118.

Ichimura, Y., Kirisako, T., Takao, T., Satomi, Y., Shimonishi, Y., Ishihara, N., et al. (2000). A ubiquitin-like system mediates protein lipidation. *Nature*, *408*(6811), 488–492.

Kaiser, S. E., Qiu, Y., Coats, J. E., Mao, K., Klionsky, D. J., & Schulman, B. A. (2013). Structures of Atg7–Atg3 and Atg7–Atg10 reveal noncanonical mechanisms of E2 recruitment by the autophagy E1. *Autophagy*, *9*(5), 778–780.

Kim, M., Sandford, E., Gatica, D., Qiu, Y., Liu, X., Zheng, Y., et al. (2016). Mutation in ATG5 reduces autophagy and leads to ataxia with developmental delay. *eLife*, *5*, e12245.

Klionsky, D. J., & Emr, S. D. (2000). Autophagy as a regulated pathway of cellular degradation. *Science*, *290*(5497), 1717–1721.

Kobayashi, S. (2015). Choose delicately and reuse adequately: The newly revealed process of autophagy. *Biological & Pharmaceutical Bulletin*, *38*(8), 1098–1103.

Lenz, H. D., Vierstra, R. D., Nurnberger, T., & Gust, A. A. (2011). ATG7 contributes to plant basal immunity towards fungal infection. *Plant Signaling & Behavior*, *6*(7), 1040–1042.

Metlagel, Z., Otomo, C., Ohashi, K., Takaesu, G., & Otomo, T. (2014). Structural insights into E2-E3 interaction for LC3 lipidation. *Autophagy*, *10*(3), 522–523.

Metlagel, Z., Otomo, C., Takaesu, G., & Otomo, T. (2013). Structural basis of ATG3 recognition by the autophagic ubiquitin-like protein ATG12. *Proceedings of the National Academy of Sciences of the United States of America*, *110*(47), 18844–18849.

Moloughney, J. G., Monken, C. E., Tao, H., Zhang, H., Thomas, J. D., Lattime, E. C., et al. (2011). Vaccinia virus leads to ATG12–ATG3 conjugation and deficiency in autophagosome formation. *Autophagy*, 7(12), 1434–1447.

Nakatogawa, H., Ichimura, Y., & Ohsumi, Y. (2007). Atg8, a ubiquitin-like protein required for autophagosome formation, mediates membrane tethering and hemifusion. *Cell*, *130*(1), 165–178.

Nath, S., Dancourt, J., Shteyn, V., Puente, G., Fong, W. M., Nag, S., et al. (2014). Lipidation of the LC3/GABARAP family of autophagy proteins relies on a membrane-curvature-sensing domain in Atg3. *Nature Cell Biology*, *16*(5), 415–424.

Noda, N. N., Fujioka, Y., Hanada, T., Ohsumi, Y., & Inagaki, F. (2013). Structure of the Atg12–Atg5 conjugate reveals a platform for stimulating Atg8-PE conjugation. *EMBO Reports*, *14*(2), 206–211.

Ohsumi, Y., & Mizushima, N. (2004). Two ubiquitin-like conjugation systems essential for autophagy. *Seminars in Cell & Developmental Biology*, *15*(2), 231–236.

Otomo, C., Metlagel, Z., Takaesu, G., & Otomo, T. (2013). Structure of the human ATG12~ATG5 conjugate required for LC3 lipidation in autophagy. *Nature Structural & Molecular Biology*, *20*(1), 59–66.

Puri, P., & Chandra, A. (2014). Autophagy modulation as a potential therapeutic target for liver diseases. *Journal of Clinical and Experimental Hepatology*, *4*(1), 51–59.

Qiu, Y., Hofmann, K., Coats, J. E., Schulman, B. A., & Kaiser, S. E. (2013). Binding to E1 and E3 is mutually exclusive for the human autophagy E2 Atg3. *Protein Science*, *22*(12), 1691–1697.

Reed, M., Morris, S. H., Owczarczyk, A. B., & Lukacs, N. W. (2015). Deficiency of autophagy protein Map1-LC3b mediates IL-17-dependent lung pathology during respiratory viral infection via ER stress-associated IL-1. *Mucosal Immunology*, *8*(5), 1118–1130.

Sakoh-Nakatogawa, M., Matoba, K., Asai, E., Kirisako, H., Ishii, J., Noda, N. N., et al. (2013). Atg12–Atg5 conjugate enhances E2 activity of Atg3 by rearranging its catalytic site. *Nature Structural & Molecular Biology*, *20*(4), 433–439.

Sou, Y. S., Tanida, I., Komatsu, M., Ueno, T., & Kominami, E. (2006). Phosphatidylserine in addition to phosphatidylethanolamine is an in vitro target of the mammalian Atg8 modifiers, LC3, GABARAP, and GATE-16. *The Journal of Biological Chemistry*, *281*(6), 3017–3024.

Taherbhoy, A. M., Tait, S. W., Kaiser, S. E., Williams, A. H., Deng, A., Nourse, A., et al. (2011). Atg8 transfer from Atg7 to Atg3: A distinctive E1-E2 architecture and mechanism in the autophagy pathway. *Molecular Cell*, *44*(3), 451–461.

Tanida, I., Ueno, T., & Kominami, E. (2004a). LC3 conjugation system in mammalian autophagy. *The International Journal of Biochemistry & Cell Biology*, *36*(12), 2503–2518.

Tanida, I., Ueno, T., & Kominami, E. (2004b). Human light chain 3/MAP1LC3B is cleaved at its carboxyl-terminal Met121 to expose Gly120 for lipidation and targeting to autophagosomal membranes. *The Journal of Biological Chemistry*, *279*(46), 47704–47710.

Tanida, I., Ueno, T., & Kominami, E. (2008). LC3 and autophagy. *Methods in Molecular Biology*, *445*, 77–88.

Tindwa, H., Jo, Y. H., Patnaik, B. B., Noh, M. Y., Kim, D. H., Kim, I., et al. (2015). Depletion of autophagy-related genes ATG3 and ATG5 in Tenebrio molitor leads to decreased survivability against an intracellular pathogen, Listeria monocytogenes. *Archives of Insect Biochemistry and Physiology*, *88*(1), 85–99.

Weissmann, F., Petzold, G., VanderLinden, R., Huis In 't Veld, P. J., Brown, N. G., Lampert, F., et al. (2016). biGBac enables rapid gene assembly for the expression of large multisubunit protein complexes. *Proceedings of the National Academy of Sciences of the United States of America*, *113*(19), E2564–E2569.

Wileman, T. (2013). Autophagy as a defence against intracellular pathogens. *Essays in Biochemistry*, *55*, 153–163.

Wu, D. H., Jia, C. C., Chen, J., Lin, Z. X., Ruan, D. Y., Li, X., et al. (2014). Autophagic LC3B overexpression correlates with malignant progression and predicts a poor prognosis in hepatocellular carcinoma. *Tumour Biology: The Journal of the International Society for Oncodevelopmental Biology and Medicine*, *35*(12), 12225–12233.

Xi, Y., Dhaliwal, J. S., Ceizar, M., Vaculik, M., Kumar, K. L., & Lagace, D. C. (2016). Knockout of Atg5 delays the maturation and reduces the survival of adult-generated neurons in the hippocampus. *Cell Death & Disease*, *7*, e2127.

Xue, L. Y., Chiu, S. M., & Oleinick, N. L. (2010). Atg7 deficiency increases resistance of MCF-7 human breast cancer cells to photodynamic therapy. *Autophagy*, *6*(2), 248–255.

CHAPTER EIGHT

Investigating Structure and Dynamics of Atg8 Family Proteins

O.H. Weiergräber*, M. Schwarten*, B. Strodel*, D. Willbold*,†,1
*Institute of Complex Systems, ICS-6 (Structural Biochemistry), Forschungszentrum Jülich, Jülich, Germany
†Institut für Physikalische Biologie und BMFZ, Heinrich-Heine-Universität Düsseldorf, Düsseldorf, Germany
[1]Corresponding author: e-mail address: d.willbold@fz-juelich.de

Contents

1. Introduction 116
2. X-Ray Crystallography 117
   2.1 Basics 117
   2.2 Structure Determination 118
   2.3 Practical Notes 120
3. NMR Spectroscopy 122
   3.1 Basics 122
   3.2 Structure Determination 122
   3.3 Investigating Dynamics 124
   3.4 Mapping Interactions 125
4. MD Simulations 126
   4.1 Linking Theory and Experiments 126
   4.2 Force Fields 127
   4.3 Numerical Integration of Newton's Second Law of Motion 128
   4.4 Enhanced Sampling Techniques in MD Simulations 129
5. Applications 130
   5.1 Determining Structures of Nonlipidated Atg8 Proteins 130
   5.2 Investigating Lipidated Atg8 Proteins 133
   5.3 Assessing Dynamic Properties 134
6. Future Prospects 138
References 139

Abstract

Atg8 family members were the first autophagy-related proteins to be investigated in structural detail and continue to be among the best-understood molecules of the pathway. In this review, we will first provide a concise outline of the major methods that are being applied for structural characterization of these proteins and the complexes they are involved in. This includes a discussion of the strengths and limitations associated with each method, along with guidelines for successful adoption to a specific problem. Subsequently, we will present examples illustrating the application of these techniques, with a particular focus on the complementarity of information they provide.

Methods in Enzymology, Volume 587
ISSN 0076-6879
http://dx.doi.org/10.1016/bs.mie.2016.09.056

1. INTRODUCTION

In eukaryotic cells, disposal of cytoplasmic constituents is accomplished by a number of mechanisms, the most prominent pathways being proteasomal degradation and macroautophagy. The latter is of particular importance both for basal turnover of long-lived proteins and organelles and for maintenance of homeostasis under conditions of stress. Of note, dysregulation of autophagy has been implicated in several human diseases, in particular neurodegenerative disorders and neoplasias (Mizushima, Levine, Cuervo, & Klionsky, 2008).

Macroautophagy is characterized by the appearance of specialized organelles termed autophagosomes; these are double-membrane vesicles formed by expansion and closure of cup-shaped precursors (phagophores). Conceptually, autophagosomes serve as containers sequestering their cargo for degradation in the lysosomal compartment. Genetic studies in yeast led to the identification of a core machinery of autophagy-related (Atg) proteins orchestrating the process, and most of these were found to be conserved in mammalian cells (Nakatogawa, Suzuki, Kamada, & Ohsumi, 2009). For instance, autophagosome formation critically depends on the ubiquitin-like protein Atg8, which is enzymatically conjugated to membrane lipids by an enzyme cascade reminiscent of the ubiquitin processing system. Following proteolytic cleavage of the C-terminus by Atg4 and adenylation of the newly exposed terminal glycine residue, Atg8 is transferred to a cysteine side chain of the E1-like enzyme Atg7, forming a thioester. Via a second thioester intermediate with the E2-like Atg3, the Atg8 molecule is finally transferred to phosphatidylethanolamine (Ichimura et al., 2000); this conjugation effectively results in membrane attachment, which can be reversed by the action of Atg4. Whereas *Saccharomyces cerevisiae* expresses only a single Atg8, human cells contain several orthologs, which are encoded by GABARAP (GABA type A receptor associated protein)-type (GABARAP, GABARAPL1, GABARAPL2, and GABARAPL3) and MAP1LC3 (microtubule-associated protein 1 light chain 3)-type genes (MAP1LC3A, MAP1LC3B, MAP1LC3B2, and MAP1LC3C). Note, however, that a GABARAPL3 translation product has not been detected in cells. The diversity of the mammalian Atg8 repertoire is further increased by the existence of splicing variants.

Despite intense research, the cellular functions of Atg8 family proteins are still incompletely understood (for recent reviews, see Lee & Lee, 2016;

Slobodkin & Elazar, 2013). Their most well-established role consists in recruitment of specific cargo to emerging phagophores, which is mediated by their reversible linkage to autophagic membranes on the one hand and a unique type of protein–protein interaction on the other. The latter is based on the affinity of Atg8 proteins for so-called LIR (LC3-interacting region) motifs found on autophagy receptors (covered in depth in chapter "Methods for studying interactions between Atg8/LC3/GABARAP and LIR-containing proteins" by Johansen et al. in this volume), which in turn associate with the actual cargo. In addition, the Atg8 family as a whole appears to fulfill several crucial tasks along the macroautophagy pathway, often with a certain prevalence of one subfamily over the other; for instance, they provide scaffolds for the Atg1/ULK (unc-51 like autophagy activating kinase) complex, enable elongation and closure of the phagophore, and support autophagosome–lysosome fusion. The precise molecular mechanisms underlying these activities still need to be established.

A wealth of data concerning structure and dynamics of Atg8 family proteins has accumulated over the past 15 years, owing to contributions both from experimentalists using, e.g., X-ray crystallography and/or NMR spectroscopy, and theoreticians applying molecular dynamics (MD) simulations. Introduction of these methods and casuistic discussion of their application to Atg8 proteins will be the focus of the present chapter.

2. X-RAY CRYSTALLOGRAPHY

2.1 Basics

X-ray crystallography is the prevailing method for determination of three-dimensional structures of proteins and other biological macromolecules (for a thorough introduction, refer to Giacovazzo et al., 2011; Rupp, 2009), accounting for 89% of all entries in the Protein Data Bank (PDB) in April 2016. The method is based on the interaction of X-ray photons with crystalline samples: the periodic structure of a crystal gives rise to a characteristic diffraction pattern containing discrete signals, which are recorded and quantitatively evaluated. The building block of this periodic structure is the *unit cell*, the content of which repeats regularly in space, thus defining a three-dimensional lattice. A very simple formalism to predict the appearance of a diffraction spot from a given crystal lattice is provided by Bragg's law:

$$2d\sin\theta = n\lambda. \tag{1}$$

Here, d denotes the spacing in a family of lattice planes that are pictured to reflect X-rays of wavelength λ at an angle θ (n being a positive integer). In addition to unit-cell (translational) symmetry, many crystals contain extra symmetry elements which effectively subdivide the unit cell into symmetry-equivalent volumes. The smallest volume element required to reconstruct the unit cell (and thus the entire crystal) via crystallographic symmetry operations is called the *asymmetric unit*; it is the content of this volume that is usually covered by coordinate files deposited in the PDB.

Scattering of X-ray photons by matter is chiefly mediated by electrons. It is instructive to realize that the crystal diffraction pattern corresponds to a Fourier transform of the electron density (structure factor equation):

$$\boldsymbol{F}(\boldsymbol{h}) = \int_{cell} \rho(\boldsymbol{x}) e^{2\pi i \boldsymbol{hx}} dv = V \int_{x=0}^{1} \int_{y=0}^{1} \int_{z=0}^{1} \rho(xyz) e^{2\pi i(hx+ky+lz)} dx dy dz \quad (2)$$

with $\boldsymbol{F}(\boldsymbol{h})$ a single *structure factor* identified by index vector $\boldsymbol{h}$ (hkl), $\rho(\boldsymbol{x})$ the electron density at fractional position $\boldsymbol{x}$ (xyz), and integration is performed over the entire unit cell with volume V. It follows that each diffraction signal carries information about a periodic modulation of electron density in a particular direction and with a particular spatial frequency. While the quantitative weight of each term, i.e., the *amplitude* of $\boldsymbol{F}(\boldsymbol{h})$, is related to the respective intensity recorded on the X-ray detector, its *phase angle*, i.e., the shift of the periodic function defining the actual positions of its peaks, needs to be assessed indirectly (see later).

2.2 Structure Determination

Inversion of the structure factor Eq. (2) leads to a Fourier summation which is known as the electron density equation:

$$\rho(\boldsymbol{x}) = \frac{1}{V} \sum_{\boldsymbol{h}} \boldsymbol{F}(\boldsymbol{h}) e^{-2\pi i \boldsymbol{hx}} = \frac{1}{V} \sum_{h} \sum_{k} \sum_{l} \boldsymbol{F}(hkl) e^{-2\pi i(hx+ky+lz)} \quad (3)$$

By virtue of (3), amplitude and phase information making up $\boldsymbol{F}(\boldsymbol{h})$ can be utilized to reconstruct the electron density distribution indicating the arrangement of atoms in space. Note that the phases required for this purpose need to be derived by indirect means. Methods of *experimental phase determination* rely on the introduction of marker atoms distinguished by either an exceptional number of electrons or the existence of an absorption edge close to the applied X-ray wavelength. These properties give rise to measurable alterations in the diffraction amplitudes of the macromolecular crystal (isomorphous and anomalous differences, respectively), from which the positions of the marker atoms can be derived. This knowledge enables calculation of the complete scattering contribution of these marker atoms and finally, together with recorded amplitude data, deduction of phase information for all diffraction signals.

In the general case, the resulting electron density is used (mostly after *density modification*; Cowtan, 2010) to build an atomic model of the macromolecule from scratch, which is subsequently refined with respect to the measured diffraction data. Availability of a related structure, e.g., from the PDB, offers an alternative (and often faster) path to phase information. In this *molecular replacement* technique, a starting model can be generated by finding the correct positioning (in terms of rotation and translation) of the search molecule in the asymmetric unit of the crystal under investigation.

Refinement of a crystallographic model typically yields, for each atom in the structure, updated coordinates as well as atomic displacement parameters. The latter parameterize different types of motion (or disorder) relative to the mean positions and can thus be useful to identify regions of enhanced conformational freedom. Irrespective of the method applied for obtaining the first model of the new structure, refinement is usually based on a target function which incorporates (i) the fit of the model to the data, i.e., residual differences between observed structure factor amplitudes (F_o) and those calculated from the model (F_c), and (ii) self-consistency, i.e., the deviation of selected stereochemical parameters from ideal values. As a measure of (i), the crystallographic residual R is reported for every refined structure:

$$R = \frac{\sum_{\boldsymbol{h}} |F_o(\boldsymbol{h}) - F_c(\boldsymbol{h})|}{\sum_{\boldsymbol{h}} F_o(\boldsymbol{h})} \quad (4)$$

It is complemented by a similar quantity (R_{free}) calculated for cross-validation purposes (Brünger, 1997), using a small fraction of reflections excluded from refinement.

Using X-ray crystallography, atomic coordinates of proteins can be determined with high accuracy and precision, with estimates of mean coordinate error typically in the 0.1–0.3 Å range for well-refined structures. While results obtained by this method are strictly valid for the crystalline state, they have been used as the primary source of structural information to describe macromolecules in their physiological condition, which usually is a crowded solution. The underlying assumption that the structure is not largely altered upon crystallization is generally considered valid. It should be noted, however, that lattice formation may be selective by favoring a single conformer out of an ensemble. In general terms, investigation of dynamic properties by crystallography is limited by the very nature of the sample: the crystalline lattice may dampen conformational dynamics or polymorphisms to a varying extent, depending on the localization of lattice contacts. In contrast to NMR spectroscopy, where flexibility (or coordinate uncertainty) is reflected by the spread within a family of models, crystallographic refinement conventionally uses measures of displacement from a set of mean atomic positions, which represent an average over the time of the experiment as well as the extent of the coherently scattering lattice. If alternative conformations are sufficiently populated to appear in the electron density, they can be modeled separately and their relative occupancies refined.

2.3 Practical Notes

Since preparation, crystallization, and crystallographic structure determination of Atg8 family proteins largely follow standard protocols, they will only be outlined briefly. These procedures need to be tuned individually to the respective protein of interest.

Atg8 proteins of the GABARAP and MAP1LC3 lineages can usually be expressed in *Escherichia coli* with high yields. For efficient purification, fusion constructs containing an N-terminal affinity tag (glutathione *S*-transferase or oligohistidine) have proven particularly convenient. It should be noted that enzymatic cleavage of such tags often leaves a few amino acids at the respective terminus which are not part of the native sequence; if such cloning artifacts are suspected to interfere with the question to be addressed, e.g., when investigating terminal dynamics, alternative strategies for tag-free expression may be chosen. For crystallization purposes, proteins should be purified to the highest achievable degree of homogeneity. In this context, homogeneity does not only refer to chemical purity but also to the conformational state and degree of oligomerization. Monodispersity can conveniently be verified by means of solution scattering methods such as dynamic light scattering. Nonlipidated Atg8 proteins are usually found not to form stable oligomers in solution even at high concentrations; transient self-interaction, however, has been detected in case of GABARAP by cross-linking

(Nymann-Andersen, Wang, & Olsen, 2002) as well as DOSY (diffusion-ordered spectroscopy) NMR (Pacheco et al., 2010). A-priori assessment of conformational homogeneity is less straightforward, but some indications can be derived from two-dimensional NMR experiments (see later).

Crystallization of soluble macromolecules is achieved by addition of precipitating agents to a reasonably concentrated solution. Since conditions yielding well-diffracting samples need to be found and optimized anew for every crystallization project, extensive screening is usually required, the most wide-spread setup being the vapor diffusion experiment. As modern instrumentation allows the experiment to be scaled down to <100 nL of protein solution per tested condition, a very large number of screening solutions can be evaluated using a moderate amount of sample. Generally, the stable globular fold of Atg8 family proteins renders them relatively easy to crystallize. We have recently observed human GABARAPL2 (also known as Golgi-associated ATPase enhancer of 16 kDa, GATE-16) to crystallize in a concentrated solution lacking conventional precipitants (Ma et al., 2015). If cocrystallization of an Atg8 ortholog with a peptide containing a LIR motif is desired, the comparably low affinities usually require a moderate excess of ligand to achieve quantitative occupancy of the binding site; as an alternative strategy to obtain a stoichiometric complex, covalent fusion of the peptide to the Atg8 protein may be considered (Suzuki et al., 2014).

In order to reduce radiation damage, the actual diffraction experiment is commonly performed at cryogenic temperature (around 100 K), requiring flash-cooling of the sample. Atg8 proteins crystallized with solutions of polymers or certain salts can often be processed directly in their mother liquor, which tends to give the best results. In other cases, treatment with cryoprotectants or replacement of aqueous buffer with oils may be necessary to prevent the formation of ice in the liquid surrounding the crystal.

Diffraction data collection and processing as well as crystallographic refinement are usually straightforward with Atg8 family proteins and are thus not described in detail. Since a large number of related structures have been deposited in the PDB, initial phases can be routinely determined via molecular replacement. When investigating (putative) Atg8-LIR cocrystals, care should be taken not to prematurely assign ambiguous difference density to the peptide ligand; interpretation of weak features by a human subject is easily biased by expectations.

3. NMR SPECTROSCOPY

3.1 Basics

NMR spectroscopy is, next to X-ray crystallography, one of the two main techniques to investigate protein structures on an atomic level. Further it gives information on the dynamics of the protein at different time scales (for a thorough introduction, refer to Cavanagh, Fairbrother, Palmer, & Skelton, 1996; Wüthrich, 1986). NMR spectroscopy relies on different nuclei possessing an intrinsic angular momentum or short a spin, which can interact with an external magnetic field. The overall spin depends on the spin quantum number I, which is ½ for ^{1}H, ^{13}C, and ^{15}N, the nuclei relevant for NMR spectroscopy of proteins. The orientation of the spin in a magnetic field is quantized and can adopt $2I+1$ different states. Thus for spin ½ nuclei two different states are possible: one oriented parallel and one antiparallel to the external field. These two states differ in their energy levels:

$$\Delta E = \gamma \cdot \hbar \cdot B_0 = h \cdot \nu \tag{5}$$

where γ is the gyromagnetic ratio, an isotope specific constant, $\hbar$ is the Planck constant h divided by 2π, B_0 is the external magnetic field, and ν is the Larmor (resonance) frequency. Thus, the resonance frequency for isolated ^{1}H nuclei is 900 MHz in a B_0 field of 21.1 T. The exact frequency of an individual ^{1}H nucleus depends on its chemical environment. To compare the frequencies at different field strengths, the chemical shift δ was defined as the fractional difference to the resonance frequency of a reference compound; this quantity is usually given in ppm (parts per million):

$$\delta = \frac{\nu_{\text{sample}} - \nu_{\text{reference}}}{\nu_{\text{reference}}} \tag{6}$$

3.2 Structure Determination

Determining the structure of a protein by NMR spectroscopy requires several steps: (i) sample preparation, (ii) spectra acquisition, (iii) resonance assignment, (iv) collection of structure defining parameters, and (v) structure calculation and validation.

For triple resonance experiments proteins enriched in ^{13}C and ^{15}N are necessary. Owing to its efficiency and ease of handling, *E. coli* is the most popular host for production of proteins for NMR spectroscopy. Incorporation of the required isotopes is achieved by using minimal media (such as M9) containing ^{15}N-ammonium chloride as nitrogen source and/or ^{13}C-glucose as carbon source (Muchmore, McIntosh, Russell, Anderson, & Dahlquist, 1989). With this protein at hand multidimensional heteronuclear correlation experiments can be performed. These experiments are possible because different nuclei are able to interact with each other, either indirectly via spin–spin coupling, also called *J*-coupling or scalar coupling, or directly via dipolar interactions.

One of the most commonly used NMR experiments for proteins is the two-dimensional ^{1}H–^{15}N correlation experiment, where a proton frequency is correlated with the frequency of a directly bound ^{15}N nucleus (Bodenhausen & Ruben, 1980). This experiment gives information about the amide groups in the backbone and side chains of the protein and is therefore also often referred to as fingerprint of a protein. Here already, first information can be gained about the homogeneity and dynamics of the protein (discussed later). After the experimental conditions have been optimized, based on the two-dimensional ^{1}H–^{15}N correlation experiments, further experiments have to be recorded that allow a sequence-specific resonance assignment and the collection of structural restraints. Several triple resonance experiments have been developed which correlate each amide proton with further defined nuclei within the same amino acid residue (intraresidual) and/or within neighboring residues (sequential). In case of GABARAP, for instance, HNCACB, CBCA(CO)NH, and HNCO experiments have been used to obtain the backbone resonance assignments and HC(CO)NH-TOCSY, HCCH-TOCSY, and HCCH-COSY experiments for the side chains (Stangler, Mayr, Dingley, Luge, & Willbold, 2001). Based on resonance assignments, first structural information can be gained from the so-called secondary chemical shifts; these are defined as the differences between the observed chemical shifts and the random coil values of the respective amino acids. These values are characteristic of the secondary structure environment and can be evaluated with programs like TALOS (Shen & Bax, 2013). The principal structural information is derived from NOESY spectra. These spectra provide restraints for distances between pairs of atoms and are based on the nuclear Overhauser effect (NOE). The intensity of an NOE is strongly dependent on the interatomic distance *r*:

$$I_{\mathrm{NOE}} \propto r^{-6} \tag{7}$$

For a typical protein, several thousand distances can be obtained from the NOESY spectra. Normally, three- or even four-dimensional heteronuclear edited NOESY spectra are recorded to reduce spectral overlap. This was also true for GABARAP, where a ^{15}N-edited NOESY-HSQC, an aliphatic ^{13}C-edited NOESY-HSQC as well as an aromatic ^{13}C-edited NOESY-HSQC have resulted in more than 4500 distance restraints. The actual protein structure is determined by a computer-based approach searching for protein conformations which are in agreement with the NMR-derived restraints and also with general constraints that ensure the proper covalent structure of the protein. Often simulated annealing with restrained MD protocols are used to fulfill this task.

3.3 Investigating Dynamics

In addition to structure determination, NMR spectroscopy is also a unique tool to assess the dynamics of a protein. This is very useful as the function of a protein cannot be completely explained by a static three-dimensional structure, but often critically relies on dynamics. These fluctuations cover different time scales from picoseconds up to several seconds or even days. That Atg8 proteins are dynamic became clear very early, when the first structures were determined.

To characterize protein dynamics on the pico- to nanosecond time scale, ^{15}N-spin relaxation experiments are very valuable; these typically imply determination of rate constants R_1 and R_2 as well as heteronuclear $\{^1H\}$-^{15}N nuclear Overhauser effects (hetNOEs). The hetNOEs can already be used for a fast estimation of subnanosecond dynamics, as hetNOE values around 0.8 are expected for rigid residues and reduced values correlate with faster motions. Note that robust mobility analysis requires these parameters to be measured at different field strengths. Based on these data, it is possible to describe the spectral density function, which is the Fourier transform of the autocorrelation function describing stochastic motions of the molecule. The final determination of the protein backbone dynamics is carried out using the so-called Lipari-Szabo model-free analysis (Lipari & Szabo, 1982a, 1982b), which yields information about the diffusion tensor, internal correlation times as well as the global rotation correlation time and the order parameter S^2. This order parameter is used to describe the motion of the N–H bond vector, where an order parameter of 1 would correspond to a totally rigid vector and 0 to complete mobility.

3.4 Mapping Interactions

Finally, NMR provides different possibilities to study the interaction of proteins with ligands. These can range in size from small molecules over peptides to whole proteins. Two types of NMR experiments are particularly important: (i) saturation transfer difference (STD) NMR spectroscopy and (ii) chemical shift perturbation (CSP) experiments using two-dimensional ^{1}H–^{15}N correlation spectra.

STD experiments are especially well suited to investigate the interaction of a protein with a small ligand where the dissociation constant K_D is between the high nanomolar and the millimolar range. Isotope enrichment of the interaction partners is not required. Typically, two spectra of a protein solution in the presence of a large ligand excess are recorded in an interleaved manner: one without saturation, which will dominantly show the signals of the ligand, and a second one after selective saturation of the protein. In case of interaction, saturation will be transferred to the ligand, causing its signals to be reduced. The difference spectrum will then only contain signals of the interacting ligand.

In addition to STD experiments, which mainly address the question whether a ligand interacts with a protein, CSP experiments further allow for determination of the binding site and (dependent on the exchange rates) also of the K_D. To this end, increasing amounts of the ligand are titrated to a ^{15}N-labeled protein solution. At each titration step a two-dimensional ^{1}H–^{15}N HSQC spectrum is recorded. As the ligand will change the chemical environment of the nuclei close to the binding site, their chemical shifts are perturbed. If resonance assignments are available, one can map the CSP to the amino acid sequence and/or the three-dimensional structure. Resonance assignments of the complex are easily achieved if its dissociation is fast on the NMR time scale, because then only one signal is detected for each nucleus, and the chemical shift is a population-weighted average of the free and bound state. Thus, during a titration, the signals of residues that are involved in the interaction move gradually from the free to the bound state position. These changes can be evaluated as a function of the ligand concentration to determine the K_D. Up to now, we experienced this fast exchange for some of the GABARAP–ligand interactions, including indole derivatives (Thielmann et al., 2008), whereas for the interaction of GABARAP with the high-affinity artificial peptide N1, for instance, an intermediate exchange rate was observed (Mohrlüder et al., 2007). The latter is characterized by a broadening of signals during the titration and the appearance of a new set of signals upon saturation of the complex. This behavior is often associated with a lower K_D. Titration experiments with very high-affinity ligands, which typically show slow exchange rates, are characterized by a decrease of the signal intensity of the free state, while a new signal appears whose intensity increases during titration. As most of the Atg8–ligand interactions are rather weak, this behavior has not been observed yet.

4. MD SIMULATIONS

4.1 Linking Theory and Experiments

MD simulations are a valuable tool to bring biomolecular structures "alive," providing insight into the protein dynamics on different time scales. The investigation of protein dynamics using experiments often faces a challenging difficulty: despite well-established theoretical foundations, modern experimental techniques to study biomolecules give only indirect access to protein dynamics and can be difficult to interpret. Moreover, often multiple scenarios can explain an observation equally well. At worst, observations from different experimental techniques might even seem to contradict each other. Here, MD simulations come into play as they allow connecting with experimental data through the calculation of experimental observables from MD data. This enables to link all pieces of information, providing a coherent picture of protein dynamics at the atomistic scale.

It is now almost 40 years since the first MD simulation of a protein was published, describing the dynamics of a folded globular protein consisting of 58 amino acids in vacuum for 8.8 ps (McCammon, Gelin, & Karplus, 1977). Since then the improved parallelization of widely used MD codes such as GROMACS (Hess, Kutzner, van der Spoel, & Lindahl, 2008), the advent of special purpose computer clusters (Shaw et al., 2008), and the possibility to perform MD simulations on graphical processing units (Friedrichs et al., 2009) have enabled MD simulations to become considerably larger in size and cover significantly longer time scales. For example, millisecond all-atom MD simulations have been reported, enabling the study of biological molecules on a timescale at which many critically important phenomena are known to occur (Shaw et al., 2009). Furthermore, biological systems are now modeled more realistically by inclusion of the surroundings, such as aqueous solvent, ions, or lipid membranes, and even the description of crowded, cell-like environments is no longer out of reach.

4.2 Force Fields

In a typical biomolecular MD simulation, the positions and velocities of the atoms composing the system are calculated using classical mechanics. The time evolution of the system is governed by Newton's second law:

$$d^2\boldsymbol{r}_i/dt^2 = \frac{\boldsymbol{f}_i}{m_i} \quad (i = 1, 2, \ldots, N) \tag{8}$$

where $\boldsymbol{f}_i = \partial U/\partial \boldsymbol{r}_i$ is the potential force acting on the i-th atom with mass m_i. In the classical representation, the potential energy $U(\boldsymbol{r})$ of the system depends on the coordinates $\boldsymbol{r} = \{\boldsymbol{r}_1, \boldsymbol{r}_2, \ldots, \boldsymbol{r}_N\}$ of N atoms constituting the protein(s) and the solvent. In practice, protein modeling is performed with one of the available empirical potentials, which are usually called force fields, such as AMBER (Cornell et al., 1995), CHARMM (MacKerell et al., 1998), and OPLS (Jorgensen, Maxwell, & Tirado-Rives, 1996). For example, the AMBER force field has the following "minimalistic" functional form:

$$\begin{aligned} U(\boldsymbol{r}) = & \sum_{\text{bonds}} K_b (l - l_0)^2 + \sum_{\text{angles}} K_a (\theta - \theta_0)^2 \\ & + \sum_{\text{dihedr.}} V_n/2(1 + \cos(n\varphi - \delta)) + U_{\text{LJ}} + U_{\text{E}} \end{aligned} \tag{9}$$

with the Lennard-Jones potential defined as

$$U_{\text{LJ}} = \sum_{i<j} \left(A_{ij}/r_{ij}{}^{12} - B_{ij}/r_{ij}{}^{6} \right) \tag{10}$$

and electrostatic interactions between atoms with charges q_i and q_j:

$$U_{\text{E}} = \sum_{i<j} q_i q_j / 4\pi\epsilon_0 r_{ij} \tag{11}$$

Here, l is a bond length, θ is a bond angle, φ a dihedral angle, r_{ij} is the distance between atoms i and j, and ε_0 is the electrical permittivity of space. All the other variables are parameters of the model, which differ between the different AMBER parameter sets, such as AMBERff99SB-ILDN (Lindorff-Larsen et al., 2010) or AMBERff03 (Duan et al., 2003). Force fields are usually developed by fitting their parameters to quantum mechanical calculations and experimental properties of small molecules, and are becoming continuously better at representing protein conformations (Beauchamp, Lin, Das, & Pande, 2012; Lindorff-Larsen et al., 2012).

4.3 Numerical Integration of Newton's Second Law of Motion

During an MD simulation, the system moves by discrete steps occurring during small time steps Δt. The moves are performed by an algorithm such as the Verlet algorithm (Verlet, 1967), which allows the integration of Eq. (8), and thus calculates the coordinates $\boldsymbol{r}_i$ and velocities $d\boldsymbol{r}_i/dt$ at time $t+\Delta t$, provided that these quantities are known at the initial time t_0. For numerical stability and accuracy in solving Newton's equations of motion, one typically needs to pick a time step that is considerably smaller than the fastest time scale in the system. If Δt is chosen too large, the MD simulation can become unstable, which is caused by devastating atomic collisions that occur when a large time step propagates the positions of two atoms to be nearly overlapping, causing both the energy and forces to rapidly rise. In a classical system, the fastest motion is that of bond vibrations, which happens on the time scale of femtoseconds. Therefore, $\Delta t = 1$ fs is often used as time step for biomolecular MD simulations. To improve efficiency, one can fix the lengths of the covalent bonds during MD simulations, thereby removing the fastest motion in the system, which enables a time step of $\Delta t = 2$ fs. The next fastest motions are changes of bond angles involving hydrogens and rotations of $-CH_3$ and $-NH_2$ groups. If one removes these motions during MD simulations, time steps of up to 4–5 fs become possible (Feenstra, Hess, & Berendsen, 1999).

In MD simulations, it is nowadays common practice to use periodic boundary conditions to ensure that the system does not have an abrupt border with a vacuum. To this end, the system is assigned to a volume of an appropriate shape, typically a box or other polyhedron, which is replicated throughout space to form an infinite system. This in turn requires the accurate treatment of long-range electrostatic interactions, which can be achieved using the particle-mesh Ewald summation method (Darden, Perera, Li, & Pedersen, 1999).

The Newtonian dynamics in Eq. (8) implies that the system conserves its total energy, i.e., it evolves in the *NVE* ensemble where the number of atoms (*N*), the volume (*V*), and energy (*E*) are constant. However, experimental systems typically involve some coupling to the external environment, causing the total energy to fluctuate within a certain distribution characterized by the temperature. In a canonical or *NVT* ensemble, the temperature (*T*) is kept constant, which can be modeled using algorithms that slightly perturb the atom velocities during the simulation to provide the energy fluctuations. The simplest approach is to rescale the velocities at each time step Δt so that the kinetic energy of the system corresponds to the desired temperature, which is typically room temperature (298 K) or a physiological temperature (e.g., 310 K). In experiments, not only the temperature but also the pressure of the environment is usually fixed. Therefore, it is common practice to also control the pressure (P) during an MD simulation by adjusting the volume of the simulation box so that the desired pressure, typically 1 bar, is maintained. Various algorithms for thermostats and barostats are available to enable MD simulations in the *NPT* ensemble.

4.4 Enhanced Sampling Techniques in MD Simulations

Even though it is now possible to simulate larger systems for longer times, depending on the question one wishes to answer, protein simulations can still be hard to converge. Hence, enhanced sampling techniques are often used to accelerate the convergence of these simulations. Many different enhancement techniques are available for biomolecular simulations (Klenin, Strodel, Wales, & Wenzel, 2011) and we limit ourselves here to a short description of replica exchange molecular dynamics (REMD) simulations as we applied this technique to study the dynamics of GATE-16 (Ma et al., 2015). In REMD, a certain number k of MD simulations, also called replicas, are carried out simultaneously and information is exchanged between these simulations. Depending on the REMD methodology, the replicas have identical energy functions $U(\mathbf{r})$ but different temperatures from a fixed set, $T_1 < T_2 < \cdots < T_k$ (Sugita & Okamoto, 1999), or they are run at the same temperature but using different energy functions $U(\mathbf{r})$ known as Hamiltonians (Jang, Shin, & Pak, 2003; Liu, Kim, Friesner, & Berne, 2005). These methods are called T-REMD and H-REMD, respectively. In T-REMD, replicas at high temperatures enable easier crossing of energy barriers, thereby enhancing the sampling of the biomolecular energy surface, whereas low-temperature replicas are used to sample the system at a relevant temperature. The idea behind H-REMD is that the system can be trapped in a local minimum because of strong bonded and nonbonded interactions. Thus, if these interactions are weakened by applying different scaling factors $\lambda_i < 1$ ($i = 2, \ldots, k$) with $\lambda_1 = 1$ for the target replica, the energy landscape is sampled more efficiently. An advantage of H-REMD over T-REMD is that fewer replicas are required if only the energy of the protein but not the solvent is modified, whereas in T-REMD always the whole system and not just the protein has to be heated (Liu et al., 2005). An important aspect of REMD is that it provides interactions between replicas by periodically exchanging configurations between replicas, where the exchange probability between two replicas is based on the Metropolis criterion. One should keep in mind that, due to the replica exchanges, the true dynamics of the system is not conserved. Nonetheless, the correct canonical distribution of protein conformations can be calculated for the target replica from the REMD data.

5. APPLICATIONS

5.1 Determining Structures of Nonlipidated Atg8 Proteins

In the course of their life cycle, Atg8 family proteins are subject to reversible C-terminal lipidation, i.e., conjugation to phosphatidylethanolamine. This type of modification has been demonstrated for both Atg8 subfamilies found in mammalian cells (GABARAP-type and MAP1LC3-type proteins) even though the two groups differ in their biological roles along the autophagosome biogenesis pathway. Since attachment of a phospholipid moiety results in a dramatic loss of solubility, structural investigations of Atg8 orthologs have been largely carried out using nonlipidated variants. The basic three-dimensional fold of Atg8 family proteins was deciphered in 2000, when Paz et al. published the crystal structure of bovine GABARAPL2/GATE-16 (Paz, Elazar, & Fass, 2000). Since templates suitable for molecular replacement were unavailable at the time, phase information was acquired by a multiple-wavelength anomalous dispersion experiment using selenomethionine-containing protein. Two years later, the structure of GABARAP, a second mammalian Atg8 family member, was determined simultaneously by five groups including ours, using X-ray crystallography (Bavro et al., 2002; Coyle, Qamar, Rajashankar, & Nikolov, 2002; Knight et al., 2002) or NMR spectroscopy (Kouno et al., 2002; Stangler, Mayr, & Willbold, 2002). These studies revealed a number of crucial properties of Atg8 orthologs, and thereby laid the foundation for understanding their function on a molecular level. They are $\alpha+\beta$ class proteins consisting of a single ellipsoid domain (Fig. 1A and B). Intriguingly, the C-terminal portion turned out to contain a β-grasp fold, which is a hallmark of the ubiquitin superfamily. This fold is defined by a four-stranded mixed β-sheet (topology 2143) with an α-helix connecting strands β2 and β3, corresponding to the minimal motif β_2-α-β_2. In Atg8 proteins, this wide-spread structure is modified in characteristic ways: first, the long loop connecting strands β3 and β4 often contains an additional short α-helix, and second, all members feature an N-terminal extension typically comprising two more helices, which attaches to the convex face of the central β-sheet.

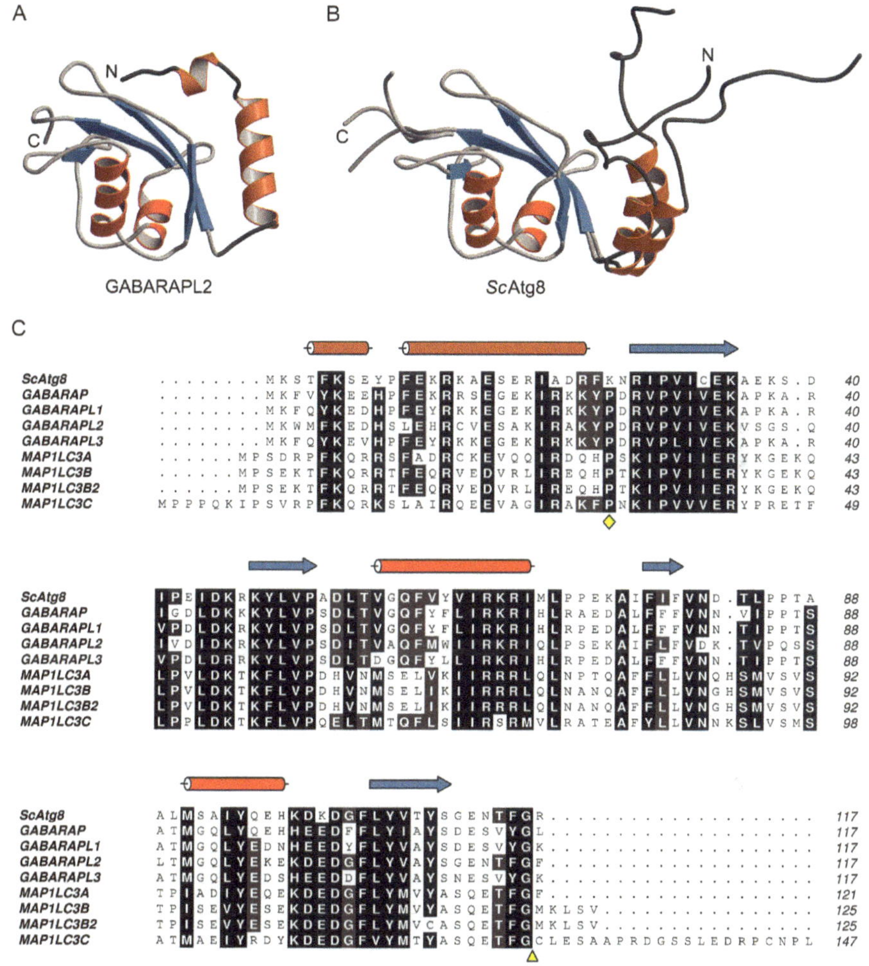

Fig. 1 Three-dimensional structures of Atg8 family proteins. (A, B) The crystal structure of human GABARAPL2/GATE-16 (PDB ID 4CO7) is shown along with the NMR structure of Atg8 from *S. cerevisiae* (PDB ID 2KQ7, terminal segments in three representative conformations). In both cases, the N-terminal extension to the β-grasp fold is distinguished by *coloring*. (C) Multiple sequence alignment of yeast Atg8 and its human orthologs, with sequence similarity encoded by shading. The secondary structure assignment provided on *top* has been adopted from GABARAPL2. A *diamond* marks the proline residue conserved in the α2–β1 loop of all human family members; the scissile bond that is cleaved prior to lipid conjugation is indicated by a *triangle*.

As anticipated based on sequence similarities (Fig. 1C), the same basic fold has been found in all Atg8 family members investigated to date, including mammalian GABARAPL1, MAP1LC3A, MAP1LC3B, and MAP1LC3C, as well as yeast Atg8. For these proteins, Table 1 provides information on

Table 1 Sequence and Structure Data for Atg8 Family Members of Yeast (*Sc*) and Human Origin

Protein Name	Length [aa]	Uniprot ID	X-Ray Structures	NMR Structures
*Sc*Atg8	117	P38182		2KQ7, 2KWC
GABARAP	117	O95166	1GNU, 1KJT	1KOT, 1KLV
GABARAPL1	117	Q9H0R8	2R2Q	
GABARAPL2	117	P60520	1EO6, 4CO7	
GABARAPL3	117	Q9BY60		
MAP1LC3A	121	Q9H492	3ECI, 3WAL	
MAP1LC3B	125	Q9GZQ8	3VTU	1V49
MAP1LC3B2	125	A6NCE7		
MAP1LC3C	147	Q9BXW4	3WAM	2NCN

The list of PDB codes excludes entries containing complexes with peptides or other proteins.

three-dimensional structures determined by X-ray crystallography, NMR spectroscopy, or both. The latest addition to the list is the NMR structure of human MAP1LC3C determined in our laboratory. This protein is distinguished from other family members by the presence of extensions at its N- and C-termini (Fig. 1C) and is thought to fulfill a unique function during elimination of intracellular pathogens. In addition to structure calculation, we were able to directly track the effects of phosphorylation at Ser18 of MAP1LC3C by protein kinase A in two-dimensional NMR spectra. Our results indicate that this posttranslational modification not only favors a conformational transition in the vicinity of Ser18 but may also change the behavior of the long helix α2 (C. Krichel, P. Neudecker, et al., unpublished observations). Indeed, it is important to realize that, notwithstanding their conserved three-dimensional fold, individual Atg8 family members appear to undergo dynamics to different extents. Indications of such intrinsic fluctuation, as obtained by various methods, are discussed later.

The biological functions of Atg8 orthologs are thought to largely rely on protein–protein interactions, involving a variety of binding partners; notable examples are the enzymes of the lipidation cascade, cargo-recruiting autophagy receptors, and components of the Atg1/ULK complex. For a thorough discussion of biochemical and biophysical characterization of the respective complexes, the reader is referred to previous reviews (Thielmann, Weiergräber, Mohrlüder, & Willbold, 2009b; Weiergräber,

Mohrlüder, & Willbold, 2013) as well as chapters "Methods for studying interactions between Atg8/LC3/GABARAP and LIR-containing proteins" by Johansen et al. and "Crystallographic characterization of ATG proteins and their interacting partners" by Qiu et al. in this volume. It is worth noting that the three-dimensional structures mentioned earlier already suggested a likely site of interaction: it was recognized that strand β2 on the edge of the central sheet is solvent-accessible and flanked by highly conserved hydrophobic patches (termed hp1 and hp2). After phage display data had indicated an affinity of GABARAP for tryptophan-containing peptides, NMR and crystallography experiments conducted in the authors' laboratory revealed distinct binding modalities for this protein (Mohrlüder et al., 2007; Thielmann et al., 2008; Thielmann, Weiergräber, Mohrlüder, & Willbold, 2009a; Weiergräber et al., 2008). Similar results have been obtained for other orthologs (for recent reviews, see Birgisdottir, Lamark, & Johansen, 2013; Noda, Ohsumi, & Inagaki, 2010). Another molecular interaction site which has been hypothesized based on sequence data and three-dimensional structures is a tubulin-binding region on the surface of the N-terminal subdomain, which has received experimental support mainly from truncation experiments (Kouno et al., 2005; Wang, Bedford, Brandon, Moss, & Olsen, 1999).

5.2 Investigating Lipidated Atg8 Proteins

Although major biological activities of Atg8 family members require C-terminal lipidation, conjugated forms of these proteins have not been extensively studied in vitro due to their poor solubility. To overcome this problem, we have explored the utility of nanodiscs as a model membrane system. In essence, a nanodisc represents a small lipid bilayer patch (approx. 10 nm in diameter) that is bounded by an amphiphilic scaffold protein derived from apolipoprotein A–I. The latter can be engineered to tailor the size distribution of the assembly to specific applications. After introduction of a cysteine residue at the C-terminus of GABARAP, the protein was chemically coupled to nanodiscs containing a thiol-reactive lipid (Ma et al., 2010). By virtue of two-dimensional NMR spectroscopy, we were able to show that this conjugated GABARAP retains both the overall fold of the soluble protein and its ligand binding capacity, while displaying changes in the environment of several residues, predominantly those facing the membrane. To some extent, these results provide a-posteriori justification for the wide-spread use of nonlipidated Atg8 proteins in biophysical studies; on the other hand, they do indicate a significant local impact of C-terminal conjugation. The nanodisc system should be generally applicable to Atg8 family proteins, facilitating their biophysical characterization in a specific functional state.

5.3 Assessing Dynamic Properties

Three-dimensional structures determined for Atg8 proteins did not only reveal the overall fold of this family but have also provided ample evidence supporting their dynamic behavior. Naturally, indications differ depending on the method employed. The first two-dimensional NMR spectra of GABARAP recorded in our laboratory, for instance, showed line broadening or signal splitting of several resonances, which are indications of conformational exchange on a millisecond to second time scale. This observation mainly concerned the N-terminal subdomain, i.e., helices α1 and α2, as well as part of the spatially adjacent α4–β4 loop (Stangler et al., 2002). It is interesting to note that one of the published crystal structures of GABARAP features an unusual conformation, where the N-terminal nine amino acids have detached from the core domain to engage in a lattice contact, a rearrangement that was accomplished by a hinge-like motion about P10 in the α1–α2 loop (Coyle et al., 2002). While the biological significance of this transition has remained unclear, it does support the notion of enhanced conformational freedom in the N-terminal segment of GABARAP. In case of MAP1LC3C, our NMR experiments indicated that the segment preceding helix α2 is largely unstructured or at least polymorphic; this implies the absence of short helix α1, in contrast to available crystal structures of LC3C. As mentioned earlier, structure and dynamics of the N-terminal region are probably subject to modulation by phosphorylation at a site located in the α1–α2 loop. The prototypical member of the family, yeast Atg8, appears to undergo even more extensive N-terminal dynamics. Indeed, NMR measurements of wild-type Atg8 yielded resonances that were strongly broadened or even undetectable for a number of residues (Schwarten, Stoldt, Mohrlüder, & Willbold, 2010). The resulting sparsity of distance restraints corresponds to the position of helix α2 being poorly defined in the structure ensemble (Fig. 1B). A similar mobility of the entire N-terminal extension has not been described for any other Atg8 protein, and its functional role is still unknown. Interestingly, it seems to be related to the identity of residue 26, which is a lysine in Atg8 and a proline in all human orthologs (Fig. 1C). Replacing K26 with P in yeast Atg8 was found to fix the position of helix α2, resulting in improved spectral resolution for N-terminal residues (Kumeta et al., 2010).

Similarly, in the original X-ray structure of GABARAPL2/GATE-16 (PDB ID 1EO6), the spatial distribution of crystallographic B factors featured marked peaks in the terminal segments, i.e., those preceding helix α1 and following strand β4, as well as certain internal loop regions, indicating enhanced mobility about mean atomic positions. Notably, the crystal structure contains two copies of GATE-16 in the asymmetric unit, which are nonequivalent by definition and may thus differ to a certain extent. The most significant difference concerns the C-terminus (res. 112–117), which is positioned close to the body of the β-grasp domain in chain B, whereas it is detached in chain A, making contact with the hydrophobic LIR binding surface of a neighboring molecule in the lattice (Paz et al., 2000). While this particular interaction is clearly a result of crystal packing, it does indicate that the C-terminus might be intrinsically flexible, with possible implications for biological function (discussed later). We have recently investigated GATE-16 crystals grown under different conditions, yielding an altered pattern of lattice contacts (Ma et al., 2015). While the new crystal form again displayed two chains per asymmetric unit, these were much less divergent than in the original structure; most importantly, both C-termini were found in a "closed" conformation, forming a number of hydrophobic and polar contacts with the remainder of the molecule. This observation prompted us to investigate the dynamics of the C-terminus in more detail. Specifically, we performed an NMR relaxation analysis where the R_1, R_2, and hetNOE values were determined. Using these data together with the three-dimensional structure, we performed a model-free analysis based on the Lipari-Szabo approach (Lipari & Szabo, 1982a, 1982b). While our data indicate that GATE-16 is well-ordered overall, several regions were found to exhibit enhanced flexibility, most notably the N- and C-termini as well as part of the β1–β2 loop.

To obtain additional information on the conformational substates populated by the termini, we performed an H-REMD simulation of GATE-16 in solution, applying the AMBERff99SB-ILDN force field and 20 replicas of 500 ns length, amounting to an accumulated H-REMD simulation time of 10 μs (Ma et al., 2015). This simulation revealed the C-terminal residues Thr114–Phe117 as the most flexible part of the protein, which is in agreement with the S^2 order parameters obtained from NMR experiments. We can classify the C-terminal dynamics based on the distance between the N- and C-termini and on the number of Phe115 side-chain atoms that are in contact with the hydrophobic pocket (hp) formed by Met1, Val36, Phe79, and the aliphatic portion of Lys82. We denote these two observables as $d_{\text{N-C}}$ and $N_{115\text{-hp}}$ and calculate the free energy ΔG at $T = 300$ K along these two observables,

$$\Delta G\left(N_{115-\text{hp}}, d_{\text{N-C}}\right) = -k_{\text{B}} T\left[\ln P\left(N_{115-\text{hp}}, d_{\text{N-C}}\right) - \ln P_{\max}\right] \qquad (12)$$

where k_B is the Boltzmann constant and $P(N_{115\text{-hp}}, d_{N-C})$ is the probability of GATE-16 to have a conformation with specific values of $N_{115\text{-hp}}$ and d_{N-C}. The maximum of this probability, P_{max}, is subtracted to ensure $\Delta G = 0$ for the free energy minimum. The resulting free energy surface (FES) is shown in Fig. 2 along with four representative GATE-16 conformations. The most stable structure is the closed conformation where the salt bridge between Met1 and Phe117 is formed, i.e., $d_{N-C} \leq 0.4$ nm, and Phe115 resides in a position where almost all of its 14 side-chain atoms are in contact with the side chains forming the hydrophobic pocket (structure 1 in Fig. 2). An alternative closed conformation exists, which is also stabilized by the Met1–Phe117 salt bridge, yet Phe115 is replaced by Phe117 in the hydrophobic pocket (structure 2 in Fig. 2). In general, the salt bridge between the termini has about equal probability of being open or closed, while the

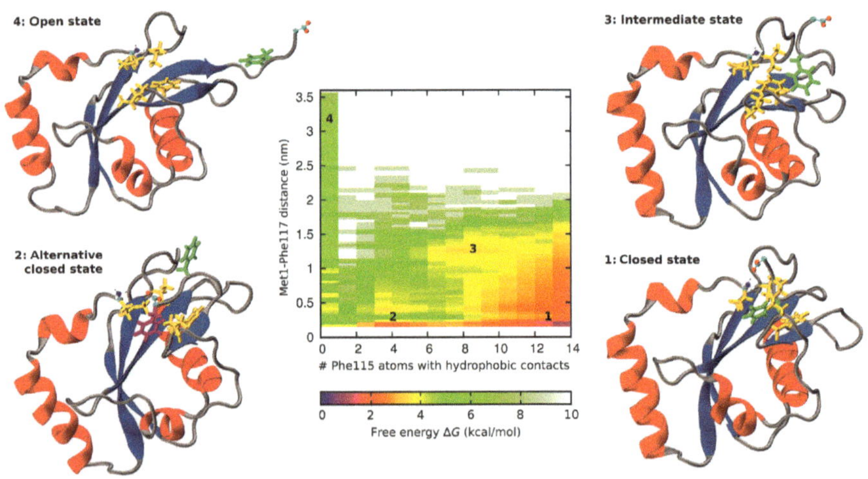

Fig. 2 The free energy ΔG, as calculated from an H-REMD simulation, is shown as surface plot for the motion of the C-terminus of GATE-16 at 300 K. ΔG is plotted as a function of the number of Phe115 side-chain atoms that are in contact with the hydrophobic pocket and the distance between the N atom of Met1 and the closest of the two carboxylate oxygens of Phe117. In addition, four representative GATE-16 structures are illustrated and their positions in the FES plot indicated by the corresponding number. The GATE-16 structures are shown in cartoon, the termini are highlighted by showing the atoms of the C_α–NH_3^+ group of Met1 and of the C_α–CO_2^- group of Phe117, the side chain of Phe115 is highlighted in *green*, and the hydrophobic pocket is indicated by showing the side chains of Met1, Val36, Phe79, and Lys82 in *yellow*. In structure 2, the side chain of Phe117 is also represented (in *magenta*) as in this structure Phe117 resides in the hydrophobic pocket.

hydrophobic contact between Phe115 and the hydrophobic pocket is more stable as it is present in about 79% of all sampled conformations. In the FES in Fig. 2, the latter conformations are characterized by a low free energy, $\Delta G < 5$ kcal/mol, and $N_{115\text{-hp}} \geq 8$, which is a suitable cutoff to decide whether or not Phe115 is in the hydrophobic pocket. One can see that Phe115 remains in the hydrophobic groove even if the Met1-Phe117 salt bridge breaks and the distance between the termini becomes rather large, reaching values of 1.5 nm and higher. One such conformation is shown in Fig. 2 (structure 3), which represents the local free energy minimum centered at $N_{115\text{-hp}} = 8$ and $d_{N\text{–}C} = 1.3$ nm. In this structure, the C-terminus starts to move away from the core of the protein, yet its complete detachment is impeded by the remaining hydrophobic contacts of Phe115. We expect this conformation to be an intermediate structure leading to a state where the C-terminus has completely detached from the core of the protein with $d_{N\text{–}C} \geq 2$ nm and both Phe115 and Phe117 being exposed to the solvent (structure 4 in Fig. 2). The FES shows that the different C-terminal conformations of GATE-16 can easily transform into each other as there are no large free energy barriers between them. Even the completely extended C-terminal conformation is accessible on the nanosecond time scale as the free energy barriers are below 8 kcal/mol at 300 K. It should be mentioned that at 300 K it takes, on average, about 800 ns to overcome a free energy barrier of 8 kcal/mol, while barriers of 7 and 6 kcal/mol can be crossed within 150 and 30 ns, respectively. In fact, we also observed the formation of the completely extended C-terminus during regular 1-μs MD simulations at 300 K. The flexibility of the C-terminus is correlated to the motion of the β3–α4 loop including Lys82, which acts like a lid for the hydrophobic pocket formed by Met1, Val36, Phe79, and thus enables Phe115 to move in and out of that pocket. Nonetheless, due to the solvent exposure of Phe115 and Phe117 we do not expect that structure 4 is a favorable state for an isolated GATE-16 molecule in solution, which is supported by the fact that ΔG is higher for this conformation compared to structures 1–3. On the other hand, to allow proteolytic cleavage of their scissile bonds (a prerequisite for lipidation), the C-termini of Atg8 family proteins need to protrude into the catalytic center of their cognate ATG4 proteases. Thus, the extended C-terminal conformation can be stabilized by the enzyme, as was observed for LC3B in complex with ATG4B (Satoo et al., 2009), leading us to conclude that the swing-out movement of the C-terminus is a built-in capability of the GATE-16 molecule and that the extended conformation found in the ATG4 complex is mostly selected, rather than induced, by the protease.

6. FUTURE PROSPECTS

Current developments in the fields of structure determination as well as simulation are likely to improve our understanding of Atg8 family proteins in the future, in particular regarding their dynamic properties. In X-ray crystallography, efforts to utilize the diffuse (non-Bragg) scattering component are particularly noteworthy. Diffuse scattering can be observed in all regions of reciprocal space and reflects deviations from the time- and space-averaged electron density encoded in the Bragg spots. Consequently, these features can result either from thermal fluctuations (*thermal diffuse scattering*) or from lattice disorder (*disorder diffuse scattering*), the former usually showing marked temperature dependence. The specific properties of diffuse scattering patterns are related to the coherence lengths of the underlying displacements and may thus prove helpful to validate models of rigid-body motion commonly used in X-ray structure refinement (Van Benschoten et al., 2016). Note, however, that temporal resolution in the diffraction data themselves can only be achieved if motions can be synchronized between unit cells, which is not applicable when studying equilibrium conformational dynamics.

To determine the conformation of a sparsely populated state that is often important for ligand binding, a new NMR method has recently become popular: the so-called Carr-Purcell-Meiboom-Gill (CPMG)-relaxation dispersion (RD) experiment. This technique allows the characterization of exchange processes if two states, the ground state A and an excited state B, are in fast exchange or if the population of the excited state is low. In such situations, it is not normally possible to observe the excited state directly by NMR. The CPMG-RD experiments allow the determination of an effective relaxation rate. With this, it is possible to evaluate the population of the excited state, the rate of exchange and also the differences in the chemical shifts of the two states (Neudecker, Lundström, & Kay, 2009). The method may therefore facilitate conclusions about the nature of functionally relevant structural fluctuations in Atg8 proteins.

The ongoing advances of computer hardware and molecular simulation techniques enable and will continue to enable elucidation of conformational transitions of proteins that occur on the microsecond time scale and possibly even slower. The application of the latest simulation techniques and usage of high-performance computing facilities will thus allow to study the slower dynamics of the Atg8 family proteins. For instance, it would be interesting to investigate the N-terminal motion of α1 and α2 of yeast Atg8, for which our NMR measurements predict considerable mobility, using molecular simulations. This would help to elucidate the role of specific residues such as K26 and may hint toward the functional role of N-terminal motions. In addition, the molecular simulation techniques nowadays available would also allow intermolecular interactions of Atg8 family proteins to be studied, be it interactions with ligands, self-interactions as present in dimerization, or interactions with lipid membranes. Finally, future simulations could directly include observables obtained from experiments, leading to simulations that are guided or restrained by experimental results, and thus allow the relevant protein conformational space to be more efficiently sampled.

Owing to recent technical advances, electron cryo-microscopy (CryoEM) is currently undergoing a "resolution revolution" and has yielded atomic models of large proteins or assemblies that are difficult to investigate by other experimental methods. Molecules as small as Atg8 family proteins pose significant challenges in both detection and alignment, which currently prevent high-resolution CryoEM reconstructions competitive with X-ray or NMR structures; investigating these proteins in the context of larger complexes may help alleviate these problems. Future developments are likely to push the limits of the method further toward smaller objects as well as higher resolution.

REFERENCES

Bavro, V. N., Sola, M., Bracher, A., Kneussel, M., Betz, H., & Weissenhorn, W. (2002). Crystal structure of the $GABA_A$-receptor-associated protein, GABARAP. *EMBO Reports*, *3*, 183–189.

Beauchamp, K. A., Lin, Y.-S., Das, R., & Pande, V. S. (2012). Are protein force fields getting better? A systematic benchmark on 524 diverse NMR measurements. *Journal of Chemical Theory and Computation*, *8*, 1409–1414.

Birgisdottir, Å. B., Lamark, T., & Johansen, T. (2013). The LIR motif—Crucial for selective autophagy. *Journal of Cell Science*, *126*, 3237–3247.

Bodenhausen, G., & Ruben, D. J. (1980). Natural abundance nitrogen-15 NMR by enhanced heteronuclear spectroscopy. *Chemical Physics Letters*, *69*, 185–189.

Brünger, A. T. (1997). Free R value: Cross-validation in crystallography. *Methods in Enzymology*, *277*, 366–396.

Cavanagh, J., Fairbrother, W. J., Palmer, A. G., III, & Skelton, N. J. (1996). *Protein NMR spectroscopy*. San Diego: Academic Press.

Cornell, W. D., Cieplak, P., Bayly, C. I., Gould, I. R., Merz, K. M., Ferguson, D. M., et al. (1995). A second generation force field for the simulation of proteins, nucleic acids, and organic molecules. *Journal of the American Chemical Society*, *117*, 5179–5197.

Cowtan, K. (2010). Recent developments in classical density modification. *Acta Crystallographica Section D, Biological Crystallography*, *66*, 470–478.

Coyle, J. E., Qamar, S., Rajashankar, K. R., & Nikolov, D. B. (2002). Structure of GABARAP in two conformations: Implications for $GABA_A$ receptor localization and tubulin binding. *Neuron*, *33*, 63–74.

Darden, T., Perera, L., Li, L., & Pedersen, L. (1999). New tricks for modelers from the crystallography toolkit: The particle mesh Ewald algorithm and its use in nucleic acid simulations. *Structure*, *7*, R55–R60.

Duan, Y., Wu, C., Chowdhury, S., Lee, M. C., Xiong, G. M., Zhang, W., et al. (2003). A point-charge force field for molecular mechanics simulations of proteins based on condensed-phase quantum mechanical calculations. *Journal of Computational Chemistry*, *24*, 1999–2012.

Feenstra, K. A., Hess, B., & Berendsen, H. J. C. (1999). Improving efficiency of large time-scale molecular dynamics simulations of hydrogen-rich systems. *Journal of Computational Chemistry*, *20*, 786–798.

Friedrichs, M. S., Eastman, P., Vaidyanathan, V., Houston, M., Legrand, S., Beberg, A. L., et al. (2009). Accelerating molecular dynamic simulation on graphics processing units. *Journal of Computational Chemistry*, *30*, 864–872.

Giacovazzo, C., Monaco, H. L., Artioli, G., Viterbo, D., Milaneso, M., Ferraris, G., et al. (2011). Fundamentals of crystallography. In *IUCr Texts on Crystallography*, Vol. 15 (3rd ed.). New York: Oxford University Press.

Hess, B., Kutzner, C., van der Spoel, D., & Lindahl, E. (2008). GROMACS 4: Algorithms for highly efficient, load-balanced, and scalable molecular simulation. *Journal of Chemical Theory and Computation*, *4*, 435–447.

Ichimura, Y., Kirisako, T., Takao, T., Satomi, Y., Shimonishi, Y., Ishihara, N., et al. (2000). A ubiquitin-like system mediates protein lipidation. *Nature*, *408*, 488–492.

Jang, S., Shin, S., & Pak, Y. (2003). Replica-exchange method using the generalized effective potential. *Physical Review Letters*, *91*, 058305.

Jorgensen, W. L., Maxwell, D. S., & Tirado-Rives, J. (1996). Development and testing of the OPLS all-atom force field on conformational energetics and properties of organic liquids. *Journal of the American Chemical Society*, *118*, 11225–11236.

Klenin, K., Strodel, B., Wales, D. J., & Wenzel, W. (2011). Modelling proteins: Conformational sampling and reconstruction of folding kinetics. *BBA–Proteins and Proteomics*, *1814*, 977–1000.

Knight, D., Harris, R., McAlister, M. S., Phelan, J. P., Geddes, S., Moss, S. J., et al. (2002). The X-ray crystal structure and putative ligand-derived peptide binding properties of gamma-aminobutyric acid receptor type A receptor-associated protein. *The Journal of Biological Chemistry*, *277*, 5556–5561.

Kouno, T., Miura, K., Kanematsu, T., Shirakawa, M., Hirata, M., & Kawano, K. (2002). ^{1}H, ^{13}C and ^{15}N resonance assignments of GABARAP, GABA$_A$ receptor associated protein. *Journal of Biomolecular NMR*, *22*, 97–98.

Kouno, T., Mizuguchi, M., Tanida, I., Ueno, T., Kanematsu, T., Mori, Y., et al. (2005). Solution structure of microtubule-associated protein light chain 3 and identification of its functional subdomains. *The Journal of Biological Chemistry*, *280*, 24610–24617.

Kumeta, H., Watanabe, M., Nakatogawa, H., Yamaguchi, M., Ogura, K., Adachi, W., et al. (2010). The NMR structure of the autophagy-related protein Atg8. *Journal of Biomolecular NMR*, *47*, 237–241.

Lee, Y. K., & Lee, J. A. (2016). Role of the mammalian ATG8/LC3 family in autophagy: Differential and compensatory roles in the spatiotemporal regulation of autophagy. *BMB Reports*, *49*, 424–430.

Lindorff-Larsen, K., Maragakis, P., Piana, S., Eastwood, M. P., Dror, R. O., & Shaw, D. E. (2012). Systematic validation of protein force fields against experimental data. *PloS One*, *7*, e32131.

Lindorff-Larsen, K., Piana, S., Palmo, K., Maragakis, P., Klepeis, J. L., Dror, R. O., et al. (2010). Improved side-chain torsion potentials for the Amber ff99SB protein force field. *Proteins*, *78*, 1950–1958.

Lipari, G., & Szabo, A. (1982a). Model-free approach to the interpretation of nuclear magnetic-resonance relaxation in macromolecules. 1. Theory and range of validity. *Journal of the American Chemical Society*, *104*, 4546–4559.

Lipari, G., & Szabo, A. (1982b). Model-free approach to the interpretation of nuclear magnetic-resonance relaxation in macromolecules. 2. Analysis of experimental results. *Journal of the American Chemical Society*, *104*, 4559–4570.

Liu, P., Kim, B., Friesner, R. A., & Berne, B. J. (2005). Replica exchange with solute tempering: A method for sampling biological systems in explicit water. *Proceedings of the National Academy of Sciences of the United States of America*, *102*, 13749–13754.

Ma, P., Mohrlüder, J., Schwarten, M., Stoldt, M., Singh, S. K., Hartmann, R., et al. (2010). Preparation of a functional GABARAP-lipid conjugate in nanodiscs and its investigation by solution NMR spectroscopy. *Chembiochem*, *11*, 1967–1970.

Ma, P., Schillinger, O., Schwarten, M., Lecher, J., Hartmann, R., Stoldt, M., et al. (2015). Conformational polymorphism in autophagy-related protein GATE-16. *Biochemistry*, *54*, 5469–5479.

MacKerell, A. D., Jr., Bashford, D., Bellott, M., Dunbrack, R. L., Jr., Evanseck, J. D., Field, M. J., et al. (1998). All-atom empirical potential for molecular modeling and dynamics studies of proteins. *The Journal of Physical Chemistry. B*, *102*, 3586–3616.

McCammon, J. A., Gelin, B. R., & Karplus, M. (1977). Dynamics of folded proteins. *Nature*, *267*, 585–590.

Mizushima, N., Levine, B., Cuervo, A. M., & Klionsky, D. J. (2008). Autophagy fights disease through cellular self-digestion. *Nature*, *451*, 1069–1075.

Mohrlüder, J., Stangler, T., Hoffmann, Y., Wiesehan, K., Mataruga, A., & Willbold, D. (2007). Identification of calreticulin as a ligand of GABARAP by phage display screening of a peptide library. *The FEBS Journal*, *274*, 5543–5555.

Muchmore, D. C., McIntosh, L. P., Russell, C. B., Anderson, D. E., & Dahlquist, F. W. (1989). Expression and nitrogen-15 labeling of proteins for proton and nitrogen-15 nuclear magnetic resonance. *Methods in Enzymology*, *177*, 44–73.

Nakatogawa, H., Suzuki, K., Kamada, Y., & Ohsumi, Y. (2009). Dynamics and diversity in autophagy mechanisms: Lessons from yeast. *Nature Reviews. Molecular Cell Biology*, *10*, 458–467.

Neudecker, P., Lundström, P., & Kay, L. E. (2009). Relaxation dispersion NMR spectroscopy as a tool for detailed studies of protein folding. *Biophysical Journal*, *96*, 2045–2054.

Noda, N. N., Ohsumi, Y., & Inagaki, F. (2010). Atg8-family interacting motif crucial for selective autophagy. *FEBS Letters*, *584*, 1379–1385.

Nymann-Andersen, J., Wang, H., & Olsen, R. W. (2002). Biochemical identification of the binding domain in the $GABA_A$ receptor-associated protein (GABARAP) mediating dimer formation. *Neuropharmacology*, *43*, 476–481.

Pacheco, V., Ma, P., Thielmann, Y., Hartmann, R., Weiergräber, O. H., Mohrlüder, J., et al. (2010). Assessment of GABARAP self-association by its diffusion properties. *Journal of Biomolecular NMR*, *48*, 49–58.

Paz, Y., Elazar, Z., & Fass, D. (2000). Structure of GATE-16, membrane transport modulator and mammalian ortholog of autophagocytosis factor Aut7p. *The Journal of Biological Chemistry*, *275*, 25445–25450.

Rupp, B. (2009). *Biomolecular crystallography: Principles, practice, and application to structural biology*. New York: Garland Science.

Satoo, K., Noda, N. N., Kumeta, H., Fujioka, Y., Mizushima, N., Ohsumi, Y., et al. (2009). The structure of Atg4B-LC3 complex reveals the mechanism of LC3 processing and delipidation during autophagy. *The EMBO Journal*, *28*, 1341–1350.

Schwarten, M., Stoldt, M., Mohrlüder, J., & Willbold, D. (2010). Solution structure of Atg8 reveals conformational polymorphism of the N-terminal domain. *Biochemical and Biophysical Research Communications*, *395*, 426–431.

Shaw, D. E., Deneroff, M. M., Dror, R. O., Kuskin, J. S., Larson, R. H., Salmon, J. K., et al. (2008). Anton, a special-purpose machine for molecular dynamics simulation. *Communications of the ACM, 51*, 91–97.

Shaw, D. E., Dror, R. O., Salmon, J. K., Grossman, J. P., Mackenzie, K. M., Bank, J. A., et al. (2009). Millisecond-scale molecular dynamics simulations on Anton. In *Proceedings of the conference on high performance computing, networking, storage and analysis (SC09)*.

Shen, Y., & Bax, A. (2013). Protein backbone and sidechain torsion angles predicted from NMR chemical shifts using artificial neural networks. *Journal of Biomolecular NMR, 56*, 227–241.

Slobodkin, M. R., & Elazar, Z. (2013). The Atg8 family: Multifunctional ubiquitin-like key regulators of autophagy. *Essays in Biochemistry, 55*, 51–64.

Stangler, T., Mayr, L. M., Dingley, A. J., Luge, C., & Willbold, D. (2001). Sequence-specific ^{1}H, ^{13}C and ^{15}N resonance assignments of human GABA receptor associated protein. *Journal of Biomolecular NMR, 21*, 183–184.

Stangler, T., Mayr, L. M., & Willbold, D. (2002). Solution structure of human $GABA_A$ receptor-associated protein GABARAP: Implications for biological function and its regulation. *The Journal of Biological Chemistry, 277*, 13363–13366.

Sugita, Y., & Okamoto, Y. (1999). Replica-exchange molecular dynamics method for protein folding. *Chemical Physics Letters, 314*, 141–151.

Suzuki, H., Tabata, K., Morita, E., Kawasaki, M., Kato, R., Dobson, R. C., et al. (2014). Structural basis of the autophagy-related LC3/Atg13 LIR complex: Recognition and interaction mechanism. *Structure, 22*, 47–58.

Thielmann, Y., Mohrlüder, J., Koenig, B. W., Stangler, T., Hartmann, R., Becker, K., et al. (2008). An indole-binding site is a major determinant of the ligand specificity of the GABA type A receptor-associated protein GABARAP. *Chembiochem, 9*, 1767–1775.

Thielmann, Y., Weiergräber, O. H., Mohrlüder, J., & Willbold, D. (2009a). Structural framework of the GABARAP-calreticulin interface–implications for substrate binding to endoplasmic reticulum chaperones. *The FEBS Journal, 276*, 1140–1152.

Thielmann, Y., Weiergräber, O. H., Mohrlüder, J., & Willbold, D. (2009b). Structural characterization of GABARAP-ligand interactions. *Molecular BioSystems, 5*, 575–579.

Van Benschoten, A. H., Liu, L., Gonzalez, A., Brewster, A. S., Sauter, N. K., Fraser, J. S., et al. (2016). Measuring and modeling diffuse scattering in protein X-ray crystallography. *Proceedings of the National Academy of Sciences of the United States of America, 113*, 4069–4074.

Verlet, L. (1967). Computer "experiments" on classical fluids. I. Thermodynamical properties of Lennard-Jones molecules. *Physical Review, 159*, 98–103.

Wang, H., Bedford, F. K., Brandon, N. J., Moss, S. J., & Olsen, R. W. (1999). $GABA_A$-receptor-associated protein links $GABA_A$ receptors and the cytoskeleton. *Nature, 397*, 69–72.

Weiergräber, O. H., Mohrlüder, J., & Willbold, D. (2013). Atg8 family proteins—Autophagy and beyond. In Y. Bailly (Ed.), *Autophagy—A double-edged sword—Cell survival or death?* (pp. 13–45). Rijeka: InTech.

Weiergräber, O. H., Stangler, T., Thielmann, Y., Mohrlüder, J., Wiesehan, K., & Willbold, D. (2008). Ligand binding mode of $GABA_A$ receptor-associated protein. *Journal of Molecular Biology, 381*, 1320–1331.

Wüthrich, K. (1986). *NMR of proteins and nucleic acids*. New York: John Wiley & Sons.

CHAPTER NINE

Methods for Studying Interactions Between Atg8/LC3/GABARAP and LIR-Containing Proteins

T. Johansen*[,1], Å.B. Birgisdottir*, J. Huber[†], A. Kniss[†], V. Dötsch[†], V. Kirkin[‡], V.V. Rogov[†,1]
*Molecular Cancer Research Group, Institute of Medical Biology, University of Tromsø—The Arctic University of Norway, Tromsø, Norway
[†]Institute of Biophysical Chemistry and Center for Biomolecular Magnetic Resonance, Goethe University, Frankfurt am Main, Germany
[‡]Merck KGaA, Darmstadt, Germany
[1]Corresponding author: e-mail address: terje.johansen@uit.no; rogov@bpc.uni-frankfurt.de

Contents

1. Introduction 144
2. Discovering LIR-Containing Proteins and Defining LIRs 146
   2.1 Identification of Atg8/LC3/GABARAP-Interacting Proteins by Yeast-Two-Hybrid Screening 147
   2.2 Use of GST Pulldown Assay to Test for Atg8/LC3/GABARAP Binding 151
   2.3 Use of Peptide Arrays for Identification and Narrowing of the LIR 155
3. Characterization of Interactions Between Atg8/LC3/GABARAP and LIR-Containing Peptides 158
   3.1 Expression and Purification of LIR-Containing Peptides 159
   3.2 Determination of Affinity and Driving Forces of the Interaction by ITC 162
4. Summary and Future Outlook 165
Acknowledgments 166
References 166

Abstract

LC3/GABARAP proteins (LC3/GABARAPs) are mammalian orthologues of yeast Atg8, small ubiquitin (Ub)-like proteins (UBLs) whose covalent attachment to lipid membranes is crucial for the growth and closure of the double membrane vesicle called the *autophagosome*. In the past decade, it was demonstrated that Atg8/LC3/GABARAPs are also required for autophagic degradation of cargos in a selective fashion. Cargo selectivity is ensured by receptor proteins, such as p62/SQSTM1, NBR1, Cue5, Atg19, NIX, Atg32, NCOA4, and FAM134B, which simultaneously bind Atg8/LC3/GABARAPs and the cargo together, thereby linking the core autophagic machinery to the target structure: a protein, an organelle, or a pathogen. *LC3-interacting regions (LIRs)* are short linear motifs within selective autophagy receptors and some other structural and

Methods in Enzymology, Volume 587
ISSN 0076-6879
http://dx.doi.org/10.1016/bs.mie.2016.10.023

signaling proteins (e.g., ULK1, ATG13, FIP200, and Dvl2), which mediate binding to Atg8/LC3/GABARAPs. Identification and characterization of LIR-containing proteins have provided important insights into the biology of the autophagy pathway, and studying their interactions with the core autophagy machinery represents a growing area of autophagy research. Here, we present protocols for the identification of LIR-containing proteins, i.e., by yeast-two-hybrid screening, glutathione S-transferase (GST) pulldown experiments, and peptide arrays. The use of two-dimensional peptide arrays also represents a powerful method to identify the residues of the LIR motif that are critical for binding. We also describe a biophysical method for studying interactions between Atg8/LC3/GABARAP and LIR-containing proteins and a protocol for preparation and purification of LIR peptides.

1. INTRODUCTION

Selective forms of macroautophagy (hereafter referred to as *autophagy*) allow targeted destruction of bulky substrates (aggrephagy, mitophagy, and reticulophagy), mobilization of metabolic substrates (lipophagy and glycophagy), and activation of the innate immune system (xenophagy), thereby playing critical roles in the maintenance of cellular homeostasis (reviewed in Johansen & Lamark, 2011; Khaminets, Behl, & Dikic, 2016; Okamoto, 2014; Rogov, Dötsch, Johansen, & Kirkin, 2014; Stolz, Ernst, & Dikic, 2014). In the selective autophagy pathway (reviewed in Khaminets et al., 2016; Kirkin, McEwan, Novak, & Dikic, 2009), the cargo is earmarked with ubiquitin (Ub) and recognized by Ub-binding domain (UBD)-containing autophagy receptors, p62/SQSTM1 (Pankiv et al., 2007), NBR1 (Kirkin, Lamark, et al., 2009), OPTN (Wild et al., 2011), NDP52 (Thurston, Ryzhakov, Bloor, von Muhlinen, & Randow, 2009), Tax1BP1 (Newman et al., 2012), and Cue5 (Lu, Psakhye, & Jentsch, 2014). Alternatively, the cargo can directly associate with selective autophagy receptors, e.g., Atg19 (Scott, Guan, Hutchins, Kim, & Klionsky, 2001), Atg34 (Suzuki, Kondo, Morimoto, & Ohsumi, 2010), NIX (Novak et al., 2010), FUNDC1 (Liu et al., 2012), FAM134B (Khaminets et al., 2015), NCOA4 (Dowdle et al., 2014; Mancias, Wang, Gygi, Harper, & Kimmelman, 2014), Atg30 (Farre, Manjithaya, Mathewson, & Subramani, 2008), Atg32 (Kanki, Wang, Cao, Baba, & Klionsky, 2009; Okamoto, Kondo-Okamoto, & Ohsumi, 2009), and Atg36 (Motley, Nuttall, & Hettema, 2012).

All selective autophagy receptors identified to date contain at least one LC3-interacting region (LIR), also known as *Atg8-family interacting motif*

(*AIM*). This is a short linear motif of up to 13 amino acids whose core sequence conforms to the generic formula Θ-X-X-Γ—where Θ is an aromatic amino acid (W/F/Y), Γ is a hydrophobic amino acid (L/I/V), and X can be any amino acid (Birgisdottir, Lamark, & Johansen, 2013; Noda et al., 2008; Rogov et al., 2014; Thielmann, Weiergraber, Mohrluder, & Willbold, 2009). The LIR allows selective autophagy receptors to interact with small Ub-like proteins (UBLs) from the Atg8/LC3/GABARAP family, which are covalently attached to the lipid phosphatidylethanolamine (PE) that is enriched in the autophagosomal membrane (Mizushima, Yoshimori, & Ohsumi, 2011). By bridging the cargo with membrane-associated Atg8/LC3/GABARAP proteins, LIR-containing receptors allow cargo to gain access to the phagophore-associated autophagic machinery consisting of the core signaling complexes, such as ULK1-ATG13-FIP200-ATG101 and VPS34-VPS15-Beclin 1-ATG14L1 in mammalian cells (reviewed in Mizushima et al., 2011).

The first, so-called canonical, LIR was described in p62/SQSTM1 (Ichimura et al., 2008; Pankiv et al., 2007) and shown to be critical for the ability of this selective autophagy receptor to target ubiquitinated proteins for autophagic degradation. Furthermore, the presence of the LIR in p62/SQSTM1 ensures that it itself is constantly turned over by autophagy, which is widely used as a basis for measuring autophagic flux in cells and tissues (Bjorkoy et al., 2009). Canonical LIRs, similar to that found in p62/SQSTM1, can be conveniently predicted using the web-based application iLIR (Kalvari et al., 2014). However, they require extensive experimental testing and validation. Furthermore, a growing number of proteins contain LIRs that do not conform to the aforementioned consensus sequence and therefore cannot be predicted by the software. Their LIRs either lack the conserved aromatic residue or show a different spacing between the aromatic and the aliphatic amino acids within the core LIR sequence, such as in NDP52 (von Muhlinen et al., 2012), Atg12 (Kaufmann, Beier, Franquelim, & Wollert, 2014), and UBA5 (Habisov et al., 2016). These noncanonical LIRs show a preference in their interactions with LC3/GABARAP proteins, which form three subfamilies: LC3, GABARAP, and GATE-16/GABARAPL-2 (Shpilka, Weidberg, Pietrokovski, & Elazar, 2011). In addition, they have the potential to bind other UBLs, such as UFM1, which interacts with LIR/UFIM (combined LIR/UFM1-interacting motif) of UBA5 (Habisov et al., 2016).

In this chapter, we provide concise protocols that can be used for discovering novel LIR-containing proteins (yeast-two-hybrid method), characterization of their interaction with members of the Atg8/LC3/GABARAP family [glutathione S-transferase (GST) pulldown assay], as well as experimental identification and mutational analyses of LIR sequences (peptide array assay). We further describe design and preparation of LIR peptides for use in biochemical and biophysical assays. Finally, a biophysical method [isothermal titration calorimetry (ITC)] that is useful for quantification of the interaction between LIRs and Atg8/LC3/GABARAP proteins is also described.

2. DISCOVERING LIR-CONTAINING PROTEINS AND DEFINING LIRs

While modern bioinformatics (Kalvari et al., 2014; Popelka & Klionsky, 2015) can predict LIRs and hence LIR-containing proteins, experimental validation is required to prove or falsify the prediction. On the other hand, unbiased biochemical or genetic methods can be applied to both discover and validate Atg8/LC3/GABARAP interactors. Proteomics approaches, such as affinity purification followed by mass spectrometry (Dunham, Mullin, & Gingras, 2012), offer an efficient way to identify protein–protein interactions. Yet, they detect partners that bind both directly and indirectly (via larger protein complexes), so that additional validation for direct interaction is required. In contrast, yeast-two-hybrid (Y2H) methods mandate that two proteins necessarily interact directly in yeast cells in vivo. One important limitation of the standard Y2H system is its unsuitability for studying interactions of integral membrane proteins. In case binding partners of membrane-bound proteins need to be identified, the split-Ub-based Y2H method (Snider, Kittanakom, Curak, & Stagljar, 2010) can be recommended. The Y2H screening protocol presented below is based on the original Y2H system by Fields and Song that makes use of the split transcription factor Gal4 (Fields & Song, 1989). It has been successfully applied to identify important LC3/GABARAP-binding proteins, such as NBR1, NIX, and FAM134B (Khaminets et al., 2015; Kirkin, Lamark, et al., 2009; Novak et al., 2010).

Following identification of Atg8/LC3/GABARAP-binding proteins, a GST pulldown assay can be used to confirm and narrow down the LIR in the protein of interest. Finally, modern peptide arrays offer a simple and powerful technique to pinpoint and limit the LIR sequence and to probe the importance of single residues of the LIR motif for binding.

2.1 Identification of Atg8/LC3/GABARAP-Interacting Proteins by Yeast-Two-Hybrid Screening

In the Y2H system presented next, the N-terminal DNA binding domain (BD) of Gal4 is encoded by a vector (e.g., pYTH9 (Hecker, Rabiller, Haglund, Bayer, & Dikic, 2006)) as an N-terminal fusion to an Atg8/LC3/GABARAP family protein (bait), in which the exposed C-terminal glycine of the processed form is removed to prevent any artifactual conjugation to the partner protein. The Gal4 C-terminal activation domain (AD) is encoded by a vector library [e.g., pACT2 (Clontech)], containing several million cDNAs derived from a tissue of interest and fused to the C-terminus of Gal4AD (prey). To perform Y2H screening, an appropriate strain of yeasts, such as *Saccharomyces cerevisiae* Y190, is used, in which reconstituted, active Gal4 is required to drive transcription of marker genes (e.g., *HIS3* encoding imidazoleglycerol-phosphate dehydratase that catalyzes the sixth step in histidine biosynthesis, and *lacZ* encoding β-galactosidase). Of note, the Y190 strain is leaky for *HIS3* expression, and 25 m*M* 3-AT (3-amino-1,2,4-triazole) is required to suppress the background growth of this strain on SD (synthetic defined)/-His medium (Clontech). The pYTH9 vector encodes *TRP1* marker, and pACT2 encodes *LEU2* marker, so that Y190 yeast transformants that harbor both pYTH9 and pACT2 can grow in SD/-Trp/-Leu double-dropout medium. Importantly, reconstitution of the functional Gal4 via the interacting couple (encoded by AD and BD fusions) allows the yeast to grow in SD/-His medium, so that in case of a positive interaction the Y190 strain can grow on SD/-His/-Trp/-Leu triple-dropout medium.

2.1.1 Preparation of Yeast for Transformation

Yeast culture is available as a glycerol stock in YPD medium (Sigma-Aldrich), supplemented with 25% glycerol and stored at −70°C, or as a short-term culture on YPD agar plates stored at 4°C. To prepare working plates, streak a small portion of the yeast stock onto an YPD agar plate and incubate at 30°C until yeast colonies reach about 2 mm in diameter (3–5 days). The sealed plate can be stored at 4°C and replated at 1–2-month intervals.

2.1.2 Transformation of Yeast With the Bait Construct

1. Inoculate a fresh yeast culture into 20–25 mL YPD medium and incubate overnight at 30°C with shaking at 200 rpm.

2. On the next day, determine the titer of yeast cells in the medium by determining optical density $(OD)_{600}$ of a 1:10 dilution in water (OD_{600} of 0.1 of the undiluted yeast culture corresponds to 1×10^6 cells/mL).
3. Add 2.5×10^8 cells to fresh YPD, prewarmed to 30°C, to obtain 50 mL yeast culture in a 250-mL flask (final yeast cell concentration of 5×10^6 cells/mL). Incubate the diluted culture for a further 3–4 h at 30°C with shaking at 200 rpm.
4. Once the cells reach the concentration of $1–2 \times 10^7$ cells/mL, pellet them by centrifugation at 3000 *g* for 5 min.
5. Wash the pellet with 25 mL of water once by resuspending the yeast and renewed centrifugation at 3000 *g* for 5 min. Resuspend the washed pellet in 900 μL sterile water, transfer to a 1.5-mL microcentrifuge tube, and centrifuge for 1 min at 15,000 *g*. Finally, resuspend yeast cells in 700 μL of 100 m*M* of lithium acetate and incubate at 30°C for 10 min.
6. For each transformation, transfer a 100-μL aliquot of the suspension to a fresh 1.5-mL microcentrifuge tube and pellet by centrifugation at 15,000 *g* for 1 min.
7. After removal of the supernatant, add 240 μL of 50% PEG [molecular weight (mw) 350; Sigma-Aldrich], 36 μL of 1 *M* lithium acetate, 50 μL single-stranded (ss)DNA (2 mg/mL; Sigma-Aldrich), 0.5–1 μg plasmid DNA, and 34 μL water (minus the volume of the plasmid DNA) to the pellet. Vortex vigorously and incubate at 30°C for 30 min.
8. Heat-shock the yeast at 42°C for 30 min.
9. Pellet the cells at 15,000 *g* for 1 min and resuspend by gentle pipetting in 250 μL sterile water.
10. Plate the culture onto an appropriate agar plate (e.g., SD/-Trp in the case of pYTH9 that encodes the *TRP1* marker) and incubate at 30°C for 2–4 days until transformants appear. *Note*: It is useful to make several identical transformations at this step for preparing a stock culture, testing growth on control plates (SD with additional dropouts), and testing bait self-activation (by β-galactosidase testing, as discussed in Section 2.1.5).

2.1.3 Testing Expression of the Bait in Yeast Culture

1. Grow a single yeast colony in 5 mL of SD/-Trp for 3 days.
2. Pellet yeast by centrifugation at 550 *g* for 2–3 min, resuspend in 300 μL of lysis buffer [0.1 *M* ethylenediaminetetraacetic acid (EDTA)

pH 8.0, 0.9 *M* sorbitol, 50 m*M* dithiothreitol (DTT), and 200 units (U) lyticase], transfer into a microcentrifuge tube, and incubate at 37°C for 3–12 h.

3. Pellet the resulting spheroplasts by centrifugation at 550 *g* for 3 min and resuspend in 300 μL protein extraction buffer (100 m*M* Tris–HCl pH 7.5, 200 m*M* NaCl, 20% glycerol, 5 m*M* EDTA, 0.1% mercaptoethanol, and protease inhibitors). Incubate lysates on ice for 20 min.
4. Clear the lysate by centrifugation at 15,000 *g* for 15 min and assess on an SDS-PAGE for protein expression.

2.1.4 Transformation of Yeast With the cDNA Library

Transform the yeast harboring pYTH9 vector (encoding the bait) with 200–250 μg of suitable cDNA library following the steps given in Section 2.1.2. The only difference is that this time, the yeasts are grown in a SD/-Trp medium and three times more cells are used in the transformation protocol. In the final step, the large cell pellet is resuspended in 20 mL of sterile water, and 1.25 mL of the suspension is plated on each dry square (1640 × 150 mm) plate with solid SD/-Trp/-Leu/-His/+3-AT medium (16 plates in total). The plates are incubated at 30°C for 4–7 days until colonies (which should express both BD- and AD-fusion proteins) appear. These colonies can be picked and streaked onto fresh SD/Trp/-Leu/-His/+3-AT plates for scoring clones positive for Atg8/LC3/GABARAP interaction with the prey proteins.

2.1.5 Identification of Colonies Positive for the Bait:Prey Interaction by β-Galactosidase Assay

To identify colonies positive for the bait:prey interaction (i.e., BD and AD domains of Gal4 are brought into a close proximity due to the bait:prey interaction; Fig. 1), the β-galactosidase assay is performed.

1. Presoak a sterile Whatman paper filter in 5 mL of Z buffer/5-bromo-4-chloro-3-indolyl-β-D-galactopyranoside (X-gal) solution [100 mL of Z buffer (60 m*M* Na_2HPO_4, 40 m*M* NaH_2PO_4, 0.075% KCl, 0.025% $MgSO_4$), supplemented with 0.27 mL of β-mercaptoethanol (β-ME), 1.67 mL X-gal stock solution] and place it in a clean 150-mm plate.
2. Place a separate dry sheet of Whatman filter on top of the colonies to be assayed. Gently lift it to make a print of the colonies and snap-freeze in liquid nitrogen.

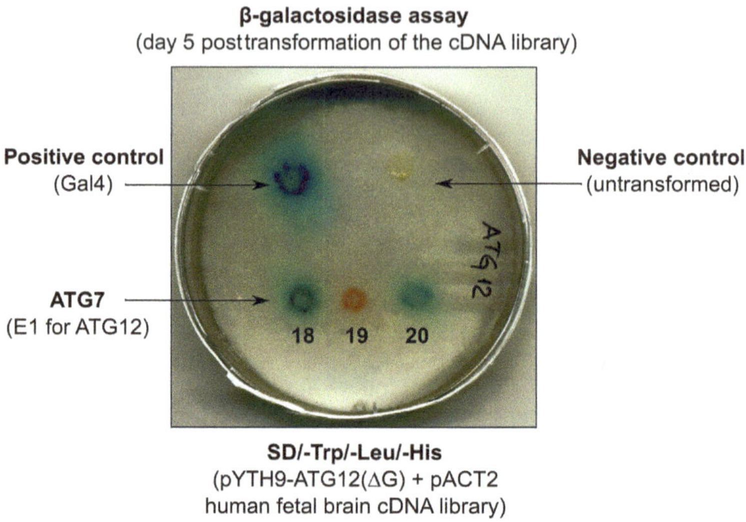

Fig. 1 Example of an outcome of Y2H screening for ATG12-interacting proteins as assessed by the β-galactosidase assay. Outgrowing clones of Y190 yeast on SD/-Trp/-Leu/-His plates suggest successful interaction of the bait ATG12(ΔG) with the prey (cDNA encoded in the pACT2-based human fetal brain cDNA library). Reconstitution of a functional Gal4 transcription factor by the interaction drives transcription of the *lacZ* gene, encoding β-galactosidase. The *blue* product of the β-galactosidase activity marks the yeast clone that is positive for the interaction (false positive clones, such as Clone 19, remain uncolored). In this Y2H screen, ATG7, the established E1 enzyme for ATG12 is expressed by Clone 18. Clone 19 is negative for the interaction with ATG12. Clone 20 is another undisclosed positive clone for the interaction with ATG12.

3. Allow the filter with colonies to thaw at room temperature and place it colony side up on top of the presoaked filter, avoiding trapping air bubbles inside.
4. Incubate filters at 30°C for 30 min to 8 h to detect blue color development. Colonies positive for the interaction will become blue.

2.1.6 Determination of Prey Identity

To determine the identity of the prey, solubilize a yeast colony in a small volume (20 μL) of 0.25% SDS and heat at 95°C for 5 min to lyse cells. This preparation of yeast can be used directly in a PCR. Alternatively, DNA can be isolated from the yeast spheroplasts (prepared as in Section 2.1.3) using a MiniPrep Kit (Qiagen). The identity of the prey can be determined by sequencing with primers specific for the pACT2 plasmid.

2.2 Use of GST Pulldown Assay to Test for Atg8/LC3/GABARAP Binding

GST pulldown assay is an in vitro assay for identification of protein–protein interactions between a GST-fusion protein and target proteins in solution. Here, a cDNA, encoding a GST-Atg8/LC3/GABARAP-fusion protein, is cloned into an isopropyl-β-D-thiogalactopyranoside (IPTG)-inducible plasmid vector that is expressed in bacteria. The GST-Atg8/LC3/GABARAP-fusion protein from the bacterial cell lysate is then captured and immobilized using glutathione-sepharose beads. The target protein, either a recombinant protein, a protein as a component of a cell lysate, or an in vitro-translated protein labeled with ^{35}S-methionine, is incubated together with GST-Atg8/LC3/GABARAP bound to glutathione-sepharose beads. Complexes recovered from the beads are resolved by running an SDS-PAGE gel. Bound target proteins are visualized by Western blot or autoradiography. Most often, we use the GST pulldown assay with ^{35}S-methionine-labeled in vitro-translated proteins as input. This format enables reliable quantification of the binding based on the radioactive signal and is especially useful when comparing the effects of mutations on the binding. For LIR motifs that are modified by phosphorylation, we use cellular extracts as inputs. When novel LIR motifs are being studied, it is recommended to use both sources of inputs. Especially when it is possible that the LIR is modified by posttranslational modifications. The protocol using in vitro-translated ^{35}S-methionine-labeled target proteins (Fig. 2) is described next.

2.2.1 Preparation of Lysates From Bacteria Expressing GST-Atg8/LC3/GABARAP

1. Subclone cDNAs encoding Atg8/LC3/GABARAPs in-frame with the pDest15 IPTG-inducible expression vector (Invitrogen), harboring an N-terminal GST tag downstream of the T7 promoter, and use the vector to transform a suitable strain of *Escherichia coli* [e.g., BL21 STAR (DE3) pLysS cells (Invitrogen)].
2. Pick a single colony of transformed *E. coli*, transfer into a sterile tube with 5 mL of 2x YT medium [16% Bacto-tryptone (w/v), 10% yeast-extract (w/v), 5% NaCl (w/v)] with ampicillin (100 μg/mL), and incubate overnight at 37°C on a horizontal shaker (250 rpm).

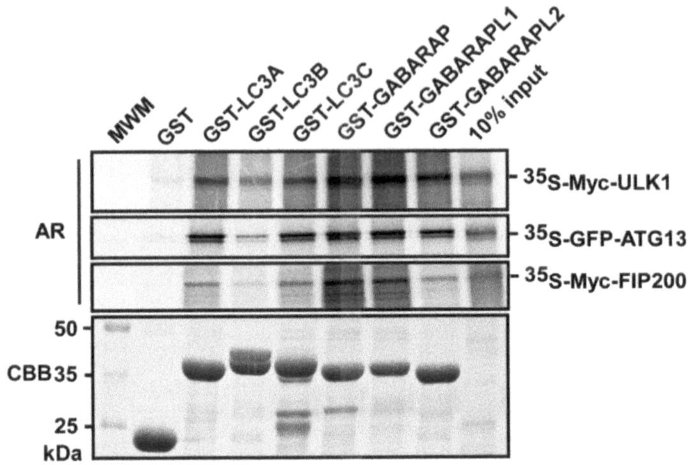

Fig. 2 Examples of GST pulldown assays showing that ULK1, ATG13, and FIP200 all preferentially bind to the GABARAP subfamily proteins. The indicated proteins were translated in vitro in the presence of [^{35}S]-methionine and tested in GST pulldown assays for interaction with the indicated GST-fused recombinant proteins. Autoradiograph (AR, *top panels*) and Coomassie-stained immobilized GST or GST-fusion proteins (CBB, *bottom panel*) are shown. *This figure was originally published in Alemu, E. A., Lamark, T., Torgersen, K. M., Birgisdottir, A. B., Larsen, K. B., Jain, A., … Johansen, T. (2012). ATG8 family proteins act as scaffolds for assembly of the ULK complex: Sequence requirements for LC3-interacting region (LIR) motifs.* Journal of Biological Chemistry, *287(47), 39275–39290 © the American Society for Biochemistry and Molecular Biology.*

3. Dilute the overnight culture 1:20 to a final volume of 100 mL of 2x YT medium with ampicillin (100 μg/mL), and grow further under same conditions until the cells reach OD_{600} of 0.9 (about 2–3 h).
4. Induce expression of the target gene by adding IPTG to a final concentration of 0.5 m*M*. Allow bacteria to grow for another 3–4 h at room temperature on a horizontal shaker.
5. After induction, transfer the bacterial culture into a sterile 250-mL plastic flask and centrifuge in a device such as the JLA 16.250 fixed angle rotor (Beckman Coulter), at 3000 *g* at 4°C for 10 min.
6. Remove the supernatant completely and resuspend the recovered pellet in 4 mL ice-cold lysis buffer made fresh with lysozyme and DTT (50 m*M* Tris–HCl pH 8.0, 250 m*M* NaCl, 1 m*M* DTT, 0.35 mg/mL lysozyme).
7. Keep the lysate on ice for 20 min and then add Triton X-100 to 1% (v/v).
8. Distribute the lysate into three cryotubes (about 1.5 mL per tube) and store at −70°C.

2.2.2 Purification and Immobilization of GST-Atg8/LC3/GABARAP Proteins on Glutathione-Sepharose Beads

1. Prepare glutathione-sepharose beads (Glutathione Sepharose™ Fast Flow; GE Healthcare, 17-5132-01) by washing three times with ice-cold phosphate-buffered saline (PBS) buffer by brief centrifugation (a few seconds) in a microcentrifuge at room temperature. After the final wash, keep the beads in a 50% solution in PBS and store at 4°C.
2. Slowly thaw one cryotube of the bacterial lysate (prepared and stored as described in Section 2.2.1) and sonicate on ice (four times at 40% amplitude and 20 kHz using a VCX 130 Sonic Vibra-Cell ultrasonic processor, for example). Divide the lysate into 1.5-mL microfuge tubes and centrifuge at 15,000 *g* at 4°C for 10 min.
3. Divide the cleared supernatant, representing the crude protein extract, into two 1.5-mL microfuge tubes, each containing 100 μL of washed GST beads (50% solution in PBS).
4. Allow binding of GST-Atg8/LC3/GABARAP from the bacterial lysates to the beads by incubating the mixture for at least 1 h on a rotating wheel at 4°C.
5. Wash the beads with bound protein at least three times with ice-cold PBS by brief centrifugation (lasting a few seconds) in a microcentrifuge at room temperature.
6. After the last wash, collect the protein-bound beads from both tubes in a clean 1.5-mL microfuge tube and adjust the volume to 50% (v/v) in ice-cold PBS.
7. Store the beads at 4°C until use.

2.2.3 Testing the Amount of Immobilized GST-Atg8/LC3/GABARAP on the Beads

1. Mix equal amounts of GST-Atg8/LC3/GABARAP-bound beads (50% solution) and 2 × SDS gel loading buffer (125 m*M* Tris–HCl pH 7.4, 4% SDS, 0.04% bromophenol blue, 8% sucrose, and 200 m*M* DTT, added immediately before use), e.g., 10 μL of each, and boil for 3–5 min.
2. Run half the mixture on a 10% SDS-PAGE gel using a protein size marker.
3. Fix the gel and stain with a dye such as Coomassie brilliant blue in order to visualize the bound proteins.
4. Assess the amount of each bound protein to use in the GST pulldown assay based on the intensity of bands.

2.2.4 In Vitro Translation of a Target Protein

To produce an in vitro-translated ^{35}S-labeled protein, the TNT T7 Quick Coupled Transcription/Translation System (Promega) and [^{35}S]-methionine (Amersham Biosciences) are used.

1. Mix 500 ng of plasmid, encoding the desired protein (under the control of the T7 promoter) with 16 μL of assay buffer [12.5 μL TNT reticulocyte-lysate, 1 μL TNT buffer, 0.5 μL amino acid mixture (minus methionine), 1 μL [^{35}S]-methionine (may be diluted 1:1 with cold methionine of same concentration), 0.5 μL RNAsin, and 0.5 μL TNT T7 polymerase (all solutions are thawed on ice before use)]. The amount of assay buffer is calculated by multiplying the number of 25-μL reactions needed plus one.
2. Adjust the volume of the reaction with water to 25 μL and incubate at 30°C in a heat block placed in a hood (marked for radioactive work) for 90 min. A 25-μL in vitro translation reaction is sufficient for two GST pulldown reactions.
3. Input controls for GST pulldown are prepared by mixing 3 μL of the in vitro-translated target protein with 15 μL of a gel loading buffer solution (6 μL water, 7.2 μL 2 × SDS loading buffer, 1.8 μL 1 *M* DTT). In this way, the in vitro-translated protein is sixfold diluted (1 μL in vitro-translated protein corresponds to 6 μL of this solution).

2.2.5 Preclearing of In Vitro-Translated Target Protein

1. To preclear the in vitro-translated target protein, distribute 20 μL of empty glutathione-sepharose beads (in a 50% solution with PBS) into microfuge tubes and wash twice with 800 μL ice-cold NET-N buffer (20 m*M* Tris–HCl pH 8.0, 125 m*M* NaCl, 0.5% Nonidet P-40, 1 m*M* EDTA); an alternative to reduce background binding is to use 6 m*M* of EDTA and 6 m*M* of ethylene glycol-bis(β-aminoethyl ether) (EGTA) by brief centrifugation (lasting a few seconds) in a microcentrifuge at room temperature. By varying the concentration of NaCl or detergent, binding conditions can be optimized for different proteins, if necessary. This can also be done during washing of the bead-bound protein (as described next).
2. Remove supernatant from each microfuge tube with empty beads and add ice-cold 100 μL NET-N buffer containing a Complete Mini EDTA-free protease inhibitor mixture (Roche).
3. Add the in vitro-translated protein (5–20 μL). Incubate the mixture on a rotating wheel at 4°C for 30 min.

4. Finally, collect the beads by brief centrifugation to give a supernatant of a precleared, in vitro-translated protein.

2.2.6 GST Pulldown, Gel Electrophoresis, and Visualization of Binding

The amount of GST-Atg8/LC3/GABARAP-loaded beads (stored as 50% solution in PBS) needs to be adjusted in order to use a similar amount of GST-fusion protein in each GST pulldown reaction. Furthermore, it is recommended to use a similar total amount of sepharose beads in each GST pulldown reaction. Therefore, empty beads should be added to tubes with less protein-loaded beads.

1. Prewash the beads loaded with GST-Atg8/LC3/GABARAP proteins two times with 500 μL NET-N buffer.
2. Add 100 μL of precleared in vitro-translated protein and incubate on a rotating wheel at 4°C for 60 min.
3. Wash the beads five times with 500 μL NET-N buffer to remove unbound protein. Remove the supernatant after the last wash and add 12 μL of NET-N buffer, 18 μL of 2× SDS gel loading buffer (without DTT), and 3 μL of 1 *M* DTT to each tube.
4. Elute the bound proteins from the beads by boiling the mixture for 3–5 min.
5. Collect the beads by brief centrifugation and store the reaction at −20°C (or run on a 10% SDS-PAGE gel the same day).
6. Run the whole GST pulldown reaction (30 μL) on a 10% SDS-PAGE gel together with the input control (10% input) from the in vitro transcription/translation reaction. After gel electrophoresis, transfer the gel onto Whatman paper (3 mm CHR; GE Healthcare, product number 3030917) and vacuum-dry the gel.
7. Expose the dried gel to a Fujifilm imaging plate in a Fujifilm cassette overnight.
8. Visualize ^{35}S-labeled proteins that have bound to the GST-Atg8/LC3/GABARAP-loaded beads on the Fujifilm imaging plate using a Fujifilm bioimaging analyzer (e.g., BAS-5000).

2.3 Use of Peptide Arrays for Identification and Narrowing of the LIR

Peptide arrays enable identification of interaction sites in a protein of interest and are ideally suited to provide information about the minimal size of each identified LIR motif. They are also ideal for probing each position of the LIR by amino acid substitutions using two-dimensional arrays

(Alemu et al., 2012). To search for candidate LIR motifs, an array of 20-mer peptides covering the entire sequence of the protein of interest is synthesized and immobilized on cellulose membrane [e.g., by the SPOT method (Frank, 2002)]. Each peptide is shifted three amino acids relative to the previous peptide. To perform two-dimensional peptide array analyses, each position of an 18-mer sequence with the core LIR motif in the center is substituted with all 19 alternative amino acids (Alemu et al., 2012). The arrays are probed with recombinant GST or GST-Atg8/LC3/GABARAP, and binding is detected with anti-GST antibodies (Fig. 3). The peptide arrays work very well for identification of candidate LIR motifs, but care should be taken in interpreting the data when studying proteins with extensive hydrophobic regions, like WD40 or KELCH repeat proteins, where the chance of detecting false positives may be high due to the presence of many aromatic and hydrophobic residues. Some of these residues may be appropriately spaced to be displayed similarly to genuine LIR motifs when arrayed as 20-mer peptides. The results should always be verified by other methods, like GST pulldown assays with and without mutations of the putative core LIR aromatic and hydrophobic residues.

2.3.1 Synthesis of Peptide Arrays

Synthesis of peptide arrays on cellulose membranes can be performed using a MultiPep automated peptide synthesizer (INTAVIS Bioanalytical Instruments AG, Germany).

2.3.2 Production and Purification of GST-Atg8/LC3/GABARAPs

It is recommended to purify the GST-Atg8/LC3/GABARAP the same day or the day before the overlay assay. The purified protein can be kept at 4°C for brief storage.

1. Slowly thaw lysates of corresponding *E. coli* prepared as described in Section 2.2.1.
2. Sonicate on ice (four times at 40% amplitude and 20 kHz using a VCX 130 Sonic Vibra-Cell ultrasonic processor, for example).
3. Transfer to a microfuge tube and centrifuge at 15,000 *g* at 4°C for 10 min.
4. Filter the supernatant by passing through a 0.45-μm sterile filter and apply on a GST-binding column (GSTrap™ FF 1 mL; GE Healthcare).
5. Elute proteins from the column with elution buffer (1 × PBS pH 8.0, 1 m*M* reduced glutathione, the buffer is pH adjusted, sterile filtered,

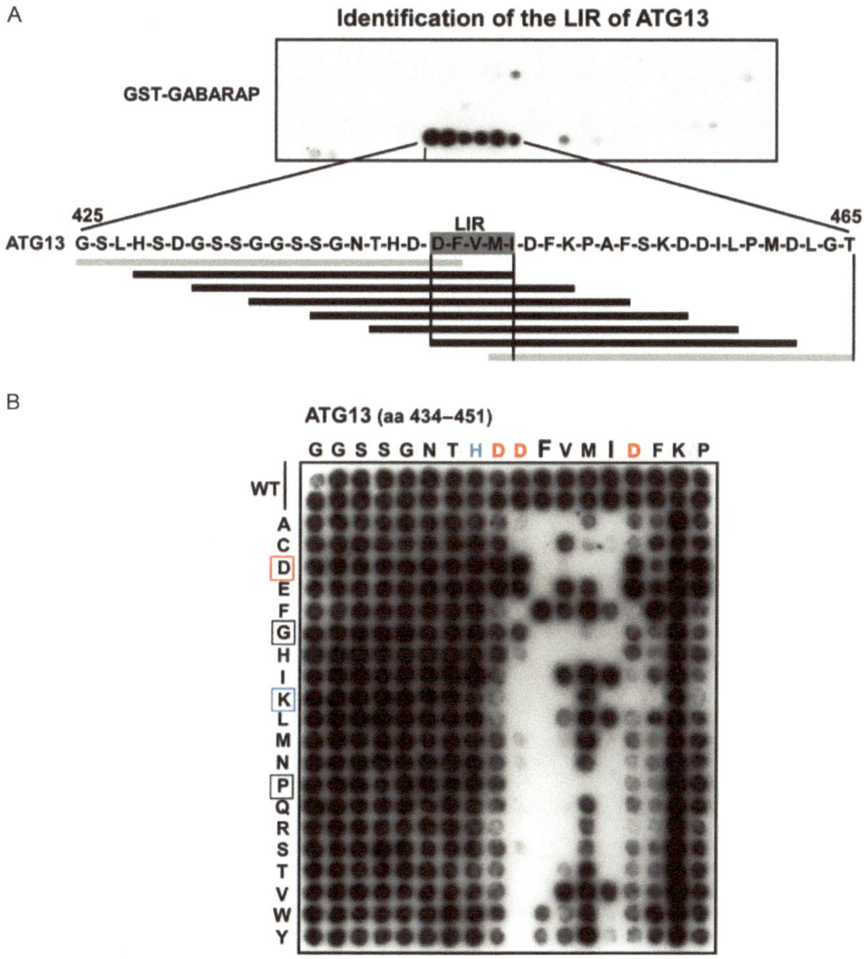

Fig. 3 Examples of a peptide array scan to identify a candidate LIR motifs and a two-dimensional peptide array to probe critical residues in the ATG13 LIR. (A) Identification of the LIR motif in ATG13 using peptide arrays. Array of 20-mer peptides covering full-length ATG13 were synthesized on cellulose membranes. Each peptide was shifted three amino acids relative to the previous peptide. Shown are sequences of the GABARAP-interacting peptides. Interacting peptides are in *black*, whereas noninteracting peptides are in *gray*. (B) Two-dimensional peptide array scan analyzing the effect of single amino acid substitutions at all positions of the indicated 18-mer peptide from human ATG13 (amino acids 434–451). Each position of the 18-mer peptides was replaced with each of the alternative 19 amino acids. The arrays were probed with 1 μg/mL of GST-GABARAP for 2 h. Binding to GST-GABARAP was detected with anti-GST antibodies. *This figure was originally published in Alemu, E. A., Lamark, T., Torgersen, K. M., Birgisdottir, A. B., Larsen, K. B., Jain, A., … Johansen, T. (2012). ATG8 family proteins act as scaffolds for assembly of the ULK complex: Sequence requirements for LC3-interacting region (LIR) motifs.* Journal of Biological Chemistry, 287*(47), 39275–39290 © the American Society for Biochemistry and Molecular Biology.*

and degassed before use) at the speed of 1 mL/min (e.g., using an ÄKTAprime chromatography system; GE Healthcare).
6. Use the Bradford protein assay to estimate protein concentration in the fractions where the protein is eluted.

2.3.3 Peptide Overlay Assay

1. Stain the peptide array membrane with Ponceau S for 1 min.
2. Wash several times with water to allow visualization of peptide spots.
3. Dry the membrane overnight.
4. On the day of the assay, activate the membrane by incubating in 100% methanol for 10 s.
5. Wash two times, at 5 min each, in TBS-T buffer (10 m*M* Tris–HCl pH 7.4, 150 m*M* NaCl, and 0.1% Tween-20).
6. Block nonspecific labeling by incubating the membrane in TBS-T containing 5% (w/v) nonfat dry milk powder for 1 h at room temperature.
7. To test the binding to GST-Atg8/LC3/GABARAP, incubate the membrane with the purified GST-Atg8/LC3/GABARAP protein at 1–2 μg/mL concentration in TBS-T buffer (the volume is adjusted according to the size of the membrane) for 2 h at room temperature.
8. After binding, wash the membrane five times in TBS-T.
9. Incubate overnight at 4°C with HRP-conjugated anti-GST (clone RPN1236; GE Healthcare) at 1:5000 dilution in 5% milk.
10. Wash the membrane five times in TBS-T.
11. Visualize the bound protein by chemiluminescence captured by a charge-coupled device (CCD) camera or by exposure to an X-ray film.

3. CHARACTERIZATION OF INTERACTIONS BETWEEN ATG8/LC3/GABARAP AND LIR-CONTAINING PEPTIDES

The short LIR motif provides the basis for both affinity and specificity for the interaction with Atg8/LC3/GABARAP proteins (Rogov et al., 2014). Therefore, purified LIR-containing peptides can be used as a model for studying all aspects of recognition and binding of a LIR-containing protein to Atg8/LC3/GABARAP. We propose a simple and effective method for expression and purification of LIR-containing peptides for biophysical studies. The expression and purification protocols for the Atg8/LC3/GABARAP proteins are published (Habisov et al., 2016; Stadel et al., 2015). We will therefore further focus on ITC, which is a simple and reliable method to determine the thermodynamic parameters of an interaction, such as the driving forces (ΔH, ΔS), as well as the strength of the LIR interaction with Atg8/LC3/GABARAPs (K_A/K_D values).

3.1 Expression and Purification of LIR-Containing Peptides

For efficient expression, isolation, and purification of peptides, the correct peptide sequence should be selected. Peptides should consist of 18–22 amino acids (smaller peptides will be difficult to handle, and longer peptides can be involved in some unspecific interactions) and preferably lack lysines and arginines (as signals of unspecific degradations) as well as cysteines (as a source of unwanted disulfide bridges) and prolines (can undergo *cis–trans* isomerization). The desired codon-optimized peptide can be cloned as a fusion with a protein of choice linked via a specific protease recognition sequence for subsequent release of the peptide in a proteolytic reaction. We use a modified Ub-tag (Rogov et al., 2012) with Nco1–BamH1 restriction sites. This tag implements a solubility-enhancing leader (modified Ub), a His-tag (affinity tag for metal-chelating chromatography comprising 6 or 10 histidines in a row), and a tobacco etch virus protease (TEV protease) cleavage site.

1. Design a codon-optimized cDNA encoding the LIR peptide flanked by a Nco1 site and a stop-codon prior to a BamH1 site. Usually, designed fragments are less than 100 nucleotides long and can be prepared by annealing the corresponding oligonucleotides (Fig. 4A).
2. To anneal, dissolve each oligonucleotide in a buffer containing 50 m*M* of Tris–HCl, pH 7.5 and 150 m*M* of NaCl, mix both solutions so that it has a molar ratio of 1:1, heat in a water bath to 95°C, and slowly cool it to room temperature.
3. Digest an aliquot (usually 0.5 μg of final double-stranded oligonucleotide) with Nco1 and BamH1 enzymes and directly use the reaction mixture in a ligation reaction with the vector that was restricted using the same enzymes and dephosphorylated.
4. After ligation, transform LIR peptide-coding plasmid DNA isolated from a sequence-verified clone into New England Biolabs (NEB) T7 Express cells and inoculate into 50 mL of lysogeny broth (LB) medium for overnight preculture.
5. Inoculate overnight preculture into 1 L of LB media and grow until OD_{600} reaches 0.8.
6. Induce expression by adding IPTG to the final concentration of 1 m*M*, incubate no longer than 3 h, harvest the *E. coli* cells by centrifugation,

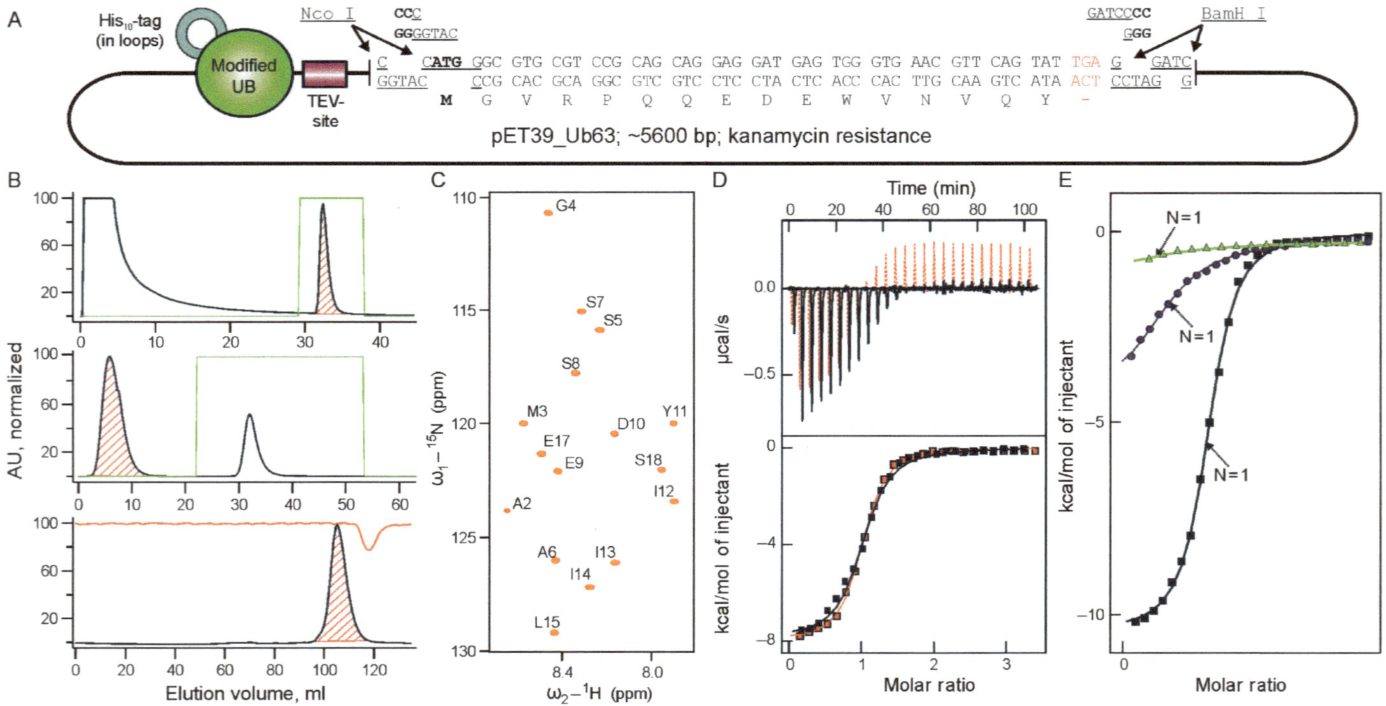

Fig. 4 See legend on opposite page.

and resuspend in buffer containing 50 m*M* of Tris–HCl pH 7.9, 200 m*M* of NaCl, 5% glycerol, 0.1 mg/mL of DNAse A, and 4 m*M* of protease inhibitor cocktail (at this stage, cells can be frozen at −80°C).

7. Lyse bacteria either by sonication or applying a French press. Add imidazole to a final concentration of 20 m*M* and remove cellular debris by centrifugation (30,000 *g*, 30 min). Apply the clear cell lysate on a nickel-nitrilotriacetic acid (Ni-NTA) column equilibrated with loading buffer (50 m*M* Tris–HCl pH 7.9, 250 m*M* NaCl, 1% glycerol, and 20 m*M* imidazole). Elute the His-tagged fusion Ub-peptide constructs with 400 m*M* imidazole in the same buffer (Fig. 4B, upper plot).
8. Process the fusion construct with TEV by supplementing the eluted protein with 1 m*M* EDTA, 10 m*M* β-ME, and 1–5 μg TEV protease (Sigma-Aldrich or self-made), and incubating the reaction overnight at 16°C.

Fig. 4 Cloning, expression, purification, and quality check of a LIR peptide and characterization of its binding to LC3B by ITC. (A) Synthetic codon-optimized cDNA encoding a LIR peptide (here PLEKHM1 LIR) was cloned into the pET39_Ub63 expression vector with a Ub63-fusion protein sequence. (B) Isolation and purification of the PLEKHM1 LIR peptide. *Top plot*—Ni-NTA chromatography of the Ub63-PLEKHM1 LIR fusion from an *E. coli* cell lysate; *middle plot*—reverse Ni-NTA chromatography of the PLEKHM1 LIR peptide after the TEV-protease treatment of the fractions above; *lower plot*—size-exclusion chromatography of the PLEKHM1 LIR peptide from fractions above. UV absorption at 280 nm is shown in *black* (normalized values in percentage of maxima are shown); concentration of elution buffer is in *green*; conductivity in *red*; and elution fractions containing PLEKHM1 LIR peptide are indicated by *red stripes*. (C) NMR characterization of the isolated LIR peptide. Section of a ^{1}H–^{15}N HCQC spectrum of the NBR1 LIR peptide obtained by the same method is shown. All assigned backbone HN resonances for the NBR1 LIR are given. (D) ITC titration of LC3B with the isolated PLEKHM1 LIR peptide (*black contours*) and Ub63-fused PLEKHM1 LIR peptide (*red contours*) in 50 m*M* of sodium phosphate, pH 7.0, at 25°C. The experiments consisted of total 26 injections (first—5 μL in 10 s spaced with 180 s, and then 25 injections of 10 μL, each spaced with 180 s). Concentration of the PLEKHM1 LIR peptide amounts to 496 μ*M*, Ub63-fused PLEKHM1 LIR peptide—510 μ*M*. The sample cell (volume 1.4567 mL) containing 21 μ*M* LC3B was stirred at 307 rpm. The *upper panel* represents raw differential heating power over time recorded in the experiment; the *lower panel* displays the integrated heat per titration over molar ratio (LC3B:LIR-containing peptide) with correction to heat of dilutions and to heat of (possible) unspecific interaction between Ub63 and LC3B. *Squares* represent experimentally obtained points; *lines* represent best fit with assumption of 1:1 interaction. (E) The integrated heat per titration over molar ratio for the LC3B:LIR interaction with different affinities. *Black squares and lines* show LC3B:p62 LIR, K_D of 1.5 μ*M*, ΔH −10.5 kcal/mol (Rozenknop et al., 2011). *Blue circles and lines* show GABARAPL1:UBA5 LIR/UFIM, K_D of 6.4 μ*M*, ΔH −3.5 kcal/mol (Habisov et al., 2016). *Green triangles and lines* indicate LC3B:NIX LIR, K_D of 91 μ*M*, ΔH −2.7 kcal/mol (Novak et al., 2010). The number of binding sites (N) for each interaction is shown by a *dashed arrow*.

9. Purify the LIR-containing peptide by reverse Ni-NTA chromatography. Therefore, change buffer of the reaction mix to the loading buffer by dialysis in 3 kDa cut-off dialyzing tubes in order to remove high concentration of imidazole, EDTA, and β-ME (3-h dialysis at room temperature). Pass the reaction mix through the Ni-NTA column, the peptide is in the flow-through (Fig. 4B, middle plot).
10. Purify the peptide from remaining Ub-tag by size-exclusion chromatography (SEC) on a Superdex 75 16 × 60 column (Fig. 4B, lower plot).
11. Concentrate purified peptides in Amicon concentrators (Merck Millipore) with a 3 kDa cut-off (>90% retention).

The resulting peptide is of >95% purity and in sufficient amount (3–15 mg/L of LB media) to perform biophysical experiments such as ITC, nuclear magnetic resonance (NMR) spectroscopy, or X-ray crystallography for detailed characterization of the Atg8/LC3/GABARAP:LIR interactions. The topics of NMR and X-ray crystallography are covered in this edition by other experts (see chapters "Investigating structure and dynamics of Atg8 family proteins" by Weiergräber et al. and "Crystallographic characterization of ATG proteins and their interacting partners" by Qiu et al.). Next, we provide a protocol for ITC-based characterization of the interactions.

3.2 Determination of Affinity and Driving Forces of the Interaction by ITC

ITC provides information on the affinity (in form of directly determined K_A and K_D values) and thermodynamic parameters (directly measured enthalpy ΔH and calculated free energy ΔG and entropy ΔS) of the interaction. The driving forces (ΔH and T*ΔS contributions) can be evaluated as well. In addition, in a single experiment one can evaluate the number of binding sites N, and via performing ITC titrations at different temperatures—the heat capacity increment ΔC_P of the interaction. This information can help to further characterize interactions between LIR peptides and the different Atg8/LC3/GABARAP orthologues, as well as to obtain additional knowledge about the differences between the binding mechanisms of canonical vs noncanonical LIR motifs. The method of ITC in application to protein–protein and protein–ligand interactions has been described in detail (Freyer & Lewis, 2008; Leavitt & Freire, 2001; Velazquez-Campoy, Ohtaka, Nezami, Muzammil, & Freire, 2004). Here, we present a short summary of the important experimental procedures and setups specifically for the Atg8/LC3/GABARAP proteins and LIR-containing peptides using the MicroCal

VP-ITC microcalorimeter (Malvern Instruments Ltd). Compared to other available techniques for studying protein–protein interactions, ITC allows label- and immobilization-free studies under "physiological" buffer conditions. It has no molecular weight limitation and is therefore a universal method for studying protein–protein, protein–peptide, and protein–ligand interactions. However, it requires considerable amounts of protein and peptides for a single experiment (for the VP-ITC, it is in the milligram range), and thus poor solubility could become an issue, especially for low-affinity interactions. High-affinity interactions result in a rectangular-shaped binding isotherm, making an accurate fit challenging. Measuring these interactions requires very low protein concentrations, which could be limited by the sensitivity of the instrument or displacement titrations. Also, it is necessary to have a measurable change in ΔH during the titration, and ΔH values close to zero are not suitable for analysis of intermolecular interactions. However, ΔH is temperature-dependent, and changes of the experimental temperature of about 5–10 degrees might result in a higher amplitude of ΔH. Consequently, solely entropy-driven reactions with small ΔC_P cannot be studied efficiently by ITC.

1. Prepare the protein and peptide samples in the same buffer (therefore, perform the gel filtration step already in the required buffer). In our previous studies, good results were obtained with 50 m*M* of Na_2HPO_4 pH 7.0, 100 m*M* of NaCl (with optional 0.5 m*M* of TCEP if the protein, such as LC3A or GABARAPL2, or LIR peptides contain cysteine residues). For LIR peptides with decreased solubility (high content of hydrophobic residues), experiments can be performed with the Ub-fused construct in the same buffer (Rozenknop et al., 2011). In this case, perform control experiments with only the Ub-fusion tag and ATG8/LC3/GABARAP proteins to rule out any unspecific interactions (Fig. 4D; compare the black solid lines—free LIR peptide titrated into GABARAP L1, and the red dashed lines—Ub63-fused peptide titrated to GABARAP L1).
2. Determine the concentration of both interaction partners on a standard UV-spectrometer by measuring the absorption at 280 nm and calculate the concentrations by using the respective molar extinction coefficient (ε), derived from protein/peptide sequence. Usually, LIR peptides contain tryptophan ($\varepsilon = 5500\ cm^{-1}/M$), tyrosine ($\varepsilon = 1490\ cm^{-1}/M$), or phenylalanine ($\varepsilon = 195\ cm^{-1}/M$ at 275.5 nm) residues. Optionally, it is also possible to add tryptophan or tyrosine residues to the very C-terminus of the designed peptide.

3. For the first time, perform the ITC at 25°C with Atg8/LC3/GABARAP concentrations between 20 and 40 μ*M*. The concentration of the LIR peptide should be at least 20–30 times higher than the UBL protein in order to obtain significant signals for weaker interactions. As a standard titration scheme, the first injection should be unique and contain only 5 μL (0.5 μL/s), which is not included in the data analysis, followed by 25 injections of 10 μL (0.5 μL/s) each. The spacing between each injection should be at least 180 s to allow reestablishment of the baseline.
4. If the affinity of the LIR peptide to the Atg8/LC3/GABARAP protein is optimal (K_D 0.1–10 μ*M*), the resulting ITC curve (integrated heat vs molar ratio) should be sigmoidal with a distinct initial and final plateau (Fig. 4E, black contours). The curve can directly be fitted with the respective program (e.g., MicroCal ITC-Origin analysis software, implemented in Origin), providing the enthalpy (ΔH), the entropy (ΔS), the association constant (K_A), and the number of binding sites (N). The values for ΔG and the dissociation constant K_D can be calculated by the following formulas:

$$\Delta G = -RT\ln(K_A) = RT\ln(K_D) = \Delta H - T*\Delta S$$

$$K_A = \frac{1}{K_D}$$

For the interactions with a lower affinity (K_D 10–50 μ*M*; Fig. 4E, blue contours), it is still possible to detect N, but with less precision. For the interactions with a very low affinity ($K_D > 50$ μ*M*, Fig. 4E, green contours), N cannot be derived from the experiment. Since all known LIR interactions occur at a 1:1 molar ratio, such a resulting curve has to be analyzed in a "one-site binding mode" with the N value fixed to 1. The strongest interaction of a validated LIR peptide to Atg8/LC3/GABARAP proteins observed so far by ITC showed K_D of 1.5 μ*M*, ΔH of −10.5 kcal/mol, and ΔS of −8.7 cal/mol/K; while the weakest one showed K_D of 91 μ*M*, ΔH of −2.7 kcal/mol, and ΔS of +9.4 cal/mol/K (Novak et al., 2010). Note that the lowest ΔH was detected for the interaction of OPTN LIR to LC3B: ΔH −0.7 kcal/mol, K_D 67 μ*M*, and ΔS +17 cal/mol/K (Wild et al., 2011). No positive ΔH values were observed for the natural LIRs.
5. If enthalpy of the interaction is either too low or too high, both the concentrations and the titration scheme have to be adjusted. If enthalpy of the interaction is low, increase the concentration of both the Atg8/LC3/GABARAP and the LIR peptide, or increase the volume of the injection

ligand during the titration step. It is strongly recommended to shift the measurement temperature by 5–10 degrees to gain a better ITC signal and to shift ΔH from the temperatures where its values are close to zero (changes will be proportional to ΔC_P of binding, which describes the temperature dependence of the ΔH, δΔH/δT). Usually, lower temperatures result in a more positive ΔH value, whereas higher temperatures give a more negative ΔH value. If the interaction is strong ($K_D < 1$ μ*M*), concentrations can be reduced, and less ligand can be injected during each titration step.

6. Weak interactions ($K_D > 50$ μ*M*) can (and should) be assessed by alternative methods (such as NMR). For example, parameters of OPTN LIR interaction with LC3B mentioned previously were confirmed by corresponding NMR titration experiments (also see chapters "Investigating structure and dynamics of Atg8 family proteins" by Weiergräber et al. and "Crystallographic characterization of ATG proteins and their interacting partners" by Qiu et al.). Very strong interactions ($K_D < 0.01$ μ*M*) might require a competitive ITC approach (Velazquez-Campoy & Freire, 2006).

The affinities and driving forces of the interaction of a specific LIR with Atg8/LC3/GABARAP proteins obtained by ITC might serve as a working model for further studies, in which the biological relevance of observed interactions has to be demonstrated.

4. SUMMARY AND FUTURE OUTLOOK

The methods presented in this chapter will allow researchers to identify and validate LIR-containing proteins, define and narrow their LIRs, purify LIR-containing peptides, and study characteristics of the interactions between LIRs and Atg8/LC3/GABARAP proteins. Additional methods can be used to support the study of LIR-containing proteins, such as mass spectrometry for identification of protein complexes involved in interactions with Atg8/LC3/GABARAP proteins (Behrends, Sowa, Gygi, & Harper, 2010), coimmunoprecipitation (Bjorkoy et al., 2009), and Förster resonance energy transfer (FRET) (Kraft & Kenworthy, 2012) for testing interactions between LIR-containing proteins and Atg8/LC3/GABARAPs in vivo. Use of NMR and X-ray crystallography is recommended to obtain structural details for the Atg8/LC3/GABARAP:LIR interaction. With the growing recognition of the complexity of the autophagy pathway, studying LIR-containing proteins using a multidisciplinary approach will further our understanding of basic principles of LIR recognition and binding. This will in turn stimulate refinement of existing and development of new methods to study protein–protein interactions involved in the initiation, progression, and regulation of the selective autophagy pathway.

ACKNOWLEDGMENTS

T.J. received grants from the FRIBIOMED program of the Norwegian Research Council (grant number 214448), and the Norwegian Cancer Society (grant number 71043-PR-2006-0320). J.H., A.K., V.D., and V.V.R. appreciate financial support from the German Cancer Consortium (DKTK); the LOEWE program of the State of Hesse, Germany; and the SFB 1177 "Molecular and Functional Characterization of Selective Autophagy," Germany.

REFERENCES

Alemu, E. A., Lamark, T., Torgersen, K. M., Birgisdottir, A. B., Larsen, K. B., Jain, A., … Johansen, T. (2012). ATG8 family proteins act as scaffolds for assembly of the ULK complex: Sequence requirements for LC3-interacting region (LIR) motifs. *The Journal of Biological Chemistry*, *287*(47), 39275–39290. http://dx.doi.org/10.1074/jbc.M112.378109.

Behrends, C., Sowa, M. E., Gygi, S. P., & Harper, J. W. (2010). Network organization of the human autophagy system. *Nature*, *466*(7302), 68–76. http://dx.doi.org/10.1038/nature09204.

Birgisdottir, A. B., Lamark, T., & Johansen, T. (2013). The LIR motif—Crucial for selective autophagy. *Journal of Cell Science*, *126*(Pt. 15), 3237–3247. http://dx.doi.org/10.1242/jcs.126128.

Bjorkoy, G., Lamark, T., Pankiv, S., Overvatn, A., Brech, A., & Johansen, T. (2009). Monitoring autophagic degradation of p62/SQSTM1. *Methods in Enzymology*, *452*, 181–197. http://dx.doi.org/10.1016/S0076-6879(08)03612-4.

Dowdle, W. E., Nyfeler, B., Nagel, J., Elling, R. A., Liu, S., Triantafellow, E., … Murphy, L. O. (2014). Selective VPS34 inhibitor blocks autophagy and uncovers a role for NCOA4 in ferritin degradation and iron homeostasis in vivo. *Nature Cell Biology*, *16*(11), 1069–1079. http://dx.doi.org/10.1038/ncb3053.

Dunham, W. H., Mullin, M., & Gingras, A. C. (2012). Affinity-purification coupled to mass spectrometry: Basic principles and strategies. *Proteomics*, *12*(10), 1576–1590. http://dx.doi.org/10.1002/pmic.201100523.

Farre, J. C., Manjithaya, R., Mathewson, R. D., & Subramani, S. (2008). PpAtg30 tags peroxisomes for turnover by selective autophagy. *Developmental Cell*, *14*(3), 365–376. http://dx.doi.org/10.1016/j.devcel.2007.12.011.

Fields, S., & Song, O. (1989). A novel genetic system to detect protein-protein interactions. *Nature*, *340*(6230), 245–246. http://dx.doi.org/10.1038/340245a0.

Frank, R. (2002). The SPOT-synthesis technique. Synthetic peptide arrays on membrane supports—Principles and applications. *Journal of Immunological Methods*, *267*(1), 13–26.

Freyer, M. W., & Lewis, E. A. (2008). Isothermal titration calorimetry: Experimental design, data analysis, and probing macromolecule/ligand binding and kinetic interactions. *Methods in Cell Biology*, *84*, 79–113. http://dx.doi.org/10.1016/S0091-679X(07)84004-0.

Habisov, S., Huber, J., Ichimura, Y., Akutsu, M., Rogova, N., Loehr, F., … Kirkin, V. (2016). Structural and functional analysis of a novel interaction motif within UFM1-activating enzyme 5 (UBA5) required for binding to ubiquitin-like proteins and ufmylation. *The Journal of Biological Chemistry*, *291*, 9025–9041. http://dx.doi.org/10.1074/jbc.M116.715474.

Hecker, C. M., Rabiller, M., Haglund, K., Bayer, P., & Dikic, I. (2006). Specification of SUMO1- and SUMO2-interacting motifs. *The Journal of Biological Chemistry*, *281*(23), 16117–16127. http://dx.doi.org/10.1074/jbc.M512757200.

Ichimura, Y., Kumanomidou, T., Sou, Y. S., Mizushima, T., Ezaki, J., Ueno, T., … Komatsu, M. (2008). Structural basis for sorting mechanism of p62 in selective autophagy. *The Journal of Biological Chemistry*, *283*(33), 22847–22857. http://dx.doi.org/10.1074/jbc.M802182200.

Johansen, T., & Lamark, T. (2011). Selective autophagy mediated by autophagic adapter proteins. *Autophagy*, 7(3), 279–296.

Kalvari, I., Tsompanis, S., Mulakkal, N. C., Osgood, R., Johansen, T., Nezis, I. P., & Promponas, V. J. (2014). iLIR: A web resource for prediction of Atg8-family interacting proteins. *Autophagy*, *10*(5), 913–925. http://dx.doi.org/10.4161/auto.28260.

Kanki, T., Wang, K., Cao, Y., Baba, M., & Klionsky, D. J. (2009). Atg32 is a mitochondrial protein that confers selectivity during mitophagy. *Developmental Cell*, *17*(1), 98–109. http://dx.doi.org/10.1016/j.devcel.2009.06.014.

Kaufmann, A., Beier, V., Franquelim, H. G., & Wollert, T. (2014). Molecular mechanism of autophagic membrane-scaffold assembly and disassembly. *Cell*, *156*(3), 469–481. http://dx.doi.org/10.1016/j.cell.2013.12.022.

Khaminets, A., Behl, C., & Dikic, I. (2016). Ubiquitin-dependent and independent signals in selective autophagy. *Trends in Cell Biology*, *26*(1), 6–16. http://dx.doi.org/10.1016/j.tcb.2015.08.010.

Khaminets, A., Heinrich, T., Mari, M., Grumati, P., Huebner, A. K., Akutsu, M., … Dikic, I. (2015). Regulation of endoplasmic reticulum turnover by selective autophagy. *Nature*, *522*(7556), 354–358. http://dx.doi.org/10.1038/nature14498.

Kirkin, V., Lamark, T., Sou, Y. S., Bjorkoy, G., Nunn, J. L., Bruun, J. A., … Johansen, T. (2009). A role for NBR1 in autophagosomal degradation of ubiquitinated substrates. *Molecular Cell*, *33*(4), 505–516. http://dx.doi.org/10.1016/j.molcel.2009.01.020.

Kirkin, V., McEwan, D. G., Novak, I., & Dikic, I. (2009). A role for ubiquitin in selective autophagy. *Molecular Cell*, *34*(3), 259–269. http://dx.doi.org/10.1016/j.molcel.2009.04.026.

Kraft, L. J., & Kenworthy, A. K. (2012). Imaging protein complex formation in the autophagy pathway: Analysis of the interaction of LC3 and Atg4B(C74A) in live cells using Forster resonance energy transfer and fluorescence recovery after photobleaching. *Journal of Biomedical Optics*, *17*(1), 011008. http://dx.doi.org/10.1117/1.JBO.17.1.011008.

Leavitt, S., & Freire, E. (2001). Direct measurement of protein binding energetics by isothermal titration calorimetry. *Current Opinion in Structural Biology*, *11*(5), 560–566.

Liu, L., Feng, D., Chen, G., Chen, M., Zheng, Q., Song, P., … Chen, Q. (2012). Mitochondrial outer-membrane protein FUNDC1 mediates hypoxia-induced mitophagy in mammalian cells. *Nature Cell Biology*, *14*(2), 177–185. http://dx.doi.org/10.1038/ncb2422.

Lu, K., Psakhye, I., & Jentsch, S. (2014). Autophagic clearance of polyQ proteins mediated by ubiquitin-Atg8 adaptors of the conserved CUET protein family. *Cell*, *158*(3), 549–563. http://dx.doi.org/10.1016/j.cell.2014.05.048.

Mancias, J. D., Wang, X., Gygi, S. P., Harper, J. W., & Kimmelman, A. C. (2014). Quantitative proteomics identifies NCOA4 as the cargo receptor mediating ferritinophagy. *Nature*, *509*(7498), 105–109. http://dx.doi.org/10.1038/nature13148.

Mizushima, N., Yoshimori, T., & Ohsumi, Y. (2011). The role of Atg proteins in autophagosome formation. *Annual Review of Cell and Developmental Biology*, *27*, 107–132. http://dx.doi.org/10.1146/annurev-cellbio-092910-154005.

Motley, A. M., Nuttall, J. M., & Hettema, E. H. (2012). Pex3-anchored Atg36 tags peroxisomes for degradation in Saccharomyces cerevisiae. *The EMBO Journal*, *31*(13), 2852–2868. http://dx.doi.org/10.1038/emboj.2012.151.

Newman, A. C., Scholefield, C. L., Kemp, A. J., Newman, M., McIver, E. G., Kamal, A., & Wilkinson, S. (2012). TBK1 kinase addiction in lung cancer cells is mediated via autophagy of Tax1bp1/Ndp52 and non-canonical NF-kappaB signalling. *PLoS One*, 7(11). e50672. http://dx.doi.org/10.1371/journal.pone.0050672.

Noda, N. N., Kumeta, H., Nakatogawa, H., Satoo, K., Adachi, W., Ishii, J., … Inagaki, F. (2008). Structural basis of target recognition by Atg8/LC3 during selective autophagy. *Genes to Cells*, *13*(12), 1211–1218. http://dx.doi.org/10.1111/j.1365-2443.2008.01238.x.

Novak, I., Kirkin, V., McEwan, D. G., Zhang, J., Wild, P., Rozenknop, A., … Dikic, I. (2010). Nix is a selective autophagy receptor for mitochondrial clearance. *EMBO Reports*, *11*(1), 45–51. http://dx.doi.org/10.1038/embor.2009.256.

Okamoto, K. (2014). Organellophagy: Eliminating cellular building blocks via selective autophagy. *The Journal of Cell Biology*, *205*(4), 435–445. http://dx.doi.org/10.1083/jcb.201402054.

Okamoto, K., Kondo-Okamoto, N., & Ohsumi, Y. (2009). Mitochondria-anchored receptor Atg32 mediates degradation of mitochondria via selective autophagy. *Developmental Cell*, *17*(1), 87–97. http://dx.doi.org/10.1016/j.devcel.2009.06.013.

Pankiv, S., Clausen, T. H., Lamark, T., Brech, A., Bruun, J. A., Outzen, H., … Johansen, T. (2007). p62/SQSTM1 binds directly to Atg8/LC3 to facilitate degradation of ubiquitinated protein aggregates by autophagy. *The Journal of Biological Chemistry*, *282*(33), 24131–24145. http://dx.doi.org/10.1074/jbc.M702824200.

Popelka, H., & Klionsky, D. J. (2015). Analysis of the native conformation of the LIR/AIM motif in the Atg8/LC3/GABARAP-binding proteins. *Autophagy*, *11*(12), 2153–2159. http://dx.doi.org/10.1080/15548627.2015.1111503.

Rogov, V., Dötsch, V., Johansen, T., & Kirkin, V. (2014). Interactions between autophagy receptors and ubiquitin-like proteins form the molecular basis for selective autophagy. *Molecular Cell*, *53*(2), 167–178. http://dx.doi.org/10.1016/j.molcel.2013.12.014.

Rogov, V. V., Rozenknop, A., Rogova, N. Y., Lohr, F., Tikole, S., Jaravine, V., … Dötsch, V. (2012). A universal expression tag for structural and functional studies of proteins. *Chembiochem*, *13*(7), 959–963. http://dx.doi.org/10.1002/cbic.201200045.

Rozenknop, A., Rogov, V. V., Rogova, N. Y., Lohr, F., Guntert, P., Dikic, I., & Dötsch, V. (2011). Characterization of the interaction of GABARAPL-1 with the LIR motif of NBR1. *Journal of Molecular Biology*, *410*(3), 477–487. http://dx.doi.org/10.1016/j.jmb.2011.05.003.

Scott, S. V., Guan, J., Hutchins, M. U., Kim, J., & Klionsky, D. J. (2001). Cvt19 is a receptor for the cytoplasm-to-vacuole targeting pathway. *Molecular Cell*, 7(6), 1131–1141.

Shpilka, T., Weidberg, H., Pietrokovski, S., & Elazar, Z. (2011). Atg8: An autophagy-related ubiquitin-like protein family. *Genome Biology*, *12*(7), 226. http://dx.doi.org/10.1186/gb-2011-12-7-226.

Snider, J., Kittanakom, S., Curak, J., & Stagljar, I. (2010). Split-ubiquitin based membrane yeast two-hybrid (MYTH) system: A powerful tool for identifying protein-protein interactions. *Journal of Visualized Experiments*. *36*. http://dx.doi.org/10.3791/1698.

Stadel, D., Millarte, V., Tillmann, K. D., Huber, J., Tamin-Yecheskel, B. C., Akutsu, M., … Behrends, C. (2015). TECPR2 cooperates with LC3C to regulate COPII-dependent ER export. *Molecular Cell*, *60*(1), 89–104. http://dx.doi.org/10.1016/j.molcel.2015.09.010.

Stolz, A., Ernst, A., & Dikic, I. (2014). Cargo recognition and trafficking in selective autophagy. *Nature Cell Biology*, *16*(6), 495–501. http://dx.doi.org/10.1038/ncb2979.

Suzuki, K., Kondo, C., Morimoto, M., & Ohsumi, Y. (2010). Selective transport of alpha-mannosidase by autophagic pathways: Identification of a novel receptor, Atg34p. *The Journal of Biological Chemistry*, *285*(39), 30019–30025. http://dx.doi.org/10.1074/jbc.M110.143511.

Thielmann, Y., Weiergraber, O. H., Mohrluder, J., & Willbold, D. (2009). Structural framework of the GABARAP-calreticulin interface—Implications for substrate binding to endoplasmic reticulum chaperones. *The FEBS Journal*, *276*(4), 1140–1152. http://dx.doi.org/10.1111/j.1742-4658.2008.06857.x.

Thurston, T. L., Ryzhakov, G., Bloor, S., von Muhlinen, N., & Randow, F. (2009). The TBK1 adaptor and autophagy receptor NDP52 restricts the proliferation of ubiquitin-coated bacteria. *Nature Immunology*, *10*(11), 1215–1221. http://dx.doi.org/10.1038/ni.1800.

Velazquez-Campoy, A., & Freire, E. (2006). Isothermal titration calorimetry to determine association constants for high-affinity ligands. *Nature Protocols*, *1*(1), 186–191. http://dx.doi.org/10.1038/nprot.2006.28.

Velazquez-Campoy, A., Ohtaka, H., Nezami, A., Muzammil, S., & Freire, E. (2004). Isothermal titration calorimetry. *Current Protocols in Cell Biology*. http://dx.doi.org/10.1002/0471143030.cb1708s23. chapter 17, unit 17.8.

von Muhlinen, N., Akutsu, M., Ravenhill, B. J., Foeglein, A., Bloor, S., Rutherford, T. J., ... Randow, F. (2012). LC3C, bound selectively by a noncanonical LIR motif in NDP52, is required for antibacterial autophagy. *Molecular Cell*, *48*(3), 329–342. http://dx.doi.org/10.1016/j.molcel.2012.08.024.

Wild, P., Farhan, H., McEwan, D. G., Wagner, S., Rogov, V. V., Brady, N. R., ... Dikic, I. (2011). Phosphorylation of the autophagy receptor optineurin restricts Salmonella growth. *Science*, *333*(6039), 228–233. http://dx.doi.org/10.1126/science.1205405.

CHAPTER TEN

Assessment of Posttranslational Modifications of ATG proteins

Y. Xie*,†, R. Kang*, D. Tang*,[1]
*University of Pittsburgh, Pittsburgh, PA, United States
†Xiangya Hospital, Central South University, Changsha, China
[1]Corresponding author: e-mail address: tangd2@upmc.edu

Contents

1. Introduction 172
2. Monitoring PTMs of ATG Proteins Using Western Blotting 173
   2.1 Overview 173
   2.2 Protocol 174
3. Monitoring PTMs of ATG Proteins Using Simple Western™ Assays 175
   3.1 Overview 175
   3.2 Protocol 175
4. Monitoring PTMs of ATG Proteins Using Immunoprecipitation Assays 181
   4.1 Overview 181
   4.2 Protocol 182
5. Monitoring PTMs of ATG Proteins Using Immunofluorescence Assays 183
   5.1 Overview 183
   5.2 Protocol 186
6. Concluding Remarks 186
Acknowledgments 187
References 187

Abstract

Posttranslational modifications (PTMs) increase the functional diversity of proteins and play a key role in many cellular processes. Macroautophagy (hereafter simply referred to as *autophagy*) is an evolutionarily conserved, lysosome-dependent degradation pathway. This process is finely regulated by autophagy-related (ATG) genes widely conserved among eukaryotes from yeast to mammals. Various PTMs of ATG proteins such as phosphorylation, ubiquitination, and acetylation have been theorized to play a critical role in modulating autophagic processes and activity. In this chapter, we introduce several antibody-based tools (e.g., Western blot, Simple Western™, immunofluorescence, and immunoprecipitation) that are widely used to assess the PTMs of ATG proteins in mammalian cells.

Methods in Enzymology, Volume 587
ISSN 0076-6879
http://dx.doi.org/10.1016/bs.mie.2016.09.057

1. INTRODUCTION

Macroautophagy (hereafter simply referred to as *autophagy*), discovered more than a half-century ago, is an intracellular catabolic process involved in the formation and maturation of multiple membrane structures (Mizushima, Yoshimori, & Levine, 2010; Yang & Klionsky, 2010). It generally starts with the formation of a double-membrane sequestering compartment termed the *phagophore,* which matures into an autophagosome and ultimately fuses with the lysosomal compartment to degrade the sequestered material by the autolysosome. A number of highly conserved autophagy-related (ATG) genes and associated proteins contribute to the dynamic arrangements of membrane structures in organisms ranging from yeast to mammalian cells (Feng, He, Yao, & Klionsky, 2014).

Posttranslational modifications (PTMs) are known as essential mechanisms used by eukaryotic cells to diversify their protein functions. PTMs are also important for modulating ATG protein function in autophagy (Popelka & Klionsky, 2015; Wani et al., 2015; Xie et al., 2015). For example, phosphorylation-mediated regulation of the ULK1 (also termed *Atg1* in yeast) complex initiates autophagosome formation via interaction with the mechanistic targets of rapamycin complex 1 (MTORC1), protein kinase A (PRKAC), AMP-activated protein kinase (AMPK), and AKT (v-akt murine thymoma viral and oncogene homolog) in response to starvation, energetic exhaustion, and insulin treatment, respectively (Bach, Larance, James, & Ramm, 2011; Dorsey et al., 2009; Kim, Kundu, Viollet, & Guan, 2011; Shang et al., 2011). In addition, nucleation of the phagophore during autophagosome formation requires phosphorylation of class III phosphatidylinositol 3-kinase (PtdIns3K) complex components. PtdIns3K complex in mammalian cells is composed of three core components, namely, PIK3C3/VPS34, PIK3R4/p150, and BECN1. In addition to ULK1, other kinases including death-associate protein kinase, Rho-associated, coiled-coil containing protein kinase 1 (ROCK), AKT, AMPK, and epidermal growth factor receptor also regulate the phosphorylation of PtdIns3K complex components (Gurkar et al., 2013; Kim et al., 2013; Russell et al., 2013; Wang et al., 2012; Wei et al., 2013; Zalckvar et al., 2009). Two ubiquitin-like (Ubl) conjugation systems, including the ATG12-ATG5 and Ubl microtubule-associated protein 1 light chain 3 (LC3) conjugation systems, are essential for phagophore elongation (Ichimura et al., 2000; Mizushima et al., 1998). Notably, LC3 lipidation determines the size of the autophagosome, which also participates in cargo recognition (Xie, Nair, & Klionsky, 2008). However, PRKAC-mediated phosphorylation of LC3 inhibits its lipidation and therefore inhibits autophagy (Cherra et al., 2010), whereas mitogen-activated protein kinase 15 (MAPK15) prevents this inhibitory phosphorylation of LC3 by PRKAC (Colecchia et al., 2012). In addition, TRAF6 (TNF receptor-associated factor 6, E3 ubiquitin protein ligase), AMBRA1 (autophagy/BECN1 regulator 1), NEDD4 (neural precursor cell expressed, developmentally downregulated 4, E3 ubiquitin protein ligase), USP10 (ubiquitin specific peptidase 10), and USP13 (ubiquitin specific peptidase 13)-mediated ubiquitination or deubiquitination of BECN1 or ULK1 are essential for cargo recognition and protein degradation (Liu et al., 2011; Platta, Abrahamsen, Thoresen, & Stenmark, 2012; Shi & Kehrl, 2010; Sowa, Bennett, Gygi, & Harper, 2009; Xia et al., 2013). Once formation of the autophagosome is completed and it fuses with the lysosome, the sequestered cargo is degraded and broken down by hydrolyses. In this chapter, we introduce methods to detect the PTMs of ATG proteins in cultured mammalian cells.

2. MONITORING PTMs OF ATG PROTEINS USING WESTERN BLOTTING

2.1 Overview

Western blotting (WB, also known as *immunoblotting* or *protein blotting*) is a widely used laboratory method to detect specific protein characteristics, including abundance, apparent molecular mass, PTMs, and splice variants, in a sample of tissue homogenate or cellular extract (Jensen, 2012; Mahmood & Yang, 2012). Gel electrophoresis is used in WB to separate native or denatured proteins by the length of the polypeptide (denaturing conditions) or by the 3-D structure of the protein (native/nondenaturing conditions). The proteins are then transferred to nitrocellulose (NC) or polyvinylidene difluoride (PVDF) membrane and are recognized with primary antibodies specific to the target protein. This is followed by a labeled secondary antibody conjugated to enzymes such as horseradish peroxidase (HRP) and alkaline phosphatase that is directed against the primary antibody. The most sensitive detection methods use a chemiluminescent substrate that, when combined with the enzyme, produces light as a by-product. This signal can be captured with traditional film or a digital imaging system.

2.2 Protocol

A modified WB protocol (based on a typical cell culture extract) from Cell Signaling Technology (http://www.cellsignal.com/common/content/content.jsp?id=western) is detailed as follows:

1. Place the cell culture dish in ice and wash the cells with ice-cold 1 × Tris-buffered saline (TBS, 137 m*M* NaCl, 20 m*M* Tris, pH 7.6).
2. Lyse cells by adding 1 × sodium dodecyl sulfate (SDS) sample buffer (500 μL for a 10-cm dish) with a protease inhibitor cocktail (Sigma, St. Louis, MO, USA) and phosphatase inhibitor cocktail (Sigma). Immediately scrape the cells off the plate using a cold plastic cell scraper and gently transfer the extract to a precooled 1.5-mL microcentrifuge tube.
3. Keep on ice for 30 min and shake three times for 5 s every 10 min.
4. If necessary, sonicate one time for 10–15 s to complete cell lysis and shear DNA to reduce sample viscosity.
5. Spin at 15,000 × *g* for 10 min at 4°C in precooled centrifuge.
6. Transfer the supernatant to a fresh tube and keep on ice. The pellet is discarded.
7. Determine the protein concentration in the supernatant using the bicinchoninic acid assay method.
8. Take 20–40 μg of each sample, add a 2 × Laemmli sample buffer (4% SDS, 20% glycerol, 10% 2-mercaptoethanol, 0.004% bromphenol blue, and 0.125 *M* Tris–HCl, pH 6.8), and boil at 95°C for 3–5 min.
9. Load equal amounts of proteins into the wells of midi-format SDS polyacrylamide gel electrophoresis (SDS-PAGE) gel, along with molecular weight markers.
10. Run the gel for 5 min at 50 V and then increase the voltage to 150 V to finish the run in about 40 min.
11. Transfer the protein from the gel to NC or PVDF membrane in a cold room at a constant current of 10 mA.
12. After transfer, wash the membrane with 25 mL of TBS for 5 min at room temperature.
13. Incubate the membrane in 25 mL of 3% bovine serum albumin (BSA) in TBST; 137 m*M* NaCl, 20 m*M* Tris, 0.1% Tween-20, pH 7.6) for 1 h at room temperature.
14. Wash three times for 5 min each with 15 mL of TBST.
15. Incubate the membrane and primary antibody (at the appropriate dilution recommended in the product data sheet) in 10 mL of primary antibody dilution buffer with gentle agitation overnight at 4°C. The

commercially available primary antibodies for the detection of PTMs (mainly phosphorylation) of ATG proteins are listed in Table 1.

16. Wash three times for 5 min each with 15 mL of TBST.
17. Incubate with antibiotin or HRP-linked secondary antibody (1:1000–1:5000) for 1 h at room temperature. The secondary antibody is diluted in 5% milk in TBST.
18. Wash three to five times for 5 min each with 15 mL of TBST.
19. Apply the chemiluminescent substrate to the blot according to the manufacturer's recommendations.
20. Capture the chemiluminescent signals using film or a digital imaging system.
21. Use image analysis software to read the band intensity and normalize the target protein levels.

3. MONITORING PTMs OF ATG PROTEINS USING SIMPLE WESTERN™ ASSAYS

3.1 Overview

The ProteinSimple Company (http://www.proteinsimple.com/) has developed Simple Western™ assays for protein sizing and quantitative immunodetection as an alternative to traditional WB analysis. Simple Western™ is an automated and capillary-based immunoassay. No gels, no transfer devices, no blots, no film, and no manual analysis are required for Simple Western™ assays (Harris, 2015). Assays are performed on an instrument that integrates and automates all manual operations associated with WB. Currently, the ProteinSimple Company provides four different detection apparatuses, including Wes™, Peggy Sue™, Sally Sue™, and NanoPro 1000™. The Simple Western™ autoassay process includes six major steps: loading matrices, loading the sample, separation, immobilization, immunoprobing, and quantitation. R&D Systems (https://www.rndsystems.com/?gclid=CIv1mIHVvswCFRJZhgodcPwMIA) and Novus Biologicals (http://www.novusbio.com/) provide certified antibodies, including PTMs of ATGs for Simple Western™ assays (Table 2). Compared with traditional WB analysis, Simple Western™ assays require few samples.

3.2 Protocol

Simple Western analyses (based on a typical cell culture extract) performed with Wes™ according to the ProteinSimple user manual

Table 1 Commercial ATG Phosphorylation Antibodies

Target Protein	PTM	Site	Activated by	Supplier (Cat. #)	Reactivity	Clonality	Host	Application	Function
ULK1	Phosphorylation	Ser555	AMPK	CST (5869)	H, M	Monoclonal	R	WB, IP	Induces starvation-induced autophagy
				Millipore (ABC124)	H, M, R, B, X	Polyclonal	R	WB, PIA	
		Ser757	MTORC1	CST (14202)	H, M, R	Monoclonal	R	WB, IP	Inhibits binding of ULK1 to AMPK and subsequent autophagy
				CST (6888)	H, M, Mk	Polyclonal	R	WB	
		Ser317	AMPK	CST (12753)	H, M	Monoclonal	R	WB	Inhibits MTORC1 and promotes glucose starvation-induced autophagy
		Ser467	AMPK	CST (4634)	M	Polyclonal	R	WB	Promotes selective autophagy (mitophagy)
				Abcam (ab179458)	H	Monoclonal	R	WB	
		Ser638	MTORC1	CST (14205)	H, M, Mk	Monoclonal	R	WB	Disrupts the interaction between ULK1 and AMPK and inhibits autophagy
				Abcam (ab133766)	H	Monoclonal	R	WB	
		Ser556	AMPK	Bioss (bs-3464R)	H, M, Rat	Polyclonal	R	WB, IHC, IF	
				Abcam (ab133747)	H, M	Monoclonal	R	WB, ICC, IF	

		Thr574	AMPK	Millipore (ABC212)	H	Polyclonal	R	WB, PIA	Promotes mitophagy
		Ser758	MTORC1	Millipore (ABC112)	H	Polyclonal	R	WB, PIA	Inhibits autophagy with nutrient
ATG3	Phosphorylation	Tyr18		Bioss (bs-5204R)	H, M, Rat	Polyclonal	R	WB, IF, IHC	
BECN1	Phosphorylation	Ser295	AKT	MyBioSource (MBS502207) LLL	H	Polyclonal	R	WB	Forms autophagy-inhibitory BECN1/14-3-3/vimentin complex
		Ser234	AKT	Rockland (600-401-W96)	H	Polyclonal	R	WB, ELISA	Forms autophagy-inhibitory BECN1/14-3-3/vimentin complex
				Abcam (ab183335)	H	Polyclonal	R	WB	
		Ser93	AMPK	CST (14717)	H	Monoclonal	R	WB, IP	Promotes glucose-starvation-induced autophagy
		Ser15	ULK1	CST (84966)	H, M	Monoclonal	R	WB	Activates the autophagy-specific VPS34 complex and autophagy induction

Continued

Table 1 Commercial ATG Phosphorylation Antibodies—cont'd

Target Protein	PTM	Site	Activated by	Supplier (Cat. #)	Reactivity	Clonality	Host	Application	Function
LC3	Phosphorylation	Ser12	PRKAC	Millipore (ABC466)	H, M	Polyclonal	R	WB, PIA	
				Abgent (AP3301a)	H	Polyclonal	R	WB, DB, ELISA	
				BioVision (6951-100)	H	N.A.	R	WB, DB	
ATG9A/B	Phosphorylation	Ser735		Bioss (bs-5199R)	H, Rat	Polyclonal	R	WB, IHC	
ATG13	Phosphorylation	Ser318	ULK1	Abnova (PAB199481)	H	Polyclonal	R	WB, ELISA	Promotes ULK1-ATG13-mediated mitophagy
				Rockland (600-401-C49)	H	Polyclonal	R	WB, ELISA, FC	
				Novus (NBP2-19127)	H	Polyclonal	R	WB, ELISA	
		Ser355	ULK1	CST (43533)	H	Polyclonal	R	WB	
ATG14	Phosphorylation	Ser29	ULK1	CST (13155)	H, M	Polyclonal	R	WB	Promotes autophagosome formation
ATG18/WIPI2	Phosphorylation	Ser413		CST (13517)	H, M, R	Polyclonal	R	WB, IP	

B, bovine; *DB*, dot blot; *ELISA*, enzyme-linked immunosorbent assay; *FC*, flow cytometry; *H*, human; *ICC*, immunocytochemistry; *IHC*, immunohistochemistry; *M*, mouse; *Mk*, monkey; *PIA*, peptide inhibition assay; *R*, rabbit; *X*, xenopus.

Table 2 Commercial ATG Antibodies for Simple Western™ Assays

Protein Target	Antibody Type	Vendor	Product Number	Protein Isoform	Species	Cell Models	Separated by Size or Charge	Dilution	kDa on Simple Western
ATG5	Primary	Novus Biologicals	NB110-53818	ATG5	R	Hela lysate 0.5 mg/mL	Size-Wes, Sally Sue/Peggy Sue	1/50	60
ATG5	Primary	R&D Systems	MAB5294	ATG5	M	Hela human cervical epithelial carcinoma cell line and PC12 rat adrenal pheochromocytoma cell line	Size	25 μg/mL	51
ATG7	Primary	R&D Systems	MAB6608	ATG7	M	Hela human cervical epithelial carcinoma cell line and HepG2 human hepatocellular carcinoma cell line	Size-Wes, Sally Sue/Peggy Sue	20 μg/mL	66
BECN1/ATG6	Primary	R&D Systems	MAB5295	BECN1/ATG6	M	Human brain (cerebellum) tissue and mouse brain tissue	Size	1 μg/mL	58
BECN1	Primary	Cell Signaling	3738	BECN1	R	Mouse liver	Size-Wes	1/100	59

Continued

Table 2 Commercial ATG Antibodies for Simple Western™ Assays—cont'd

Protein Target	Antibody Type	Vendor	Product Number	Protein Isoform	Species	Cell Models	Separated by Size or Charge	Dilution	kDa on Simple Western
LC3	Primary	Novus Biologicals	NB100-2331	LC3	R	Neuro2A lysate 0.5 mg/mL	Size-Wes, Sally Sue/Peggy Sue	1/50	16
LC3	Primary	Novus Biologicals	NBP1-19167	LC3	R	Neuro2A lysate 0.5 mg/mL	Size-Wes, Sally Sue/Peggy Sue	1/40	17
LC3	Primary	Novus Biologicals	NB100-2220	LC3	R	Neuro2A lysate 0.5 mg/mL	Size-Wes, Sally Sue/Peggy Sue	1/50	
LC3B	Primary	Novus Biologicals	NB600-1384	LC3B	R	Neuro2A lysate 0.5 mg/mL	Size-Wes, Sally Sue/Peggy Sue	1/100	

M, mouse; R, rabbit.

(http://www.proteinsimple.com/wes.html) are detailed later. All reagents are obtained from the Wes kit from ProteinSimple™.

1. Dilute cell lysates to 1 mg/mL using 0.1 × sample buffer.
2. Add 4 μL of diluted lysate to 1 μL of 5 × master mix, with a final concentration = 0.8 mg/mL.
3. Heat samples at 95°C for 5 min.
4. The samples, blocking reagent, primary antibodies (1:50 or 1:100 dilution), HRP-conjugated secondary antibodies (1:200 dilution), chemiluminescent substrate, and separation and stacking matrices are dispensed to designated wells in the plate. The commercially available ATG antibodies tested with Simple Western™ are listed in Table 2 (refer to ProteinSimple's antibody database at http://www.proteinsimple.com/antibody/antibodies.html).
5. After plate loading, the separation electrophoresis and immunodetection steps take place in the capillary system and are fully automated. Capillaries are first filled with separation matrix (200 s), followed by stacking matrix (15 s) and loading approximately 40 nL of sample (9 s). During electrophoresis, proteins are separated on the basis of molecular weight through the stacking and separation matrices at 375 V for 25 min and then immobilized on the capillary wall using proprietary photoactivated capture chemistry. The matrices are then washed out (230 s). Capillaries are next incubated with a blocking reagent for 30 min, and target proteins are immunoprobed with primary antibodies (30 min), followed by HRP-conjugated secondary antibodies (30 min). Luminol and peroxide are then added to generate chemiluminescence, which is captured by a charge-coupled device camera (15–480 s).
6. The digital image is analyzed with Compass software (ProteinSimple), and the quantified data of the detected protein are reported as molecular weight and signal/peak intensity. Compass software also gives the option to view the data as an electropherogram or as a lane view.

4. MONITORING PTMs OF ATG PROTEINS USING IMMUNOPRECIPITATION ASSAYS

4.1 Overview

Immunoprecipitation or coimmunoprecipitation is used to enrich a specific protein or protein complex from a tissue homogenate, cell lysate, or culture supernatant (Kaboord & Perr, 2008). This enrichment is accomplished by binding the protein of interest with a specific antibody or epitope-tagged plasmid with a specific tag antibody. The antibody/antigen complex is then pulled out of the sample using Protein A/G-coupled agarose beads. The precipitated immune complexes are then denatured and resolved using SDS-PAGE and further analyzed with tools that include WB or protein mass spectrometry. Immunoprecipitation can be used to isolate and concentrate a particular protein from a protein complex, determine the concentration of low protein levels, identify protein–protein interactions, or evaluate protein PTMs. The immunoprecipitation assay process includes five major steps: sample preparation, preclearing, antibody incubation/formation of antibody-antigen complexes, precipitation and washing, and analysis using SDS-PAGE and other methods. Immunoprecipitation assays combined with site mutation are widely used to study the effects of PTMs (e.g., ubiquitination and phosphorylation) on ATG protein complex formation.

4.2 Protocol

A modified immunoprecipitation protocol (based on a typical cell culture extract assay) from Abcam (http://www.abcam.com/protocols/immunoprecipitation-protocol-1) is detailed later:

1. Place the cell culture dish in ice and wash the cells with ice-cold 1 × PBS.
2. Drain the PBS, and then add ice-cold cell lysis buffer (20 m*M* Tris, pH 7.5, 150 m*M* NaCl, 1 m*M* EDTA, 1 m*M* EGTA, 1% Triton X-100, 2.5 m*M* sodium pyrophosphate, 1 m*M* β-glycerophosphate, 1 m*M* Na3VO4, 1 μg/mL leupeptin, and 1 m*M* PMSF) under nondenaturing conditions (1 mL per 10^7 cells/100 mm^2 dish).
3. Scrape adherent cells off the dish using a cold plastic cell scraper, then gently transfer the cell suspension into a precooled microcentrifuge tube.
4. Maintain constant agitation for 30 min at 4°C.
5. Spin at 15,000 × *g* for 10 min at 4°C in precooled centrifuge.
6. Transfer the supernatant to a fresh tube and keep on ice. The pellet is discarded.
7. Take 1 mL of cell lysate and add to either Protein A or G agarose beads (100 μL of 50% bead slurry). Incubate for 1 h at 4°C with gentle agitation.
8. Spin 15,000 × *g* for 10 min at 4°C in precooled centrifuge.
9. Discard the bead pellet and keep the supernatant for immunoprecipitation.

10. On ice, add 10–50 μg of cell lysate plus the recommended amount of antibody (1–2 μg) in a microcentrifuge tube. The commercially available ATG antibodies for immunoprecipitation are listed in Table 3.
11. Incubate the sample with the antibody for 12 h at 4°C with gentle agitation.
12. Add either Protein A or G agarose beads (20 μL of 50% bead slurry). Incubate with gentle rocking for 1–3 h at 4°C. If using a monoclonal antibody, choose Protein G-coupled Sepharose beads. If using a polyclonal antibody, Protein A-coupled Sepharose beads are usually suitable.
13. Microcentrifuge for 30 s at 4°C. Wash the pellet five times with 500 μL of 1 × cell lysis buffer.
14. Resuspend the pellet in 20 μL 3 × SDS sample buffer (187.5 m*M* Tris–HCl, pH 6.8, 6% w/v SDS, 30% glycerol, 150 m*M* DTT, and 0.03% w/v bromophenol blue). Vortex, and then microcentrifuge for 30 s.
15. Heat the sample to 95–100°C for 2–5 min and microcentrifuge for 1 min at 15,000 × *g*.
16. Load the sample (15–30 μL) into SDS-PAGE gel as described earlier.
17. Use total phosphoserine, ubiquitin, acetylation, or other PTM antibody to test.

5. MONITORING PTMs OF ATG PROTEINS USING IMMUNOFLUORESCENCE ASSAYS

5.1 Overview

Immunofluorescence (IF) or cell imaging techniques rely on the use of antibodies to label a specific target antigen with a fluorescent dye such as fluorescein isothiocyanate or cyanine dye. IF can be used in combination with nonantibody methods of fluorescent staining (e.g., labeling DNA using DAPI). The fluorescent dye allows visualization of the target distribution in the samples (e.g., tissue sections or cultured cells) under a fluorescent microscope (e.g., epifluorescence and confocal microscopes). Direct IF uses a single antibody directed against the target of interest. The primary antibody is directly conjugated to a fluorophore. In contrast, indirect IF uses two antibodies. The primary antibody is unconjugated and a fluorophore-conjugated secondary antibody directed against the primary antibody is used for detection. In addition to antibodies, fluorescent protein tagging can be used to observe the subcellular distribution of ATG proteins, as well as PTMs in autophagy in vitro or in vivo.

Table 3 Commercial ATG Antibodies for Immunoprecipitation Assays

Target Protein	Ubiquited by	Supplier (Cat. #)	Reactivity	Clonality	Host	Function
ULK1	TRAF6 AMBRA1	Santa Cruz (sc-390904)	H	Monoclonal	M	Stabilizes ULK1 and promotes autophagy induction
		Santa Cruz (sc-33182)	H, M, rat	Polyclonal	R	
		CST (8054)	H, M, rat	Monoclonal	R	
		CST (6439)	H	Monoclonal	R	
BECN1	TRAF6 AMBRA1 NEDD4 USP10 USP13	GeneTex (GTX61619)	H	Monoclonal	R	Promotes starvation-induced autophagy by TRAF and AMBRA1-mediated ubiquitination Promotes BECN1 proteasomal degradation; Promotes autophagy by USP10 and USP13-mediated deubiquitination on BECN1
		Proteintech Group (11306-1-AP)	H, M, rat	Polyclonal	R	
		Santa Cruz (sc-11427)	H, M, rat, cow	Polyclonal	R	
		Santa Cruz (sc-10087 Sc-48341)	H, M, rat	Polyclonal	G	
		Santa Cruz (sc-10086)	H, M, rat	Monoclonal	M	
		Santa Cruz (sc-48381)	H, M	Monoclonal	M	
		Invitrogen (PA1-16857)	H, M, rat	Polyclonal	R	
		US Biological (B0981-23E)	H	Polyclonal	R	
		Millipore (MABN16)	H	Monoclonal	R	
		LifeSpan Biosciences (LS-C363502)	H	Polyclonal	R	

	LifeSpan Biosciences (LS-C363501 LS-C363496 LS-C363495 LS-C363494 LS-C363489 LS-C289303 LS-C289301 LS-C142267 LS-C142266 LS-C142265 LS-C121120 LS-C83852 LS-C83851 LS-C74776)	H	Polyclonal	R
	LifeSpan Biosciences (LS-B7113)	H	Monoclonal	R
	Novus Biologicals (NB500-249)	H, M, rat	Polyclonal	R
	Novus Biologicals (NBP1-00085)	H, M, rat, cow	Monoclonal	M
	Bethyl (A310-935A A302-566A A302-567A)	H, M	Polyclonal	R
	Sigma (B6186)	H, M, rat	Polyclonal	R
	Sigma (B6061)	H	Polyclonal	R
	CST (4122)	H	Monoclonal	M
	CST (3738)	H, M, rat	Polyclonal	R

H, human; *M*, mouse; *R*, rabbit.

5.2 Protocol

A modified IF protocol (based on a typical cell culture) from the University of Pittsburgh Center for Biologic Imaging (http://www.cbi.pitt.edu/) is detailed later. Cells should be grown, treated, fixed, and stained directly in multiwell plates, chamber slides, or on coverslips.

1. Remove the medium and wash the cells with ice-cold 1 × PBS.
2. Fix the sample in 2% paraformaldehyde made in PBS for 15 min at room temperature.
3. Aspirate the fixative and rinse three times in 1 × PBS for 5 min each.
4. Permeabilize cells with 0.1% Triton X-100 made in PBS solution for 10 min.
5. Wash the monolayer three times with PBS.
6. Wash the monolayer five times with a PBS+ 0.5% BSA (PBB).
7. Block with 2% BSA for 45 min.
8. Wash the monolayer five times with PBB.
9. Primary antibody: Dilute to the desired concentration in PBB. Gently drop 70–100 μL of antibody solution over the section (so that the section is covered). Incubate overnight at 4°C.
10. Wash the monolayer five times with PBB.
11. Incubate the specimen in fluorochrome-conjugated secondary antibody diluted in PBB for 1 h at room temperature in the dark.
12. Rinse three times in 1 × PBS for 5 min each.
13. Hoescht stain for 30 s. Hoechst (Sigma, #B-2883) is made at a concentration of 1 mg/100 mL in dH_2O and is stored in the dark at 4°C when not in use.
14. Wash the monolayer three times with PBS.
15. Coverslip slides with Prolong® Gold Antifade Reagent (Cell Signaling Technology).

6. CONCLUDING REMARKS

Commercial antibodies are extensively used in basic and clinical research, although their qualities and performances remain a challenge in applications. In this chapter, we summarized several methods for detecting the PTMs of ATG proteins using commercial antibodies. These antibody-based methods are important techniques used in cell and molecular biology. In addition, PTMs can be analyzed using nonantibody methods such as mass spectrometry. New developments in instrumentation (e.g., Simple Western™)

and new methods may have a profound effect on autophagy research in the future (Klionsky et al., 2016).

ACKNOWLEDGMENTS

We thank Christine Heiner (Department of Surgery, University of Pittsburgh) for her critical reading of the manuscript. This work was supported by the National Institutes of Health (R01CA160417 and R01GM115366 to D.T.) and a Research Scholar Grant from the American Cancer Society (RSG-16-014-01-CDD to D.T.).

REFERENCES

Bach, M., Larance, M., James, D. E., & Ramm, G. (2011). The serine/threonine kinase ULK1 is a target of multiple phosphorylation events. *The Biochemical Journal*, *440*, 283–291.

Cherra, S. J., 3rd, Kulich, S. M., Uechi, G., Balasubramani, M., Mountzouris, J., Day, B. W., et al. (2010). Regulation of the autophagy protein LC3 by phosphorylation. *The Journal of Cell Biology*, *190*, 533–539.

Colecchia, D., Strambi, A., Sanzone, S., Iavarone, C., Rossi, M., Dall'Armi, C., et al. (2012). MAPK15/ERK8 stimulates autophagy by interacting with LC3 and GABARAP proteins. *Autophagy*, *8*, 1724–1740.

Dorsey, F. C., Rose, K. L., Coenen, S., Prater, S. M., Cavett, V., Cleveland, J. L., et al. (2009). Mapping the phosphorylation sites of Ulk1. *Journal of Proteome Research*, *8*, 5253–5263.

Feng, Y., He, D., Yao, Z., & Klionsky, D. J. (2014). The machinery of macroautophagy. *Cell Research*, *24*, 24–41.

Gurkar, A. U., Chu, K., Raj, L., Bouley, R., Lee, S. H., Kim, Y. B., et al. (2013). Identification of ROCK1 kinase as a critical regulator of Beclin1-mediated autophagy during metabolic stress. *Nature Communications*, *4*, 2189.

Harris, V. M. (2015). Protein detection by Simple Western analysis. *Methods in Molecular Biology*, *1312*, 465–468.

Ichimura, Y., Kirisako, T., Takao, T., Satomi, Y., Shimonishi, Y., Ishihara, N., et al. (2000). A ubiquitin-like system mediates protein lipidation. *Nature*, *408*, 488–492.

Jensen, E. C. (2012). The basics of western blotting. *Anatomical Record (Hoboken, N.J. 2007)*, *295*, 369–371.

Kaboord, B., & Perr, M. (2008). Isolation of proteins and protein complexes by immunoprecipitation. *Methods in Molecular Biology*, *424*, 349–364.

Kim, J., Kim, Y. C., Fang, C., Russell, R. C., Kim, J. H., Fan, W., et al. (2013). Differential regulation of distinct Vps34 complexes by AMPK in nutrient stress and autophagy. *Cell*, *152*, 290–303.

Kim, J., Kundu, M., Viollet, B., & Guan, K. L. (2011). AMPK and mTOR regulate autophagy through direct phosphorylation of Ulk1. *Nature Cell Biology*, *13*, 132–141.

Klionsky, D. J., Abdelmohsen, K., Abe, A., Abedin, M. J., Abeliovich, H., Acevedo Arozena, A., et al. (2016). Guidelines for the use and interpretation of assays for monitoring autophagy (3rd edition). *Autophagy*, *12*, 1–222.

Liu, J., Xia, H., Kim, M., Xu, L., Li, Y., Zhang, L., et al. (2011). Beclin1 controls the levels of p53 by regulating the deubiquitination activity of USP10 and USP13. *Cell*, *147*, 223–234.

Mahmood, T., & Yang, P. C. (2012). Western blot: Technique, theory, and trouble shooting. *North American Journal of Medical Sciences*, *4*, 429–434.

Mizushima, N., Noda, T., Yoshimori, T., Tanaka, Y., Ishii, T., George, M. D., et al. (1998). A protein conjugation system essential for autophagy. *Nature*, *395*, 395–398.

Mizushima, N., Yoshimori, T., & Levine, B. (2010). Methods in mammalian autophagy research. *Cell*, *140*, 313–326.

Platta, H. W., Abrahamsen, H., Thoresen, S. B., & Stenmark, H. (2012). Nedd4-dependent lysine-11-linked polyubiquitination of the tumour suppressor Beclin 1. *The Biochemical Journal*, *441*, 399–406.

Popelka, H., & Klionsky, D. J. (2015). Post-translationally-modified structures in the autophagy machinery: An integrative perspective. *The FEBS Journal*, *282*, 3474–3488.

Russell, R. C., Tian, Y., Yuan, H., Park, H. W., Chang, Y. Y., Kim, J., et al. (2013). ULK1 induces autophagy by phosphorylating Beclin-1 and activating VPS34 lipid kinase. *Nature Cell Biology*, *15*, 741–750.

Shang, L., Chen, S., Du, F., Li, S., Zhao, L., & Wang, X. (2011). Nutrient starvation elicits an acute autophagic response mediated by Ulk1 dephosphorylation and its subsequent dissociation from AMPK. *Proceedings of the National Academy of Sciences of the United States of America*, *108*, 4788–4793.

Shi, C. S., & Kehrl, J. H. (2010). TRAF6 and A20 regulate lysine 63-linked ubiquitination of Beclin-1 to control TLR4-induced autophagy. *Science Signaling*, *3*, ra42.

Sowa, M. E., Bennett, E. J., Gygi, S. P., & Harper, J. W. (2009). Defining the human deubiquitinating enzyme interaction landscape. *Cell*, *138*, 389–403.

Wang, R. C., Wei, Y., An, Z., Zou, Z., Xiao, G., Bhagat, G., et al. (2012). Akt-mediated regulation of autophagy and tumorigenesis through Beclin 1 phosphorylation. *Science*, *338*, 956–959.

Wani, W. Y., Boyer-Guittaut, M., Dodson, M., Chatham, J., Darley-Usmar, V., & Zhang, J. (2015). Regulation of autophagy by protein post-translational modification. *Laboratory Investigation; A Journal of Technical Methods and Pathology*, *95*, 14–25.

Wei, Y., Zou, Z., Becker, N., Anderson, M., Sumpter, R., Xiao, G., et al. (2013). EGFR-mediated Beclin 1 phosphorylation in autophagy suppression, tumor progression, and tumor chemoresistance. *Cell*, *154*, 1269–1284.

Xia, P., Wang, S., Du, Y., Zhao, Z., Shi, L., Sun, L., et al. (2013). WASH inhibits autophagy through suppression of Beclin 1 ubiquitination. *The EMBO Journal*, *32*, 2685–2696.

Xie, Y., Kang, R., Sun, X., Zhong, M., Huang, J., Klionsky, D. J., et al. (2015). Posttranslational modification of autophagy-related proteins in macroautophagy. *Autophagy*, *11*, 28–45.

Xie, Z., Nair, U., & Klionsky, D. J. (2008). Atg8 controls phagophore expansion during autophagosome formation. *Molecular Biology of the Cell*, *19*, 3290–3298.

Yang, Z., & Klionsky, D. J. (2010). Eaten alive: A history of macroautophagy. *Nature Cell Biology*, *12*, 814–822.

Zalckvar, E., Berissi, H., Mizrachy, L., Idelchuk, Y., Koren, I., Eisenstein, M., et al. (2009). DAP-kinase-mediated phosphorylation on the BH3 domain of beclin 1 promotes dissociation of beclin 1 from Bcl-XL and induction of autophagy. *EMBO Reports*, *10*, 285–292.

CHAPTER ELEVEN

Tagged ATG8-Coding Constructs for the In Vitro and In Vivo Assessment of ATG4 Activity

C. López-Otín*, G. Mariño[†,1]

*Instituto Universitario de Oncología (IUOPA), de Bioquímica y Biología Molecular, Universidad de Oviedo, Oviedo, Spain

[†]Instituto para la investigación sanitaria del Principado de Asturias, Universidad de Oviedo, Oviedo, Spain

[1]Corresponding author: e-mail address: marinoguillermo@uniovi.es

Contents

1. Introduction 190
2. The Atg4–Atg8 System and Its Unexpected Evolutionarily Acquired Complexity 191
3. Monitoring Atg8-Like Proteins Processing in the Context of Autophagic Flux Analysis 194
4. Construction of Mammalian Expression Vectors Including Tagged Atg8 Proteins for Cleavage Assays 195
5. Transfection of Cultured Cells for the Analysis of Atg4-Like Activity 197
6. Hydrodynamic Delivery-Based Expression of Atg8-Like Tagged Constructs for the Analysis of Atg4-Like Activity in Live Animals 200
7. Concluding Remarks 203

Acknowledgments 204

References 204

Abstract

Autophagy is a catabolic pathway, which mediates the degradation of cytoplasmic components and sustains many essential cellular functions. More than 30 genes have been involved in different aspects of this essential process in simple eukaryotes as yeast. Among these genes, those coding for members of the Atg4–Atg8 proteolytic system have acquired a high degree of complexity throughout evolution. Contrasting with the situation in unicellular eukaryotes, in which the system is composed by just a single protease (Atg4) and a single substrate (Atg8), evolution has led to the presence of several members for both Atg4 and Atg8 families in multicellular organisms. In human cells, there are four Atg4 proteases and six Atg8 substrates, which have probably evolved to cope with specific requirements for autophagic pathway in more complex scenarios. Despite these considerations, the reasons for the evolutionarily acquired complexity of this proteolytic system are still not completely understood. In this work, we describe two different applications of a relatively simple but useful technique to analyze protease–substrate specificity of this system in mammalian cells. By using the described

Methods in Enzymology, Volume 587
ISSN 0076-6879
http://dx.doi.org/10.1016/bs.mie.2016.11.001

technique, it is possible to determine the cellular efficiency in the initial cleavage for each of the Atg8 family members in diverse experimental settings both in cultured cells and live laboratory mice.

1. INTRODUCTION

Autophagy (derived from Greek, meaning "self-eating") was originally identified as an adaptive response triggered by fasting and other types of stressors in simple unicellular eukaryotes, as yeast (Reggiori & Klionsky, 2013). Apart from its essential role, autophagy was later described as an evolutionarily conserved pathway for maintaining cellular homeostasis in higher eukaryotes. Through this catabolic route, cytoplasmic components (including damaged organelles, misfolded proteins, or even intracellular pathogens) are degraded by lysosomal hydrolases (Choi, Ryter, & Levine, 2013; Klionsky et al., 2016). There are at least three major autophagic pathways: macroautophagy, microautophagy, and chaperone-mediated autophagy, which primarily differ in the method by which the cargo is transferred to the lysosome (Marino & Lopez-Otin, 2004). Macroautophagy (herein referred to as autophagy) involves the formation and elongation of a membrane sac called the phagophore that develops into a double-membrane vesicle, the autophagosome, in which the cytoplasmic cargo is sequestered. Once mature, autophagosomes eventually fuse with lysosomes, allowing for the degradation of the cargo and the autophagosome's inner membrane. Once degraded, the content of the autophagosome is returned to the cytoplasm as basic building blocks (mainly amino acids, lipids, sugars, and nucleotides) that can be reused by the cell to synthesize new biomolecules (Singh & Cuervo, 2011). Autophagy execution requires the activation of a complex molecular machinery that includes regulatory signaling cascades, the formation and completion of autophagosomes, the transport of autophagosomes, and final fusion of autophagosomes with lysosomes (Yang & Klionsky, 2010). Although the intricate molecular machinery of this multistep pathway is not yet fully understood, some of its components have been identified as essential effectors required for a correct autophagic response. This is the case of proteins such as Atg5, Atg7, the proteases from the Atg4 family, or the ubiquitin-like proteins from the Atg8 family and their conjugation system, which is indispensable for the correct expansion and closure of the preautophagosomal double membrane.

2. THE ATG4–ATG8 SYSTEM AND ITS UNEXPECTED EVOLUTIONARILY ACQUIRED COMPLEXITY

Proteases are key components of the autophagic pathway. Some proteolytic enzymes are known to indirectly regulate the autophagic response through processing of autophagy-essential proteins. For instance, cleavage of the autophagy regulator Beclin 1 by different caspases abolishes its proautophagic activity (Djavaheri-Mergny, Maiuri, & Kroemer, 2010), and a switch from autophagy to apoptosis can also be observed when ATG5 is processed by calpain proteases in different cell types (Yousefi et al., 2006). Another example is the activity of caspase-3, a protease that can target Crohn's disease-associated T300A variant of the autophagic protein ATG16L1 (Murthy et al., 2014). Moreover, acidic lysosomal (or vacuolar) proteases are required for the degradation of autophagosomal content during the latest steps of the autophagic process (Repnik, Stoka, Turk, & Turk, 2012). Apart from the aforementioned examples, there is one family of proteases involved in autophagy, whose specific activity is essential for the initiation and execution of this process: the Atg4 family of cysteine proteinases (Kaminskyy & Zhivotovsky, 2012).

Most of our knowledge about the molecular mechanisms that take place during the autophagic response initially originated from a large number of genetic screens done in autophagy-defective mutants of *Saccharomyces cerevisiae*. More than 30 genes specifically involved in autophagy (termed Atg (AuTophaGy-related) genes) (Klionsky et al., 2003) have been described (Feng, He, Yao, & Klionsky, 2014). These genes are essential for different steps of the autophagic process, and deletion of any of them leads to autophagic impairment in yeast, as functional redundancy for this pathway is absent in unicellular eukaryotes. In more complex organisms, such as mammals, some essential genes for autophagy execution still present a single ortholog, including Atg3, Atg5, or Atg7. Hence, mutations or deletions of any of these genes lead to a total autophagic impairment either in simple or more complex systems. This particularity has facilitated the study of the pathophysiological roles of autophagy in diverse model organisms, including worms, flies, or laboratory mice (Mizushima, 2007; Mizushima & Komatsu, 2011). In these settings, autophagy has been successfully abolished by the targeted deletion of one of these essential genes, either by classic whole-body approaches or in a tissue-specific or time-dependent fashion. Most of the results obtained by using these strategies have revealed prominent roles for autophagy pathway in maintaining both cellular integrity and tissue-specific functions, illustrating the relevance of this degradative route as a housekeeping mechanism in multicellular organisms. As a caveat, given the housekeeping nature of basal autophagy, its total inhibition usually leads to dramatic consequences, which may hide new essential functions of the pathway as an inducible antistress response.

Contrasting with the simple scenario in which one single gene is responsible for an essential function in autophagy execution from yeast to mammals, there are some other Atg genes that have undergone amplification during evolution and present several orthologs in higher eukaryotes. This is the case of the Atg4–Atg8 system, which in yeast consists of a single protease—Atg4—and its substrate—Atg8—but has gained a considerable complexity in multicellular organism (Fernandez & Lopez-Otin, 2015).

In mammalian cells, the Atg4–Atg8 system consists of four different ATG4 proteases, and a variable number of ATG8-like substrates (Fig. 1A), depending on the analyzed species. In humans, we can find six Atg8 orthologs, which are distributed in two subfamilies: the LC3 (microtubule-associated

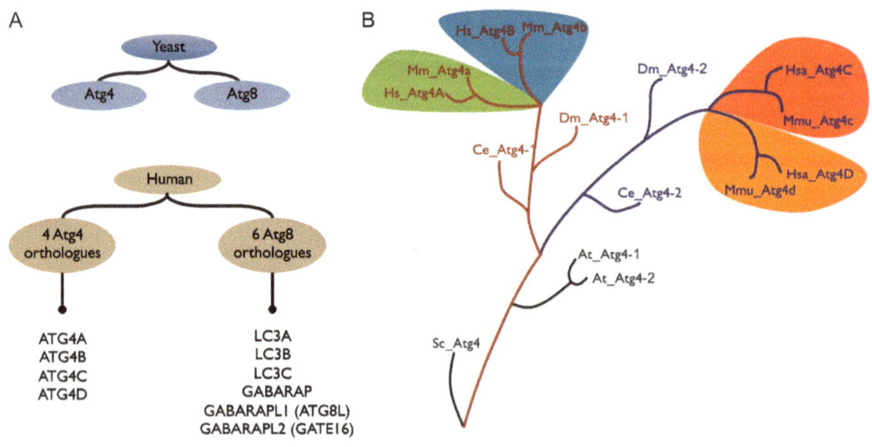

Fig. 1 Evolutionarily acquired complexity in the Atg4 and Atg8 families of autophagy-related proteins. (A) Simple organisms, as yeast, present only one Atg4 protease and a single substrate. By contrast, in more complex organisms, as humans, these two families consist of several members (four proteases, Atg4A–D and six substrates, LC3A–C, GABARAP, GABARAPL1–2). This illustrates the high degree of complexity acquired by the processes in which these two families of proteins are involved, either in the context of autophagic degradation or in new possible unknown pathways. (B) Phylogenetic tree showing Atg4 gene amplification through evolution (*Sc, Saccharomyces cerevisiae; At, Arabidopsis thaliana; Ce, Caenorhabditis elegans; Dm, Drosophila melanogaster; Mm, Mus musculus; Hs, Homo sapiens*).

protein 1 light chain 3) subfamily, which consists of LC3A–C, and the GABARAP (GABAA receptor-associated protein) subfamily, including GABARAP, GABARAPL1/Atg8L/GEC1, and GABARAPL2/GATE-16/GEF2 (Le Grand et al., 2011). This points to a complex network of enzyme–substrate interactions likely with diverse spatiotemporal modulation, although the exact in vivo specificity for the mammalian Atg4–Atg8 system is currently not totally understood. Although the existing redundancy for Atg4 and Atg8 functions complicates our whole understanding of the system, it can also represent an interesting chance to understand and dissect newly acquired functions for autophagy pathway in diverse pathophysiological contexts. Prompted by this idea, we have previously generated and characterized mutant mice deficient for different members of the Atg4 family. Contrary to mice deficient for unique components of the route, such as ATG5, ATG3, or ATG7, which share a common phenotype characterized by perinatal lethality (Mizushima & Komatsu, 2011), deficiency in components of the Atg4 family is compatible with normal development until adulthood. This has helped to unveil new physiological functions for this degradative route in mammalian cells, as well as specific functions for the members of the Atg4–Atg8 system (Cabrera et al., 2013; Marino et al., 2010, 2007).

In this context, it is of special importance to characterize protease–substrate specificity in order to understand specific and more general functions played by mammalian Atg4 proteases and their substrates. Recently, several interesting studies have shed some light into this question by analyzing the activity of recombinant Atg4 proteases toward some of the mammalian Atg8 substrates in vitro. Out of these studies, it seems that Atg8-like substrates are differentially cleaved by each Atg4 protease, being ATG4B the most potent and promiscuous of them in terms of substrate specificity (Li et al., 2011). ATG4A can also cleave a wide range of ATG8 orthologs, and it effectively targets GABARAPL2 (Scherz-Shouval, Sagiv, Shorer, & Elazar, 2003), but cannot cleave LC3B. Meanwhile, ATG4C and ATG4D show little to no catalytic activity against these substrates in these types of assays. This is not surprising, as sequence analysis of the Atg4 family shows that two subfamilies diverged through evolution: ATG4A/ATG4B and ATG4C/ATG4D (Fig. 1B). Although the results yielded by these studies are important, the fact that they rely on recombinant protein expression and in vitro cleavage assays somehow limits the scope of their conclusions.

In the course of our studies to characterize murine mice deficient for members of the Atg4 family of proteases, we have relied on a simple method, which enables unambiguous characterization of Atg4–Atg8 system specificity and also allows detecting functional redundancy and compensation events in the Atg4 protease family. This method is based on the addition of N- and C-terminal epitopes to Atg8-like proteins, which will be then used to detect the amount of uncleaved proteins (positive for the C-terminal tag) and refer it to the total pool of the expressed tagged Atg8 form (positive for the N-terminal tag). Although there are some limitations for the use of this technique, as the determination of enzymatic parameters (K_{cat}, K_M, and related parameters), it offers many advantages to characterize Atg4–Atg8 system specificity both in live cells and laboratory mice (Marino et al., 2010; Tanida, Sou, Minematsu-Ikeguchi, Ueno, & Kominami, 2006).

3. MONITORING ATG8-LIKE PROTEINS PROCESSING IN THE CONTEXT OF AUTOPHAGIC FLUX ANALYSIS

In the last years, due to the growing importance of autophagy pathway in a growing number of disciplines, it has become increasingly necessary to monitor autophagic activity in diverse experimental settings. Although there are many articles in which autophagy-related parameters are evaluated, there are still some published studies in which autophagic activity is not addressed properly. To date, several methodological articles have described in detail how to assess the so-called autophagic flux that allows determining not only the amount of autophagosomes present in a cell or tissue at a determined moment but also autophagosome clearance rate. This technique mainly relies on the quantification of the amount of membrane-bound form of LC3B (the most studied member of the Atg8 family in mammalian cells) in the absence or presence of lysosomal inhibitors. The amount of this membrane-bound form of LC3B, known as LC3B-II, is proportional to the amount of cellular autophagosomes in a determined moment. Briefly, by comparing the amount of LC3B-II in both presence/absence of lysosomal inhibitors, we can quantify the "autophagic flux" in a determined cell population or tissue (Kuma, Matsui, & Mizushima, 2007; Rubinsztein et al., 2009; Tanida, Minematsu-Ikeguchi, Ueno, & Kominami, 2005).

Very few studies take into account the possibility that inhibition of components of the autophagic molecular machinery involved in LC3B maturation (Atg4 proteases) might yield experimental artifacts that can be misinterpreted. LC3B and its paralogs are synthesized as an inactive protein that has to be proteolytically cleaved to become functional in the context of autophagic degradation. This initial cleavage is generally constitutive, and in normal conditions the half-life of uncleaved LC3B is significantly brief. In certain specific cases, however, Atg4 proteases activity can be inhibited, and the unprocessed form of LC3B (pro-LC3B) can be accumulated. In fact, inhibition of Atg4A and Atg4B activity by reactive oxygen species has been shown to block starvation-induced autophagy (Perez-Perez, Zaffagnini, Marchand, Crespo, & Lemaire, 2014; Scherz-Shouval et al., 2007), suggesting that the cysteine protease nature of Atg4 can be regulated by redox events.

This situation can lead to experimental artifacts that can be very difficult to detect and interpret. The electrophoretical mobilities of Atg8-like proteins in SDS-PAGE analyses do not always correspond with their predicted behavior and are more related to the sequence of the last carboxyl-terminal amino acids than to the protein's molecular weight. Indeed, it has been recently reported that the electrophoretic mobility of uncleaved human pro-LC3B is almost undistinguishable from its membrane-bound lipidated LC3B-II form, which may have led to misinterpretation of some results in the past (Wang, Chen, Billiar, Stang, & Gao, 2013). Altogether, the particularities of the mammalian Atg4–Atg8 system make the analysis of the processing status of Atg8-like proteins a good companion for the analysis of autophagic flux by the currently accepted methods based on the inhibition of lysosomal activity.

In the following sections, we will describe how the processing status of the different Atg8-like proteins can be determined. These analyses can be applied to diverse experimental contexts, from cultured cells to live laboratory mice.

4. CONSTRUCTION OF MAMMALIAN EXPRESSION VECTORS INCLUDING TAGGED ATG8 PROTEINS FOR CLEAVAGE ASSAYS

Rationale: Mammalian expression vectors coding for either human/murine Atg8 orthologs flanked by different epitopes at the N- and C-terminal regions of the proteins are constructed for either transfection or hydrodynamic delivery of the corresponding constructs in order to analyze initial cleavage of Atg8 proteins in different experimental settings.

Methods: cDNAs of the murine and human Atg8 orthologs (Atg8s) are either subcloned from another plasmid or generated from RNA extracted from either MEFs (or a different murine cell line) or HEK-293 cells (or a different human cell line), respectively. The different Atg8's cDNAs are cloned into a recombinant vector able to yield sufficient amounts of protein expression in mammalian cells, as pcDNA3.1 or derived vectors. To assure easy detection of the epitopes, three consecutive tags are inserted in frame with the corresponding Atg8 protein in the N- and C-terminal regions. There are many commercial vectors containing 3xFLAG, 3xHA, 3xMyc, or different protein tags in triplicate flanking either N- or C-terminal regions of the vector's multicloning site (MCS), which can be used for this purpose. Some of these vectors can be obtained from the Addgene nonprofit repository. In any case, examples of optimized DNA sequences for each of these three epitopes are listed below.

3xHA:5′TACCCATACGATGTTCCAGATTACGCTGGC
TATCCCTATGACGTCCCGGACTATGCAGGA
TCCTATCCATATGACGTTCCAGATTACGCT
GTT-

3xMyc:5′GAGCAGAAACTCATCTCTGAAGAAGATCTG
GAACAAAAGTTGATTTCAGAAGAAGATCTG
GAACAGAAGCTCATCTCTGAGGAAGATCTG
-3′

3xFlag:5′GACTACAAAGACCATGACGGTGATTATAAA
GATCATGACATCGATTACAAGGATGACGAT
GACAAGC-3′

These DNA fragments can be chemically synthesized, with additional nucleotides containing the convenient restriction enzyme recognition sequence(s).

(1) *Vector MCS and Atg8-like cDNA sequence analysis.* By analyzing the specific mammalian Atg8-like cDNA sequence (ENSMUSG000000 31812; CCDS22726.1), we can exclude the use of restriction enzymes whose recognition sequence is present in the chosen Atg8-like cDNA sequence. As an example, in the case of murine LC3B, many conventionally used restriction enzymes, as for example, *Eco*RI, *Bam*HI, or *Hin*dIII, can be used, as they do not present any consensus recognition sequence in murine LC3B cDNA. A list of cutting and noncutting commercial enzymes can be easily obtained by analyzing a determined sequence with either appropriate software (as an example, free SnapGene® Viewer application, www.snapgene.com) or by using a web-based tool for restriction analysis (as an example, New England Biolabs® freely accessible NebCutter V2.0, http://nc2.neb.com/NEBcutter2/).

(2) *Oligonucleotide design.* There are several approaches to add the 3x epitopes to N- and C-terminal regions of the cDNA of interest. In this case, we recommend the design specific oligonucleotides containing the chosen 3x epitope, flanked by restriction enzyme compatible ends, which will be annealed and then subcloned in frame to the corresponding cDNA.

(3) *Atg8s subcloning.* We may either choose to amplify cDNA directly from RNA, from another vector in which the cDNA is inserted, or subclone it without PCR amplification in the case that retrieval of the cDNA by specific digestion with restriction enzymes leaves an in-frame cDNA sequence when added to that from the 3x epitopes in the vector of choice.

(4) Once the desired vector constructed and confirmed by DNA sequence, it is recommended to large-scale purify it with specific endotoxin-free kits, specially for in vivo hydrodynamic delivery-based approaches.

5. TRANSFECTION OF CULTURED CELLS FOR THE ANALYSIS OF ATG4-LIKE ACTIVITY

Rationale: Cultured cells will be transfected with the generated plasmids, and cell extracts will be prepared and analyzed by Western blotting (WB) against the epitopes present both at the N- and C-terminal regions of the Atg8-like proteins. By using this method, we will be able to determine the efficiency of the initial activating cleavage for Atg8-like proteins by proteases from the Atg4 family in mammalian cells. This assay can be used to obtain information on the efficiency of a determined approach aimed at disrupting or reducing the activity of proteases from the Atg4 family.

(1) *Cell transfection.* Depending on the cell line of interest, a transfection method should be selected in order to obtain the best transfection efficiency. A moderate expression is recommended to avoid experimental artifacts due to overexpression, which may cause protein aggregates that will probably not be accessible for tag cleavage by Atg4 proteases.

(2) *Cell extracts preparation.* 24 or 48 h after transfection, cell extracts should be prepared from either treated or untreated cells. This step should be as quick as possible to avoid experimental artifacts. A conventional RIPA buffer contains 50 m*M* Tris (alternatively, 10 m*M* sodium phosphate may be used instead), pH 7.5; 150 m*M* NaCl; 0.1% SDS (optional); 0.5% sodium deoxycholate; and 1% Triton X 100 or NP-40. It is

essential to add protease inhibitors to RIPA buffer to avoid protein degradation and experimental artifacts due to potential nonspecific epitope cleavage during manipulation. Several companies sell prepared protease inhibitor cocktails which can be added to RIPA buffer before use (i.e., Complete™ Protease Inhibitor Cocktail from Roche®). Although it is not strictly necessary for this assay, it is recommended to also include phosphatase inhibitors to RIPA buffer.

Adherent cells

6-mm polystyrene tissue culture plates will be used. Cells should at least at 80% confluency when lysed.

Aspirate or decant media and keep plates on ice for all steps.

Wash cell monolayer gently one time with 10 mL ice-cold PBS. Aspirate excess PBS.

Add RIPA lysis buffer with inhibitors to culture plates and distribute buffer homogeneously. If multiple plates for the same condition are used, the lysis buffer can be employed to sequentially lyse several plates; this yields higher concentration of protein in the final extract. The volume of lysis buffer should be empirically determined for each cell type to ensure efficient lysis and optimal final protein concentration. As an indication, 200–500 μL per 100-mm plate should be adequate.

Using a cell scraper or silicone spatula, scrape the cells and transfer the lysate to a 15-mL conical tube. Incubate the lysate on ice for 15 min.

Sonicate the lysate to reduce extract viscosity if required.

Incubate the lysate an additional 15 min.

Centrifuge at 13,000 × *g* for 5 min at 4°C and collect the supernatant (avoiding the pellet) into new microtubes.

Cells in suspension

Cells should be cultured to a density of 1–2 million cells/mL.

Centrifuge cells at 500 × *g* for 5 min at room temperature.

Aspirate supernatant and keep cells on ice for all steps. Wash pellets with ice-cold PBS.

Repeat centrifugation. Decant PBS and aspirate the supernatant. The volume of lysis buffer should be empirically determined for each cell type to ensure efficient lysis and optimal final protein concentration. As an indication, 100 μL of lysis buffer per each million of cells should give an adequate protein concentration. Incubate the lysate on ice for 15 min.

From here on, the lysis protocol is the same as for adherent cells.

(3) Cell extracts are quantified, and protein concentration is adjusted to perform WB. Due to the size of the tagged Atg8-like proteins, a 12–15% polyacrylamide concentration is recommended for SDS-PAGE.

(4) Detection of the tag epitopes present both at the N- and C-terminal regions is performed by WB. As shown in Fig. 2, the N-tag epitope should be always detected and can be used as transfection + loading control. The C-terminal tag, by contrast, should only be detected when Atg4's cellular activity is totally or partially impaired, as shown for *Case 2* in Fig. 2.

(5) Quantification of cellular Atg4-like activity against a specific Atg8-like protein can be easily performed by densitometry analyses. Briefly, by

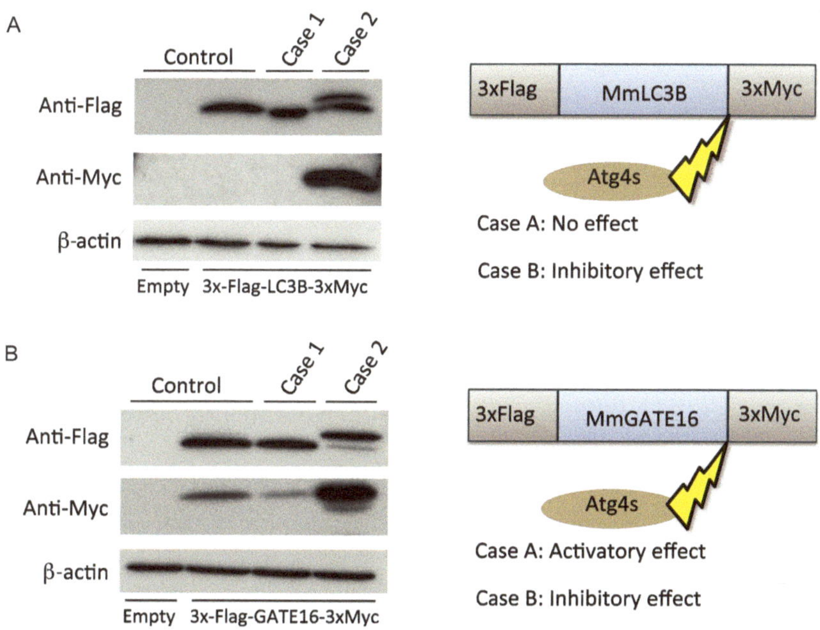

Fig. 2 Example of LC3B and GATE16 initial cleavage analysis in transfected cultured cells. Mouse embryonic fibroblasts (MEFs) were transfected with mammalian expression vectors coding for 3xFlag, 3xMyc epitopes for either murine LC3B (A) or GATE16 (B). Initial processing of LC3B and GATE16 is analyzed in two experimental settings (Cases A and B) by Western blot-mediated analysis of Flag (N-terminal) and Myc (C-terminal) epitopes. (A) For LC3B, no significant alteration in the initial cleavage in found in Case 1 (because endogenous activity seems to be already maximal). By contrast, Case 2 shows an example of inhibition of this proteolytic event. (B) As for GATE16, an increment on the initial cleavage efficiency for this protein is detected for Case 1, whereas a stronger inhibition than the observed for LC3B is detected in Case 2.

calculating the ratio of the signal from C-terminal epitope relative to the N-terminal tag, we should have an estimation of the cleavage efficiency. This ratio can be compared with that obtained in control conditions, in case that a significant signal for the uncleaved protein is detected. Cleavage ratio = OD_C-term-tag/OD_N-term-tag.

6. HYDRODYNAMIC DELIVERY-BASED EXPRESSION OF ATG8-LIKE TAGGED CONSTRUCTS FOR THE ANALYSIS OF ATG4-LIKE ACTIVITY IN LIVE ANIMALS

Rationale: The plasmids whose generation has been previously described will be used to transfect liver cells in live laboratory mice by hydrodynamic gene delivery. By using this method, we will be able to determine the efficiency of the initial activating cleavage for Atg8-like proteins by proteases from the Atg4 family in vivo, in hepatic cells.

There are a wide variety of methods for the transient and stable expression of exogenous DNA in live laboratory animals (Zhang, Satterlee, & Huang, 2012). Although virus-based methods or electroporation-based approaches may also be used for this same purpose, here we will describe the hydrodynamic gene delivery approach, which we have successfully used for the analysis of Atg4-like activity in vivo and represent a noninvasive, safe, and simple strategy to recombinant expression of naked DNA in hepatocytes of live laboratory mice.

Hydrodynamic delivery of naked cDNAs can be applied to evaluate Atg8-like proteins efficiency in different experimental settings in vivo. Detailed theoretical basis and extensive protocols can be found in the literature (Ahlen et al., 2016; Kovacsics & Raper, 2014).

(1) *Preparation of the solution to be injected through the tail vein.* As previously mentioned, it is important to use endotoxin-free DNA. For each mice, 15–50 μg of plasmid DNA should be injected. As a starting point, 10 μg of DNA per 10 g of mouse weight can be injected initially. The exact DNA amounts should be empirically determined depending of the type of promoter and the intensity of expression it would yield in liver cells.

DNA solutions should be prepared in saline (0.9% sodium chloride). The total volume of injected solution should correspond 10% of body weight. Thus, for a 25-g mouse, 25 μg of endotoxin-free DNA should be diluted in 2.5 mL of saline.

For injection, 21-gauge needles are recommended. Syringes with higher gauges could make difficult fast injection of the required volumes. When filling the syringes, special attention should be put to avoid neither introducing air bubbles nor letting the uncapped needle to touch nonsterile surfaces.

(2) *Mice preparation for hydrodynamic delivery.* Temperature of mice is an essential factor for this technique. Thus, it should be avoided performing the procedure in rooms in which room temperature is too cold due to low temperature-induced constriction of the tail vein. Similarly, anesthesia normally leads to vein constriction. To avoid these inconveniences, mice can be positioned on a heating pad at 37°C. Additionally, mice tail can be warmed with a heat lamp to dilate blood vessels. However, caution should be taken to avoid overheating of the mouse.

(3) *Hydrodynamic DNA tail vein injection.* Mice should be placed into a restrainer. Special attention should be put to guarantee that the mouse is breathing freely. Also avoid severe stress of the animal. Tail should be wiped with an alcohol swab.

Lateral caudal veins can be identified by visualization of the back of the animal: the blue-colored vessels at both sides of the tail are the lateral caudal veins, whereas the red vessel in the middle of the tail is an artery. For injection, the syringe should be placed in parallel to the tail. The bevel of the syringe should be facing up for best results. Once injection is performed, the needle should slide with virtually no resistance if the injection is correctly carried out. Moving of the needle should be avoided during the procedure. Also avoid full insertion of the syringe into the vein, as this may lead to vessel shearing.

Injection should be performed in a single and continuous manner. In order to achieve good transfection efficiency, full volume of liquid should be injected in less than 10 s.

(4) Once injection is successfully performed, mouse is removed from the restrainer and placed into a recovery cage, ideally heated at 37–38°C. This step is important and may be critical for survival of the mouse if room temperature is not too warm. Injection leads to temporary arrhythmia, so animals can remain immobile for several minutes. Recovery should be evident in 5 min.

(5) Liver tissue extraction for preparing tissue extracts should be performed 24–48 h after injection. Animals should be euthanized painlessly, according to ethic principles for work with laboratory mice. Once extracted, liver tissue is rinsed with PBS and flash-frozen immediately in liquid nitrogen.

For liver tissue protein extracts preparation, a small portion of liver tissue (25–50 mg) should be placed in around 1 mL of ice-cold tissue homogenization buffer (50 m*M* Tris pH 7.5; 150 m*M* NaCl; 0.1% SDS (optional); 0.5% sodium deoxycholate; 1% Triton X-100 or NP-40; 300 m*M* sucrose and protease, and phosphatase inhibitor cocktail). Tissues should be homogenized on ice by means of a mechanical grinder. Tissue homogenates are let 30 min on ice and then centrifuged at 13,000 × *g* for 5 min at 4°C and collect the supernatant (avoiding the pellet) into new microtubes.

Tissue extracts should be next analyzed by WB against N- and C-terminal tags as described previously for transfected cells (Fig. 3).

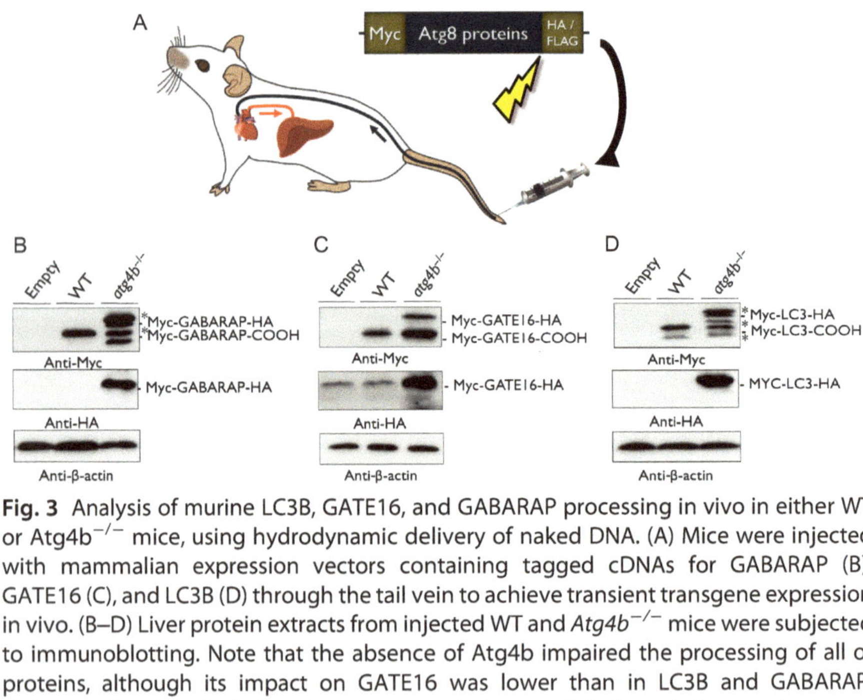

Fig. 3 Analysis of murine LC3B, GATE16, and GABARAP processing in vivo in either WT or Atg4b$^{-/-}$ mice, using hydrodynamic delivery of naked DNA. (A) Mice were injected with mammalian expression vectors containing tagged cDNAs for GABARAP (B), GATE16 (C), and LC3B (D) through the tail vein to achieve transient transgene expression in vivo. (B–D) Liver protein extracts from injected WT and *Atg4b*$^{-/-}$ mice were subjected to immunoblotting. Note that the absence of Atg4b impaired the processing of all of proteins, although its impact on GATE16 was lower than in LC3B and GABARAP. Unknown bands, which likewise corresponded to partial degradation products, are denoted by *asterisks*. β-Actin was used as a loading control. *Taken from Marino, G., Fernandez, A. F., Cabrera, S., Lundberg, Y. W., Cabanillas, R., Rodriguez, F., et al. (2010). Autophagy is essential for mouse sense of balance.* The Journal of Clinical Investigation, 120, *2331–2344.*

7. CONCLUDING REMARKS

Among the essential activities required for autophagy execution, the Atg4–Atg8 system has acquired an unexpected complexity during evolution. This fact illustrates that although autophagy pathway plays a wide variety of essential functions either as a constitutive housekeeping process or as an inducible antistress response, it has also acquired new functions that we are currently starting to elucidate. It is very likely that a fine-tuning of the autophagic process, in terms of substrate recognition or protein–protein interactions, would mediate the development of these possible new acquired functions. The analysis of the Atg4-like enzymatic activity as well as the proteolytic status of Atg8 in different experimental settings would be required to unveil these new potential aspects of autophagy pathway. In this sense, the use of tagged ATG8-coding constructs might be of special interest in animal or cellular models deficient for one or several Atg4 proteins. In fact, the analysis of the proteolytic status of Atg8-like proteins when one or several Atg4 proteases are absent will likely contribute to the understanding of the Atg4–Atg8 system in higher eukaryotes. As a caveat, it has to be noted that Atg4 proteases also cleave the membrane-bound forms of Atg8 proteins once autophagosome–lysosome fusion has occurred, being this is an important event for the development of the autophagic process. Unfortunately, free Atg8 proteins generated by this second cleavage cannot be distinguished from the rest of the cytosolic pool either by the recognition of N-terminal epitopes or that of the endogenous protein. Thus, additional methods should be used in combination with the described techniques, in order to specifically evaluate this second cleavage.

In this work, we have described two different applications of a relatively simple but useful technique to measure proteolytic status of Atg8-like proteins. Given the specific aspects of the Atg4–Atg8 system, the use of these methods in combination with classic autophagic flux analyses might help to faithfully address the status of autophagy pathway both in cultured cells and live laboratory animals. Moreover, in certain occasions, Western blot detection of the lipidated forms of Atg8-like proteins may be problematic. This may be caused either by a reduced presence of the membrane-bound forms of some Atg8 orthologs, or by a reduced reactivity of certain antibodies against the autophagosome-bound forms. In the latter case, Western blot analysis of tagged Atg8-like proteins (by detection of the N-terminal epitope) may constitute an alternative approach, as the affinity against cytosolic and lipidated forms of Atg8-like proteins should be comparable for this experimental setting.

ACKNOWLEDGMENTS

We thank A.F. Fernández for help with figure preparation and helpful comments. G.M. is funded by the Ramon y Cajal Program (RYC-2013-12751) and supported by Spain's Ministerio de Economía y Competitividad, BFU2015-68539 and the BBVA foundation, SV-15-FBBVA-2.

Conflicts of Interest: There are no conflicts of interest for this work to disclose.

REFERENCES

Ahlen, G., Frelin, L., Holmstrom, F., Smetham, G., Augustyn, S., & Sallberg, M. (2016). A targeted controlled force injection of genetic material in vivo. *Molecular Therapy. Methods & Clinical Development*, *5*, 16016.

Cabrera, S., Fernandez, A. F., Marino, G., Aguirre, A., Suarez, M. F., Espanol, Y., et al. (2013). ATG4B/autophagin-1 regulates intestinal homeostasis and protects mice from experimental colitis. *Autophagy*, *9*, 1188–1200.

Choi, A. M., Ryter, S. W., & Levine, B. (2013). Autophagy in human health and disease. *The New England Journal of Medicine*, *368*, 651–662.

Djavaheri-Mergny, M., Maiuri, M. C., & Kroemer, G. (2010). Cross talk between apoptosis and autophagy by caspase-mediated cleavage of Beclin 1. *Oncogene*, *29*, 1717–1719.

Feng, Y., He, D., Yao, Z., & Klionsky, D. J. (2014). The machinery of macroautophagy. *Cell Research*, *24*, 24–41.

Fernandez, A. F., & Lopez-Otin, C. (2015). The functional and pathologic relevance of autophagy proteases. *The Journal of Clinical Investigation*, *125*, 33–41.

Kaminskyy, V., & Zhivotovsky, B. (2012). Proteases in autophagy. *Biochimica et Biophysica Acta*, *1824*, 44–50.

Klionsky, D. J., Abdelmohsen, K., Abe, A., Abedin, M. J., Abeliovich, H., Acevedo Arozena, A., et al. (2016). Guidelines for the use and interpretation of assays for monitoring autophagy (3rd edition). *Autophagy*, *12*, 1–222.

Klionsky, D. J., Cregg, J. M., Dunn, W. A., Jr., Emr, S. D., Sakai, Y., Sandoval, I. V., et al. (2003). A unified nomenclature for yeast autophagy-related genes. *Developmental Cell*, *5*, 539–545.

Kovacsics, D., & Raper, J. (2014). Transient expression of proteins by hydrodynamic gene delivery in mice. *Journal of Visualized Experiments: JoVE*, *87*, e51481.

Kuma, A., Matsui, M., & Mizushima, N. (2007). LC3, an autophagosome marker, can be incorporated into protein aggregates independent of autophagy: Caution in the interpretation of LC3 localization. *Autophagy*, *3*, 323–328.

Le Grand, J. N., Chakrama, F. Z., Seguin-Py, S., Fraichard, A., Delage-Mourroux, R., Jouvenot, M., et al. (2011). GABARAPL1 (GEC1): Original or copycat? *Autophagy*, *7*, 1098–1107.

Li, M., Hou, Y., Wang, J., Chen, X., Shao, Z. M., & Yin, X. M. (2011). Kinetics comparisons of mammalian Atg4 homologues indicate selective preferences toward diverse Atg8 substrates. *The Journal of Biological Chemistry*, *286*, 7327–7338.

Marino, G., Fernandez, A. F., Cabrera, S., Lundberg, Y. W., Cabanillas, R., Rodriguez, F., et al. (2010). Autophagy is essential for mouse sense of balance. *The Journal of Clinical Investigation*, *120*, 2331–2344.

Marino, G., & Lopez-Otin, C. (2004). Autophagy: Molecular mechanisms, physiological functions and relevance in human pathology. *Cellular and Molecular Life Sciences*, *61*, 1439–1454.

Marino, G., Salvador-Montoliu, N., Fueyo, A., Knecht, E., Mizushima, N., & Lopez-Otin, C. (2007). Tissue-specific autophagy alterations and increased tumorigenesis in mice deficient in Atg4C/autophagin-3. *The Journal of Biological Chemistry*, *282*, 18573–18583.

Mizushima, N. (2007). Autophagy: Process and function. *Genes & Development*, *21*, 2861–2873.

Mizushima, N., & Komatsu, M. (2011). Autophagy: Renovation of cells and tissues. *Cell*, *147*, 728–741.

Murthy, A., Li, Y., Peng, I., Reichelt, M., Katakam, A. K., Noubade, R., et al. (2014). A Crohn's disease variant in Atg16l1 enhances its degradation by caspase 3. *Nature*, *506*, 456–462.

Perez-Perez, M. E., Zaffagnini, M., Marchand, C. H., Crespo, J. L., & Lemaire, S. D. (2014). The yeast autophagy protease Atg4 is regulated by thioredoxin. *Autophagy*, *10*, 1953–1964.

Reggiori, F., & Klionsky, D. J. (2013). Autophagic processes in yeast: Mechanism, machinery and regulation. *Genetics*, *194*, 341–361.

Repnik, U., Stoka, V., Turk, V., & Turk, B. (2012). Lysosomes and lysosomal cathepsins in cell death. *Biochimica et Biophysica Acta*, *1824*, 22–33.

Rubinsztein, D. C., Cuervo, A. M., Ravikumar, B., Sarkar, S., Korolchuk, V., Kaushik, S., et al. (2009). In search of an "autophagomometer". *Autophagy*, *5*, 585–589.

Scherz-Shouval, R., Sagiv, Y., Shorer, H., & Elazar, Z. (2003). The COOH terminus of GATE-16, an intra-Golgi transport modulator, is cleaved by the human cysteine protease HsApg4A. *The Journal of Biological Chemistry*, *278*, 14053–14058.

Scherz-Shouval, R., Shvets, E., Fass, E., Shorer, H., Gil, L., & Elazar, Z. (2007). Reactive oxygen species are essential for autophagy and specifically regulate the activity of Atg4. *The EMBO Journal*, *26*, 1749–1760.

Singh, R., & Cuervo, A. M. (2011). Autophagy in the cellular energetic balance. *Cell Metabolism*, *13*, 495–504.

Tanida, I., Minematsu-Ikeguchi, N., Ueno, T., & Kominami, E. (2005). Lysosomal turnover, but not a cellular level, of endogenous LC3 is a marker for autophagy. *Autophagy*, *1*, 84–91.

Tanida, I., Sou, Y. S., Minematsu-Ikeguchi, N., Ueno, T., & Kominami, E. (2006). Atg8L/Apg8L is the fourth mammalian modifier of mammalian Atg8 conjugation mediated by human Atg4B, Atg7 and Atg3. *The FEBS Journal*, *273*, 2553–2562.

Wang, W., Chen, Z., Billiar, T. R., Stang, M. T., & Gao, W. (2013). The carboxyl-terminal amino acids render pro-human LC3B migration similar to lipidated LC3B in SDS-PAGE. *PLoS One*, *8*. e74222.

Yang, Z., & Klionsky, D. J. (2010). Mammalian autophagy: Core molecular machinery and signaling regulation. *Current Opinion in Cell Biology*, *22*, 124–131.

Yousefi, S., Perozzo, R., Schmid, I., Ziemiecki, A., Schaffner, T., Scapozza, L., et al. (2006). Calpain-mediated cleavage of Atg5 switches autophagy to apoptosis. *Nature Cell Biology*, *8*, 1124–1132.

Zhang, Y., Satterlee, A., & Huang, L. (2012). In vivo gene delivery by nonviral vectors: Overcoming hurdles? *Molecular Therapy*, *20*, 1298–1304.

CHAPTER TWELVE

Measurement of the Activity of the Atg4 Cysteine Proteases

M. Li*[,1], Y. Fu*, Z. Yang*, X.-M. Yin[†,1]
*School of Pharmaceutical Sciences, Sun Yat-Sen University, Guangzhou, Guangdong, China
[†]Indiana University School of Medicine, Indianapolis, IN, United States
[1]Corresponding authors: e-mail address: limin65@mail.sysu.edu.cn; xmyin@iupui.edu

Contents

1. Introduction 208
2. The Structure and Regulatory Machinery of Atg4 210
3. Overview of the Methods to Detect the Atg4 Activity In Vitro and Ex Vivo 211
4. Expression and Purification of Atg4 and Atg8 Proteins for In Vitro Assays 213
   4.1 Construct Design for Recombinant Atg4s and Atg8s 213
   4.2 Protein Expression and Purification 214
   4.3 Design, Expression, and Purification of FRET Substrates 214
5. Measurement of Atg4 Activity In Vitro and Ex Vivo 215
   5.1 Gel-Based Assay Under in vitro Conditions 215
   5.2 FRET-Based Assay Under in vitro Conditions 216
   5.3 FRET-Based HTS Assay 217
   5.4 Detection of Atg4 Activity in ex vivo Condition 219
6. Kinetics Analysis of Atg4 Enzymes 219
   6.1 Rationale 219
   6.2 Kinetics Measurement With the Gel-Based Assay 220
   6.3 Kinetics Measurement With the FRET-Based Assay 221
7. Summary 222
Acknowledgments 222
References 223

Abstract

While only one Atg4 is present in yeast, there are four Atg4 homologues in human and in mouse with different substrate specificities and catalytic efficiencies. The molecule Atg4 is a type of cysteine protease, and is known for its crucial roles in cleavage of the Atg8 family proteins before they can be conjugated to phospholipids, and also in cleavage of the conjugated Atg8 molecules from the membrane, a process known as deconjugation. Both processes are required for the maximal efficiency in autophagosome biogenesis. Atg4 could thus be a target for intervention of the autophagy process. It is thus important to measure Atg4 activity to determine and to modulate the autophagy function. Here, we review the catalytic functions and regulatory mechanisms of human Atg4 proteases and discuss the methodology for analyzing Atg4 activity in details.

Methods in Enzymology, Volume 587
ISSN 0076-6879
http://dx.doi.org/10.1016/bs.mie.2016.10.024

1. INTRODUCTION

Autophagy is an evolutionarily conserved, multistep process, whereby cellular components and damaged organelles are sequestered within autophagosomes for lysosomal degradation. Among them, autophagosome biogenesis requires two ubiquitin-like conjugation systems: the Atg12–Atg5 and the Atg8–phosphatidylethanolamine (PE) systems. Atg4 is a cysteine protease of the C54 family and plays an important role in the Atg8/LC3 lipid conjugation system (Marino et al., 2003).

Atg4 was first found to physically interact with Atg8 in the yeast *Saccharomyces cerevisiae* in 1998 (Lang et al., 1998). There is only one single member of Atg4 proteins in yeast, and deletion of Atg4 impairs the autophagy process (Kirisako et al., 2000). In mammals, there are four Atg4 homologues, Atg4A, Atg4B, Atg4C, and Atg4D (Marino et al., 2003). There are eight human Atg8 homologues belonging to two subfamilies: the LC3 (microtubule-associated protein 1 light chain 3) subfamily, which is comprised of LC3A (isoform a and b), LC3B, and LC3C; and the GABARAP ($GABA_A$ receptor-associated protein) subfamily, including GABARAP, GABARAPL1/Atg8L/GEC1, GABARAPL2/GATE-16/GEF2, and GABARAPL3 (Le Grand et al., 2011). All substrates have a conserved cleavage site for Atg4 (Fig. 1). The substrate specificity of different Atg4 homologues is not identical. Studies have shown Atg4B is able to cleave most of the human Atg8 homologues tested so far (Hemelaar, Lelyveld, Kessler, & Ploegh, 2003; Kabeya et al., 2004; Li et al., 2011; Tanida et al., 2004). Atg4A is a potent protease for the GABARAP subfamily, but not the LC3 subfamily, whereas Atg4C and Atg4D seem to possess marginal activities until they are activated by the cleavage of their N-termini via a caspase (Li et al., 2011; Scherz-Shouval, Sagiv, Shorer, & Elazar, 2003).

Atg4 cleaves Atg8 at the peptide bond on the glycine residue at the C-terminus, thus allowing the conjugation of Atg8 to phosphatidylethanolamine (PE) with the participation of other autophagy molecules. Atg4 can also serve as a deconjugating enzyme, which cleaves the amide bond of the conjugated Atg8 and releases it from the autophagosomal membrane. The latter step is important for reusing Atg8 upon autophagy completion. Moreover, this deconjugation process may also occur on isolation membrane, which would positively or negatively affect membrane formation (Nair et al., 2012; Scherz-Shouval et al., 2007; Yu et al., 2012). Therefore,

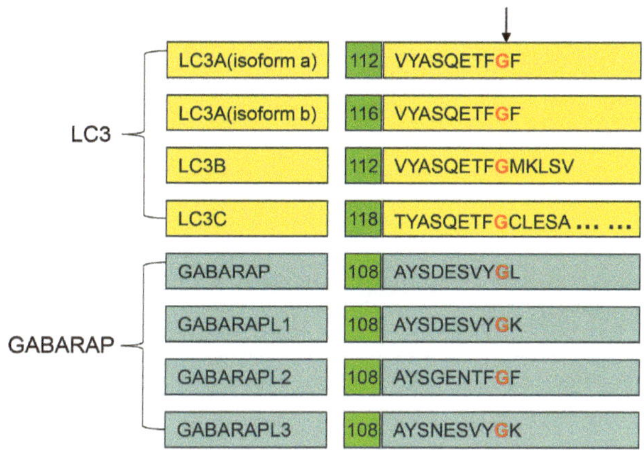

Fig. 1 The cleavage site of Atg4 substrates. The sequences around the Atg4 cleavage sites are listed for the LC3B subfamily substrates (LC3Aa, LC3Ab, LC3B, LC3C) and the GABARAP subfamily substrates (GABARAP, GABARAPL1/Atg8L/GEC1, GABARAPL2/GATE-16/GEF2, and GABARAPL3). The sequences are aligned at the cleavage site next to the glycine residue (indicated by the *arrow*) with the sequence location indicated.

Atg4-mediated deconjugation of Atg8 from the membrane may regulate the autophagy process.

Autophagic activity is diminished in both Atg4-deficient yeast and Atg4B-deficient mice (Marino et al., 2010; Nakatogawa, Ishii, Asai, & Ohsumi, 2012; Yu et al., 2012), while deletion of Atg4C in mice displayed minimal impact on autophagy (Marino et al., 2007), indicating that Atg4B is likely the principal mammal Atg4 homolog. Inhibition of Atg4A and Atg4B activity by reactive oxygen species has been shown in one study to block starvation-induced autophagy (Scherz-Shouval et al., 2007), suggesting that the cysteine protease nature of Atg4 can be regulated by redox events. A more recent study found that the yeast Atg4 could be regulated by thioredoxin on the reduction of the disulfide bond formed by Cys338 and Cys394 (Perez-Perez, Zaffagnini, Marchand, Crespo, & Lemaire, 2014). Mutation of the two cysteines resulted in an increased recruitment of Atg8 to the phagophore assembly site. In addition, Akin et al. have shown that Atg4B knockdown in osteosarcoma cell line Saos-2 and breast cancer cell line MDA-MB468 reduced starvation-induced autophagy and that Saos-2 cells lacking Atg4B failed to survive in amino acid-starvation conditions and failed to grow as xenografted tumors in mice (Akin et al., 2014). Atg4 proteins may thus have a role in cancer development. Finally, some studies have suggested a potential link between Atg4D and apoptosis, indicating a putative role for Atg4D at the interface between autophagy and apoptosis (Kaminskyy & Zhivotovsky, 2014; Norman, Cohen, & Bampton, 2010).

It is therefore important to measure Atg4 activity in order to study autophagy function and to modulate the activity of autophagy.

2. THE STRUCTURE AND REGULATORY MACHINERY OF ATG4

Atg4 proteases range from 393 to 474 amino acids in size and possess several structural features of cysteine proteases. The crystal structure of Atg4A (PDB ID: 2P82) has been resolved (Fig. 2A). Atg4A shares a similar catalytic triad (Cys77/Asp279/His281) with Atg4B (Marino et al., 2003). Structure studies have shown that all the residues of Atg4B that interact

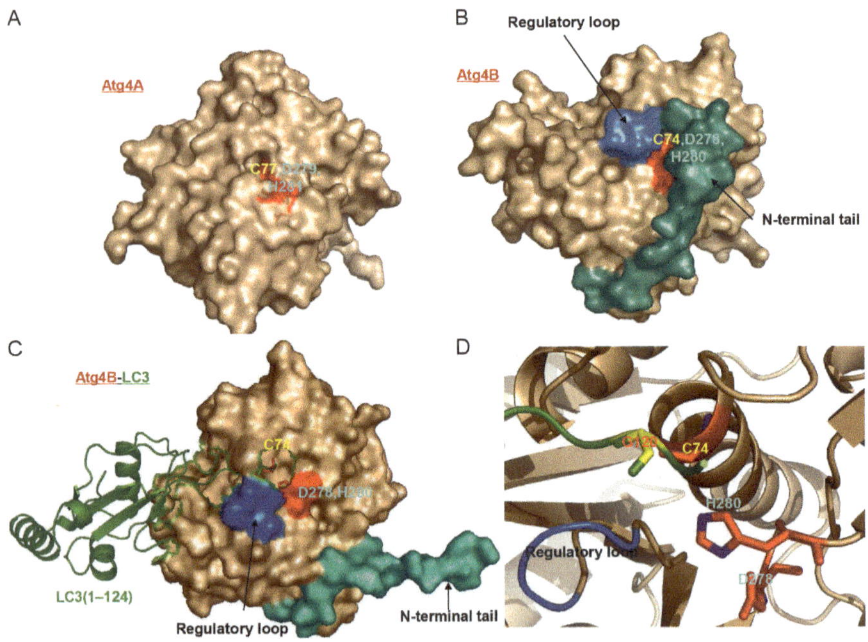

Fig. 2 Structures of human Atg4A and Atg4B. (A) Structure of human Atg4A protein with labeled amino acids in the catalytic pocket. (B) Structure of human Atg4B protein with labeled amino acids in the catalytic pocket. (C) Structure of human Atg4B binding to LC3B. (D) Enlarged catalytic pocket of Atg4B interacting with LC3B. The crucial sites, regulatory loop, free N-terminus of Atg4B are highlighted. All structure models were generated using PyMOL.

with LC3B are conserved in Atg4A except Leu232. Atg4A possesses Ile233 instead of Leu232 at the corresponding position. When Ile233 is changed to Leu, the mutated Atg4A acquires a notable ability to cleave LC3B (Satoo et al., 2009).

The crystal structure of human Atg4B (PDB ID: 2CY7) has also been resolved (Fig. 2B). The structure of Atg4B contains a papain-like fold and a small α/β-fold domain, which is thought to be the binding sites for Atg8 homologues. The active site of Atg4B is composed of Cys74, Asp278, and His280. Mutation of these sites is associated with the complete loss of its catalytic activity (Sugawara et al., 2005). The active site of free Atg4B is masked by a regulatory loop (residues 259–262) (Satoo et al., 2009). A large conformational change of Atg4B is induced in the regulatory loop and the N-terminal tail (residues 1–24) when Atg4B interacts with LC3 (Fig. 2C and D). In this process, the regulatory loop masking the entrance of the active site of Atg4B is lifted by LC3 Phe119, resulting in the formation of a groove, into which the LC3 tail can enter, gaining the access to the active site. Besides, the N-terminal tail is originally positioned at the back of the active sites and undergoes a large conformational change as well upon interaction with LC3, which could affect the exit of the cleaved substrates. Thus, deletion of this N-terminal region has been found to increase Atg4B activity (Satoo et al., 2009).

Although the crystal structures of Atg4C and Atg4D are not available, their three dimensional structures can be acquired by homology modeling, using the structure of Atg4B as a template (Zhang, Li, Ouyang, Liu, & Cheng, 2016). In that model, the catalytic triad is conserved (Cys110/Asp345/His347 for Atg4C and Cys134/Asp356/His358 for Atg4D). It is of note that Atg4C and Atg4D have longer sequences than Atg4A and Atg4B, and the catalytic triad may be spatially positioned differently, rendering access to the substrate difficult. Interestingly, deletion of the N-terminal 63 amino acids of Atg4D stimulated its catalytic activity toward Atg8L (Betin & Lane, 2009), indicating that structural changes of Atg4D could improve substrate access to the active sites.

3. OVERVIEW OF THE METHODS TO DETECT THE ATG4 ACTIVITY IN VITRO AND EX VIVO

The classical method to measure Atg4 activity is based on electrophoretic separation of cleaved Atg8 substrates with a C-terminal tag after the cleavage site (Kabeya et al., 2000; Kirisako et al., 2000; Sugawara et al., 2005). This method is described in details later. By measuring the amount of cleaved band of Atg8 migrating at the expected position on SDS-PAGE, one can determine the activity of Atg4. Compared with the SDS-PAGE-based assay, adoption of specific fluorogenic substrates can make the process more automated and convenient. One example is the use of a site-specific fluorogenic tetrapeptide (Shu et al., 2010), such as acetyl-Gly-Thr-Phe-Gly-AFC (Ac-GTFG-AFC), which is close to the cleavage site (Fig. 1). Release of the fluorogenic AFC following Atg4 cleavage can be detected with a fluorescence spectrometer. A concern on the use of the short 4-amino acid peptide is that the cleavage efficiency by Atg4 is quite low (Shu et al., 2010; Vezenkov et al., 2015) (M. Li and X.-M. Yin, unpublished observations). Although different peptide substrates have been optimized with self-immolative linker, which might be convenient for single point read assays, the K_M of such peptides are still 10-fold higher than that measured with full-length LC3 (Vezenkov et al., 2015).

Another in vitro system uses the enzyme phospholipase A_2, which is inactivated as the result of fusion to the C-terminus of LC3B adjacent to the cleavage site (Ketteler & Seed, 2008; Shu et al., 2010). When the fusion protein is cleaved by Atg4B, PLA_2 is released to regain its activity, which can be determined using a separate substrate for PLA_2. Although this method is sensitive, quantitative, and suitable for high-throughput screening (HTS), the Atg4 activity is indirectly measured based on PLA_2 activity and thus could be affected by any agents that interfere with the PLA_2 reaction.

Methods to measure ex vivo Atg4 activity have been developed. One early study fused *Gaussia princeps* luciferase (GLUC) and β-actin to the N- and C-terminus of LC3, respectively (Tannous, Kim, Fernandez, Weissleder, & Breakefield, 2005). When expressed in cells, the fusion protein is anchored to the cytoskeleton via β-actin and the soluble cytosol contains few GLUC activity. The increased GLUC activity in the cytosol represents that of Atg4B, which cleaves the fusion protein, releasing it from the cytoskeleton. This method does not interrupt normal metabolism in cells with the overexpression of the substrate. As this is an indirect method, any factors that affect β-actin interaction with the substrates or affect GLUC activity can lead to assay instability, and require multiple parallel controls.

Simple one-reaction designs for ex vivo measurement have been reported, which considerably reduce the complexity and interferences. In one assay (Choi et al., 2011), FITC-labeled Atg4-substrate peptides are conjugated to polymeric nanoparticles, which are highly permeable to cells. Once inside the cells, the substrates are cleaved by Atg4, releasing and dequenching the labeled fluorescent dye. The fluorescence is measured and the intensity reflects the Atg4 activity. In another assay, two cell penetrating peptides are used (Ni et al., 2015). One peptide named as AU4S contains the Atg4-recognizable motif "GTFG" and the fluorophore FITC. The other peptide named as AU4R is used as a control, and is not recognized and cleaved by Atg4. The fluorescence ratio defined as "*F–D* value" between AU4S and AU4R changes as the Atg4 activity. The concern on the ex vivo use of the short 4-amino acid peptide is the same as in the in vitro case, that is, the cleavage efficiency of Atg4 on peptides is quite low (Shu et al., 2010; Vezenkov et al., 2015) (M. Li and X.-M. Yin, unpublished observations). Thus, the use of full-length substrate would be preferred to better reflect Atg4 activity. Toward that end a fluorescence resonance energy transfer (FRET)-based ex vivo assay has been developed (Li, Chen, Ye, Vogt, & Yin, 2012), which is detailed in the following section.

4. EXPRESSION AND PURIFICATION OF ATG4 AND ATG8 PROTEINS FOR IN VITRO ASSAYS

In vitro analysis using purified Atg4 and/or Atg8 substrates is necessary for the study of enzymatic mechanisms and kinetics, and for determining structure and activity relationship. The assay can also be adapted into the high-throughput version for rapid screening of potential chemical modulators that inhibit or enhance the activity of Atg4. In this and subsequent sections, the focus is the methodology and only limited data are shown for illustrative purposes. The readers are referred to the original publications for detailed experimental findings.

4.1 Construct Design for Recombinant Atg4s and Atg8s

4.1.1 Rationale

Recombinant human Atg8 homologues are prepared for in vitro enzymatic cleavage assays with either purified recombinant Atg4 proteins or cell lysates that contain Atg4 activity. The methods are adapted from Li et al. (2011).

4.1.2 Methods

The open reading frames of the human Atg4 homologues (Atg4s) and Atg8 homologues (Atg8s) are amplified using RNA extracted from HEK-293A cells. Atg4A (NP_443168), Atg4B (NP_037457), and Atg4C (NP_116241) are cloned into pET-28a (+) in fusion with the N-terminal 6 × His tag. Atg4D (NP_116274) is cloned into pGEX-6P-1 in fusion with the N-terminal GST tag for a better expression. For some reasons, Atg4D in fusion with the 6 × His tag did not express well in our hands.

Representatives of the human Atg8 homologues, LC3B (NP_073729), GATE16 (NP_009216), GABARAP (NP_009209), and Atg8L (NP_113600) belonging to the two subfamilies are cloned into pET-28a (+) in fusion with the 6 × His tag at the N-terminus. The DNA fragment encoding GST is then amplified from the pGEX-6P-1 plasmid and inserted in frame at the C-termini of the Atg8s in the above constructs. The catalytically defective mutant Atg4B^{C74S} is constructed using the ExSite in vitro site-directed mutagenesis system. Atg4B with the deletion of the N-terminal first 24 amino acids (Atg4B$^{\Delta 1-24}$) is constructed by PCR-based amplification of the corresponding Atg4B sequence. All DNA constructs are confirmed by sequencing.

4.2 Protein Expression and Purification

4.2.1 Methods

The plasmids are transformed into the BL21 (DE3) strain of *Escherichia coli*. Protein expression is induced by 0.2–0.5 m*M* isopropyl β-D-1-thiogalactopyranoside (IPTG) at 16°C for 16 h. Atg4D is purified by affinity chromatography using glutathione Sepharose, whereas all other Atg4s and Atg8s are purified by affinity chromatography using Ni^{2+}-NTA resin and eluted with a 20–250 m*M* imidazole gradient buffer containing150 m*M* NaCl, 10 m*M* β-mercaptoethanol, and 20 m*M* Tris (pH 8.0). All preparations are desalted with a PD-10 desalting column. The purity and the apparent molecular weight of the purified proteins are verified by SDS-PAGE followed by Coomassie Brilliant Blue (CBB) staining (Fig. 3), which also allows for quantification by densitometry.

4.3 Design, Expression, and Purification of FRET Substrates

4.3.1 Rationale

Atg8 substrates flanked with CFP and YFP are used in FRET-based assay for a better sensitivity and the ability to be modified for automatic analysis. The methods are adapted from Li et al. (2012). The illustrated examples are LC3B and GATE-16, representing the LC3 and GABARAP subfamilies of the Atg8-like proteins, respectively.

4.3.2 Methods

Full-length LC3B or GATE-16 is first cloned into pET-28a (+) between the BamHI and HindIII restriction sites. The DNA fragments of CFP and YFP

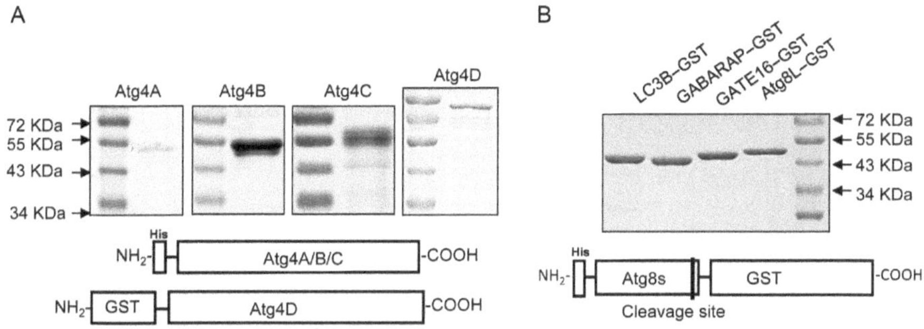

Fig. 3 Expression of Atg4 homologues and Atg8 homologues. (A) His-tagged human Atg4A, Atg4B, and Atg4C, and GST-tagged Atg4D were expressed in *E. coli*, and purified using Ni^{2+}-NTA resins, or glutathione sepharose, respectively. (B) Four human Atg8 homologues were fused with the His tag at the N-terminus and GST at the C-terminus, expressed in *E. coli* and purified using Ni^{2+}-NTA resins. Recombinant proteins were separated by SDS-PAGE and stained with CBB for quantification by densitometry.

are amplified from the commercial vectors pECFP-C1 and pEYFP-C1. CFP is inserted at the N-terminal side of LC3B or GATE-16 at the NdeI and BamHI sites, whereas YFP is inserted at the C-terminal side at the HindIII and XhoI sites.

The expression and purification of the FRET substrates are essentially the same as the non-FRET substrates as described earlier, except that the *E. coli* strain used is BL21 (DE3)-Codon Plus followed by 0.5 m*M* IPTG induction at 16°C for 16 h.

5. MEASUREMENT OF ATG4 ACTIVITY IN VITRO AND EX VIVO

5.1 Gel-Based Assay Under in vitro Conditions

1. Purified Atg4s (0.25 μg for Atg4A and Atg4B, 0.5 μg for Atg4C and Atg4D) and Atg8s–GST (5 μg) are admixed with Buffer A (150 m*M* NaCl, 1 m*M* EDTA, 1 m*M* DTT, and 50 m*M* Tris–HCl, pH 7.2) in a total volume of 20 μL. Atg4C and Atg4D are weaker in catalytic efficiency than Atg4A and Atg4B, and more are used in the assay. The mixture is incubated at 37°C for 0–60 min. Reactions are stopped by adding the 5 × Laemmli sample buffer (5 μL). Samples are boiled at 95°C for 5 min and resolved by SDS-PAGE. Recombinant GST alone will be run as a quantification control.

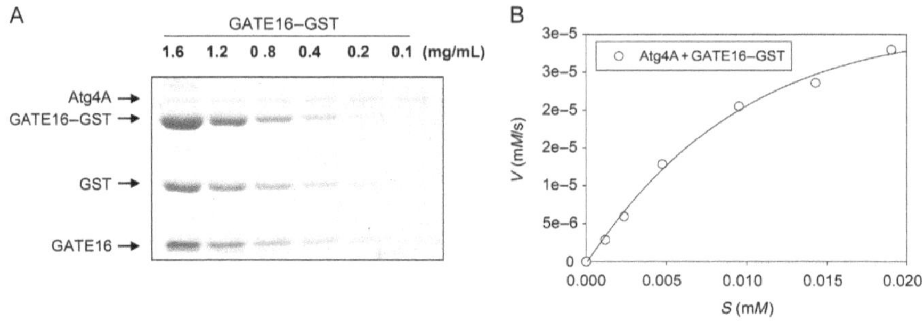

Fig. 4 Determination of the kinetics of Atg4A toward GATE-16 using the gel-based assay. (A) Different concentrations (mg/mL) of GATE-16-GST were incubated with Atg4A (0.25 μg) in a 20-μL volume for 1 min. The reaction was then stopped and the proteins were separated by SDS-PAGE, followed by staining with CBB for quantification by standard densitometry. GATE16–GST was quickly cleaved by Atg4A into two fragments representing GST and GATE16. (B) The initial velocity (*V*, *y*-axis) is defined as the change of the concentration of GST/s ([m*M*]/s), which is then plotted against the concentration of the substrate (S [m*M*], *x*-axis). The curves are fitted using the nonlinear regression method.

2. The amounts of substrates (Atg8s–GST) and the cleaved products (Atg8s and GST) (Fig. 4A) were quantified by densitometry based on the standard curve of GST proteins. Higher Atg4 activities will result in a fewer amount of intact Atg8s–GST but a higher amount of cleavage products: GST and Atg8s. The values of the densitometry measurement can be compared to determine the relative activity of Atg4s, and can be used in kinetics analysis (Fig. 4B), which is discussed in Section 6.

5.2 FRET-Based Assay Under in vitro Conditions

5.2.1 Rationale

Although gel-based method is intuitive and accurate, the sensitivity is limited and the procedure can be labor-intensive. A larger amount of substrates are needed for the assay, and it is not feasible for HTS. A number of fluorescence or luminescence-based assays have been developed as discussed earlier. Provided here in detail is the FRET-based assay that we have developed (Li et al., 2012).

The structures of LC3B and GATE-16 suggest that the N-terminus and the C-terminus of these proteins are spatially close to each other (Kouno et al., 2005; Paz, Elazar, & Fass, 2000). When LC3B and GATE-16 are fused with CFP and YFP at the N- and C-terminus, respectively, the fusion structure would allow FRET to occur, corresponding to a higher fluorescence at the emission wavelength of 527 nm (YFP signals). When cleavage occurs, the YFP moiety is separated from the CFP-Atg8 part and FRET signal is lost, leading to a higher fluorescence at the emission wavelength of 477 nm (CFP signals). The fluorescence signal ratio of 527 nm/477 nm is high before the substrates are cleaved, and is low after cleavage.

The FRET-based assay needs a fewer amount of substrate as it is more sensitive than the gel-based assay. The assay is not prone to interferences as the signals are directly coupled with the Atg4 activity without the involvement of a secondary reaction. Small molecules with autofluorescence may still interfere with the FRET signal in a screening assay, but these can be filtered out easily. The FRET-based assay delivers time- and dose-dependent quantitative responses and has a broad dynamic range sufficient for quantifying and discriminating Atg4 activities under different conditions. The use of a full-length molecule, instead of a peptide, as the substrate can ensure maximal specificity and hydrolysis efficiency.

5.2.2 Methods

Purified FRET-LC3B, FRET-LC3B^{G120A}, and FRET-GATE-16 (5 μg) are mixed with 0.5 μg of purified Atg4A, Atg4B, or Atg4B^{C74S} in Buffer B (150 m*M* NaCl, 1 m*M* EDTA, 50 m*M* Tris–HCl, pH 7.5). The assay can be laid out on a 96-well plate in a total volume of 100 μL. The plate is placed on a multiwavelength fluorescence spectrometer (such as TECAN infinite M200pro) at 37°C and assessed at the excitation wavelength of 434 nm and emission wavelength of 477 and 527 nm. The fluorescence ratio of 527 nm/477 nm reflects the relative strength of Atg4, with a lower ratio correlating with a higher Atg4 activity (Fig. 5A).

The cleavage can be verified independently by SDS-PAGE. The full-length protein (CFP-LC3B-YFP or CFP-GATE-16-YFP) would be cleaved into CFP-LC3B or CFP-GATE-16, plus YFP by the Atg4A or Atg4B, respectively (Fig. 5B).

5.3 FRET-Based HTS Assay

HTS assays for Atg4B inhibitors have been conducted (Li et al., 2012; Shu, Liu, & Huang, 2012; Xu et al., 2016). The method described by Shu et al. is based on the Atg4B cleavage of LC3-PLA2 followed by the analysis of PLA2 activity (see Section 3). This HTS assay has a Z' factor of >0.7 and has been validated in pilot screening with the 1280-compound Lopac™ library and 2000-compound Spectrum™ library, with a hit rate of 0.23% and 0.70%, respectively. The HTS developed by Xu et al. (2016) is based on a time-resolved fluorescence resonance energy transfer (TR-FRET) assay for

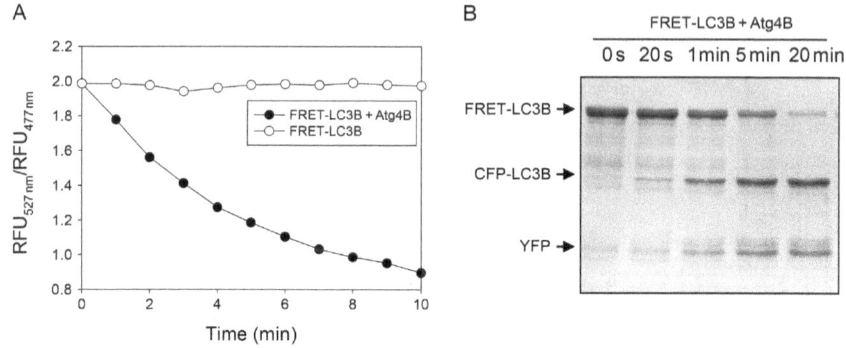

Fig. 5 Cleavage of FRET-LC3B by Atg4B using FRET-based assay. (A) FRET-LC3B (5 μg) was incubated with or without Atg4B (0.5 μg) in a volume of 100 μL. The ratio of relative fluorescence unit (RFU) at 477 and 527 nm with excitation at 434 nm was recorded at different time. (B) FRET-LC3B (5 μg) was incubated with or without Atg4B (0.25 μg) in a 20-μL volume. The reaction was stopped at different times using the SDS-PAGE sample buffer, and separated by SDS-PAGE. Proteins were visualized with CBB staining. Two bands, CFP-LC3B (donor) and YFP (acceptor), were produced after cleavage.

Atg4B. This HTS assay has an average Z' factor of approximately 0.9. They validated the assay with a 57,000-compound Roche focus library, which yielded a hit rate of 0.49%.

We have developed the FRET-based HTS for Atg4A and Atg4B with excellent Z' factors of 0.78 and 0.81, respectively (Li et al., 2012). Pilot screening using FRET-GATE-16 and FRET-LC3B on a 10,000-compound ChemDiv library yielded a hit rate of 0.67%, and 0.7% for Atg4A, and Atg4B, respectively (M. Li and X.-M. Yin, unpublished results). The compounds are found to be able to modulate Atg4A and/or Atg4B activity. The basic setup of the HTS assay can be summarized as follows. An amount of 0.5 μg of Atg4A or 0.1 μg of Atg4B are first incubated with different chemicals (final concentration is 10 μ*M*) for 30 min at 37°C in 384-well plates. The stock of compounds is prepared in DMSO at 10 m*M*. The final concentration of DMSO at the screening is thus 0.1%, which has no effects on Atg4 activity. More Atg4A proteins are used in the assay than Atg4B because Atg4A is weaker in catalytic efficiency. FRET-GATE-16 or FRET-LC3B (3 μg) is then added to a total volume of 50 μL. After 30 min, the fluorescence intensity is recorded. The inhibition effect by a given chemical is calculated as: percentage of inhibition $(\%) = (RFU_X - RFU_{min})/(RFU_{max} - RFU_{min}) \times 100\%$. RFU_X is the RFU ratio of 527 nm/477 nm in the presence of a given chemical, RFU_{min} is the ratio in the absence of the chemical, and RFU_{max} is the ratio in the absence of Atg4 proteins.

5.4 Detection of Atg4 Activity in ex vivo Condition

5.4.1 Rationale

Endogenous Atg4 activity in cells can be analyzed by incubating cell lysates with the recombinant Atg8 substrates. To study the regulatory mechanism of Atg4, it is also possible to express a relevant Atg4 homologous in cells and prepare the lysates (or the immunoprecipitates of the specific Atg4 molecules) after proper treatment, which can then be incubated with the substrates to assess any changes in the Atg4 activity. Below is a published example using the gel-based assay (Li et al., 2012). However, by using FRET substrate, the assay can be run with the FRET detection method.

5.4.2 Methods

HEK-293A cells are transfected with Atg4s tagged with FLAG and lysed 24 h later in Buffer C (150 m*M* NaCl, 1 m*M* EDTA, 0.1% Triton-100, 50 m*M* Tris–HCl, pH 8.0) with freshly prepared protease inhibitors. About 20–30 μg of the supernatants with overexpressed Atg4s are incubated with 5 μg of Atg8s–GST. The reactions are stopped by SDS-PAGE sample loading buffer for 3 min. The samples are boiled and resolved by SDS-PAGE.

The cleavage assay is characterized for the generation of the cleaved products, GST (Li et al., 2012). Cell lysates with transfected Atg4B can result in increased activities against all four substrates, whereas transfection of Atg4A can result in an increased activity mainly against GATE-16. Transfection of Atg4C, Atg4D, and Atg4B^{C74S} can lead to no more activities than the endogenous background level.

6. KINETICS ANALYSIS OF ATG4 ENZYMES

6.1 Rationale

One of the most important goals of in vitro assay is to determine the enzymatic kinetics in order to understand the catalytic mechanism and catalytic efficiency. This knowledge will become quite useful in comparing the different Atg4 homologues for the substrate specificity and catalytic efficiency, and in analyzing how chemical modulators work in inhibiting or enhancing the Atg4 enzymes, thus facilitating the better design of such molecules. There are a number of different ways to measure kinetic parameters, depending largely on how enzyme activity is measured (Eleanore Seibert, 2014). Below are the methods we have used when Atg4 enzyme activity is measured by the SDS-PAGE method (Li et al., 2011) or by the FRET method (Li et al., 2012).

Based on these assays, Atg4B is the most potent enzyme among the four Atg4 homologues with a catalytic efficiency ranging from 78,900 to 107,000 (mol^{-1} L s^{-1}) for LC3B–GST, GATE-16–GST, GABARAP–GST, and Atg8L–GST in the gel-based assay. The results from the gel-based assay and from the FRET-based assay are quite comparable with the same approach for calculation. A different calculation method had been used, which resulted in somewhat different numbers (Shu et al., 2010). However, those data still lead to the same conclusions that GATE-16-GST is the best substrate for both Atg4B and Atg4A and that Atg4B has a broad substrate specificity than Atg4A. In fact, a more detailed analysis indicates that Atg4A has a low activity toward the substrates other than GATE-16 (Li et al., 2011). In addition, Atg4C and Atg4D possess the lowest activity toward the four Atg8 homologues. Furthermore, the catalytic triad of Atg4s has an influence on the catalytic efficiency. The catalytic mutant (C74S) of Atg4B had dramatically reduced the ability of Atg4B to cleave all four substrates. Finally, the deletion of 24 amino acids at N-terminus of Atg4B can increase the catalytic efficiency (Li et al., 2011).

6.2 Kinetics Measurement With the Gel-Based Assay

1. Purified Atg4s (0.25 μg for Atg4A and Atg4B, 0.5 μg for Atg4C and Atg4D) and Atg8s–GST (5 μg) are admixed together in Buffer A (150 m*M* NaCl, 1 m*M* EDTA, 1 m*M* DTT, and 50 m*M* Tris–HCl, pH 7.2) in a volume of 20 μL at 37°C for 0–60 min. Reactions are stopped by adding the Laemmli sample buffer and resolved by SDS-PAGE. An example is illustrated in Fig. 4A.
2. The amounts of uncleaved substrates (Atg8s–GST) and the cleaved products (Atg8s and GST) are quantified by densitometry (OD) based on the standard curve of GST proteins.
3. The percentage of noncleaved substrates is equal to $OD_{Atg8s-GST}/(OD_{Atg8s-GST} + OD_{GST} + OD_{Atg8s}) \times 100\%$. This value is then plotted against the reaction time. The optimal time is determined, at which the reaction is linear in rate under the given experiment conditions.
4. Atg8s–GST at different amounts (0.1–32 μg) are incubated with a particular type of Atg4 enzyme (0.25–0.5 μg) for an optimal period of time as determined in step 3 earlier.
5. The initial velocity is defined as the change in the concentration ([S] in the unit of m*M*) of GST, one of the cleavage products of GST–Atg8s, per second ([S]/s).

6. The velocity (mM/s) is plotted against the concentration of the substrate ([S] in the unit of mM) (Fig. 4B). The curve is then fitted using the nonlinear regression method, from which the V_{max} (mM/s) and K_M (Michaelis constant, mM) for each enzyme–substrate reaction are derived using the Michaelis–Menten formula.
7. The catalytic constant (k_{cat}) is determined by dividing V_{max} by the concentration of the enzyme used in the assay. Catalytic efficiency is defined as k_{cat}/K_M (mol^{-1} L s^{-1}).

6.3 Kinetics Measurement With the FRET-Based Assay

1. The Atg4 enzyme (0.05 μg for Atg4A, 0.02 μg for Atg4B) is incubated with FRET substrates in different amounts (0.1–20 μg) in Buffer B (150 mM NaCl, 1 mM EDTA, 50 mM Tris–HCl, pH 7.5) in a total volume of 20 μL at 37°C for the designated times.
2. The initial velocity of the cleavage reaction is defined as $V = [S]_t/t$, in which $[S]_t$ is the substrate concentration ([S]) in the unit of mM at a given time point (t) (in the unit of second). $[S]_t$ is estimated based on the fluorescence change of the substrates at 527 nm ($[S]_{t-527nm}$) as well as on the fluorescence change of the substrates at 477 nm ($[S]_{t-477nm}$), so that $[S]_t = \left([S]_{t-527\,nm} + [S]_{t-477\,nm}\right)/2$. $[S]_{t-527nm}$ was calculated as $[S] \times (RFU_{max-527\,nm} - RFU_{t-527\,nm})/(RFU_{max-527\,nm} - RFU_{min-527\,nm})$, whereas $[S]_{t-477\,nm} = [S] \times (RFU_{t-477\,nm} - RFU_{min-477\,nm})/(RFU_{max-477\,nm} - RFU_{min-477\,nm})$, in which [S] is the initial substrate concentration. RFU_{max} and RFU_{min} are the fluorescence intensity at the maximal and minimal levels across the entire reaction period at the corresponding wavelength, and RFU_t is the fluorescence intensity at a given time (t).
3. The initial velocity (V) is plotted against the concentration of the substrate at the given time points ($[S]_t$) and the curve is fitted using the ligand-binding method, from which the V_{max} ([S]/s, or mM/s) and K_M (Michaelis constant, [S] or mM) for each enzyme–substrate reaction are derived based on the Michaelis–Menten formula. The curve-fitting of FRET-GATE-16 cleaved by Atg4B is illustrated (Fig. 6)
4. The catalytic constant (k_{cat}, s^{-1}) is determined by dividing V_{max} by the concentration of the enzyme. The catalytic efficiency is defined as k_{cat}/K_M (mol^{-1} L s^{-1}).

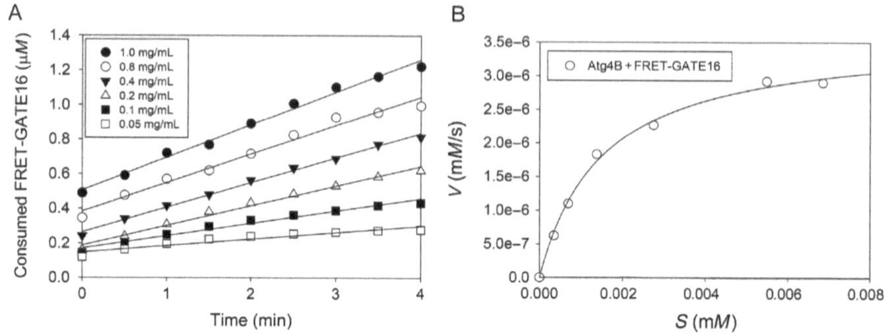

Fig. 6 Determination of the kinetics of Atg4B toward GATE-16 using the FRET-based assay. (A) FRET-GATE-16 at different concentrations as indicated by the legends were incubated with 0.02 μg of Atg4B in a 20-μL volume. The consumed substrates were calculated and plotted against the incubation time. The first 4 min were used for initial velocity calculation, which was defined as the increment of the consumed substrate per second. (B) The initial velocity (*V*, *y*-axis) was plotted against the concentration of the substrate (*S* [m*M*], *x*-axis). The curve was fitted using the ligand-binding method.

7. SUMMARY

Atg4 proteases play a crucial role in preparing Atg8 for conjugation to lipid membranes and for the deconjugation of Atg8 from the autophagosomes (Till & Subramani, 2010). Of the four proteases, Atg4B is 1500-fold more catalytically efficient for LC3B activation than the other Atg4s, whereas Atg4A is most selective toward GATE-16 (Li et al., 2011). Atg4B and LC3B are best studied in autophagy and Atg4B seems to be most potent (Nakatogawa, Ichimura, & Ohsumi, 2007; Tanida et al., 2004). These findings suggest that Atg4B could be a potential target for inhibiting autophagy. HTS assays for Atg4B inhibitors have been conducted (Li et al., 2012; Shu et al., 2012; Xu et al., 2016), but it is still a long journey to the availability of practically useful agents with sufficient potency and specificity. We hope that the methods discussed in this chapter will further stimulate studies of Atg4 enzymes and the development of Atg4 modulators.

ACKNOWLEDGMENTS

The authors and their relevant works were in part supported by National Institute of Health (R03MH083154, R01CA111456, R01CA83817, R01AA021751) (to X.-M.Y.), the Natural Science Foundation of Guangdong Province, China (S2013010015876, 2016A030313335) (to M.L.), the National Natural Science Foundation of China (31671437) (to M.L.), and the Specialized Research Fund for the Doctoral Program of Higher Education (20130171120049) (to M.L.).

The Conflict-of-Interest and Financial Disclosure Statements: There were no potential conflicts of interest and no significant financial support for this work to disclose.

REFERENCES

Akin, D., Wang, S. K., Habibzadegah-Tari, P., Law, B., Ostrov, D., Li, M., ... Dunn, W. A., Jr. (2014). A novel ATG4B antagonist inhibits autophagy and has a negative impact on osteosarcoma tumors. *Autophagy, 10*, 2021–2035.

Betin, V. M. S., & Lane, J. D. (2009). Caspase cleavage of Atg4D stimulates GABARAP-L1 processing and triggers mitochondrial targeting and apoptosis. *Journal of Cell Science, 122*, 2554–2566.

Choi, K. M., Nam, H. Y., Na, J. H., Kim, S. W., Kim, S. Y., Kim, K., ... Ahn, H. J. (2011). A monitoring method for Atg4 activation in living cells using peptide-conjugated polymeric nanoparticles. *Autophagy, 7*, 1052–1062.

Eleanore Seibert, T. S. T. (2014). Fundamentals of enzyme kinetics. *Methods in Molecular Biology, 1113*, 23–35.

Hemelaar, J., Lelyveld, V. S., Kessler, B. M., & Ploegh, H. L. (2003). A single protease, Apg4B, is specific for the autophagy-related ubiquitin-like proteins GATE-16, MAP1-LC3, GABARAP, and Apg8L. *Journal of Biological Chemistry, 278*, 51841–51850.

Kabeya, Y., Mizushima, N., Uero, T., Yamamoto, A., Kirisako, T., Noda, T., ... Yoshimori, T. (2000). LC3, a mammalian homologue of yeast Apg8p, is localized in autophagosome membranes after processing. *The EMBO Journal, 19*, 5720–5728.

Kabeya, Y., Mizushima, N., Yamamoto, A., Oshitani-Okamoto, S., Ohsumi, Y., & Yoshimori, T. (2004). LC3, GABARAP and GATE16 localize to autophagosomal membrane depending on form-II formation. *Journal of Cell Science, 117*, 2805–2812.

Kaminskyy, V. O., & Zhivotovsky, B. (2014). Free radicals in cross talk between autophagy and apoptosis. *Antioxidants & Redox Signaling, 21*, 86–102.

Ketteler, R., & Seed, B. (2008). Quantitation of autophagy by luciferase release assay. *Autophagy, 4*, 801–806.

Kirisako, T., Ichimura, Y., Okada, H., Kabeya, Y., Mizushima, N., Yoshimori, T., ... Ohsumi, Y. (2000). The reversible modification regulates the membrane-binding state of Apg8/Aut7 essential for autophagy and the cytoplasm to vacuole targeting pathway. *The Journal of Cell Biology, 151*, 263–276.

Kouno, T., Mizuguchi, M., Tanida, I., Ueno, T., Kanematsu, T., Mori, Y., ... Kawano, K. (2005). Solution structure of microtubule-associated protein light chain 3 and identification of its functional subdomains. *Journal of Biological Chemistry, 280*, 24610–24617.

Lang, T., Schaeffeler, E., Bernreuther, D., Bredschneider, M., Wolf, D. H., & Thumm, M. (1998). Aut2p and Aut7p, two novel microtubule-associated proteins are essential for delivery of autophagic vesicles to the vacuole. *The EMBO Journal, 17*, 3597–3607.

Le Grand, J. N., Chakrama, F. Z., Seguin-Py, S., Fraichard, A., Delage-Mourroux, R., Jouvenot, M., & Boyer-Guittaut, M. (2011). GABARAPL1 (GEC1): Original or copycat? *Autophagy, 7*, 1098–1107.

Li, M., Chen, X., Ye, Q. Z., Vogt, A., & Yin, X. M. (2012). A high-throughput FRET-based assay for determination of Atg4 activity. *Autophagy, 8*, 401–412.

Li, M., Hou, Y., Wang, J., Chen, X., Shao, Z. M., & Yin, X. M. (2011). Kinetics comparisons of mammalian Atg4 homologues indicate selective preferences toward diverse Atg8 substrates. *The Journal of Biological Chemistry, 286*, 7327–7338.

Marino, G., Fernandez, A. F., Cabrera, S., Lundberg, Y. W., Cabanillas, R., Rodriguez, F., ... Lopez-Otin, C. (2010). Autophagy is essential for mouse sense of balance. *Journal of Clinical Investigation, 120*, 2331–2344.

Marino, G., Salvador-Montoliu, N., Fueyo, A., Knecht, E., Mizushima, N., & Lopez-Otin, C. (2007). Tissue-specific autophagy alterations and increased tumorigenesis in mice deficient in Atg4C/autophagin-3. *The Journal of Biological Chemistry, 282*, 18573–18583.

Marino, G., Uria, J. A., Puente, X. S., Quesada, V., Bordallo, J., & Lopez-Otin, C. (2003). Human autophagins, a family of cysteine proteinases potentially implicated in cell degradation by autophagy. *The Journal of Biological Chemistry, 278*, 3671–3678.

Nair, U., Yen, W. L., Mari, M., Cao, Y., Xie, Z., Baba, M., ... Klionsky, D. J. (2012). A role for Atg8-PE deconjugation in autophagosome biogenesis. *Autophagy*, *8*, 780–793.

Nakatogawa, H., Ichimura, Y., & Ohsumi, Y. (2007). Atg8, a ubiquitin-like protein required for autophagosome formation, mediates membrane tethering and hemifusion. *Cell*, *130*, 165–178.

Nakatogawa, H., Ishii, J., Asai, E., & Ohsumi, Y. (2012). Atg4 recycles inappropriately lipidated Atg8 to promote autophagosome biogenesis. *Autophagy*, *8*, 177–186.

Ni, Z. H., Gong, Y., Dai, X. F., Ding, W., Wang, B., Gong, H. Y., ... He, F. T. (2015). AU4S: A novel synthetic peptide to measure the activity of ATG4 in living cells. *Autophagy*, *11*, 403–415.

Norman, J. M., Cohen, G. M., & Bampton, E. T. W. (2010). The in vitro cleavage of the hAtg proteins by cell death proteases. *Autophagy*, *6*, 1042–1056.

Paz, Y., Elazar, Z., & Fass, D. (2000). Structure of GATE-16, membrane transport modulator and mammalian ortholog of autophagocytosis factor Aut7p. *Journal of Biological Chemistry*, *275*, 25445–25450.

Perez-Perez, M. E., Zaffagnini, M., Marchand, C. H., Crespo, J. L., & Lemaire, S. D. (2014). The yeast autophagy protease Atg4 is regulated by thioredoxin. *Autophagy*, *10*, 1953–1964.

Satoo, K., Noda, N. N., Kumeta, H., Fujioka, Y., Mizushima, N., Ohsumi, Y., & Inagaki, F. (2009). The structure of Atg4B-LC3 complex reveals the mechanism of LC3 processing and delipidation during autophagy. *The EMBO Journal*, *28*, 1341–1350.

Scherz-Shouval, R., Sagiv, Y., Shorer, H., & Elazar, Z. (2003). The COOH terminus of GATE-16, an intra-Golgi transport modulator, is cleaved by the human cysteine protease HsApg4A. *Journal of Biological Chemistry*, *278*, 14053–14058.

Scherz-Shouval, R., Shvets, E., Fass, E., Shorer, H., Gil, L., & Elazar, Z. (2007). Reactive oxygen species are essential for autophagy and specifically regulate the activity of Atg4. *The EMBO Journal*, *26*, 1749–1760.

Shu, C. W., Drag, M., Bekes, M., Zhai, D., Salvesen, G. S., & Reed, J. C. (2010). Synthetic substrates for measuring activity of autophagy proteases: Autophagins (Atg4). *Autophagy*, *6*, 936–947.

Shu, C. W., Liu, P. F., & Huang, C. M. (2012). High throughput screening for drug discovery of autophagy modulators. *Combinatorial Chemistry & High Throughput Screening*, *15*, 721–729.

Sugawara, K., Suzuki, N. N., Fujioka, Y., Mizushima, N., Ohsumi, Y., & Inagaki, F. (2005). Structural basis for the specificity and catalysis of human Atg4B responsible for mammalian autophagy. *Journal of Biological Chemistry*, *280*, 40058–40065.

Tanida, I., Sou, Y. S., Ezaki, J., Minematsu-Ikeguchi, N., Ueno, T., & Kominami, E. (2004). HsAtg4B/HsApg4B/autophagin-1 cleaves the carboxyl termini of three human Atg8 homologues and delipidates microtubule-associated protein light chain 3- and GABAA receptor-associated protein-phospholipid conjugates. *The Journal of Biological Chemistry*, *279*, 36268–36276.

Tannous, B. A., Kim, D. E., Fernandez, J. L., Weissleder, R., & Breakefield, X. O. (2005). Codon-optimized Gaussia luciferase cDNA for mammalian gene expression in culture and in vivo. *Molecular Therapy*, *11*, 435–443.

Till, A., & Subramani, S. (2010). A balancing act for autophagin. *Journal of Clinical Investigation*, *120*, 2273–2276.

Vezenkov, L., Honson, N. S., Kumar, N. S., Bosc, D., Kovacic, S., Nguyen, T. G., ... Young, R. N. (2015). Development of fluorescent peptide substrates and assays for the key autophagy-initiating cysteine protease enzyme, ATG4B. *Bioorganic & Medicinal Chemistry*, *23*, 3237–3247.

Xu, D., Xu, Z., Han, L., Liu, C., Zhou, Z., Qiu, Z., … Ding, H. (2016). Identification of new ATG4B inhibitors based on a novel high-throughput screening platform. *Journal of Biomolecular Screening*, pii: 1087057116639202. [Epub ahead of print]. https://www.ncbi.nlm.nih.gov/pubmed/26993319.
Yu, Z. Q., Ni, T., Hong, B., Wang, H. Y., Jiang, F. J., Zou, S., … Xie, Z. (2012). Dual roles of Atg8-PE deconjugation by Atg4 in autophagy. *Autophagy*, *8*, 883–892.
Zhang, L., Li, J., Ouyang, L., Liu, B., & Cheng, Y. (2016). Unraveling the roles of Atg4 proteases from autophagy modulation to targeted cancer therapy. *Cancer Letters*, *373*, 19–26.

CHAPTER THIRTEEN

Crystallographic Characterization of ATG Proteins and Their Interacting Partners

Y. Qiu*[,1], Y. Zheng*[,†,1], A.M. Taherbhoy*[,2], S.E. Kaiser*[,3], B.A. Schulman*[,†,‡,4]

*St. Jude Children's Research Hospital, Memphis, TN, United States
†University of Tennessee Health Sciences Center, Memphis, TN, United States
‡Howard Hughes Medical Institute, St. Jude Children's Research Hospital, Memphis, TN, United States
[4]Corresponding author: e-mail address: brenda.schulman@stjude.org

Contents

1. Introduction 228
2. Domain Structures of Atg Proteins in the UBL Conjugation Pathway 230
3. Proteins That Have Been Crystallized in the UBL Conjugation System in Autophagy 231
4. Crystallization of *Saccharomyces cerevisiae* Atg7 Alone or in Complex With Atg3 or Atg10 232
   4.1 Crystallization of *S. cerevisiae* Atg7^{NTD} and Atg7^{CTD} 232
   4.2 Crystallization of *S. cerevisiae* Atg7^{NTD}–Atg3FR^{pep} 236
   4.3 Crystallization of *S. cerevisiae* Atg7–Atg3 and Atg7–Atg10 237
5. Crystallization of Human ATG5–ATG16L1 (1–69) E122D Disease-Associated Mutant 239
   5.1 Baculovirus/Insect Cell Expression System 239
   5.2 Transposition 240
   5.3 Bacmid Preparation 240
   5.4 Transfection of Insect Cells (P1 Production) 241
   5.5 P2/P3 Amplification 242
   5.6 Infection of Suspension Hi5 Insect Cells for Protein Expression 242
   5.7 Purification of the ATG5–ATG16L1 (1–69) Complex 243
   5.8 Crystallization of the ATG5–ATG16L1 (1–69) Complex 244

Acknowledgment 244
References 244

---

[1] These authors contributed equally.
[2] Current position: Nurix, Inc., San Francisco, CA 94158, United States.
[3] Current position: Pfizer, Inc., San Diego, CA 92121, United States.

Methods in Enzymology, Volume 587
ISSN 0076-6879
http://dx.doi.org/10.1016/bs.mie.2016.09.058

Abstract

Autophagosome formation and specific substrate recruitment during autophagy require ligation of the ubiquitin-like protein (UBL) Atg8 to the head group of the lipid phosphatidylethanolamine. Atg8 lipidation is mediated by distinctive UBL cascades involving autophagy-specific E1, E2, and E3 enzymes that differ substantially in sequence from components of other UBL conjugation cascades. Structural studies are important for elucidating the roles of Atg proteins that regulate multiple steps involved in autophagy. This chapter describes methods to prepare and crystallize selected proteins and complexes involved in autophagy UBL conjugation pathways, as a guide for strategies for structural and biochemical characterization of Atg proteins.

1. INTRODUCTION

Autophagy is characterized by the formation of double-membrane structures called autophagosomes, which sequester the cellular components to be degraded (Mizushima, 2007). This process comprises phagophore initiation and vesicle nucleation, vesicle elongation, fusion with lysosome/vacuole, and degradation (Kraft & Martens, 2012; Levine & Klionsky, 2004). Each step is tightly regulated by numerous proteins, including at least 15 distinct autophagy-related (Atg) proteins that are core components of many forms of autophagy (Hurley & Schulman, 2014; Klionsky & Schulman, 2014; Mizushima, Yoshimori, & Ohsumi, 2011). Of these, Atg12 and Atg8 are ubiquitin-like proteins (UBLs) required for autophagosome formation and recruitment of specific substrates (e.g., protein aggregates and damaged organelles) to autophagomes (Geng & Klionsky, 2008). In mammals, six subfamilies of Atg8, including LC3, GABARAP, and GATE-16 proteins (for simplicity, referred to as Atg8 hereafter), have been identified (Geng & Klionsky, 2008; Ohsumi & Mizushima, 2004; Schulman & Harper, 2009). Similar to ubiquitin, Atg12 and Atg8 are ligated to their targets via E1–E2–E3 enzyme cascades. Atg12 is unusual in that after activation by the E1 enzyme Atg7, it is transferred to its E2 enzyme Atg10 and is then directly ligated to its protein substrate Atg5 without the involvement of an E3 enzyme (Fig. 1A–C). Unlike Atg12, Atg8 is processed by the cysteine protease Atg4 to expose its C-terminal glycine, activated by the E1 enzyme Atg7, and then transferred to its E2 enzyme Atg3. This results in a transient Atg3~Atg8 intermediate in which the C-terminus of Atg8 is linked by a covalent but reactive thioester bond to the catalytic cysteine of Atg3 ("~" represents a covalent

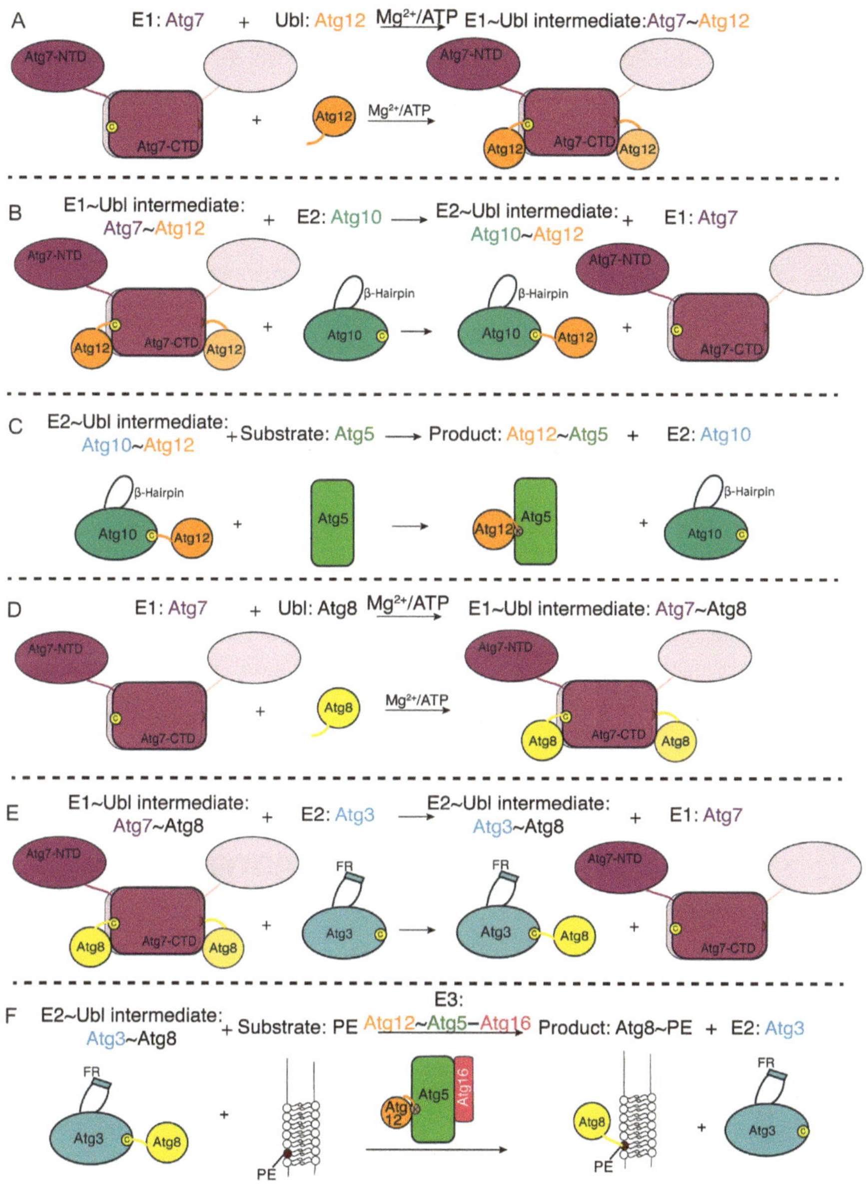

Fig. 1 Schematic view of Atg12 and Atg8 conjugation cascades. (A) Atg7 (E1) mediated Atg12 (Ubl) activation, (B) transthiolation reaction transfers Atg12 (Ubl) from Atg7 (E1) to Atg10 (E2), (C) Atg12 (Ubl) ligation to its substrate Atg5 from Atg10 (E2), (D) Atg7 (E1) mediated Atg8 (Ubl) activation, (E) Transthiolation reaction transfers Atg8 (Ubl) from Atg7 (E1) to Atg3 (E2), and (F) Atg8 (Ubl) ligation to its substrate lipid PE from Atg3 (E2) catalyzed by Atg12~Atg5–Atg16 (E3).

bond). Atg8 is then transferred from Atg3 to the head group of the lipid phosphatidylethanolamine (PE). This reaction is catalyzed by a multiprotein E3 enzyme composed of the product of the Atg12 ligation reaction bound to Atg16 (Atg12~Atg5–Atg16) (Fig. 1D–F) (Geng & Klionsky, 2008; Nakatogawa, Suzuki, Kamada, & Ohsumi, 2009; Ohsumi & Mizushima, 2004). Mutations of the genes involved in these cascades are observed in human diseases such as ataxia, asthma, and Crohn's disease (Barrett et al., 2008; Hampe et al., 2007; Kim et al., 2016; Martin et al., 2012; Poon, Eidelman, Laprise, & Hamid, 2012; Rioux et al., 2007).

Given the important role of the UBL conjugation system in autophagy and its close association with human health, the modulation of protein function in the Atg8 lipidation pathway through chemical inhibition of the E1–E2–E3 cascade has become an area of significant interest. Several promising autophagy inhibitors have been used in clinical trials for cancer (Jiang & Mizushima, 2014), and a recent study showed that an intestinal tissue-specific knockout of Atg7 prevents tumor initiation (Levy et al., 2015).

Structures of enzymes in this pathway have greatly enhanced our understanding of the molecular mechanisms underlying autophagy and may facilitate the development of approaches to manipulate autophagy to treat diseases in the future. In this chapter, we present the strategy and detailed methodology for crystallizing several Atg proteins and their interacting partners involved in the autophagy UBL conjugation pathway.

2. DOMAIN STRUCTURES OF ATG PROTEINS IN THE UBL CONJUGATION PATHWAY

Seven proteins or protein conjugates are involved in the UBL conjugation cascade for autophagy: Atg7 (E1), Atg3 (E2), Atg10 (E2), Atg12~Atg5–Atg16 (E3), Atg12 (UBL), Atg8 (UBL), and Atg4 (deconjugation enzyme). These E1, E2, and E3 enzymes are "noncanonical," because they differ in many respects from the "canonical" E1, E2, and E3 enzymes such as those for ubiquitin, SUMO, and NEDD8. Canonical E1 enzymes contain three domains: a pseudosymmetric UBL adenylation domain for Ubl activation, a separate catalytic cysteine-containing domain for Ubl thioester formation, and a ubiquitin-fold domain that recruits the E2 for transfer of the UBL from the catalytic cysteine on the E1 to that on the E2. In contrast, Atg7 is a symmetric homodimer, also defined by three distinct domains: a C-terminal domain (CTD) that houses both the adenylation domain and the catalytic cysteine; a flexible extreme CTD (ECTD) extension; and a unique N-terminal

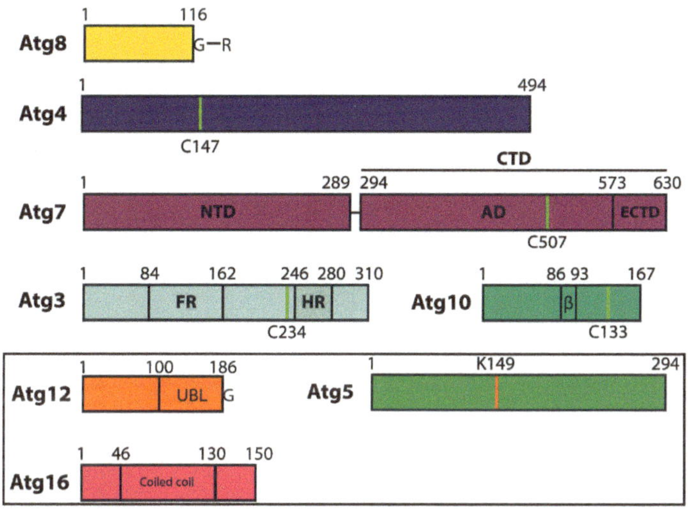

Fig. 2 Scheme showing domain structures of ATG proteins involved in the UBL conjugation pathway in autophagy. *NTD*, N-terminal domain; *AD*, adenylation domain; *ECTD*, extreme C-terminal domain; *CTD*, C-terminal domain; *FR*, flexible region; *HR*, handle region; *β*, β-hairpin.

domain (NTD) responsible for recruiting the E2s. Atg7 homodimerization is mediated via its CTD (Fig. 2). Atg3 and Atg10 contain unique catalytic residues around the catalytic cysteine, and they have additional secondary structures inserted within the E2 fold (Fig. 2). Atg12∼Atg5–Atg16 is a multiprotein complex (Fig. 2) that functions as an E3 enzyme to ligate Atg8 to the head group of PE. This is in contrast to other canonical E3 enzymes that ligate their UBLs to a lysine residue of a protein substrate. The Atg12∼Atg5–Atg16 complex does not share sequence similarity with any known UBL E3 ligases. Given these differences between the autophagy and canonical UBL conjugation systems, it is important to structurally characterize each component in the autophagy UBL conjugation pathway in order to understand the mechanisms that control the ligation of Atg8 to PE (Fig. 3).

3. PROTEINS THAT HAVE BEEN CRYSTALLIZED IN THE UBL CONJUGATION SYSTEM IN AUTOPHAGY

Table 1 lists existing crystal structures for several components and protein complexes in the UBL conjugation system.

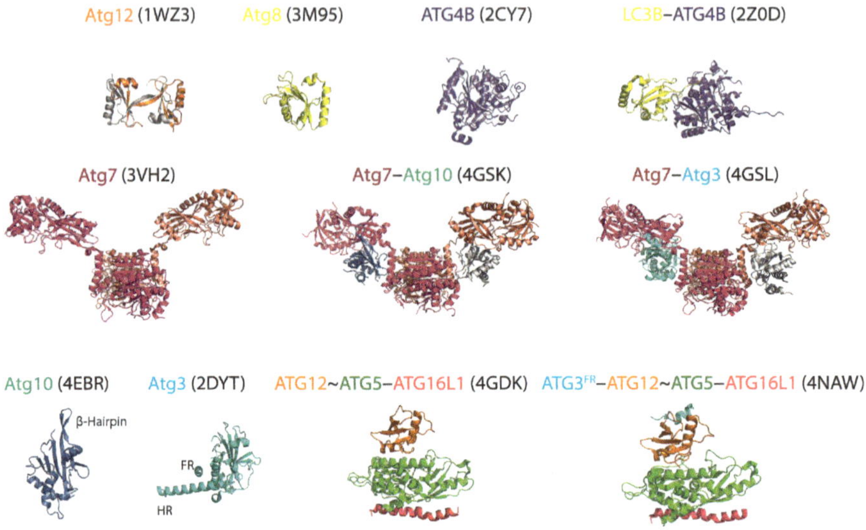

Fig. 3 Representative structures of ATG proteins in the UBL conjugation pathway in autophagy.

4. CRYSTALLIZATION OF *SACCHAROMYCES CEREVISIAE* ATG7 ALONE OR IN COMPLEX WITH ATG3 OR ATG10

4.1 Crystallization of *S. cerevisiae* Atg7^{NTD} and Atg7^{CTD}

Atg7 is composed of two domains (NTD, residues 1–289; and CTD, residues 294–630) separated by a short linker. We previously crystallized the NTD and CTD separately (Taherbhoy et al., 2011) (Fig. 3). Atg7^{NTD} and Atg7^{CTD} coding sequences were each cloned into the pGEX-4T1 vector (GE Healthcare Life Sciences, Marlborough, MA) and transformed into BL21 Gold (DE3) cells (Novagen).

4.1.1 Expression

1. Inoculate 1-L LB medium supplemented with 100 μg/mL ampicillin with 5-mL LB culture containing Atg7^{NTD} or Atg7^{CTD}. Grow the 1-L culture in a 2-L flask at 37°C until OD_{600nm} reaches 0.6–0.8.
2. Transfer the flask to a 23°C shaker and add 1 mL of 0.6 *M* isopropyl β-D-1-thiogalactopyranoside to the 1-L culture to induce protein expression, and continue to grow the culture for 16–20 h. Harvest the bacteria by centrifuging at 4°C.

Table 1 ATG Proteins Crystallized in the UBL Conjugation Pathway in Autophagy

Protein or Protein Complex	Organism	PDB ID	References
Atg7 (1–595)	*Saccharomyces cerevisiae*	3VH1	Noda et al. (2011)
Atg7 (1–613)	*Saccharomyces cerevisiae*	3VH2	Noda et al. (2011)
Atg7 NTD	*Kluyveromyces marxianus*	3VX6	Yamaguchi et al. (2012)
Atg7 NTD	*Saccharomyces cerevisiae*	3RUJ, 3T7F	Hong et al. (2011) and Taherbhoy et al. (2011)
Atg7 CTD	*Saccharomyces cerevisiae*	3T7E	Taherbhoy et al. (2011)
Atg7 NTD and Atg3FR complex	*Saccharomyces cerevisiae*	3T7G	Taherbhoy et al. (2011)
Atg7 NTD and Atg3 complex	*Arabidopsis thaliana*	3VX8	Yamaguchi et al. (2012)
Atg7 NTD and Atg10 complex	*Kluyveromyces marxianus*	3VX7	Yamaguchi et al. (2012)
Atg7 CTD and Atg8 complex	*Saccharomyces cerevisiae*	3VH3, 3VH4, 3RUI	Hong et al. (2011) and Noda et al. (2011)
Atg7 (1–613) and Atg3 complex	*Saccharomyces cerevisiae*	4GSL	Kaiser et al. (2012)
Atg7 (1–613) and Atg10 complex	*Saccharomyces cerevisiae*	4GSK	Kaiser et al. (2012)
Atg3	*Saccharomyces cerevisiae*	2DYT	Yamada et al. (2007)
Atg10	*Saccharomyces cerevisiae*	4EBR	Hong, Kim, Kim, and Song (2012)
ATG12~ATG5–ATG16L1	*Homo sapiens*	4GDK, 4GDL	Otomo, Metlagel, Takaesu, and Otomo (2013)
Atg12~Atg5–Atg16	*Saccharomyces cerevisiae*	3W1S	Noda, Fujioka, Hanada, Ohsumi, and Inagaki (2013)
ATG12~ATG5–ATG16L1 and ATG3FR complex	*Homo sapiens*	4NAW	Metlagel, Otomo, Takaesu, and Otomo (2013)

Continued

Table 1 ATG Proteins Crystallized in the UBL Conjugation Pathway in Autophagy—cont'd

Protein or Protein Complex	Organism	PDB ID	References
Atg12	*Arabidopsis thaliana*	1WZ3	Suzuki, Yoshimoto, Fujioka, Ohsumi, and Inagaki (2005)
Atg8	*Bombyx mori*	3M95	Hu, Zhang, Teng, Hu, and Li (2010)
Atg8	*Trypanosoma brucei*	3H9D	Koopmann, Muhammad, Perbandt, Betzel, and Duszenko (2009)
LC3	*Homo sapiens*	1UGM	Sugawara et al. (2004)
LC3A	*Homo sapiens*	3WAL	Suzuki et al. (2014)
LC3B	*Homo sapiens*	3VTU	Rogov et al. (2013)
LC3C	*Homo sapiens*	3WAM	Suzuki et al. (2014)
GABARAP	*Homo sapiens*	1KJT, 1GNU	Bavro et al. (2002) and Knight et al. (2002)
GABARAPL1	*Homo sapiens*	2R2Q	
GABARAPL2	*Homo sapiens*	1EO6, 4CO7	Ma et al. (2015)
Atg8 and Atg3 peptide complex	*Plasmodium falciparum*	4EOY	Hain et al. (2012)
ATG4A	*Homo sapiens*	2P82	
ATG4B	*Homo sapiens*	2CY7, 2D1L	Kumanomidou et al. (2006) and Sugawara et al. (2005)
ATG4B and LC3 complex	*Homo sapiens*	2Z0D, 2Z0E	Satoo et al. (2009)

PDB, Protein Data Bank identifier.

4.1.2 Purification

1. Vortex to fully homogenize the cell paste in ice-cold lysis buffer containing 25 m*M* Tris, pH 7.6, 200 m*M* NaCl, 10 m*M* dithiothreitol (DTT), and 2.5 m*M* PMSF (freshly added).
2. Disrupt the bacteria by using a microfluidizer (Avestin) or by sonication.

3. Centrifuge the lysate at 20,000 × *g* for at least 40 min and collect the supernatant.
4. Load the supernatant onto preequilibrated Glutathione Sepharose 4B beads (GE Healthcare Life Sciences, Marlborough, MA).
5. Wash the beads with 12 column volumes (CVs) of lysis buffer and then elute the protein with 5 CVs of elution buffer containing 25 m*M* Tris, pH 7.6, 200 m*M* NaCl, 5 m*M* β-mercaptoethanol (BME), and 10 m*M* glutathione.
6. Collect all five elution fractions and check the purity of each fraction by SDS-PAGE.
7. Pool the fractions containing the pure target protein and add thrombin to cleave the GST-tag from $Atg7^{NTD}$ or $Atg7^{CTD}$ with a mass ratio of 1:150. Dialyze overnight into buffer containing 25 m*M* Tris, pH 7.6, 200 m*M* NaCl, and 5 m*M* BME at 4°C.
8. Check the cleavage efficiency by SDS-PAGE, then pass the cleavage reactions back over preequilibrated Glutathione Sepharose 4B beads (GE Healthcare Life Sciences, Marlborough, MA) to remove GST.
9. Collect the flow-through, and use 2 CV of buffer to wash out protein remaining in the column bed volume.
10. Further purify the $Atg7^{NTD}$ or $Atg7^{CTD}$ protein by anion-exchange chromatography (HiTrap Q HP, GE Healthcare Life Sciences, Marlborough, MA) and gel filtration (Superdex 200 10/300 GL, GE Healthcare Life Sciences, Marlborough, MA) in a final buffer containing 25 m*M* Tris, pH 7.6, 150 m*M* NaCl, and 10 m*M* DTT; concentrate to 30 mg/mL for $Atg7^{NTD}$ and 15 mg/mL for $Atg7^{CTD}$; divide into aliquots; flash freeze in liquid nitrogen; and store at −80°C.

4.1.3 Crystallization

1. Add 1 mL of the reservoir solution in each well of a 24-well VDXm Plate Greased™ (Hampton Research).
2. To obtain the initial $Atg7^{NTD}$ crystal, mix 1 μL of 30 mg/mL $Atg7^{NTD}$ in 25 m*M* Tris, pH 7.6, 150 m*M* NaCl, and 10 m*M* DTT with 1 μL of the well solution containing 9.2% isopropanol, 0.1 *M* potassium thiocyanate, and 0.1 *M* citrate buffer (pH 4.0) by using hanging-drop vapor diffusion at 4°C.
3. To improve crystal quality, use streak seeding. Set up drops under conditions that contain 60–80% of the precipitant concentration for a normal condition that produces crystals (in order to avoid the formation of crystal nuclei) and equilibrate the drop overnight at 4°C.

4. To make the seed stock, crush the crystals into microcrystals as best possible within the drop and then mix the drop containing microcrystals with 100 μL of well solution and vortex.
5. Dip a clean cat whisker or seeding tool (Hampton) into the seed stock, and run the cat whisker or seeding tool through the preequilibrated drops. Use the loaded cat whisker or seeding tool for three more drops in order to transfer progressively fewer seeds to subsequent drops.
6. After crystal growth, cryoprotect the crystals in well solution supplemented with 6% glycerol, 6% ethylene glycol, and 6% sucrose.
7. To obtain crystals of $Atg7^{CTD}$, mix 1 μL of 15 mg/mL $Atg7^{CTD}$ in 25 m*M* Tris, pH 7.6, 150 m*M* NaCl, and 10 m*M* DTT with 1 μL of 12% polyethylene glycol 20k (PEG-20k), 0.1 *M* Tris, pH 8.5, and 10 m*M* TCEP and incubate at 4°C. For harvesting, soak crystals in 13% PEG-20k, 0.1 *M* Tris, pH 8.5, 10 m*M* TCEP, 7.5% ethylene glycol, and 7.5% glycerol before cryoprotection in 13% PEG-20k, 0.1 *M* Tris, pH 8.5, 10 m*M* TCEP, 15% ethylene glycol, and 15% glycerol.

4.2 Crystallization of *S. cerevisiae* $Atg7^{NTD}$–$Atg3FR^{pep}$

To understand the molecular mechanism underlying the recognition of Atg3 by Atg7, we have crystallized $Atg7^{NTD}$ in complex with $Atg3FR^{pep}$ (Fig. 3) by the following procedure:

1. Synthesize $Atg3FR^{pep}$ (Atg3 residue 128–144: SIDDIDELIQDMEIKEE) with N-terminal acetylation and an additional NH_2 group at the C-terminus.
2. Dissolve $Atg3FR^{pep}$ in 25 m*M* Tris, pH 7.6, 150 m*M* NaCl, 10 m*M* DTT. Adjust the pH to 7.6 with NaOH.
3. To prepare the sample for crystallization, mix 30 mg/mL $Atg7^{NTD}$ with an eightfold molar excess of the $Atg3FR^{pep}$ peptide.
4. Grow crystals of $Atg7^{NTD}$–$Atg3FR^{pep}$ at room temperature by mixing 1 μL of the $Atg7^{NTD}$–$Atg3FR^{pep}$ sample and 1 μL of the well solution containing 1.8 *M* ammonium sulfate, 0.1 *M* Bis-Tris, pH 5.5, and 5 m*M* DTT.
5. After soaking the crystals in the well solution supplemented with 6.25% glycerol and 6.25% ethylene glycol, cryoprotect crystals in the well solution supplemented with 12.5% glycerol and 12.5% ethylene glycol.

4.3 Crystallization of *S. cerevisiae* Atg7–Atg3 and Atg7–Atg10

Atg7 is the only E1 enzyme that activates the two UBLs Atg8 and Atg12 and transfers them to two distinct E2 enzymes—Atg3 and Atg10, respectively. Crystallization of the Atg7–Atg3 and Atg7–Atg10 complexes provided insights into how a single E1 enzyme recognizes the two different E2s and how active sites are juxtaposed for these noncanonical E1–E2 complexes to promote autophagy. To visualize the juxtaposition of E1 and E2 catalytic cysteines and to facilitate the crystallization process, cross-linking is performed with the homobifunctional sulfhydryl cross-linker bismaleimidoethane (BMOE) by using the following strategy. Versions of Atg3 and Atg10 containing only a single cysteine at the active site are prepared, BMOE is reacted with each E2, and excess cross-linker is removed by desalting. For Atg3, Atg7 (1–613) is added. For Atg10, a version of Atg7 (1–613) with C39S C195S C375A mutations is added to reduce background cross-linking. By this strategy, we obtain crystals of Atg7–BMOE–Atg3 and Atg7–BMOE–Atg10 (hereafter referred to as Atg7–Atg3 and Atg7–Atg10, respectively) (Fig. 3). The detailed methodology is as follows.

4.3.1 Protein Preparation

1. Prepare Atg7 (1–613) and Atg7 (1–613) C39S C195S C375A by the same method as that used for $Atg7^{NTD}$ or $Atg7^{CTD}$.
2. Clone Atg3 C41A C76A C83A (Atg3 C234only) and Atg10 C26S C137S (Atg10 C133only) into the pGEX-4T1 vector (GE Healthcare Life Sciences, Marlborough, MA), and transform into RIL codon-enhanced BL21 (DE3) (Agilent Technologies).
3. For expression, dilute small-scale culture in 1 L of LB medium containing 100 μg/mL ampicillin and 25 mg/mL chloramphenicol in a 2-L flask and grow at 37°C until OD_{600nm} reaches 0.6–0.8.
4. Transfer the flask to shaker equilibrated at 16°C and add 1 mL of 0.6 *M* IPTG (isopropyl β-D-1-thiogalactopyranoside) to the 1-L culture to induce protein expression. Grow the culture for 16–20 h.
5. Use the same purification procedure as that used for $Atg7^{NTD}$ or $Atg7^{CTD}$.

4.3.2 Atg7–Atg3 Cross-Linking

1. Reduce Atg3 C234only and Atg7 (1–613) with 10 m*M* DTT and purify by using a Superdex 200 10/300 GL column (GE Healthcare Life

Sciences, Marlborough, MA) equilibrated with buffer containing 20 m*M* HEPES, pH 7.0, and 150 m*M* NaCl.

2. Mix approximately 0.5 m*M* Atg3 C234only with a 5 × molar excess of BMOE (Pierce) on ice for 1 h, followed by desalting on a PD-10 column (GE Healthcare Life Sciences, Marlborough, MA) to remove unreacted cross-linker.
3. Mix Atg3 C234only~BMOE (3 μ*M*) and Atg7 (1–613) (1.5 μ*M*) for 20 s, and quench the cross-linking reaction with 50 m*M* DTT.
4. Purify Atg7–Atg3 C234only by anion-exchange chromatography (5 mL HiTrap Q HP) and gel filtration (Superdex 200 10/300 GL) into a final buffer of 50 m*M* HEPES, pH 7.0, 150 m*M* NaCl, and 10 m*M* DTT; concentrate to 12.5 mg/mL; divide into aliquots; flash freeze in liquid nitrogen; and store at −80°C.

4.3.3 Atg7–Atg10 Cross-Linking

Prepare Atg10 C133only cross-linked to Atg7 (1–613) C39S C195S C375A by the same procedure as that used for Atg7–Atg3 cross-linking, except that a 5 × molar excess of BMOE is added to approximately 0.2 m*M* Atg10 C133only. Perform cross-linking with 1.2 μ*M* Atg7 (1–613) C39S C195S C375A and 2.4 μ*M* Atg10 C133only in the final buffer of 25 m*M* HEPES, pH 7.0, 100 m*M* NaCl, and 5 m*M* DTT. Concentrate Atg7–Atg10 to 7 mg/mL.

4.3.4 Crystallization of Atg7–Atg3

1. Grow crystals of Atg7–Atg3 by microseeding at room temperature sitting drops of 2 μL of Atg7 (1–613)~BMOE~Atg3 C234only with 1.4 μL of the well solution of 100 m*M* sodium citrate, pH 5.8, 13.5% PEG 4000, 10% isopropanol, 5% dioxane, and 88 m*M* ammonium sulfate.
2. To prepare the seed stock, mix drops containing microcrystals with the well solution.
3. After vortexing, serially dilute the seed stock from 10^{-1}- to 10^{-6}-fold and add 200 nL to 2 μL of protein and 1.8 μL of the well solution mixture.
4. Cryoprotect the crystals in the well solution supplemented with 15% glycerol and 15% ethylene glycol before data collection.

4.3.5 Crystallization of Atg7–Atg10

1. Grow crystals of Atg7–Atg10 by at room temperature microseeding sitting drops containing 400 nL of Atg7 (1–613) C39S C195S

C375A~BMOE~Atg10 C133only in 25 m*M* HEPES, pH 7.0, 100 m*M* NaCl, and 5 m*M* DTT mixed with 400 nL of a well solution containing 76.5 m*M* Na/K phosphate, pH 6.5, 9 m*M* Tris, pH 8.5, 153 m*M* NaCl, 100 m*M* glycine, 72 m*M* Na/K tartrate, 19.125% PEG 1000, 0.045% PEG 5000MME, and 4.5% dioxane.

2. To prepare the seeding solution, mix drops containing microcrystals with the well solution.
3. After vortexing, serially dilute the seeding solution from 10^{-1}- to 10^{-6}-fold and add a 20 nL of seed stock dilution to 500 nL of protein and 500 nL of the well solution mixture.
4. Cryoprotect the crystals in well solution supplemented with 30% glycerol before data collection.

5. CRYSTALLIZATION OF HUMAN ATG5–ATG16L1 (1–69) E122D DISEASE-ASSOCIATED MUTANT

5.1 Baculovirus/Insect Cell Expression System

To obtain wild-type or mutant version of the ATG5–ATG16L1 (1–69) complex, a baculovirus/insect cell expression system is used. Specifically, ATG5 wild type or the ATG5 E122D mutant is cloned into the pFastBac-HTb vector, which contains an N-terminal His-tag fusion with ampicillin resistance (Life Technologies). ATG16L1 (1–69) is cloned into the pFastBac-hisMBP vector, which is modified by insertion of MBP between the His-tag and the TEV cleavage site from pFastBac-HTb. After amplification by DH5alpha cells, the ATG5 containing pFastBac-HTb plasmid and ATG16L1 (1–69) containing plasmid are transformed into DH10-EmBacY cells. DH10-EmBacY cells are DH10B *Escherichia coli* containing a plasmid encoding the transposase and tetracycline resistance and the bacmid encoding the baculovirus genome, yellow fluorescent protein, kanamycin resistance, and ampicillin resistance. At this step, the *ATG5* gene or *ATG16L1* (1–69) gene, as well as a gentamicin marker, is transferred from the pFastBac-HTb plasmid into a bacmid encoding the baculovirus genome via transposase-mediated recombination, or transposition, in *E. coli*. When the *ATG5* or *ATG16L1* (1–69) gene is transposed into the bacmid, it gets inserted into the *lacZα* gene. The *lacZα* gene provides β-galactosidase complementation and turns colonies blue on plates containing X-gal. Therefore, the desired recombinant bacmids produce white colonies, whereas undesirable transformants that contain nonrecombinant bacmids are blue when grown on plates containing X-gal. To verify blue/white selection, white colonies need to be restreaked on selective plates with X-gal before proceeding to the next steps.

The detailed methodology is as follows.

5.2 Transposition

1. Thaw competent DH10-EmBacY cells on ice. Spread 40 μL of 40 mg/mL X-gal (dissolved in dimethylformamide) and 40 μL of 100 m*M* IPTG on plates containing 50 μg/mL kanamycin, 7 μg/mL gentamicin, 10 μg/mL tetracycline, and 100 μg/mL ampicillin.
2. Add 1 μg of ATG5 containing pFastBac-HTb plasmid to DH10-EmBacY competent cells. Incubate on ice for 30 min.
3. Heat shock the competent cells at 42°C for 30 s. Then, allow cells to recover on ice for 2 min.
4. Add 900 μL of the SOC medium, transfer to a culture tube, and incubate at 37°C with shaking for 4–8 h.
5. Spread 500 μL of cells on LB-agar plates containing the four antibiotics and incubate at 37°C until blue color develops. This step takes at least 24 h.
6. Restreak several white colonies on a fresh plate with X-gal, IPTG, and the four antibiotics. Include a blue colony as a control for color development. Incubate overnight at 37°C until blue color develops. This step usually takes less than 24 h.

5.3 Bacmid Preparation

1. Select 4–8 colonies and grow the culture in a 37°C shaking incubator overnight.
2. Transfer 1.5 mL of bacterial culture to a 1.5-mL microcentrifuge tube and centrifuge at 6800 × *g* for 3 min.
3. Remove the supernatant by vacuum aspiration or by pouring and resuspend the cell pellet in 0.3 mL of P1. Gently vortex or pipet up and down to resuspend.
4. Add 0.3 mL of P2 and gently mix. Incubate at room temperature for 5 min.
5. Slowly add 0.3 mL of 3 *M* potassium acetate, pH 5.5 (also known as P3), mixing gently during addition. A thick white precipitate of protein and *E. coli* genomic DNA will form. Place the sample on ice for 5–10 min.
6. Centrifuge for 10 min at 16,000 × *g* at room temperature.

7. Gently transfer the supernatant to a microcentrifuge tube containing 0.8 mL of isopropanol. Do not transfer any white precipitate. Invert the tube a few times to mix and place on dry ice for 15 min.
8. Centrifuge the sample for 20 min at 16,000 × *g* at room temperature.
9. Carefully remove the supernatant, taking care not to disturb the pellet.
10. Add 0.5 mL of 70% ethanol to the pellet. When adding the ethanol, first cover the pellet and then continue adding ethanol around the edges of the tube to get as much isopropanol as possible. Invert the tube carefully to remove isopropanol completely. Centrifuge for 5–10 min at 16,000 × *g*, remove most of the ethanol gently with a pipette and then aspirate the remaining ethanol carefully, taking care not to disturb the pellet.
11. Air-dry the pellet for 5–10 min at room temperature.
12. Dissolve the DNA pellet in 40 μL of 1 × TE buffer, pH 8.0. To avoid shearing, do not mechanically resuspend the DNA. Allow the solution to sit in the tube, with occasional gentle tapping of the bottom of the tube.
13. Store the DNA at 4°C.

    Note: We do not recommend storing the purified bacmid DNA at −20°C, because repeated freezing and thawing might shear the DNA.
14. Perform PCR screening of the recombinant bacmids for the insert gene.

5.4 Transfection of Insect Cells (P1 Production)

The initial virus (P1) is made by transfecting SF9 cells in plates with FuGENE HD (Promega) and harvesting the supernatant 3–5 days later, as described later:

1. Measure cell density of the SF9 culture with a hemocytometer or other cell counter.
2. Seed 5×10^6 cells into the well of a 6-well plate (Corning Costar 3516, Sigma-Aldrich). Allow cells to attach to the plate by placing in an incubator at 27°C for approximately 30 min.
3. Verify cell attachment by using a microscope. Cells should appear very dense.
4. Dilute 2 μg of bacmid DNA into 100 μL of medium warmed to 27°C in a sterile, polystyrene tube.
5. Add 10 μL of FuGENE HD and incubate at 27°C for 30 min.
6. Aspirate the medium from the cells in the plate.

7. Add 0.9 mL of medium to the DNA–FuGENE mixture and transfer to cells in the plate.
8. Incubate cells for 5–6 h at 27°C with occasional rocking.
9. Add another 1 mL of medium to the cells and incubate at 27°C for 4 days.
10. Pass the medium (containing the P1 virus) through a 0.22-μm syringe filter to remove detached cells and debris.
11. Store the virus in the dark at 4°C.

5.5 P2/P3 Amplification

Later-generation viruses (P2 and P3) are produced in the same manner by infecting a plate of SF9 cells and harvesting the medium 3 days later as follows:

1. Measure cell density of the SF9 culture with a hemocytometer or other cell counter.
2. Seed 25–30 $\times$ 10^6 cells in a 150-mm plate (Corning 430599) and allow cells to attach by incubating at 27°C for approximately 30 min.
3. Verify cell attachment with a microscope. Cells should appear 75–90% confluent.
4. Adjust the volume of medium on the cells to 10 mL by aspirating or adding additional medium, as needed.
5. Add 0.25 mL of virus to the cells and gently rock the plate, mixing thoroughly. Incubate at 27°C for 1 h.
6. Add 12 mL of medium to the plate and incubate for 3–4 days at 27°C, gently rocking or shaking the plate each day to agitate the medium.
7. Remove the medium (containing the virus) from the cells. To remove detached cells and debris, centrifuge the medium in a sterile 50-mL conical tube at 500 $\times$ *g* for 10 min. Store in the dark at 4°C.

5.6 Infection of Suspension Hi5 Insect Cells for Protein Expression

Hi5 cells need to be maintained in HyQ-SFX serum-free medium, and the culture should be in the mid-log phase.

1. Count the cells (ideal cell count 1×10^6–2×10^6 cells/mL) and resuspend the expression cultures at a final density of 2×10^6 cells/mL. For expression in 1-L scale, 2×10^9 cells are needed.
2. Spin the cells down at 500 $\times$ *g* for 5 min.
3. Mix 10 mL each of the P3 virus for ATG5 and ATG16L1 (1–69).

4. After the cells have been centrifuged, aspirate and discard the supernatant. Resuspend the cells at a density of 1×10^7 cells/mL (200 mL in total). Return the cells to the growth flasks. Add the virus mixture and incubate for 1–2 h at 27°C, with shaking at 75–100 rpm.
5. After 1–2 h, add medium to bring the cells to a density of 2×10^6 cells/mL and incubate at 27°C, with shaking at 155 rpm.
6. After 3 days of expression, harvest the cells by centrifugation and isolate the protein.

5.7 Purification of the ATG5–ATG16L1 (1–69) Complex

1. Resuspend the cell pellets in 80 mL (per liter culture) of the lysis buffer containing 25 m*M* Tris, pH 8.0, 500 m*M* NaCl, 5 m*M* BME, 20 mg/L aprotinin, and10 mg/L leupeptin. Add PMSF after resuspension to a final concentration of 2.5 m*M*.
2. Transfer the suspension into Falcon tubes with 15–20 mL in each tube. Sonicate each tube two times for 15 s each for insect cells.
3. After sonication, transfer the lysate into cold Oakridge tubes for clarification. Centrifuge for 40 min at 32,000 × *g*.
4. Equilibrate 8 mL of Ni NTA beads (Sigma) with the lysis buffer.
5. Load the supernatant onto the preequilibrated NTA beads slowly at 4°C.
6. Wash the beads twice with 10 CV of wash buffer containing 25 m*M* Tris, pH 8.0, 500 m*M* NaCl, and 5 m*M* BME.
7. Elute the protein with 6 CV of elution buffer containing 25 m*M* Tris, pH 8.0, 500 m*M* NaCl, 200 m*M* imidazole, and 5 m*M* BME. Collect elution fractions for every CV and check the purity of each fraction by SDS-PAGE.
8. Pool the fractions containing target protein and add TEV to cleave the His-tag from ATG5 and HisMBP-tag from ATG16L1 (1–69) with a mass ratio of 1:50. Then, dialyze overnight into buffer containing 25 m*M* Tris, pH 8.0, 200 m*M* NaCl, 5 m*M* BME at 4°C.
9. Check cleavage efficiency by SDS-PAGE. The cleaved protein should be further purified by anion-exchange chromatography (5 mL HiTrap Q HP, GE Healthcare Life Sciences, Marlborough, MA) and gel filtration (Superdex 200 10/300 GL, GE Healthcare Life Sciences, Marlborough, MA) into a final buffer of 20 m*M* Tris, pH 8.5, 50 m*M* NaCl, and 10 m*M* DTT.
10. Concentrate the complex to 18.5 mg/mL, divide into aliquots, flash freeze, and store at −80°C until further use.

5.8 Crystallization of the ATG5–ATG16L1 (1–69) Complex

1. Crystals of a wild-type complex of ATG5–ATG16L1 were reported elsewhere (Kim et al., 2015). To crystallize the version with the ATG5 E122D patient-derived mutations (Kim et al., 2016), add 1 mL of the well solution containing 37.5 m*M* MES, pH 5.2–5.8, 0.2 *M* sodium tartrate, and 11–13% PEG 3350 to each well of the 24-well plate.
2. To obtain the initial ATG5 E122D–ATG16L1 (1–69) crystal, mix 1 μL of 18.5 mg/mL ATG5 E122D–ATG16L1 (1–69) complex in buffer containing 20 m*M* Tris, pH 8.5, 50 m*M* NaCl, 10 m*M* DTT with 1 μL of the well solution by using hanging-drop vapor diffusion at 4°C.
3. To improve crystal quality, use microseeding. Mix drops containing microcrystals by using the well solution as the seeding solution.
4. After vortexing, serially dilute the seeding solution from 10^{-1}- to 10^{-6}-fold, and add 500 nL to the preequilibrated mixture of 1 μL of ATG5 E122D–ATG16L1 (1–69) and 1 μL of the well solution of 40 m*M* MES, pH 5.5, 0.2 *M* sodium tartrate, and 8.5% PEG 3350.
5. Cryoprotect the crystals in the well solution supplemented with 25% xylitol and flash freeze in liquid nitrogen before data collection.

ACKNOWLEDGMENT

This study was supported by ALSAC, HHMI, NIH R01-GM077053, and American Heart Association 14POST19890021 (Y.Q.).

REFERENCES

Barrett, J. C., Hansoul, S., Nicolae, D. L., Cho, J. H., Duerr, R. H., Rioux, J. D., et al. (2008). Genome-wide association defines more than 30 distinct susceptibility loci for Crohn's disease. *Nature Genetics*, *40*, 955–962.

Bavro, V. N., Sola, M., Bracher, A., Kneussel, M., Betz, H., & Weissenhorn, W. (2002). Crystal structure of the GABA(A)-receptor-associated protein, GABARAP. *EMBO Reports*, *3*, 183–189.

Geng, J., & Klionsky, D. J. (2008). The Atg8 and Atg12 ubiquitin-like conjugation systems in macroautophagy. 'Protein modifications: Beyond the usual suspects' review series. *EMBO Reports*, *9*, 859–864.

Hain, A. U., Weltzer, R. R., Hammond, H., Jayabalasingham, B., Dinglasan, R. R., Graham, D. R., et al. (2012). Structural characterization and inhibition of the Plasmodium Atg8-Atg3 interaction. *Journal of Structural Biology*, *180*, 551–562.

Hampe, J., Franke, A., Rosenstiel, P., Till, A., Teuber, M., Huse, K., et al. (2007). A genome-wide association scan of nonsynonymous SNPs identifies a susceptibility variant for Crohn disease in ATG16L1. *Nature Genetics*, *39*, 207–211.

Hong, S. B., Kim, B. W., Kim, J. H., & Song, H. K. (2012). Structure of the autophagic E2 enzyme Atg10. *Acta Crystallographica Section D, Biological Crystallography*, *68*, 1409–1417.

Hong, S. B., Kim, B. W., Lee, K. E., Kim, S. W., Jeon, H., Kim, J., et al. (2011). Insights into noncanonical E1 enzyme activation from the structure of autophagic E1 Atg7 with Atg8. *Nature Structural & Molecular Biology, 18*, 1323–1330.

Hu, C., Zhang, X., Teng, Y. B., Hu, H. X., & Li, W. F. (2010). Structure of autophagy-related protein Atg8 from the silkworm Bombyx mori. *Acta Crystallographica Section F, Structural Biology and Crystallization Communications, 66*, 787–790.

Hurley, J. H., & Schulman, B. A. (2014). Atomistic autophagy: The structures of cellular self-digestion. *Cell, 157*, 300–311.

Jiang, P., & Mizushima, N. (2014). Autophagy and human diseases. *Cell Research, 24*, 69–79.

Kaiser, S. E., Mao, K., Taherbhoy, A. M., Yu, S., Olszewski, J. L., Duda, D. M., et al. (2012). Noncanonical E2 recruitment by the autophagy E1 revealed by Atg7-Atg3 and Atg7-Atg10 structures. *Nature Structural & Molecular Biology, 19*, 1242–1249.

Kim, J. H., Hong, S. B., Lee, J. K., Han, S., Roh, K. H., Lee, K. E., et al. (2015). Insights into autophagosome maturation revealed by the structures of ATG5 with its interacting partners. *Autophagy, 11*, 75–87.

Kim, M., Sandford, E., Gatica, D., Qiu, Y., Liu, X., Zheng, Y., et al. (2016). Mutation in ATG5 reduces autophagy and leads to ataxia with developmental delay. *eLife, 5*, e12245.

Klionsky, D. J., & Schulman, B. A. (2014). Dynamic regulation of macroautophagy by distinctive ubiquitin-like proteins. *Nature Structural & Molecular Biology, 21*, 336–345.

Knight, D., Harris, R., McAlister, M. S., Phelan, J. P., Geddes, S., Moss, S. J., et al. (2002). The X-ray crystal structure and putative ligand-derived peptide binding properties of gamma-aminobutyric acid receptor type A receptor-associated protein. *The Journal of Biological Chemistry, 277*, 5556–5561.

Koopmann, R., Muhammad, K., Perbandt, M., Betzel, C., & Duszenko, M. (2009). Trypanosoma brucei ATG8: Structural insights into autophagic-like mechanisms in protozoa. *Autophagy, 5*, 1085–1091.

Kraft, C., & Martens, S. (2012). Mechanisms and regulation of autophagosome formation. *Current Opinion in Cell Biology, 24*, 496–501.

Kumanomidou, T., Mizushima, T., Komatsu, M., Suzuki, A., Tanida, I., Sou, Y. S., et al. (2006). The crystal structure of human Atg4b, a processing and de-conjugating enzyme for autophagosome-forming modifiers. *Journal of Molecular Biology, 355*, 612–618.

Levine, B., & Klionsky, D. J. (2004). Development by self-digestion: Molecular mechanisms and biological functions of autophagy. *Developmental Cell, 6*, 463–477.

Levy, J., Cacheux, W., Bara, M. A., L'Hermitte, A., Lepage, P., Fraudeau, M., et al. (2015). Intestinal inhibition of Atg7 prevents tumour initiation through a microbiome-influenced immune response and suppresses tumour growth. *Nature Cell Biology, 17*, 1062–1073.

Ma, P., Schillinger, O., Schwarten, M., Lecher, J., Hartmann, R., Stoldt, M., et al. (2015). Conformational polymorphism in autophagy-related protein GATE-16. *Biochemistry, 54*, 5469–5479.

Martin, L. J., Gupta, J., Jyothula, S. S., Butsch Kovacic, M., Biagini Myers, J. M., Patterson, T. L., et al. (2012). Functional variant in the autophagy-related 5 gene promotor is associated with childhood asthma. *PloS One*, 7, e33454.

Metlagel, Z., Otomo, C., Takaesu, G., & Otomo, T. (2013). Structural basis of ATG3 recognition by the autophagic ubiquitin-like protein ATG12. *Proceedings of the National Academy of Sciences of the United States of America, 110*, 18844–18849.

Mizushima, N. (2007). Autophagy: Process and function. *Genes & Development, 21*, 2861–2873.

Mizushima, N., Yoshimori, T., & Ohsumi, Y. (2011). The role of Atg proteins in autophagosome formation. *Annual Review of Cell and Developmental Biology, 27*, 107–132.

Nakatogawa, H., Suzuki, K., Kamada, Y., & Ohsumi, Y. (2009). Dynamics and diversity in autophagy mechanisms: Lessons from yeast. *Nature Reviews Molecular Cell Biology*, *10*, 458–467.

Noda, N. N., Fujioka, Y., Hanada, T., Ohsumi, Y., & Inagaki, F. (2013). Structure of the Atg12-Atg5 conjugate reveals a platform for stimulating Atg8-PE conjugation. *EMBO Reports*, *14*, 206–211.

Noda, N. N., Satoo, K., Fujioka, Y., Kumeta, H., Ogura, K., Nakatogawa, H., et al. (2011). Structural basis of Atg8 activation by a homodimeric E1, Atg7. *Molecular Cell*, *44*, 462–475.

Ohsumi, Y., & Mizushima, N. (2004). Two ubiquitin-like conjugation systems essential for autophagy. *Seminars in Cell & Developmental Biology*, *15*, 231–236.

Otomo, C., Metlagel, Z., Takaesu, G., & Otomo, T. (2013). Structure of the human ATG12~ATG5 conjugate required for LC3 lipidation in autophagy. *Nature Structural & Molecular Biology*, *20*, 59–66.

Poon, A., Eidelman, D., Laprise, C., & Hamid, Q. (2012). ATG5, autophagy and lung function in asthma. *Autophagy*, *8*, 694–695.

Rioux, J. D., Xavier, R. J., Taylor, K. D., Silverberg, M. S., Goyette, P., Huett, A., et al. (2007). Genome-wide association study identifies new susceptibility loci for Crohn disease and implicates autophagy in disease pathogenesis. *Nature Genetics*, *39*, 596–604.

Rogov, V. V., Suzuki, H., Fiskin, E., Wild, P., Kniss, A., Rozenknop, A., et al. (2013). Structural basis for phosphorylation-triggered autophagic clearance of Salmonella. *The Biochemical Journal*, *454*, 459–466.

Satoo, K., Noda, N. N., Kumeta, H., Fujioka, Y., Mizushima, N., Ohsumi, Y., et al. (2009). The structure of Atg4B-LC3 complex reveals the mechanism of LC3 processing and delipidation during autophagy. *The EMBO Journal*, *28*, 1341–1350.

Schulman, B. A., & Harper, J. W. (2009). Ubiquitin-like protein activation by E1 enzymes: The apex for downstream signalling pathways. *Nature Reviews Molecular Cell Biology*, *10*, 319–331.

Sugawara, K., Suzuki, N. N., Fujioka, Y., Mizushima, N., Ohsumi, Y., & Inagaki, F. (2004). The crystal structure of microtubule-associated protein light chain 3, a mammalian homologue of Saccharomyces cerevisiae Atg8. *Genes to Cells*, *9*, 611–618.

Sugawara, K., Suzuki, N. N., Fujioka, Y., Mizushima, N., Ohsumi, Y., & Inagaki, F. (2005). Structural basis for the specificity and catalysis of human Atg4B responsible for mammalian autophagy. *The Journal of Biological Chemistry*, *280*, 40058–40065.

Suzuki, H., Tabata, K., Morita, E., Kawasaki, M., Kato, R., Dobson, R. C., et al. (2014). Structural basis of the autophagy-related LC3/Atg13 LIR complex: Recognition and interaction mechanism. *Structure*, *22*, 47–58.

Suzuki, N. N., Yoshimoto, K., Fujioka, Y., Ohsumi, Y., & Inagaki, F. (2005). The crystal structure of plant ATG12 and its biological implication in autophagy. *Autophagy*, *1*, 119–126.

Taherbhoy, A. M., Tait, S. W., Kaiser, S. E., Williams, A. H., Deng, A., Nourse, A., et al. (2011). Atg8 transfer from Atg7 to Atg3: A distinctive E1-E2 architecture and mechanism in the autophagy pathway. *Molecular Cell*, *44*, 451–461.

Yamada, Y., Suzuki, N. N., Hanada, T., Ichimura, Y., Kumeta, H., Fujioka, Y., et al. (2007). The crystal structure of Atg3, an autophagy-related ubiquitin carrier protein (E2) enzyme that mediates Atg8 lipidation. *The Journal of Biological Chemistry*, *282*, 8036–8043.

Yamaguchi, M., Matoba, K., Sawada, R., Fujioka, Y., Nakatogawa, H., Yamamoto, H., et al. (2012). Noncanonical recognition and UBL loading of distinct E2s by autophagy-essential Atg7. *Nature Structural & Molecular Biology*, *19*, 1250–1256.

CHAPTER FOURTEEN

Dynamics of Atg5–Atg12–Atg16L1 Aggregation and Deaggregation☆

P. Kharaziha, T. Panaretakis[1]
Cancer Centrum Karolinska, Karolinska Institutet and University Hospital, Stockholm, Sweden
[1]Corresponding author: e-mail address: theoharis.panaretakis@ki.se

Contents

1. Introduction 248
   1.1 *Atg5–Atg12–Atg16L1* System 248
2. Experimental Conditions 250
   2.1 Model Systems 250
   2.2 Generation of Stable Cell Lines 250
3. *Atg5–Atg12–Atg16L1* Reporter Systems 251
   3.1 Immunoprecipitation 251
   3.2 Proximity Ligation Assays 253
4. Conclusion 254
References 255

Abstract

Macroautophagy is a physiological process that is implicated in various pathological conditions, including neurodegenerative diseases and cancer. The execution of canonical autophagy is regulated by a core signaling cascade and it involves two well-characterized, ubiquitin-like conjugation systems—the *Atg5/Atg12/Atg16L1* and the *Atg8*-phosphatidyl ethanolamine (PE), which are both catalyzed by *Atg7*. The conjugation of *Atg5–Atg12* and the subsequent interaction with the positive regulator Atg16L1 are essential for the conjugation of *Atg8* to PE and the subsequent formation of autophagosomes. The interaction between *Atg5–Atg12* complex and *Atg16L1* is highly dynamic, induced upon activation of the autophagic process, and required for the recruitment of the *At5–Atg12* complex to sites of autophagosome formation. Monitoring the *Atg5–Atg12–Atg16L1* aggregation and deaggregation may be used not only as means to study the dynamics of autophagy, but in another important point, it may provide important insights on the basic molecular mechanisms of autophagy in physiological and pathological settings. In this chapter, we describe methods of monitoring the *Atg5–Atg12–Atg16L1* aggregation and deaggregation, with emphasis on prostate cancer.

☆The authors declare no conflict of interest.

Methods in Enzymology, Volume 587
ISSN 0076-6879
http://dx.doi.org/10.1016/bs.mie.2016.09.059

ABBREVIATIONS

Baf. A1 bafilomycin A1
BSA bovine serum albumin
CQ chloroquine
DAPI 4',6-diamidino-2-phenylindole
DTT 1,4-dithiothreitol
EBSS Earle's balanced salt solution
EGF epidermal growth factor
FCS fetal calf serum
G418 geneticin
GFP green fluorescent protein
HCl Hydrochloric acid
HRP horseradish peroxidase
LDS lithium dodecyl sulfate
PBS phosphate-buffered saline
PE phosphatidyl ethanolamine
PMSF phenylmethylsulfonyl fluoride
PVDF Polyvinylidene difluoride
RIPA buffer Radioimmunoprecipitation assay buffer
ROS Reactive oxygen species
RPMI 1640 Roswell Park Memorial Institute medium
SDS sodium dodecyl sulfate
SFM serum-free medium
TBS Tris-buffered saline

1. INTRODUCTION

1.1 *Atg5–Atg12–Atg16L1* System

Autophagy is a physiological process that facilitates the removal of damaged organelles and protein aggregates (Boya, Reggiori, & Codogno, 2013). Molecular or chemical perturbation of this signaling cascades leads to the development of several pathological conditions, including neurodegenerative and inflammatory diseases and carcinogenesis (Levine & Kroemer, 2008). Particularly under conditions of stress, metabolic, bioenergetic, hypoxic, or cytotoxicity, autophagy serves as a cell defense mechanism to alleviate the insult and promote the survival of the cell and the organism. In cancer, autophagy plays both a tumor-suppressing and a tumor-promoting role (Galluzzi et al., 2015). In the early stages of carcinogenesis, autophagy functions as a tumor-suppressing mechanism, aiming to remove putative mutagenic damage (e.g., ROS generated by damaged mitochondria). In the later stages of tumor growth, when nutrient depletion due to deficient vascularization of the tumor, the ensuing metabolic and bioenergetic stress are counteracted by the autophagic degradation and recycling of intracellular organelles (Galluzzi et al., 2015).

Autophagosomes are the key organelles executing the autophagic program (Boya et al., 2013). Formation of the autophagosomes that either engulfs damaged organelles, protein aggregates, or cytoplasmic content destined for recycling is carefully orchestrated and regulated. The execution of autophagy requires the activity of two ubiquitin-like conjugation systems. Both systems require the E1 activation enzyme *Atg7* and the E2 conjugation enzymes *Atg3* and *Atg10*. The first ubiquitin-like conjugation system catalyzes the conjugation of phosphatidyl ethanolamine (PE) to *Atg8* and in the second, the conjugation of *Atg12–Atg5* takes place. Both events are critical for efficient phagophore formation. The formation of *Atg8*-PE is a key signaling event in the formation of the autophagosomes and is critical for the tethering of *Atg8* to the phagophore membrane and the formation of stable autophagosomes in the canonical autophagy (Hamasaki, Shibutani, & Yoshimori, 2013; Mizushima, Yoshimori, & Ohsumi, 2011). In addition, *Atg8* functions as an adaptor for substrates that are targeted for degradation in the autophagolysomes. The complex *Atg5–Atg12* exhibits E3 ligase activity and the structure of the complex is critical for *Atg8* lipidation in autophagy (Otomo, Metlagel, Takaesu, & Otomo, 2013). The activity of this complex is regulated by a noncovalent interaction with *Atg16L1* and this triad is required for the localization of this complex in the phagophore (Hamasaki et al., 2013; Mizushima et al., 2011). The main function of *Atg12–Atg5–Atg16L1* is to facilitate the conjugation of PE to the activated *Atg8* in vivo (Carlsson & Simonsen, 2015). In addition, this complex determines the site of interaction of the *Atg8*-PE with the phagophore membrane.

In this chapter, we describe two alternative protocols for monitoring the dynamics of *Atg5–Atg12–Atg16L1* aggregation in cancerous and noncancerous cell lines. We used PZ-HPV-7 (ATCC® CRL-2221™) as the noncancer cell line and DU145 (ATCC® HTB-81™), 22Rv1 (ATCC® CRL-2505™), and PC-3 (ATCC® CRL-1435™) as prostate cancer cell lines. The DU145 cell line has no endogenous *Atg5*, so it is a good model system to compare the complex with and without *Atg5*. We use Earle's balanced salt solution (EBSS) and rapamycin for induction of autophagy and chloroquine (CQ), and bafilomycin A1 (Baf. A1) and leupeptin as inhibitors of autophagy. We monitor the dynamics of *Atg5–Atg12–Atg16L1* aggregation/deaggregation by using two different approaches—namely, immunoprecipitation and proximity ligation assays.

2. EXPERIMENTAL CONDITIONS

2.1 Model Systems

1. Warm the required media to 37°C in a water bath: complete RPMI growth medium (catalog number: 11875085; Thermo Fisher Scientific, 10% FCS, 1 × Pen-Strep), complete Keratinocyte-SFM (catalog number: 17005-042; Thermo Fisher Scientific, supplemented with 25 μg/mL bovine pituitary extract and 0.15 ng/mL EGF), phosphate-buffered saline (PBS) and 1 × trypsin– EDTA solution.
2. Use complete RPMI for DU145, PC-3, and 22Rv1.
3. Use complete Keratinocyte-SFM to culture PZ-HPV-7.
4. Remove the growth medium.
5. Wash the cells with 10 mL of warm PBS.
6. Trypsinize the cells for 1 min in 2.5 mL of trypsin–EDTA solution.
7. Monitor the detachment of cells with an inverted microscope.
8. Add 5 mL of complete RPMI to inactivate trypsin.
9. Count the cells and put 1×10^6 cells in 75-cm^2 tissue culture flasks.
10. When the confluency of cells reaches 90%, repeat these steps.

2.2 Generation of Stable Cell Lines

In this protocol, we describe a method to establish DU145 expressing *Atg5*, and DU145 expressing *Atg5-GFP*.

1. Put 2×10^5 DU145 cells per well in a six-well plate.
2. After 24 h, the cells are usually ready for transfection.
3. Dilute 2 μg of the required plasmid (e.g., pEGFP-C1-hApg5) in 200 μL of Opti-MEM™ (catalog number: 31985062; Thermo Fisher Scientific).
4. Dilute 5 μL of Lipofectamine 2000 (catalog number: 11668030; Thermo Fisher Scientific) in 200 μL of Opti-MEM™.
5. Add the diluted plasmid to the Lipofectamine solution and wait 5 min.
6. Remove the RPMI media from the wells and add 400 μL of solution 5 to each well.
7. Wait 30 s and add 2 mL of complete RPMI to each well and incubate at 37°C.
8. Change the RPMI media with fresh RPMI media containing 200 μg/mL G418 after 24 h.
9. Replace media daily for 1 week with RPMI media containing 200 μg/mL G418 and monitor cell growth.
10. Put one cell per well in a 96-well plate and mark wells containing only one cell. Mark at least 50 wells.

11. Monitor the marked wells after 72 h, and if the cells have grown to 70–80% confluency, transfer them to a 24-well plate.
12. When the cells reach 90% confluence, lyse half of them and transfer the rest to a six-well plate. Perform this for at least 20 clones.
13. Lyse the cells using RIPA Lysis Buffer System (catalog number: sc-24948; Santa Cruz Biotechnology) containing 1:100 of PMSF solution, 1:100 sodium orthovanadate solution, and 1:100 protease inhibitor cocktail solution.
14. Measure protein concentration.
15. Load 30 μg of protein for each clone, run PC-3 cell lysate as a positive control, and DU145 cell lysate as a negative control on Bis–Tris gel.
16. Use anti-*Atg5* antibody (catalog number: 2630; Cell Signaling technology).
17. Choose the appropriate, stably transfected clones, make a stock of at least five vials, and store them in the liquid nitrogen tank.

3. *ATG5–ATG12–ATG16L1* REPORTER SYSTEMS

3.1 Immunoprecipitation

1. Put 3×10^5 PZ-HPV-7, DU145 (wildtype, *Atg5* null), DU145 (stable expressing *Atg5*), PC-3, and 22Rv1 cells in each well of a six-well plate and add 2 mL of the appropriate culture medium to each well.
2. Remove the media after 24 h and add 2 mL of the appropriate medium to the control well, 2 mL of EBSS, and 2 mL of the appropriate medium containing either rapamycin (catalog number: R8781; Sigma-Aldrich) at 1 μ*M*, CQ (catalog number: PHR1258; Sigma-Aldrich) at 10 μ*M*, Baf. A1 at 50 n*M* (catalog number: B1793; Sigma-Aldrich) and leupeptin (catalog number: L5793; Sigma-Aldrich).
3. Harvest cells by adding trypsin–EDTA after 24 h and then inactivate them by adding media containing 10% FBS. Work on ice.
4. Centrifuge at 1500 rpm for 5 min at 4°C and remove the supernatant.
5. Wash the cells once with ice-cold PBS.
6. Lyse the cells by adding 100 μL of low-salt lysis buffer [1% Igepal CA630 (NP-40) and 50 m*M* Tris–HCl (pH 8.0), containing a 1:100 protein inhibitor]. Keep on ice for 30 min.
7. Centrifuge for 10 min at 12,000 × *g* at 4°C to the sediment, remove the cell debris, and store the supernatant.
8. Measure the protein concentration of the supernatant.
9. Shake the μMACS™ Protein A/G MicroBeads bottles well.

10. Use 100 μL of beads for each sample.
11. Wash beads with 0.5 mL of low-salt lysis buffer once for 10 min at 2000 × *g* at 4°C.
12. Add 200 μg protein to 50 μL of beads for each sample and shake it for 1 h at 250 rpm at room temperature.
13. Centrifuge samples for 10 min at 2000 × *g* at 4°C.
14. Keep the supernatant and discard the beads (prewash step).
15. Add 2 μg of the *Atg5* antibody to 50 μL of beads and add it to the supernatant from step 14.
16. Incubate overnight on a shaking platform at 250 rpm, at 4°C.
17. Place a 20-μ column in the μMACS™ separator and a suitable waste container under the 20-μ column. Rinse the 20-μ column by 200 μL of low-salt lysis buffer.
18. Apply the cell lysate onto the column and let the lysate run through. Magnetically labeled protein is retained in the 20-μ columns.
19. Rinse columns four times with 250 μL of low-salt lysis buffer.
20. Rinse columns once with 100 μL of low-salt wash buffer and wait for the buffer to go through each column.
21. Detach each column from the magnet and put it in a clean tube.
22. Elute the retained proteins in the columns by adding 50 μL of preheated (95°C) 1 × sodium dodecyl sulfate (SDS) gel loading buffer (1 ×): 50 m*M* Tris–HCl (pH 6.8), 50 m*M* DTT, 1% SDS, 0.005% bromphenol blue, and 10% glycerol.
23. Mix 20 μL of elute with 2.5 μL of reducing buffer and 6.25 μL of LDS sample buffer, and then boil the mixture for 10 min at 95°C.
24. Run it on a 12% Bis–Tris gel with an MES running buffer.
25. Transfer the proteins to PDVF membrane with 40 V, 75 min.
26. Block the membrane using 5% bovine serum albumin (BSA).
27. Incubate the membrane overnight at 4°C with 1 μg/mL of mouse anti-*Atg16L1* antibody (ab90442, Abcam) diluted in BSA.
28. Wash the membrane three times with TBS–Tween 20.
29. Incubate the membrane with anti-mouse horseradish peroxidase (HRP)-conjugated antibody for 1 h at room temperature on a shaker at 200 rpm.
30. Incubate the membrane overnight at 4°C with 1 μg/mL goat anti-*Atg12* antibody (AB0083-200, Sicgen) diluted in BSA and after washing with an anti-goat HRP-conjugated antibody.
31. Develop the Western blotting with X-ray film.

3.2 Proximity Ligation Assays

1. Sterilize the coverslips with a steam autoclave.
2. Grow 25,000 DU145 cells, which stably express *Atg5*-GFP on the sterile coverslips overnight in a 24-well plate.
3. After 24 h, remove the media and add 1 mL of complete RPMI to the control wells, 1 mL of EBSS, and 1 mL of the appropriate media containing either rapamycin (Sigma-Aldrich) at 1 μ*M*, CQ (Sigma-Aldrich) at 10 μ*M*, Baf. A1 at 50 n*M* (Sigma-Aldrich), and leupeptin at 10 μ*M*.
4. Remove the media after 24 h.
5. Wash the coverslips three times with PBS for 5 min, at room temperature on a shaker at 100 rpm. Be careful not to detach the cells from the coverslips.
6. Add 250 μL of freshly made 4% paraformaldehyde to PBS and incubate for 10 min at room temperature.
7. Wash the coverslips three times with PBS for 5 min, at room temperature on a shaker at 100 rpm.
8. Add digitonin diluted in PBS (10 ng/mL) for 10 min to permeabilize the cells.
9. Add 200 μL of blocking buffer (blocking buffer: 2% BSA, 0,2% Tween 20, 5% glycerol, 0.02% sodium azide in PBS, followed by filtration) and incubate for 1 h at RT.
10. Remove the blocking buffer by using an aspirator.
11. Add a goat anti-*Atg12* antibody (AB0083-200; Sicgen) and a mouse anti-*Atg16L1* antibody (ab90442, Abcam) diluted 1/100 in the blocking buffer and incubate for 1 h at RT.
12. Calculate for 40 μL of total reaction for each sample. Use the same blocking buffer to mix and dilute the Duolink® In Situ PLA® Probe Anti-Goat MINUS and Duolink® In Situ PLA® Probe Anti-Mouse PLUS probes at a 1:5 ratio. Incubate the mixture for 20 min at room temperature. For example, use 8 μL of PLA probe MINUS stock, 8 μL of PLA probe PLUS stock, and 24 μL of the blocking buffer for a 40-μL reaction.
13. Aspirate the primary antibody from the wells of the 24-well plate.
14. Wash the coverslips three times with PBS for 5 min, at room temperature on a shaker at 100 rpm.
15. Add the diluted PLA probe solution.
16. Incubate the slides in a humidity chamber at 37°C for 1 h.

17. Dilute the Ligation stock at a 1:5 ratio in high-purity water.
18. Aspirate the PLA probe solution from the slides.
19. Wash the coverslips three times with PBS for 5 min, at room temperature on a shaker at 100 rpm.
20. Calculate and add 40 μL of ligase solution for each sample. Add 8 μL of the 5 × Ligation stock and 31 μL of high-purity water, plus 1 μL of Ligase for each 40-μL reaction. Use Duolink® In Situ Detection Reagents FarRed.
21. Incubate the slides in a humidity chamber at 37°C for 1 h.
22. Dilute the Amplification stock to 1:5 in high-purity water and mix.
23. Aspirate the Ligation solution from the slides.
24. Wash the slides twice in wash buffer A for 2 min under gentle agitation. Aspirate the wash solution after the last washing.
25. Calculate and add 40 μL of ligase solution for each sample. Add 8 μL of the 5 × Amplification stock and 31.5 μL of high-purity water, plus 0.5 μL of polymerase for each 40 μL.
26. Incubate the slides in a humidity chamber for 100 min at 37°C.
27. Aspirate the amplification-polymerase solution from the slides and wash the slides twice in 1 × wash buffer B for 10 min.
28. Wash the slides in 0.01 × wash buffer B for 1 min.
29. Dry the slides at room temperature in the dark.
30. Mount the slides using Duolink In Situ Mounting Medium with DAPI.
31. Wait for 15 min and capture the image by using a confocal fluorescence microscope.
32. Complex of *Atg5–Atg12* and *Atg16L1* will appear in yellow. *Atg12–Atg16L1* without *Atg5* will appear in red.

4. CONCLUSION

In this chapter, we explain a couple of experimental techniques that can be used to measure the *Atg5–Atg12–Atg16L1* aggregation–deaggregation process during the autophagic program. Measuring the dynamics of the formation and the dissociation of this complex may provide important insights into the regulation of autophagy and also allow the discovery of additional molecules that are important for the execution of the autophagic process under physiological and pathological conditions. Furthermore, these high-throughput methods may be used during drug screening programs for the identification of new molecules that may disrupt the autophagic process and thereby inhibit it. It is expected that as more knowledge is acquired about the autophagic process, additional technological advancements will allow the live detection of this important regulatory step in the execution of the autophagic program.

REFERENCES

Boya, P., Reggiori, F., & Codogno, P. (2013). Emerging regulation and functions of autophagy. *Nature Cell Biology*, *15*(7), 713–720.

Carlsson, S. R., & Simonsen, A. (2015). Membrane dynamics in autophagosome biogenesis. *Journal of Cell Science*, *128*(2), 193–205.

Galluzzi, L., Pietrocola, F., Bravo-San Pedro, J. M., Amaravadi, R. K., Baehrecke, E. H., Cecconi, F., et al. (2015). Autophagy in malignant transformation and cancer progression. *The EMBO Journal*, *34*(7), 856–880.

Hamasaki, M., Shibutani, S. T., & Yoshimori, T. (2013). Up-to-date membrane biogenesis in the autophagosome formation. *Current Opinion in Cell Biology*, *25*(4), 455–460.

Levine, B., & Kroemer, G. (2008). Autophagy in the pathogenesis of disease. *Cell*, *132*(1), 27–42.

Mizushima, N., Yoshimori, T., & Ohsumi, Y. (2011). The role of Atg proteins in autophagosome formation. *Annual Review of Cell and Developmental Biology*, *27*, 107–132.

Otomo, C., Metlagel, Z., Takaesu, G., & Otomo, T. (2013). Structure of the human ATG12∼ATG5 conjugate required for LC3 lipidation in autophagy. *Nature Structural & Molecular Biology*, *20*(1), 59–66.

CHAPTER FIFTEEN

Fluorescent FYVE Chimeras to Quantify PtdIns3P Synthesis During Autophagy

S.M.S. Yakhine-Diop[*,†], G. Martínez-Chacón[*,†], R.A. González-Polo[*,†], J.M. Fuentes[*,†,1], M. Niso-Santano[*,†,1]

[*]Centro de Investigación Biomédica en Red en Enfermedades Neurodegenerativas (CIBERNED), Universidad de Extremadura, Cáceres, Spain

[†]Facultad de Enfermería y Terapia Ocupacional, Universidad de Extremadura, Cáceres, Spain

[1]Corresponding authors: e-mail address: jfuentes@unex.es; mnisosan@unex.es

Contents

1. Introduction 258
2. Cell Culture Conditions 261
3. Transient Transfection of Plasmid 262
4. Autophagy Assay Conditions 264
5. Detection by Fluorescence Microscopy 264
6. Quantification of PtdIns3P Puncta Formation 265
7. Concluding Remarks 265
8. Notes 266

Acknowledgments 267

References 267

Abstract

Autophagy is the major cellular process of degradation and is modulated by several signaling pathways. Phosphatidylinositol 3-kinase (PtdIns3K) class III (Vps34) and PtdIns3K class I regulate the autophagy pathway positively and negatively, respectively. Both classes of PtdIns3K participate in the synthesis of phosphatidylinositol 3-phosphate (PtdIns3P), which plays a crucial role in autophagosome biogenesis and membrane traffic. PtdIns3P is a membrane phospholipid that is associated with endogenous FYVE domain-containing proteins. Indeed, such interactions facilitate autophagosome fusion with lysosomes and subsequent cargo degradation. During starvation-induced autophagy, the expression of FYVE domain-containing proteins increases, and their binding to PtdIns3P is strengthened. Nonetheless, not all FYVE domain proteins are related to the induction of autophagy. This method report presents the quantification of PtdIns3P synthesis by using cells either transiently transfected with or stably expressing FYVE-dsRed.

Methods in Enzymology, Volume 587
ISSN 0076-6879
http://dx.doi.org/10.1016/bs.mie.2016.09.060

1. INTRODUCTION

Phosphoinositides (PtdIns) are phosphorylated derivatives of phosphatidylinositol that regulate essential cellular functions including signaling, intracellular membrane trafficking, cytoskeleton regulation and motility, and cell proliferation (Di Paolo & De Camilli, 2006). Seven PtdIns have been identified with distinct cellular localization and regulatory functions; for instance, phosphatidylinositol 4-phosphate (PtdIns4P) and phosphatidylinositol-4,5-bisphosphate (PtdIns(4,5)P_2) are enriched at the plasma membrane, on the Golgi membrane and on lysosomes, whereas phosphatidylinositol 3-phosphate (PtdIns3P) are concentrated in endosomes and in autophagic structures (Balla, 2013; Di Paolo & De Camilli, 2006; Mayinger, 2012). All these PtdIns exert their functions through the recruitment of cytosolic proteins to specific membrane compartments. PtdIns are recognized by a variety of specific domains that are present in numerous cytosolic and membrane proteins.

PtdIns3P and its effectors play a fundamental role in autophagosome biogenesis during starvation-induced autophagy (Burman & Ktistakis, 2010; Dall'Armi, Devereaux, & Di Paolo, 2013; Noda, Matsunaga, Taguchi-Atarashi, & Yoshimori, 2010; Simonsen & Tooze, 2009). Autophagy is an intracellular degradation process characterized by the formation of vesicles called autophagosomes that recycle organelles or cytosolic cargo for delivery to the lysosomes for degradation (Mizushima, 2007). The biogenesis of autophagosomes is a critical feature of autophagy, which is inducible in response to nutrient deprivation or various other forms of cellular stress (Mizushima, Levine, Cuervo, & Klionsky, 2008; Rubinsztein, Shpilka, & Elazar, 2012). The formation of autophagosomal membranes requires PtdIns3P, which serves as a platform to recruit PtdIns3P-binding proteins that control membrane docking and fusion in the final steps of autophagosome biogenesis (Rubinsztein et al., 2012; Shibutani & Yoshimori, 2014; Walker, Chandra, Manifava, Axe, & Ktistakis, 2008).

Two different pathways are involved in PtdIns3P synthesis: the phosphorylation of phosphatidylinositol at the 3-position of the inositol ring by class II or class III PtdIns 3-kinases and the dephosphorylation of PtdIns(3,4)P_2 by PtdIns3P phosphatases after phosphatidylinositol 3-kinase (PtdIns3K) class I activation (Balla, 2013; Raiborg, Schink, & Stenmark, 2013). The mammalian class I PtdIns3K has been found to regulate autophagic activity through the activation of Akt/PKB, which activates the kinase mammalian target of rapamycin (mTOR), the major negative regulator of autophagy (Mehrpour, Esclatine, Beau, & Codogno, 2010; Zoncu, Efeyan, & Sabatini, 2011).

The synthesis of PtdIns3P by class III PtdIns3K (Vps34) is crucial for the initial stage of autophagosome biogenesis (Proikas-Cezanne, Takacs, Donnes, & Kohlbacher, 2015; Stack, Horazdovsky, & Emr, 1995). Vps34 is the principal source of PtdIns3P in yeast (Schu et al., 1993; Stack et al., 1995); however, in mammalian cells, PtdIns3P can be synthesized through multiple pathways (Devereaux et al., 2013) (Fig. 1). Although Vps34 has been considered a critical regulator of autophagy, the existence of Vps34-independent autophagy has been reported (Codogno, Mehrpour, & Proikas-Cezanne, 2012). In addition to autophagosome biogenesis, PtdIns3P regulates many other aspects of autophagy, including maturation and intracellular movement of autophagosomes (Dall'Armi et al., 2013). PtdIns3P controls maturation steps of autophagy via specific recruitment of protein effectors that mediate the fusion of autophagosomes with endosomes and lysosomes. Moreover, PtdIns3P is involved in the recruitment of the cargo into the autophagosome through a scaffold protein called Alfy (Autophagy-linked FYVE) that targets cytosolic protein aggregates for autophagic degradation (Clausen et al., 2010; Filimonenko et al., 2010; Simonsen et al., 2004). PtdIns3P can also facilitate the retrograde movement of autophagosomes formed at the periphery through microtubule plus-end-directed transport (Pankiv et al., 2010).

Therefore, PtdIns3P is recognized and mediates its physiological effects at membranes via the recruitment and activation of specific protein effectors containing PtdIns3P-binding domains such as FYVE (*F*ab1p, *Y*OTB,

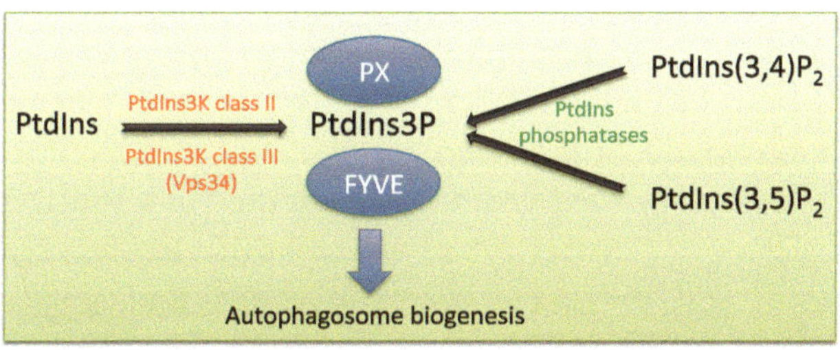

Fig. 1 Signaling pathway of PtdIns3P synthesis. The phosphorylation of PtdIns is catalyzed by class II or III PtdIns3K (in *red*). The production of PtdIns3P is regulated through the dephosphorylation of PtdIns(3,4)P$_2$ or PtdIns(3,5)P$_2$ by PtdIns phosphatases (in *green*). PtdIns3P effectors bind to PtdIns3P through PX and FYVE domains. FYVE domain-containing proteins play a role in the biogenesis of the autophagosome.

*V*acp1, *E*EA1) and PX domains (Di Paolo & De Camilli, 2006). The PX domain is not specific to PtdIns3P alone but also shows affinity for PtdIns5P, PtdIns(4,5)P_2, and PtdIns(3,4,5)P_3 (Di Paolo & De Camilli, 2006). The FYVE domain preferentially associates with PtdIns3P and contributes to endosomal or autophagosomal protein localization (Stenmark, Aasland, & Driscoll, 2002). The FYVE domain, originally identified in the proteins Fab1p, YOTB, Vac1p, and EEA1, is a zinc finger-containing protein domain composed of approximately 60–80 amino acid residues with eight conserved cysteine residues (Burd & Emr, 1998; Stenmark et al., 2002). In contrast to other membrane-binding domains, FYVE domains are well conserved. All FYVE domains harbor three conserved sequences that are crucial for PtdIns3P association: the amino terminal WxxD motif, the central HHCR motif, and the carboxy terminal RVC motif (Stenmark et al., 2002). Five FYVE domain-containing proteins have been identified in yeast and approximately 30 FYVE domain-containing proteins have been identified in human as regulating membrane dynamics and protein trafficking (Moravcevic, Oxley, & Lemmon, 2012; Stenmark et al., 2002).

Several PtdIns3P-binding FYVE domain proteins have been associated with autophagy initiation. For example, DFCP1 (double FYVE domain-containing protein 1) is localized in the endoplasmic reticulum and Golgi. During nutrient starvation, DFCP1 binds to PtdIns3P through its FYVE domains, showing the site of omegasome formation (Axe et al., 2008). However, DFCP1 is not essential for autophagy. Alfy is a FYVE domain protein that binds to PtdIns3P-containing autophagosome membranes to engulf cytoplasmic protein aggregates for their autophagic degradation. Alfy is not involved in starvation-induced autophagy (Clausen et al., 2010; Filimonenko et al., 2010; Simonsen et al., 2004). FYCO1 (FYVE and coiled-coil domain-containing protein 1), another FYVE domain protein, interacts with LC3 and Rab7 to mediate plus-end-directed microtubule transport of the autophagosome to late endosomes or lysosomes (Gutierrez, Munafo, Beron, & Colombo, 2004; Jager et al., 2004; Pankiv et al., 2010).

A FYVE domain-derived probe has been used as a PtdIns3P sensor to show the localization of intracellular PtdIns3P to endosomes, vacuoles, and autophagosomes and to characterize the contribution of the type III PtdIns3K (Axe et al., 2008; Gillooly et al., 2000; Zhang et al., 2007). To improve the sensitivity and threshold of intracellular PtdIns3P detection, several other constructs have been generated. In particular, tandem FYVE domain 2 × FYVE and 4 × FYVE constructs show high affinity through an avidity effect (Bohdanowicz, Cosio, Backer, & Grinstein, 2010; Gillooly et al., 2000). However, a high expression level of the 2 × FYVE domain can sequester PtdIns3P and disrupt PtdIns3P-dependent signaling (Byfield, Murray, & Backer, 2005).

Here, we present a protocol for quantifying the formation of PtdIns3P puncta in human cell lines using fluorescence microscopy and a PtdIns3P probe based on the FYVE domain.

2. CELL CULTURE CONDITIONS

1. We used the following human cell lines:
   a. Human neuroglioma H4 cells.
   b. Human neuroblastoma SH-SY5Y cells.
   c. Human osteosarcoma U2OS cells stably expressing FYVE-dsRed reporter (Fig. 2) were a kind gift of Dr. Junying Yuan, Harvard Medical School.
2. Cells were maintained in T150 flasks (150 cm^2) at 37°C with 5% CO_2 in Dulbecco's modified Eagle's medium (DMEM, Sigma-Aldrich, St. Louis, MO) supplemented with 10% fetal bovine serum (Sigma-Aldrich, St. Louis, MO), 2 m*M* L-glutamine (Sigma-Aldrich, St. Louis, MO), 100 U/mL Penicillin, 100 μg/mL Streptomycin (Hyclone, Logan, UT), and/or 500 μg/mL G418 (see *Note 1*).
3. Once cultured cells reached 70–85% confluence, the culture medium was discarded, and the cells were washed with prewarmed phosphate-buffered

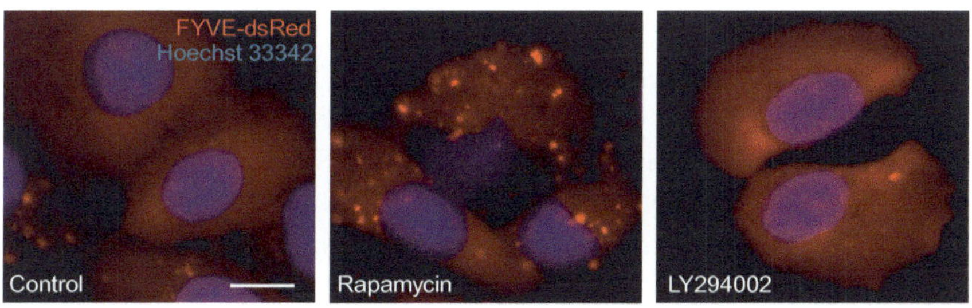

Fig. 2 PtdIns3P synthesis in U2OS cells stably expressing FYVE-dsRed. Human osteosarcoma U2OS cells stably expressing FYVE-dsRed were seeded in 24-well plates. One day later, cells were treated with or without rapamycin (10 μ*M*) or LY294002 (10 μ*M*) for 2 h. Representative immunofluorescence images were processed using an Olympus IX51 fluorescence microscope. Autophagy induction increases the expression of FYVE-dsRed puncta. Nuclei were stained with Hoechst 33342, and the scale bar represents 10 μm. Original magnification: 40×.

saline (PBS) (see *Notes 2 and 3*) and detached with trypsin/EDTA solution (Sigma-Aldrich, St. Louis, MO) for 2–5 min at room temperature (RT). The trypsin reaction was stopped by adding fresh culture medium, and cells were centrifuged for 5 min at $300 \times g$ at RT. Supernatant was removed and discarded. Cells were then resuspended in fresh medium and reseeded in a new T150 flask (see *Note 4*).

3. TRANSIENT TRANSFECTION OF PLASMID

1. One day before transfection, 5×10^4 cells were seeded on sterile coverslips (see *Note 5*) in a 24-well plate in DMEM without antibiotics (see *Note 6*). When cells reached 70–80% confluence, we performed the transient transfection of FYVE-dsRed plasmid (a kind gift of Dr. Junying Yuan of Harvard Medical School) by using two different transfection reagents (see *Notes 7–9*).
   - **a.** Lipofectamine 2000 (Invitrogen Life Technologies, Carlsbad, CA): 2 μL of Lipofectamine was diluted into 100 μL Opti-MEM (Gibco®-Thermo Fisher Scientific, Carlsbad, CA) and incubated for 15 min at RT. In parallel, we diluted 2.5 μg DNA in 100 μL Opti-MEM in a new tube. These two tubes were then mixed and incubated for 10 min at RT to form the DNA/Lipofectamine complex. Then, we replaced the culture medium with fresh medium without antibiotics and added 50 μL of the DNA/Lipofectamine complex to each well (see *Note 10*). Cells were incubated for 12–18 h at 37°C and 5% CO_2.
   - **b.** Neuromag transfection reagent (OZBioscience, San Diego, CA): DNA (2.5 μg) was diluted in 200 μL Opti-MEM. The diluted DNA was added to 2 μL Neuromag and incubated for 20 min at RT after homogenization (see *Note 11*). Subsequently, the culture medium was replaced, and 50 μL of DNA/Neuromag complex was added onto the cells and incubated for 20 min on a magnetic plate at 37°C. We removed the culture plate from the magnetic plate and incubated the cells overnight at 37°C and 5% CO_2.
2. For both transfection methods, the medium was replaced 24 h after transfection, and cells were treated at 48 h after the plasmid transfection to ensure the optimal expression of FYVE-dsRed reporter (Fig. 3A). Transfection efficiency was greater than 50% with each transfection reagent (see *Note 12*).

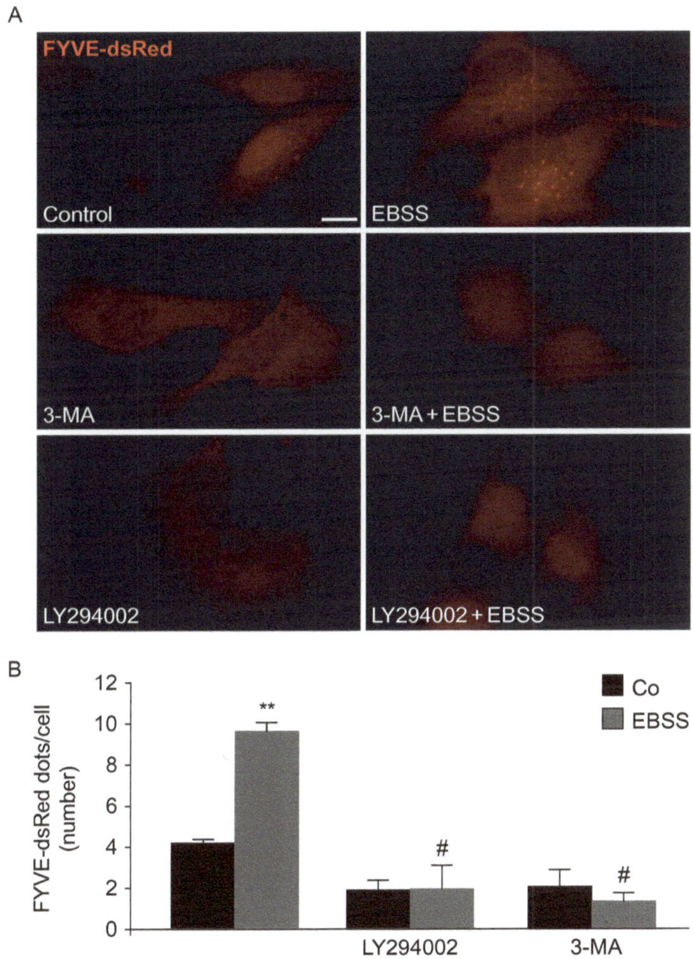

Fig. 3 Quantification of FYVE-dsRed puncta in H4 cells. (A) Representative immunofluorescence images of FYVE-dsRed dots. Neuroglioma H4 cells grown on coverslips in 24-well plate were transiently transfected with FYVE-dsRed plasmid as described in Section 3. Cells were subsequently incubated for 4 h with or without EBSS (starvation), LY294002 (10 μ*M*) (class I PI3K inhibitor), 3-MA (10 m*M*) (class III PI3K inhibitor), EBSS + 3-MA or EBSS + LY294002. (B) Quantification of dots was performed using Ifdotmeter® software (at least 200 cells per condition), and data are expressed as the mean values ± SD. $^{**}p < 0.01$ vs cells maintained in control conditions (in conformity with Student's *t* test). $^{\#}p < 0.05$ vs cells treated with EBSS (in conformity with Student's *t* test). Scale bar: 10 μm. Original magnification: 40 ×.

4. AUTOPHAGY ASSAY CONDITIONS

We carried out autophagy assays in 24-well plates for 2–4 h. Culture medium was removed, and cells were washed with PBS and treated with the following autophagy inducers and/or inhibitors (see *Note 13*):

1. Rapamycin (Sigma-Aldrich, St. Louis, MO) treatment: Rapamycin is an autophagy inducer that binds to FKBP1/FKBP2 and inhibits mTORC1. Rapamycin was dissolved in DMSO (see *Note 14*), and a final concentration of 10 μ*M* was used to induce autophagy.
2. Earle's balanced salt solution (EBSS, Sigma-Aldrich, St. Louis, MO) treatment: EBSS is a nutrient-free medium that induces classical autophagy.
3. LY294002 (Tocris Bioscience, Bristol, UK) administration: LY294002 is a potent inhibitor of PI3Ks that acts on the ATP-binding site of the enzyme. LY294002 inhibits macroautophagy (see *Notes 15 and 16*). LY294002 was dissolved in ethanol and was added at a final concentration of 10 μ*M*.
4. 3-Methyladenine (3-MA) treatment: 3-MA inhibits an early stage of autophagy through the inhibition of class III PtdIns3K. 3-MA is normally used to inhibit starvation- or rapamycin-induced autophagy (see *Notes 17 and 18*). 3-MA was dissolved in culture medium and prepared just before use at a final concentration of 10 m*M*.

5. DETECTION BY FLUORESCENCE MICROSCOPY

1. Cells were plated on glass coverslips (see *Note 19*), transfected with FYVE-dsRed plasmid, and treated with autophagy inhibitors and inducers (see Sections 3 and 4).
2. Culture medium was removed, and cells were washed once with PBS and fixed with 4% PFA supplemented with 5–10 μg/mL Hoechst 33342 for 20 min (see *Note 20*).
3. The fixation reagent was discarded, and cells were washed twice with PBS.
4. Glass coverslips were mounted onto glass slides using Fluoromount G (Southern Biotech, Birmingham, AL) (see *Note 21*).
5. Cells were visualized with a fluorescence microscope (Olympus IX51, Center Valley, PA) using a fluorophore-specific filter (see *Note 22*).

6. QUANTIFICATION OF PtdIns3P PUNCTA FORMATION

Digital images for the quantification of intracellular PtdIns3P were obtained on a fluorescence microscope and photographed with an Olympus DP71 camera (Olympus, Center Valley, PA).

To quantify dots, we used Ifdotmeter® (Rodriguez-Arribas et al., 2016) software because it is easy to use for counting dots. Several other programs, such as ImageJ or CellProfiler, may be used, depending on the user. First of all, it is very important to take images with the same settings across the treatment conditions to minimize subsequent variation analysis or misinterpretation. Further, the threshold should be selected to differentiate the background fluorescence signal, caused by the overexpression, from the FYVE-dsRed dots (see *Notes 23 and 24*). Above the threshold, dots can be selected by establishing a minimum size for pixel clusters. With Ifdotmeter®, dots are considered as groups. As dot size varies substantially between treatment conditions, it is important to determine the number of pixels that best represents the different sizes of dots in all conditions. Finally, the results can be represented either as dots per cell or as dots per area, based strictly on the spreadsheet generated from the data (Fig. 3B). The cell number can be determined based on saved images of the nuclei labeled with DAPI or Hoechst stains. For each condition, 200–500 cells were evaluated.

7. CONCLUDING REMARKS

Fluorescent FYVE chimera's probes are widely used for visualizing and quantifying intracellular PtdIns3P synthesis by fluorescence microscopy. This approach provides a rapid and simple method for general assessment of PtdIns3P production. However, it is important to note that FYVE domain proteins not only associate with PtdIns3P located on the autophagosomal membranes but also associate with PtdIns3P on membranes of other organelles such as endosomes and vacuoles. Indeed, high expression levels of the PtdIns3P probes can block PtdIns3P-dependent signaling. Therefore, additional complementary assays, such as FYVE colocalization with autophagy markers, should be conducted in order to confirm the PtdIns3P synthesis and the localization in autophagy.

8. NOTES

1. G418 is provided by Invitrogen under the name Geneticin™. G418 was added as a selection agent in U2OS FYVE-dsRed medium.
2. All reagents for cell culture were prewarmed at 37°C.
3. This step removes general debris and also facilitates cell detachment.
4. Cells should not be used for experiments within 24 h after trypsinization, to avoid cellular stress.
5. We used precleaned 12-mm round glass coverslips (Thermo Fisher Scientific, Carlsbad, CA).
6. Antibiotics may reduce the transfection efficiency.
7. The optimal ratio of transfection reagent to DNA depends on cell type and plasmid size, among other factors. The ratio should be optimized for every new cell type and plasmid combination used.
8. Transfection reagents should be added directly into the medium without touching the sides of the microtubes.
9. All the dilutions for transfection assays should be gently mixed.
10. The DNA/transfection reagent complex is added slowly dropwise.
11. Incubation time was according to the manufacturer's protocol.
12. Transfected cells were distinguished from untransfected cells by the fluorescence intensity of the fluorescent reporter protein dsRed.
13. In control conditions, cells were directly incubated with fresh medium.
14. The effects of autophagy inducers can vary significantly depending on the cell type. For instance, rapamycin may fail to activate autophagy in cultured primary neurons.
15. Most autophagy inhibitors are not specific. Appropriate inhibitory concentrations should be determined for each cell type.
16. LY294002 is a PtdIns3K inhibitor that is less potent but more stable than wortmannin.
17. 3-MA was dissolved in prewarmed DMEM medium.
18. It has been reported that 3-MA may promote autophagy by inhibition of class I PtdIns3P.
19. Coat with poly-D-lysine, if cells are not very adherent.
20. It is not necessary to mix PFA and Hoechst 33342 solutions because they can be prepared separately.
21. Wait for at least 60 min prior to storing samples at 4°C.
22. It is recommended to take images promptly because FYVE-dsRed probes tend to lose fluorescence intensity over time.

23. It is important to highlight that pH variation, due to treatment conditions or medium storage conditions, may affect the fluorescence intensity and number of PtdIns3P dots.
24. High expression levels of the probe may also disrupt the association of PtdIns3P with endogenous FYVE domain-containing proteins.

ACKNOWLEDGMENTS

J.M.F. received research support from the Ministerio de Economia y Competitividad, Spain, CIBERNED (CB06/05/004 and PI12/02280). R.A.G.-P. was supported by a "Contrato destinado a la retención y atracción del talento investigador, TA13009" from Junta de Extremadura, Spain, and she received research support from Ministerio de Economía y Competitividad, Spain (PI14/00170). M.N.-S. was supported by "Contrato Juan de la Cierva" (JCI-2012-14383) from Ministerio de Economia y Competitividad, Spain. This work was supported also by "Fondo Europeo de Desarrollo Regional" (FEDER) from European Union. The authors also thank FUNDESALUD for helpful assistance.

REFERENCES

Axe, E. L., Walker, S. A., Manifava, M., Chandra, P., Roderick, H. L., Habermann, A., ... Ktistakis, N. T. (2008). Autophagosome formation from membrane compartments enriched in phosphatidylinositol 3-phosphate and dynamically connected to the endoplasmic reticulum. *The Journal of Cell Biology*, *182*(4), 685–701. http://dx.doi.org/10.1083/jcb.200803137.

Balla, T. (2013). Phosphoinositides: Tiny lipids with giant impact on cell regulation. *Physiological Reviews*, *93*(3), 1019–1137. http://dx.doi.org/10.1152/physrev.00028.2012.

Bohdanowicz, M., Cosio, G., Backer, J. M., & Grinstein, S. (2010). Class I and class III phosphoinositide 3-kinases are required for actin polymerization that propels phagosomes. *The Journal of Cell Biology*, *191*(5), 999–1012. http://dx.doi.org/10.1083/jcb.201004005.

Burd, C. G., & Emr, S. D. (1998). Phosphatidylinositol(3)-phosphate signaling mediated by specific binding to RING FYVE domains. *Molecular Cell*, *2*(1), 157–162.

Burman, C., & Ktistakis, N. T. (2010). Autophagosome formation in mammalian cells. *Seminars in Immunopathology*, *32*(4), 397–413. http://dx.doi.org/10.1007/s00281-010-0222-z.

Byfield, M. P., Murray, J. T., & Backer, J. M. (2005). hVps34 is a nutrient-regulated lipid kinase required for activation of p70 S6 kinase. *The Journal of Biological Chemistry*, *280*(38), 33076–33082. http://dx.doi.org/10.1074/jbc.M507201200.

Clausen, T. H., Lamark, T., Isakson, P., Finley, K., Larsen, K. B., Brech, A., ... Johansen, T. (2010). p62/SQSTM1 and ALFY interact to facilitate the formation of p62 bodies/ALIS and their degradation by autophagy. *Autophagy*, *6*(3), 330–344.

Codogno, P., Mehrpour, M., & Proikas-Cezanne, T. (2012). Canonical and non-canonical autophagy: Variations on a common theme of self-eating? *Nature Reviews. Molecular Cell Biology*, *13*(1), 7–12. http://dx.doi.org/10.1038/nrm3249.

Dall'Armi, C., Devereaux, K. A., & Di Paolo, G. (2013). The role of lipids in the control of autophagy. *Current Biology*, *23*(1), R33–R45. http://dx.doi.org/10.1016/j.cub.2012.10.041.

Devereaux, K., Dall'Armi, C., Alcazar-Roman, A., Ogasawara, Y., Zhou, X., Wang, F., ... Di Paolo, G. (2013). Regulation of mammalian autophagy by class II and III PI 3-kinases through PI3P synthesis. *PLoS One*, *8*(10), e76405. http://dx.doi.org/10.1371/journal.pone.0076405.

Di Paolo, G., & De Camilli, P. (2006). Phosphoinositides in cell regulation and membrane dynamics. *Nature*, *443*(7112), 651–657. http://dx.doi.org/10.1038/nature05185.

Filimonenko, M., Isakson, P., Finley, K. D., Anderson, M., Jeong, H., Melia, T. J., … Yamamoto, A. (2010). The selective macroautophagic degradation of aggregated proteins requires the PI3P-binding protein Alfy. *Molecular Cell*, *38*(2), 265–279. http://dx.doi.org/10.1016/j.molcel.2010.04.007.

Gillooly, D. J., Morrow, I. C., Lindsay, M., Gould, R., Bryant, N. J., Gaullier, J. M., … Stenmark, H. (2000). Localization of phosphatidylinositol 3-phosphate in yeast and mammalian cells. *The EMBO Journal*, *19*(17), 4577–4588. http://dx.doi.org/10.1093/emboj/19.17.4577.

Gutierrez, M. G., Munafo, D. B., Beron, W., & Colombo, M. I. (2004). Rab7 is required for the normal progression of the autophagic pathway in mammalian cells. *Journal of Cell Science*, *117*(Pt. 13), 2687–2697. http://dx.doi.org/10.1242/jcs.01114.

Jager, S., Bucci, C., Tanida, I., Ueno, T., Kominami, E., Saftig, P., & Eskelinen, E. L. (2004). Role for Rab7 in maturation of late autophagic vacuoles. *Journal of Cell Science*, *117*(Pt. 20), 4837–4848. http://dx.doi.org/10.1242/jcs.01370.

Mayinger, P. (2012). Phosphoinositides and vesicular membrane traffic. *Biochimica et Biophysica Acta*, *1821*(8), 1104–1113. http://dx.doi.org/10.1016/j.bbalip.2012.01.002.

Mehrpour, M., Esclatine, A., Beau, I., & Codogno, P. (2010). Overview of macroautophagy regulation in mammalian cells. *Cell Research*, *20*(7), 748–762. http://dx.doi.org/10.1038/cr.2010.82.

Mizushima, N. (2007). Autophagy: Process and function. *Genes & Development*, *21*(22), 2861–2873. http://dx.doi.org/10.1101/gad.1599207.

Mizushima, N., Levine, B., Cuervo, A. M., & Klionsky, D. J. (2008). Autophagy fights disease through cellular self-digestion. *Nature*, *451*(7182), 1069–1075. http://dx.doi.org/10.1038/nature06639.

Moravcevic, K., Oxley, C. L., & Lemmon, M. A. (2012). Conditional peripheral membrane proteins: Facing up to limited specificity. *Structure*, *20*(1), 15–27. http://dx.doi.org/10.1016/j.str.2011.11.012.

Noda, T., Matsunaga, K., Taguchi-Atarashi, N., & Yoshimori, T. (2010). Regulation of membrane biogenesis in autophagy via PI3P dynamics. *Seminars in Cell & Developmental Biology*, *21*(7), 671–676. http://dx.doi.org/10.1016/j.semcdb.2010.04.002.

Pankiv, S., Alemu, E. A., Brech, A., Bruun, J. A., Lamark, T., Overvatn, A., … Johansen, T. (2010). FYCO1 is a Rab7 effector that binds to LC3 and PI3P to mediate microtubule plus end-directed vesicle transport. *The Journal of Cell Biology*, *188*(2), 253–269. http://dx.doi.org/10.1083/jcb.200907015.

Proikas-Cezanne, T., Takacs, Z., Donnes, P., & Kohlbacher, O. (2015). WIPI proteins: Essential PtdIns3P effectors at the nascent autophagosome. *Journal of Cell Science*, *128*(2), 207–217. http://dx.doi.org/10.1242/jcs.146258.

Raiborg, C., Schink, K. O., & Stenmark, H. (2013). Class III phosphatidylinositol 3-kinase and its catalytic product PtdIns3P in regulation of endocytic membrane traffic. *The FEBS Journal*, *280*(12), 2730–2742. http://dx.doi.org/10.1111/febs.12116.

Rodriguez-Arribas, M., Pizarro-Estrella, E., Gomez-Sanchez, R., Yakhine-Diop, S. M., Gragera-Hidalgo, A., Cristo, A., … Fuentes, J. M. (2016). IFDOTMETER: A new software application for automated immunofluorescence analysis. *Journal of Laboratory Automation*, *21*(2), 246–259. http://dx.doi.org/10.1177/2211068215600650.

Rubinsztein, D. C., Shpilka, T., & Elazar, Z. (2012). Mechanisms of autophagosome biogenesis. *Current Biology*, *22*(1), R29–R34. http://dx.doi.org/10.1016/j.cub.2011.11.034.

Schu, P. V., Takegawa, K., Fry, M. J., Stack, J. H., Waterfield, M. D., & Emr, S. D. (1993). Phosphatidylinositol 3-kinase encoded by yeast VPS34 gene essential for protein sorting. *Science*, *260*(5104), 88–91.

Shibutani, S. T., & Yoshimori, T. (2014). A current perspective of autophagosome biogenesis. *Cell Research*, *24*(1), 58–68. http://dx.doi.org/10.1038/cr.2013.159.

Simonsen, A., Birkeland, H. C., Gillooly, D. J., Mizushima, N., Kuma, A., Yoshimori, T., ... Stenmark, H. (2004). Alfy, a novel FYVE-domain-containing protein associated with protein granules and autophagic membranes. *Journal of Cell Science*, *117*(Pt. 18), 4239–4251. http://dx.doi.org/10.1242/jcs.01287.

Simonsen, A., & Tooze, S. A. (2009). Coordination of membrane events during autophagy by multiple class III PI3-kinase complexes. *The Journal of Cell Biology*, *186*(6), 773–782. http://dx.doi.org/10.1083/jcb.200907014.

Stack, J. H., Horazdovsky, B., & Emr, S. D. (1995). Receptor-mediated protein sorting to the vacuole in yeast: Roles for a protein kinase, a lipid kinase and GTP-binding proteins. *Annual Review of Cell and Developmental Biology*, *11*, 1–33. http://dx.doi.org/10.1146/annurev.cb.11.110195.000245.

Stenmark, H., Aasland, R., & Driscoll, P. C. (2002). The phosphatidylinositol 3-phosphate-binding FYVE finger. *FEBS Letters*, *513*(1), 77–84.

Walker, S., Chandra, P., Manifava, M., Axe, E., & Ktistakis, N. T. (2008). Making autophagosomes: Localized synthesis of phosphatidylinositol 3-phosphate holds the clue. *Autophagy*, *4*(8), 1093–1096.

Zhang, L., Yu, J., Pan, H., Hu, P., Hao, Y., Cai, W., ... Yuan, J. (2007). Small molecule regulators of autophagy identified by an image-based high-throughput screen. *Proceedings of the National Academy of Sciences of the United States of America*, *104*(48), 19023–19028. http://dx.doi.org/10.1073/pnas.0709695104.

Zoncu, R., Efeyan, A., & Sabatini, D. M. (2011). mTOR: From growth signal integration to cancer, diabetes and ageing. *Nature Reviews. Molecular Cell Biology*, *12*(1), 21–35. http://dx.doi.org/10.1038/nrm3025.

CHAPTER SIXTEEN

Quantification of Phosphatidylinositol Phosphate Species in Purified Membranes

A. Jeschke*, N. Zehethofer†, D. Schwudke†, A. Haas*,[1]
*Cell Biology Institute, University of Bonn, Bonn, Germany
†Priority Research Area Infections, Research Center Borstel, Leibniz-Center for Medicine and Biosciences, Borstel, Germany
[1]Corresponding author: e-mail address: albert.haas@uni-bonn.de

Contents

1. Introduction 272
2. Detection and Quantification of PI(3)P and PI(4)P in Lysosomes or Autophagosomes by PIP-Binding Proteins 274
   2.1 Ectopic Expression and Purification of PI(3)P-Binding Hrs 2xFYVE Domain 274
   2.2 Lysosome Preparation From J774E Macrophage-Like Cells 276
   2.3 Incubation of Lysosomes With Lipid-Binding Probes 277
   2.4 Enrichment of HeLa Cell Autophagosomes 279
   2.5 Incubation of Autophagosomes With Lipid-Binding Probes 280
   2.6 Staining of Lysosomes and Autophagosomes for Bound Lipid Probes 280
   2.7 Quantification of the Amount of Lipid-Binding Probes on Lysosomes or Autophagosomes in ImageJ 281
3. Quantification of PIPs in Lysosome Preparations by RP-HPLC-MS 282
   3.1 RP-HPLC-MS-Based Analysis of PIPs Extracted From Purified Lysosomes 282
4. Results and Discussion 285
Acknowledgments 289
References 289

Abstract

Phosphoinositide lipids (PIPs) are required for various processes during macroautophagy, such as phagophore formation and autophagosome–lysosome fusion. Hence, quantification of the seven PIP species in autophagosome membranes is an important tool to understand how these lipids govern the transition of autophagosomes into autolysosomes. Here, we describe microscopic and mass spectrometry methods which, although designed to quantify the different PIP species on purified lysosomes, can also be applied to analyze autophagosomal PIPs.

Methods in Enzymology, Volume 587
ISSN 0076-6879
http://dx.doi.org/10.1016/bs.mie.2016.10.037

1. INTRODUCTION

During macroautophagy (henceforth "autophagy"), a portion of the cytosol is sequestered by a double-membrane phagophore (isolation membrane), the edges of which fuse with each other leading to incorporation of the selected cargo into a double-membrane-bound vacuole, the autophagosome. This compartment eventually fuses with lysosomes, leading to degradation of its contents and transport of the resulting degradation products in the cytosol for reuse. Autophagy provides building blocks and energy during starvation and removes damaged or superfluous organelles (Feng, He, Yao, & Klionsky, 2014) or (detrimental) protein aggregates. Moreover, it helps to fight infections by ingestion and degradation of microbial intruders (Boya, Reggiori, & Codogno, 2013; Feng et al., 2014). Genetic studies in *Saccharomyces cerevisiae* identified 35 autophagy-related genes (Atg) (Thumm et al., 1994; Tsukada & Ohsumi, 1993), most of which are essential for autophagosome formation and possess functional homologs in mammals (ATGs) (Mizushima, Yoshimori, & Ohsumi, 2011). In addition to ATG proteins and other autophagy-relevant proteins (e.g., SNARE proteins; Moreau, Renna, & Rubinsztein, 2013), phosphoinositide lipids (PIPs) are emerging as important regulators of autophagy. PIPs are mono-, bis-, or trisphosphate derivatives of the glycerophospholipid phosphatidylinositol (PI), in which the D3-, D4-, and/or D5-hydroxyl groups of the inositol head are phosphorylated. Most of the various PIPs can be interconverted into each other by PI(P) kinases or PIP phosphatases (Sasaki et al., 2009). Due to the specific subcellular localization of these enzymes, PIPs are enriched in particular subcellular compartments (De Matteis & Godi, 2004). Although PIPs represent less than 1% of the total lipid content of a mammalian cell (Skwarek & Boulianne, 2009), they regulate numerous cellular processes, including various subreactions of autophagy. For instance, autophagosome formation requires phosphatidylinositol 3-phosphate (PI(3)P) (Dooley et al., 2014) and/or phosphatidylinositol 5-phosphate (PI(5)P) (Vicinanza et al., 2015), whereas autophagosome–lysosome fusion depends on phosphatidylinositol 4-phosphate (PI(4)P) (Wang et al., 2015).

Whether and which PIPs are required for autophagy can be determined using a combination of techniques (i) to track autophagosome formation and maturation, (ii) to manipulate, and (iii) to analyze the PIP composition of autophagosomes.

A common technique to monitor autophagosome formation relies on the detection of the covalent conjugation of LC3 (yeast Atg8) to phosphatidylethanolamine (PE). This modification is introduced at the phagophore assembly site (PAS) and persists on the autophagosome until fusion with lysosomes has occurred. Membrane-bound LC3 can be used to microscopically identify autophagosomes within cells (Mizushima, Yoshimori, & Levine, 2010). Moreover, the amount of PE-conjugated LC3 in cells (upon inhibition of lysosomal proteases or of autophagosome–lysosome fusion; Mizushima & Yoshimori, 2007) is a measure of autophagosome formation and can be assessed by immunoblotting.

Assays for autophagosome maturation analyze the degradation of PE-conjugated LC3 in lysosomes using western blotting (Mizushima & Yoshimori, 2007) or rely on microscopic evaluation of colocalization between autophagosomes and lysosome marker molecules (e.g., LAMP1; McEwan et al., 2015). Furthermore, autolysosome formation can be assessed using GFP-RFP/mCherry-LC3 double label constructs. In autophagosomes both fluorophores fluoresce, whereas in acidic autolysosomes GFP fluorescence is quenched and pH-independent red fluorescence persists (Kimura, Noda, & Yoshimori, 2007).

Autophagosome PIP compositions can be manipulated by nonselective incorporation of exogenic PIP analogs into subcellular compartment membranes (Ozaki, DeWald, Shope, Chen, & Prestwich, 2000; Vicinanza et al., 2015; Wang et al., 2015) or by overexpression, inhibition, silencing, and/or depletion of PIP-modifying enzymes. The impact of such manipulations on autophagosome PIPs can be followed using techniques to quantify PIPs in biological samples.

These techniques basically rely on the use of specific PIP-binding proteins or on separation of (deacylated) PIPs according to hydrophobicity, charge, and/or mass using TLC (thin-layer chromatography), HPLC (high-performance liquid chromatography), or mass spectrometry (Rusten & Stenmark, 2006). Whereas PIP-binding proteins can be used to detect PIPs in situ, i.e., in live or fixed cells, in purified compartments or tissue sections, TLC, HPLC, and mass spectrometry require lipid extraction. When combined with subcellular fractionation techniques, the biochemical approach is also useful to attribute PIPs to a defined membrane compartment.

The subcellular distribution of PIPs, however, has been largely analyzed using ectopic expression of PIP-binding protein domains fused to fluorescent proteins (e.g., GFP, green fluorescent protein). Although it is a valuable tool, overexpression of PIP-binding proteins can competitively inhibit PIP-dependent cellular processes (Rusten & Stenmark, 2006). For instance, overexpression of a PI(5)P-binding domain blocks autophagosome formation (Vicinanza et al., 2015). This problem can be circumvented, using PIP-binding domains or antibodies with fixed and permeabilized cells, as exemplified by the detection of PI(3)P and PI(4)P in autophagosomes using a purified PI(3)P-binding protein (Axe et al., 2008) or a PI(4)P-specific antibody (Wang et al., 2015). PIP-binding domains sometimes require additional partner molecules to bind to membranes, such as other lipids (e.g., PIPs) or proteins. This phenomenon of "coincidence detection" can bias the localization of a PIP-binding probe within cells (Carlton & Cullen, 2005; Hammond & Balla, 2015). Accordingly, it is advisable to use different binding domains to quantify the amount of a PIP in a subcellular compartment. Ideally, the presence of PIPs is confirmed using one of the above direct approaches to quantify PIPs in lipid extracts from subcellular compartment preparations. Corresponding protocols for the purification of autophagosomes are based on density gradient centrifugation (Mancias, Wang, Gygi, Harper, & Kimmelman, 2014; Marzella, Ahlberg, & Glaumann, 1982; Stromhaug, Berg, Fengsrud, & Seglen, 1998) or LC3 coimmunoprecipitation (Gao et al., 2010). A combination of the methods described earlier can be used with autophagosomes from differentially treated cells and help to quantify the various PIPs in order to understand regulation of autophagy by these lipids.

Here, we describe protocols to detect and quantify the amounts of monophosphorylated PIPs (i.e., PI(3)P or PI(4)P) in purified lysosomes by either a combinatorial reverse-phase high-performance liquid chromatography–mass spectrometry (RP-HPLC-MS) approach or by specific PIP-binding proteins, immunofluorescence microscopy, and computational image analysis. Moreover, we present data on the application of the latter method to also quantify the amounts of these PIPs in purified autophagosomes.

2. DETECTION AND QUANTIFICATION OF PI(3)P AND PI(4)P IN LYSOSOMES OR AUTOPHAGOSOMES BY PIP-BINDING PROTEINS

2.1 Ectopic Expression and Purification of PI(3)P-Binding Hrs 2xFYVE Domain

To visualize PI(3)P on autophagosomes and lysosomes, a PI(3)P-binding construct consisting of two successive copies of the FYVE (Fab1, YOTB,

Vac1, EEA1) domain of mouse Hrs (hepatocyte growth factor-regulated tyrosine kinase substrate) is used (Gillooly et al., 2000). This protein is produced in and purified from *Escherichia coli* BL21 DE3 (genotype: *fhuA2 [lon] ompT gal (λ DE3) [dcm] ΔhsdS λ DE3 = λ sBamHIo ΔEcoRI-B int::(lacI::PlacUV5::T7 gene1) i21 Δnin5*).

2.1.1 Buffers, Reagents, and Equipment

1. 50 × Protease inhibitor cocktail cOmplete™ (PIC) [Roche, 11697498001]
2. Bacterial expression plasmid encoding two successive copies of mouse Hrs FYVE domain: pGEX-5X-3_mouse_Hrs_2xFYVE (Gillooly et al., 2000)
3. Bacterial expression plasmid for fusion proteins with glutathione *S*-transferase (GST) of *Schistosoma japonicum* (GE Healthcare Life Sciences, pGEX-4T-1)
4. Elution buffer: 50 m*M* Tris, 10 m*M* glutathione, pH 7.5
5. Glutathione sepharose 4B [GE Healthcare Life Sciences, 17-0756-01]
6. Homogenization buffer (HB): 250 m*M* sucrose, 20 m*M* HEPES, 0.5 m*M* EGTA, pH 7.2
7. Lysis buffer: 140 m*M* NaCl, 10 m*M* Na_2HPO_4, 1.8 m*M* KH_2PO_4, pH 7.5

2.1.2 Procedure

1. Inoculate 20 mL of LB (lysogeny broth; 5 g/L yeast extract, 10 g/L tryptone, 5 g/L NaCl) containing 100 μg/mL ampicillin with *E. coli* BL21 DE3 transformants containing pGEX-5X-3_mouse_Hrs_2xFYVE construct or pGEX-4T-1. Incubate on a rotatory shaker for 16 h at 200 rpm and 37°C.
2. Add the resulting culture to 1 L of LB containing 100 μg/mL ampicillin and incubate until it reaches an OD_{600} (optical density at 600 nm) of 0.5.
3. Add isopropyl-β-D-galactoside from a 100 m*M* stock in distilled water to a final concentration of 0.25 m*M* and incubate for 16 h at 200 rpm and 16°C.
4. Harvest bacteria (10 min, 6000 × *g*, 4°C) and store at −20°C or process directly.
5. Resuspend bacteria in lysis buffer containing 1 × PIC.
6. Disintegrate bacteria by sonication.
7. Centrifuge homogenate (30 min, 16,100 × *g*, 4°C).
8. Mix homogenate with glutathione sepharose (0.3 mL slurry per 10 mL homogenate) and incubate for 30 min on an end-over-end shaker.
9. Wash glutathione sepharose thrice in 10 mL lysis buffer (centrifugation: 1 min, 1800 × *g*, 4°C).

10. Transfer glutathione sepharose to a 10 mL polycarbonate chromatography column and elute bound proteins from the column with elution buffer.
11. Dialyze eluted proteins with a 3-kDa molecular weight cutoff dialysis membrane against 2000 volumes HB for 16 h at 4°C, snap-freeze in liquid nitrogen, and store at −80°C.

2.2 Lysosome Preparation From J774E Macrophage-Like Cells

The J774 cell lines are derived from a murine reticulum cell sarcoma whose ascites form has macrophage properties. This cell line was selected for upregulated expression of the surface mannose receptor (Diment, Leech, & Stahl, 1987). Lysosomes in J774E macrophages are double labeled with 10 nm paramagnetic particles (ferrofluid) and a red fluorescent carboxytetramethylrhodamine conjugate of BSA (bovine serum albumin; BSA-rhodamine). Labeled lysosomes are then purified from cell homogenates using a magnet (Becken, Jeschke, Veltman, & Haas, 2010; Jeschke et al., 2015). Lysosomes prepared according to this protocol are devoid of marker molecules of the ER (i.e., calnexin, GRP-78), the Golgi apparatus (i.e., giantin, α-mannosidase), mitochondria (i.e., mitochondrial Hsp70), and of the plasma membrane (i.e., Na^+/K^+ ATPase) as revealed by immunoblotting (Becken et al., 2010; Li, Stolz, & Romero, 2005).

2.2.1 Buffers, Reagents, Cells, and Equipment

1. BSA-rhodamine: Prepare as described (Becken et al., 2010)
2. DMEM (Dulbecco's modified Eagle medium, Gibco, 31885-83)
3. Dura Grind stainless steel homogenizer (Wheaton Scientific; through Zinsser Analytic, 357572)
4. Dynal MPC-S magnetic particle concentrator (Thermo Scientific, A13346)
5. J774E cells, kindly supplied by Philip D. Stahl (Washington University, St. Louis, MO, USA)
6. FCS (fetal calf serum, Sigma Aldrich, F7524)
7. Ferrofluid (Ferrotec Europe, EMG 508)
8. GlutaMAX™ (Gibco, 35050-38)
9. HB (see Section 2.1)
10. PBS (phosphate-buffered saline)
11. PBS containing 5 m*M* EDTA
12. Protein assay dye reagent concentrate (Bio-Rad, 5000006)

2.2.2 Procedure

1. Subcultivate J774E cells at a 1:3 ratio 2 days before the experiment.
2. The day before the experiment, add fresh medium containing 50 μg/mL BSA-rhodamine to cells. Dilute 15 μL ferrofluid with 150 μL of 10 mg/mL BSA in PBS. Shake ferrofluid suspension on a rocker at 4°C for 16 h.
3. Discard medium, add 150 μL of ferrofluid suspension, diluted in 2.5 mL of DMEM/FCS per 10 cm dish, and incubate for 5 min at 37°C.
4. Wash cells twice in 3 mL PBS.
5. Add 4 mL of fresh DMEM/FCS and incubate for 120 min at 37°C.
6. Discard medium and add 2 mL of cold PBS per dish.
7. Scrape off cells from dish in PBS and harvest cells by centrifugation (160 × *g*, 4°C, 7 min).
8. Resuspend cells in 20 mL cold PBS/5 m*M* EDTA and centrifuge again (160 × *g*, 4°C, 7 min).
9. Resuspend cells in 20 mL cold HB and centrifuge again (160 × *g*, 4°C, 7 min).
10. Resuspend cells in 2 mL cold HB, 1 × PIC and rupture cells in a Dounce homogenizer using 6 strokes.
11. Prepare a postnuclear supernatant (PNS) from the homogenate by centrifugation (800 × *g*, 4°C, 5 min).
12. Dilute PNS to 10 mL by addition of HB and place tubes on a magnetic particle concentrator (30 min, 4°C).
13. Discard supernatant and harvest lysosomes by centrifugation (160 × *g*, 10 s).
14. Determine protein concentration of lysosome preparation using the Bradford protein assay.

2.3 Incubation of Lysosomes With Lipid-Binding Probes

The PIP composition of purified compartments can be analyzed directly after their isolation or after isolation and incubation with ATP and cytosol at 37°C. Direct analysis reveals the steady-state levels of PIPs in a subcellular compartment; yet since PIPs are short-lived, this analysis neglects PIPs which are dephosphorylated during sample preparation. Incubation of membranes in the presence of cytosol and ATP can regenerate PIPs lost during compartment isolation, even though the different PIP species may be generated at slightly different concentrations than present in living cells. In any instance, this procedure helps to identify the PIP-modifying enzymes involved in the biochemical shaping of these compartments (Defacque et al.,

2002; Jeschke et al., 2015). For instance, purified lysosomes contain little PI(3)P, yet accumulate this PIP during incubation with ATP and cytosol (Jeschke et al., 2015). The usefulness of this approach of purification and in vitro incubation under defined conditions was demonstrated by the identification of PI(3)P and the corresponding lipid kinase Vps34 as regulators of membrane fusion between lysosomes and phagosomes (Jeschke et al., 2015).

To visualize PI(4)P and PI(3)P in lysosome membranes, purified lysosomes are incubated in vitro in the presence or absence of ATP with cytosol and a PI(4)P-binding antibody or a PI(3)P-binding 2xFYVE domain.

2.3.1 Buffers, Reagents, and Equipment

1. 10 × ATP-regenerating system: HB containing 10 m*M* $MgCl_2$, 10 m*M* ATP, 1 mg/mL creatine kinase, and 85 m*M* creatine phosphate, pH 7.2.
2. 10 × Salt solution: HB containing 15 m*M* $MgCl_2$ and 1 *M* KCl.
3. HB containing 100 m*M* DTT.
4. Mouse anti-PI(4)P IgM (Echelon, Z-P004).
5. Purified mouse Hrs 2xFYVE domain and GST in HB (see Section 2.1).

2.3.2 Procedure

1. Prepare a reaction mixture containing 12 μg (protein) of lysosome preparation, 1 × ATP-regenerating system, 1 × salt solution, 2 mg/mL cytosolic proteins, 1 m*M* DTT, and 2 μ*M* GST, 2xFYVE, or 1:200 anti-PI(4)P IgM.
2. Incubate reaction mixtures at 37°C for 60 min.
3. Add 1 mL of cold HB/1 m*M* DTT.
4. Place tubes on magnetic particle concentrator and incubate for 10 min at 4°C.
5. Discard excess fluids, add 1 mL of cold HB/1 m*M* DTT, and incubate for 5 min at 4°C.
6. Repeat step 5 once.
7. Discard excess fluids, and harvest lysosomes by centrifugation (160 × *g*, 10 s).
8. Resuspend lysosome pellet in 500 μL of 2 mg/mL BSA in HB.
9. Centrifuge lysosome suspension onto a glass coverslip in a 24-well tissue culture plate (1800 × *g*, 4°C, 15 min).
10. Discard supernatant and add 300 μL of 4% (w/v) formaldehyde in HB per well.
11. Incubate for 16 h at 4°C.
12. Discard supernatant, add 300 μL 50 m*M* NH_4Cl in HB per well, and incubate for 15 min at ambient temperature.

2.4 Enrichment of HeLa Cell Autophagosomes

Nutrient deprivation induces autophagy which provides urgently needed building blocks and energy during starvation. PE-conjugated LC3 can be used to microscopically detect (nascent) autophagosomes within cells (Mizushima et al., 2010). Here, we detect autophagosomes in an autophagosome-enriched fraction prepared from starved HeLa cells by microscopy through a fluorescent protein fusion of LC3.

2.4.1 Buffers, Reagents, Cells, and Equipment

1. A mammalian expression plasmid encoding LC3b fused to eGFP and mCherry: pEGFP-C1-mCherry-GFP-LC3B (Hundeshagen, Hamacher-Brady, Eils, & Brady, 2011)
2. EBSS (Earle's balanced salt solution, Gibco, 24010-043)
3. HeLa cells (human cervix adenocarcinoma cells, American Type Culture Collection, Manassas, VA, CCL-2)
4. jetPRIME™ transfection buffer and reagent (Polyplus transfection, 114-7)

2.4.2 Procedure

1. Subcultivate HeLa cells in a 1:10 ratio the day before transfection.
2. Prepare a mixture of 40 μg purified pEGFP-C1-mCherry-GFP-LC3B plasmid in 2 mL of jetPRIME buffer.
3. Add 80 μL of jetPRIME reagent and mix thoroughly.
4. Incubate mixture at ambient temperature for 10 min.
5. Dropwise add 200 μL mixture per dish of HeLa cells. Incubate at 37°C for 4 h.
6. Discard medium and add 8 mL of new medium per 10-cm dish.
7. Incubate for 24 h at 37°C.
8. Discard medium and add 3 mL of EBSS to cells.
9. Incubate for 2 h at 37°C.
10. Discard medium and add 2 mL of cold PBS per dish.
11. Scrape off cells from dish and harvest cells by centrifugation (160 × *g*, 4°C, 7 min).
12. Resuspend cells in 20 mL cold PBS/5 m*M* EDTA and centrifuge again (160 × *g*, 4°C, 7 min).
13. Resuspend cells in 20 mL HB and centrifuge again (160 × *g*, 4°C, 7 min).
14. Resuspend cells in 2 mL cold HB/1 × PIC and homogenize using 8 strokes in a homogenizer.

15. Prepare a PNS by centrifugation of the homogenate (800 × *g*, 4°C, 5 min).
16. Prepare an autophagosome-enriched fraction by centrifugation of PNS (16,100 × *g*, 4°C, 10 min).
17. Resuspend pellet in 200 μL of HB and determine protein concentration using the Bio-Rad protein assay.

2.5 Incubation of Autophagosomes With Lipid-Binding Probes

To visualize PI(4)P and PI(3)P in autophagosome membranes, an autophagosome-enriched fraction is prepared from EBBS-treated HeLa cells and incubated in vitro in the presence or absence of ATP with cytosol and a PI(4)P-binding antibody or the PI(3)P-binding 2xFYVE domain.

2.5.1 Buffers, Reagents, and Equipment

See Section 2.3.1.

2.5.2 Procedure

1. Incubate 2.5 μg of autophagosome-enriched fraction with 4 μ*M* 2xFYVE domain or GST, or 1:100 anti-PI(4)P IgM for 30 min at 37°C in the presence of 1 × ATP-regenerating system, 1 × salts, 1 m*M* DTT, 2 mg/mL HeLa cell cytosol in a total reaction volume of 15 μL.
2. Add 1 mL of HB containing 5 mg/mL BSA and spin membranes onto glass coverslips (1800 × *g*, 4°C, 30 min).
3. Discard supernatant, add 300 μL of 4% formaldehyde in HB, and incubate for 30 min at ambient temperature.
4. Discard supernatant, add 300 μL of 50 m*M* NH_4Cl in HB, and incubate for 15 min at ambient temperature.

2.6 Staining of Lysosomes and Autophagosomes for Bound Lipid Probes

The amount of lipid-binding probes associated with autophagosomes or lysosomes is a measure of how much of the corresponding lipid is present in these compartments. Lipid-binding probes on autophagosomes or lysosomes can be visualized by immunofluorescence microscopy.

2.6.1 Buffers, Reagents, and Equipment

1. Goat antimouse IgG Alexa 488 (Invitrogen, A11029)
2. Goat antimouse IgG Alexa Fluor 647 (Invitrogen, A21236)
3. Goat antirabbit IgG Alexa Fluor 488 (Invitrogen, A11008)

4. Mouse anti-GST [B-14] (Santa Cruz Biotech, sc-138)
5. Rabbit anti-GST [Z-5] (Santa Cruz Biotech, sc-459)

2.6.2 *Procedure*

2.6.2.1 Lysosomes

1. Discard supernatant and incubate coverslips with 30 μL of 4% (w/v) BSA in PBS for 30 min.
2. Incubate coverslips with 30 μL of 1:200 rabbit anti-GST (sc-459) for 1 h.
   Note: For detection of mouse anti-PI(4)P IgM skip steps 2 and 3.
3. Rinse coverslips 5 × in PBS.
4. Cover coverslips with 30 μL of 1:200 goat antirabbit Alexa Fluor 488 or goat antimouse Alexa Fluor 488 in 4% BSA in PBS for 30 min.
5. Rinse coverslips 5 × in PBS.
6. Mount coverslips using Mowiol and dry for 2 h at ambient temperature.
7. Microscopic analysis. *Note*: Micrographs from same-experiment samples need to be recorded using identical microscope settings.

2.6.2.2 Autophagosomes

1. Discard supernatant and incubate coverslips with 30 μL of 4% (w/v) BSA in PBS for 30 min.
   Incubate coverslips with 30 μL of 1:100 mouse anti-GST (sc-138) for 1 h.
   Note: For detection of mouse anti-PI(4)P IgM skip steps 2 and 3.
2. Rinse coverslips 5 × in PBS.
3. Cover coverslips with 30 μL of 1:100 goat antimouse Alexa Fluor 647 in 4% (w/v) BSA in PBS for 30 min.
4. Rinse coverslips 5 × in PBS.
5. Mount coverslips using Mowiol.
6. Microscopic analysis. *Note*: Micrographs from same-experiment samples need to be recorded using identical microscope settings.

2.7 Quantification of the Amount of Lipid-Binding Probes on Lysosomes or Autophagosomes in ImageJ

To assess the relative amounts of PI(3)P or PI(4)P in autophagosome or lysosome membranes after incubation in vitro, the fluorescence signals on these compartments, resulting from immunofluorescence labeling of the PIP-binding probes, can be measured using the image analysis software ImageJ (NIH, National Institutes of Health).

1. Split micrographs to be analyzed into the respective channels, transform to grayscale, and export as tif files using Zeiss AxioVision™ software (release 4.8).
2. Define regions of interest (ROIs, lysosomes, or autophagosomes): Open tif file depicting the fluorescence of the lysosomal fluorophore or mCherry-LC3b, transform it to 8-bit (Image > Type > 8-bit), and adjust brightness and contrast (Image > Adjust > Brightness/Contrast > Auto).
3. Transform micrograph into a binary image (Process > Binary > Make binary).
4. Identify ROIs and add ROI coordinates to the "ROI manager" using the "analyze particles" function.
5. Select all ROIs added to the ROI manager and determine fluorescence intensities in ROIs using the "measure" function (Analyze > Tools > ROI manager > Measure).

3. QUANTIFICATION OF PIPs IN LYSOSOME PREPARATIONS BY RP-HPLC-MS

In addition to the use of PIP-binding protein domains, other techniques have been developed to quantify the PIP content of purified compartments, e.g., the detection of radiolabeled (deacylated) PIPs by TLC or HPLC, or of unlabeled PIPs by HPLC or mass spectrometry (Rusten & Stenmark, 2006). Radioactivity-based methods are highly sensitive, yet they only detect PIP turnover and neglect quiescent PIP pools (Kiefer et al., 2010). Nonisotopic assays measure all PIPs in a biological sample; however, detection of unlabeled PIPs by HPLC is far less sensitive than by radioactivity-based methods. Furthermore, mass spectrometry detection of unlabeled PIPs, though rather sensitive, cannot distinguish between the three mono- or the three bisphosphate PIP isomers (Idevall-Hagren & De Camilli, 2015). Importantly, a combination of HPLC separation and subsequent mass spectrometry identification of deacylated PIPs enables detection and quantification of all PIP isomers in a complex lipid mixture (Kiefer et al., 2010). The latter approach is used here to quantify the various monophosphorylated PIP isomers in lysosome preparations.

3.1 RP-HPLC-MS-Based Analysis of PIPs Extracted From Purified Lysosomes

3.1.1 Extraction of PIPs From Purified Lysosomes

3.1.1.1 Buffers, Reagents, and Equipment

Chloroform, hydrochloric acid (HCl), methanol, and water of LC-MS grade.

3.1.1.2 Procedure

PIPs are extracted using an acidified Bligh and Dyer extraction protocol (Bligh & Dyer, 1959).

1. Add 500 μL chloroform/methanol, 1:2 (by volume) to the lysosome suspension (260 μg protein) and vortex for 5 s.
2. Incubate the sample on ice for 10 min.
3. Add 300 μL of 20% (v:v) HCl and vortex for 5 s.
4. Centrifuge the suspension (20,000 × *g*, 90 s).
5. Remove the lower phase (~120 μL) and collect in a plastic tube.
   Note: Do not remove supernatant during this step.
6. Add 200 μL "synthetic lower phase" to the remaining sample and vortex for 5 s.
   Note: "Synthetic lower phase" is the lower phase obtained from a mixture of water, chloroform, methanol, and 20% (v:v) HCl (10:17:34:30, by volume) after shaking for 10 s.
7. Centrifuge the suspension (20,000 × *g*, 90 s).
8. Remove the lower phase (~200 μL) and collect it in a plastic tube.
   Note: Do not remove supernatant during this step.
9. Add 200 μL synthetic lower phase to the remaining sample and vortex for 5 s.
10. Centrifuge the suspension (20,000 × *g*, 90 s).
11. Remove the lower phase (~200 μL) and collect in plastic tube.
12. Dry the pooled organic phases in vacuum at 37°C.
13. Deacylate extracted lipids or store at −20°C.

Note: To improve the recovery of PIPs, it is recommended to perform extraction procedures in plastic tubes. However, the tubes should be tested for their usability for the extraction and deacylation of PIPs. In this respect, tubes from different brands were tested for chemical stability and recoveries. Here, polypropylene (PP) tubes were used (Sarstedt, Nümbrecht, Germany: 72.690), reaching recovery levels of approximately 50% (Zehethofer, 2013). Unless stated otherwise, all pipetting steps were performed with PP tips (Eppendorf, Hamburg, Germany: 0030 000.870/0030 000.919).

3.1.2 Deacylation of Extracted Lipids

3.1.2.1 Buffers, Reagents, and Equipment

Methylamine, *n*-butanol, and *n*-propanol of LC-MS grade.

3.1.2.2 Procedure

Deacylation of extracted PIPs is performed according to a procedure modified from Clarke and Dawson (1981).

1. Add 50 μL of methylamine solution in water/methanol/*n*-butanol (43:46:11, by volume) to dried lipids extracts using a glass pipette.
2. Incubate samples at 53°C in a water bath for 50 min.
3. Cool samples on ice, add 25 μL of cold *n*-propanol, and vortex.
4. Dry samples in vacuum at 37°C.
5. Perform LC-MS analysis of deacylated samples or store the samples at −20°C.

3.1.3 LC-MS Analysis of Deacylated PIPs

3.1.3.1 Buffers, Reagents, and Equipment

1. Solvent A: 5 m*M* DMHA (*N*,*N*-dimethylhexylamine) and 4 m*M* glacial acetic acid in water.
2. Solvent B: 5 m*M* DMHA and 4 m*M* glacial acetic acid in methanol.
3. 1100 HPLC system (Agilent, Böblingen, Germany) coupled to a Bruker Apex Qe FT-ICR MS (Bruker, Bremen, Germany) or similar.
4. Synergi Hydro C18 column (250 × 0.5 mm, 3 μm particle size) (Phenomenex, Torrance, CA) or similar.

3.1.3.2 Procedure

LC-MS analysis of deacylated PIPs is performed according to Kiefer et al. (2010).

1. Prior to the RP-HPLC-MS analysis, dissolve dried and deacylated lipids in 50 μL of solvent A.

   Note: 10 μL of the resulting solution is injected into the RP-HPLC-MS device.
2. RP-HPLC separation is performed using a gradient from 85% solvent A to 35% solvent B within 58 min at a constant flow rate 10 μL/min.
3. MS data are acquired in the negative ion mode between *m*/*z* 200 and 1000.
4. RP-HPLC-MS data processing is performed by evaluation of mass spectra in the elution time range of the deacylated PIP species of interest (mass-to-charge ratio (*m*/*z*): 413.0239). Elution time ranges of particular PIP isomers are obtained by deacylation and RP-HPLC-MS analysis of suitable standards.

Note: A high-resolution mass spectrometer is not necessarily required for this RP-HPLC-MS analysis. However, the increased specificity, which is based on accurate mass determination, is beneficial for analyses with complex sample matrices.

4. RESULTS AND DISCUSSION

PIPs are emerging as important regulators of the endocytic and autophagic pathways. To understand how these lipids govern formation and maturation of endosomes and autophagosomes, methods are required to analyze the PIP composition of these compartments. Here, we applied techniques originally designed to visualize PI monophosphates on purified lysosomes (Jeschke et al., 2015) to also characterize the PIP content of autophagosomes.

To analyze lysosomes for PI(3)P and PI(4)P, lysosomes were purified, incubated in vitro with ATP and cytosol and PI(3)P-binding 2xFYVE domain or a PI(4)P-specific antibody. Lysosomes were rinsed and assayed by immunofluorescence microscopy for associated lipid-binding probes. Both, PI(3)P and PI(4)P, were present in lysosomes as judged by colocalization of a fluorescent lysosomal content marker (i.e., BSA-rhodamine) with the lipid probes (Fig. 1A1–A6). Omission of ATP from the incubations strongly decreased the amount of lipid-binding probes on lysosomes, showing that PI(3)P and PI(4)P are generated in lysosome membranes during the incubation, likely by phosphorylation of their common precursor PI (Jeschke et al., 2015).

Presence of PI(3)P and PI(4)P in lysosomes was confirmed by RP-HPLC-MS (Fig. 2A and B): Lysosomes were purified and incubated in vitro with ATP and cytosol in the absence or presence of the PI3K inhibitor wortmannin. Lysosomes were harvested; lipids were extracted, deacylated, and analyzed by RP-HPLC-MS. PI(4)P was present in lysosomes in both conditions (Fig. 2A and B), whereas PI(3)P was not detected in samples incubated with wortmannin (Fig. 2B). The functional relevance of PI(3)P and PI(4)P in lysosomes is discussed elsewhere (Jeschke et al., 2015).

To visualize PI(3)P and PI(4)P in autophagosomes, we expressed an mCherry-fusion of LC3b ectopically in HeLa cells. Such cells were starved and an autophagosome-enriched fraction was isolated. This fraction was incubated with ATP and cytosol plus PI(3)P- or PI(4)P-binding probes as described earlier for lysosomes. Whereas about 55% of the LC3b-decorated compartments stained positive for PI(4)P, less than a tenth colocalized with PI(3)P-binding 2xFYVE domain (Fig. 1B1–B6). This is in agreement with previous data, showing that PI(4)P is present in autophagosomes and required for autophagosome–lysosome fusion (Wang et al., 2015).

Despite the key role of PI(3)P in autophagosome formation and its well-documented presence at the PAS, it is not known whether it is also required

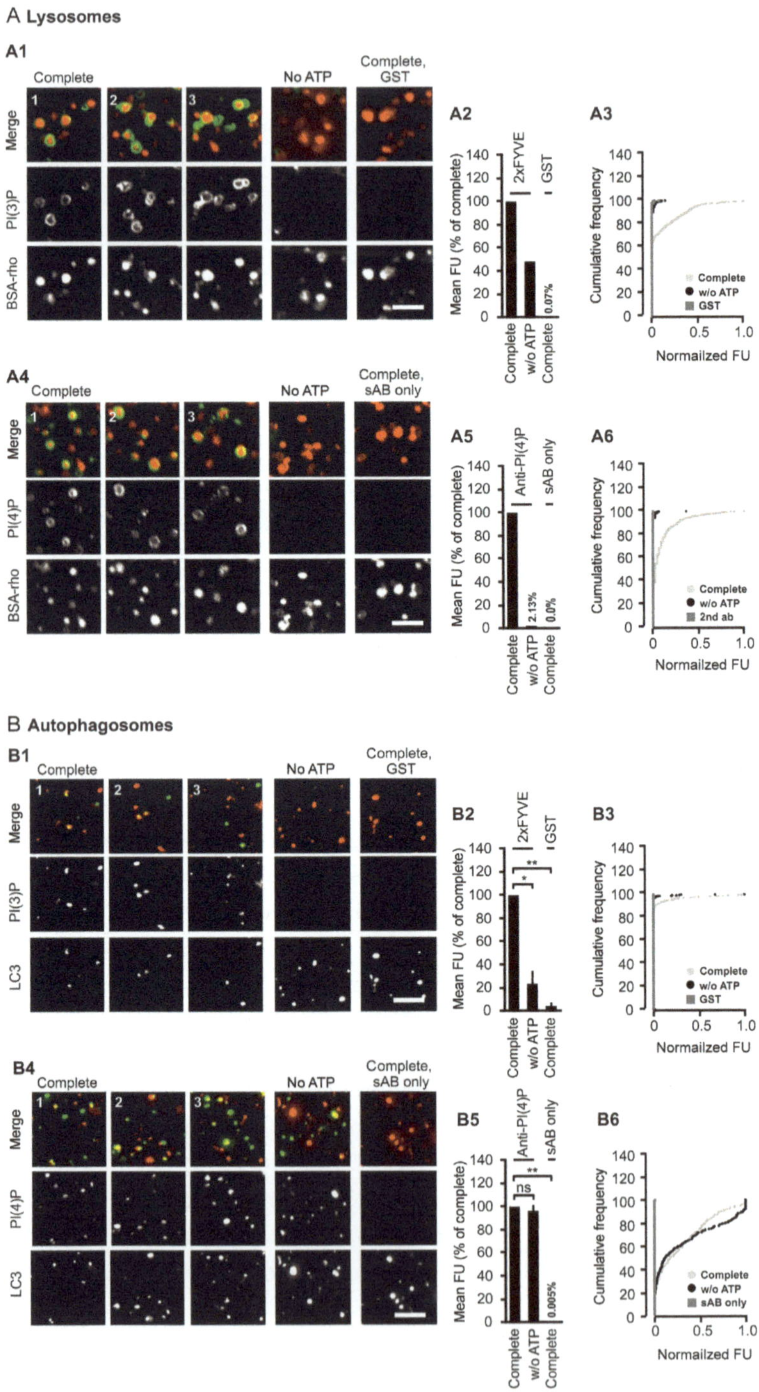

Fig. 1 See legend on opposite page.

for subsequent autophagosome maturation (Burman & Ktistakis, 2010). Here, we used the autophagosome marker LC3 (isoform LC3b) which labels autophagosomes of various maturation stages (Kabeya et al., 2000). If PI(3)P were only present in phagophores, the rare colocalization between LC3- and PI(3)P-binding 2xFYVE domain described here could be a reflection of the observation that phagophores make up only a small portion of LC3-positive compartments in starved cells (Gao et al., 2010).

Omission of ATP from the incubations virtually abolished colocalization of PI(3)P with LC3-containing compartments (Fig. 1B2), possibly reflecting the generation of this PIP by class III PI3K Vps34.

Fig. 1 Detection of PI(3)P on purified lysosomes and autophagosomes by fluorescence microscopy. (A1–A3) Purified BSA-rhodamine-loaded lysosomes were incubated with J774E cytosol, 1 m*M* DTT, 1 × salt mix, and with 2 μ*M* of PI(3)P-binding 2xFYVE domain or GST in the presence (complete) or absence of ATP (no ATP) for 60 min at 37°C. Lysosomes were washed, spun and fixed onto coverslips, and stained for associated 2xFYVE or GST. (A1) Representative fluorescence micrographs depicting fluorescence signals recorded for BSA-rhodamine (*red channel*, lysosomes) and/or 2xFYVE domain (*green channel*, PI(3)P). (A2) Mean green fluorescence per lysosome incubated under conditions specified. The mean FU in an unstained region of the "complete" sample (="background fluorescence") was set as 0 and subtracted from the mean FU of each ROI in same-experiment samples. 35.5% of the lysosomes showed mean FU (fluorescence units) > 0. (A3) Cumulative distribution plot depicting fluorescence signals of individual lysosomes from a representative experiment. (A4–A6) Purified BSA-rhodamine-loaded lysosomes were incubated as above in the presence or absence of a PI(4)P-binding antibody. Lysosomes assayed for bound anti-PI(4)P antibodies as above. 61.4% of lysosomes showed mean FU > 0. (B1–B3) Autophagosome-enriched fractions from starved, LC3b-mCherry-expressing HeLa cells were incubated with 4 μ*M* 2xFYVE or 4 μ*M* GST, HeLa cytosol, 1 m*M* DTT, 1 × salt mix, in the presence or absence of ATP for 60 min at 37°C. Compartment membranes were washed in excess buffer, spun and fixed onto coverslips, and assayed for bound 2xFYVE domain or GST by immunofluorescence microscopy. (B1) Representative micrographs depicting fluorescence signals recorded for mCherry-LC3b (*red channel*, autophagosomes) and/or 2xFYVE (*green channel*, PI(3)P). (B2) Mean green fluorescence per autophagosome incubated under conditions as indicated. Data are means ± SEM (standard error of the mean) from three independent experiments. 6.84% (±3.28% (SEM)) of autophagosomes showed mean FU > 0. (B3) Cumulative distribution plot for a representative experiment depicting fluorescence signals of individual autophagosomes. (B4–B6) Autophagosome-enriched fractions were incubated in the presence or absence of anti-PI(4)P antibodies and stained for associated anti-PI(4)P as above. (B4) Representative fluorescence micrographs (*red channel*, autophagosomes; *green channel*, PI(4)P). (B5) See B3. 54.9% ± 14.0% of autophagosomes showed mean FU > 0. Data are means ± SEM from three independent experiments. (B6) Cumulative distribution plot for one representative experiment. $^{**}p < 0.01$, $^{*}p < 0.05$ for two-tailed Student's *t*-test. Scale bar: 2 μm.

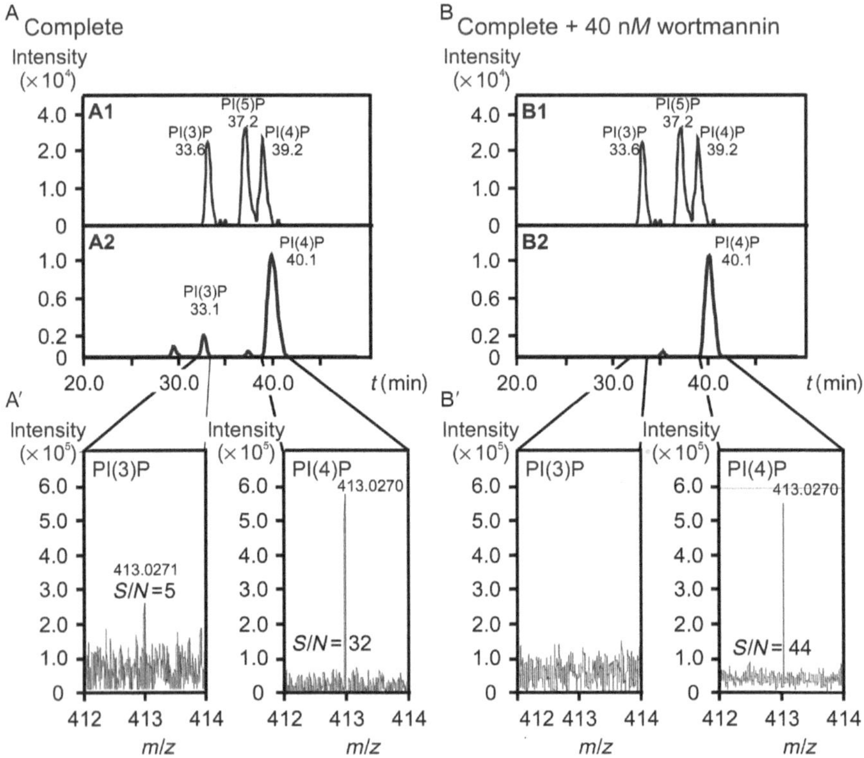

Fig. 2 Detection of PI(3)P or PI(4)P in purified lysosomes by RP-HPLC-MS. Lysosomes were purified from J774E macrophages and incubated with J774E cytosol, an ATP-regenerating system, 1 × salt mix, and 1 m*M* DTT at 37°C for 60 min in the presence of 40 n*M* of PI3K inhibitor wortmannin (B) or equal volumes of the vehicle DMSO (dimethylsulfoxide) (A). Compartments were harvested from reaction mixtures by centrifugation, and lipids were extracted, deacylated, and analyzed by RP-HPLC-MS. (A1 and B1) Extracted-ion chromatogram (EIC) depicting separation of deacylated PIP standards (i.e., PI(3)P, PI(5)P, or PI(4)P). (A2 and B2) EIC of separation of deacylated PIPs prepared from purified lysosomes after incubation under conditions specified. Retention times are indicated. (A′ and B′) Mass spectra of the elution time range of PI(3)P, PI(4)P, or PI(5)P. Signal-to-noise ratios (S/N) are indicated. *This figure is adapted from Jeschke, A., Zehethofer, N., Lindner, B., Krupp, J., Schwudke, D., Haneburger, I., et al. (2015). Phosphatidylinositol 4-phosphate and phosphatidylinositol 3-phosphate regulate phagolysosome biogenesis.* Proceedings of the National Academy of Sciences of the United States of America, 112*(15), 4636–4641.*

In sum, the methods presented here can be used to analyze the turnover of PIPs on autophagosomes and will help to identify PIPs relevant to autolysosome formation.

Note that fluorescence microscopy detection of PIPs on individual lysosomes or autophagosomes by lipid-binding proteins does not require pure compartment preparations, because the respective compartments can be readily distinguished from possible contaminants by detection of suitable marker molecules on the membranes of interest.

By contrast, analysis of lysosomal or autophagosomal PIPs via RP-HPLC-MS requires pure compartment preparations, as it cannot distinguish between PIPs in the compartment of interest and contaminant PIPs. Whereas our lysosome preparations are largely devoid of contaminants (Becken et al., 2010; Li et al., 2005), the autophagosome preparations used here are crude and require additional purification steps to be suitable for MS-based lipid analysis. Such protocols are now available as discussed earlier.

ACKNOWLEDGMENTS

We thank the Deutsche Forschungsgemeinschaft for supporting our studies through grants SPP1580 and SFB645 C2 (to A.H.). Moreover, we thank Michael Weinkauf for excellent technical assistance.

REFERENCES

Axe, E. L., Walker, S. A., Manifava, M., Chandra, P., Roderick, H. L., Habermann, A., et al. (2008). Autophagosome formation from membrane compartments enriched in phosphatidylinositol 3-phosphate and dynamically connected to the endoplasmic reticulum. *The Journal of Cell Biology*, *182*(4), 685–701.

Becken, U., Jeschke, A., Veltman, K., & Haas, A. (2010). Cell-free fusion of bacteria-containing phagosomes with endocytic compartments. *Proceedings of the National Academy of Sciences of the United States of America*, *107*(48), 20726–20731.

Bligh, E. G., & Dyer, W. J. (1959). A rapid method of total lipid extraction and purification. *Canadian Journal of Biochemistry and Physiology*, *37*(8), 911–917.

Boya, P., Reggiori, F., & Codogno, P. (2013). Emerging regulation and functions of autophagy. *Nature Cell Biology*, *15*(7), 713–720.

Burman, C., & Ktistakis, N. T. (2010). Regulation of autophagy by phosphatidylinositol 3-phosphate. *FEBS Letters*, *584*(7), 1302–1312.

Carlton, J. G., & Cullen, P. J. (2005). Coincidence detection in phosphoinositide signaling. *Trends in Cell Biology*, *15*(10), 540–547.

Clarke, N. G., & Dawson, R. M. (1981). Alkaline O leads to N-transacylation. A new method for the quantitative deacylation of phospholipids. *The Biochemical Journal*, *195*(1), 301–306.

Defacque, H., Bos, E., Garvalov, B., Barret, C., Roy, C., Mangeat, P., et al. (2002). Phosphoinositides regulate membrane-dependent actin assembly by latex bead phagosomes. *Molecular Biology of the Cell*, *13*(4), 1190–1202.

De Matteis, M. A., & Godi, A. (2004). PI-loting membrane traffic. *Nature Cell Biology*, *6*(6), 487–492.

Diment, S., Leech, M. S., & Stahl, P. D. (1987). Generation of macrophage variants with 5-azacytidine: Selection for mannose receptor expression. *Journal of Leukocyte Biology*, *42*(5), 485–490.

Dooley, H. C., Razi, M., Polson, H. E., Girardin, S. E., Wilson, M. I., & Tooze, S. A. (2014). WIPI2 links LC3 conjugation with PI3P, autophagosome formation, and pathogen clearance by recruiting Atg12-5-16L1. *Molecular Cell*, *55*(2), 238–252.

Feng, Y., He, D., Yao, Z., & Klionsky, D. J. (2014). The machinery of macroautophagy. *Cell Research*, *24*(1), 24–41.

Gao, W., Kang, J. H., Liao, Y., Ding, W.-X., Gambotto, A. A., Watkins, S. C., et al. (2010). Biochemical isolation and characterization of the tubulovesicular LC3-positive autophagosomal compartment. *The Journal of Biological Chemistry*, *285*(2), 1371–1383.

Gillooly, D. J., Morrow, I. C., Lindsay, M., Gould, R., Bryant, N. J., Gaullier, J. M., et al. (2000). Localization of phosphatidylinositol 3-phosphate in yeast and mammalian cells. *The EMBO Journal*, *19*(17), 4577–4588.

Hammond, G. R., & Balla, T. (2015). Polyphosphoinositide binding domains: Key to inositol lipid biology. *Biochimica et Biophysica Acta*, *1851*(6), 746–758.

Hundeshagen, P., Hamacher-Brady, A., Eils, R., & Brady, N. R. (2011). Concurrent detection of autolysosome formation and lysosomal degradation by flow cytometry in a high-content screen for inducers of autophagy. *BMC Biology*, *9*, 38.

Idevall-Hagren, O., & De Camilli, P. (2015). Detection and manipulation of phosphoinositides. *Biochimica et Biophysica Acta*, *1851*(6), 736–745.

Jeschke, A., Zehethofer, N., Lindner, B., Krupp, J., Schwudke, D., Haneburger, I., et al. (2015). Phosphatidylinositol 4-phosphate and phosphatidylinositol 3-phosphate regulate phagolysosome biogenesis. *Proceedings of the National Academy of Sciences of the United States of America*, *112*(15), 4636–4641.

Kabeya, Y., Mizushima, N., Ueno, T., Yamamoto, A., Kirisako, T., Noda, T., et al. (2000). LC3, a mammalian homologue of yeast Apg8p, is localized in autophagosome membranes after processing. *The EMBO Journal*, *19*(21), 5720–5728.

Kiefer, S., Rogger, J., Melone, A., Mertz, A. C., Koryakina, A., Hamburger, M., et al. (2010). Separation and detection of all phosphoinositide isomers by ESI-MS. *Journal of Pharmaceutical and Biomedical Analysis*, *53*(3), 552–558.

Kimura, S., Noda, T., & Yoshimori, T. (2007). Dissection of the autophagosome maturation process by a novel reporter protein, tandem fluorescent-tagged LC3. *Autophagy*, *3*(5), 452–460.

Li, H.-S., Stolz, D. B., & Romero, G. (2005). Characterization of endocytic vesicles using magnetic microbeads coated with signalling ligands. *Traffic*, *6*(4), 324–334.

Mancias, J. D., Wang, X., Gygi, S. P., Harper, J. W., & Kimmelman, A. C. (2014). Quantitative proteomics identifies NCOA4 as the cargo receptor mediating ferritinophagy. *Nature*, *509*(7498), 105–109.

Marzella, L., Ahlberg, J., & Glaumann, H. (1982). Isolation of autophagic vacuoles from rat liver: Morphological and biochemical characterization. *The Journal of Cell Biology*, *93*(1), 144–154.

McEwan, D. G., Popovic, D., Gubas, A., Terawaki, S., Suzuki, H., Stadel, D., et al. (2015). PLEKHM1 regulates autophagosome-lysosome fusion through HOPS complex and LC3/GABARAP proteins. *Molecular Cell*, *57*(1), 39–54.

Mizushima, N., & Yoshimori, T. (2007). How to interpret LC3 immunoblotting. *Autophagy*, *3*(6), 542–545.

Mizushima, N., Yoshimori, T., & Levine, B. (2010). Methods in mammalian autophagy research. *Cell*, *140*(3), 313–326.

Mizushima, N., Yoshimori, T., & Ohsumi, Y. (2011). The role of Atg proteins in autophagosome formation. *Annual Review of Cell and Developmental Biology*, *27*, 107–132.

Moreau, K., Renna, M., & Rubinsztein, D. C. (2013). Connections between SNAREs and autophagy. *Trends in Biochemical Sciences*, *38*(2), 57–63.

Ozaki, S., DeWald, D. B., Shope, J. C., Chen, J., & Prestwich, G. D. (2000). Intracellular delivery of phosphoinositides and inositol phosphates using polyamine carriers. *Proceedings of the National Academy of Sciences of the United States of America*, *97*(21), 11286–11291.

Rusten, T. E., & Stenmark, H. (2006). Analyzing phosphoinositides and their interacting proteins. *Nature Methods*, *3*(4), 251–258.

Sasaki, T., Takasuga, S., Sasaki, J., Kofuji, S., Eguchi, S., Yamazaki, M., et al. (2009). Mammalian phosphoinositide kinases and phosphatases. *Progress in Lipid Research, 48*(6), 307–343.

Skwarek, L. C., & Boulianne, G. L. (2009). Great expectations for PIP: Phosphoinositides as regulators of signaling during development and disease. *Developmental Cell, 16*(1), 12–20.

Stromhaug, P. E., Berg, T. O., Fengsrud, M., & Seglen, P. O. (1998). Purification and characterization of autophagosomes from rat hepatocytes. *The Biochemical Journal, 335*(Pt. 2), 217–224.

Thumm, M., Egner, R., Koch, B., Schlumpberger, M., Straub, M., Veenhuis, M., et al. (1994). Isolation of autophagocytosis mutants of Saccharomyces cerevisiae. *FEBS Letters, 349*(2), 275–280.

Tsukada, M., & Ohsumi, Y. (1993). Isolation and characterization of autophagy-defective mutants of Saccharomyces cerevisiae. *FEBS Letters, 333*(1–2), 169–174.

Vicinanza, M., Korolchuk, V. I., Ashkenazi, A., Puri, C., Menzies, F. M., Clarke, J. H., et al. (2015). PI(5)P regulates autophagosome biogenesis. *Molecular Cell, 57*(2), 219–234.

Wang, H., Sun, H.-Q., Zhu, X., Zhang, L., Albanesi, J., Levine, B., et al. (2015). GABARAPs regulate PI4P-dependent autophagosome:lysosome fusion. *Proceedings of the National Academy of Sciences of the United States of America, 112*(22), 7015–7020.

Zehethofer, N. (2013). *Development of mass spectrometry-based methods to elucidate the role of lipids in asthma.* Dissertation, Germany: University of Lübeck.

CHAPTER SEVENTEEN

Mass Assays to Quantify Bioactive PtdIns3P and PtdIns5P During Autophagic Responses

J. Viaud*, G. Chicanne*, R. Solinhac*, K. Hnia*, F. Gaits-Iacovoni*, B. Payrastre*,†,1
*INSERM, U1048 and Université Toulouse 3, I2MC, Toulouse, France
†CHU (Centre Hospitalier Universitaire) de Toulouse, Laboratoire d'Hématologie, Toulouse, France
[1]Corresponding author: e-mail address: bernard.payrastre@inserm.fr

Contents

1. Introduction 293
2. Preparation of Recombinant Proteins 298
   2.1 Preparation of Recombinant PIKfyve for PtdIns3P Mass Assay 298
   2.2 Preparation of Recombinant GST-PIP4KIIα for PtdIns5P Mass Assay 299
3. Lipid Extraction From Biological Samples 301
   3.1 Lipid Extraction From Cell Extracts 301
   3.2 Lipid Extraction From Tissues 303
4. Purification of Phosphatidylinositol Monophosphates 304
5. Quantification of PtdIns3P by Mass Assay 306
6. Quantification of PtdIns5P by Mass Assay 308
Acknowledgments 309
References 309

Abstract

Autophagy is a cellular process whereby cytoplasmic substrates are targeted for degradation in the lysosome via the membrane structures autophagosomes. This process is initiated by specific phosphoinositides, PtdIns3P and PtdIns5P, which play a key role in autophagy by recruiting effectors such as Atg18/WIPI2. Therefore, quantifying those lipids is important to better understand the assembly of the complex autophagic machinery. Herein, we describe in detail methods to quantify PtdIns3P and PtdIns5P by specific mass assays feasible in most laboratories.

1. INTRODUCTION

Autophagic processes are grouped into three major types: macroautophagy, microautophagy, and chaperone-mediated autophagy

Methods in Enzymology, Volume 587
ISSN 0076-6879
http://dx.doi.org/10.1016/bs.mie.2016.09.061

(Bento et al., 2016; de Duve, 2005; Dice, 1990). Macroautophagy (referred to as "autophagy" hereafter) is the most extensively described and quantitatively more important type of autophagy. In species from yeasts to mammals, autophagy is mediated by a specific organelle, the autophagosome (He & Klionsky, 2009), formed sequentially through the elongation of small membrane structures, called phagophores, that elongate and fuse to form a de novo double-membrane vesicle enclosing a portion of the cytosol (Fig. 1). This vesicle then acquires the proteases required to degrade the sequestered material through fusion with a lysosome to form an autolysosome. Initiation of this process requires specific phosphoinositides (PIs), which are low abundant phospholipids having the capacity to bind to specific protein domains (Ceccato et al., 2016). They regulate protein recruitment and formation of protein complexes involved in the control of cellular signaling, cytoskeleton reorganization, and vesicular trafficking (Viaud et al., 2016). Among the seven PIs, PtdIns3P has rapidly been linked to the canonical autophagy pathway following its detection in autophagosomes (Obara, Noda, Niimi, & Ohsumi, 2008). In this process, PtdIns3P, produced by the class III phosphoinositide 3-kinase VPS34 (Burman & Ktistakis, 2010), binds the autophagosome marker protein DFCP1 and Atg18 (also called WIPI2), an effector protein essential for autophagosome nucleation (Fig. 1, *middle*). Then, this structure elongates through the Atg12 and the LC3/Atg8-conjugation systems (Fig. 1, *right*) (Karanasios & Ktistakis, 2015; Obara, Sekito, Niimi, & Ohsumi, 2008). Therefore, PtdIns3P is a critical lipid that allows autophagosome nucleation and coordination of the autophagic machinery under basal and amino acid starvation conditions (Cebollero et al., 2012). Local PtdIns3P concentration has to be accurately regulated for proper control of autophagy (Noda, Matsunaga, Taguchi-Atarashi, & Yoshimori, 2010). Indeed, several PtdIns3P-phosphatases of the myotubularin family (MTMs; MTM1, MTMR3, MTMR6, MTMR8, MTMR14/jumpy) have been proposed to regulate autophagy in mammalian and nonmammalian models (Fetalvero et al., 2013; Mochizuki et al., 2013; Noda et al., 2010; Taguchi-Atarashi et al., 2010; Vergne et al., 2009; Zou et al., 2012).

Recently, a pathway of VPS34-independent initiation of autophagy has been described in response to glucose starvation. It relies on the still poorly known lipid, PtdIns5P (Viaud, Boal, Tronchere, Gaits-Iacovoni, & Payrastre, 2014), which is a mirror image of PtdIns3P (Vicinanza et al., 2015). Interestingly, PtdIns5P was able to recruit WIPI2, which is also a

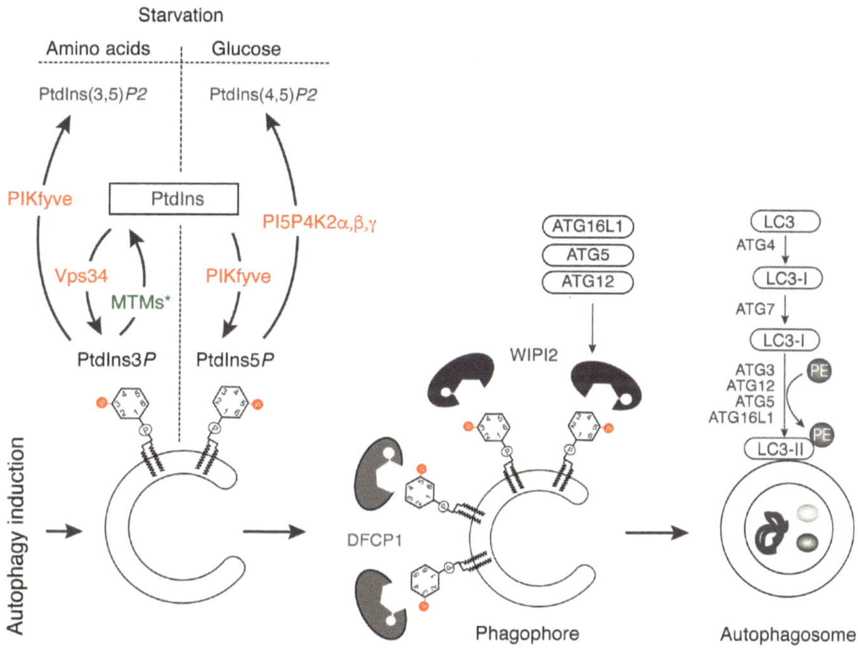

* Active MTMs (myotubularins) with reported impact on autophagy: MTM1, MTMR3, MTMR6, MTMR8, MTMR14/jumpy

Fig. 1 Role of phosphoinositides in autophagy. PtdIns3P and PtdIns5P are generated by the indicated PI-metabolizing enzymes and are localized on the phagophore (preautophagosome) membrane. PtdIns3P is mostly involved in Vps34-dependent autophagy (amino acid starvation), while PtdIns5P is likely implicated in noncanonical autophagy (glucose starvation). PtdIns3P or PtdIns5P is required to recruit and bind WIPI2, which allows phagophore elongation and formation of the autophagosome. A sequential cascade of conjugaison and deconjugaison reactions (mediated by ATG proteins) leads to the formation of the autophagosome marker LC3II-PE indicating the formation of mature autophagosome.

PtdIns3P effector, suggesting that both lipids may act interchangeably (Vicinanza et al., 2015). Moreover, PtdIns5P can functionally substitute for PtdIns3P during basal autophagy and in conventional amino acid/serum starvation-induced autophagy.

In this study, the increase in PtdIns5P levels, leading to autophagy, came from PIKfyve activation (PtdIns5P produced by either direct phosphorylation ofPtdIns or 5-dephosphorylation, by MTM family members, of PtdIns(3,5)P_2 generated by PIKfyve from PtdIns3P) or by invalidation of PI5P4K2s, which phosphorylates PtdIns5P to form PtdIns(4,5)P_2 (Fig. 1).

Quantifying changes in PtdIns3P and PtdIns5P levels in mammalian cells remains challenging due to the fact that they are present in such low quantities. The traditional techniques to separate and quantify the isomers of phosphatidylinositol monophosphates (PtdIns3P, PtdIns4P, and PtdIns5P) require reaching isotopic equilibrium state using [^{3}H]-inositol or [^{32}P]-inorganic phosphate radiolabeled cells, lipid extraction, deacylation, and finally separation and analysis by ion exchange high-pressure liquid chromatography (Munnik, 2013; Payrastre, 2004; Sarkes & Rameh, 2010). This method is appropriate to analyze PIs in cultured cells but remains radioactive and found mainly in laboratories specialized in lipid biochemistry. Moreover, this approach does not allow monitoring PI amounts in biopsies and is difficult to use to analyze isolated subcellular compartments. A combination of liquid chromatography and mass spectrometry has recently been developed to quantify PIs from lipid extracts (Clark et al., 2011; Kielkowska et al., 2014). However, at present, this very potent and promising approach does not permit separation and quantification of the different isomers of phosphatidylinositol monophosphates. While total phosphatidylinositol monophosphates can be quantified by this method, it is not possible to get information about the quantity of the variation in PtdIns3P or PtdIns5P levels, due to the fact that they are largely underrepresented in biological samples in comparison with PtdIns4P, which represents more than 80% of total phosphatidylinositol monophosphates.

As an alternative, nonradioactive methods using protein domains or antibodies selectively recognizing a lipid have been developed and some of them are commercially available. However, quantification of PtdIns3P using these kits does not always give reproducible and reliable results, possibly because the PIs extraction does not always allow to obtain a sufficient degree of purity of extracted PIs. To circumvent these drawbacks, we and others have developed specific mass assays based on the phosphorylation of a given PI such as PtdIns3P (Chicanne et al., 2012) or PtdIns5P (Morris, Hinchliffe, Ciruela, Letcher, & Irvine, 2000) using specific recombinant kinases.

In this chapter, we describe the protocols to quantify by mass assay PtdIns3P and PtdIns5P following a procedure that can be setup in most laboratories and allowing quantification of those PIs in whole cell extracts or isolated subcellular compartments (Fig. 2).

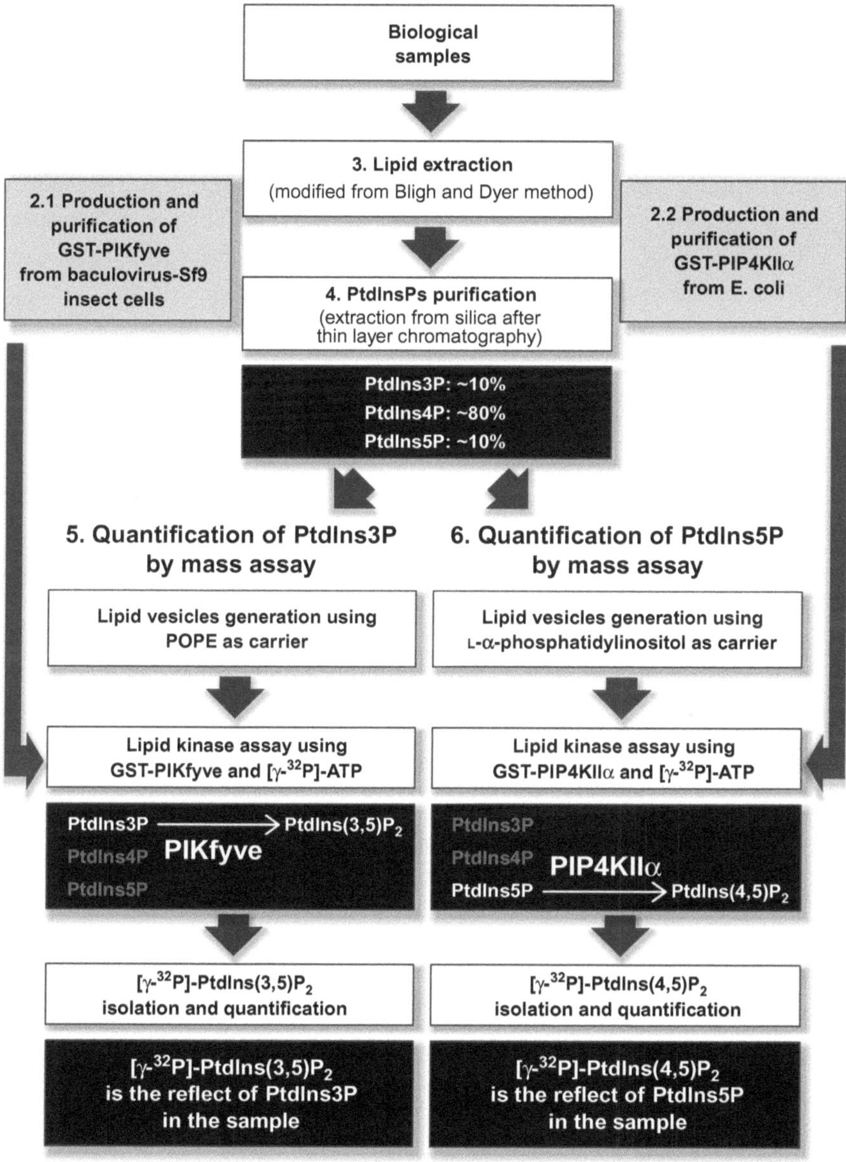

Fig. 2 Key steps for PtdIns3P and PtdIns5P quantification by mass assay.

2. PREPARATION OF RECOMBINANT PROTEINS

2.1 Preparation of Recombinant PIKfyve for PtdIns3P Mass Assay

2.1.1 Production of Recombinant PIKfyve in Sf9 Cells

For all those steps, work under a sterile hood. DO NOT FORGET to spray laboratory equipment such as flasks, pipettes, etc. with 70% ethanol and wipe dry before transferring into the hood. Sf9 cells are cultivated in an incubator shaker (27°C, 150 rpm) only dedicated to insect cell culture.

1. Grow Sf9 cells to a density of 1–2 × 10^6 cells/mL (500 mL culture/2 L flask for large-scale expression) in SF900II medium supplemented with 2 m*M* L-glutamine, 30 μg/mL sodium benzylpenicillin, and 50 μg/mL streptomycin sulfate. NOTE: store medium in the dark at 4°C. Warm medium to 27°C in order for it to be "ready to use."
2. Infect cells by the addition of 10 M.O.I. baculovirus expressing GST-PIKfyve.
3. Incubate infected cells for 36 h.
4. Recover cells by centrifugation in a conical 500 mL tubes spun at 800 × *g* for 10 min.
5. Decant supernatant back into the culture flasks and kill virus by adding bleach to reach a final 10% per volume for 10 min minimum.
6. Resuspend pellet in 30 mL PBS to wash cells and transfer to a 50-mL conical tube.
7. Resediment cells in 50 mL tubes at 800 × *g* for 10 min. Decant supernatant carefully to avoid dislodging the cell pellet.
8. Freeze the cell pellet in liquid nitrogen or store at −80°C overnight.

2.1.2 Purification of Recombinant GST-PIKfyve

Lysis buffer: 25 m*M* Hepes pH 7.4, 150 m*M* NaCl, and a cocktail of inhibitors composed of 4 m*M* Na_3VO_4, 10 μ*M* Leupeptin; 10 μ*M* Aprotinin; and 1 m*M* AEBSF.

Washing buffer: 25 m*M* Hepes pH 7.4, 150 m*M* NaCl.

Elution buffer: 25 m*M* Hepes pH 7.4, 150 m*M* NaCl, 30 m*M* reduced glutathione.

Storage buffer: 25 m*M* Hepes pH 7.4, 150 m*M* NaCl.

1. Thaw the cell pellet on ice and conduct all subsequent steps at 4°C.
2. Resuspend the cell pellet derived from 1 L of culture with 16 mL of lysis buffer, keeping cold at all times to minimize proteolysis.

3. Sonicate gently using a probe sonicator in an ice bath and monitor lysis by microscopy. NOTE: take a sample after each round of sonication. Ninety percent of the cells should be lysed before proceeding to the next step.
4. Pellet the debris by centrifugation at 25,000 × *g* for 20 min. Collect the supernatant and add sodium azide 0.02% to prevent bacterial growth. *Save 5 μL of the supernatant for SDS-PAGE and add 10 μL SDS-PAGE sample buffer. Boil 4 min.*
5. Add 1 mL of glutathione sepharose beads prewashed with washing buffer and rotated at 4°C for 2 h.
6. Pellet beads gently at 800 × *g* for 5 min and remove the depleted lysate (save in case of residual GST-PIKfyve in the lysate). *Save 5 μL of the depleted lysate for SDS-PAGE and add 10 μL SDS-PAGE sample buffer. Boil 4 min.*
7. Wash the beads with 3 × 40 mL washing buffer.
8. Transfer to a disposable column for elution and elute with six sequential elutions of 1 mL elution buffer. *Save 5 μL of each elution and add 10 μL SDS-PAGE sample buffer. Boil 4 min and load 10 μL for SDS-PAGE along with the lysate (10 μL), depleted lysate (10 μL), and the beads (10 μL from 10 μL of beads resuspended in 40 μL SDS-PAGE sample buffer and boil 4 min).*
9. Select elution fractions for pooling, pool and dialyze (cutoff: 100 kDa) vs PIKfyve storage buffer: 2 × 1 h in 1 L (at 4°C).
10. Recover dialysate, add glycerol to 10%, do a protein assay to determine yield, aliquot and snap freeze in liquid nitrogen.
11. Store at −80°C.

2.2 Preparation of Recombinant GST-PIP4KIIα for PtdIns5P Mass Assay

Although the protein has a GST-tag, a higher yield of the active kinase is recovered when eluted under denaturing conditions using 8 *M* urea instead of reduced glutathione.

Lysis buffer: PBS pH 7.4, 0.5 mg/mL lysozyme and a cocktail of inhibitors composed of 10 μ*M* Leupeptin, 10 μ*M* Aprotinin, and 1 m*M* AEBSF.

Washing buffer: PBS pH 7.4, 300 m*M* NaCl, 0.1% Tween 20 then PBS pH 7.4, 300 m*M* NaCl.

Elution buffer: 50 m*M* Tris pH 8, 8 *M* urea.

Storage buffer: 50 m*M* Tris pH 8.

2.2.1 Production of Recombinant GST-PIP4KIIα

1. Grow 500 mL ampicillin-resistant bacteria transformed with a pGEX plasmid coding for GST-PIP4KIIα at 37°C from a 20 mL bacterial starter until the optical density of the culture reaches between 0.6 and 0.8.
2. Add IPTG to a final concentration of 200 μ*M* and grow the bacterial culture for 3 h at 37°C.
3. Sediment the bacteria at ~5000 × *g* for 20 min.
4. Decant supernatant back into the culture flasks and kill virus by adding bleach to reach a final 10% per volume for 10 min minimum.
5. Resuspend the pellet in PBS to wash bacteria and transfer to a 50-mL conical tube.
6. Repellet bacteria at 3000 × *g* for 10 min and remove supernatant.
7. Freeze the bacterial pellet in liquid nitrogen or store at −80°C overnight.

2.2.2 Preparation of the Column

This step should be performed simultaneously between steps 1 and 8 of Section 2.2.3.

1. Hydrate 2 mL of Sephadex G-10 with H_2O and transfer to a thin disposable column.
2. Equilibrate the column with five volumes of 50 m*M* Tris pH 8 and add the cap when the liquid is just above the Sephadex G-10. NOTE: always keep the column hydrated.

2.2.3 Purification of Recombinant GST-PIP4KIIα

1. Thaw the cell pellet on ice and conduct all subsequent steps at 4°C.
2. Add 20 mL lysis buffer per 500 mL cell culture and resuspend well, keeping cold at all times to minimize proteolysis.
3. Sonicate using a probe sonicator in an ice bath (we usually make three cycles of 20 1-s pulses, output 50%).
4. Add 1% triton X-100 (final concentration) and rotate at 4°C during 30 min.
5. Pellet the debris by centrifugation 25,000 × *g* for 30 min in an adapted centrifuge tube.
6. Collect the supernatant in a new 50-mL conical tube. *Save 5 μL of the supernatant for SDS-PAGE and add 10 μL SDS-PAGE sample buffer. Boil 4 min.*
7. Add 0.5 mL volume of glutathione sepharose beads prewashed into washing buffer and rotate at 4°C for 2 h.

8. Pellet beads gently at 800 × *g* for 5 min and collect depleted lysate to save in case not all of the GST-PIP4KIIα was bound. *Save 5 μL of the depleted lysate for SDS-PAGE and add 10 μL SDS-PAGE sample buffer. Boil 4 min.*
9. Wash the beads twice with 30 mL washing buffer containing Tween 20 then twice with 30 mL washing buffer without Tween 20.
10. Add 0.5 mL of elution buffer containing urea to the beads and incubate for 1 min.
11. Load the beads on the top of the column prepared during steps in Section 2.2.2.
12. Elute the first 500 μL by removing the cap of the column and place back the cap before to dry the column.
13. Slowly add 50 m*M* Tris pH 8 and elute 6 × 250 μL. *Save 5 μL of each elution and add 10 μL SDS-PAGE sample buffer. Boil 4 min and load 10 μL for SDS-PAGE along with the lysate (10 μL) and the depleted lysate (10 μL).*
14. Select elution fractions for pooling, pool and dialyze (cutoff: 50 kDa) vs storage buffer: 2 × 1 h in 1 L (at 4°C).
15. Recover dialysate, add glycerol to 10%, do a protein assay to determine yield, aliquot and snap freeze in liquid nitrogen.
16. Store at −80°C.

3. LIPID EXTRACTION FROM BIOLOGICAL SAMPLES

- **Important! All work with solvents must be done under a chemical hood.**

3.1 Lipid Extraction From Cell Extracts

The number of cells needed depends on the cell type but 10 million cells are generally a good starting point. NOTE: lipid extraction can be performed directly from cells in a culture dish, in suspension, or from a cell pellet stored at −80°C (pellets can be stored up to 1 month, longer storage times lead to unreliable measurements due to alterations in the lipids).

3.1.1 Adherent Cells

The given protocol is based on a 10-cm culture dish but can be adapted to other formats.

1. Wash the cell monolayer gently one time with 10 mL ice-cold PBS.
2. Scrape the cells in 1 mL ice-cold HCl 1.2 N using a rubber policeman.

3. Collect and transfer the lysate in a 50 mL polypropylene tube.
4. Wash the dish with 1 mL of ice-cold HCl 1.2 N and transfer to the corresponding tube.
5. Rewash the dish with 2 mL CH_3OH and transfer to the corresponding tube.
6. Add 2 mL of $CHCl_3$.
7. Sonicate using a probe sonicator in an ice bath (10 s, output 30%).
8. Strongly shake the 50 mL polypropylene tube by manual agitation during 3 × 10 s. NOTE: at this step, your extract should be milky a sign of good homogenization.
9. Centrifuge at 300 × *g* for 10 min at room temperature. NOTE: at this step, two phases are distinguishable with a white interphase (proteins). If not the case, add 200 μL of $CHCl_3$ and proceed again to step 8.
10. Collect the lower phase (organic phase) with a glass pipette and transfer into a ~6 mL glass tube.
11. Dry samples under a nitrogen stream (this step should take ~20 min). NOTE: to accelerate the process, you can use a waterbath set to 37°C.
12. At this step, you can store your dried lipids at −70°C for 1–2 days. NOTE: flush the samples with nitrogen gas to limit fatty acid oxidation.
13. Resuspend the dried lipid extracts in 100 μL of $CHCl_3/CH_3OH$ (1/1; v/v). NOTE: the resuspended lipids can be stored at 4°C overnight. NOTE: flush the samples with nitrogen gas to limit fatty acid oxidation. At this step, you can control lipid recovery between samples by total lipid phosphate assay (Jones, Ramirez, Lowe, & Divecha, 2013).
14. Proceed to Section 4.

3.1.2 Suspension Cells

Suspension cells should be in 50 mL polypropylene tubes for stimulation. The given protocol is based on an 800-μL cell volume, but can be adapted to other formats. For large volumes greater than 4 mL, follow steps in Section 3.1.4.

1. To stop stimulation, add 3 mL of $CHCl_3/CH_3OH$ (1/2; v/v) to the cells. NOTE: the final ratio of $CHCl_3/CH_3OH/H_2O$ will be 1/2/0.8; v/v/v.
2. Sonicate using a probe sonicator in an ice bath (10 s, output 30%).
3. Add 1 mL of $CHCl_3$ and 1 mL of HCl 2.3 N.
4. Proceed to step 8 of Section 3.1.1.

3.1.3 Preparation of a Cell Pellet From Adherent Cells

The given protocol is based on a 10-cm culture dish but can be adapted to other formats.

1. Wash the cell monolayer gently one time with 10 mL ice-cold PBS.
2. Add 1 mL of ice-cold HCl 1 N and scrape the cells with a rubber policeman.
3. Collect and transfer the lysate into 2 mL tube.
4. Wash the dish with 1 mL of ice-cold HCl 1 N and transfer to the corresponding tube.
5. Centrifuge at 15,000 × *g* for 5 min, 4°C.
6. Remove the supernatant, snap freeze the pellet in liquid N_2, and store at −80°C.
7. For lipid extraction, resuspend the pellet in 1 mL of $CHCl_3/CH_3OH$ (1/2; v/v) and transfer in a 50 mL polypropylene tube.
8. Wash the 2 mL tube with 2 × 1 mL of $CHCl_3/CH_3OH$ (1/2; v/v) and transfer in the corresponding 50 mL polypropylene tube.
9. Add 800 μL of H_2O and sonicate using a probe sonicator in an ice bath (10 s, output 30%).
10. Add 1 mL of $CHCl_3$ and 1 mL of HCl 2.3 N.
11. Proceed to step 8 of Section 3.1.1.

3.1.4 Preparation of a Cell Pellet From Suspension Cells

Suspension cells should be in 2 mL tubes for stimulation. The given protocol is based on an 800-μL cell volume, but can be adapted for other formats.

1. Add 800 μL of ice-cold HCl 2 N and homogenate very gently by tube inversion.
2. Proceed to step 5 of Section 3.1.3.

3.2 Lipid Extraction From Tissues

The amount of tissue should be determined case by case. Based on our experience, 200–300 mg per point is needed for skeletal muscle and ~50 mg for liver.

1. Grind the tissues using a benchtop homogenizer. We currently use a FastPrep Instrument with Matrix D (MP Bio). Tissues are lysed twice during 1 min at 6.5 m/s in 1 mL of $CHCl_3/CH_3OH$ (1/2; v/v).
2. Transfer samples to 50 mL polypropylene tube and rinse sample tubes twice with 1 mL of $CHCl_3/CH_3OH$ (1/2; v/v).
3. Add 800 μL of H_2O.

4. Sonicate using a probe sonicator in an ice bath (15 s, output 40%).
5. Add 1 mL of $CHCl_3$ and 1 mL of HCl 2.3 N.
6. Proceed to steps 8–14 of Section 3.1.1.

4. PURIFICATION OF PHOSPHATIDYLINOSITOL MONOPHOSPHATES

Before migrating a TLC plate, it is necessary to saturate the tank with the corresponding solvent for ~48 h. A solvent tank can be used for a maximum of four migrations and it is very important to store them and use them at room temperature.

1. Migrate a 20 cm × 20 cm TLC silica gel 60 plate (1.05721.001, Merck-Millipore, Billerica, USA) in a mixture containing 90 mL H_2O, 60 mL CH_3OH, 3 mL EDTA 0.1 *M*, and 1.5 g potassium oxalate and let the TLC dry at room temperature. NOTE: this step should take ~12 h and a migrated plate can be stored for 1 month.
2. "Activate" the TLC plate for 30 min at 70°C in an oven.
3. On the TLC plate, with a soft pencil, draw a parallel line starting at 2 cm from the bottom edge. This will corresponds to the origin line. Each sample should be separated by 1 cm from each other and by 1.5 cm minimum from the side edge (Fig. 3A).
4. Spot the resuspended samples onto the TLC and dry them using a nitrogen stream.
5. Rinse each glass tube with 30 μL of $CHCl_3/CH_3OH$ (1/1; v/v) and spot them.
6. Spot 5 μg of natural standards PtdIns4P (840045X, Avanti Polar Lipids, Alabaster, USA) and PtdIns(4,5)P_2 (840046X, Avanti Polar Lipids) on each side (Fig. 2A).
7. Resolve the TLC plates using $CHCl_3/CH_3OH$/4.3 *M* NH_4OH (90/70/20; v/v) as solvent (time of migration ~90 min).
8. Take out the TLC plate and allow it to dry under the hood at room temperature (~30 min).
9. Protect the samples with a glass plate and reveal the natural standards using iodine vapor in a glass tank in order to localize the migration of phosphatidylinositol monophosphates in the different samples (Fig. 3B). NOTE: to prepare the iodine tank, just add a few crystals of iodine, which have a high vapor pressure and will rapidly saturate the chamber.

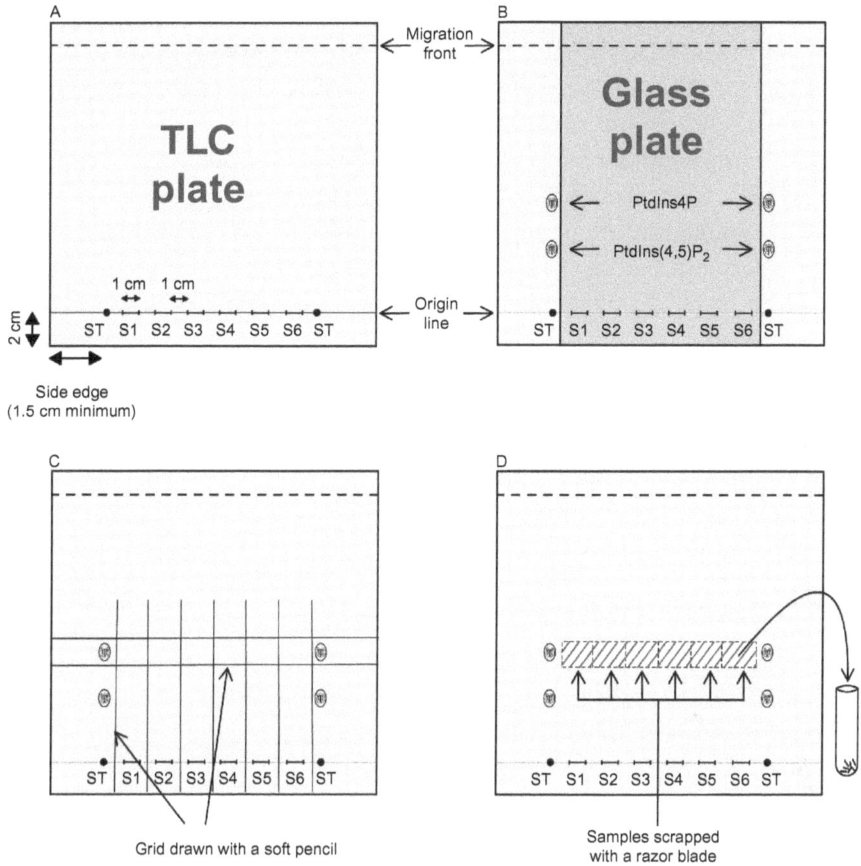

Fig. 3 Phosphatidylinositol monophosphates isolation procedure (see Section 3). (A) TLC map. (B) Standards are visualized using iodine vapor. Note that the samples are protected with a glass plate. (C) Identification of phosphoinositides migration based on standards. (D) Scrape off the phosphatidylinositol monophosphates.

10. Based on the position of the standards, identify the phosphatidylinositol monophosphates of your samples with a soft pencil and a ruler. Draw a rectangle of the same area for each sample (Fig. 2C).
11. Scrape off the silica with a razor blade and transfer into ~6 mL glass tubes (Fig. 2D).
12. For PtdIns3P mass assay, add 20 nmol of 1-palmitoyl-2-oleoyl-sn-glycero-3-phosphoethanolamine (POPE, 850757P, Avanti Polar Lipids) to each tube. For PtdIns5P mass assay, add 20 nmol of natural L-α-phosphatidylinositol (840042P, Avanti Polar Lipids) to each tube.
13. Add 1 mL of $CHCl_3/CH_3OH$ (1/1; v/v) then 0.5 mL of HCl 1.2 N.

14. Sonicate using a probe sonicator in an ice bath (2 × 7 s, output 30%).
15. Centrifuge at 300 × *g* for 10 min at room temperature.
16. Collect the lower phase (organic phase) with a glass pipette and transfer it in a new ~6 mL glass tube.
17. Dry samples under a nitrogen stream (this step should take ~20 min). NOTE: to accelerate the process, you can use a waterbath set to 37°C. The dried lipids can be stored at 4°C (flushed with nitrogen gas) for up to overnight before doing the kinase assay.
18. To prepare the standard curve, scrape four rectangles of blank silica (same size of the scraped rectangles) and transfer into ~6 mL glass tubes.
19. Add different amounts of C16-PtdIns3P (P3016, Echelon Biosciences, Salt Lake City, USA) for PtdIns3P mass assay or C16-PtdIns5P (P5016, Echelon Biosciences) for PtdIns5P mass assay from a stock solution of 1 pmol/μL: 0, 10, 50, and 100 pmol.
20. Proceed to step 12 of Section 4.

5. QUANTIFICATION OF PtdIns3P BY MASS ASSAY

Our lab has developed a mass assay that is able to accurately determine the amount of PtdIns3P in cell extracts (Chicanne et al., 2012). This assay is based on the use of recombinant PIKfyve (produced in the baculovirus-Sf9 insect cell system) and [γ-^{32}P]ATP to produce and quantify radiolabeled PtdIns(3,5)P_2 formed from PtdIns3P.

- **Important! The following steps require the use of radioactivity. Follow local rules when working with radioactivity.**

PtdIns3P Kinase Buffer: 25 m*M* Hepes pH 7.4, 150 m*M* NaCl, 5.5 m*M* β-glycerophosphate, 2.7 m*M* $MgCl_2$, 0.22 m*M* EDTA. Add extemporaneously 1.1 m*M* final DTT.

The protocol is adapted for 60 μL kinase reactions.

1. Resuspend the dried lipids (samples and standards) with 55 μL of PtdIns3P Kinase Buffer.
2. Sonicate 2 × 30 s in a cooled bath sonicator.
3. To start the kinase assay, add 5 μL of a mix to have 1 μg of recombinant GST-PIKfyve, 40 μ*M* cold ATP (A2383, Sigma-Aldrich, St. Louis, USA), and 20 μCi of [γ-^{32}P]ATP (BLU502A, Perkin Elmer, Waltham, USA) in each tube.
4. Incubate 15 min at 37°C with constant mixing and stop the reaction by adding 400 μL of $CHCl_3/CH_3OH$ (1/1; v/v).
5. To each tube, add 100 μL of H_2O, 30 μL of HCl 1.2 N, and 5 μg of Folch-extracted cow brain lipids (B1502, Sigma-Aldrich).

6. Vigorously mix using a vortex.
7. Centrifuge at 300 × *g* for 5 min, room temperature.
8. Collect the lower phase (organic phase) with a glass pipette and transfer it in a ~6 mL glass tube.
9. Dry samples under a nitrogen stream.
10. Perform a TLC according to steps 1–8 of Section 4.
11. Visualize [γ-^{32}P]-PtdIns(3,5)P_2 produced from PtdIns3P using an autoradiography film or an autoradiography scanner. PtdIns(3,5)P_2 migration is easily identifiable through the standard points (Fig. 4A).
12. Localize PtdIns(3,5)P_2 migration using a soft pencil as done for step 10 of Section 4.
13. Scrape off the silica with a razor blade and collect in liquid scintillation tubes.
14. Quantify radioactivity with a liquid scintillation counter.
15. Plot the standard curve to determine the amount of [γ-^{32}P]-PtdIns(3,5)P_2 formed. This amount of radioactivity is proportional to the PtdIns3P present in the samples (Fig. 4B).

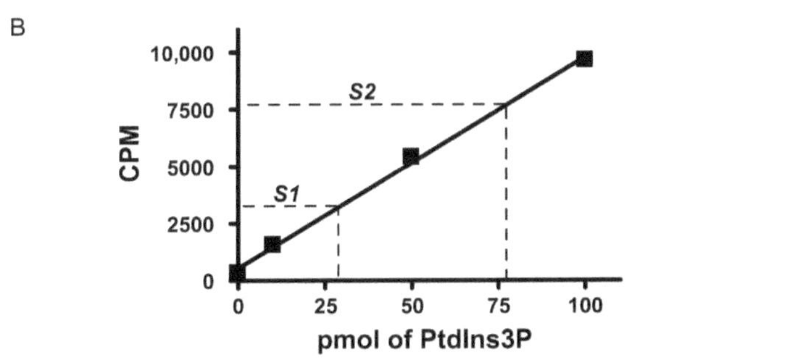

Fig. 4 Example of PtdIns3P quantification by mass assay (see Section 5). (A) Autoradiography of a TLC after a PIKfyve-kinase assay. [©-^{32}P]-PtdIns(3,5)P_2 is visualized using an autoradiography film or a phosphorimager and the resulting image is then used to identify the spots on the TLC plate to be scrape off. (B) Calibration curve obtained from standards points of panel A. This curve is then used to quantify the PtdIns3P present in the samples.

6. QUANTIFICATION OF PtdIns5P BY MASS ASSAY

This assay is based on the use of recombinant PIP4KIIα (produced in *Escherichia coli*) and [γ-^{32}P]ATP to produce and quantify radiolabeled PtdIns(4,5)P_2 formed from PtdIns5P (Morris et al., 2000).

- **Important! The following steps require the use of radioactivity. Follow local rules while working with radioactivity.**

PtdIns5P Kinase Buffer: 50 m*M* Tris pH 7.4, 100 m*M* NaCl, 5 m*M* $MgCl_2$, 1.5 m*M* DTT, 0.5 m*M* EDTA, 0.01% DOC.

1. Resuspend the dried lipids (samples and standards) with 60 μL of PtdIns5P Kinase Buffer.
2. Sonicate 2 × 30 s in a cooled bath sonicator.
3. To start the kinase assay, add a mixture of 0.3 μg of recombinant GST-PIP4KIIα, 10 μ*M* cold ATP (A2383, Sigma-Aldrich), and 20 μCi of [γ-^{32}P]ATP (BLU502A, Perkin Elmer) in each tube.
4. Incubate 15 min at 37°C with constant mixing and stop the reaction by adding 400 μL of CHCl3/CH3OH (1/1; v/v).
5. To each tube, add 100 μL of H_2O, 30 μL of HCl 1.2 N, and 5 μg of Folch-extracted cow brain lipids (B1502, Sigma-Aldrich).
6. Vigorously mix using a vortex.
7. Centrifuge at 300 × *g* for 5 min, room temperature.
8. Collect the lower phase (organic phase) with a glass pipette and transfer it in a ~6 mL glass tube.
9. Dry samples under a nitrogen stream.
10. Perform a TLC according to steps 1–8 of Section 4.
11. Visualize [γ-^{32}P]-PtdIns(4,5)P_2 produced from PtdIns5P using an autoradiography film or an autoradiography scanner. PtdIns(4,5)P_2 migration is easily identifiable through the standard points.
12. Localize PtdIns(4,5)P_2 migration using a soft pencil as done for step 10 of Section 4.
13. Scrape off the silica with a razor blade and collect in liquid scintillation tubes.
14. Quantify radioactivity with a liquid scintillation counter.
15. Plot the standard curve to determine the amount of [γ-^{32}P]-PtdIns(4,5) P_2 formed. This amount of radioactivity is proportional to the PtdIns3P present in the samples.

ACKNOWLEDGMENTS

The project was supported by grants from the Agence Nationale de la Recherche (ANR) (ANR-13-BSV2-0004-01) and ARC (Fondation ARC pour la recherche sur le cancer). B.P. is a senior member of the Institut Universitaire de France. We thank Jason S. Iacovoni for critical reading of the manuscript.

Data and materials availability: All plasmids described in the chapter are available upon request.

Conflict of interests: The authors declare that they have no competing financial interests.

REFERENCES

Bento, C. F., Renna, M., Ghislat, G., Puri, C., Ashkenazi, A., Vicinanza, M., et al. (2016). Mammalian autophagy: How does it work? *Annual Review of Biochemistry*, *85*, 685–713.

Burman, C., & Ktistakis, N. T. (2010). Regulation of autophagy by phosphatidylinositol 3-phosphate. *FEBS Letters*, *584*, 1302–1312.

Cebollero, E., van der Vaart, A., Zhao, M., Rieter, E., Klionsky, D. J., Helms, J. B., et al. (2012). Phosphatidylinositol-3-phosphate clearance plays a key role in autophagosome completion. *Current Biology*, *22*, 1545–1553.

Ceccato, L., Chicanne, G., Nahoum, V., Pons, V., Payrastre, B., Gaits-Iacovoni, F., et al. (2016). PLIF: A rapid, accurate method to detect and quantitatively assess protein-lipid interactions. *Science Signaling*, *9*, rs2.

Chicanne, G., Severin, S., Boscheron, C., Terrisse, A. D., Gratacap, M. P., Gaits-Iacovoni,-F., et al. (2012). A novel mass assay to quantify the bioactive lipid PtdIns3P in various biological samples. *The Biochemical Journal*, *447*, 17–23.

Clark, J., Anderson, K. E., Juvin, V., Smith, T. S., Karpe, F., Wakelam, M. J., et al. (2011). Quantification of PtdInsP3 molecular species in cells and tissues by mass spectrometry. *Nature Methods*, *8*, 267–272.

de Duve, C. (2005). The lysosome turns fifty. *Nature Cell Biology*, *7*, 847–849.

Dice, J. F. (1990). Peptide sequences that target cytosolic proteins for lysosomal proteolysis. *Trends in Biochemical Sciences*, *15*, 305–309.

Fetalvero, K. M., Yu, Y., Goetschkes, M., Liang, G., Valdez, R. A., Gould, T., et al. (2013). Defective autophagy and mTORC1 signaling in myotubularin null mice. *Molecular and Cellular Biology*, *33*, 98–110.

He, C., & Klionsky, D. J. (2009). Regulation mechanisms and signaling pathways of autophagy. *Annual Review of Genetics*, *43*, 67–93.

Jones, D. R., Ramirez, I. B., Lowe, M., & Divecha, N. (2013). Measurement of phosphoinositides in the zebrafish *Danio rerio*. *Nature Protocols*, *8*, 1058–1072.

Karanasios, E., & Ktistakis, N. T. (2015). Live-cell imaging for the assessment of the dynamics of autophagosome formation: Focus on early steps. *Methods*, *75*, 54–60.

Kielkowska, A., Niewczas, I., Anderson, K. E., Durrant, T. N., Clark, J., Stephens, L. R., et al. (2014). A new approach to measuring phosphoinositides in cells by mass spectrometry. *Advances in Biological Regulation*, *54*, 131–141.

Mochizuki, Y., Ohashi, R., Kawamura, T., Iwanari, H., Kodama, T., Naito, M., et al. (2013). Phosphatidylinositol 3-phosphatase myotubularin-related protein 6 (MTMR6) is regulated by small GTPase Rab1B in the early secretory and autophagic pathways. *The Journal of Biological Chemistry*, *288*, 1009–1021.

Morris, J. B., Hinchliffe, K. A., Ciruela, A., Letcher, A. J., & Irvine, R. F. (2000). Thrombin stimulation of platelets causes an increase in phosphatidylinositol 5-phosphate revealed by mass assay. *FEBS Letters*, *475*, 57–60.

Munnik, T. (2013). Analysis of D3-,4-,5-phosphorylated phosphoinositides using HPLC. *Methods in Molecular Biology*, *1009*, 17–24.

Noda, T., Matsunaga, K., Taguchi-Atarashi, N., & Yoshimori, T. (2010). Regulation of membrane biogenesis in autophagy via PI3P dynamics. *Seminars in Cell & Developmental Biology*, *21*, 671–676.

Obara, K., Noda, T., Niimi, K., & Ohsumi, Y. (2008). Transport of phosphatidylinositol 3-phosphate into the vacuole via autophagic membranes in *Saccharomyces cerevisiae*. *Genes to Cells*, *13*, 537–547.

Obara, K., Sekito, T., Niimi, K., & Ohsumi, Y. (2008). The Atg18-Atg2 complex is recruited to autophagic membranes via phosphatidylinositol 3-phosphate and exerts an essential function. *The Journal of Biological Chemistry*, *283*, 23972–23980.

Payrastre, B. (2004). Phosphoinositides: Lipid kinases and phosphatases. *Methods in Molecular Biology*, *273*, 201–212.

Sarkes, D., & Rameh, L. E. (2010). A novel HPLC-based approach makes possible the spatial characterization of cellular PtdIns5P and other phosphoinositides. *The Biochemical Journal*, *428*, 375–384.

Taguchi-Atarashi, N., Hamasaki, M., Matsunaga, K., Omori, H., Ktistakis, N. T., Yoshimori, T., et al. (2010). Modulation of local PtdIns3P levels by the PI phosphatase MTMR3 regulates constitutive autophagy. *Traffic*, *11*, 468–478.

Vergne, I., Roberts, E., Elmaoued, R. A., Tosch, V., Delgado, M. A., Proikas-Cezanne, T., et al. (2009). Control of autophagy initiation by phosphoinositide 3-phosphatase Jumpy. *The EMBO Journal*, *28*, 2244–2258.

Viaud, J., Boal, F., Tronchere, H., Gaits-Iacovoni, F., & Payrastre, B. (2014). Phosphatidylinositol 5-phosphate: A nuclear stress lipid and a tuner of membranes and cytoskeleton dynamics. *Bioessays*, *36*, 260–272.

Viaud, J., Mansour, R., Antkowiak, A., Mujalli, A., Valet, C., Chicanne, G., et al. (2016). Phosphoinositides: Important lipids in the coordination of cell dynamics. *Biochimie*, *125*, 250–258.

Vicinanza, M., Korolchuk, V. I., Ashkenazi, A., Puri, C., Menzies, F. M., Clarke, J. H., et al. (2015). PI(5)P regulates autophagosome biogenesis. *Molecular Cell*, *57*, 219–234.

Zou, J., Zhang, C., Marjanovic, J., Kisseleva, M. V., Majerus, P. W., & Wilson, M. P. (2012). Myotubularin-related protein (MTMR) 9 determines the enzymatic activity, substrate specificity, and role in autophagy of MTMR8. *Proceedings of the National Academy of Sciences of the United States of America*, *109*, 9539–9544.

CHAPTER EIGHTEEN

Fluorescence-Based Assays to Analyse Phosphatidylinositol 5-Phosphate in Autophagy

M. Vicinanza, M.J. Gratian, M. Bowen, D.C. Rubinsztein[1]
Cambridge Institute for Medical Research, Wellcome/MRC Building, Cambridge Biomedical Campus, Cambridge, United Kingdom
[1]Corresponding author: e-mail address: dcr1000@cam.ac.uk

Contents

1. PI(5)P Role in Autophagy 312
2. Microscopy-Based Detection of PI(5)P 315
   2.1 Cell Culture Conditions 316
   2.2 Autophagosomal Markers 317
   2.3 Expression of GFP-PHD3X(ING2) to Visualize PI(5)P During Autophagy by Confocal Microscopy 317
3. Manipulations of PI(5)P Levels to Visualize PI(5)P During Autophagy 318
   3.1 siRNA Transfection 319
   3.2 Genetic Perturbation of PI(5)P Levels 320
   3.3 Pharmacological Manipulations of PI(5)P 320
   3.4 Fluorescent PI(5)P Analogs 320
4. Super-Resolution Structured Illumination Microscopy (SR-SIM) to Visualize PI(5)P During Autophagy 321
   4.1 Super-Resolution Structured Illumination Microscopy (SR-SIM) 322
   4.2 Instrument Preparation 324
   4.3 Sample Preparation 325
   4.4 Sample Imaging 325
   4.5 Image Processing 326
5. Concluding Remarks 328
Acknowledgments 328
References 328

Abstract

Autophagosome formation is stimulated by VPS34-dependent PI(3)P formation and by alternative VPS34-independent pathways. We recently described that PI(5)P regulates autophagosome biogenesis and rescues autophagy in VPS34-inactivated cells, suggesting that PI(5)P contributes to canonical autophagy. Our analysis revealed a hitherto unknown functional interplay between PIKfyve and PIPK type II in controlling PI(5)P levels in the context of autophagy. Among phosphoinositides, visualization of PI(5)P in

Methods in Enzymology, Volume 587
ISSN 0076-6879
http://dx.doi.org/10.1016/bs.mie.2016.09.062

intact cells has remained difficult. While PI(5)P has been implicated in signaling pathways, chromatin organization, bacterial invasion, and cytoskeletal remodeling, our study is the first report showing PI(5)P localization on autophagosomes and early autophagosomal structures when autophagy is induced by nutrient deprivation (amino acids or glucose starvation). We provided a detailed analysis of PI(5)P distribution by the use of super-resolution structured illuminated microscopy. Here, we present a set of tools for detection of PI(5)P during autophagy by confocal microscopy, live-cell imaging, and super-resolution microscopy.

1. PI(5)P ROLE IN AUTOPHAGY

Phosphoinositides (PIs) are low-abundance lipids that cluster on cytosol-facing platforms to dynamically recruit effector proteins to specific membranes (Balla, 2013). The metabolism of PIs is locally regulated by sets of phosphoinositide kinases and phosphatases distributed in diverse intracellular compartments. Their interconvertibility enables PIs to undergo rapid changes in response to nutrient availability and cell stress, and makes PIs ideal transducers of perturbations of cellular homeostasis and metabolism into membrane trafficking events.

First identified in 1997 (Rameh, Tolias, Duckworth, & Cantley, 1997), PI(5)P has been implicated in growth factor signaling pathways (Boal et al., 2015; Liu et al., 2006), chromatin organization and gene expression (Alvarez-Venegas et al., 2006; Di Lello et al., 2005; Gelato et al., 2014; Stijf-Bultsma et al., 2015; Viiri et al., 2009), bacterial invasion and cytoskeletal remodeling (Grindheim et al., 2014; Viaud, Lagarrigue, et al., 2014). PI(5)P levels have been shown to increase in response to stimuli such as thrombin, insulin, oxidative and osmotic stress, etoposide and UV irradiation, cell cycle progression, fibroblast growth factor 1 (Viaud, Boal, Tronchere, Gaits-Iacovoni, & Payrastre, 2014). PI(5)P production is relevant to infections with *Salmonella* and *Shigella flexneri* (Niebuhr et al., 2002), T-cell activation and antiviral innate immune responses (Guittard et al., 2009, 2010), myogenic differentiation (Fugier et al., 2011; Stijf-Bultsma et al., 2015), and carcinogenesis (Viaud, Boal, et al., 2014).

Autophagy is a bulk degradation process that delivers cytosolic content to lysosomes and is induced by stress conditions like nutrient deprivation, growth factor withdrawal, and low cellular energy levels (Rubinsztein, Codogno, & Levine, 2012). The double-membraned autophagosomes are formed from cup-shaped, double-membrane precursors, called phagophores, whose edges have not yet fused. Phagophores are positive for both LC3 and ATG16L1, while autophagosomes have LC3 but no ATG16L1 (see Fig. 1A). Autophagy has been classically considered to be phosphatidylinositol 3-phosphate (PI(3)P) dependent. Genetic studies in yeast and flies and pharmacological manipulations in cells suggested that PI(3)P production by the class III phosphatidylinositol 3-kinase (VPS34) is required for autophagosome formation as this lipid enables the recruitment of proteins necessary for autophagosome biogenesis (like WIPI1-2, WD-repeat protein interacting with phosphoinositide) (Dall'Armi, Devereaux, & Di Paolo, 2013; Proikas-Cezanne, Takacs, Donnes, & Kohlbacher, 2015). However, autophagosomes can be detected in T-lymphocytes and sensory neurons from *Vps34*$^{-/-}$ mice (Zhou et al., 2010) and in glucose-depleted cells treated with VPS34 inhibitors (Codogno, Mehrpour, & Proikas-Cezanne, 2012; McAlpine, Williamson, Tooze, & Chan, 2013), suggesting that a PI(3)P independent, noncanonical autophagy pathway exists.

We recently reported a role for phosphatidylinositol 5-phosphate (PI(5)P) in autophagy (Vicinanza et al., 2015). This is compatible with previous reports showing autophagy regulatory proteins binding to PI(5)P (Baskaran, Ragusa, Boura, & Hurley, 2012; Lorenzo, Urbe, & Clague, 2005). We detected PI(5)P on nascent and mature autophagosomes. Increased cellular PI(5)P levels enhanced autophagosome formation, while depletion of PI(5)P decreased the numbers of autophagosome precursors and completed autophagosomes. We found that elevating PI(5)P levels was able to induce autophagosome formation even when VPS34 was inhibited or silenced, leading us to conclude that PI(5)P can serve as an "alternative" to PI(3)P in the autophagosome biogenesis. We confirmed that PI(5)P is involved in PI(3)P-independent autophagy pathways like glucose starvation- or resveratrol-induced autophagy, where PI(3)P is dispensable (Fig. 1A).

The levels of PIs are also important for autophagosome maturation. PI(3)P promotes retrograde movement of autophagosomes via the dynein–dynactin complex and its effector FYCO1 (FYVE and coiled-coil domain containing 1) (Pankiv et al., 2010), while the completion of autophagy requires clearance of PI(3)P (via myotubularin 3-phosphatases) enabling fusion with the lysosome (Vergne & Deretic, 2010). PI(3,5)P2 has been proposed to promote fusion events and compartment acidification at the endolysosomal network and promote recycling of autophagolysosome membrane (McCartney, Zhang, & Weisman, 2014).

A

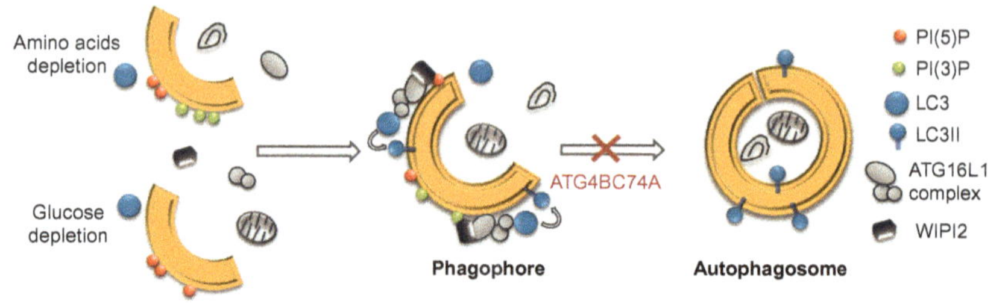

B

PI(5)P-binding proteins

Protein	PI-Binding Domain	Localization	References
WIPI1, WIPI2	FRRG motif	Autophagosome	Baskaran et al. (2012)
MTMR3	PH-GRAM	Autophagosome, Golgi	Lorenzo et al. (2005)
Dok1, Dok2, Dok4, Dok5	PH	Plasma membrane	Guittard et al. (2009, 2010)
BIN1	—	Endosomes	Fugier et al. (2011)
TOM1	VHS	Endosomes	Boal et al. (2015)
TIAM1	PH	Endosomes	Viaud, Lagarrigue et al. (2014) and Viaud, Boal, et al. (2014)
SNX5	—	Endosomes	Liu et al. (2006)
Annexin A2	—	Endosomes, plasma membrane, nucleus	Grindheim et al. (2014)
ING2	PHD	Nucleus	Gozani et al. (2003) and Bua, Martin, Binda, and Gozani (2013)
ATX1	PHD	Nucleus	Alvarez-Venegas et al. (2006)
TFIIH	PH	Nucleus	Di Lello et al. (2005)
Sap30/Sap30L	—	Nucleus	Viiri et al. (2009)
TAF3	PHD	Nucleus	Stijf-Bultsma et al. (2015)
UHRF1	PBR	Nucleus	Gelato et al. (2014)
PLEKHA5	PH	—	Yamada, Nomura, Yamano, Yamada, and Wakamatsu (2012)

C

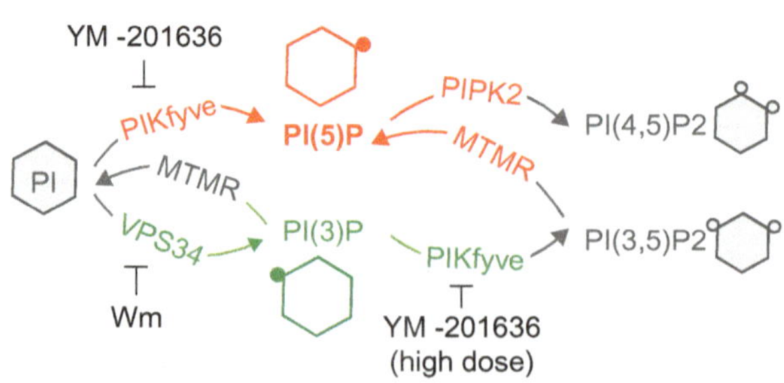

Fig. 1 See legend on opposite page.

Although PI(5)P has been reported to regulate endosome maturation, it remains undefined whether PI(5)P plays a role in autophagosome–lysosome fusion (Boal et al., 2015).

2. MICROSCOPY-BASED DETECTION OF PI(5)P

About 2% of the human genome encodes proteins carrying specific PI-binding domains (including PH, FYVE, PX, and PHD domains) (Hammond & Balla, 2015). These domains have become valuable tools in visualizing PIs in intact cells. When expressed as fusions with fluorescent probes (i.e., GFP, RFP), PI-binding domains can be used to determine the intracellular location and relative levels of PIs in live and fixed cells by wide-field or confocal fluorescent microscopy.

Note: Extreme care must be applied when PI-reporters are exogenously expressed in cells, because, depending on the level of expression and the PI affinity of the probe, problems may occur. These include limited access to certain PIs pools that are already in complex with cellular proteins or interference of exogenous PI-domains that can prevent endogenous proteins from interacting with their cognate lipid.

A major limitation in studying the intracellular localization of endogenous PI(5)P is its low abundance (estimated as approximately 1–2% of monophosphorylated PIs). Several biochemical and microscopy approaches have detected PI(5)P pools at plasma membrane, nucleus, and endosomal compartments (Jones & Smirnoff, 2006; Ramel et al., 2011; Sarkes & Rameh, 2010). Another limitation in studying PI(5)P localization is the uncertain specificity of PI(5)P bioprobes. PI(5)P-binding proteins are beginning to be identified (Fig. 1B) and include proteins carrying the PHD (plant homeodomain) finger, the PH (pleckstrin homology) domain, the PH-GRAM (glucosyl transferase, Rab-like GTPase activator and myotubularins), and PBR (PolyBasicRegion) motif (Fig. 1B). The PHD domains are zinc-finger motifs and PHD finger of inhibitor of growth

Fig. 1 (A) Autophagy pathways regulated by PI(5)P (Vicinanza et al., 2015); (B) Summary of cellular proteins shown to bind PI(5)P by different lipid interaction domains and localized in diverse subcellular compartments. *PH*, Pleckstrin homology domain; *PHD*, plant homeo-domain; *PH-GRAM*, glucosyl transferase, Rab-like GTPase activator and myotubularins; *PBR*, PolyBasicRegion; *VHS*, (Vps27-Hrs-STAM). Note that some of these proteins (i.e., WIPI1 and WIPI2) have been shown to bind PI(3)P and PI(3,5)P2. (C) Schematic representation of pathways for PI(5)P and PI(3)P synthesis/turnover and compounds targeting enzymes involved in these pathways.

protein 2 (ING2) was one of the first convincingly shown to interact with PI(5)P (Gozani et al., 2003). The PHD domain of ING2 (Bua et al., 2013; Gozani et al., 2003) and the PH domain of Dok (downstream of kinase) proteins (Guittard et al., 2009, 2010) show strong preference for PI(5)P and have been used to obtain spatial information on the distribution of PI(5)P.

We detected PI(5)P using GFP-tagged PHD3X (three tandem repeats of PHD domain of ING2) on discrete puncta, which were positive for autophagosome markers and increased in numbers when cells were starved (Vicinanza et al., 2015).

A PHD mutant defective in PI(5)P binding has been used as a control (PHD3X Znmut) (Bua et al., 2013). It is critical to avoid extreme PHD3X overexpression, as overexpression (30 h) of GFP-PHD3X (and GFP-PH-Dok-5) sequester intracellular PI(5)P and dramatically decrease the number of mature autophagosomes appearing in starved-cells.

2.1 Cell Culture Conditions

We grow HeLa, human cervix, adenocarcinoma cells (ATCC, CCL-2) at 37°C and 5% CO_2 in Dulbecco's modified Eagle's medium (DMEM) with 4500 mg/L glucose, sodium pyruvate (1 m*M*) (Gibco/Invitrogen Corporation, United Kingdom), supplemented with L-glutamine (2 m*M*), 10% fetal calf serum (FCS) (Sigma, USA), 100 U/mL penicillin, and 100 mg/mL streptomycin solution (Gibco/Invitrogen Corporation, United Kingdom). We routinely use 75 cm^2 BD Falcon culture flasks with 0.2 mm vented blue plug seal cups (BD Europe, France). For long-term storage in liquid nitrogen, we freeze $1–5 \times 10^6$ cells in 2 mL of FCS 10% dimethyl-sulfoxide (DMSO). We routinely check for the absence of mycoplasma by PCR analyses.

Generally, we seed $1–2 \times 10^5$ cells in 2 mL of culture medium on glass coverslips in each well of BD Falcon six-well tissue culture plates (BD Falcon, Europe, France). Transfection was performed with TransIT 2020 Mirus (for DNA) or LipofectAMINE 2000 (for siRNA) reagents (Invitrogen), using the manufacturer's protocol.

Autophagy induction by nutrient deprivation:

Amino acid and serum deprivation: cells incubated in Hank's balanced salt solution (HBSS) media (Invitrogen) for 1–2 h (*Note*: HBSS does not contain amino acids and serum and does contain low levels of glucose: 1 g/L of D-glucose vs 4.5 g/L glucose in DMEM); glucose deprivation: cells incubated in DMEM lacking glucose (Invitrogen, #11966) with 10% dialyzed FBS (Sigma) and glutamine for 4–6 h.

2.2 Autophagosomal Markers

Mature autophagosomes and autophagolysosomes can be detected using RFP-LC3B transiently expressed in HeLa cells for 16 h. Phagophores and prephagophore structures can be detected using mStrawberry-ATG16L1 transiently expressed in HeLa cells for 16 h (Ravikumar, Moreau, Jahreiss, Puri, & Rubinsztein, 2010). In both cases, care needs to be taken to avoid overexpression artifacts. To better visualize phagophores we use the expression of proteolytic activity-deficient mutant of ATG4B (ATG4BC74A). This is tool that freezes the phagophore stage (that in normal conditions are transient and rare structures) by preventing LC3 lipidation and autophagosome completion (Fujita et al., 2008).

2.3 Expression of GFP-PHD3X(ING2) to Visualize PI(5)P During Autophagy by Confocal Microscopy

1. HeLa cells are seeded at 1–2 × 10^5 cells in 2 mL of culture medium on glass coverslips in six-well plates. Cells are incubated overnight to 70–80% confluency at 37°C and 5% CO_2.
2. Place 200 μL of prewarmed OPTIMEM reduced-serum medium in a sterile tube under the hood.
3. Add 1 μg of GFP-PHD3X (or GFP-PHD3X Znmut as negative control) and 0.3 μg of ATG16L1-mStrawberry/0.5 μg of RFP-LC3 to 200 μL OPTIMEM by gently pipetting (*Note*: in cases of triple transfection, we use 0.8 μg of GFP-PHD3X, 0.3 μg of ATG16L1-mStrawberry, and 0.5 μg of flag-tagged ATG4B (ATG4BC74A)).
4. Briefly vortex transfection reagent TransIT 2020 Mirus (Invitrogen) at room temperature and then add 3–4 μL of Mirus to the diluted DNA mixture. (*Note*: we usually keep a DNA:Mirus ratio of 1:3.)
5. Pipette gently to mix and incubate at room temperature for 15–20 min, with mixing every 5 min.
6. Slowly apply the 200 μL Mirus–DNA complex to wells by gently pipetting into 1 mL full serum culture media (*Note*: before adding transfection complex, cells are washed in complete culture medium containing 10% serum and low levels of antibiotics).
7. Incubate cells for 16 h.
8. Aspirate off the full serum culture medium, wash twice with prewarmed HBSS (37°C) to remove residual serum and incubate with 1 mL of prewarmed HBSS (37°C) for 1 h.

9. Fix the cells adding 1 mL of prewarmed 4% paraformaldehyde (dissolved in PBS, pH 7.0 freshly made a few hours earlier from powder) to 1 mL of HBSS (final concentration is 2% paraformaldehyde) and incubate for 10 min at room temperature.
10. Wash the cells two times with PBS and once with water.
11. Coverslips are mounted with 20 μL ProLong Antifade Reagent (Molecular Probes, catalog number P-36930) and stored at 4°C in the dark.

For confocal laser scanning microscopy, we used LSM710 and LSM780 microscopes (Zeiss). All images were taken with 63 × 1.4 NA DIC Plan-Apochromat oil-immersion objectives (Zeiss, Jena). We currently use helium neon lasers for excitations at 543 or 633 nm, a diode laser for excitation at 561 nm and argon lasers for excitations at 488 nm. As a standard procedure, we take 8–10 fields, enough to collect images from about 80 cells for each condition and then we also acquire higher magnification images of five individual cells.

Any saturated images identified by using the integrated range indicator tool are discarded and acquisition settings (i.e., laser intensity) are adjusted accordingly.

ImageJ is used for further analysis (number of vesicles analysis and colocalization) and processing of confocal images.

3. MANIPULATIONS OF PI(5)P LEVELS TO VISUALIZE PI(5)P DURING AUTOPHAGY

Manipulation of intracellular PI(5)P levels by targeting enzymes relevant for its biogenesis and turnover showed that GFP-PHD3X puncta formation under autophagy induction conditions is dependent on PI(5)P (Vicinanza et al., 2015). PI(5)P biosynthesis is regulated by the type III PtdInsP 5-kinase PIKfyve, as reduced PI(5)P levels are seen in PIKfyve hypomorph and heterozygous mice and in cells silenced by siRNA, overexpressing a dominant-negative, or treated with pharmacological inhibitor of the kinase (Ikonomov et al., 2011; Sbrissa, Ikonomov, Filios, Delvecchio, & Shisheva, 2012; Sbrissa, Ikonomov, & Shisheva, 2002; Zolov et al., 2012).

PIKfyve can either directly phosphorylate PI, or make PI(3,5)P2, which is then transformed into PI(5)P by 3-phosphatases of the myotubularin family (MTMRs) (Fig. 1C). On the other hand, the major route for

PI(5)P removal is attributed to type II PIPK kinases (phosphatidylinositol 5-phosphate 4-kinases, PI5P4K) (Clarke, Wang, & Irvine, 2010; Shisheva, 2013).

Although we cannot exclude that the PHD3X domain binds other lipids (mainly PI(3)P) in addition to PI(5)P, PHD3X localized in our experiments at the nucleus and plasma membrane, as previously described for PI(5)P (Jones & Smirnoff, 2006; Sarkes & Rameh, 2010). PHD3X detected PI(5)P on autophagosomes when PI(3)P levels where diminished (VPS34 inhibition by wortmannin treatment or 3-phosphatase MTMR3 overexpression), while PHD3X-positive structures were lost in conditions expected to increase PI(3)P levels but decrease PI(5)P levels (PIKFyve and MTMR3 silencing) (Fig. 1C). These observations lead us to conclude that PHD3X detects PI(5)P independently of PI(3)P, at least in autophagy-inducing conditions (HBSS or glucose-depleted media).

3.1 siRNA Transfection

1. Routinely, two rounds of knockdown for 5 days with final siRNA concentrations of 50 or 100 n*M* are applied.
2. Seed HeLa cells at 60–70% confluency in six-well plates and incubate overnight to at 37°C and 5% CO_2.
3. Day 1: dilute 5–10 μL Lipofectamine 2000 (Invitrogen, catalog number 11668019) in 250 μL of prewarmed OPTIMEM reduced-serum medium (Invitrogen, catalog number 51985042). Dilute 5–10 μL of siRNA oligos (stock solution 20 μ*M*) in 250 μL of prewarmed OPTIMEM. Incubate for 5 min at room temperature. Combine diluted siRNA and diluted Lipofectamine 2000 and incubate 20 min at room temperature. Wash the cells with 1 mL of OPTIMEM and slowly apply the siRNA/Lipofectamine 2000 mixture (finale volume of 500 μL). Incubate for 6 h at 37°C and 5% CO_2 and then replace the transfection medium with culture medium containing serum.
4. Day 2: repeat the siRNA oligo transfection.
5. Day 3: split cells and seed 1–2 × 10^5 cells in 2 mL of culture medium on glass coverslips in six-well plates.
6. Day 4: transfect cells with GFP-PHD3X in combination with RFP-LC3 or ATG16L1-mStrawberry (see Section 2.3).
7. Day 5: incubate with 1 mL of prewarmed HBSS (37°C) for 1 h, fix and process for confocal microscopy (see Section 2.3).

3.2 Genetic Perturbation of PI(5)P Levels

Decreased PI(5)P synthesis: PIKfyve (Gene ID: 200576) and MTMR3 (Gene ID: 8897) silencing;

Decreased PI(5)P removal: PIP4K2A (Gene ID: 5305), PIP4K2B (Gene ID: 8396), and PIP4K2C (Gene ID: 79837) silencing.

See Table S1 from Vicinanza et al. (2015) for siRNA oligos used (Dharmacon).

3.3 Pharmacological Manipulations of PI(5)P

PIKfyve inhibitors: YM-201636 (Cayman Chem).

Incubate cells with 1 mL of fresh culture medium or HBSS or glucose-depleted medium containing 100 n*M* YM-201636 (*Note*: we use low doses of YM-201636, between 100 and 200 n*M*, previously shown to selectively inhibit PI(5)P synthesis via PIKfyve (Sbrissa et al., 2012)).

3.4 Fluorescent PI(5)P Analogs

Fluorescently labeled PI(5)P analogs are commercially available and include acyl-labeled BODIPY derivatives of PI(5)P. BODIPY (4,4-difluoro-3a,4a-diaza-s-indacene) is an hydrophobic fluorescent compound that, when attached to the end of one of the acyl chain (sn-1 position), can be used to monitor the distribution of PI(5)P without altering the structure of the inositol head group. BODIPY fluorescent analogs exist as FL (Excitation: 503 nm/Emission: 513 nm) and TMR (Excitation: 542 nm/Emission: 574 nm) versions (Echelon Bioscience, C-05F6a and C-05M6, respectively). BODIPY-PI(5)P analogs are loaded into cellular membranes with use of a specific carrier (Ozaki, DeWald, Shope, Chen, & Prestwich, 2000).

We use BODIPY-PI(5)P di-C6 at final concentrations of 1 or 10 μ*M*; carrier-only is used as negative control. Cells are processed for live cell imaging, since BODIPY-PI(5)P analogs are not fixable. (*Note*: PIs cannot be fixed by standard fixatives and fixed cells cannot be permeabilized with standard detergents that may extract the lipids.)

Cautionary note: use low adherent polystyrene tubes or glass tubes; reconstitute PIs and carriers in aqueous solution containing small amounts of methanol, ethanol, or *tert*-BuOH; vortex mix and briefly sonicate to completely dissolve carriers and PIs; avoid using phosphate buffers when combining carrier and lipids.

3.4.1 Addition of Exogenous PI(5)P to Cells

1. BODIPY-labeled PI(5)P di-C6 and carrier (Echelon Bioscience) are reconstituted in H_2O:*tert*-BuOH (9:1) solution.
2. Vortex for 1 min and apply 1 min of bath sonication (4°C, cold room) to reconstitute solutions.
3. Combine carrier and lipids at a 1:1 ratio in sterile low adherent tubes and incubate for 10 min at room temperature.
4. Dilute the mixture in starvation medium (HBSS) and use for short incubations (30 min to 1 h) on cells.

3.4.2 Live Cell Imaging of Cells Loaded With Fluorescent PI(5)P Analogs

HeLa cells transfected with RFP-LC3 (see Section 2.3) are exposed to HBSS containing BODIPY-labeled-PI(5)P for 1 h, followed by live cell imaging for 10 min in prewarmed buffered HBSS medium (10 m*M* Hepes).

We performed time-lapse microscopy using a Zeiss LSM 780 confocal microscope equipped with 488 nm argon, 561 nm diode, and 633 nm HeNe lasers and a 63 × 1.4NA DIC Plan-Apochromat oil immersion objective (Carl Zeiss, Jena, Germany). The microscope was operated using the ZEN black software, version 2012. To allow live cell imaging, the microscope was equipped with a PeCon XL3 environmental chamber, temperature control unit and heating unit (Zeiss, Jena).

To prepare for time-lapse microscopy, HeLa cells are grown to 70% confluency in 35 mm MatTek glass bottom dishes (MatTek Corporation, Ashland, MA), transfected with RFP-LC3 (see Section 2.3) and loaded with fluorescent PI(5)P (see Section 3.4.1).

The incubator is equilibrated to 37°C before starting the time-lapse microscopy. Images were taken at interval of 10 s for 10 min. Images series were elaborated by using ImageJ software.

4. SUPER-RESOLUTION STRUCTURED ILLUMINATION MICROSCOPY (SR-SIM) TO VISUALIZE PI(5)P DURING AUTOPHAGY

Conventional light microscopy (i.e., confocal) is useful to determine whether a given phosphoinositide is enriched on a specific organelle but does not allow discrimination between closely adjacent small membrane compartments or membrane subdomains. In this respect, electron microscopy (EM) provides a much higher resolving power than standard light microscopy; however, fixation and postfixation manipulations required can potentially affect the structure of lipids. Super-resolution microscopy provides resolution between conventional light microscopy and EM. Conventional light microscopy, including confocal, is subject to optical diffraction which places a physical limit on the resolution achievable. This limit is approximately half the wavelength of the emitted light and is often generalized to be 200 nm maximum. In recent years, various techniques and technologies have been used to circumvent this limit and thereby achieve so-called super resolution.

We carried out 3D analysis of GFP-PHD3X-labeled structures by super-resolution microscopy using a Zeiss Elyra PS1 microscope, in order to have a deeper understanding of PI(5)P localization on nascent autophagosomes in autophagy-stimulating-conditions (amino acid and serum starvation, HBSS media) (Vicinanza et al., 2015) (Fig. 3).

4.1 Super-Resolution Structured Illumination Microscopy (SR-SIM)

Super-resolution structured illumination microscopy (SR-SIM) is a technique which allows extraction of normally unresolvable high-frequency information by means of a moving pattern in the illuminating light, coupled with postacquisition computational reconstruction of multiple images into a single super-resolved image or volume. By super-imposition of a known pattern (the illumination) onto an unknown pattern (the details of the sample), followed by careful analysis of the interference between the two patterns, extraction of detail beyond the resolving power of the objective is achieved.

The structured illumination pattern in the Zeiss Elyra is formed by inserting a grid into the incident light-path which projects a regular arrangement of light and dark stripes into the sample. The pattern spacing is tailored to the wavelength(s) required and is shifted in both phase and rotation during the acquisition of multiple images.

SR-SIM offers two main advantages over confocal imaging, a twofold increase in resolution in x, y, and z dimensions and a large increase in dynamic range. For example, the 250 nm conventional resolution for GFP is increased to 125 nm laterally and around 250 nm axially. This twofold increase in x, y, and z resolution equates to an $8\times$ reduction in the size of the smallest imaging volume (Gustafsson et al., 2008) (Fig. 2A and C).

SR-SIM is a widefield technique making use of cameras (typically sCMOS type), which results in a large increase in dynamic range vs confocal microscopes using photomultiplier type detectors, therefore, more weakly

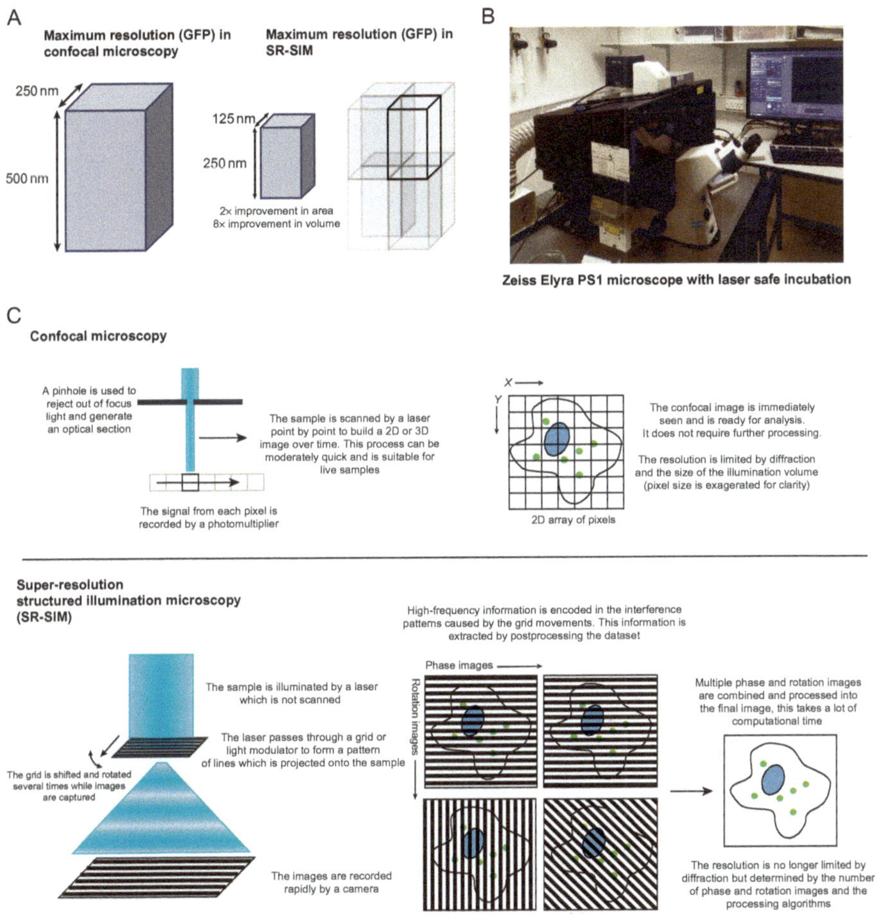

Fig. 2 (A) Maximum image resolution of super-resolution structured illuminated microscopy (SR-SIM) and conventional confocal microscopy. (B) Zeiss Elyra PS1 microscope equipped with an environment-controlled chamber. (C) Image acquisition and image processing of super-resolution structured illuminated microscopy (SR-SIM) and conventional confocal microscopy.

fluorescent structures are easily detected without the risk of saturation of brighter structures. Thus, SR-SIM images routinely demonstrate a much larger population of fluorescent structures than can be observed in a single confocal image.

The main disadvantage of SR-SIM is the speed of acquisition. In order to generate one super resolved optical section in a single color, it is necessary to collect a minimum of five grid phases and three grid rotations, i.e., a minimum of 15 images (Fig. 2C). Resolution isotropy is improved by increasing the number of grid rotations captured up to 5, resulting in 25 images per color per optical section. Imaging is therefore necessarily slow and this greatly reduces its applicability for live samples. If structures in the sample move during the acquisition of the multiple images required for capture of a single time point, then the reconstruction algorithms will give rise to artifacts. Recently, this speed issue has been addressed by the use of very rapid imaging techniques using light-sheet illumination and spatial light modulators to generate the SIM pattern. These instruments bring SR-SIM into the realm of rapid live cell imaging (Chen et al., 2014).

4.2 Instrument Preparation

For SR-SIM, we use a Zeiss Elyra PS1 (Carl Zeiss GmbH, Jena, Germany. http://www.zeiss.com/microscopy/en_de/products/superresolution-microscopy.html) (Fig. 2B) with a carefully matched 63× 1.4NA Zeiss Plan-Apochromat objective. This lens was specially selected from a batch to produce an optimally formed point-spread function on our particular instrument.

1. Stabilization: In order to maximize system stability and compensate for environmental variations, the microscope is sited on an active anti-vibration table (Vision IsoStation, Newport Corp., Irvine, CA) and is isolated with an environmental chamber (PeCon GmbH, Jena, Germany). This chamber also provides additional laser safety interlocks and is black to prevent the transmission of harmful laser radiation.
2. Temperature: We maintain the room temperature at around 18–20°C during imaging and heat the environmental chamber to precisely 23°C which produces the optimal refractive index in the immersion oil (Immersol 1.518 at 23°C, Carl Zeiss GmbH, Jena). Slides and oil are stored in the chamber so the entire optical train is at 23°C.
3. Stage equilibration/thermal alignment: The Elyra instrument is allowed to equilibrate for at least 3 h after switching on, to stabilize the focus (i.e., to minimize expansion/contraction of elements forming the optical path). Subsequent to this equilibration, the stage insert is aligned so that it is exactly perpendicular to the optical light path through the objective. This is achieved by means of two set-screw adjustments on the stage insert which are turned to super-impose two reflected images using a custom made alignment tool.
4. Chromatic aberration correction: After stage alignment and thermal equilibration, it is necessary to perform a bead alignment to correct for any chromatic aberration which may be present. Chromatic

aberration, which arises from imperfections in the optical system, would be too small an effect to be problematic in confocal imaging but is resolvable by super-resolution instruments; it manifests as a mismatch in the alignment of the different color channels. For example, different colors will be offset from one another to some degree. By imaging multicolor fluorescent beads (2076-515 Low Density MultiSpec, Carl Zeiss GmbH, Jena), this mismatch can be quantified and a table of corrections generated using an affine analysis method. These corrections, which include lateral and axial offsets, image rotations and shear in all three dimensions, are saved as an alignment calibration file and are subsequently used to correct the experimental images. Thus, we can ensure that structures which colocalize to within the resolution of the Elyra are correctly presented as colocalized.

4.3 Sample Preparation

1. HeLa cells are seeded at $1–2 \times 10^5$ cells in 2 mL of culture medium on high-precision 18 mm square number 1.5 coverslips (Zeiss Ltd.) (which have a thickness of 170 μm and a variation of less than ±5 μm) in six-well plates. Cells are incubated overnight to 70–80% confluency at 37°C and 5% CO_2.
2. Cells are transfected with 0.3 μg of ATG16L1-mStrawberry (or RFP-LC3) and 1 μg of GFP-PHD3X (see Section 2.3).
3. Coverslips are mounted with 20 μL ProLong Gold (P36934, LifeTechnologies) and allowed to cure at room temperature for 72 h in order to achieve a refractive index which matches the immersion oil at 23°C. These coverslips are mounted on conventional glass slides without frosting or adhesive labels, in the geometric center, one coverslip per slide. Note: if single side frosted slides are used, it is important to mount the coverslip on the plain side so that it will lie completely flat on the microscope stage. Slides frosted on both sides should not be used.

4.4 Sample Imaging

The slides are preheated to 23°C by placing them in the microscope chamber for 10 min before imaging.

Image acquisition: Images are acquired with the following microscope configuration. Excitation of the three channels is by 488 nm/200 mW, 560 nm/200 mW, and 640 nm/150 mW diode lasers, structured illumination grid pitches are 28, 34, and 34 μm, respectively, and emission filters are BP495-550, BP570-620, and LP655nm. We image z-stack through-focus series of images at a spacing interval of 110 nm in three colors sequentially using five grid phase shifts and five grid rotations, resulting in a raw data image sequence comprising 75 images per z-slice. A typical z-stack comprises of 20 slices covering an approximately 2 μm thickness of the sample, thus resulting in 1500 images total per single output volume. The camera used is a PCO edge sCMOS device (PCO AG, Kelheim, Germany), which is thermoelectrically cooled to 5°C. Waste heat is extracted with a liquid cooling system (Innovatak OS GmbH, Stammham, Germany) to prevent the vibrations, which occur with fan-based camera cooling. This camera has a large area sensor and is capable of imaging up 2240 × 2154 pixels but this is typically not necessary, as 1024 × 1024 or 512 × 512 usually provide an ample field area. 512 × 512 pixels equates to a field area of 52.7 μm^2. Raw data in the form of multidimensional image arrays are processed in the frequency domain using complex reconstructive algorithms (Gustafsson et al., 2008).

We have found that as long as the sample is prepared optimally and the correct materials are used to maintain consistency of refractive index, the automated structured illumination processing built into the Elyra version of ZEN software performs extremely well. Adjustment of manual processing parameters is possible for "difficult" samples such as particularly thick specimens, low signal-to-noise ratio, or (to some extent) refractive index mismatch. However, we have found that it is far better to optimize the sample preparation than to adjust the image processing parameters. If the structured illumination pattern is not visible during imaging due to inadequate matching of refractive index, no amount of postprocessing will generate an acceptable super-resolved image.

4.5 Image Processing

Reconstructed images are then corrected for spherical and chromatic aberrations using channel alignment information created earlier. The output images are then saved in the proprietary *.czi file format. The average final image resolution was calculated to be 110 nm in x and y dimensions and 240 nm in the z dimension, which represents a slightly better than twofold lateral and axial improvement in resolution, compared to conventional microscopy. Final images can be further analyzed in additional software packages. We use Volocity 6.3 software (Perkin Elmer Cell Imaging, Seer Green, UK) to generate final 3D isosurface renderings and videos of selected cropped regions of the image volumes for presentation and publication. Note that this rendering type means that vesicles positive for green and red do not look yellow but have green and red on the surface (Fig. 3).

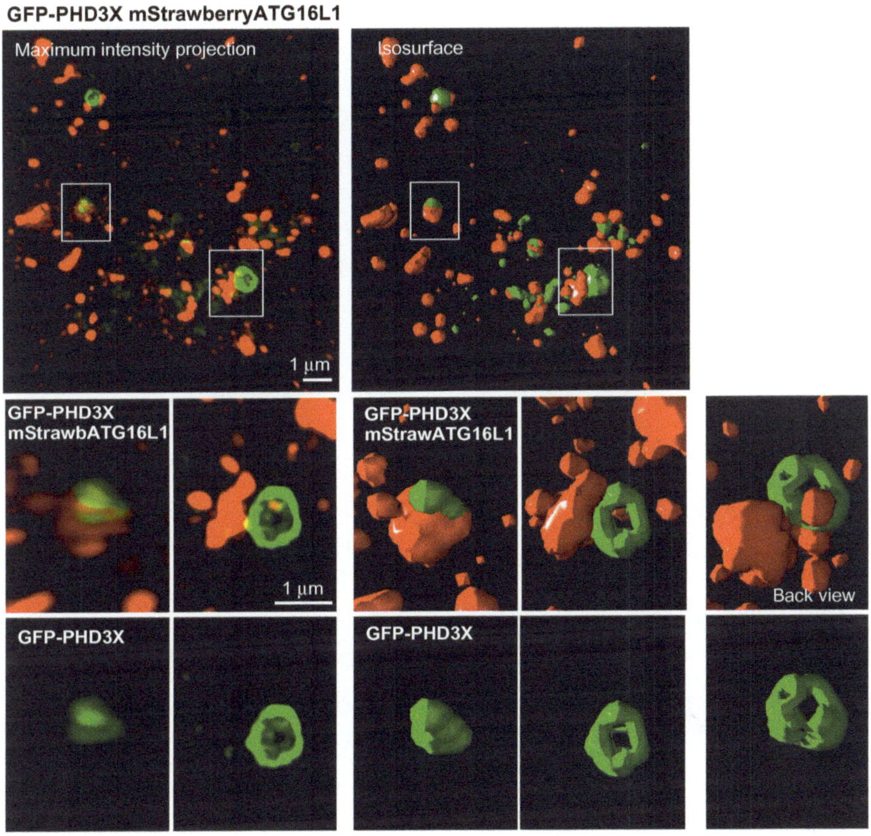

Fig. 3 HeLa cells transfected with GFP-PHD3X and Strawberry-ATG16L1 for 16 h were left in HBSS for 1 h, then fixed and imaged on Elyra super-resolution microscope (Vicinanza et al., 2015). Final visualization was performed in Volocity 6.3 Software using maximum intensity projection (*left*) or isosurface rendering (*right*) of selected cropped regions of the datasets. Note that when rendering is applied, vesicles that are positive for *green* and *red*, have *green* and *red* on the surface but do not appear *yellow* (as showed maximum intensity projection on the left). It is inherently difficult to present such 3D data in a 2D format. Usually such data are presented via maximum intensity projection (MIP) which we have also shown on the *left*. We felt that MIP rendering was insufficient in this case to properly appreciate the shapes of the structures presented. To improve matters, we chose to use isosurface rendering, which uses light and shade to better present 3D structures as solid forms. The reflected light and shadows are important visual cues, which allow a certain degree of three dimensionality in a 2D reproduction. Of course, the best way to evaluate the structure is to see a movie where the isosurface image is rotated in space.

5. CONCLUDING REMARKS

Visualizing PI(5)P and interfering with its metabolism during autophagy represent a new tool to monitor and study autophagosome formation. Although will be important to develop new and more specific PI(5)P probes as well as continue to identify additional effectors to better understand PI(5)P function in autophagy, our approach is employable for screening regulators of autophagy.

ACKNOWLEDGMENTS

We are grateful to the Wellcome Trust (Principal Research Fellowship to DCR (095317/Z/11/Z)) and a Strategic Grant to Cambridge Institute for Medical Research (100140/Z/12/Z) for funding.

REFERENCES

Alvarez-Venegas, R., Sadder, M., Hlavacka, A., Baluska, F., Xia, Y., Lu, G., et al. (2006). The Arabidopsis homolog of trithorax, ATX1, binds phosphatidylinositol 5-phosphate, and the two regulate a common set of target genes. *Proceedings of the National Academy of Sciences of the United States of America*, *103*(15), 6049–6054.

Balla, T. (2013). Phosphoinositides: Tiny lipids with giant impact on cell regulation. *Physiological Reviews*, *93*(3), 1019–1137.

Baskaran, S., Ragusa, M. J., Boura, E., & Hurley, J. H. (2012). Two-site recognition of phosphatidylinositol 3-phosphate by PROPPINs in autophagy. *Molecular Cell*, *47*(3), 339–348.

Boal, F., Mansour, R., Gayral, M., Saland, E., Chicanne, G., Xuereb, J. M., et al. (2015). TOM1 is a PI5P effector involved in the regulation of endosomal maturation. *Journal of Cell Science*, *128*(4), 815–827.

Bua, D. J., Martin, G. M., Binda, O., & Gozani, O. (2013). Nuclear phosphatidylinositol-5-phosphate regulates ING2 stability at discrete chromatin targets in response to DNA damage. *Scientific Reports*, *3*, 2137.

Chen, B. C., Legant, W. R., Wang, K., Shao, L., Milkie, D. E., Davidson, M. W., et al. (2014). Lattice light-sheet microscopy: Imaging molecules to embryos at high spatiotemporal resolution. *Science*, *346*(6208), 1257998.

Clarke, J. H., Wang, M., & Irvine, R. F. (2010). Localization, regulation and function of type II phosphatidylinositol 5-phosphate 4-kinases. *Advances in Enzyme Regulation*, *50*(1), 12–18.

Codogno, P., Mehrpour, M., & Proikas-Cezanne, T. (2012). Canonical and non-canonical autophagy: Variations on a common theme of self-eating? *Nature Reviews. Molecular Cell Biology*, *13*(1), 7–12.

Dall'Armi, C., Devereaux, K. A., & Di Paolo, G. (2013). The role of lipids in the control of autophagy. *Current Biology*, *23*(1), R33–R45.

Di Lello, P., Nguyen, B. D., Jones, T. N., Potempa, K., Kobor, M. S., Legault, P., et al. (2005). NMR structure of the amino-terminal domain from the Tfb1 subunit of TFIIH and characterization of its phosphoinositide and VP16 binding sites. *Biochemistry*, *44*(21), 7678–7686.

Fugier, C., Klein, A. F., Hammer, C., Vassilopoulos, S., Ivarsson, Y., Toussaint, A., et al. (2011). Misregulated alternative splicing of BIN1 is associated with T tubule alterations and muscle weakness in myotonic dystrophy. *Nature Medicine*, *17*(6), 720–725.

Fujita, N., Hayashi-Nishino, M., Fukumoto, H., Omori, H., Yamamoto, A., Noda, T., et al. (2008). An Atg4B mutant hampers the lipidation of LC3 paralogues and causes defects in autophagosome closure. *Molecular Biology of the Cell*, *19*(11), 4651.

Gelato, K. A., Tauber, M., Ong, M. S., Winter, S., Hiragami-Hamada, K., Sindlinger, J., et al. (2014). Accessibility of different histone H3-binding domains of UHRF1 is allosterically regulated by phosphatidylinositol 5-phosphate. *Molecular Cell*, *54*(6), 905–919.

Gozani, O., Karuman, P., Jones, D. R., Ivanov, D., Cha, J., Lugovskoy, A. A., et al. (2003). The PHD finger of the chromatin-associated protein ING2 functions as a nuclear phosphoinositide receptor. *Cell*, *114*(1), 99–111.

Grindheim, A. K., Hollas, H., Ramirez, J., Saraste, J., Trave, G., & Vedeler, A. (2014). Effect of serine phosphorylation and Ser25 phospho-mimicking mutations on nuclear localisation and ligand interactions of annexin A2. *Journal of Molecular Biology*, *426*(13), 2486–2499.

Guittard, G., Gerard, A., Dupuis-Coronas, S., Tronchere, H., Mortier, E., Favre, C., et al. (2009). Cutting edge: Dok-1 and Dok-2 adaptor molecules are regulated by phosphatidylinositol 5-phosphate production in T cells. *Journal of Immunology*, *182*(7), 3974–3978.

Guittard, G., Mortier, E., Tronchere, H., Firaguay, G., Gerard, A., Zimmermann, P., et al. (2010). Evidence for a positive role of PtdIns5P in T-cell signal transduction pathways. *FEBS Letters*, *584*(11), 2455–2460.

Gustafsson, M. G., Shao, L., Carlton, P. M., Wang, C. J., Golubovskaya, I. N., Cande, W. Z., et al. (2008). Three-dimensional resolution doubling in wide-field fluorescence microscopy by structured illumination. *Biophysical Journal*, *94*(12), 4957–4970.

Hammond, G. R., & Balla, T. (2015). Polyphosphoinositide binding domains: Key to inositol lipid biology. *Biochimica et Biophysica Acta*, *1851*(6), 746–758.

Ikonomov, O. C., Sbrissa, D., Delvecchio, K., Xie, Y., Jin, J. P., Rappolee, D., et al. (2011). The phosphoinositide kinase PIKfyve is vital in early embryonic development: Preimplantation lethality of PIKfyve −/− embryos but normality of PIKfyve +/− mice. *The Journal of Biological Chemistry*, *286*(15), 13404–13413.

Jones, M., & Smirnoff, N. (2006). Nuclear dynamics during the simultaneous and sustained tip growth of multiple root hairs arising from a single root epidermal cell. *Journal of Experimental Botany*, *57*(15), 4269–4275.

Liu, H., Liu, Z. Q., Chen, C. X., Magill, S., Jiang, Y., & Liu, Y. J. (2006). Inhibitory regulation of EGF receptor degradation by sorting nexin 5. *Biochemical and Biophysical Research Communications*, *342*(2), 537–546.

Lorenzo, O., Urbe, S., & Clague, M. J. (2005). Analysis of phosphoinositide binding domain properties within the myotubularin-related protein MTMR3. *Journal of Cell Science*, *118*(Pt 9), 2005–2012.

McAlpine, F., Williamson, L. E., Tooze, S. A., & Chan, E. Y. (2013). Regulation of nutrient-sensitive autophagy by uncoordinated 51-like kinases 1 and 2. *Autophagy*, *9*(3), 361–373.

McCartney, A. J., Zhang, Y., & Weisman, L. S. (2014). Phosphatidylinositol 3,5-bisphosphate: Low abundance, high significance. *Bioessays*, *36*(1), 52–64.

Niebuhr, K., Giuriato, S., Pedron, T., Philpott, D. J., Gaits, F., Sable, J., et al. (2002). Conversion of PtdIns(4,5)P(2) into PtdIns(5)P by the S. flexneri effector IpgD reorganizes host cell morphology. *The EMBO Journal*, *21*(19), 5069–5078.

Ozaki, S., DeWald, D. B., Shope, J. C., Chen, J., & Prestwich, G. D. (2000). Intracellular delivery of phosphoinositides and inositol phosphates using polyamine carriers. *Proceedings of the National Academy of Sciences of the United States of America*, *97*(21), 11286–11291.

Pankiv, S., Alemu, E. A., Brech, A., Bruun, J. A., Lamark, T., Overvatn, A., et al. (2010). FYCO1 is a Rab7 effector that binds to LC3 and PI3P to mediate microtubule plus end-directed vesicle transport. *The Journal of Cell Biology*, *188*(2), 253–269.

Proikas-Cezanne, T., Takacs, Z., Donnes, P., & Kohlbacher, O. (2015). WIPI proteins: Essential PtdIns3P effectors at the nascent autophagosome. *Journal of Cell Science, 128*(2), 207–217.

Rameh, L. E., Tolias, K. F., Duckworth, B. C., & Cantley, L. C. (1997). A new pathway for synthesis of phosphatidylinositol-4,5-bisphosphate. *Nature, 390*(6656), 192–196.

Ramel, D., Lagarrigue, F., Pons, V., Mounier, J., Dupuis-Coronas, S., Chicanne, G., et al. (2011). Shigella flexneri infection generates the lipid PI5P to alter endocytosis and prevent termination of EGFR signaling. *Science Signaling, 4*(191), ra61.

Ravikumar, B., Moreau, K., Jahreiss, L., Puri, C., & Rubinsztein, D. C. (2010). Plasma membrane contributes to the formation of pre-autophagosomal structures. *Nature Cell Biology, 12*(8), 747–757.

Rubinsztein, D. C., Codogno, P., & Levine, B. (2012). Autophagy modulation as a potential therapeutic target for diverse diseases. *Nature Reviews. Drug Discovery, 11*(9), 709–730.

Sarkes, D., & Rameh, L. E. (2010). A novel HPLC-based approach makes possible the spatial characterization of cellular PtdIns5P and other phosphoinositides. *The Biochemical Journal, 428*(3), 375–384.

Sbrissa, D., Ikonomov, O. C., Filios, C., Delvecchio, K., & Shisheva, A. (2012). Functional dissociation between PIKfyve-synthesized PtdIns5P and PtdIns(3,5)P2 by means of the PIKfyve inhibitor YM201636. *American Journal of Physiology Cell Physiology, 303*(4), C436–C446.

Sbrissa, D., Ikonomov, O. C., & Shisheva, A. (2002). Phosphatidylinositol 3-phosphate-interacting domains in PIKfyve. Binding specificity and role in PIKfyve. Endomenbrane localization. *The Journal of Biological Chemistry, 277*(8), 6073–6079.

Shisheva, A. (2013). PtdIns5P: News and views of its appearance, disappearance and deeds. *Archives of Biochemistry and Biophysics, 538*(2), 171–180.

Stijf-Bultsma, Y., Sommer, L., Tauber, M., Baalbaki, M., Giardoglou, P., Jones, D. R., et al. (2015). The basal transcription complex component TAF3 transduces changes in nuclear phosphoinositides into transcriptional output. *Molecular Cell, 58*(3), 453–467.

Vergne, I., & Deretic, V. (2010). The role of PI3P phosphatases in the regulation of autophagy. *FEBS Letters, 584*(7), 1313–1318.

Viaud, J., Boal, F., Tronchere, H., Gaits-Iacovoni, F., & Payrastre, B. (2014). Phosphatidylinositol 5-phosphate: A nuclear stress lipid and a tuner of membranes and cytoskeleton dynamics. *Bioessays, 36*(3), 260–272.

Viaud, J., Lagarrigue, F., Ramel, D., Allart, S., Chicanne, G., Ceccato, L., et al. (2014). Phosphatidylinositol 5-phosphate regulates invasion through binding and activation of Tiam1. *Nature Communications, 5*, 4080.

Vicinanza, M., Korolchuk, V. I., Ashkenazi, A., Puri, C., Menzies, F. M., Clarke, J. H., et al. (2015). PI(5)P regulates autophagosome biogenesis. *Molecular Cell, 57*(2), 219–234.

Viiri, K. M., Janis, J., Siggers, T., Heinonen, T. Y., Valjakka, J., Bulyk, M. L., et al. (2009). DNA-binding and -bending activities of SAP30L and SAP30 are mediated by a zinc-dependent module and monophosphoinositides. *Molecular and Cellular Biology, 29*(2), 342–356.

Yamada, K., Nomura, N., Yamano, A., Yamada, Y., & Wakamatsu, N. (2012). Identification and characterization of splicing variants of PLEKHA5 (Plekha5) during brain development. *Gene, 492*(1), 270–275.

Zhou, X., Wang, L., Hasegawa, H., Amin, P., Han, B. X., Kaneko, S., et al. (2010). Deletion of PIK3C3/Vps34 in sensory neurons causes rapid neurodegeneration by disrupting the endosomal but not the autophagic pathway. *Proceedings of the National Academy of Sciences of the United States of America, 107*(20), 9424–9429.

Zolov, S. N., Bridges, D., Zhang, Y., Lee, W. W., Riehle, E., Verma, R., et al. (2012). In vivo, Pikfyve generates PI(3,5)P2, which serves as both a signaling lipid and the major precursor for PI5P. *Proceedings of the National Academy of Sciences of the United States of America, 109*(43), 17472–17477.

CHAPTER NINETEEN

Ultrastructural Characterization of Phagophores Using Electron Tomography on Cryoimmobilized and Freeze Substituted Samples

J. Biazik*, H. Vihinen†, E. Jokitalo†, E.-L. Eskelinen*,[1]
*University of Helsinki, Helsinki, Finland
†Electron Microscopy Unit, Institute of Biotechnology, University of Helsinki, Helsinki, Finland
[1]Corresponding author: e-mail address: eeva-liisa.eskelinen@helsinki.fi

Contents

1. Introduction 332
2. Cryoimmobilization 334
3. Materials and Methods 335
   3.1 Chemical Fixation 335
   3.2 Cryoimmobilization and Freeze Substitution 335
   3.3 Electron Tomography 336
4. Results 337
   4.1 Ultrastructure of Autophagosomes in HPF–FS Samples 337
   4.2 ET Performed on Cryoimmobilized Samples: The Phagophore 337
   4.3 Chemical Fixation vs Cryoimmobilization 341
   4.4 Chemical Fixation vs Cryoimmobilization: MCSs 342
5. Discussion 342
Acknowledgments 346
References 346

Abstract

Electron tomography has significantly contributed to recent findings regarding the biogenesis of the phagophore, an organelle which initiates autophagic sequestration. The information obtained from 1.9 nm slices through the tomograms have revealed that during biogenesis the phagophore is in contact with the membranes of apposing organelles to form tubular connections and membrane contact sites (MCSs). The most reported and established tubular connections occur between the phagophore and the endoplasmic reticulum. However, as the phagophore continues to grow and expand, connections and MCSs have also been reported to occur between the phagophore and several other organelles in a possible attempt to utilize lipids for membrane expansion from alternative sources. Since the lifespan of the phagophore is only a few minutes and membrane connections and MCSs are very dynamic, capturing these two events

Methods in Enzymology, Volume 587
ISSN 0076-6879
http://dx.doi.org/10.1016/bs.mie.2016.09.063

requires precision during fixation. Up to date there is no quicker alternative for sample preservation in transmission electron microscopy than cryoimmobilization. In this report, we describe our protocol for cryoimmobilization using high-pressure freezing and freeze substitution, and report our first findings on phagophore morphology using this approach.

ABBREVIATIONS

EM electron microscopy
ER endoplasmic reticulum
ET electron tomography
FS freeze substitution
HPF high-pressure freezing
MCS membrane contact site

1. INTRODUCTION

Electron microscopy (EM) has been instrumental in the study of autophagy, which has generously benefited from technological advancements in the field of EM, including electron tomography (ET) (Biazik, Vihinen, Anwar, Jokitalo, & Eskelinen, 2015; Hayashi-Nishino et al., 2009; Yla-Anttila, Vihinen, Jokitalo, & Eskelinen, 2009a), immuno-EM (Cook et al., 2014; Eskelinen, 2005; Jager et al., 2004), and more recently serial block face EM (Biazik, Yla-Anttila, Vihinen, Jokitalo, & Eskelinen, 2015). EM, particularly ET, has been pivotal in autophagy research in elucidating the membrane relationships during the biogenesis and expansion of the phagophore. In these studies, ET was performed on chemically fixed samples revealing that the endoplasmic reticulum (ER) membranes form tubular connections with the phagophore (Hayashi-Nishino et al., 2009; Yla-Anttila et al., 2009a). In a more recent study, also performed on chemically fixed cells, membrane contact sites (MCSs) or tubular connections were also evident between the expanding phagophore and a series of other organelles, including mitochondria, ER exit sites, late endosomes/lysosomes, as well as the Golgi complex (Biazik, Yla-Anttila, et al., 2015). These studies have been informative and provided three-dimensional ultrastructural clues about the possible origins of the phagophore membrane. However, the constant scrutiny that morphologists face in the field of EM is that chemical fixation contributes to the formation of many artifacts.

The word artifact is used to refer to "man-made" structures induced by fixation or later sample preparation, which are not present in the living cell. Fixation, however, is unavoidable. It is required to prevent the alterations to the cells caused by autolysis that would occur during the later sample preparation steps. In the case of routine chemical fixation using glutaraldehyde, the fixative penetrates cells within seconds, but the cross-linking properties of glutaraldehyde are selective (McDonald & Auer, 2006). The reason for this is that glutaraldehyde mainly reacts with primary amino groups (e.g., side chains of lysines) that are rarely or not at all present in lipids, nucleic acids, and carbohydrate molecules (Buys & Pretorius, 2012; McDonald & Auer, 2006). Postfixation in osmium tetroxide gives support and contrast to lipid membranes, but osmium tetroxide mainly reacts with double bonds that are not present in all lipids (White, Andrews, Faller, & Barrnett, 1976). As a consequence, the unbound molecules may be extracted from the cells during subsequent washes and during the dehydration step. For general morphological purposes, this is not problematic and electron microscopists are familiar with these artifacts. Various researchers have identified very little change to morphology depending on the method of fixation used (Mazzone, Kornblau, & Durand, 1980), whereas some have reported small artifacts arising from chemical fixation. The reported artifacts from chemical fixation include the following: filamentous components tend to aggregate (Gulley & Reese, 1981; Rostaing et al., 2006), fibers in general are more compact and dense (Buys & Pretorius, 2012), endosomal vacuoles undergo deformation and reduction in volume (Murk et al., 2003), tubular extensions are more pronounced in endosomes (Murk et al., 2003), membranes which are clearly perforated tend to reseal (Szczesny, Walther, & Muller, 1996), and finally, specimens with impermeable cell walls or cuticles such as plant pollen and *Caenorhabditis elegans* are almost impossible to preserve (Studer, Humbel, & Chiquet, 2008). At the other end of the spectrum, some researchers have noticed very large changes in ultrastructural morphology depending on the mode of fixation used. One study suggests that the two layers of the basement membrane, lamina lucida and lamina densa, are an artifact of chemical fixation and that only one layer, the lamina densa is present in the basement membrane after the sample has been cryoimmobilized (Chan & Inoue, 1994).

Therefore, we aimed to assess whether phagophore morphology is different depending on whether cryoimmobilization is used instead of conventional chemical fixation.

2. CRYOIMMOBILIZATION

The alternative method to chemical fixation employed in this study is cryoimmobilization with high-pressure freezing (HPF). This is followed by freeze substitution (FS) and chemical fixation carried out at low temperature, and finally by resin embedding (Studer et al., 2008). This method is used in many studies, particularly when complimented with ET (Murk et al., 2004; Seemann, Kurth, & Entzeroth, 2012; Tomova et al., 2009). The goal of cryoimmobilization is the removal of heat at an ultra-rapid rate so that the water molecules in the sample form vitreous ice (Dubochet et al., 1988; McDonald & Auer, 2006) instead of crystalline cubic ice which would cause damage to the sample. This step is performed in a HPF machine which exerts 210 MPa (2048 bar) pressure on the sample while rapidly cooling it. The high pressure works against the expansion of water molecules as the sample cools, hence preventing ice crystal formation. Once the cells are frozen, there is no thermal energy left in the sample to allow lipids and proteins to move (Kellenberger, 1987; McDonald & Auer, 2006), and as long as the specimen is kept below –90°C, these macromolecules remain immobilized and there is also no risk of any of the vitreous ice reverting to cubic ice. The sample is then transferred into a FS machine for chemical cross-linking to be carried out without altering the cellular ultrastructure. Here, the sample is dehydrated in a solvent such as acetone during the initial incubation step below –90°C, followed by cross-linking with osmium tetroxide, uranyl acetate, or glutaraldehyde. During the cross-linking step, the FS chamber is slowly and gradually warmed as different components of the FS medium cross-link macromolecules at different temperatures, e.g., osmium tetroxide begins to cross-link carbon double bonds at –70°C (White et al., 1976). Recently, it was also concluded that adding 5–10% water to the FS medium improved contrast in the sample (Biazik, Vihinen, et al., 2015; Knoops et al., 2008), whereby the presence of water was critical for the retention of structure at temperatures around –60°C (Buser & Walther, 2008). After gradual warming to 0°C or room temperature, the sample is embedded in resin and remains stable allowing it to be manipulated for either immuno-EM or ET.

The HPF–FS method has been successfully used to fix and process difficult samples including *C. elegans* and *Drosophila melanogaster* (Hohenberg, Mannweiler, & Muller, 1994; McDonald & Morphew, 1993; Rostaing, Weimer, Jorgensen, Triller, & Bessereau, 2004). This method has also improved the preservation of synapses in the rat brain (Frotscher, Zhao, Graber, Drakew, & Studer, 2007), preserved the ultrastructure of biofilms (Palsdottir et al., 2009), helped to identify MCS between the ER and the intracellular parasite *Toxoplasma gondii* (Tomova et al., 2009), and helped to elucidate the three-dimensional (3D) structure of multilaminar lysosomes (Murk et al., 2004). It was also shown that the amount of lipids lost during HPF–FS was minimal (Pfeiffer et al., 2000). Immunoreactivity within the sample is also well preserved using this method (Rostaing et al., 2004). Sharp et al. coupled superior close to native state cryoimmobilization with immuno-EM and demonstrated that gold particles were evenly spaced at 60 nm intervals which coincided with the predicted distance between antibody epitopes in the bipolar kinesin molecule (Sharp et al., 1999).

To date, most ET studies describing phagophore morphology have mainly relied on sample preparation with routine chemical fixation (Biazik, Yla-Anttila, et al., 2015; Hayashi-Nishino et al., 2009; Yla-Anttila et al., 2009a). We aimed to clarify whether similar phagophore morphology is observed when the cells are fixed in the near-native state using cryoimmobilization with HPF followed by FS, resin embedding, and ET. The growth and elongation phase of the phagophore is a very short lived and dynamic event. Therefore, we reasoned that it requires a rapid immobilization to capture the 3D morphology, including membrane connections and MCSs.

3. MATERIALS AND METHODS

3.1 Chemical Fixation

Adherent cells were grown on 3 cm dishes containing glass cover slips to subconfluency. The cells were fixed in 2% glutaraldehyde in 0.1 *M* sodium cacodylate buffer, pH 7.4, postfixed in reduced osmium tetroxide, and flat embedded in Epon as described (Yla-Anttila, Vihinen, Jokitalo, & Eskelinen, 2009b).

3.2 Cryoimmobilization and Freeze Substitution

1. Human retinal pigment epithelium (ARPE-19) cells on normal rat kidney (NRK-52E) cells were cultured on 1.2- or 6-mm sapphire discs or directly on poly-L-lysine-coated gold HPF carriers (Leica Microsystems, Vienna, Austria).

2. The cells were overlaid with 20% bovine serum albumin in phosphate buffer, and immediately high-pressure frozen using a Leica EM HPM100 apparatus (Leica Microsystems).
3. The samples were transferred under liquid nitrogen to the automated freeze substitution apparatus (Leica EM AFS) into a solution containing 2% osmium tetroxide, 0.3% uranyl acetate, and 10% water in acetone. To make the mixture, the water was added first and frozen, based on a method modified from Knoops et al. (2008) and as detailed in Biazik, Vihinen, et al. (2015).
4. Samples were maintained at –95°C for 4 h, slowly warmed to –60°C (5 degree per hour) and maintained for 2 h, slowly warmed to –30°C (5° per hour) and maintained for 2 h, and slowly warmed to 0°C (5 degree per hour).
5. Two washes with cold acetone were carried out at 0°C and finally the cells were infiltrated in TAAB resin at room temperature and polymerized at +60°C.
6. For routine transmission EM, 70-nm sections were cut with a diamond knife and stained using uranyl acetate and lead citrate.

3.3 Electron Tomography

1. Semithick 200- or 250-nm sections were cut with a diamond knife and picked up on formvar and carbon-coated single-slot grids.
2. Colloidal gold particles, 10 nm in diameter, were placed on the top and bottom of the sections to serve as fiducial markers for alignment of the tomograms as described previously (Biazik, Vihinen, et al., 2015; Biazik, Yla-Anttila, et al., 2015).
3. Dual axis tilt series were acquired using SerialEM software running on a Tecnai FEG 20 microscope (FEI, Comp) operating at 200 kV with a nominal magnification of 19,000 ×, giving a 2 × binned pixel size of 1.15 nm.
4. Images were recorded at 1 degree intervals for the *a*-axis and 1.5 degree intervals for the *b*-axis with a 4 × 4 k CCD camera (Gatan Inc., USA) over a tilt range of ±62 degrees.
5. IMOD software was used to create 3D reconstructions from the tilt series and to trace the membranes in order to create 3D models of the membranes (Biazik, Yla-Anttila, et al., 2015; Kremer, Mastronarde, & McIntosh, 1996).

4. RESULTS

4.1 Ultrastructure of Autophagosomes in HPF–FS Samples

We used 70-nm thin sections to observe the morphology or autophagosomes in the samples prepared using HPF–FS. The morphological preservation varied between different cells on the same sample, or even within different regions of the same cell. This was likely due to uneven freezing speed, or uneven penetration of the solvent and fixative during FS. This is a common finding in FS samples, irrespective of the laboratory and FS protocol used. However, we were able to find a sufficient number of cells with good preservation. While the contrast of endosome membranes was good that of ER was poor, and in some cases ER could only be identified by the presence of ribosomes in rows at regular intervals (Fig. 1A). Mitochondria were very electron dense (Fig. 1A). Three types of autophagosome morphologies were observed. One type showed a round shape, with an empty space between the outer and inner limiting membrane (Fig. 1A). Another type showed a wavy limiting membrane, with the two limiting membranes tightly attached to each other (Fig. 1B). The third type showed a round shape, with the two limiting membranes tightly attached to each other (Fig. 1C). It is possible that these morphologies represent different stages of autophagosome biogenesis. Another possibility is that the differences are due to different preservation of the native morphology during the HPF–FS process. Further studies are needed to resolve this issue. Further, we also noted that microtubules (arrows) and small vesicles (v) were frequently observed in the vicinity of the autophagosomes (Fig. 1A–C). Although these two structures are also visible in conventionally fixed cells, they were considerably more frequent in the HPF–FS samples.

4.2 ET Performed on Cryoimmobilized Samples: The Phagophore

In 70-nm thin sections, it is not always possible to know whether the putative autophagosome is closed, or still open, i.e., whether it is a phagophore or an autophagosome. In ET, this distinction is easier to make since the organelles are seen as 3D objects. We used ET to identify phagophores and to study their morphology and relation to the surrounding organelles in samples prepared using HPF–FS. Phagophores (indicated by P, and depicted

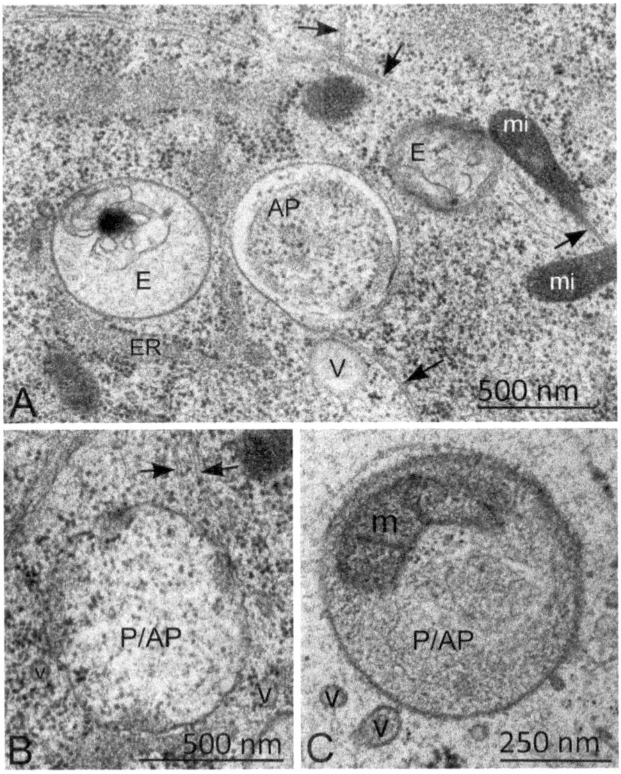

Fig. 1 Autophagosome morphology in 70 nm sections of NRK-52E cells (A, B) and ARPE-19 cells (C) processed with HPF–FS. (A) An autophagosome (AP) with two endosomes (E) next to it. There is an empty space between the limiting membranes of this autophagosome. Note the microtubule below the autophagosome (*arrow*) that seems to touch the autophagosome membrane. Other microtubules (*arrows*) are visible further away from the autophagosome. Mitochondria (mi) are very electron dense. One vesicle (V) is located next to the autophagosome. Membranes of rough ER have low contrast, but can be identified by the presence of ribosomes (the *dark dots*) at regular intervals. (B) A phagophore or an autophagosome (P/AP), with the two limiting membrane tightly attached to each other. In addition, the limiting membrane shows a wavy appearance. Two microtubules (*arrows*) are visible next to the phagophore/autophagosome. (C) A P/AP, with the two limiting membrane tightly attached to each other. This P/AP contains a mitochondrion (m). Small vesicles (v) are visible next to the P/AP.

in green color) were well contrasted, and more importantly, all phagophores showed their two limiting membranes tightly attached to each other (Figs. 2A and D, and 3A–C). Similar to earlier results with chemical fixation (Biazik, Yla-Anttila, et al., 2015; Hayashi-Nishino et al., 2009; Yla-Anttila et al., 2009a), we observed tubular membrane connections between the

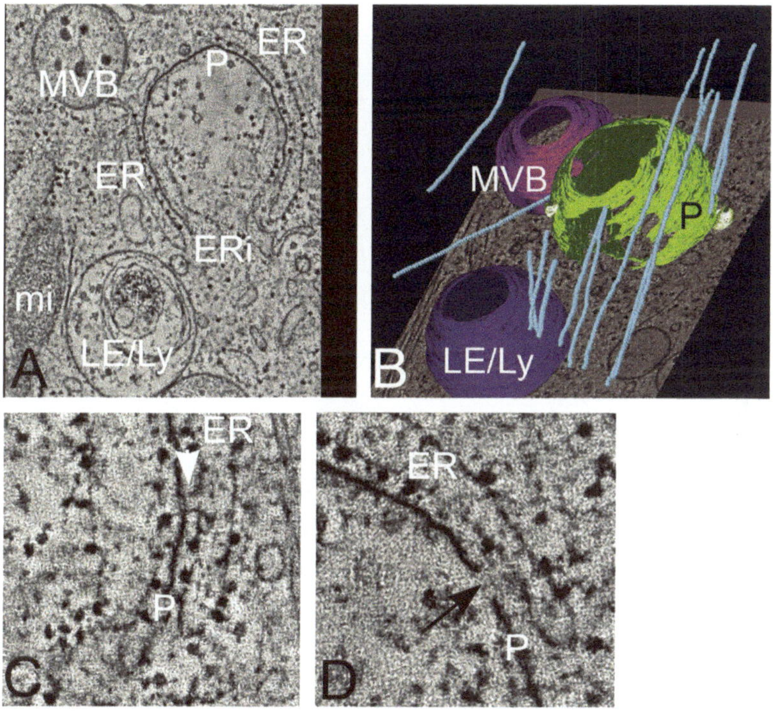

Fig. 2 Tomographic slices (approximately 1.9 nm thick) (A, B, D) and a 3D model (B) from an ARPE-19 cell prepared using HPF–FS. (A) A phagophore (P) is lined by ER cisternae on both inside and outside. A late endosome or lysosome (LE/Ly) and a multivesicular body (MVB) are located close to the phagophore. mi, mitochondrion. While the contrast of ER membrane is poor, rough ER can be identified due to the presence of ribosomes (the *dark dots*) at regular intervals. (B) 3D model of the phagophore (P, *green*), late endosome/lysosome (LE/Ly, *blue*), and multivesicular body (MVB, *purple*). Two small vesicles are shown next to the phagophore in *white*. Microtubules are shown in *light blue*. (C, D) Enlarged views of tomographic slices through the same phagophore as in (A and B). (C) The *arrowhead* indicates a short tubular membrane extension of the rough ER, connecting the ER, and the phagophore (P). (D) The *arrow* indicates a discontinuity in the phagophore (P) membrane. Movie 1 (http://dx.doi.org/10.1016/bs.mie.2016.09.063) shows all the tomographic slices, combined with the 3D model.

phagophore and ER (Fig. 2C). In addition, we observed discontinuities in the phagophore membrane (Fig. 2D, Movie 1; http://dx.doi.org/10.1016/bs.mie.2016.09.063). Many phagophores were seen in close proximity to late endosomes or lysosomes (LE/Ly, blue) and multivesicular bodies (MVB, purple) (Figs. 2A and B, and 3A and F; Movies 1 and 2; http://dx.doi.org/10.1016/bs.mie.2016.09.063). Interestingly, the open end of the phagophore typically directed toward the endosome/lysosome

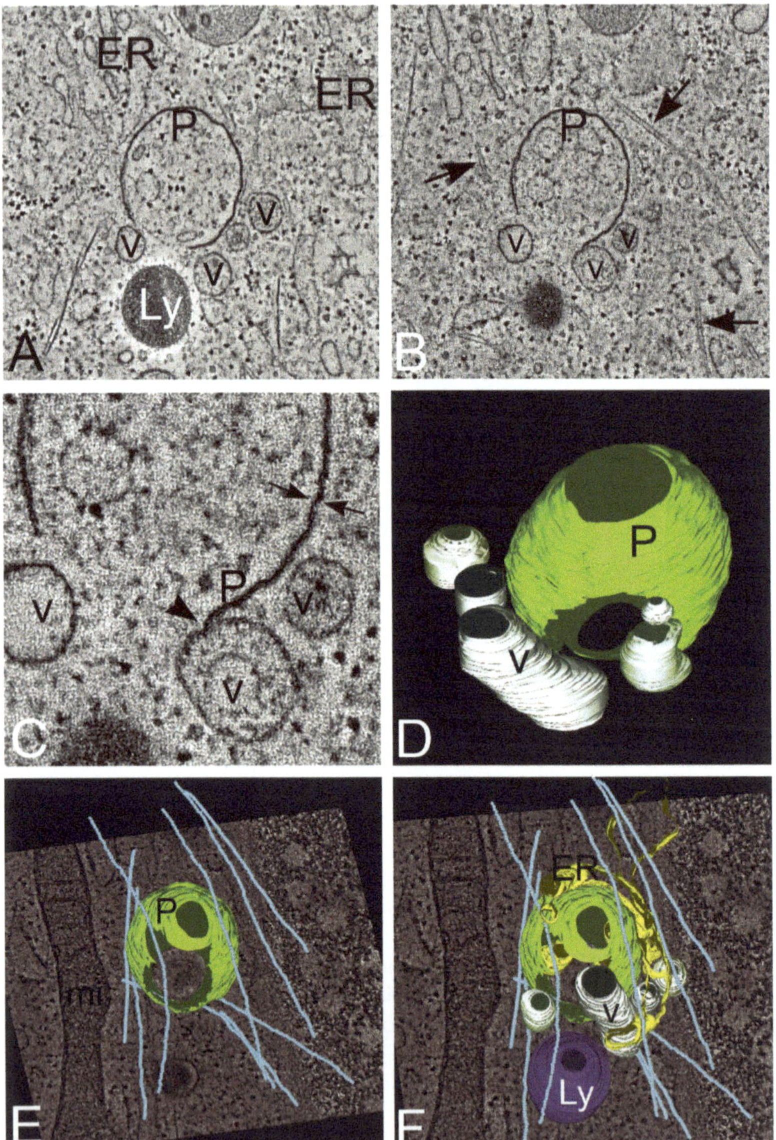

Fig. 3 Tomographic slices (approximately 1.9-nm thick) (A–C) and 3D models (D–F) from an ARPE-19 cell prepared using HPF–FS. (A) A phagophore (P) is surrounded by vesicles or tubules (V), cisternae of rough ER, and a lysosome (Ly). (B) The *arrows* indicate microtubules which are well preserved and contrasted in the HPF–FS samples. (C) The *arrowhead* indicates that the edge of the phagophore (P) is in contact with the vesicle/tubule (V). The two *arrows* indicate the phagophore membrane, where the two bilayers are tightly attached to each other. (D) A 3D model of the phagophore (*green*) and the vesicles/tubules (*white*) next to it. (E and F) 3D models of the phagophore (P, *green*), vesicles/tubules (V, *white*), lysosome (Ly, *blue*), part of the ER (*yellow*), and microtubules (*light blue*). *mi*, mitochondrion. Movie 2 shows all the tomographic slices, combined with the 3D model.

(Figs. 2A and B, and 3A and F). Further, an array of microtubules, approximately 25 in diameter (light blue) was observed in close vicinity of the phagophore (Figs. 2B and 3B, E, and F; Movies 1 and 2; http://dx.doi.org/10.1016/bs.mie.2016.09.063). These microtubules were mainly running parallel with each other, forming a tunnel-like structure around the phagophore. The open end of the phagophore was oriented toward one "end" of the tunnel (Figs. 2B and 3F). We also observed vesicles or tubules (V, white) in close proximity to the phagophore (Figs. 2B and 3A–D, F; Movies 1 and 2; http://dx.doi.org/10.1016/bs.mie.2016.09.063). One of these vesicles/tubules was observed to be in contact with the edge of the phagophore (Fig. 3C). The vesicles/tubules were mainly observed in areas void of ER (yellow), so that the vesicles were in direct proximity to the phagophore membrane and not separated from the phagophore by the ER (Figs. 1A and C, 2B, and 3B and C).

4.3 Chemical Fixation vs Cryoimmobilization

ET studies elucidating phagophore morphology and MCSs, performed using chemically fixed samples, have been published by us and others (Biazik, Vihinen, et al., 2015; Biazik, Yla-Anttila, et al., 2015; Hayashi-Nishino et al., 2009; Yla-Anttila et al., 2009a). The first obvious difference between chemically fixed samples and those immobilized by HPF–FS is that the contrast in the chemically fixed samples is better than in the HPF–FS samples, especially in ER membranes. Our experience with HPF–FS has shown that, somewhat counterintuitively, when the vitrification is very good, the contrast of the ER membranes is poor. Further, there are typically prominent thicker regions of the phagophore membrane in chemically fixed samples, which we did not observe in HPF–FS samples. In the HPF–FS samples there was generally good contrast in the phagophore membrane, and their membranes were uniform in thickness along the entire contour of the phagophore without any thickening. Similar to our observations with chemically fixed cells postfixed in reduced osmium tetroxide, in the HPF–FS samples the two bilayers of the phagophore membrane appeared as one thick membrane with no empty cleft between them (Fig. 3C, arrows).

Late endosomes, lysosomes, and multivesicular bodies in the HPF–FS samples had good contrast and were round and uniform in shape (Figs. 1A, 2A, and 3A; Movies 1 and 2; http://dx.doi.org/10.1016/bs.mie.2016.09.063). Numerous microtubules were observed in the tomograms from HFP–FS samples (Figs. 1A and B; 2B; and 3B, E, and F). In contrast to this, our aldehyde-fixed tomograms showed considerably fewer microtubules and not all microtubules were found throughout the whole volume of the tomograms. Vesicles and ribosomes as well as Golgi complex cisternae were well preserved and contrasted in HPF–FS samples.

4.4 Chemical Fixation vs Cryoimmobilization: MCSs

Using chemical fixation, we have previously observed membrane connections and MCSs between the phagophore and various surrounding organelles in our tomograms (Biazik, Vihinen, et al., 2015; Biazik, Yla-Anttila, et al., 2015; Yla-Anttila et al., 2009a). The phagophore membrane was observed to have connections with organelles including the ER, ER exit sites, mitochondria, Golgi complex, as well as membranes originating from late endosomes or lysosomes. The connections were detected either as tubular membrane bridges between the two organelles or as touching points (MCSs) between the two adjacent membranes. Membrane continuity was also observed between the phagophore and ER, and the phagophore and a late endosomal or lysosomal membrane. Membrane continuity was defined as an event where the phagophore membrane is continuous with the membrane of another organelle.

After cryoimmobilization using the HPF–FS protocol, we detected tubular connections between the phagophore and the ER (Fig. 2C), and contacts between the phagophore and vesicles/tubules (Fig. 3C). Because of the low number of phagophores we have studied so far using ET with HPF–FS samples, it is possible that connections between phagophores and other organelles will be revealed in the future.

5. DISCUSSION

At present, even with the most recent HPF and FS equipment, the reproducibility of HPF–FS is still lower than in the routine chemical fixation and plastic embedding protocols. This means that more time and effort has to be used to find cells with good ultrastructural preservation that contain the structures of interest. However, compared to the earlier cryoimmobilization methods including metal mirror and plunge freezing, HPF represents an improvement in the depth and reproducibility of vitrification (Dahl & Staehelin, 1989; Dubochet, 1995; McDonald, 1999; Studer, Michel, & Muller, 1989). However, similar to conventional plastic embedding protocols, HPF–FS protocols utilize heavy metals (osmium tetroxide and uranyl acetate) to generate contrast to the sample. Because of this, the images are not a true presentation of the cellular structures in their native state. Osmication, for example, has been proposed to cause ion redistribution artifacts (Edelmann, 1986). However, since in FS these heavy metal treatments are performed at low temperature, the artifacts are likely to be less destructive than in conventional plastic embedding protocols carried out at room temperature. For instance, osmium has been shown to behave differently at room temperature and below –30°C (White et al., 1976). Artifacts caused by sectioning the 70- or 200-nm sections, and beam damaged caused by imaging in the EM are naturally the same in conventional plastic-embedded samples and in HPF–FS samples.

There are only few published images of mammalian phagophores and autophagosomes prepared using cryoimmobilization and FS (Kovacs, Palfia, Rez, Vellai, & Kovacs, 2007; Kovacs, Rez, Palfia, & Kovacs, 2000). In these earlier studies, the samples were frozen with metal mirror freezing, and the ultrastructure of phagophores and autophagosomes was similar to our findings in the present study. However, unlike in our study, ER membranes had good contrast. The outcome of cryoimmobilization and FS depends on the water and solvent content of the sample. Therefore, one possible explanation for the difference in ER contrast is that the tissues used in the earlier studies were more suitable for cryoimmobilization and FS than the cultured cells lines used in our studies. Further, our present experience with HPF–FS suggests that the better the vitrification, the lower the contrast in the ER. In other words, it is possible that when ER is perfectly vitrified in HPF, it does not bind the contrasting agents (osmium and uranyl acetate) during freeze substitution. Further studies are needed to solve this problem.

Despite these shortcomings, HPF–FS and ET have helped answer many biological questions, particularly when ultrastructural details have needed confirmation or clarification, and when transient structures have been captured. Our aim was to utilize this approach to confirm the findings made using conventional chemical fixation and to determine whether novel information can be attained about phagophore morphology by using this method.

The shape, size, and morphology of the phagophore as well as the morphology of the other membrane-bound organelles were well preserved in the HPF–FS samples. Even though the level of ER membrane contrast was compromised in the HPF–FS samples (Murk et al., 2003), we still managed to obtain new knowledge from this data. Firstly, HPF–FS confirmed our observations with chemical fixation followed by reduced osmium tetroxide postfixation (Biazik, Yla-Anttila, et al., 2015; Yla-Anttila et al.,

2009a): there was no evidence of an empty cleft between the two bilayers of the phagophores, in the tomographic slices after HPF–FS protocol. This suggests that the cleft is indeed an artifact of chemical fixation as has been previously reported by Kovacs and colleagues using a different type of cryoimmobilization followed by FS (Kovacs et al., 2007, 2000). In another study, Murk and colleagues reported the changes that aldehyde fixation had on the morphology of endosomes and lysosomes when comparing them to cryoimmobilized samples (Murk et al., 2003). Although membranes of various organelles including the ER, Golgi, and mitochondria were often poorly contrasted after cryoimmobilization, Murk et al. concluded that aldehyde fixation caused significant deformation and reduction of endosomal volume without affecting the membrane length and that extensive tubular extensions were present in endosomes that were aldehyde fixed. We observed that the phagophore membrane in aldehyde fixed samples showed regional thickening in some parts of the membrane (Biazik, Yla-Anttila, et al., 2015; Yla-Anttila et al., 2009a) which was not evident in HPF–FS samples. This finding suggests that dynamic membranes such as the phagophore benefit from the rapid cryoimmobilization technique. Further, the preservation of lipid content in the membrane is also suggested to be maximized when using cryoimmobilization (Murk et al., 2003), another favorable feature when examining phagophores that are rich in lipids and poor in membrane-spanning proteins.

Phagophore membrane discontinuities were also observed in the tomography samples prepared using the HPF–FS protocol. In some phagophores there were several discontinuities spanning the contour of the membrane (Fig. 2B and D; Movie 1; http://dx.doi.org/10.1016/bs.mie.2016.09.063). Of note, we have occasionally observed similar discontinuities in phagophore membranes in tomograms from chemically fixed cells (unpublished observation).

It has been thought that microtubules can be preserved when samples are fixed at room temperature. However, we found that cryoimmobilized cells had a very rich distribution of microtubules which were considerably fewer in our tomograms from aldehyde fixed cells. Microtubular trajectories were well contrasted and easily tracked in the tomographic slices of HPF–FS samples by using the slicer tool in IMOD. The data showed that the microtubules predominantly ran in close proximity to the phagophore, forming a tunnel-like cage around the phagophore. In most cases the open end of the phagophore was oriented along the tunnel and was facing a late endosome or lysosome. These findings are in agreement with previous reports which imply that microtubules are important for autophagosome biogenesis and maturation (Fass, Shvets, Degani, Hirschberg, & Elazar, 2006; Jahreiss, Menzies, & Rubinsztein, 2008; Kochl, Hu, Chan, & Tooze, 2006; Kouno et al., 2005; Mann & Hammarback, 1994). Earlier studies using microtubule drugs concluded that microtubules are not necessary for autophagosome formation (Aplin, Jasionowski, Tuttle, Lenk, & Dunn, 1992; Rez, Laszlo, Fellinger, Kovacs, & Kovacs, 1990; Seglen et al., 1996). However, newer studies show that autophagosome formation is enhanced by microtubules (Fass et al., 2006; Kochl et al., 2006) and requires the dynamic microtubule population, while sealed autophagosomes bind the stable microtubules (Geeraert et al., 2010). Further, microtubules have been reported by several groups to assist in autophagosome maturation (reviewed by Monastyrska, Rieter, Klionsky, & Reggiori, 2009). Other studies showed that phagophores are immobile (Fass et al., 2006), probably due to their initial close relationship with the ER during biogenesis. Autophagosomes on the other hand are mobile structures (Fass et al., 2006; Jahreiss et al., 2008; Kochl et al., 2006) which form in the cellular periphery and move bi-directionally along microtubules en route for degradation (Fass et al., 2006) and tend to be distributed with a bias toward the perinuclear microtubule-organizing center (Jahreiss et al., 2008). The microtubule-associated protein (MAP) 1 light chain 3 (MAP1-LC3) associates with phagophores, autophagosomes (in both inner and outer membranes) and autolysosomes (Jahreiss et al., 2008) and is a widely used marker for these organelles (Yu et al., 2010). LC3 was shown to be required for the microtubule- and dynein-dependent centripetal movement of autophagosomes toward lysosomes (Kimura, Noda, & Yoshimori, 2008). Our tomographic data generated by HPF–FS showed a clear ultrastructural relationship between the phagophore and microtubules, which has not been reported before. Furthermore, since the microtubules tend to run parallel and toward a late endosome/lysosome, it is tempting to speculate that once the phagophore membrane seals over, the newly formed mobile autophagosome will then potentially travel along the microtubular tunnel in the direction of the late endosome/lysosome in order for organelle fusion to occur.

Collectively, ET data generated from cryoimmobilized samples using the HPF–FS technique have uncovered ultrastructural detail about phagophores, particularly on their spatial relationship with microtubules, which has not been reported using ET on aldehyde fixed samples. ET on cells prepared using HPF–FS revealed that (i) no empty cleft between the two phagophore membranes is evident, (ii) the phagophore membrane is uniform in thickness, and (iii) microtubules are very well preserved and surround the forming phagophores in a defined orientation. Our data suggest that microtubules arrange themselves in a parallel orientation that surrounds the phagophore, where the open end of the phagophore faces a late endosome/lysosome. Further, we confirmed previous findings from chemically fixed samples, by showing that the phagophore membrane has tubular connections with the ER. Finally, we also observed that phagophores can have contacts with vesicles/tubules, which has not been reported in aldehyde-fixed samples. Further studies are needed to clarify the identity of these vesicles.

ACKNOWLEDGMENTS

This study was supported by the Academy of Finland and Biocenter Finland.

REFERENCES

Aplin, A., Jasionowski, T., Tuttle, D. L., Lenk, S. E., & Dunn, W. A. (1992). Cytoskeletal elements are required for the formation and maturation of autophagic vacuoles. *Journal of Cellular Physiology*, *152*, 458–466.

Biazik, J. M., Vihinen, H., Anwar, T., Jokitalo, E., & Eskelinen, E. L. (2015). The versatile electron microscope: An ultrastructural overview of autophagy. *Methods*, *75*, 44–53. http://dx.doi.org/10.1016/j.ymeth.2014.11.013.

Biazik, J. M., Yla-Anttila, P., Vihinen, H., Jokitalo, E., & Eskelinen, E. L. (2015). Ultrastructural relationship of the phagophore with surrounding organelles. *Autophagy*, *11*(3), 439–451. http://dx.doi.org/10.1080/15548627.2015.1017178.

Buser, C., & Walther, P. (2008). Freeze-substitution: The addition of water to polar solvents enhances the retention of structure and acts at temperatures around −60 degrees C. *Journal of Microscopy*, *230*(Pt. 2), 268–277. http://dx.doi.org/10.1111/j.1365-2818.2008.01984.x.

Buys, A. V., & Pretorius, E. (2012). Comparing different preparation methods to study human fibrin fibers and platelets using TEM. *Microscopy Research and Technique*, *75*(6), 801–806. http://dx.doi.org/10.1002/jemt.21129.

Chan, F. L., & Inoue, S. (1994). Lamina lucida of basement membrane: An artefact. *Microscopy Research and Technique*, *28*(1), 48–59. http://dx.doi.org/10.1002/jemt.1070280106.

Cook, K. L., Soto-Pantoja, D. R., Abu-Asab, M., Clarke, P. A., Roberts, D. D., & Clarke, R. (2014). Mitochondria directly donate their membrane to form autophagosomes during a novel mechanism of parkin-associated mitophagy. *Cell & Bioscience*, *4*(1), 16. http://dx.doi.org/10.1186/2045-3701-4-16.

Dahl, R., & Staehelin, L. A. (1989). High-pressure freezing for the preservation of biological structure: Theory and practice. *Journal of Electron Microscopy Technique*, *13*(3), 165–174. http://dx.doi.org/10.1002/jemt.1060130305.

Dubochet, J. (1995). High-pressure freezing for cryoelectron microscopy. *Trends in Cell Biology*, *5*(9), 366–368.

Dubochet, J., Adrian, M., Chang, J. J., Homo, J. C., Lepault, J., McDowall, A. W., et al. (1988). Cryo-electron microscopy of vitrified specimens. *Quarterly Reviews of Biophysics*, *21*(2), 129–228.

Edelmann, L. (1986). Freeze-dried embedded specimens for biological microanalysis. *Scanning Electron Microscopy*, (Pt. 4), 1337–1356.

Eskelinen, E. L. (2005). Maturation of autophagic vacuoles in mammalian cells. *Autophagy*, *1*(1), 1–10.

Fass, E., Shvets, E., Degani, I., Hirschberg, K., & Elazar, Z. (2006). Microtubules support production of starvation-induced autophagosomes but not their targeting and fusion with lysosomes. *The Journal of Biological Chemistry*, *281*(47), 36303–36316.

Frotscher, M., Zhao, S., Graber, W., Drakew, A., & Studer, D. (2007). New ways of looking at synapses. *Histochemistry and Cell Biology*, *128*(2), 91–96. http://dx.doi.org/10.1007/s00418-007-0305-7.

Geeraert, C., Ratier, A., Pfisterer, S. G., Perdiz, D., Cantaloube, I., Rouault, A., et al. (2010). Starvation-induced hyperacetylation of tubulin is required for the stimulation of autophagy by nutrient deprivation. *Journal of Biological Chemistry*, *285*(31), 24184–24194. http://dx.doi.org/10.1074/jbc.M109.091553.

Gulley, R. L., & Reese, T. S. (1981). Cytoskeletal organization at the postsynaptic complex. *The Journal of Cell Biology*, *91*(1), 298–302.

Hayashi-Nishino, M., Fujita, N., Noda, T., Yamaguchi, A., Yoshimori, T., & Yamamoto, A. (2009). A subdomain of the endoplasmic reticulum forms a cradle for autophagosome formation. *Nature Cell Biology*, *11*(12), 1433–1437.

Hohenberg, H., Mannweiler, K., & Muller, M. (1994). High-pressure freezing of cell suspensions in cellulose capillary tubes. *Journal of Microscopy*, *175*(Pt. 1), 34–43.

Jager, S., Bucci, C., Tanida, I., Ueno, T., Kominami, E., Saftig, P., et al. (2004). Role for Rab7 in maturation of late autophagic vacuoles. *Journal of Cell Science*, *117*(Pt. 20), 4837–4848. http://dx.doi.org/10.1242/jcs.01370.

Jahreiss, L., Menzies, F. M., & Rubinsztein, D. C. (2008). The itinerary of autophagosomes: From peripheral formation to kiss-and-run fusion with lysosomes. *Traffic*, *9*(4), 574–587. http://dx.doi.org/10.1111/j.1600-0854.2008.00701.x.

Kellenberger, E. (1987). The response of biological macromolecules and supramolecular structures to the physics of specimen cryopreparation. In R. A. Steinbrecht & K. Zierold (Eds.), *Cryotechniques in biological electron microscopy* (pp. 35–63). Berlin, Heidelberg: Springer-Verlag.

Kimura, S., Noda, T., & Yoshimori, T. (2008). Dynein-dependent movement of autophagosomes mediates efficient encounters with lysosomes. *Cell Structure and Function*, *33*(1), 109–122. http://dx.doi.org/10.1247/csf.08005.

Knoops, K., Kikkert, M., Worm, S. H., Zevenhoven-Dobbe, J. C., van der Meer, Y., Koster, A. J., et al. (2008). SARS-coronavirus replication is supported by a reticulovesicular network of modified endoplasmic reticulum. *PLoS Biology*, *6*(9), e226. http://dx.doi.org/10.1371/journal.pbio.0060226.

Kochl, R., Hu, X. W., Chan, E. Y., & Tooze, S. A. (2006). Microtubules facilitate autophagosome formation and fusion of autophagosomes with endosomes. *Traffic*, 7(2), 129–145.

Kouno, T., Mizuguchi, M., Tanida, I., Ueno, T., Kanematsu, T., Mori, Y., et al. (2005). Solution structure of microtubule-associated protein light chain 3 and identification of its functional subdomains. *The Journal of Biological Chemistry*, *280*(26), 24610–24617. http://dx.doi.org/10.1074/jbc.M413565200.

Kovacs, A. L., Palfia, Z., Rez, G., Vellai, T., & Kovacs, J. (2007). Sequestration revisited: Integrating traditional electron microscopy, de novo assembly and new results. *Autophagy*, *3*(6), 655–662.

Kovacs, A. L., Rez, G., Palfia, Z., & Kovacs, J. (2000). Autophagy in the epithelial cells of murine seminal vesicle in vitro. Formation of large sheets of nascent isolation membranes, sequestration of the nucleus and inhibition by wortmannin and 3-ethyladenine. *Cell and Tissue Research*, *302*, 253–261.

Kremer, J. R., Mastronarde, D. N., & McIntosh, J. R. (1996). Computer visualization of three-dimensional image data using IMOD. *Journal of Structural Biology, 116*(1), 71–76. http://dx.doi.org/10.1006/jsbi.1996.0013.

Mann, S. S., & Hammarback, J. A. (1994). Molecular characterization of light chain 3. A microtubule binding subunit of MAP1A and MAP1B. *The Journal of Biological Chemistry, 269*(15), 11492–11497.

Mazzone, R. W., Kornblau, S., & Durand, C. M. (1980). Shrinkage of lung after chemical fixation for analysis of pulmonary structure-function relations. *Journal of Applied Physiology: Respiratory, Environmental and Exercise Physiology, 48*(2), 382–385.

McDonald, K. (1999). High-pressure freezing for preservation of high resolution fine structure and antigenicity for immunolabeling. *Methods in Molecular Biology, 117*, 77–97. http://dx.doi.org/10.1385/1-59259-201-5:77.

McDonald, K. L., & Auer, M. (2006). High-pressure freezing, cellular tomography, and structural cell biology. *Biotechniques, 41*(2). 137, 139, 141, 143 passim.

McDonald, K., & Morphew, M. K. (1993). Improved preservation of ultrastructure in difficult-to-fix organisms by high pressure freezing and freeze substitution: I. Drosophila melanogaster and Strongylocentrotus purpuratus embryos. *Microscopy Research and Technique, 24*(6), 465–473. http://dx.doi.org/10.1002/jemt.1070240603.

Monastyrska, I., Rieter, E., Klionsky, D. J., & Reggiori, F. (2009). Multiple roles of the cytoskeleton in autophagy. *Biological Reviews, 84*(3), 431–448. http://dx.doi.org/10.1111/j.1469-185X.2009.00082.x.

Murk, J. L., Lebbink, M. N., Humbel, B. M., Geerts, W. J., Griffith, J. M., Langenberg, D. M., et al. (2004). 3-D Structure of multilaminar lysosomes in antigen presenting cells reveals trapping of MHC II on the internal membranes. *Traffic, 5*(12), 936–945. http://dx.doi.org/10.1111/j.1600-0854.2004.00235.x.

Murk, J. L., Posthuma, G., Koster, A. J., Geuze, H. J., Verkleij, A. J., Kleijmeer, M. J., et al. (2003). Influence of aldehyde fixation on the morphology of endosomes and lysosomes: Quantitative analysis and electron tomography. *Journal of Microscopy, 212*(Pt. 1), 81–90.

Palsdottir, H., Remis, J. P., Schaudinn, C., O'Toole, E., Lux, R., Shi, W., et al. (2009). Three-dimensional macromolecular organization of cryofixed Myxococcus xanthus biofilms as revealed by electron microscopic tomography. *Journal of Bacteriology, 191*(7), 2077–2082. http://dx.doi.org/10.1128/jb.01333-08.

Pfeiffer, S., Vielhaber, G., Vietzke, J. P., Wittern, K. P., Hintze, U., & Wepf, R. (2000). High-pressure freezing provides new information on human epidermis: Simultaneous protein antigen and lamellar lipid structure preservation. Study on human epidermis by cryoimmobilization. *The Journal of Investigative Dermatology, 114*(5), 1030–1038. http://dx.doi.org/10.1046/j.1523-1747.2000.00966.x.

Rez, G., Laszlo, L., Fellinger, E., Kovacs, A. L., & Kovacs, J. (1990). Time course of the quantitative changes in the autophagic-lysosomal and secretory granule compartments of murine liver cells under the influence of vinblastine. *Virchows Archiv B, Cell Pathology Including Molecular Pathology, 58*(3), 189–197.

Rostaing, P., Real, E., Siksou, L., Lechaire, J. P., Boudier, T., Boeckers, T. M., et al. (2006). Analysis of synaptic ultrastructure without fixative using high-pressure freezing and tomography. *The European Journal of Neuroscience, 24*(12), 3463–3474. http://dx.doi.org/10.1111/j.1460-9568.2006.05234.x.

Rostaing, P., Weimer, R. M., Jorgensen, E. M., Triller, A., & Bessereau, J. L. (2004). Preservation of immunoreactivity and fine structure of adult C. elegans tissues using high-pressure freezing. *The Journal of Histochemistry and Cytochemistry, 52*(1), 1–12.

Seemann, E., Kurth, T., & Entzeroth, R. (2012). Insight into the ultrastructural organisation of sporulated oocysts of Eimeria nieschulzi (Coccidia, Apicomplexa). *Parasitology Research, 111*(5), 2143–2147. http://dx.doi.org/10.1007/s00436-012-3066-8.

Seglen, P. O., Berg, T. O., Blankson, H., Fengsrud, M., Holen, I., & Stromhaug, P. E. (1996). Structural aspects of autophagy. *Advances in Experimental Medicine and Biology, 389*, 103–111.

Sharp, D. J., McDonald, K. L., Brown, H. M., Matthies, H. J., Walczak, C., Vale, R. D., et al. (1999). The bipolar kinesin, KLP61F, cross-links microtubules within interpolar microtubule bundles of Drosophila embryonic mitotic spindles. *The Journal of Cell Biology, 144*(1), 125–138.

Studer, D., Humbel, B. M., & Chiquet, M. (2008). Electron microscopy of high pressure frozen samples: Bridging the gap between cellular ultrastructure and atomic resolution. *Histochemistry and Cell Biology, 130*(5), 877–889. http://dx.doi.org/10.1007/s00418-008-0500-1.

Studer, D., Michel, M., & Muller, M. (1989). High pressure freezing comes of age. *Scanning Microscopy Supplement, 3*, 253–268. discussion 268–259.

Szczesny, P. J., Walther, P., & Muller, M. (1996). Light damage in rod outer segments: The effects of fixation on ultrastructural alterations. *Current Eye Research, 15*(8), 807–814.

Tomova, C., Humbel, B. M., Geerts, W. J., Entzeroth, R., Holthuis, J. C., & Verkleij, A. J. (2009). Membrane contact sites between apicoplast and ER in Toxoplasma gondii revealed by electron tomography. *Traffic, 10*(10), 1471–1480. http://dx.doi.org/10.1111/j.1600-0854.2009.00954.x.

White, D. L., Andrews, S. B., Faller, J. W., & Barrnett, R. J. (1976). The chemical nature of osmium tetroxide fixation and staining of membranes by x-ray photoelectron spectroscopy. *Biochimica et Biophysica Acta, 436*(3), 577–592.

Yla-Anttila, P., Vihinen, H., Jokitalo, E., & Eskelinen, E. L. (2009a). 3D tomography reveals connections between the phagophore and endoplasmic reticulum. *Autophagy, 5*(8), 1180–1185.

Yla-Anttila, P., Vihinen, H., Jokitalo, E., & Eskelinen, E. L. (2009b). Monitoring autophagy by electron microscopy in mammalian cells. *Methods in Enzymology, 452*, 143–164.

Yu, L., McPhee, C. K., Zheng, L., Mardones, G. A., Rong, Y., Peng, J., et al. (2010). Termination of autophagy and reformation of lysosomes regulated by mTOR. *Nature, 465*(7300), 942–946. http://dx.doi.org/10.1038/nature09076. nature09076.

CHAPTER TWENTY

A Simple Cargo Sequestration Assay for Quantitative Measurement of Nonselective Autophagy in Cultured Cells

M. Luhr, P. Szalai, F. Sætre, L. Gerner, P.O. Seglen, N. Engedal[1]

Centre for Molecular Medicine Norway, Nordic EMBL Partnership, University of Oslo, Oslo, Norway

[1]Corresponding author: e-mail address: k.n.engedal@ncmm.uio.no

Contents

1. Introduction 352
2. Measuring Nonselective Autophagic Sequestration of Cytosol in Cultured Cells 353
3. Concluding Remarks 362
Acknowledgments 363
References 363

Abstract

Autophagy (self-eating) is a common term for various processes by which cellular components are transferred to lysosomes for degradation. In macroautophagy, intracellular membrane structures termed "phagophores" expand to encapsulate autophagic cargo into sealed, double-membrane vacuoles termed "autophagosomes," which subsequently may fuse with endosomes to form intermediary vacuoles called "amphisomes," and finally with lysosomes to have their contents degraded and recycled. Autophagy is frequently analyzed by monitoring phagophore- and autophagosome-associated markers such as LC3. Although useful, it is becoming increasingly clear that very few, if any, of these marker proteins are entirely specific to the autophagic process. Moreover, phagophore/autophagosome markers cannot be used to measure autophagic activity since they are part of the autophagic machinery, or "cart," rather than autophagic cargo. Thus, there is a great need for functional assays in autophagy research. Here, we describe a method that quantitatively measures the nonselective autophagic sequestration of endogenous cytosolic cargo. The method is based on a crude separation of sedimentable cellular material from cytosol and a subsequent measurement of the fraction of a cytosolic enzyme activity transferred to the sedimentable fraction by autophagic sequestration. The original assay was first developed in 1990, but during the last few years we have systematically downscaled and simplified the method into the time- and cost-efficient procedure presented here, which can be performed with standard laboratory equipment and is suitable for any cell type.

Methods in Enzymology, Volume 587
ISSN 0076-6879
http://dx.doi.org/10.1016/bs.mie.2016.09.064

1. INTRODUCTION

The machinery involved in phagophore assembly and maturation has been extensively studied, and the roles of several autophagy-related (Atg) proteins taking part in these processes have been characterized. Proteins of the Atg8 family become anchored to the phagophore membrane by lipidation at a late step in phagophore assembly. One of the Atg8 proteins, LC3B, has become the most frequently used marker for autophagy in mammalian cells. It is, however, of paramount importance to be aware that assessing alterations in autophagic markers or in components of the autophagic machinery cannot be used as a measure of autophagic activity, as the latter requires direct monitoring of *cargo* flux through the autophagic pathway. Attesting to this, whereas both we and others have found that calcium-modulating compounds increase cellular levels of lipidated LC3, we could also demonstrate that such compounds actually block autophagic flux prior to phagophore closure (Engedal et al., 2013; Gordon, Holen, Fosse, Rotnes, & Seglen, 1993; Sætre, Hagen, Engedal, & Seglen, 2015). Furthermore, we recently showed that autophagy of cytosolic cargo can occur in the complete absence of autophagic–lysosomal LC3 flux, and that efficient simultaneous silencing of the LC3 protein family affects neither autophagic sequestration of cytosolic cargo (Szalai et al., 2015) nor long-lived protein degradation (our unpublished results), strongly implying that LC3 is not required for bulk autophagic flux. It is increasingly being realized that most Atgs have autophagy-independent functions besides their role in autophagy (Bestebroer, V'Kovski, Mauthe, & Reggiori, 2013; Klionsky et al., 2016; Subramani & Malhotra, 2013). Thus, altered expression levels, posttranslational modifications, and changes in intracellular localization of Atgs do not provide direct evidence of changes in autophagic activity, but may, alternatively, be related to their involvement in other processes. Altogether, this underlines the importance of including functional assays in order to assess autophagic activity.

Autophagic sequestration assays measure the transfer of cargo from cytosol to autophagic vacuoles (Seglen et al., 2015). Bulk autophagy includes the nonselective sequestration of cytosol and can consequently be monitored as the sequestration of soluble cytosolic molecules. In principle, any highly abundant cytosolic protein that is degraded by the autophagic–lysosomal pathway can be used as a cargo marker for bulk autophagic sequestration. In fact, it was originally reported that seven cytosolic enzymes with different half-lives were sequestered at the same rate during amino acid starvation in hepatocytes (Kopitz, Kisen, Gordon, Bohley, & Seglen, 1990). Recently, we have in particular focused on refining, optimizing, and validating the assay that measures autophagic sequestration of the cytosolic protein lactate dehydrogenase (LDH) (Fig. 1, and protocol later) (Seglen et al., 2015). The advantage of using LDH as a cargo marker is that (i) the amount of sequestered protein can easily be quantified by an enzymatic assay, (ii) LDH is ubiquitously expressed, meaning that it can be used as an endogenous cargo marker in any cell type without the need for transfection, and (iii) LDH is exclusively degraded by the autophagic–lysosomal pathway in a nonselective manner (Kopitz et al., 1990).

While the method was originally used in primary hepatocytes (Kopitz et al., 1990), we recently adapted and optimized it for cultured cells (Engedal et al., 2013; Seglen et al., 2015; Szalai et al., 2015). This has reduced the amount of material required to carry out experiments, as well as enabled the convenient use of a table-top centrifuge for separation of soluble LDH from sequestered LDH. Since then, we have simplified and streamlined the method even further and validated it in numerous cell lines, including LNCaP, VCaP, PC3, DU145, PNT2, HEK293, HeLa, U2OS, HEC1A, MCF7, MEFs, Huh7, G361, and BJ. First, we have been able to omit several washing steps as compared to previous protocols. Second, we have eliminated the need for a density cushion in order to separate soluble LDH from sequestered LDH inside sedimentable autophagic vacuoles. Third, until recently, we could only achieve the crucial, selective electrodisruption of the plasma membrane using a homemade device (Gordon & Seglen, 1982; Seglen et al., 2015). We have now overcome this limitation by identifying settings that achieve effective and reproducible plasma membrane electrodisruption with a commercially available electroporation apparatus. Here, we present our current, optimized protocol for measuring nonselective autophagic sequestration of the endogenous cytosolic cargo protein LDH.

2. MEASURING NONSELECTIVE AUTOPHAGIC SEQUESTRATION OF CYTOSOL IN CULTURED CELLS

1. Seed cells in a 12-well plate.

   Note: The method can in principle be carried out in any format as long as the LDH activity (measured as the decline in the UV absorbance of NADH) is in the linear range. With the current protocol, the assay

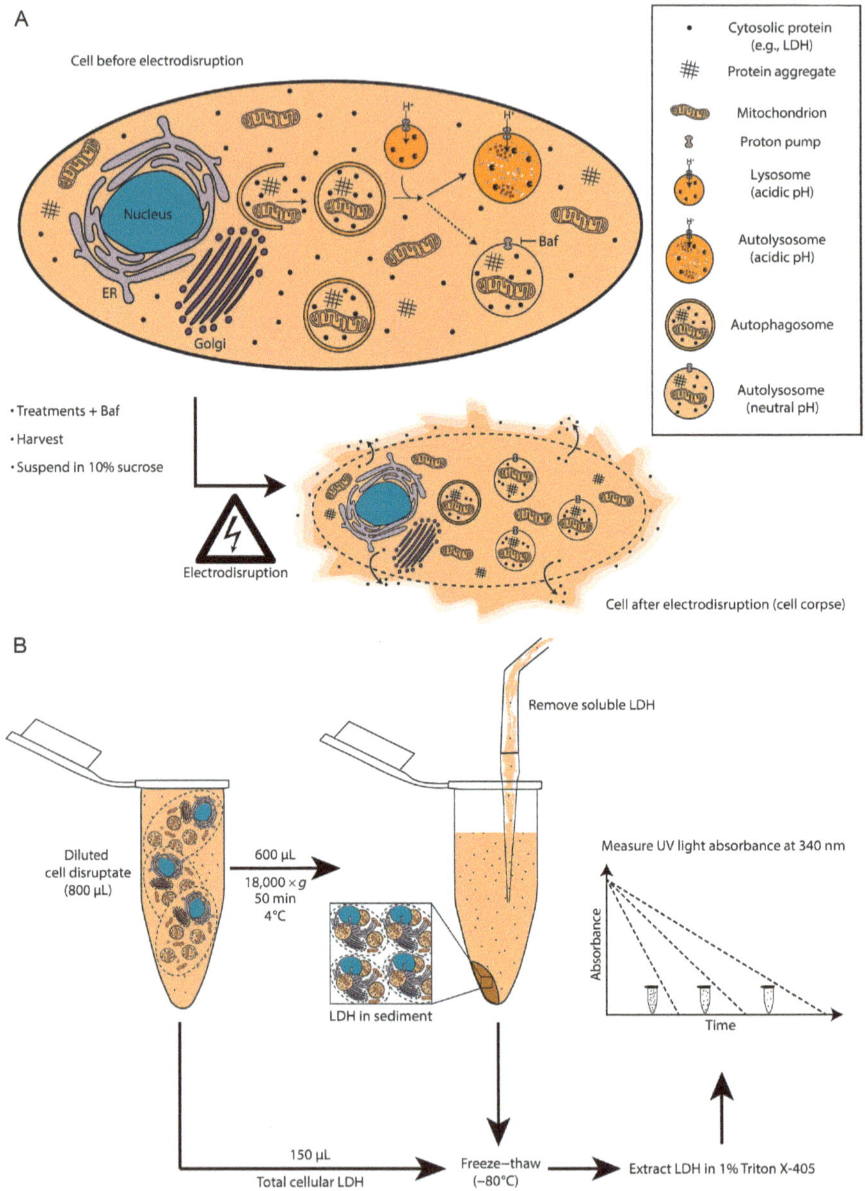

Fig. 1 Measuring macroautophagic sequestration using LDH as cargo marker. (A) During bulk autophagy, cytosolic proteins (*black dots*) such as LDH will be non-selectively sequestered alongside other cargo. Incubation with a degradation inhibitor acting at a postsequestration step (e.g., Baf) will result in the retention and accumulation of intact LDH inside autophagic vacuoles during the period of inhibitor presence. After the cells have been harvested, the plasma membrane is selectively disrupted by an electric pulse in an isotonic buffer, resulting in LDH leakage from the cytoplasm, but not

can be performed in 12-well plates for most cell types. The method can be used for both adherent and nonadherent cells, the only difference being the harvest step (step 4). We usually perform the assay with three replicate wells per treatment condition.

2. Administer treatments when the cells have reached the desired confluency. In order to prevent proteolytic cleavage of sequestered LDH, include a postsequestration step inhibitor of autophagic–lysosomal degradation, the addition of which defines the start of the actual measurement period during which sequestered LDH accumulates.

   Note: The duration of the inhibitor treatment should be as short as possible to minimize secondary effects caused by the block in autophagic–lysosomal degradation. Generally, the inhibitor is added only during the last 2–4 h of treatment. We routinely include 100–200 n*M* of the vacuolar-type H^+-ATPase inhibitor bafilomycin A1 (Baf) for 3 h to block proton pumping and thus acidity-dependent degradation of sequestered material. Other proton pump inhibitors, like concanamycin A, protease inhibitors like leupeptin or E-64/E-64d, or lysosomotropic agents such as chloroquine, propylamine, or ammonium chloride may, in principle, be used for the same purpose (Seglen, 1983; Seglen, Grinde, & Solheim, 1979; Seglen et al., 2015). The efficacy of the inhibitors may vary depending on the cell type, concentration, and duration of the treatment; for example, most of the cell lines we have tested respond poorly to leupeptin, in contrast to primary rat hepatocytes (Kopitz et al., 1990). It can be useful to perform the assay also in the absence of degradation inhibitors to determine whether a particular treatment or genetic alteration may block autophagy at a postsequestration step. For example, we have recently used this approach to confirm that RAB7A acts at an autophagic postsequestration step prior to lysosomal

from sealed autophagic vacuoles. (B) Following electrodisruption, nonsequestered (cytosolic) LDH is separated from sequestered LDH inside sedimentable autophagic vacuoles by centrifugation. Part of the diluted cell disruptate is subjected to a freeze–thaw cycle without centrifugation in order to measure the total cellular amount of LDH in the sample. Following a freeze–thaw cycle, total cellular LDH and sequestered LDH are extracted by 1% Triton X-405 and the amount of LDH in both fractions is measured by an enzymatic assay as the decline in NADH absorbance at 340 nm. *nm*, nanometer; *Baf*, bafilomycin A1; *LDH*, lactate dehydrogenase; *NADH*, nicotinamide adenine dinucleotide.

degradation (our unpublished results). Moreover, it is possible to detect some LDH *en route* through the autophagic–lysosomal pathway under certain conditions, e.g., during amino acid starvation in LNCaP cells (Engedal et al., 2013) or primary rat hepatocytes (Kopitz et al., 1990).

3. Incubate three wells in complete medium (the same nutrient-rich growth medium that is used to culture the cells) without the addition of vehicle control to define the background percentage of LDH in the sediment, referred to as "LDH in sediment $(\%)_{\mathrm{INITIAL}}$," to be subtracted from the percentage of LDH in the sediment after the measurement period, referred to as "LDH in sediment $(\%)_{\mathrm{FINAL}}$." We refer to the text and figures later for a full explanation of how to calculate and present LDH sequestration data. A typical experimental set-up is outlined in Table 1.
4. To harvest adherent cells, aspirate medium from the wells and wash once with 500 μL 37°C PBS (pH 7.4). Remove the PBS, add 200 μL Accumax (Sigma A7089) to the cells, and incubate at 37°C until the cells have detached from the plate.

   Note: Other agents such as trypsin–EDTA may also be used to harvest the cells and works well with most cell types. The advantage of

Table 1 A Typical Initial Experimental Set-up with Controls to Determine How a Given Treatment (treatment X) Impacts on Macroautophagic Sequestration

Sample Number	Treatment
1–3	Untreated cells
4–6	Cells subjected to vehicle (e.g., DMSO)
7–9	Cells subjected to Baf
10–12	Cells subjected to Baf + sequestration inhibitor (e.g., 3MA)
13–15	Cells subjected to positive control treatment (e.g., Torin1) + Baf
16–18	Cells subjected to positive control treatment + Baf + sequestration inhibitor (e.g., 3MA)
19–21	Cells subjected to "treatment X"
22–24	Cells subjected to "treatment X" + Baf

For treatments that are found to enhance autophagic sequestration in the presence of Baf, it will in subsequent experiments be relevant to include sequestration inhibitors such as 3MA and thapsigargin (Engedal et al., 2013; Sætre et al., 2015) as controls. In this example, all cells are kept in complete medium (CM). *3MA*, 3-methyladenine; *Baf*, bafilomycin A1; DMSO, dimethyl sulfoxide.

using Accumax is that it contains DNase, which degrades DNA that may leak out of cells that are damaged during harvest. When extracellular DNA is not degraded, the cells may become "sticky," making it harder to obtain single-cell suspensions prior to electrodisruption. Accumax is a trademark of Innovative Cell Technologies, Inc. (CA, USA). Trypsin–EDTA can also be supplemented with DNase in order to degrade DNA that has leaked out of cells during harvest.

5. Add 800 μL of 37°C PBS/5% BSA, resuspend gently until the cells are near a single-cell suspension, and transfer to a microcentrifuge tube on ice.
6. Centrifuge at 300 × *g* for 5 min at 4°C.
7. Aspirate the supernatant, add 400 μL ice-cold 10% sucrose to the tube, and place it on ice.

   Note: Whereas previous protocols included two washing steps with 10% sucrose (Seglen, Øverbye, & Sætre, 2009; Seglen et al., 2015), we have recently found these washing steps to be unnecessary in the current, downscaled protocol.
8. Resuspend the cells gently until they are near a single-cell suspension, transfer to a 0.4 cm electroporation cuvette (Bio-Rad 1652088), and place the cuvette in an electroporation chamber. Prior to electrodisruption, the cells should be Trypan blue negative, as shown in Fig. 2A, confirming that their plasma membrane has remained intact throughout the harvest and washing steps.

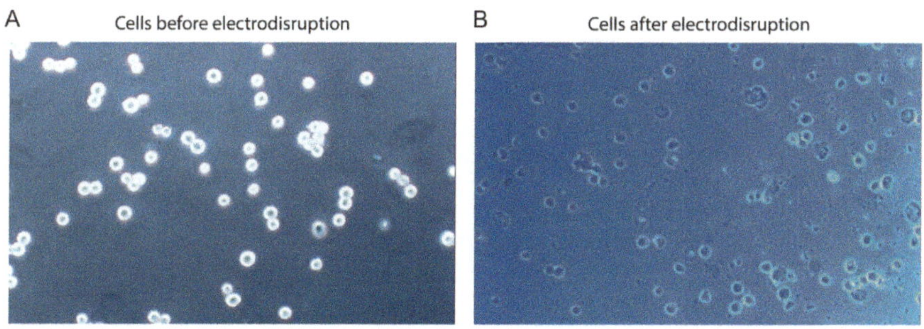

Fig. 2 Trypan blue staining of cells before and after electrodisruption. LNCaP cells were stained with Trypan blue prior to (A), or following (B) electrodisruption (800 V, 25 μF, and 400 Ω with a BTX Harvard EMC 630 apparatus, producing a ~8 ms pulse). Trypan blue selectively stains cells with permeabilized membranes. All cells remained Trypan blue positive for >30 min, indicating that the plasma membranes were electrodisrupted, and thus unable to reseal. *μF*, microfarad; *ms*, millisecond; *V*, volt.

9. Apply a single electric pulse of 800 V, 25 μF, and 400 **Ω** (settings used with a BTX Harvard EMC 630 electroporation apparatus, which produces a pulse of approximately 8 ms) to selectively disrupt the plasma membrane. The plasma membrane-disrupted cells (cell corpses) should now be Trypan blue positive, as shown in Fig. 2B.

Note: Remarkably, this method of plasma membrane electrodisruption does not affect intracellular structure (Gordon & Seglen, 1982), and all organelles including autophagic vacuoles are well preserved (Kopitz et al., 1990). Selective electrodisruption of the plasma membrane is simple, efficient, and reproducible. We routinely also use a homemade device for electrodisruption (2000 V and 1.2 μF in a 1 × 1 × 5 cm electrode chamber), which has been extensively validated (Seglen et al., 2009). We are able to obtain equally low background levels of LDH and the same sequestration rates with the commercially available electroporator as with our custom-made instrument (Fig. 3). Codogno and colleagues have successfully used glass/teflon homogenization instead of electrodisruption (Pattingre, Petiot, & Codogno, 2004). Whichever method is preferred, it is important to validate that the plasma membrane does not reseal after it has been disrupted and that the autophagic vacuoles remain intact. At suboptimal electric pulses we have observed that even though all cells are Trypan blue positive initially (indicating full electropermeabilization), after some time (10–30 min) in the continued presence of Trypan blue at room temperature, many of the cells become Trypan blue negative, indicating that the cells can reseal their plasma membrane and pump out the dye. This phenomenon does not occur in cells subjected to optimal electric pulses. The Trypan blue assay can thus be used as a test for sustained plasma membrane electrodisruption. Validation that the electrodisruption has not been too harsh, and thus has not affected intracellular organelles and autophagic vacuoles can be done by electron microscopy (Kopitz et al., 1990), as well as by testing the purity of the cytosolic fraction after centrifugation of the electrodisrupted cells through a density cushion (Seglen et al., 2015). The resulting cytosolic fraction should, for instance, not contain the lysosomal enzyme Cathepsin B (Sætre et al., 2015). Satisfactory electrodisruption can be further tested empirically by the use of positive and negative controls in the LHD sequestration assay, for example, by observing sequestration of LDH in cells incubated with Torin1 or in amino acid-free medium, in combination with bafilomycin A1, and validating that this

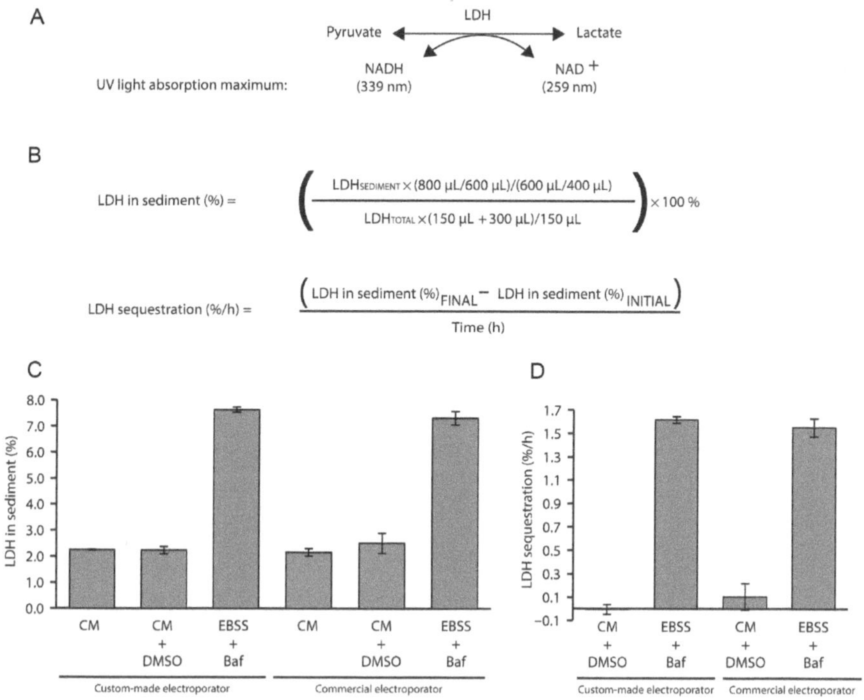

Fig. 3 Measurement of LDH activity and calculation of LDH sequestration (%/h), and example of results obtained in LNCaP cells upon use of our custom-made electroporator vs a commercially available electroporator. (A) When pyruvate is in excess, LDH catalyzes its conversion to lactate, and in this process consumes NADH. The amount of LDH in the samples is measured as the rate of decrease in NADH absorbance at 340 nm. (B) To account for dilutions in step 10, 12, 13, 17, and 18, multiply measured values for LDH_{TOTAL} by (150 + 300 μL)/150 μL and $LDH_{SEDIMENT}$ by (800/600 μL)/(600/400 μL). Multiply the thus calculated $LDH_{SEDIMENT}/LDH_{TOTAL}$ ratio by 100%, as shown in the upper panel, to express the values as "LDH in sediment (%)." Subtract the background percentage of LDH obtained in the sediment of untreated cells ("LDH in sediment $(\%)_{INITIAL}$") from the percentage of LDH in the sediment of experimentally treated cells ("LDH in sediment $(\%)_{FINAL}$") and divide by hours incubation with degradation inhibitor (the measurement period), as shown in the lower panel, to express the values as "LDH sequestration (%/h)." The autophagic cargo sequestration rate is the net accumulation of sequestered LDH (initial background subtracted) per hour of incubation with degradation inhibitor, expressed as percent of total cellular LDH/h. (C–D) LNCaP cells were subjected to the indicated treatments for 3 h and harvested according to the protocol described in this text. Cells were either electrodisrupted with our custom-made electroporator (2000 V and 1.2 μF in a 1 × 1 × 5 cm electrode chamber) or the commercial electroporator BTX Harvard EMC 630 (800 V, 25 μF, and 400 Ω). In (C) the results are expressed as "LDH in sediment (%)." In (D), background LDH has been subtracted and the resulting values have been divided by time (3 h) according to (B), in order to express the values as "LDH sequestration (%/h)." Plasma membrane electrodisruption with the commercial electroporator yields equally low background and virtually identical results as when using the original, custom-made apparatus. Mean values from triplicates are shown, with *error bars* representing standard deviation. Variability between values from treatment replicates can be expected to be up to 0.1–0.2%/h. *μL*, microliter; *μF*, microfarad; *cm*, centimeter; *h*, hours; *nm*, nanometer; *LDH*, lactate dehydrogenase; *NADH*, nicotinamide adenine dinucleotide; *Ω, ohm; V*, volt.

increase is sensitive to sequestration inhibitors like 3-methyladenine (3MA) and thapsigargin, and/or to knockdown/knockout of critical autophagy-related genes that act at presequestration steps. We have found that LDH sequestration is in general inhibited by 3MA and thapsigargin (Engedal et al., 2013; Gordon et al., 1993; Kopitz et al., 1990; Sætre et al., 2015; Seglen et al., 2015), and is strongly reduced upon RNAi-mediated silencing of FIP200 (Seglen et al., 2015) or the GABARAP Atg8 subfamily (Szalai et al., 2015), or upon knockout of Atg5 (Szalai et al., 2015). Note that even though the pulse discharge obtained with our homemade device has proven appropriate for all cell types we have tested thus far, the optimal pulse settings for selective plasma membrane disruption of different cell types may vary. It is also important to note that different methods of plasma membrane disruption may affect the cells in different ways, and this may drastically alter the centrifugation conditions that are needed for sedimentation (step 14). We recommend using electrodisruption because it is extremely efficient and reproducible from sample to sample, as well as across various cell types, at the same time that it has minimal effects on intracellular structures.

10. Transfer the 400 μL cell disruptate from the electroporation cuvette to a new microcentrifuge tube containing 400 μL ice-cold phosphate-buffered sucrose (100 m*M* sodium monophosphate, 2 m*M* dithiothreitol (DTT), 2 m*M* EDTA, and 1.75% sucrose, pH 7.5) to a total volume of 800 μL and store on ice.
11. Repeat steps 8–10 with the remaining samples, one sample at a time.
12. Transfer 150 μL of the diluted cell disruptate to a new microcentrifuge tube and subject it to one freeze–thaw cycle at −80°C (generally we keep it overnight at −80°C). This sample is hereafter referred to as "LDH_{TOTAL}."
13. Transfer 600 μL of the diluted cell disruptate to a microcentrifuge tube containing 900 μL ice-cold resuspension buffer (50 m*M* sodium monophosphate, 1 m*M* EDTA, 1 m*M* DTT) supplemented with 0.5% BSA and 0.01% Tween 20. Mix briefly by pipetting up-and-down, and store on ice.
14. Centrifuge at 18,000 × *g* for 50 min at 4°C in a normal table-top centrifuge (e.g., Beckman Coulter 368831).

    Note: Centrifugation at 18,000 × *g* for 50 min at 4°C is routinely used but may be reduced for some cell lines. It is, for example, sufficient to centrifuge at 5000 × *g* for 7 min at 4°C with MEFs. Too forceful centrifugation results in dense pellets, making it strenuous to solubilize the autophagic vacuoles in 1% Triton X-405 (step 17). Too weak centrifugation results in incomplete sedimentation of the cell corpses.

15. Aspirate the supernatant and subject the pellet to one freeze–thaw cycle at –80°C (generally we keep it overnight at –80°C). This is hereafter referred to as "$LDH_{SEDIMENT}$."

    Note: Use suction to thoroughly aspirate all liquid, leaving the pellet as dry as possible.

16. The following day/after one freeze–thaw cycle, thaw all samples on ice.
17. Vortex the $LDH_{SEDIMENT}$ sample until the pellet detaches from the microcentrifuge tube wall, add 400 μL ice-cold resuspension buffer supplemented with 1% Triton X-405, and resuspend until the pellet is dissolved.
18. Dilute the total cellular LDH sample (see step 12) by adding 300 μL resuspension buffer supplemented with 1.5% Triton X-405 and mix until the solution is homogenous.
19. Centrifuge all samples at 18,000 × *g* for 5 min at 4°C to sediment debris.
20. Measure LDH enzymatic activity in the LDH_{TOTAL} and $LDH_{SEDIMENT}$ samples as the decline in NADH absorbance at 340 nm (Fig. 3A). Fig. 3B illustrates how to express the measured values as "LDH sequestration (%/h)." First, the percentage of LDH in the sediment is calculated for all experimental samples. Next, the LDH sequestration rate during experimental treatments is calculated by subtracting the "LDH in sediment $(\%)_{INITIAL}$" from the "LDH in sediment $(\%)_{FINAL}$" (see step 3 for definitions) divided by the time of experimental treatment with the inhibitor of LDH degradation (e.g., Baf). Fig. 3C shows typical "LDH in sediment (%)" values that we obtain in LNCaP cells. Note that the background levels (~2.2%) and the levels observed upon subjection to EBSS medium (starvation) + Baf for 3 h (~7–7.5%) were in the same range whether we used our custom-made electroporator or a commercially available electroporator for plasma membrane electrodisruption (Fig. 3C). Fig. 3D shows "LDH sequestration (%/h)" values obtained after subtracting the "LDH in sediment $(\%)_{INITIAL}$" from the "LDH in sediment $(\%)_{FINAL}$" and dividing by the time of experimental treatment (equation shown in Fig. 3B). This is the preferred, general way of illustrating results from the LDH sequestration assay. Fig. 3D shows the sequestration rate one can typically expect in LNCaP cells under amino

acid starvation (~1.6%/h), using the protocol described herein. Note that the obtained sequestration rate was similar whether we used our custom-made electroporator or a commercially available electroporator for plasma membrane electrodisruption (Fig. 3D). The background percentages of LDH in the sediment as well as the sequestration rates obtained with the downscaled and simplified method presented herein (Fig. 3C and D) are very similar to those obtained using the original method (Engedal et al., 2013). Interestingly, macroautophagic sequestration rates vary substantially between cell types. For example, whereas primary rat hepatocytes, which have an exceptionally high macroautophagic capacity, sequester LDH at a rate of 3–5% per hour under amino acid starvation conditions (Kisen et al., 1993; Kopitz et al., 1990; Szalai et al., 2015), LNCaP, MEF, HeLa, PC3, and U2OS sequester LDH at a rate of approximately 1.7, 1.2, 1.0, 0.9, and 0.6% per hour, respectively, as measured under amino acid starvation conditions in the presence of Baf (Seglen et al., 2015).

Note: We measure LDH activity with a multianalyzer (MaxMat PL-II, Erba Diagnostics) in combination with an LDH-assay kit (RM LADH0126V, Erba Diagnostics). The LDH kit consists of two reagents, R1 (65 m*M* imidazol and 0.75 m*M* pyruvate, pH 7.5) and R2 (65 m*M* imidazol and 0.9 m*M* NADH). The working solution (65 m*M* imidazol, 0.6 m*M* pyruvate, and 0.18 m*M* NADH, pH 7.5) is prepared by mixing four volumes of the reagent R1 with one volume of the reagent R2. The working solution is stable for up to 15 days when stored at 2–8°C and 5 days when stored at 15–25°C. LDH activity can in principle be measured with any spectrophotometer/plate reader and LDH-assay kit.

3. CONCLUDING REMARKS

The autophagy field is expanding and so is the array of methods available for studying different aspects of this process. It is, therefore, increasingly important to critically evaluate which methods are best suited to answer a given scientific question related to autophagy (Klionsky et al., 2016). An important distinction is between autophagic *markers* (the cart) and autophagic *cargo*. If the aim is to examine whether a compound and/or genetic manipulation alters autophagic activity, it is essential to employ functional assays that monitor autophagic *cargo*. Here, we have presented a method for measuring autophagic sequestration of the cytosolic protein LDH. The advantage of using LDH as a cargo marker is that it is ubiquitously expressed, it is exclusively degraded by the autophagic–lysosomal pathway (Kopitz et al., 1990), and the amount of sequestered protein can easily be quantified by an enzymatic assay. This method represents one of very few assays that are available to analyze autophagic sequestration, whereas other methods such as fluorescence and electron microscopy cannot accurately determine whether, or to which degree, structures with characteristics of autophagosomes are sealed or not. The process of phagophore closure remains one of the least understood steps in the macroautophagic pathway. Sequestration assays, such as the one described here, may be one of the most useful tools to uncover regulators and mechanisms of phagophore closure. Our data obtained by employing the LDH sequestration assay indicate that the final steps in autophagosome formation are sensitive to alterations in intracellular calcium homeostasis as well as to the phosphatidylinositol 3-kinase inhibitor 3-methyladenine (Engedal et al., 2013; Sætre et al., 2015). The LDH sequestration method, originally developed for primary hepatocytes (Kopitz et al., 1990) has now been adapted to cultured cells (Engedal et al., 2013; Seglen et al., 2015; Szalai et al., 2015). Importantly, we have been able to increase the throughput of the method by eliminating several washing and centrifugation steps, without reducing the quality of the data. We have also, for the first time, been successful in applying a commercial electroporator for selective plasma membrane disruption, so that our current optimized method can now be carried out using standard laboratory equipment.

ACKNOWLEDGMENTS

This work was generously supported by grants from the Norwegian Research Council, the University of Oslo, the Norwegian Cancer Society, the Nansen Foundation, and the Legacy in the memory of Henrik Homan.

REFERENCES

Bestebroer, J., V'Kovski, P., Mauthe, M., & Reggiori, F. (2013). Hidden behind autophagy: The unconventional roles of ATG proteins. *Traffic*, *14*(10), 1029–1041. http://dx.doi.org/10.1111/tra.12091.

Engedal, N., Torgersen, M. L., Guldvik, I. J., Barfeld, S. J., Bakula, D., Sætre, F., … Mills, I. G. (2013). Modulation of intracellular calcium homeostasis blocks autophagosome formation. *Autophagy*, *9*(10), 1475–1490. http://dx.doi.org/10.4161/Auto.25900.

Gordon, P. B., Holen, I., Fosse, M., Rotnes, J. S., & Seglen, P. O. (1993). Dependence of hepatocytic autophagy on intracellularly sequestered calcium. *Journal of Biological Chemistry*, *268*(35), 26107–26112.

Gordon, P. B., & Seglen, P. O. (1982). Autophagic sequestration of [14C]sucrose, introduced into rat hepatocytes by reversible electro-permeabilization. *Experimental Cell Research*, *142*(1), 1–14.

Kisen, G. Ø., Tessitore, L., Costelli, P., Gordon, P. B., Schwarze, P. E., Baccino, F. M., & Seglen, P. O. (1993). Reduced autophagic activity in primary rat hepatocellular carcinoma and ascites hepatoma cells. *Carcinogenesis*, *14*(12), 2501–2505.

Klionsky, D. J., Abdelmohsen, K., Abe, A., Abedin, M. J., Abeliovich, H., Acevedo Arozena, A., … Zughaier, S. M. (2016). Guidelines for the use and interpretation of assays for monitoring autophagy (3rd edition). *Autophagy*, *12*(1), 1–222. http://dx.doi.org/10.1080/15548627.2015.1100356.

Kopitz, J., Kisen, G. O., Gordon, P. B., Bohley, P., & Seglen, P. O. (1990). Nonselective autophagy of cytosolic enzymes by isolated rat hepatocytes. *Journal of Cell Biology*, *111*(3), 941–953.

Pattingre, S., Petiot, A., & Codogno, P. (2004). Analyses of Galpha-interacting protein and activator of G-protein-signaling-3 functions in macroautophagy. *Methods in Enzymology*, *390*, 17–31. http://dx.doi.org/10.1016/S0076-6879(04)90002-X.

Sætre, F., Hagen, L. K., Engedal, N., & Seglen, P. O. (2015). Novel steps in the autophagic-lysosomal pathway. *The FEBS Journal*, *282*(11), 2202–2214. http://dx.doi.org/10.1111/febs.13268.

Seglen, P. O. (1983). Inhibitors of lysosomal function. *Methods in Enzymology*, *96*, 737–764.

Seglen, P. O., Grinde, B., & Solheim, A. E. (1979). Inhibition of the lysosomal pathway of protein degradation in isolated rat hepatocytes by ammonia, methylamine, chloroquine and leupeptin. *European Journal of Biochemistry*, *95*(2), 215–225.

Seglen, P. O., Luhr, M., Mills, I. G., Sætre, F., Szalai, P., & Engedal, N. (2015). Macroautophagic cargo sequestration assays. *Methods*, *75*, 25–36. http://dx.doi.org/10.1016/j.ymeth.2014.12.021.

Seglen, P. O., Øverbye, A., & Sætre, F. (2009). Sequestration assays for mammalian autophagy. *Methods in Enzymology*, *452*, 63–83. http://dx.doi.org/10.1016/S0076-6879(08)03605-7.

Subramani, S., & Malhotra, V. (2013). Non-autophagic roles of autophagy-related proteins. *EMBO Reports*, *14*(2), 143–151. http://dx.doi.org/10.1038/embor.2012.220.

Szalai, P., Hagen, L. K., Sætre, F., Luhr, M., Sponheim, M., Øverbye, A., … Engedal, N. (2015). Autophagic bulk sequestration of cytosolic cargo is independent of LC3, but requires GABARAPs. *Experimental Cell Research*, *333*(1), 21–38. http://dx.doi.org/10.1016/j.yexcr.2015.02.003.

CHAPTER TWENTY-ONE

In Vitro Reconstitution of Autophagosome–Lysosome Fusion

J. Diao*[,1], L. Li[†,‡], Y. Lai[§], Q. Zhong[†,‡,1]
*University of Cincinnati College of Medicine, Cincinnati, OH, United States
[†]Center for Autophagy Research, University of Texas Southwestern Medical Center, Dallas, TX, United States
[‡]University of Texas Southwestern Medical Center, Dallas, TX, United States
[§]Stanford University, Stanford, CA, United States
[1]Corresponding authors: e-mail address: jiajie.diao@uc.edu; qing.zhong@utsouthwestern.edu

Contents

1. Introduction 366
2. SNARE Protein Purification 367
   2.1 Syntaxin 17 and VAMP8 368
   2.2 SNAP-29 369
3. Protein Reconstitution 370
   3.1 Preparation of Lipid-Mixing Vesicles 370
   3.2 Preparation of Content-Mixing Vesicles 371
   3.3 Reconstitution of SNAREs 371
4. Fluorescent Measurement 372
5. Single-Vesicle Assay 373
6. Summary 374
Acknowledgments 375
References 375

Abstract

SNARE (soluble *N*-ethylmaleimide-sensitive factor attachment protein receptors) proteins are a highly regulated class of membrane proteins lying in the center of membrane fusion. In conjunction with accessory proteins, SNAREs drive efficient merger of two distinct lipid bilayers into one interconnected structure. This chapter describes our fluorescence resonance energy transfer (FRET)-based proteoliposome fusion assays for the roles of various SNARE proteins, accessory proteins, and effects of different lipid compositions on membrane fusion involved in autophagy.

Methods in Enzymology, Volume 587
ISSN 0076-6879
http://dx.doi.org/10.1016/bs.mie.2016.09.065

1. INTRODUCTION

Autophagy is a crucial catabolic pathway conserved from yeast to humans. In this pathway, double membrane vesicles called autophagosomes sequester and engulf cytosol, protein aggregates, damaged organelles, and invading pathogens. Small autophagosomes expand into large autophagosomes, later fusing with lysosomes to degrade cargo. The membrane fusion between autophagosomes and lysosomes is the critical step in the maturation of autolysosomes. Although over 30 different autophagy-specific genes have been discovered in *S. cerevisiae* (Levine & Klionsky, 2004), the autophagy-specific factors and mechanism responsible for the membrane fusion activity of autophagosomes remain poorly understood.

Membrane fusion is an important cellular process by which two initially distinct lipid bilayers merge to form one interconnected structure. Intracellular trafficking, egg fertilization, and communication between neurons are a few among many processes that rely on some forms of fusion (Jahn, Lang, & Sudhof, 2003). Membrane fusion in eukaryotic cells is mediated by a conserved family of proteins called SNAREs (soluble *N*-ethylmaleimide-sensitive factor attachment protein receptors) (Chernomordik & Kozlov, 2008; Wickner & Schekman, 2008). All SNARE systems consist of two cognate pairs, v- and t-SNAREs (vesicular- and target-), which are anchored in two sides of membranes. They can form a four-alpha-helix-bundle via association of their "SNARE motifs" that are composed of 15 hydrophobic core layers (Sutton, Fasshauer, Jahn, & Brunger, 1998). The SNAREs required for autophagy in *S. cerevisiae* and human cells have recently been identified (Itakura, Kishi-Itakura, & Mizushima, 2012; Nair et al., 2011). In particular, the yeast SNAREs containing Sso1, Snc2, and Sec9 have been implicated in early autophagosome formation (Nair et al., 2011). Conversely, syntaxin 17, VAMP8 (vesicle-associated membrane protein 8), and SNAP-29 (synaptosome-associated protein 29 kDa) have been implicated in autophagosome maturation in human cells (Itakura et al., 2012).

Ensemble fluorescence (Förster) resonance energy transfer (FRET) for protein-reconstituted proteoliposomes has been used as a standard in vitro tool for studying membrane fusion (Struck, Hoekstra, & Pagano, 1981), which could be mediated by SNAREs (Weber et al., 1998). In this assay, *N*-(7-nitro-2,1,3-benzoxadiazole-4-yl)-phosphatidylethanolamine (NBD-PE)- and *N*-(lissamine rhodamine B sulfonyl)-phosphatidylethanolamine

(Rh-PE)-containing vesicles reconstituted with v-SNAREs (synaptobrevin/VAMP) fuse with unlabeled vesicle containing t-SNAREs (syntaxin and SNAP), resulting in the dequenching of NBD fluorescence. To study membrane fusion involved in autophagy, v- and t-SNARE proteins are reconstituted into two independent populations of vesicles containing acceptor and donor fluorophore-labeled lipophilic molecules, respectively. Upon fusion, lipid mixing results in a decrease of the average distance between donors and acceptors, and the degree of fusion can be quantified from the FRET efficiency, which is a measure of energy transfer between donor and acceptor fluorophores and is commonly approximated as the ratio of the acceptor intensity and the sum of the donor and acceptor intensities (Fig. 1A). Prequenched dye molecules, such as sulforhodamine B, can be used as content indicators (Fig. 1B) (Lai et al., 2014).

2. SNARE PROTEIN PURIFICATION

For syntaxin 17 and VAMP8 purification, two types of detergents were applied. Dodecylmaltoside (DDM) was used for solubilizing membranes

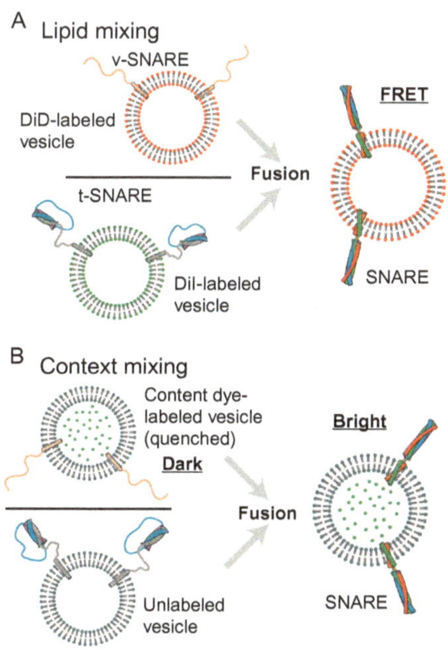

Fig. 1 Ensemble vesicle fusion assays of lipid mixing (A) and content mixing (B).

efficiently and replaced in the following wash step by octyl glucoside (OG) for reconstitution.

2.1 Syntaxin 17 and VAMP8

1. Suspend 8 L of autoinduced C43 cells (low-expression derivative of BL-21 *E. coli* cells) in 500 mL of Buffer A (10 m*M* sodium phosphate, 2.7 m*M* KCl, 137 m*M* NaCl, 5 m*M* EDTA, pH 7.4) supplemented with 10 EDTA-free Complete Protease Inhibitor tablets (Roche, Basel, Switzerland) and 1 m*M* PMSF.
2. After homogenizing thoroughly with a blender, lyse the cells by passing the cell suspension through the Emulsiflex C5 homogenizer (Avestin, Ottawa, Canada) for three passes with a pressure of at least 15,000 psi.
3. Centrifuge cell lysate in a JA-14 rotor (Beckman Coulter, Brea, CA) at 10,000 rpm for 30 min to remove inclusion bodies and cell debris. Collect supernatant carefully and try not to dislodge the pellet.
4. Distribute the supernatant into Ti-45 tubes (Beckman Coulter, Brea, CA), isolate the membrane by centrifugation at 40,000 rpm for 2 h in the Ti-45 rotor (Beckman Coulter, Brea, CA) at 4°C.
5. The membrane will be in the pellet, pour out the supernatant, and drain tubes thoroughly by placing inverted on a paper towel for 2 min.
6. Homogenize the membrane pellet in Buffer B (20 m*M* HEPES, 500 m*M* NaCl, 2 m*M* DTT, 10 w/v% glycerol, pH 7.5) supplemented with 1 EDTA-free Complete Protease Inhibitor tablet for every 50 mL and 1 m*M* PMSF. Measure the concentration of total membrane proteins by a Bradford assay.
7. Extract the membrane proteins by solubilizing the membranes with 2% DDM (Anatrace, Maumee, OH) at 5 mg/mL membrane protein in Buffer B. Stir gently with a magnetic stir bar, allow membranes to solubilize for 1 h in a cold room.
8. Spin the sample in Ti-45 rotor at 40,000 rpm for 35–60 min at 4°C. While the sample is centrifuging, prepare a gravity flow column with 1 mL of GST resin (GE Healthcare, Uppsala, Sweden) and wash with 20 mL of Buffer B to equilibrate it.
9. Incubate the supernatant with the equilibrated GST resin, stir gently with a magnetic stir bar, and allow the protein to bind overnight in a cold room.
10. Collect the resin through the filter on the polypropylene gravity flow column by pouring the supernatant–resin mixture through the column.

11. Wash the column with 50 mL of Buffer C (20 m*M* HEPES, 300 m*M* NaCl, 2 m*M* DTT, 10 w/v% glycerol, 110 m*M* OG, pH 7.5).
12. Suspend the resin in 2 mL of Buffer C, digest with 0.2 mg/mL TEV protease to cleave the GST tag, and incubate overnight on a shaker in a cold room.
13. Collect the fractions containing the protein.
14. Apply the sample to a Superdex 200 (10/300 GL) column (GE Healthcare, Uppsala, Sweden), preequilibrated with Buffer D (20 m*M* HEPES, 300 m*M* NaCl, 1 m*M* TCEP, 10 w/v% glycerol, 110 m*M* OG, pH 7.5). Collect the fractions containing the protein peak.
15. Measure the concentration and freeze in liquid nitrogen, store in −80°C freezer. Purified proteins could be stored for at least 1 month.

2.2 SNAP-29

1. Suspend 4 L of autoinduced C43 cells (low-expression derivative of BL-21 *E. coli* cells) in 200 mL of Buffer AA (50 m*M* Tris, 300 *M* NaCl, 2 m*M* DTT, 20 m*M* imidazole, pH 8.0) supplemented with 4 EDTA-free Complete Protease Inhibitor tablets and 1 m*M* PMSF.
2. After homogenizing thoroughly with a blender, lyse the cells by passing the cell suspension through the Emulsiflex C5 homogenizer for three passes with a pressure of at least 15,000 psi.
3. The lysate was clarified by centrifugation in the Ti-45 rotor for 1.5 h at 40,000 rpm at 4°C. While the sample is centrifuging, prepare a gravity flow column with 2 mL Ni-NTA resin (Qiagen, Hilden, Germany) and wash with 40 mL of Buffer AA to equilibrate it.
4. Incubate the supernatant with the equilibrated Ni-NTA resin, stir gently with a magnetic stir bar, and allow protein to bind for 2 h in a cold room.
5. Collect the resin through the filter on the polypropylene gravity flow column by pouring the supernatant–resin mixture through the column.
6. Wash the column with 100 mL of Buffer BB (50 m*M* Tris, 300 m*M* NaCl, 2 m*M* DTT, 60 m*M* imidazole, pH 8.0). Elute the protein with Buffer CC (50 m*M* Tris, 300 m*M* NaCl, 2 m*M* DTT, 500 m*M* imidazole, pH 8.0).
7. Pool the protein fractions and digest twice with 0.2 mg/mL TEV protease for 30 min at ambient temperature to cleave the hexa-histidine tag; the TEV protease can be removed by centrifugation at 5000 rpm for 10 min since it is not stable in high salt buffer.

8. Apply the sample to a Superdex 200 (10/300 GL) column preequilibrated with Buffer DD (20 m*M* HEPES, 300 m*M* NaCl, 1 m*M* TCEP, 10 w/v% glycerol, pH 7.5). Collect the fractions containing the protein peak.
9. Measure the concentration and freeze in liquid nitrogen, store in −80°C freezer. Purified proteins could be stored for at least 1 month.

3. PROTEIN RECONSTITUTION

Standard and direct methods are two major assays for SNARE protein reconstitution. The standard method refers to a comicellization method whereby proteins and lipids are initially cosolubilized with detergent. The direct method is a preparation process whereby detergent-solubilized proteins are incorporated into preformed vesicles, which is used in our study.

3.1 Preparation of Lipid-Mixing Vesicles

1. Combine the following lipid solution in a glass tube for t- and v-vesicles, respectively (POPC:POPE = 78:20 mol%, 2 mol% DiI or DiD dyes). The composition of the lipid can be changed according to the purpose of experiments. For instance, 0.1 mol% biotinylated phosphatidylethanolamine (biotin-PE) is required to supplement POPC in DiD vesicles for surface immobilization in single-vesicle tethering assay.
2. Evaporate the organic solvent carefully under a gentle stream of nitrogen. Tilt and rotate the tube continuously until a thin film is formed at the bottom of the tube.
3. Wrap the tube with a piece of foil with a small hole on the top for chloroform evaporation and place the tube into a vacuum desiccator for at least 4 h for further drying. Shield the samples from exposure to the ambient light by aluminum foil or a box to minimize photobleaching of dyes.
4. Hydrate the lipid film in 500 μL of vesicle buffer (20 m*M* HEPES, 90 m*M* NaCl) by vortexing until it is completely suspended in the buffer (~30 min).
5. Freeze and thaw the lipid solution using liquid nitrogen and 30–40°C water bath for at least five cycles.
6. Extrude the lipid solution through a 50-nm polycarbonate membrane (25 cycles) according to the manufacturer's instruction. Detailed mini-extruder assembly and extrusion information is available from

Avanti Polar Lipids, Inc. (http://www.avantilipids.com). The vesicles may be stored at 4°C and should be used within 2 weeks.

3.2 Preparation of Content-Mixing Vesicles

1. Combine the following lipid solution in a glass tube for both t- and v-vesicles (POPC:POPE = 80:20 mol%). The composition of the lipid can be changed according to the purpose of experiments.
2. Evaporate the organic solvent carefully under a gentle stream of nitrogen. Tilt and rotate the tube continuously until a thin film is formed at the bottom of the tube.
3. Wrap the tube with a piece of foil with a small hole on the top for chloroform evaporation and place the tube into a vacuum desiccator for at least 4 h for further drying.
4. For t-vesicles, hydrate the lipid film in vesicle buffer (20 m*M* HEPES, 90 m*M* NaCl), making the final lipid concentration at 10 m*M*; for v-vesicles, hydrate the lipid film in vesicle buffer (20 m*M* HEPES, 90 m*M* NaCl) plus 50 m*M* sulforhodamine B, also making the final lipid concentration at 10 m*M*, and vortex the glass tube until it is completely suspended in the buffer (~10 min).
5. Freeze and thaw the lipid solution using liquid nitrogen and 30–40°C water bath for at least five cycles.
6. Extrude the lipid solution through a 50-nm polycarbonate membrane (25 cycles) according to the manufacturer's instruction. The vesicles may be stored at 4°C and should be used within 2 weeks.

3.3 Reconstitution of SNAREs

1. For t-SNARE vesicle sample, mix syntaxin 17 and SNAP-29 (1:2 molar ratio of syntaxin 17/SNAP-29) in vesicle buffer containing 1 wt% OG by gentle agitation for 0.5–1 h at 25°C to form t-SNARE complex.
2. Reconstitute t-SNARE (syntaxin 17/SNAP-29) and v-SNARE (VAMP8) proteins with lipid vesicles containing lipophilic dyes at 1:50–1:200 protein to lipid ratio in buffers containing 0.8 wt% OG detergent solution. For content-mixing sample (only v-SNARE vesicles), 50 m*M* sulforhodamine B was dissolved into VAMP8 protein solution and mixed with v-vesicles containing 50 m*M* sulforhodamine B at 1:50–1:200 protein to lipid ratio in buffers containing 0.8 wt% OG detergent solution. Keep proteins and vesicles on ice. Protect the vesicles from exposure to light.

3. Wrap the tubes with foil and rotate them in a cold room (4°C) for 30 min.
4. For lipid-mixing samples, dilute the vesicle/protein mixture twofold with vesicle buffer. For content-mixing sample (only v-SNARE vesicles), dilute the vesicle/protein mixture twofold with vesicle buffer containing 50 m*M* sulforhodamine B. The vesicle harboring SNARE protein is formed as the OG concentration is reduced below the critical micelle concentration.
5. Move the solutions into prehydrated dialysis tubes and dialyze against 2 L dialysis buffer (20 m*M* HEPES, 90 m*M* NaCl, 2 m*M* DTT, pH 7.4) for 10–12 h, followed by another 2 L dialysis buffer for 4 h. The dialysis should be performed at 4°C. Bio-beads SM-2 (Bio-Rad, Hercules, CA) could be used for removing OG. For content-mixing sample (only v-SNARE vesicles), before the dialysis step, the v-vesicles subsequently reformed and free sulforhodamine B dye was removed by size exclusion chromatography using a Sepharose CL-4B column.
6. Collect the dialyzed vesicles into microfuge tubes. Measure the volume to calculate the lipid concentration. The concentration of the lipid may change due to the change in the volume during dialysis. The protein-reconstituted vesicle samples may be stored at 4°C and should be used within 3 days.

4. FLUORESCENT MEASUREMENT

For the content-mixing experiments, CL-4B column is essential to clean free sulforhodamine B. Moreover, it is critical to perform measurement at the 25°C to minimize false signal from vesicle leakages.

1. Set the fluorescence spectrophotometer to a continuing model (5 s per measurement for 2000 s) with the following parameters: an excitation wavelength of 530 nm and emission wavelengths of 570 and 670 nm and 35°C for the cuvette holder. For content-mixing measurement, an excitation wavelength of 530 nm and emission wavelength of 570 nm and 25°C for the cuvette holder. The slip sizes for excitation and emission could be variable for a suitable intensity.
2. Prepare 50 μL mixture of t-SNARE and v-SNARE vesicles with 1:1 molar ratio. And then add vesicle buffer or accessory proteins to make a total volume of 100 μL in the quartz cuvette.
3. Put the quartz cuvette into the cuvette holder for the fluorescence intensity measurement.

4. For data analysis, set the lowest stable fluorescence intensity point of 670 nm emission as the starting point for the lipid mixing, which could cancel out the unbalanced initial stage.

5. SINGLE-VESICLE ASSAY

The custom software for single-particle data acquisition and analysis is available for download from Taekjip Ha's group Web site.

1. Flow in 30 μL of 0.2 mg/mL NeutrAvidin into each empty flow channel and incubate for 5 min. Preparation of PEGylated quartz slides and assembly steps of the flow chamber have been described in our previous publications (Diao et al., 2012; Kyoung, Zhang, Diao, Chu, & Brunger, 2013).
2. Wash out the excess NeutrAvidin with 400 μL buffer (20 m*M* HEPES, 90 m*M* NaCl, pH 7.4).
3. Flow in 100 μL of ~1 μ*M* (lipid concentration) DiD vesicles or with reconstituted SNARE proteins for fusion and incubate for 30 min at ambient temperature.
4. Wash out the excess DiD vesicles with 400 μL buffer (20 m*M* HEPES, 90 m*M* NaCl, pH 7.4). Avoid introduction of any air bubble which can destroy vesicles upon contact.
5. Flow in 100 μL of solution containing 100–200 n*M* DiI vesicles or with reconstituted SNARE proteins for fusion and ATG14 proteins.
6. Incubate the flow chamber at 37°C inside a hydrated box (e.g., empty pipette tip box with water) for 30–120 min.
7. Wash out the excess DiI vesicles with 400 μL buffer (20 m*M* HEPES, 90 m*M* NaCl, pH 7.4).
8. Place the flow cell on a total internal reflection fluorescence microscope setup and acquire short movies from at least 15 random locations within the flow channel of interest under 532 nm laser excitation. A higher number of short movies may be necessary for samples with low tethering efficiencies.
9. Analyze data for the average number of tethered DiI vesicles per imaging area for tethering effect. Alternatively, for single-vesicle fusion, extract the donor (I_d) and acceptor (I_a) intensities from short movies for fitted molecules. The relative FRET efficiency (E) correlated to the level of fusion could be calculated by $E = I_a/(I_d + I_a)$.

6. SUMMARY

This simple and sensitive fluorescence assay is useful in membrane fusion research and associated molecular study. In the field of SNARE studies, it has led to the finding of minimal membrane fusion machinery in 1998 (Weber et al., 1998). FRET-based lipid-mixing assays have been widely used for a long time to study a variety of membrane fusion processes; among these are fusion events in neurobiology (Ma, Su, Seven, Xu, & Rizo, 2013), autophagy research (Diao et al., 2015; Nair et al., 2011), and other cell biology processes (Gorvel, Chavrier, Zerial, & Gruenberg, 1991).

Meanwhile, the lipid mixing is still blind to the fusion pore opening and expansion, the final critical step of the full-collapse fusion pathway. In 1999, the fusion assay based on duplex formation of oligonucleotides has been employed to research the integrity of liposomes in SNARE-mediated membrane fusion (Nickel et al., 1999), which leads to the conclusion that SNAREs can mediate complete membrane fusion shown by mixing of luminal contents of vesicles. However, in this assay, the content-mixing signal was measured after vesicles were lysed, which left a possibility that docked vesicles without fusion could still contribute to the final readout. A recent study showed that lipid mixing can occur without content mixing (Chan, van Lengerich, & Boxer, 2009), strongly suggesting that protein-reconstituted vesicle fusion experiments have to employ content-mixing indicators.

Despite its extensive usage, the ensemble assays cannot clearly distinguish different stages of the fusion process such as docking, hemifusion, and full fusion due to ensemble averaging. Since the rate of ensemble assays is mainly limited by the docking step (Cypionka et al., 2009), ensemble assays can misinterpret factors with a significant effect on docking rather than fusion (Diao, Ishitsuka, & Bae, 2011). In order to overcome these limitations of bulk lipid-mixing assay, new techniques are needed for observing different fusion steps rather than (or in addition to) lipid mixing at the single-vesicle level (Brunger, Cipriano, & Diao, 2015). As a demonstration, a single-vesicle assay (Diao, Yoon, Su, Shin, & Ha, 2009; Diao et al., 2013) has been successfully used to study the membrane tethering effect of Atg14(L) on protein-free liposomes (Fig. 2) (Diao et al., 2015). Dissecting membrane fusion at distinct steps at the single-vesicle level will greatly facilitate our understanding of this sophisticated and delicate process.

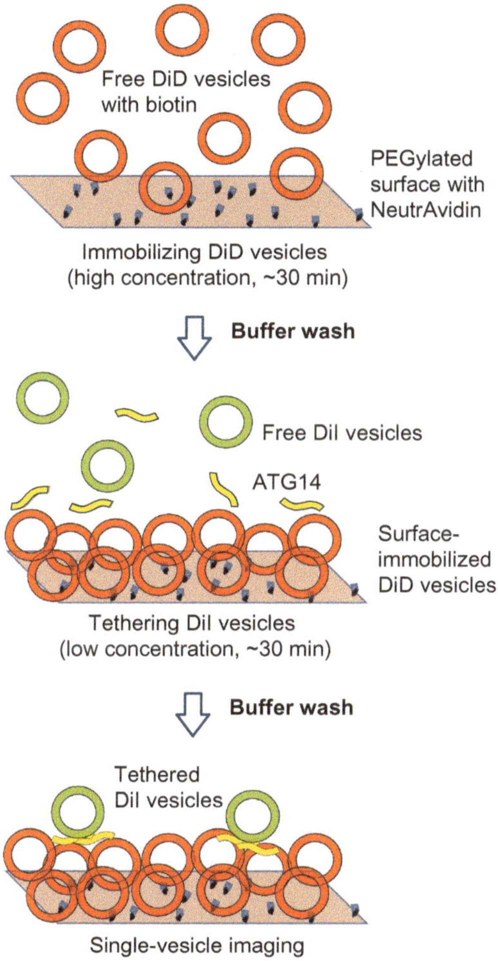

Fig. 2 Experimental scheme of single-vesicle assay.

ACKNOWLEDGMENTS

We want to thank Dr. Qiangjun Zhou and Mr. Richard A. Pfuetzner for help on protein purification protocol. This work was supported by grants to Q.Z. from the Welch Foundation (I-1864), the Cancer Prevention & Research Institute of Texas (RP140320), and NIH R01 (GM116908).

REFERENCES

Brunger, A. T., Cipriano, D. J., & Diao, J. (2015). Towards reconstitution of membrane fusion mediated by SNAREs and other synaptic proteins. *Critical Reviews in Biochemistry and Molecular Biology*, *50*, 231–241.

Chan, Y. H. M., van Lengerich, B., & Boxer, S. G. (2009). Effects of linker sequences on vesicle fusion mediated by lipid-anchored DNA oligonucleotides. *Proceedings of the National Academy of Sciences of the United States of America, 106*(4), 979–984.

Chernomordik, L. V., & Kozlov, M. M. (2008). Mechanics of membrane fusion. *Nature Structural & Molecular Biology, 15*(7), 675–683.

Cypionka, A., et al. (2009). Discrimination between docking and fusion of liposomes reconstituted with neuronal SNARE-proteins using FCS. *Proceedings of the National Academy of Sciences of the United States of America, 106*(44), 18575–18580.

Diao, J., Ishitsuka, Y., & Bae, W. R. (2011). Single-molecule FRET study of SNARE-mediated membrane fusion. *Bioscience Reports, 31*(6), 457–463.

Diao, J., Yoon, T. Y., Su, Z. L., Shin, Y. K., & Ha, T. (2009). C2AB: A molecular glue for lipid vesicles with a negatively charged surface. *Langmuir, 25*(13), 7177–7180.

Diao, J., et al. (2012). A single vesicle-vesicle fusion assay for in vitro studies of SNAREs and accessory proteins. *Nature Protocols*, 7(5), 921–934.

Diao, J. J., et al. (2013). Complexin-1 enhances the on-rate of vesicle docking via simultaneous SNARE and membrane interactions. *Journal of the American Chemical Society, 135*(41), 15274–15277.

Diao, J., et al. (2015). ATG14 promotes membrane tethering and fusion of autophagosomes to endolysosomes. *Nature, 520*(7548), 563–566.

Gorvel, J.-P., Chavrier, P., Zerial, M., & Gruenberg, J. (1991). rab5 controls early endosome fusion in vitro. *Cell, 64*(5), 915–925.

Itakura, E., Kishi-Itakura, C., & Mizushima, N. (2012). The hairpin-type tail-anchored SNARE syntaxin 17 targets to autophagosomes for fusion with endosomes/lysosomes. *Cell, 151*(6), 1256–1269.

Jahn, R., Lang, T., & Sudhof, T. C. (2003). Membrane fusion. *Cell, 112*(4), 519–533.

Kyoung, M. J., Zhang, Y. X., Diao, J. J., Chu, S., & Brunger, A. T. (2013). Studying calcium-triggered vesicle fusion in a single vesicle-vesicle content and lipid-mixing system. *Nature Protocols, 8*(1), 1–16.

Lai, Y., et al. (2014). Complexin inhibits spontaneous release and synchronizes Ca^{2+}—triggered synaptic vesicle fusion by distinct mechanisms. *eLife, 3*, e03756.

Levine, B., & Klionsky, D. J. (2004). Development by self-digestion: Molecular mechanisms and biological functions of autophagy. *Developmental Cell, 6*(4), 463–477.

Ma, C., Su, L., Seven, A. B., Xu, Y., & Rizo, J. (2013). Reconstitution of the vital functions of Munc18 and Munc13 in neurotransmitter release. *Science, 339*(6118), 421–425.

Nair, U., et al. (2011). SNARE proteins are required for macroautophagy. *Cell, 146*(2), 290–302.

Nickel, W., et al. (1999). Content mixing and membrane integrity during membrane fusion driven by pairing of isolated v-SNAREs and t-SNAREs. *Proceedings of the National Academy of Sciences of the United States of America, 96*(22), 12571–12576.

Struck, D. K., Hoekstra, D., & Pagano, R. E. (1981). Use of resonance energy transfer to monitor membrane fusion. *Biochemistry, 20*(14), 4093–4099.

Sutton, R. B., Fasshauer, D., Jahn, R., & Brunger, A. T. (1998). Crystal structure of a SNARE complex involved in synaptic exocytosis at 2.4 angstrom resolution. *Nature, 395*(6700), 347–353.

Weber, T., et al. (1998). SNAREpins: Minimal machinery for membrane fusion. *Cell, 92*, 759–772.

Wickner, W., & Schekman, R. (2008). Membrane fusion. *Nature Structural & Molecular Biology, 15*(7), 658–664.

CHAPTER TWENTY-TWO

In Vitro Reconstitution of Atg8 Conjugation and Deconjugation

D. Fracchiolla, B. Zens, S. Martens[1]
Max F. Perutz Laboratories, University of Vienna, Vienna Biocenter, Vienna, Austria
[1]Corresponding author: e-mail address: sascha.martens@univie.ac.at

Contents

1. Introduction 378
2. Methods for In Vitro Atg8 Lipidation 379
   2.1 Expression and Purification of Recombinant Proteins of the Atg8 Conjugation System 379
   2.2 Conjugation/Lipidation of Atg8 to SUVs 384
   2.3 Conjugation of mGFP-Atg8 to GUVs 387
References 388

Abstract

Macroautophagy, hereafter autophagy, is a major degradation pathway in eukaryotic systems that allows the removal of large intracellular structures such as entire organelles or protein aggregates, thus contributing to the homeostasis of cells and tissues. Autophagy entails the de novo formation of an organelle termed *autophagosome*, where a cup-shaped structure called *isolation membrane* nucleates in proximity of a cytoplasmic cargo material. Upon elongation and closure of isolation membranes, the mature autophagosome delivers the sequestered cargo into the lysosomal system for degradation. Among the factors for autophagosome formation are the autophagy-related (Atg) proteins belonging to the Atg8 conjugation system. In this system, the ubiquitin-like Atg8 protein is conjugated to the membrane lipid phosphatidylethanolamine present in autophagosomal membranes. Atg8 can also be removed from membranes by Atg4-mediated deconjugation. Here, we describe in vitro systems that recapitulate the enzymatic reactions occurring in vivo by presenting expression and purification strategies for all the components of the *Saccharomyces cerevisiae* Atg8 conjugation system. We also present protocols for in vitro Atg8 conjugation and deconjugation reactions employing small and giant unilamellar vesicles.

Methods in Enzymology, Volume 587
ISSN 0076-6879
http://dx.doi.org/10.1016/bs.mie.2016.09.066

1. INTRODUCTION

Autophagy is a central quality control pathway that mediates the delivery of cytoplasmic cargo material into the lysosomal system for degradation (Ge, Baskaran, Schekman, & Hurley, 2014; Zaffagnini & Martens, 2016). This is achieved by the sequestration of the cargo within double-membrane vesicles termed *autophagosomes*. Autophagosomes form in an inducible manner and initially appear as small membrane structures referred to as isolation membranes or phagophores. While isolation membranes expand, they gradually surround the cargo material. After closure of the isolation membranes, the cargo is isolated from the rest of the cell and subsequently degraded upon fusion of autophagosomes with lysosomes or vacuoles in yeast.

The conjugation of the ubiquitin-like Atg8 proteins to the membrane lipid phosphatidylethanolamine (PE) (Ichimura et al., 2000) is important for cargo selectivity (Noda et al., 2008) and also has a role during the expansion of isolation membranes (Xie, Nair, & Klionsky, 2008) (Fig. 1). *Saccharomyces cerevisiae* Atg8 (and its human homologs, for example, LC3B) are synthesized as precursors with a C-terminal extension. For *S. cerevisiae* Atg8, this extension consists of a single arginine residue (R117). This arginine is removed by the cysteine protease Atg4 exposing the penultimate glycine residue (G116), which in turn becomes conjugated to the headgroup of PE (Ichimura et al., 2000; Kirisako et al., 2000) (Fig. 1). During this ubiquitin-like conjugation reaction, the E1-like Atg7 enzyme activates Atg8 under consumption of ATP. This reaction requires Mg^{2+} as a cofactor. From Atg7, Atg8 is transferred to the E2-like enzyme Atg3, which transfers Atg8 to PE (Ichimura et al., 2000). This last step is massively accelerated by an E3-like enzyme composed of the Atg12–Atg5 conjugate and Atg16 (Hanada et al., 2007; Mizushima et al., 1998; Romanov et al., 2012). Atg8 is deconjugated from PE by the Atg4 protein, the same enzyme that activated Atg8 by removing the C-terminal arginine (Kirisako et al., 2000) (Fig. 1).

Since the formation of Atg8–PE is central to autophagosome formation, it is important to understand the mechanism of its conjugation to PE and how this reaction is regulated. This requires the reconstitution of the reaction in order to determine which factors or modifications are necessary and sufficient for a given step of this reaction and/or its regulation. Indeed, reconstitution approaches for Atg8 conjugation have provided unique insights into the role of Atg8 during autophagosome formation (Hanada et al., 2007; Kaufmann, Beier, Franquelim, & Wollert, 2014;

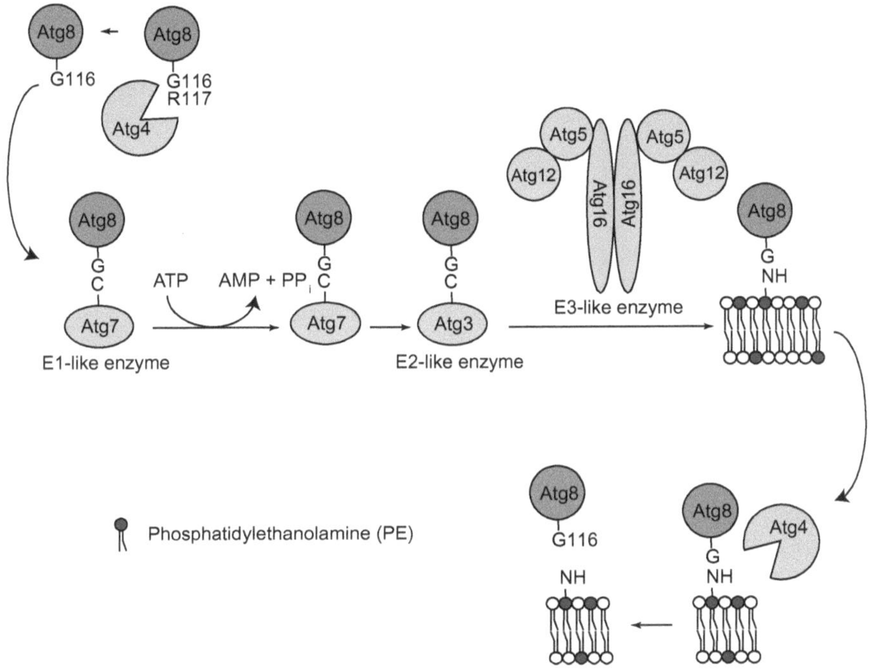

Fig. 1 Scheme of the Atg8 conjugation and deconjugation reactions. Atg8 is synthesized as precursor form with a C-terminal arginine (R117), which is proteolytically removed by Atg4. The resulting C-terminal glycine (G116) is subsequently transferred to a cysteine residue in Atg7. This step requires ATP and Mg^{2+} as a cofactor. Next, Atg8 is transferred to Atg3 and from there to the headgroup of PE. This last step is promoted by the Atg12–Atg5-Atg16 complex. Atg8–PE can be deconjugated from PE by Atg4.

Knorr et al., 2014; Nakatogawa, Ichimura, & Ohsumi, 2007; Nath et al., 2014; Romanov et al., 2012; Sawa-Makarska et al., 2014; Turco & Martens, 2016; Weidberg et al., 2011; Zens, Sawa-Makarska, & Martens, 2015). Here, we describe protocols for the in vitro conjugation and deconjugation of Atg8 from PE using small and giant unilamellar vesicles (SUVs and GUVs).

2. METHODS FOR IN VITRO ATG8 LIPIDATION

2.1 Expression and Purification of Recombinant Proteins of the Atg8 Conjugation System

In vitro reconstitution systems represent a reductionist approach to study complex biological processes and are useful means to identify the essential components needed for the studied reactions. The first step toward a reconstituted system is to obtain the components at a high purity. In this paragraph, we describe protocols used to express and purify active enzymes involved in the ubiquitin-like conjugation reaction of the *S. cerevisiae* Atg8 to PE on artificial membranes.

2.1.1 Expression and Purification of S. cerevisiae Atg8 and mGFP-Atg8

Recombinant Atg8 (NP_009475) and monomeric eGFP (mGFP)-Atg8 with a N-terminal 6xHis-tag followed by a TEV protease cleavage site are expressed from pETDuet-1 in *Escherichia coli* Rosetta pLysS. The coding sequence of both proteins are lacking the C-terminal arginine (R117) to mimic the activating proteolytic cleavage mediated by Atg4. Cells are grown in conical flasks shaking at 37°C in lysogeny broth (LB) to an OD_{600} of 0.8 and protein expression was induced with 50 μ*M* isopropyl β-D-1-thiogalactopyranoside (IPTG) at 18°C. Glucose is added to a final concentration of 0.1% 30 min after induction to increase the yield. Cells are further grown for 16 h at 18°C. Subsequently, cells are pelleted and resuspended in lysis buffer containing 50 m*M* HEPES pH 7.5, 300 m*M* NaCl, 10 m*M* imidazole, 1 m*M* $MgCl_2$, 2 m*M* β-mercaptoethanol, complete protease inhibitors (Roche), and DNAse I (Sigma) and are lysed by freeze thawing and brief sonication for 30 s, 50% cycles. The lysate is centrifuged at 40,000 rpm (Beckman Ti45 rotor) for 40 min at 4°C. The supernatant is loaded on a HisTrap column (GE Healthcare, Buckinghamshire, UK) and the protein is eluted in a step gradient of imidazole (50–75–100–150–200–300 m*M* in 50 m*M* HEPES, pH 7.5, 300 m*M* NaCl, 2 m*M* β-mercaptoethanol). Fractions containing 6xHis-(mGFP)-Atg8 are pooled and the 6xHis-tag is cleaved off with TEV protease (tobacco etch virus nuclear-inclusion-a endopeptidase) overnight at 4°C. The Atg8 protein is subsequently concentrated and diluted to reach a final imidazole concentration of 40 m*M* and the TEV protease containing a N-terminal His-tag is removed by incubation with nickel beads. The supernatant is concentrated to 2 mL using an Amicon Ultra centrifugal filter (MW cut-off 3 kDa) and run on a 16/60 Superdex S75 size exclusion column in 50 m*M* HEPES, pH 7.5, 150 m*M* NaCl, and 1 m*M* DTT buffer. Fractions containing the pure Atg8 (or mGFP-Atg8) protein are pooled, concentrated, flash frozen in liquid N_2 and kept at −80°C for long-term storage. mGFP-Atg8 can be diluted 1:1 in glycerol and stored at −20°C in the dark.

2.1.2 Expression and Purification of S. cerevisiae Atg7

Full-length Atg7 (NP_012041.1) is expressed as a N-terminally 6xHis-tagged protein from pOPTHrsTEV in *E. coli* Rosetta pLysS. Cells are grown at 37°C in terrific broth (TB) medium to an OD_{600} of 0.8, induced with 50 μ*M* IPTG and grown for a further 16 h at 18°C. Cells are pelleted and resuspended in lysis buffer containing 50 m*M* HEPES, pH 7.5, 300 m*M* NaCl, 10 m*M* imidazole, 1 m*M* $MgCl_2$, 2 m*M* β-mercaptoethanol, complete protease inhibitors (Roche), and DNAse I (Sigma). Cells are lysed by freeze thawing and brief sonication for 30 s, 50% cycles. The lysate is centrifuged at 40,000 rpm (Beckman Ti45 rotor) for 40 min at 4°C. The supernatant is then loaded on a 5 mL HisTrap column (GE Healthcare) and eluted in a step gradient of imidazole (50–75–100–150–200–300 m*M* in 50 m*M* HEPES, pH 7.5, 300 m*M* NaCl, 2 m*M* β-mercaptoethanol). The purest Atg7 fractions usually elute at 150 m*M* imidazole. The 6xHis-tag is cleaved off with TEV protease overnight at 4°C. During the overnight cleavage protein aggregates may form, and hence the solution is centrifuged for 10 min at 4°C at 4000 × *g*. The supernatant containing the nonaggregated Atg7 is diluted with 50 m*M* HEPES and 1 m*M* DTT to a final concentration of 50 m*M* HEPES, pH 7.5, 150 m*M* NaCl, and 1 m*M* DTT. To avoid further formation of protein aggregates during the dilution step, the supernatant containing TEV-cleaved Atg7 is stirred gently at 4°C and the dilution buffer is added very slowly drop by drop. Once the final buffer concentration has been reached, the solution is again centrifuged for 10 min at 4°C at 4000 × *g* and subsequently loaded onto a 10/300 Q-Sepharose anion exchange column (GE Healthcare). Atg7 is eluted with a linear gradient reaching from 150 m*M* to 1 *M* NaCl in 50 m*M* HEPES, pH 7.5, and 1 m*M* DTT. The protein is expected to elute at a NaCl concentration of 400–500 m*M*. Fractions containing Atg7 are pooled, concentrated to 2 mL using an Amicon Ultra centrifugal filter (MW cut-off 50 kDa), and run on a 16/60 Superdex 200 size exclusion column in 50 m*M* HEPES, pH 7.5, 150 m*M* NaCl, and 1 m*M* DTT. Fractions containing pure Atg7 are pooled, concentrated, and aliquots are directly flash frozen in liquid N_2, and kept at −80°C for long-term storage.

2.1.3 Expression and Purification of S. cerevisiae Atg3

Full-length Atg3 (NP_014404) is expressed as a N-terminal GST fusion protein from pGEX4T1 in *E. coli* Rosetta pLysS cells. Cells are grown at 37°C to an OD_{600} of 0.8, induced with 50 μ*M* IPTG, and further grown at 18°C for 16 h. To increase the yield, glucose is added 30 min after induction to a final concentration of 0.1%. Cells are pelleted and resuspended in a buffer containing 50 m*M* HEPES, pH 7.5, 300 m*M* NaCl, 2 m*M* $MgCl_2$, 1 m*M* DTT, complete protease inhibitors (Roche), and DNAse I (Sigma). Cells are lysed by freeze thawing and brief sonication for 30 s, 50% cycles. The lysate is centrifuged at 40,000 rpm (Beckman Ti45 rotor) for 40 min at 4°C. The supernatant is incubated with 3 mL/1 L culture glutathione beads (GE Healthcare) for 90 min at 4°C. Beads are washed five times with 50 m*M* HEPES, pH 7.5, 300 m*M* NaCl, and 1 m*M* DTT followed by two washes with 50 m*M* HEPES, pH 7.5, 1 *M* NaCl, 1 m*M* DTT, and two washes with 50 m*M* HEPES, pH 7.5, 300 m*M* NaCl, and 1 m*M* DTT. The GST-tag is cleaved off overnight at 4°C by incubation with thrombin protease (Serva). The supernatant containing the untagged Atg3 protein is then concentrated and run on a 16/60 S75 size exclusion column in 50 m*M* HEPES, pH 7.5, 150 m*M* NaCl, and 1 m*M* DTT. Fractions containing pure Atg3 are pooled and concentrated. Aliquots are directly flash frozen in liquid N_2 and kept at −80°C for long-term storage.

2.1.4 Expression and Purification of S. cerevisiae Atg12–Atg5

The Atg12–Atg5 conjugate (Atg12: NP_009776.1; Atg5: NP_015176.1) is produced in *E. coli* Rosetta pLysS by co-expression of Atg12, 6xHis-tagged Atg5 (both from pETDuet-1), Atg7, and Atg10 (both from pCOLADuet-1). Atg7 and Atg10 are required for the Atg12–Atg5 conjugate formation. Cells are grown at 37°C in TB medium to an OD_{600} of 0.8, induced with 1 m*M* IPTG, and grown for another 4 h at 37°C. Harvested cells are resuspended in the lysis buffer (50 m*M* HEPES, pH 7.5, 300 m*M* NaCl, 10 m*M* imidazole, 2.5 m*M* Pefabloc (Roth), 1 m*M* $MgCl_2$, 2 m*M* β-mercaptoethanol, DNase I) and disrupted by freeze thawing and brief sonication for 30 s, 50% cycles. The yield of full-length and active protein conjugate is increased when all steps are strictly done at 4°C and the purification is conducted as fast as possible. The lysate is centrifuged at 25,000 rpm (Beckman Ti45 rotor) for 40 min at 4°C. The cleared lysate is applied to a 5 mL HisTrap column (GE Healthcare) and the protein is eluted with a step-wise imidazole gradient (50–75–100–150–200–300 m*M* imidazole in 50 m*M* HEPES, pH 7.5, 300 m*M* NaCl, 2 m*M* β-mercaptoethanol). The Atg12–Atg5 conjugate elutes in fractions containing 100–150 m*M* imidazole. The protein is concentrated using an Amicon Ultra centrifugal filter (MW cut-off 30 kDa) and further purified using a 16/60 S200 size exclusion column (GE Healthcare). The protein is eluted with a buffer containing 50 m*M* HEPES, pH 7.5, 150 m*M* NaCl, 1 m*M* DTT; concentrated, flash frozen in liquid N_2 and kept at −80°C until usage.

2.1.5 *Expression and Purification of* S. cerevisiae *Atg16 and Atg16-mCherry*

GST-tagged Atg16 (NP_013882.1) is expressed in *E. coli* Rosetta pLysS from pOPTG-Atg16. Cells are grown in LB medium at 37°C to an OD_{600} of 0.8, induced with 1 m*M* IPTG and grown for another 4 h. Cells are disrupted by freeze thawing and brief sonication for 30 s, 50% cycles. The cleared lysate is incubated with glutathione beads (Glutathione Sepharose 4B, GE Healthcare). The protein tag is cleaved off with TEV protease overnight at 4°C, concentrated using an Amicon Ultra centrifugal filter (MW cut-off 10 kDa) and applied to a 16/60 S75 size exclusion column (GE Healthcare). The protein is finally eluted with 50 m*M* HEPES, pH 7.5, 300 m*M* NaCl, and 1 m*M* DTT buffer.

Atg16-mCherry with a N-terminal 6xHis-tag followed by a TEV cleavage site is expressed from pETDuet-1 in *E. coli* Rosetta pLysS. Cells are grown in LB medium at 37°C to an OD_{600} of 0.6 and induced with 0.1 m*M* IPTG. Protein expression is carried out for 16 h at 18°C. Cells are pelleted and resuspended in a buffer containing 50 m*M* HEPES, pH 7.5, 300 m*M* NaCl, 10 m*M* imidazole, 1 m*M* $MgCl_2$, 2.5 m*M* β-mercaptoethanol; complete protease inhibitors (Roche); and DNAse I (Sigma). Cells are lysed by freeze thawing followed by brief 30 s sonication, 50% cycle, and the lysate is centrifuged at 40,000 rpm (Beckman Ti45 rotor) for 40 min at 4°C. The supernatant is applied to a 5 mL HisTrap column (GE Healthcare) and eluted via a step-wise imidazole gradient (50–75–100–150–200–300 m*M* imidazole in 50 m*M* HEPES, pH 7.5, 300 m*M* NaCl, 2 m*M* β-mercaptoethanol). Atg16-mCherry-containing fractions are pooled and subjected to overnight cleavage with TEV protease at 4°C in the dark. The cleaved protein is concentrated using an Amicon Ultra centrifugal filter (MW cut-off 10 kDa) applied to a Superdex 200 column (16/60 prep grade, GE Healthcare) and eluted with a buffer containing 25 m*M* HEPES, pH 7.5, 500 m*M* NaCl and 1 m*M* DTT. Fractions containing the purified protein are pooled, concentrated, frozen in liquid N_2 and stored at −80°C.

2.1.6 *In Vitro Formation of the* S. cerevisiae *Atg12–Atg5-Atg16 Complex*

The Atg12–Atg5-Atg16 as well as the Atg12–Atg5-Atg16–mCherry are generated by mixing the previously purified (see earlier) Atg12–Atg5 with Atg16 or Atg16–mCherry in a molar ratio of 1:1. It is recommended to mix the two proteins very slowly in pre-pipetted buffer (50 m*M* HEPES, pH 7.5, 300 m*M* NaCl, and 1 m*M* DTT) under constant and gentle mixing. At this stage it is important to monitor the transparency of the solution because the coiled-coil Atg16 molecule has a tendency to precipitate in low salt buffers. Incubate the mix on ice for 30 min (in the dark in case of mCherry-labeled Atg16). After incubation the protein mix is centrifuged at 16,000 × *g* in order to clear the solution from any protein precipitates. The resulting Atg12–Atg5-Atg16 or Atg12–Atg5-Atg16–mCherry complexes are concentrated using an Amicon Ultra centrifugal filter (MW cut-off 50 kDa) and further purified by size exclusion chromatography on Superdex S200 (16/60 prep grade, GE Healthcare) in 50 m*M* HEPES, pH 7.5, 150 m*M* NaCl and 1 m*M* DTT.

2.1.7 Expression and Purification of S. cerevisiae *Atg4*

Atg4 (NP_014176.2) is expressed and purified as a N-terminal GST fusion protein from pGEX4T3 in *E. coli* Rosetta pLysS cells. Cells are grown at 37°C in LB medium until OD_{600} of 0.8. Protein expression is induced by addition of IPTG to a final concentration of 0.1 m*M* and the protein is expressed at 18°C for 16 h. 30 min after induction glucose is added to a final concentration of 0.1% to increase the yield. The cell pellets are resuspended in 50 m*M* HEPES, pH 7.5, 300 m*M* NaCl, 1 m*M* DTT, 1 m*M* $MgCl_2$, DNAse I, complete protease inhibitors (Roche), 1 m*M* Pefabloc (Roche). Cells are disrupted by freeze thawing and brief sonication of 30 s, 50% cycles. The lysate is cleared by centrifugation at 40,000 rpm (Beckman Ti45 rotor) for 40 min at 4°C, and the supernatant is loaded on a 5 mL GSTrap HP column (GE Healthcare) and eluted with an isocratic elution with 5 m*M* HEPES, pH 7.5, 300 m*M* NaCl, 1 m*M* DTT from 0, 20 m*M* L-glutathione. Peak fractions containing Atg4 are pooled and cleaved with thrombin protease (Serva) overnight at 4°C. The cleaved protein is concentrated using an Amicon Ultra centrifugal filter (MW cut-off 30 kDa) and run on a Superdex 200 16/60 column (GE Healthcare) and the fractions containing Atg4 are pooled and concentrated. Aliquots are flash frozen in liquid N_2 and stored at −80°C for long-term storage.

2.2 Conjugation/Lipidation of Atg8 to SUVs

The lipid composition of the autophagosomal membrane has not yet been characterized. However, it is clear that PE is an important factor because it is the substrate for Atg8 conjugation to the autophagosomal membranes (Ichimura et al., 2004). Atg8 lipidation efficiency is increased in the presence of negatively charged lipids (Ichimura et al., 2004). An optimal lipid composition was also characterized that allows membrane binding of the

Atg12–Atg5-Atg16 complex, which promotes the conjugation reaction (Romanov et al., 2012). Based on these considerations, a lipid mixture composed of 39% POPC, 35% POPS, 21% POPE, and 5% PI3P is used to reconstitute Atg8 lipidation on artificial membranes. However, it was shown that Atg8 can also be conjugated to phosphatidylserine and that this conjugation is resistant to deconjugation by Atg4 (Sou, Tanida, Komatsu, Ueno, & Kominami, 2006). Due to this a lipid mixture composed of 30% DOPE, 65% DOPC, and 5% PI/5% PI3P (phosphatidylinositol/phosphatidylinositol 3-phosphate) is used for reconstitution systems designed to follow the activity of Atg4 in Atg8 delipidation. When dioleoyl- (DO) instead of palmitoyl-oleoyl (PO)-based lipids are used, the rate of Atg8 conjugation to the artificial vesicles and in particular to GUVs is increased.

2.2.1 Preparation of SUVs

SUVs are composed of 39% POPC (Avanti Polar Lipids, Inc., 850457C, 10 mg/mL), 35% POPS (Avanti Polar Lipids, Inc., 840034C, 10 mg/mL), 21% POPE (Avanti Polar Lipids, Inc., 850757C, 10 mg/mL), and 5% PI3P (Avanti Polar Lipids, Inc., 850150P). Alternatively, 30% DOPE (Avanti Polar Lipids, Inc., 850725C, 10 mg/mL), 65% DOPC (Avanti Polar Lipids, Inc., 850375C, 10 mg/mL), and 5% PI/5% PI3P (Avanti Polar Lipids, Inc., 850149/850150, 1 mg/mL) are used. Corresponding amounts of the lipid stocks are transferred into a glass vial and mixed before they are dried under an Argon stream. They are then further dried for 1 h in a desiccator. The dried lipids are rehydrated with buffer containing 25 m*M* HEPES, pH 7.5, 150 m*M* NaCl, and 1 m*M* DTT for 15 min, prewarmed at room temperature to allow better resuspension of the lipids. The lipids are resuspended by tapping and gently sonicated for 2 min in a water bath sonicator. The resuspended SUVs are then extruded 21 times through a 0.4-μm membrane followed by extrusion through a 0.1-μm membrane for 21 times (Whatman, Nucleopore) using the Mini Extruder from Avanti Polar Lipids Inc. The final SUV suspension has a concentration of 1 mg lipids/mL buffer. SUVs are stable for approximately 1 week when stored at 4°C.

2.2.2 Conjugation Reaction

A buffer containing 25 m*M* HEPES, 150 m*M* NaCl, and 1 m*M* DTT is used for the conjugation reaction and it is warmed to 30°C together with SUVs prior to the addition of proteins. Proteins are thawed on ice, gently mixed, and centrifuged at 16,000 × *g* for 10 min at 4°C to remove protein aggregates. The protein concentration should be measured after spinning with a spectrophotometer at A_{280} or with the Bradford method. Before newly purified proteins are used for the first time, they should additionally be checked by SDS-gel electrophoresis and staining with Coomassie Brilliant Blue to ensure the measured concentration is accurate. Proteins are added to the reaction at the following final concentrations: Atg7 and Atg3 are used at 1 μ*M*, while Atg8 is used at 5 μ*M*. Atg12–Atg5 (or Atg12–Atg5-Atg16) is used at a final concentration of 0.5 μ*M*. $MgCl_2$, a cofactor of the reaction, is used at a final concentration of 1 m*M*, while ATP is used at a final concentration of 100 μ*M*. ATP is added as the last component and immediately starts the reaction. The reaction is carried out at 30°C and stopped by the addition of SDS-loading buffer. Under these conditions more than 80% of the substrate Atg8 is converted into its lipidated form within the first 5 min after addition of ATP (Fig. 2A). Successful conjugation of Atg8 to the SUVs results in increased turbidity of the reaction mixture. Lower temperatures or lower concentrations of the enzymes can be employed to slow down the reaction, enabling the comparison of different mutants of proteins under investigation or the effect of other factors on the conjugation rate.

2.2.3 Detection of Conjugated Atg8

The kinetics of the reaction can be monitored by sampling at defined time points and subjecting the samples to SDS-PAGE electrophoresis. SDS-loading dye is added to stop the reaction followed by heating at 60°C for 10 min. These samples are then loaded on 11% SDS-polyacrylamide gels

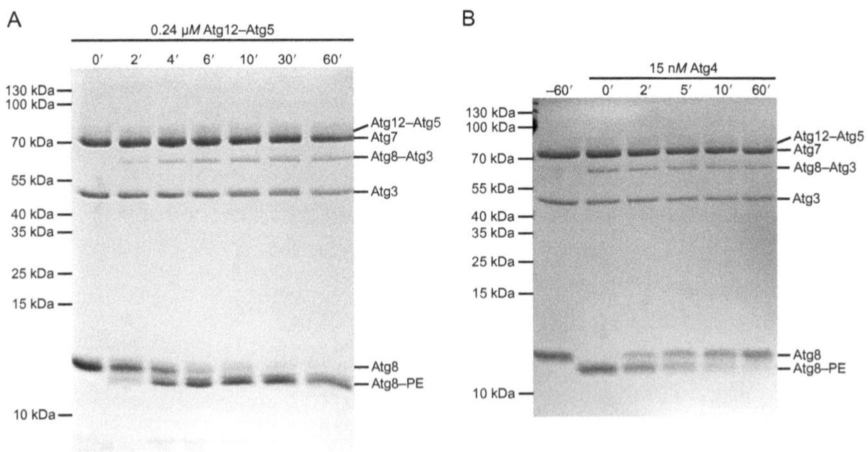

Fig. 2 Coomassie stained gels of typical conjugation (A) and deconjugation (B) reactions. Atg8–PE is detected as faster migrating band when urea (here 4.5 *M*) is added to the gels. The numbers above the gels indicate the reaction times in minutes.

containing 4.5 *M* urea in the separating gel. The presence of a faster migrating band indicates the conjugated form of Atg8 (Atg8–PE) (Fig. 2). Gels are stained with Coomassie Brilliant Blue staining solution (10% acetic acid, 40% methanol, 0.2% Coomassie Brilliant Blue) and destained with destaining solution (10% acetic acid, 40% methanol). A quantitative analysis can be done by measuring the relative amounts of unconjugated and conjugated Atg8.

2.2.4 Deconjugation Reaction With Atg4

The deconjugation reaction is performed after complete conjugation of Atg8 to the SUVs, which occurs within 1 h after addition of ATP as described earlier. After conjugation, 1 unit of calf intestinal alkaline phosphatase (CIP; NEB) per 100 μL reaction is added to dephosphorylate the remaining ATP. This abolishes the conjugation reaction by consuming the ATP molecules present in the mix and ensures that only the deconjugation reaction is observed. Without this step, deconjugated Atg8 will be reconjugated to the SUVs masking the deconjugation effect to some degree. The CIP-mediated dephosphorylation reaction is conducted by incubation for 30 min at 30°C. Subsequently, Atg4 is added to a final concentration of 0.01–0.025 μ*M* (0.015 μ*M* in Fig. 2B). At this concentration, the deconjugation rate is low enough to enable the determination of the reaction kinetics. The speed of the reaction can be fine-tuned with higher or lower concentrations of Atg4. As each batch of purified Atg4 might show some variation in enzymatic activity, it is recommended to titrate each new batch singly to ensure optimal conditions. The reaction is performed at 30°C and stopped by the addition of SDS-loading buffer. With increasing deconjugation of Atg8 from the SUVs, the reaction mixture becomes less turbid. Samples are analyzed by SDS-PAGE as described in Section 2.2.3 (Fig. 2B).

2.3 Conjugation of mGFP-Atg8 to GUVs

GUVs offer more stringent conditions for the lipidation reaction as their curvature is lower than for SUVs. In addition, they offer the possibility to follow reaction in a spatiotemporal manner by light microscopy.

2.3.1 Preparation of GUVs

Detailed protocols of the formation of GUVs and optimization approaches have been described and summarized elsewhere (Romanov et al., 2012; Schmid, Richmond, & Fletcher, 2015). For preparation of GUVs employed in the in vitro Atg8 lipidation assay described here, the following lipid composition was used: 39% DOPC, 35% DOPS, 20% DOPE, 5% PI3P.

1% Marina Blue or Rhodamine (Invitrogen) at the expense of DOPC is added to the mixture for membrane labeling. GUVs are prepared by electroformation: 3 μL of a lipid mixture (10 mg/mL in chloroform/methanol (3:1)) are applied onto the surface of indium-tin-oxide-coated glass slides in a dropwise manner and desiccated for at least 3 h under vacuum. Subsequently, electroformation chambers are assembled using silicon gaskets and the chambers are filled with a solution containing 300 m*M* sucrose. The electroformation protocol contains three phases. Phase 1: 30 min Sine wave with frequency $f=10$ Hz. Peak-to-peak amplitude ramps up linearly from 0.05 to 1.41 V. Phase 2: 120 min Sine wave with frequency $f=10$ Hz. Peak-to-peak amplitude held constant at 1.41 V. Phase 3: 30 min Square wave with frequency $f=4.5$ Hz. Peak-to-peak amplitude held constant at 2.12 V. The electroformation has to be conducted above the transition temperature of the lipids used for the mixture. The GUVs are then gently diluted in 15 m*M* HEPES, pH 7.5, 135 m*M* NaCl, and 1 m*M* DTT.

2.3.2 Conjugation of mGFP-Atg8 to GUVs and Its Detection

Electroformed GUVs are diluted 1:2 or 1:4 in a buffer containing 15 m*M* HEPES, pH 7.5, 135 m*M* NaCl, and 1 m*M* DTT, and gently transferred to a 96-well glass-bottom microplate (Greiner Bio-One). Prior to transferring the GUVs, the wells are incubated with 5 mg/mL BSA in 50 m*M* Tris–HCl pH 7.4, 150 m*M* NaCl for 1 h, and subsequently washed once with 15 m*M* HEPES, pH 7.5, 135 m*M* NaCl buffer. To conjugate Atg8 to PE containing GUVs, the conjugation reaction is performed at the following final proteins concentrations: 400 n*M* mGFP-Atg8 and Atg12–Atg5-Atg16, 80 n*M* Atg3 and Atg7. ATP and $MgCl_2$ are added to final concentrations of 1 and 0.5 m*M*, respectively. All components required for Atg8 lipidation should be gently pipetted directly to the 96-well glass-bottom microplate (Greiner Bio-One) containing GUVs, since the GUVs are very fragile and tend to burst. The reaction is carried out at 30°C in the dark. Microscopic images can be acquired using a fluorescence microscope with the appropriate filters.

REFERENCES

Ge, L., Baskaran, S., Schekman, R., & Hurley, J. H. (2014). The protein-vesicle network of autophagy. *Current Opinion in Cell Biology*, *29*, 18–24. http://dx.doi.org/10.1016/j.ceb.2014.02.005.

Hanada, T., Noda, N. N., Satomi, Y., Ichimura, Y., Fujioka, Y., Takao, T., … Ohsumi, Y. (2007). The Atg12-Atg5 conjugate has a novel E3-like activity for protein lipidation in autophagy. *The Journal of Biological Chemistry*, *282*(52), 37298–37302. http://dx.doi.org/10.1074/jbc.C700195200.

Ichimura, Y., Imamura, Y., Emoto, K., Umeda, M., Noda, T., & Ohsumi, Y. (2004). In vivo and in vitro reconstitution of Atg8 conjugation essential for autophagy. *The Journal of Biological Chemistry*, *279*(39), 40584–40592. http://dx.doi.org/10.1074/jbc.M405860200.

Ichimura, Y., Kirisako, T., Takao, T., Satomi, Y., Shimonishi, Y., Ishihara, N., … Ohsumi, Y. (2000). A ubiquitin-like system mediates protein lipidation. *Nature*, *408*(6811), 488–492. http://dx.doi.org/10.1038/35044114.

Kaufmann, A., Beier, V., Franquelim, H. G., & Wollert, T. (2014). Molecular mechanism of autophagic membrane-scaffold assembly and disassembly. *Cell*, *156*(3), 469–481. http://dx.doi.org/10.1016/j.cell.2013.12.022.

Kirisako, T., Ichimura, Y., Okada, H., Kabeya, Y., Mizushima, N., Yoshimori, T., … Ohsumi, Y. (2000). The reversible modification regulates the membrane-binding state of Apg8/Aut7 essential for autophagy and the cytoplasm to vacuole targeting pathway. *The Journal of Cell Biology*, *151*(2), 263–275.

Knorr, R. L., Nakatogawa, H., Ohsumi, Y., Lipowsky, R., Baumgart, T., & Dimova, R. (2014). Membrane morphology is actively transformed by covalent binding of the protein Atg8 to PE-lipids. *PloS One*, *9*(12), e115357. http://dx.doi.org/10.1371/journal.pone.0115357.

Mizushima, N., Noda, T., Yoshimori, T., Tanaka, Y., Ishii, T., George, M. D., … Ohsumi, Y. (1998). A protein conjugation system essential for autophagy. *Nature*, *395*(6700), 395–398. http://dx.doi.org/10.1038/26506.

Nakatogawa, H., Ichimura, Y., & Ohsumi, Y. (2007). Atg8, a ubiquitin-like protein required for autophagosome formation, mediates membrane tethering and hemifusion. *Cell*, *130*(1), 165–178. http://dx.doi.org/10.1016/j.cell.2007.05.021.

Nath, S., Dancourt, J., Shteyn, V., Puente, G., Fong, W. M., Nag, S., … Melia, T. J. (2014). Lipidation of the LC3/GABARAP family of autophagy proteins relies on a membrane-curvature-sensing domain in Atg3. *Nature Cell Biology*, *16*(5), 415–424. http://dx.doi.org/10.1038/ncb2940.

Noda, N. N., Kumeta, H., Nakatogawa, H., Satoo, K., Adachi, W., Ishii, J., … Inagaki, F. (2008). Structural basis of target recognition by Atg8/LC3 during selective autophagy. *Genes to Cells*, *13*(12), 1211–1218. http://dx.doi.org/10.1111/j.1365-2443.2008.01238.x.

Romanov, J., Walczak, M., Ibiricu, I., Schuchner, S., Ogris, E., Kraft, C., & Martens, S. (2012). Mechanism and functions of membrane binding by the Atg5-Atg12/Atg16 complex during autophagosome formation. *The EMBO Journal*, *31*(22), 4304–4317. http://dx.doi.org/10.1038/emboj.2012.278.

Sawa-Makarska, J., Abert, C., Romanov, J., Zens, B., Ibiricu, I., & Martens, S. (2014). Cargo binding to Atg19 unmasks additional Atg8 binding sites to mediate membrane-cargo apposition during selective autophagy. *Nature Cell Biology*, *16*(5), 425–433. http://dx.doi.org/10.1038/ncb2935.

Schmid, E. M., Richmond, D. L., & Fletcher, D. A. (2015). Reconstitution of proteins on electroformed giant unilamellar vesicles. *Methods in Cell Biology*, *128*, 319–338. http://dx.doi.org/10.1016/bs.mcb.2015.02.004.

Sou, Y. S., Tanida, I., Komatsu, M., Ueno, T., & Kominami, E. (2006). Phosphatidylserine in addition to phosphatidylethanolamine is an in vitro target of the mammalian Atg8 modifiers, LC3, GABARAP, and GATE-16. *The Journal of Biological Chemistry*, *281*(6), 3017–3024. http://dx.doi.org/10.1074/jbc.M505888200.

Turco, E., & Martens, S. (2016). Insights into autophagosome biogenesis from in vitro reconstitutions. *Journal of Structural Biology*, *196*(1), 29–36. http://dx.doi.org/10.1016/j.jsb.2016.04.005.

Weidberg, H., Shpilka, T., Shvets, E., Abada, A., Shimron, F., & Elazar, Z. (2011). LC3 and GATE-16 N termini mediate membrane fusion processes required for autophagosome biogenesis. *Developmental Cell*, *20*(4), 444–454. http://dx.doi.org/10.1016/j.devcel.2011.02.006.

Xie, Z., Nair, U., & Klionsky, D. J. (2008). Atg8 controls phagophore expansion during autophagosome formation. *Molecular Biology of the Cell*, *19*(8), 3290–3298. http://dx.doi.org/10.1091/mbc.E07-12-1292.

Zaffagnini, G., & Martens, S. (2016). Mechanisms of selective autophagy. *Journal of Molecular Biology*, *428*(9, Part A), 1714–1724. http://dx.doi.org/10.1016/j.jmb.2016.02.004.

Zens, B., Sawa-Makarska, J., & Martens, S. (2015). In vitro systems for Atg8 lipidation. *Methods*, *75*, 37–43. http://dx.doi.org/10.1016/j.ymeth.2014.11.004.

CHAPTER TWENTY-THREE

Study of ULK1 Catalytic Activity and Its Regulation

B. Stork*[,1], J. Dengjel†[,1]

*Institute of Molecular Medicine I, Medical Faculty, Heinrich-Heine-University, Düsseldorf, Germany
†University of Freiburg, Freiburg, Germany
[1]Corresponding authors: e-mail address: bjoern.stork@uni-duesseldorf.de; joern.dengjel@unifr.ch

Contents

1. Introduction 392
2. Detection of Phospho-ULK1 Variants by Immunoblotting 393
   2.1 Phosphorylation of ULK1 at Ser556, Ser638, and Ser758 393
3. Analysis of ULK1 Phosphorylation by Mass Spectrometry 397
4. Analysis of ULK1 Inhibitors by In Vitro Kinase Assays 400
5. Summary 402
Acknowledgments 403
References 403

Abstract

During the last decade, the molecular mechanisms controlling the initiation of (macro-) autophagy have been extensively studied. Two macromolecular kinase complexes are central for the initiation of autophagy: the protein kinase unc-51-like kinase 1 (ULK1) complex and the lipid kinase VPS34/Beclin 1 complex. The serine/threonine kinase ULK1 represents the mammalian ortholog of yeast autophagy-related (Atg) protein 1 (Atg1). ULK1 is regulated by upstream nutrient- and energy-sensing kinases, and transmits these signals to the core autophagic machinery. To date, the analysis of ULK1 activation and/or activity is an effective tool to investigate autophagy pathways. As described in this chapter, this can be performed by immunoblotting with phosphosite-specific antibodies against ULK1 and/or ULK1 substrates, by mass spectrometry, or by in vitro kinase assays. Furthermore, the recent design and development of ULK1-specific inhibitors established this kinase as an attractive therapeutic target in settings, where the inhibition of autophagy is desired.

ABBREVIATIONS

mTOR mammalian target of rapamycin
AMPK AMP-activated protein kinase
ATG autophagy-related
FIP200 focal adhesion kinase family interacting protein of 200 kDa
ULK1/2 unc-51-like autophagy activating kinase 1/2

Methods in Enzymology, Volume 587
ISSN 0076-6879
http://dx.doi.org/10.1016/bs.mie.2016.09.067

1. INTRODUCTION

The unc-51-like kinase 1 (ULK1) complex is centrally involved in the initiation of autophagy. Next to the catalytic component ULK1, the core complex is composed of the components autophagy-related (ATG) protein 13 (ATG13), focal adhesion kinase family interacting protein of 200 kDa (FIP200; alternatively termed retinoblastoma-associated protein 1-inducible coiled-coil protein 1, RB1CC1), and ATG101 (Alers, Löffler, Wesselborg, & Stork, 2012a, 2012b; Chan & Tooze, 2009; Mizushima, 2010; Wesselborg & Stork, 2015; Wong, Puente, Ganley, & Jiang, 2013). ULK1 consists of an N-terminal serine/threonine protein kinase domain, followed by a proline/serine (P/S)-rich domain and a conserved C-terminal domain (CTD). So far five ULK family members have been identified, i.e., ULK1-4 and STK36. Of these, especially ULK1 and ULK2 have been implicated in the initiation of autophagy (Alers et al., 2012a, 2012b; Chan & Tooze, 2009; Mizushima, 2010; Wesselborg & Stork, 2015; Wong et al., 2013). The ULK1 complex is regulated by upstream nutrient- and energy-sensing kinases, of which the mammalian target of rapamycin (mTOR), AMP-activated protein kinase (AMPK), and Akt are the best-characterized examples. Accordingly, the ULK1 complex represents the central node between sensing kinases and the core autophagic machinery. To date, several downstream substrates of ULK1 have been identified, including both canonical ATG proteins and proteins with no obvious function for autophagic pathways (Wesselborg & Stork, 2015). Additionally, it has to be noted that autophagy pathways independent of ULK1/2 exist (Alers et al., 2011; Cheong, Lindsten, Wu, Lu, & Thompson, 2011).

In recent years, it has become obvious that autophagy and/or its dysregulation contribute to the outcome of several diseases or pathophysiological conditions. Dependent on the context, either the induction or the inhibition of autophagy might be desirable. With regard to the inhibition of autophagy, currently several clinical trials rely on the usage of the antimalarial drugs chloroquine or hydroxychloroquine. These compounds inhibit lysosome-dependent proteolysis by raising lysosomal pH, and are thus not specific for autophagic pathways. In the recent past, novel inhibitors have been designed and developed, which target the catalytic activities of the autophagy-inducing ULK1 and VPS34/Beclin 1 complexes and are presumably more specific for autophagy pathways (see also Section 4).

2. DETECTION OF PHOSPHO-ULK1 VARIANTS BY IMMUNOBLOTTING

2.1 Phosphorylation of ULK1 at Ser556, Ser638, and Ser758

Several groups have investigated the regulatory phosphorylation of ULK1 by upstream kinases. Collectively, a rather complex "ULK1-phospho-barcode" emerged (Wesselborg & Stork, 2015; Wong et al., 2013). Of the several identified phospho-acceptor serine- or threonine residues, three ULK1 phosphosites were reported by at least three independent groups, i.e., Ser556, Ser638, and Ser758 of human ULK1 (corresponding to murine Ser555, Ser637, and Ser757, respectively).

Using a SILAC-based approach, Shang et al. compared the ULK1 phospho-states under fed and starvation conditions, respectively (Shang et al., 2011). They identified 13 phosphosites within ULK1, and two of these sites revealed a more than 10-fold decrease in phosphorylation level upon amino acid starvation, i.e., Ser638 and Ser758. Both sites became dephosphorylated by rapamycin treatment or mTOR knockdown, and accordingly the authors suggested that mTOR mediates the phosphorylation of these sites (Shang et al., 2011). Phospho-Ser758 was also identified by Dorsey et al. (2009), and its mTOR-dependent phosphorylation was confirmed by Kim, Kundu, Viollet, and Guan (2011). AMPK-catalyzed Ser638 phosphorylation of ULK1 was reported by Egan et al. (2011) and Mack, Zheng, Asara, and Thomas (2012). The latter group reported that Ser638 is the predominant AMPK-dependent phosphosite in ULK1 both in vitro and in vivo (Mack et al., 2012). Notably, Shang et al. showed that Ser638 can be phosphorylated by both mTOR and AMPK, respectively (Shang et al., 2011).

Although Ser758 has been characterized as authentic mTOR-dependent ULK1 phosphosite by several groups (Dorsey et al., 2009; Kim et al., 2011; Shang et al., 2011), physiological consequences are discussed controversially. Kim et al. reported that phospho-Ser758 prevents association of AMPK and ULK1, and thus the AMPK-mediated activation of ULK1. Accordingly, mTOR inhibition would result in the association of AMPK and ULK1 (Kim et al., 2011). In contrast, Shang et al. suggest that phospho-Ser758 promotes the interaction between ULK1 and AMPK. Accordingly, mTOR inhibition would result in the dissociation of AMPK from ULK1 (Shang et al., 2011). Notably, the ULK1 mutant S758A initiates starvation-induced autophagy faster at an early time point, although it does not change the maximum autophagic capacity. This observation let the authors speculate that the AMPK–ULK1 interaction serves as "sequestering reservoir," which retains ULK1 under nutrient-rich conditions (Shang et al., 2011).

Finally, AMPK-dependent Ser556 phosphorylation was reported by Egan et al. (2011). They screened for proteins that bind to recombinant 14-3-3 proteins in wild-type but not in AMPK-deficient cells and identified ULK1 (Egan et al., 2011). Ser556 became phosphorylated upon treatment with the AMPK activator phenformin, and this result was confirmed by an in vitro kinase assay (Egan et al., 2011). Ser556 phosphorylation was also confirmed by the SILAC approach of Shang et al. (2011). Interestingly, this site revealed an approximately sevenfold decrease in phosphorylation level upon starvation, which is the third strongest change after that of Ser638 and Ser758 (Shang et al., 2011). Two additional studies demonstrated that phospho-Ser556 represents a major site for AMPK-dependent 14-3-3 binding (Bach, Larance, James, & Ramm, 2011; Mack et al., 2012).

The different ULK1 phosphorylation patterns might also be due to different proautophagic stimuli. Kim et al. observed that AMPK-dependent phosphorylation and activation of ULK1 is stimulated by glucose starvation (Kim et al., 2011). In contrast, Shang et al. reported AMPK sites in ULK1 which become dephosphorylated upon amino acid starvation, suggesting that these sites are rather inhibiting autophagy (Shang et al., 2011). As already stated by Wong et al., these differences might reflect the distinct autophagy-regulatory functions of AMPK when sensing different triggers (Wong et al., 2013). With regard to the sites Ser556, Ser638, and Ser758, in murine embryonic fibroblasts, they appear to become dephosphorylated upon amino acid starvation (Fig. 1A). In contrast, glucose starvation does not particularly affect ULK1 phosphorylation in this cellular system (Fig. 1B).

Next to the detection of ULK1 phosphorylation, the analysis of phosphorylation of ULK1 substrates is another suitable approach to investigate ULK1 activation and/or activity. Several autophagy-relevant proteins have been reported to become phosphorylated by ULK1, including ATG13, ATG101, FIP200, ATG14, Beclin 1, and VPS34 (Alers et al., 2011; Chan, Longatti, McKnight, & Tooze, 2009; Chang & Neufeld, 2009; Egan et al., 2015; Ganley et al., 2009; Hosokawa et al., 2009; Joo et al., 2011; Jung et al., 2009; Park et al., 2016; Russell et al., 2013). In 2011, Joo et al. described that ULK1 phosphorylates ATG13 at serine 318 (isoform 2, Uniprot identifier O75143-2; corresponding to serine 355 in

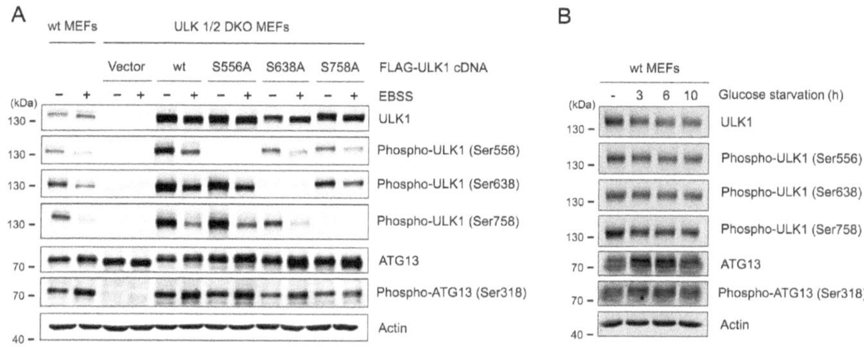

Fig. 1 Phosphorylation of ULK1 at Ser556, Ser638, and Ser758 upon autophagy induction. (A) Wild-type murine embryonic fibroblasts (MEFs) or ULK1/2 double-knockout (DKO) MEFs reconstituted with either empty vector or cDNA encoding FLAG-tagged wild-type ULK1, S556A mutant, S638A mutant, or S757A mutant were incubated in full medium (DMEM) or starvation medium (EBSS) for 2 h. Cells were lysed and cleared cellular lysates (CCLs) were subjected to immunoblotting for ULK1, phospho-ULK1 (Ser556), phospho-ULK1 (Ser638), phospho-ULK1 (Ser758), ATG13, phospho-ATG13 (Ser318), and actin. (B) Wild-type MEFs were incubated in full medium (DMEM) or starvation medium (glucose-free DMEM) for indicated times. Cells were lysed and CCLs were subjected to immunoblotting for proteins indicated in (A). (A and B) For near-infrared immunofluorescent detection, IRDye® 800CW goat antimouse IgG (H+L) or IRDye® 600RD goat antirabbit IgG (H+L) secondary antibodies from LI-COR Biosciences were used. Signals were detected with an Odyssey® Infrared Imaging system (LI-COR Biosciences). Separate immunoblots were performed for the anti-ULK1 and anti-ATG13 antibodies. The actin-loading control refers to the total ULK1 and ATG13 immunoblots. Positions of molecular weight markers (kDa) are indicated on the *left*.

isoform 1, Uniprot identifier O75143-1) (Joo et al., 2011). They observed that this phosphorylation leads to the release of ATG13 from ULK1 and the subsequent recruitment of ATG13 to damaged mitochondria. Finally, overexpression of an ATG13 S318A mutant impaired mitophagy but not basal or starvation-induced autophagy (Joo et al., 2011). We observe that ATG13 phosphorylation at Ser318 increases during amino acid starvation-induced autophagy (Hieke et al., 2015; Fig. 1A), but not during glucose starvation-induced autophagy (Fig. 1B).

To analyze the phosphorylation status of ULK1 at Ser556, Ser638, and Ser758 or of ATG13 at Ser318 by immunoblotting, the following protocol can be used (the procedure is described for mouse embryonic fibroblasts (MEFs), but can easily be adapted to any other cell line):

1. MEFs are seeded in a density of 1×10^6 cells in 3 mL full medium (DMEM high glucose, 10% FCS, 1% penicillin/streptomycin) on 60-mm cell culture dishes the day before treatment.

2. Cells are grown in full medium at 37°C and 5% CO_2 until the confluence reaches 80–90%.
3. Cells are washed once with PBS.
4. Cells are incubated in full medium or starvation medium (EBSS with Ca^{2+} and Mg^{2+} for amino acid starvation, and glucose-free DMEM, 10% dialyzed FCS, 1% penicillin/streptomycin for glucose starvation) at 37°C and 5% CO_2 for desired times. Due to the dynamics of phosphorylation processes, generally different time points should be investigated.
5. For harvesting, cells are abraded and transferred into 1.5-mL Eppendorf tubes and centrifuged 5 min at 4°C and 1200 rpm. The supernatants are discarded and the remaining pellets are washed with 1 mL PBS.
6. Then cell pellets are lysed with 50 μL lysis buffer (20 m*M* Tris/HCl, pH 7.5, 150 m*M* NaCl, 1% (v/v) Triton X-100, 0.5 m*M* EDTA, 10 μ*M* Na_3VO_4, 10 m*M* NaF, 2.5 m*M* $Na_4P_2O_7$, 1× protease inhibitor) and lysates are incubated on ice for 30 min.
7. Lysates are centrifuged at 4°C and 14,000 rpm for 15 min and the cleared cellular lysates (CCLs) are transferred to fresh tubes.
8. The protein amounts of CCLs are assessed by an appropriate assay.
9. SDS sample buffer is added to CCLs and lysates are heated to 95°C for 5 min.
10. 25 μg of protein are separated by 8% SDS-PAGE.
11. Immunoblotting is performed according to standard techniques. The following primary antibodies are used:
    a. Rabbit anti-ULK1, clone D8H5, Cell Signaling Technology (Danvers, MA, USA) #8054 (1:2000)
    b. Rabbit anti-phospho-ULK1 (Ser555),[a] clone D1H4, Cell Signaling Technology #5869 (1:1000)
    c. Rabbit anti-phospho-ULK1 (Ser638), clone D8K9O, Cell Signaling Technology #14205 (1:1000)
    d. Rabbit anti-phospho-ULK1 (Ser757),[a] Cell Signaling Technology #6888 (1:1000)
    e. Rabbit anti-ATG13, Sigma-Aldrich (St. Louis, MO, USA) # SAB4200100 (1:1000)
    f. Rabbit anti-phospho-ATG13 (Ser318), Rockland Immunochemicals (Limerick, PA, USA) #600-401-C49 (1:1000)
12. As mentioned earlier, mTOR, AMPK, and Akt are central kinases regulating ULK1. Next to ULK1 or ULK1 substrate phosphorylation,

[a] Murine amino acid sequence positions; correspond to human Ser556 and Ser758, respectively.

we recommend to simultaneously analyze activation and/or activity of these kinases by immunoblotting. This can be done by phospho-specific antibodies directly targeting these kinases (e.g., anti-phospho-mTOR Ser2448 [Cell Signaling Technology #2971], anti-phospho-mTOR Ser2481 [Cell Signaling Technology #2974], anti-phospho-AMPKα Thr172 [Cell Signaling Technology #2535], anti-phospho-Akt Thr308 [Cell Signaling Technology #4056], anti-phospho-Akt Ser473 [Cell Signaling Technology #9271]), or by phospho-specific antibodies targeting other substrates of these kinases (e.g., anti-phospho-p70 S6 Kinase Thr389 [Cell Signaling Technology #9206], anti-phospho-4E-BP1 Thr37/46 [Cell Signaling Technology #9459], or anti-phospho-PRAS40 Ser183 [Cell Signaling Technology #5936] for mTORC1 activity; anti-phospho-acetyl-CoA-carboxylase Ser79 [Cell Signaling Technology #3661] or anti-phospho-Raptor Ser792 [Cell Signaling Technology #2083] for AMPK activity; and anti-phospho-GSK-3β Ser9 [Cell Signaling Technology #9336], anti-phospho-PRAS40 Thr246 [Cell Signaling Technology #2997] or anti-phospho-FoxO1 Thr24/FoxO3a Thr32 [Cell Signaling Technology #9464] for Akt activity). All phospho-signals should be normalized to total kinase or kinase substrate levels.

3. ANALYSIS OF ULK1 PHOSPHORYLATION BY MASS SPECTROMETRY

Next to the analysis of specific ULK1 phospho-acceptor sites, we recommend the analysis of global ULK1 phosphorylation patterns. As described earlier, several groups characterized ULK1 phosphorylation sites by mass spectrometry, both under nutrient-rich and starvation conditions. Of note, the phospho-acceptor sites are distributed over the entire protein, i.e., within the kinase domain, the PS-rich domain, and the CTD. These proteomic screens revealed that some of the sites are constitutively phosphorylated, whereas others show a dependency on the nutritional conditions. Furthermore, it appears that different autophagy-inducing conditions result in different phosphorylation patterns. In order to analyze global ULK1 phosphorylation, affinity purification of ULK1 should be coupled to a mass spectrometry readout. If one plans to compare different nutritional conditions, we suggest to use stable isotope labeling by amino acids in cell culture (SILAC) (Rigbolt et al., 2014) prior affinity purification of ULK1 and phosphosite mapping.

A. Stable isotope labeling by amino acids in cell culture.

1. Concentration of SILAC amino acids (L-arginine hydrochloride (Arg0), L-arginine-$^{13}C_6$ hydrochloride (Arg6), L-arginine-$^{13}C_6$, $^{15}N_4$ hydrochloride (Arg10), L-lysine hydrochloride (Lys0), L-lysine-4,4,5,5-D_4 hydrochloride (Lys4), L-lysine-$^{13}C_6$, $^{15}N_2$ hydrochloride (Lys8)) and proline in cell culture medium is cell line dependent and needs to be adjusted. The following recipe is used for MCF7 cells:

   Dissolve amino acids (final concentration of 42 mg/L L-arginine HCl, 73 mg/L L-lysine HCl, and 26 mg/L L-proline) in SILAC medium and sterile filter the solution.
2. Complement SILAC-DMEM with 10% dialyzed FCS, 2 m*M* L-glutamine and 1% Pen/Strep. Cells need to be cultured in SILAC medium containing light (Lys0 and Arg0), medium-heavy (Lys4 and Arg6), or heavy (Lys8 and Arg10) amino acids. To gain full incorporation of labeled amino acids, cells should be cultured for at least five cell doublings in SILAC medium. Incomplete incorporation could lead to quantification inaccuracies. The same number of cells should be used for each condition to ensure accurate quantification. One to two 70% confluent 15 cm dishes should be used per condition (approx. 2×10^7 cells, corresponding to 2–4 mg of protein per SILAC condition).

B. Affinity purification (Fig. 2).

1. Remove medium from the culture dishes, abrade cells, and transfer into 15 mL reaction tubes combining differentially SILAC-labeled cells. Centrifuge for 5 min at 4°C and 300 × *g*. Discard supernatants and wash remaining pellets with 1 mL PBS. Lyse cell pellets with three volumes of lysis buffer (20 m*M* Tris/HCl, pH 7.5, 150 m*M* NaCl, 1% (v/v) Triton X-100; 0.5 m*M* EDTA; 10 μ*M* Na_3VO_4; 10 m*M* NaF; 2.5 m*M* $Na_4P_2O_7$; 1 × protease inhibitor) and incubate on ice for 30 min. Afterward remove cell debris by centrifugation.
2. Add protein A sepharose to preclear lysates of "sticky" proteins. Rotate minimally 30 min at 4°C and remove protein A sepharose by passing lysate through a disposable Bio-Spin column.
3. Add immobilized anti-ULK1 antibody (Santa Cruz Biotechnology, Dallas, TX, USA; #SC-33182, minimally 30 μg; this antibody has previously been shown to be well suited for immunopurification of endogenous ULK1; Jung et al., 2009; Löffler et al., 2011) into the mixed lysates and rotate for 4–6 h at 4°C. Wash beads three times

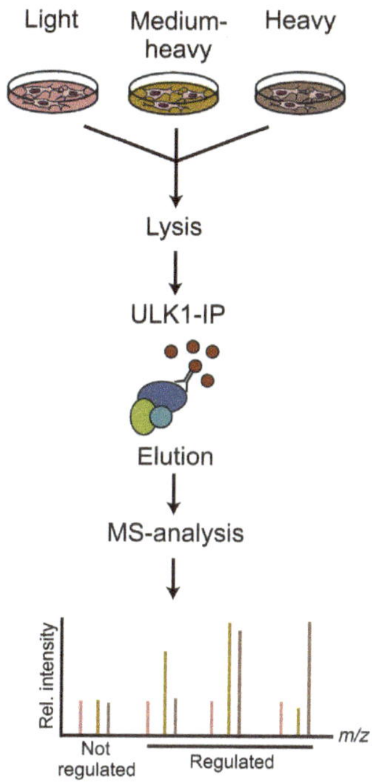

Fig. 2 Affinity purification-mass spectrometry-based analysis of ULK1 phosphorylation. SILAC-labeled cells are differentially treated and ULK1 protein complexes are purified. Purified complexes are processed as outlined and generated peptide mixtures are analyzed by mass spectrometry. XICs of ULK1 phosphopeptide are compared to identify stimulus responsive phosphosites. Nonstimulated cells were light labeled serving as control. Nonregulated phosphopeptides are present in a 1:1:1 ratio. Regulated phosphopeptides show ratios that differ from one. Depending on stimulus and phosphosite, sites might be coregulated or respond only to one stimulus.

with five volumes of ice-cold lysis buffer, resuspend beads in SDS-loading buffer, and elute purified ULK1 complexes by heating at 75°C for 10 min. Repeat the elution twice and combine eluates.

C. GeLC–MS/MS.

1. Use 1 m*M* DTT and 5.5 m*M* iodoacetamide for the sequential reduction and alkylation of proteins prior to gel electrophoresis. Use precast gradient gels, e.g., 4–12% Novex® Wedge Well gels for separation.
2. After staining with Colloidal Blue to visualize proteins, excise the gel area corresponding to ULK1 forms. For optimal sequence coverage and phosphosite mapping we suggest to split the sample and use minimally two proteolytic enzymes for in gel digestion. Trypsin (Promega, Madison, WI, USA; #V5113) is the standard protease

which cleaves after lysine and arginine residues if not followed by proline, and is complementary to the described SILAC approach. Different enzymes can be used for the second sample, e.g., chymotrypsin, GluC, or AspN. We made good experiences with elastase (Promega, #V1891), which cleaves rather unspecific yielding good sequence coverage. It should be kept in mind, however, that database searches using "unspecific" as digestion mode are significantly longer.

3. Extract peptides from gel pieces using 100 μL digestion buffer and two rounds of 100 μL ethanol. Combine extracts and concentrate peptide solutions to less than 50 μL in a speedvac to remove ethanol.
4. Prepare STAGE tips for desalting of peptide solutions (Rappsilber, Mann, & Ishihama, 2007). Punch two 0.5 mm discs out of a C18 Empore disc and pack them tightly into a 200 μL pipette tip. Wash the tip first with 50 μL buffer B (80% acetonitrile, 0.5% acetic acid in deionized water) to remove impurities followed by two washing steps with 50 μL buffer A (0.5% acetic acid in deionized water) for equilibration. Add 200 μL of buffer A/A* [3:1; buffer A*: (3% acetonitrile, 0.3% TFA in deionized water)] to the peptide solution and load the mixture onto the C18 STAGE tip. Wash the tip with 100 μL buffer A and elute the peptides in 50 μL buffer B into a new reaction tube. Concentrate the sample to less than 5 μL to remove acetonitrile and add 10 μL of buffer A/A* (3:1).
5. The samples are ready to be analyzed by mass spectrometry. Comparing the extracted ion currents (XICs) of nonmodified ULK1 peptides allows checking of the mixing ratio. All peptides should have a ratio of 1:1:1. Regulated phosphopeptides should have ratios different from 1 (Fig. 2). Triple SILAC allows comparing directly nonstimulated with two differentially stimulated cells either comparing different time points of stimulation or different stimuli.

4. ANALYSIS OF ULK1 INHIBITORS BY IN VITRO KINASE ASSAYS

Next to the analysis of ULK1 phosphorylation by immunoblotting or mass spectrometry, ULK1 activity and its regulation can be assessed by in vitro kinase assays. As kinase, both endogenous ULK1 or overexpressed affinity tag-labeled ULK1 can be used for in vitro kinase assays (Löffler et al., 2011), although there might exist different substrate specificities caused by

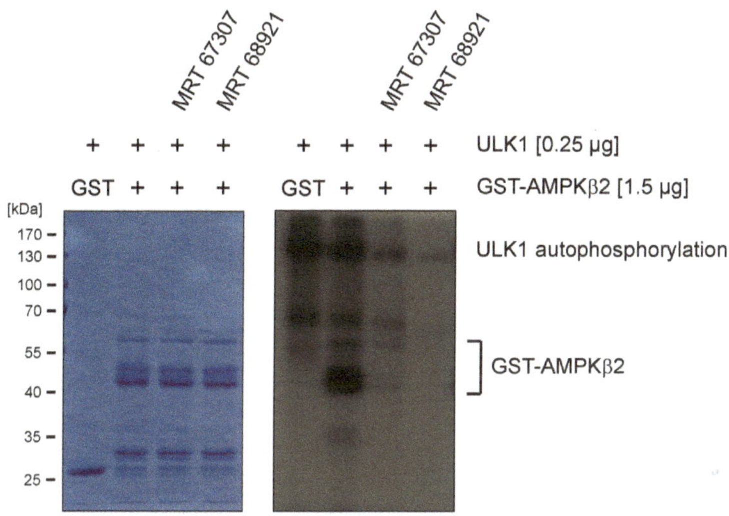

Fig. 3 Analysis of ULK1 kinase inhibitors by in vitro kinase assays. Activated ULK1 (Sigma-Aldrich, St. Louis, MO, USA; #SRP 0252) was left untreated or incubated with 10 μ*M* of MRT 67307 (Selleck Chemicals, Munich, Germany; #S7948) or MRT 68921 (Selleck Chemicals, #S7949) for 30 min at 30°C. Then the samples were employed for in vitro phosphorylation of GST or GST-AMPKβ2 with [^{32}P]ATP. The reactions were subjected to SDS-PAGE. After Coomassie staining of the gel, autoradiography was performed. Positions of molecular weight markers (kDa) are indicated on the *left*.

the presence or absence of native interaction partners (Papinski et al., 2014; own unpublished observations). As substrate, different ULK1 target proteins have been used for in vitro kinase assays, including ATG13, FIP200, Beclin 1, and VPS34 (Chan et al., 2009; Egan et al., 2015; Ganley et al., 2009; Hosokawa et al., 2009; Jung et al., 2009; Russell et al., 2013). In the protocol described below, we make use of recombinant AMPK-subunit β2 as ULK1 substrate (Fig. 3). We have previously observed that all three subunits of AMPK serve as ULK1 substrates (Löffler et al., 2011).

So far, most of the clinical trials targeting autophagy pathways employ the antimalarial drugs chloroquine or hydroxychloroquine, respectively. Accordingly, there is an urgent need for autophagy-specific inhibitors for therapeutic interventions. Recently, several inhibitors for the autophagy-inducing ULK1 complex have been reported, e.g., aminoquinazoline or aminopyrimidine class of compounds, MRT67307 and MRT68921, or SBI-0206965, and the efficacy of these compounds within cells has been assessed by the analysis of ATG13 phosphorylation (see Section 2 of this chapter), LC3 turnover, or LC3 puncta formation (Egan et al., 2015;

Lazarus, Novotny, & Shokat, 2015; Lazarus & Shokat, 2015; Petherick et al., 2015). However, the efficacy of these or novel compounds can also be assessed by ULK1 in vitro kinase assays (Fig. 3). As describe earlier, ULK1 associates with several other Ser/Thr kinases. In order to ensure ULK1 specificity, one should include either ULK1 inhibitors or a kinase-dead version of ULK1, e.g., a mutant of the Mg^{2+}-binding $D^{165}FG$ motif of the kinase domain, D165A (Löffler et al., 2011).

ULK1 in vitro kinase assays can be performed according to the following protocol:

1. ULK1 is purified from cells expressing endogenous ULK1 or affinity tag-labeled wild-type or kinase-dead ULK1. Alternatively, commercially available recombinant ULK1 can be used.
2. Immunopurified ULK1 is washed three times with lysis buffer.
3. Immunopurified ULK1 is resuspended with buffer A (50 m*M* Tris/HCl, pH 7.5, 0.1 m*M* EGTA, 1 m*M* DTT, 5 m*M* $Mg(CH_3COO)_2$) to a final volume of 10 μL (*Mix A*). Add 10 μCi [^{32}P]ATP to *Mix A*. If desired, include ULK1 inhibitors and incubate at 30°C for 30 min.
4. 1–2 μg of substrate and 2 μ*M* cold ATP are resuspended in 10 μL H_2O (*Mix B*)
5. Add *Mix A* to *Mix B* and incubate at 30°C for 30–45 min.
6. The kinase reaction is stopped by the addition of SDS sample buffer and boiling at 95°C for 5 min.
7. The reaction is subjected to SDS-PAGE. After Coomassie staining of the gel, autoradiography is performed.

5. SUMMARY

In the current protocol, we summarize three different methodological approaches to analyze ULK1 activity and regulation based on the analysis of phosphorylation changes of ULK1 itself and known downstream targets. Whereas phosphosite-specific antibodies allow robust and quick testing of regulation of known phosphosites, mass spectrometry-based proteomics generates a comprehensive picture also allowing discovery of new sites and regulation mechanisms. In vitro kinase assays may also be analyzed by mass spectrometry generating site-specific data compared to radioisotope-based assays. As deregulation of autophagy has been linked to several classes of diseases, e.g., cancer, neurodegeneration, or metabolic diseases, modulation of ULK1 activity holds great promises for therapeutic approaches. The described assays can be used to monitor ULK1 activity modulation and to study underlying molecular mechanisms.

ACKNOWLEDGMENTS

We thank Wenxian Wu, Nora Wallot-Hieke, Alexandra Ziemski, and Jana Deitersen for assistance with the preparation of figures. We thank Kathrin Thedieck for helpful comments. We thank Tullia Lindsten for providing wild-type and ULK1/2 DKO MEFs. B.S. is supported by Grants from the Deutsche Forschungsgemeinschaft (STO 864/3-1 and STO 864/4-1) and the Research Committee of the Medical Faculty of the Heinrich-Heine-University Düsseldorf 22/2015. J.D. is supported by the Swiss National Science Foundation, Grant 31003A-166482/1.

REFERENCES

Alers, S., Löffler, A. S., Paasch, F., Dieterle, A. M., Keppeler, H., Lauber, K., … Stork, B. (2011). Atg13 and FIP200 act independently of Ulk1 and Ulk2 in autophagy induction. *Autophagy*, 7(12), 1423–1433.

Alers, S., Löffler, A. S., Wesselborg, S., & Stork, B. (2012a). The incredible ULKs. *Cell Communication and Signaling: CCS*, *10*(1), 7.

Alers, S., Löffler, A. S., Wesselborg, S., & Stork, B. (2012b). Role of AMPK-mTOR-Ulk1/2 in the regulation of autophagy: Cross talk, shortcuts, and feedbacks. *Molecular and Cellular Biology*, *32*(1), 2–11.

Bach, M., Larance, M., James, D. E., & Ramm, G. (2011). The serine/threonine kinase ULK1 is a target of multiple phosphorylation events. *The Biochemical Journal*, *440*(2), 283–291.

Chan, E. Y., Longatti, A., McKnight, N. C., & Tooze, S. A. (2009). Kinase-inactivated ULK proteins inhibit autophagy via their conserved C-terminal domains using an Atg13-independent mechanism. *Molecular and Cellular Biology*, *29*(1), 157–171.

Chan, E. Y., & Tooze, S. A. (2009). Evolution of Atg1 function and regulation. *Autophagy*, *5*(6), 758–765.

Chang, Y. Y., & Neufeld, T. P. (2009). An Atg1/Atg13 complex with multiple roles in TOR-mediated autophagy regulation. *Molecular Biology of the Cell*, *20*(7), 2004–2014.

Cheong, H., Lindsten, T., Wu, J., Lu, C., & Thompson, C. B. (2011). Ammonia-induced autophagy is independent of ULK1/ULK2 kinases. *Proceedings of the National Academy of Sciences of the United States of America*, *108*(27), 11121–11126.

Dorsey, F. C., Rose, K. L., Coenen, S., Prater, S. M., Cavett, V., Cleveland, J. L., & Caldwell-Busby, J. (2009). Mapping the phosphorylation sites of Ulk1. *Journal of Proteome Research*, *8*(11), 5253–5263.

Egan, D. F., Chun, M. G., Vamos, M., Zou, H., Rong, J., Miller, C. J., … Shaw, R. J. (2015). Small molecule inhibition of the autophagy kinase ULK1 and identification of ULK1 substrates. *Molecular Cell*, *59*(2), 285–297.

Egan, D. F., Shackelford, D. B., Mihaylova, M. M., Gelino, S., Kohnz, R. A., Mair, W., … Shaw, R. J. (2011). Phosphorylation of ULK1 (hATG1) by AMP-activated protein kinase connects energy sensing to mitophagy. *Science*, *331*(6016), 456–461.

Ganley, I. G., Lam du, H., Wang, J., Ding, X., Chen, S., & Jiang, X. (2009). ULK1.ATG13. FIP200 complex mediates mTOR signaling and is essential for autophagy. *The Journal of Biological Chemistry*, *284*(18), 12297–12305.

Hieke, N., Loffler, A. S., Kaizuka, T., Berleth, N., Bohler, P., Driessen, S., … Stork, B. (2015). Expression of a ULK1/2 binding-deficient ATG13 variant can partially restore autophagic activity in ATG13-deficient cells. *Autophagy*, *11*(9), 1471–1483.

Hosokawa, N., Hara, T., Kaizuka, T., Kishi, C., Takamura, A., Miura, Y., … Mizushima, N. (2009). Nutrient-dependent mTORC1 association with the ULK1-Atg13-FIP200 complex required for autophagy. *Molecular Biology of the Cell*, *20*(7), 1981–1991.

Joo, J. H., Dorsey, F. C., Joshi, A., Hennessy-Walters, K. M., Rose, K. L., McCastlain, K., ... Kundu, M. (2011). Hsp90-Cdc37 chaperone complex regulates Ulk1- and Atg13-mediated mitophagy. *Molecular Cell*, *43*(4), 572–585.

Jung, C. H., Jun, C. B., Ro, S. H., Kim, Y. M., Otto, N. M., Cao, J., ... Kim, D. H. (2009). ULK-Atg13-FIP200 complexes mediate mTOR signaling to the autophagy machinery. *Molecular Biology of the Cell*, *20*(7), 1992–2003.

Kim, J., Kundu, M., Viollet, B., & Guan, K. L. (2011). AMPK and mTOR regulate autophagy through direct phosphorylation of Ulk1. *Nature Cell Biology*, *13*(2), 132–141.

Lazarus, M. B., Novotny, C. J., & Shokat, K. M. (2015). Structure of the human autophagy initiating kinase ULK1 in complex with potent inhibitors. *ACS Chemical Biology*, *10*(1), 257–261.

Lazarus, M. B., & Shokat, K. M. (2015). Discovery and structure of a new inhibitor scaffold of the autophagy initiating kinase ULK1. *Bioorganic & Medicinal Chemistry*, *23*(17), 5483–5488.

Löffler, A. S., Alers, S., Dieterle, A. M., Keppeler, H., Franz-Wachtel, M., Kundu, M., ... Stork, B. (2011). Ulk1-mediated phosphorylation of AMPK constitutes a negative regulatory feedback loop. *Autophagy*, *7*(7), 696–706.

Mack, H. I., Zheng, B., Asara, J. M., & Thomas, S. M. (2012). AMPK-dependent phosphorylation of ULK1 regulates ATG9 localization. *Autophagy*, *8*(8), 1197–1214.

Mizushima, N. (2010). The role of the Atg1/ULK1 complex in autophagy regulation. *Current Opinion in Cell Biology*, *22*(2), 132–139.

Papinski, D., Schuschnig, M., Reiter, W., Wilhelm, L., Barnes, C. A., Maiolica, A., ... Kraft, C. (2014). Early steps in autophagy depend on direct phosphorylation of Atg9 by the Atg1 kinase. *Molecular Cell*, *53*(3), 471–483.

Park, J. M., Jung, C. H., Seo, M., Otto, N. M., Grunwald, D., Kim, K. H., ... Kim, D. H. (2016). The ULK1 complex mediates MTORC1 signaling to the autophagy initiation machinery via binding and phosphorylating ATG14. *Autophagy*, *12*(3), 547–564.

Petherick, K. J., Conway, O. J., Mpamhanga, C., Osborne, S. A., Kamal, A., Saxty, B., & Ganley, I. G. (2015). Pharmacological inhibition of ULK1 kinase blocks mammalian target of rapamycin (mTOR)-dependent autophagy. *The Journal of Biological Chemistry*, *290*(18), 11376–11383.

Rappsilber, J., Mann, M., & Ishihama, Y. (2007). Protocol for micro-purification, enrichment, pre-fractionation and storage of peptides for proteomics using StageTips. *Nature Protocols*, *2*(8), 1896–1906.

Rigbolt, K. T., Zarei, M., Sprenger, A., Becker, A. C., Diedrich, B., Huang, X., ... Dengjel, J. (2014). Characterization of early autophagy signaling by quantitative phosphoproteomics. *Autophagy*, *10*(2), 356–371.

Russell, R. C., Tian, Y., Yuan, H., Park, H. W., Chang, Y. Y., Kim, J., ... Guan, K. L. (2013). ULK1 induces autophagy by phosphorylating Beclin-1 and activating VPS34 lipid kinase. *Nature Cell Biology*, *15*(7), 741–750.

Shang, L., Chen, S., Du, F., Li, S., Zhao, L., & Wang, X. (2011). Nutrient starvation elicits an acute autophagic response mediated by Ulk1 dephosphorylation and its subsequent dissociation from AMPK. *Proceedings of the National Academy of Sciences of the United States of America*, *108*(12), 4788–4793.

Wesselborg, S., & Stork, B. (2015). Autophagy signal transduction by ATG proteins: From hierarchies to networks. *Cellular and Molecular Life Sciences: CMLS*, *72*(24), 4721–4757.

Wong, P. M., Puente, C., Ganley, I. G., & Jiang, X. (2013). The ULK1 complex: Sensing nutrient signals for autophagy activation. *Autophagy*, *9*(2), 124–137.

CHAPTER TWENTY-FOUR

Evaluating the mTOR Pathway in Physiological and Pharmacological Settings

S. Hong*, K. Inoki*,†,1
*Life Sciences Institute, University of Michigan, Ann Arbor, MI, United States
†University of Michigan Medical School, Ann Arbor, MI, United States
[1]Corresponding author: e-mail address: inokik@umich.edu

Contents

1. Introduction 406
2. Functional Readouts and Inhibitors for the mTOR Pathway 409
3. Methods 410
   3.1 Cell Culture and Treatments 410
   3.2 Transfection 412
   3.3 Protein Extraction for Western Blotting 415
   3.4 Assessing Cellular mTOR Activity Using Phosphospecific Antibodies 415
   3.5 Accessing Levels of mTOR Complexes and Their Activities by Western Blotting 417
   3.6 Membrane Stripping and Reprobing Membrane 418
   3.7 Coimmunoprecipitation of mTORC1 and mTORC2 418
   3.8 In Vitro Kinase Assay 419
   3.9 mTOR Immunofluorescence Staining 421
   3.10 GTP Loading Assay of Small GTPase 421
4. Concluding Remarks 424
Acknowledgments 425
References 425

Abstract

Mammalian/mechanistic target of rapamycin (mTOR) is an evolutionarily conserved genuine protein kinase, which phosphorylates serine/threonine in response to growth factors and nutrients. It functions as a catalytic core in two distinct multiprotein complexes: mTOR complex 1 (mTORC1) and mTOR complex 2 (mTORC2). mTORC1 promotes cell growth and proliferation by positively regulating translation, transcription, and lipid biosynthesis in response to growth factors and amino acids, whereas it inhibits autophagy, an essential degradation and recycling pathway. mTORC2 regulates cell survival and cytoskeleton organization. Mechanistic insights into the function and regulation of mTOR complexes have been provided in various experimental settings and monitoring mTOR activity has been a most valuable way to judge whether levels of

Methods in Enzymology, Volume 587
ISSN 0076-6879
http://dx.doi.org/10.1016/bs.mie.2016.09.068

environmental cues such nutrients and growth factors can satisfy cellular needs for cell growth, proliferation, and autophagic response. Here, we describe useful methods to access mTOR activity in different experimental settings.

1. INTRODUCTION

Mammalian/mechanistic target of rapamycin (mTOR) is a master kinase that regulates autophagy, cell growth, proliferation, and survival in response to growth factors and nutrients such as amino acids. It forms two structurally and functionally distinct multiprotein kinase complexes named mTORC complex 1 (mTORC1) and mTORC complex 2 (mTORC2) (Dibble & Cantley, 2015; Guertin & Sabatini, 2007; Wullschleger, Loewith, & Hall, 2006). While mTORC1 consists of mTOR, Raptor (regulatory-associated protein of mTOR), mLST8 (mammalian lethal with SEC Thirteen 8), PRAS40 (proline-rich Akt substrate of 40 kDa), and Deptor (DEP domain-containing mTOR-interacting protein), mTORC2 comprises of mTOR, Rictor (rapamycin-insensitive companion of mTOR), SIN1 (SAPK-interacting 1), mLST8, Protor (protein observed with rictor), and Deptor (Laplante & Sabatini, 2012). In mTORC1, Raptor functions as a scaffold for several specific mTORC1 substrates, including S6 kinase (S6K), eIF4E binding protein (4EBP), and ULK1 (Unc-51-like kinase 1), as well as for tethering mTORC1 to the endosomal membrane for its activation (Hara et al., 2002; Kim et al., 2002; Sancak et al., 2008). In contrast, PRAS40 and Deptor negatively regulate the activity of mTORC1 (Peterson et al., 2009; Sancak et al., 2007; Vander Haar, Lee, Bandhakavi, Griffin, & Kim, 2007). In mTORC2, Rictor, SIN1, and mLST8 play an essential role in the activity of mTORC2 to phosphorylate its substrates, including Akt, PKCα, and SGK1 (Jacinto et al., 2006; Su & Jacinto, 2011; Yang, Inoki, Ikenoue, & Guan, 2006).

Bacteria-produced rapamycin is a macrolide and its pharmaceutical derivatives are drugs approved by the FDA for organ transplantation, coronary artery stenosis, and several types of cancer (Cargnello, Tcherkezian, & Roux, 2015; Geissler, 2015). mTORC1 is defined as the rapamycin-sensitive complex, whereas mTORC2 is insensitive. Rapamycin forms a complex with FKBP12 to interact with the FKBP12–rapamycin-binding (FRB) domain of mTOR kinase in mTORC1 and allosterically suppresses the mTOR kinase activity by blocking the accessibility of substrates to the active site of mTOR kinase and ultimately disrupts the formation of mTORC1 (Brown et al., 1994; Hara et al., 2002; Kim et al., 2002). It has been demonstrated that the FRB domain of mTOR in mTORC2 is hindered by Rictor once mTORC2 is established (Gaubitz et al., 2015). However, prolonged rapamycin treatment often decreases the expression of mTORC2 and inhibits its functions (Sarbassov et al., 2006). Thus, it is likely that the rapamycin–FKBP12 complex may gain access to newly synthesized mTOR and prevent mTOR from forming mTORC2. Recently, several specific mTOR kinase inhibitors have been synthesized and now are commercially available. These second-generation mTOR inhibitors function as an ATP-competitive inhibitor for mTOR kinase that potently inhibits the kinase activity of both mTORC1 and mTORC2 (Feldman et al., 2009; Hsieh et al., 2012; Thoreen et al., 2009).

Two important environmental cues have long been studied in the regulation of mTORC1 activation: growth factors such as insulin and nutrients such as amino acids. The activity of mTORC1 is stimulated by growth factors and nutrients through two distinct Ras-related small guanosine triphosphatases (GTPases): monomeric ras homolog enriched in the brain (Rheb) and the heterodimeric Rag complex, respectively, on the lysosomal membrane (Dibble & Manning, 2013; Jewell, Russell, & Guan, 2013; Sancak et al., 2010).

Growth factors activate Rheb by inhibiting the trimeric TSC1/TSC2/TBC1D7 complex (hereafter called the *TSC complex*), a well-known tumor suppressor and the specific GTPase-activating protein (GAP) for Rheb, through inhibitory phosphorylation of TSC2 by Akt (Dibble et al., 2012; Garami et al., 2003; Inoki, Li, Xu, & Guan, 2003; Inoki, Li, Zhu, Wu, & Guan, 2002; Manning, Tee, Logsdon, Blenis, & Cantley, 2002; Potter, Pedraza, & Xu, 2002; Zhang et al., 2003). Akt-dependent TSC2 phosphorylation induces the dissociation of the TSC complex from the lysosomal membrane, thereby maintaining lysosomal active Rheb, which directly activates mTORC1 (Demetriades, Doumpas, & Teleman, 2014; Menon et al., 2014).

Amino acids, especially leucine and arginine, activate mTORC1 through the activation of Rag GTPases (Kim, Goraksha-Hicks, Li, Neufeld, & Guan, 2008; Sancak et al., 2008). There are four mammalian Rag proteins, which form obligate heterodimers. RagA and RagB are functionally redundant and form heterodimeric complexes with either RagC or RagD (Nakashima, Noguchi, & Nishimoto, 1999; Sekiguchi, Hirose, Nakashima, Ii, & Nishimoto, 2001). Intriguingly, when Rag complexes are active, RagA and RagB are GTP-bound forms, whereas RagC and RagD are GDP-bound forms (Kim et al., 2008; Sancak et al., 2008). Another unique feature of Rags is their lack of a lipid moiety, even though they reside on lysosomes. The lysosomal expression of Rags is dependent on the lysosome-anchored Ragulator (Bar-Peled, Schweitzer, Zoncu, & Sabatini, 2012; Sancak et al., 2010). The Ragulator is a pentameric protein complex consisting of five subunits, p18 (LAMTOR1), p14 (LAMPTOR2), MP1 (LAMTOR3), C7orf59 (LAMTOR4), and HBXIP (LAMTOR5). Importantly, Ragulator functions as not only a scaffold but also a guanine nucleotide exchange factor (GEF) for RagA and RagB (Bar-Peled et al., 2012). In response to amino acids, Ragulator is activated through v-ATPase on the lysosomal membrane, thereby stimulating the activity of Rags. In contrast, GATOR1, a trimetric protein complex consisting of DEPDC5, NPRL2, and NPRL3, functions as a GAP for both RagA and RagB. GATOR1 is inhibited by another pentameric protein complex, GATOR2, which interacts with Sestrins and CASTORs (Bar-Peled et al., 2013; Chantranupong et al., 2014, 2016; Kim et al., 2015; Parmigiani et al., 2014). Importantly, recent studies have revealed that Sestrin2 and CASTOR1 directly interact with leucine and arginine, respectively (Chantranupong et al., 2016; Wolfson et al., 2016). Both leucine-binding to Sestrin2 and arginine-binding to CASTOR1 are required for leucine and arginine to activate mTORC1 through GATOR2 activation. Although it has been proposed that lysosomal v-ATPase transmits lumenal amino acid signal to the Rag complexes through Ragulator (Wolfson et al., 2016), newly identified cytosolic amino acid sensors such as Sestrins and CASTORs are not expressed on the lysosomal membrane (Chantranupong et al., 2016; Wolfson et al., 2016). Therefore, these observations suggest that essential amino acids are sensed at the lysosome and cytosol for mTORC1 activation.

In addition to the role of amino acids in recruiting mTORC1 to lysosomal membranes, recent studies have revealed other roles of amino acids in the regulation of the TSC complex. Under amino acid starvation conditions, TSC2 interacts with the inactive form of RagA on lysosomes, and this interaction is required for complete inactivation of mTORC1 upon amino acid starvation (Demetriades et al., 2014). Furthermore, amino acids, especially arginine, disrupt the interaction between TSC2 and Rheb, which enhances the accessibility of mTORC1 to active Rheb on the lyososomal membrane (Carroll et al., 2016). Presumably, this is why the activity of mTORC1 is relatively insensitive to amino acid depletion in cells lacking a functional TSC complex.

Compared to mTORC1, molecular mechanisms of mTORC2 activation have not been clearly shown, although a recent study proposed that ribosomes are required for mTORC2 activation (Zinzalla, Stracka, Oppliger, & Hall, 2011). mTORC2 interacts with ribosomal proteins (including RpL26) in a manner dependent on the activity of PI3K, and reduction of ribosomal proteins mitigates cellular mTORC2 activity.

Given that mTOR, especially mTORC1, plays a critical role in suppressing the induction of autophagy, monitoring cellular mTOR activity is a valuable tool to determine the status of cellular autophagic activity. Here, we summarize established methods for monitoring the activity of mTOR, its subcellular localization, and the activity of Rheb to determine cellular mTOR activity.

2. FUNCTIONAL READOUTS AND INHIBITORS FOR THE mTOR PATHWAY

There are a plethora of substrates that have been shown to be phosphorylated by the mTOR complexes. These substrates include S6K1, 4EBP1, PRAS40, and ULK1 (UNC-51-like kinase 1) as mTORC1 substrates, and serum- and glucocorticoid-induced kinase 1 (SGK1) and Akt (Laplante & Sabatini, 2012). For monitoring the activity of mTORC1 in vivo and in vitro, levels of S6K1 phosphorylation on Thr389 (hydrophobic motif) and 4EBP1 phosphorylation on Thr37/Thr46 and Ser65 have been widely used. These substrates are known to play essential roles in mTORC1-dependent mRNA translation (Fig. 1) (Moschetta, Reale, Marasco, Vacca, & Carratu, 2014). In addition, ULK1 phosphorylation on Ser757 can be monitored to determine cellular mTORC1 activity in the regulation of autophagy (Kim, Kundu, Viollet, & Guan, 2011). Along with the abovementioned biochemical approaches, monitoring lysosomal localization of mTOR has begun to be accepted as a new biological method for assessing mTORC1 activation.

Akt phosphorylation on Ser473 (hydrophobic motif) has been widely used for monitoring mTORC2 activity both in vivo and in vitro (Fig. 1) (Sarbassov, Guertin, Ali, & Sabatini, 2005).

To assess functions of mTOR in a variety of experimental settings, two types of mTOR inhibitors have been well used; allosteric inhibitors such as rapalogs, and ATP-competitive inhibitors, including Torin1, PP242, and INK128. Since ATP-competitive inhibitors directly inhibit the kinase activity of mTOR, a catalytic core in both mTORC1 and mTORC2, these inhibitors completely block the activity of both mTORC1 and mTORC2

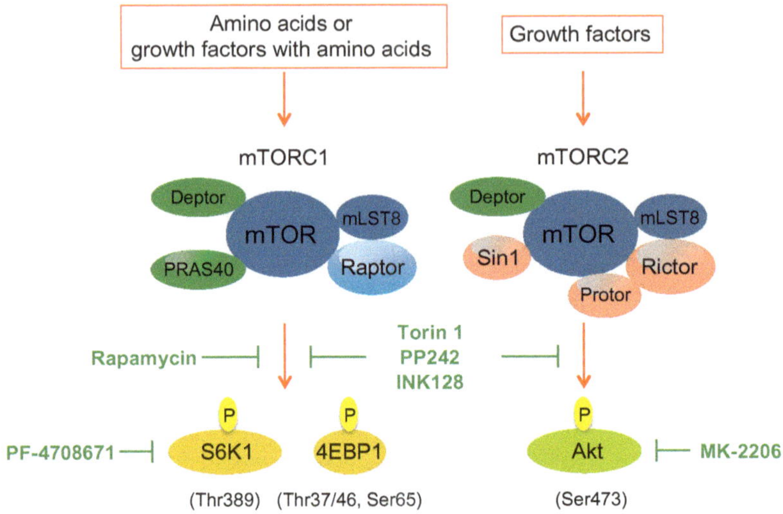

Fig. 1 mTORC1 and mTORC2 and mTOR pathway inhibitors. Schematic illustration shows the key components of mTORC1 and mTORC2, essential cellular cues activating these complexes, inhibitors for the mTOR pathway; rapamycin for mTORC1, Torin1, PP242, and INK128 for mTORC1 and mTORC2, PF-470861 for S6K1, and MK-2206 for Akt. Major phosphorylation sites of mTORC1 and mTORC2 substrates are indicated.

(Figs. 1 and 3). While rapamycin and its derivatives, rapalogs are well-known mTORC1 inhibitors, and the effect of these allosteric inhibitors on mTORC1 inhibition is widely acknowledged, they do not completely suppress the phosphorylation of some mTORC1 substrates, such as 4EBP1 and ULK1. In addition, S6K1 and Akt, which are downstream kinases and substrates of mTORC1 and mTORC2, respectively, can be inhibited by PF-4708671 (S6K1 inhibitor) and MK-2206 (Akt inhibitor) (Figs. 1 and 3).

mTORC1 receives at least two essential signals from growth factors and amino acids for its activation. These two signals impinge on the lysosomal membrane for mTORC1 activation through Rheb and Rag small GTPases. Thus, monitoring the active status of these two small GTPases provides important information for the molecular mechanisms by which mTORC1 regulators stimulate mTORC1 activity (Fig. 2).

3. METHODS

3.1 Cell Culture and Treatments

The signal transductions from growth factors such as insulin and amino acids to mTOR are well conserved in mammalian cells. Representative

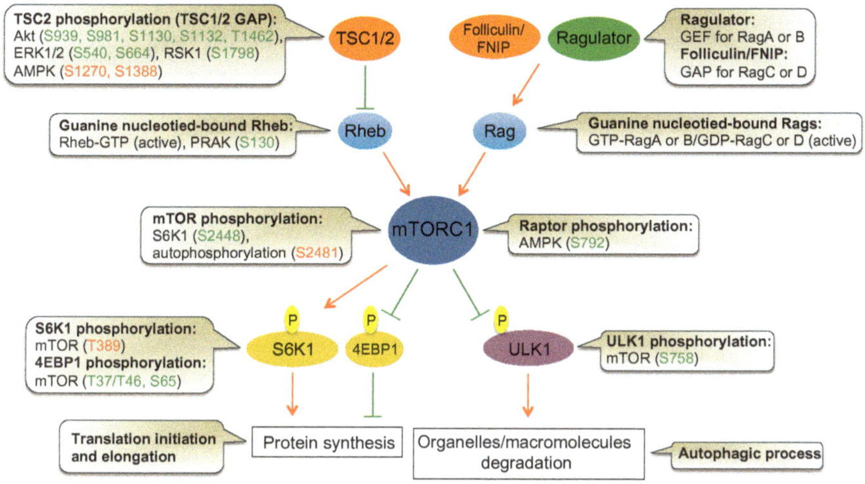

Fig. 2 Diagnostic posttranslational modifications in the mTORC1 pathway. Diagnostic phosphorylation events and critical regulators such as small GTPases and their GAPs and GEF are shown. The phosphorylation sites depicted in red have a positive role, whereas those in *green* play a negative role for the phosphorylated proteins.

mammalian cells widely used in the research of mTOR signaling include HEK293T (human embryonic kidney 293T), MEFs (mouse embryonic fibroblasts), and some cancer cell lines such as HeLa cells. These cells can be cultured in standard culture media such as DMEM (Dulbecco's modified eagle medium) with fetal bovine serum (FBS). However, it is noteworthy that the mTOR pathway in HEK293T and some cancer cells is less sensitive to growth factor stimulation or depletion, in part due to a lack of the activity of phosphatase and tensin homolog (PTEN), a key lipid phosphatase that removes the phosphate in the D3 position of inositol rings from a variety of phophstidylinositols. In addition, HeLa cells that lack the expression of serine/threonine-protein kinase STK11 (LKB1), a master kinase for the T-loop of AMPK family of proteins, show less sensitive to glucose stimulation or depletion in the regulation of the mTOR pathway (Lizcano et al., 2004; Shaw et al., 2004).

In order to inhibit mTOR activity by suppressing upstream inputs, cells need to be starved with growth factor, amino acids, or both by culturing growth factor-free DMEM, PBS containing dialyzed FBS, and PBS containing calcium, magnesium, and glucose (DPBS, ThermoFisher, Waltham, MA, USA, cat# 14040216), respectively. By depleting growth factors in medium, the activity of both mTORC1 and mTORC2 is inhibited. By depleting amino acids in medium, the activity of mTORC1 but not mTORC2 is greatly inhibited. Media lacking a specific amino acid such as leucine or glutamine are also commercially available. To ensure complete removal of growth factors or amino acids, cells should be washed at least once with growth factor- or amino acid-free media before culturing with the starvation media. To inhibit mTORC1 activity by suppressing Akt activity, specific and potent pan-Akt inhibitors such as MK-2206 (IC_{50} = 8/12/65 n*M* for Akt1/2/3, respectively), AZD5363 (IC_{50} = 3/8/8 n*M*), and GSK690693 (IC_{50} = 2/13/9) are commercially available.

In order to inhibit mTORC1 activity directly, cells are treated with rapamycin (sirolimus) or RAD001 (everolimus). To inhibit both mTORC1 and mTORC2, ATP-competitive inhibitors such as Torin 1 (IC_{50} = 2–10 n*M*, 1000-fold selectivity for mTOR than PI3K), KU-006379 (IC_{50} = ~10 n*M*) AZD8055 (IC_{50} = ~1 n*M*, 1000-fold selectivity), INK128 (IC_{50} = 1 n*M*, 200-fold selectivity), and Torkinib (PP242) (IC_{50} = 8 n*M*, 10- to 100-fold selectivity) can be used (Figs. 1 and 3).

3.2 Transfection

Treatments with physiological cues including mitogens, growth factors, and nutrients, or pharmacological compounds such as inhibitors generally produce their effects in all of cultured cells. Therefore, it is able to examine the mTOR signaling by analyzing endogenous proteins. However, in order to determine the role of an exogenous protein in the regulation of mTORC1 or mTORC2, coexpression of the exogenous protein with mTORC1 substrate (e.g., S6K1) or mTORC2 substrate (i.e., Akt) as a reporter helps to analyze its effects on the mTOR pathway in cells that have low transfection efficiency.

Liposome-mediated transfection is a common and efficient method for introducing negatively charged nucleic acid molecules, including cDNA and RNA. Lipofectamine (Invitrogen, Carlsbad, CA, USA) and fugene (Promega, Madison, WI, USA) are two representative commercially available transfection reagents with less cytotoxicity and high efficiency of cDNA transfection. However, due to the cost of these transfection reagents, calcium phosphate transfection is a more attractive method, especially for large-scale transfections such as shRNA-expressing virus production. In addition, PEI (polyethylenamine) can be used as a transfection reagent.

3.2.1 Calcium Phosphate Transfection

All the solutions should be warmed at room temperature before transfection.

1. Grow cells to 70% confluent in 10-cm plates: less confluent cells can die due to cytotoxicity of transfection mixtures.

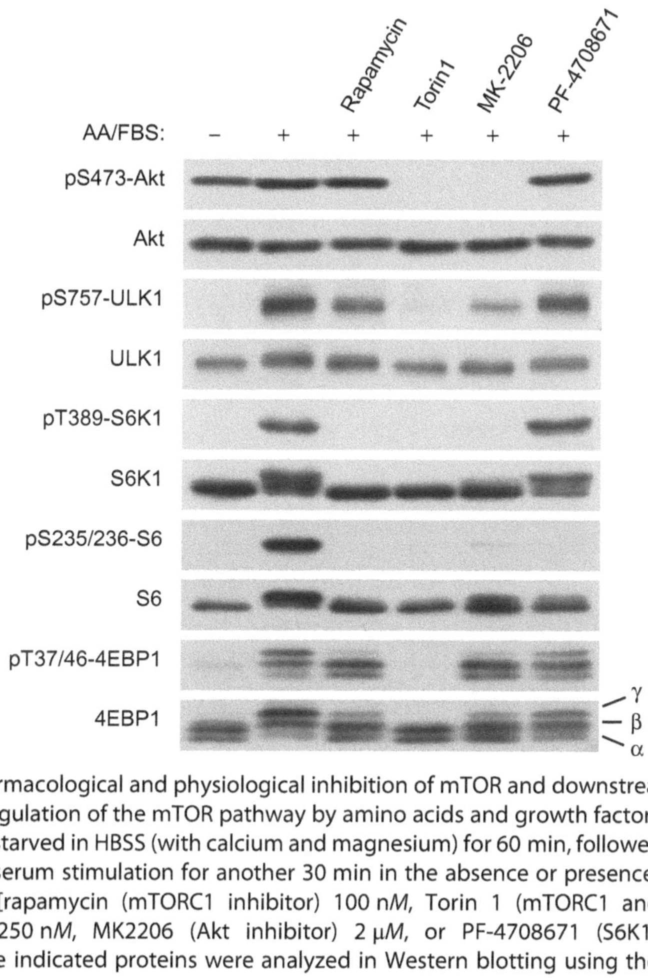

Fig. 3 Pharmacological and physiological inhibition of mTOR and downstream effector kinases. Regulation of the mTOR pathway by amino acids and growth factors. HEK293T cells were starved in HBSS (with calcium and magnesium) for 60 min, followed by amino acids and serum stimulation for another 30 min in the absence or presence of various inhibitors [rapamycin (mTORC1 inhibitor) 100 n*M*, Torin 1 (mTORC1 and mTORC2 inhibitor) 250 n*M*, MK2206 (Akt inhibitor) 2 μ*M*, or PF-4708671 (S6K1 inhibitor) 20 μ*M*]. The indicated proteins were analyzed in Western blotting using the indicated antibodies. α, β, and γ denote non/hypo-, less-, and hyperphosphorylated forms of 4E-BP1, respectively.

2. Change with fresh growth media 3 h before transfection.
3. Add 10–50 μg of DNA into a 50-mL disposable tube and add autoclaved water to 1095 μL.
4. Add 155 μL of 0.22-μm-filtered 2 *M* calcium chloride (stored at 4°C) and mix by gentle swirling.
5. Add 1250 μL of 2 × HBS dropwise within 1 min to evenly form calcium phosphate particles: 2 × HBS: For 500 mL, 8 g NaCl, 0.2 g Na_2HPO_4, 6.5 g HEPES, pH 7.0 stored at –80°C.

6. After 12 h, remove the media, wash once with media to remove calcium phosphates and add fresh growing media.
7. Cells can be harvested and expression can be assessed 36–48 h after transfection.

3.2.2 Transfection Using Liposome (Lipofectamine)

1. Grow cells to 60% confluent in one well of a six-well plate.
2. Before transfection, wash and change media with 0.85 mL of Opti-MEM (ThermoFisher, Waltham, MA, USA, cat# 31985062).
3. In a microcentrifuge tube, add 150 μL of Opti-MEM and DNA constructs.
4. Mix by gentle vortex.
5. Add 5 μL of lipofectamine into the same tube. (This step is different from the manufacturer's instructions.)
6. Mix by gentle vortex and incubate at room temperature for 30 min.
7. Add the transfection mixture from step 6 dropwise and do not disturb attached cells.
8. Incubate cells in the CO_2-humidified incubator for 4–12 h.
9. Change with growing media.
10. Cells can be harvested and levels of an exogenous protein can be assessed 36–48 h after transfection.

3.2.3 Transfection Using PEI

1. Grow cells to reach 90% confluent in a 10-cm plate.
2. Change media with 10 mL of serum-free DMEM (without antibiotics and FBS).
3. In a microcentrifuge tube, add 800 μL of serum-free DMEM, and DNA constructs (less than 50 μg) and mix by vortex.
4. In another microcentrifuge tube, add 800 μL of serum-free DMEM and 60 μL of PEI (DNA: PEI ratio is 1:3) and vortex; 2 mg/mL PEI pH 7.0 stored at −80°C.
5. Mix the abovementioned two tubes and incubate at room temperature for 20 min.
6. Add the PEI transfection mixture dropwise and incubate for 6–8 h.
7. Change media with growing media.
8. Cells can be harvested and levels of an exogenous protein expression can be assessed 36–48 h after transfection.

3.3 Protein Extraction for Western Blotting

Since mTORC1 senses intracellular nutrient levels, we recommend that cells are lysed promptly with lysis buffer without washing with nutrient-free solution such as PBS buffer.

1. For six-well plates, remove media completely by aspiration and immediately add 300 μL of cold NP-40 lysis buffer (10 m*M* Tris–Cl pH 7.5, 2 m*M* EDTA, 150 m*M* NaCl, 1% NP-40, 10 m*M* pyrophosphate, 10 m*M* glycerophosphate, 50 m*M* NaF, and EDTA-free protease inhibitors [Roche]). For a 10-cm plate, add 1 mL of lysis buffer.
2. Incubate on ice for 15 min with occasional tapping.
3. Transfer suspension into 1.5-mL microcentrifuge tubes.
4. Spin at maximum rpm for 15 min at 4°C.
5. Transfer 150 μL of supernatant into a new tube.
6. Add 50 μL of 4 × SDS sample buffer (200 m*M* Tris–Cl pH 6.8, 8% SDS, 40% glycerol, 20% β-mercaptoethanol, and 0.4% bromophenol blue) and vortex briefly.
7. Denature proteins in a 95°C heating block for 5 min.
8. Keep the samples at room temperature for 10 min.
9. Use 10 μL of each sample to apply to a well of SDS-PAGE, followed by Western blot analysis. Samples can be stored at −20°C for several years.

3.4 Assessing Cellular mTOR Activity Using Phosphospecific Antibodies

Since mTORC1 is sensitive to nutrients and growth factors (Figs. 1 and 3), washing cells with buffers without them is not desirable, which may lower the activity of mTORC1 during washing or harvesting. If it is necessary to remove unwanted components in media from cell surfaces, cells should be rinsed quickly. As described earlier, the starvation of nutrients such as amino acids inhibits the activity of mTORC1 within 30 min in most cells. However, the inhibition of mTORC1 and mTORC2 caused by growth factor starvation varies among cells. For instance, the activity of mTORC2 can be inhibited by serum depletion in MEFs within 60 min, whereas it takes much longer in HEK293T or certain cancer cells with diminished PTEN or TSC2 activity. Both the mTORC1 and mTORC2 activity suppressed by nutrient- and growth factor-starvation can be regained within 15 min by the replenishment of nutrients and growth factors (Fig. 3).

To determine the activity of mTORC1 in vivo, Western blot analysis with phosphospecific antibodies against mTORC1 substrates is the most accurate and straightforward method. In addition, monitoring mobility shift caused by protein phosphorylation in SDS-PAGE is an alternative way when the phosphospecific antibody is not available. However, the appropriate percentage of separating gel in SDS-PAGE needs to be empirically determined to obtain a clear mobility shift of target proteins by phosphorylation. For example, to detect the mobility shift of S6K1 or 4E-BP1, generally 8% or 13% SDS-PAGE, respectively, is considered as the ideal setting for obtaining clear mobility shift of these proteins. 4EBP1 is a representative protein that can be phosphorylated as multiple residues and detected as three major bands (α, β, and γ form) in Western blotting. The α, β, and γ 4EBP1 correspond to non/hypo-, less-, and hyperphosphorylated form, respectively (Fig. 3).

To monitor cellular mTORC2 activity, levels of Akt phosphorylation on serine 473 (hydrophobic site) can be determined by Western blotting with phospho-Ser473 Akt antibody. In this section, we describe methods to monitor the activity of mTORC1.

1. Cells grow 70% confluent in a 10-cm plate.
2. Wash with HBSS with calcium and magnesium twice.
3. Add 10 mL of HBSS with calcium and magnesium and incubate in the humidified CO_2 incubator for 1 h to inhibit mTORC1 activity completely.
4. Remove HBSS by aspiration and add 10 mL of DMEM media containing amino acids and 10% FBS for maximum activation of mTORC1. For only amino acid stimulation, add DMEM without FBS.
5. Incubate for additional 15–30 min.
6. Extract proteins as mentioned earlier: For the Western blotting, NP-40 lysis buffer is preferred.
7. In Western blotting, levels of phosphorylation of S6K1 and its total protein can be determined using phosphospecific S6K1 (phospho-T389) and S6K1 antibody, respectively. Levels of S6 phosphorylation and its total protein can also be monitored for determining cellular S6K activity using phospho-S6 (phospho-S240/244) and S6 antibodies. Note that Ser235/236 phosphorylation of S6 can be induced by not only mTORC1-S6K pathway, but also other kinases, including RSK. The activity of mTORC1 or S6K1 can be determined by the ratio of pS6K1/S6K or pS6/S6, respectively.

3.5 Accessing Levels of mTOR Complexes and Their Activities by Western Blotting

Although Western blotting is one of the most widely used techniques, the task of obtaining a clear band of high-molecular-weight proteins such as mTOR (288 kDa) and the components of mTOR complexes by Western blotting needs some extra effort. To detect mTOR, Raptor, and Rictor, approximately 8% SDS-PAGE can be used.

1. During SDS-PAGE, briefly rinse a PVDF membrane with water.
2. Activate a membrane with MeOH until it becomes transparent: this takes a couple of seconds.
3. Remove MeOH and incubate with transfer buffer. The transfer buffer can be prepared a day before the experiment; keep it at 4°C to enhance transfer efficiency (for 4 L, Tris 12.1 g, glycine 57.6 g, MeOH 800 mL).
4. Incubate the membrane in transfer buffer for at least 5 min at room temperature by shaking.
5. Transfer proteins to the membrane at 4°C for 180 min at fixed 350 mA for mTOR, Raptor, S6K1, Rictor, and Akt in a 8% SDS-PAGE gel, and for 120 min at the same mA for S6 and 4EBP1 in a 13% SDS-PAGE gel.
6. After transfer, briefly wash the membrane three times with TBST.
7. Block the membrane with blocking buffer (5% nonfat dry milk in TBST) for 20–60 min.
8. Wash the membrane as in step 6 to remove excess milk from the membrane.
9. Incubate the membrane in antibody solution overnight at 4°C with gentle rocking: primary antibodies are diluted in 10 mL of TBST containing 5% BSA and 0.02% sodium azide.
10. The next day, collect primary antibodies in polystyrene tubes to recycle these antibodies and keep them at −20°C.
11. Wash the membrane three times with TBST for 30 min.
12. Incubate the membrane with secondary antibodies conjugated with HRP (1:5000) in blocking solution and incubate for 2 h at room temperature with rocking.
13. Wash the membrane with TBST four times for 40 min (longer than 40 min washing is also acceptable).
14. Visualize target protein bands using ECL mixture onto X-ray film.

3.6 Membrane Stripping and Reprobing Membrane

1. After visualizing proteins by ECL reagents, recover the membrane and wash with water several times to remove TBST and ECL: if membrane is dried, activate with MeOH and wash with water.
2. Wash the membrane with stripping buffer (25 m*M* glycine, 1% SDS, pH 2) for 40 min (each 10 min × four times).
3. Briefly wash the membrane with water several times to remove any trace of SDS in the stripping buffer and wash three times with TBST for 30 min.
4. Repeat Western blotting from the blocking stage to probe other proteins.

3.7 Coimmunoprecipitation of mTORC1 and mTORC2

mTOR immunoprecipitation using mTOR antibodies can pull-down all the key components of mTORC1 and mTORC2, since mTOR is a common component in both mTORC1 and mTORC2. If specific isolation of mTORC1 or mTORC2 is desired, immunoprecipitation of Raptor for mTORC1 or Rictor for mTORC2 is necessary. For the coimmunoprecipitation (co-IP) of mTOR complexes, CHAPS lysis buffer must be used because other nonionic detergents such as NP-40 and TX-100 disrupt the integrity of these complexes.

1. Grow cells 70% confluent in 10-cm plates.
2. Remove media completely by aspiration and add 1 mL of CHAPS lysis buffer [40 m*M* HEPES pH 7.5, 120 m*M* NaCl, 1 m*M* EDTA, 0.6% CHAPS 10 m*M* pyrophosphate, 10 m*M* glycerophosphate, 50 m*M* NaF, and EDTA-free protease inhibitors (Roche)] immediately.
3. Incubate on ice for 15 min with occasional tapping.
4. Collect suspension and spin at maximum rpm for 15 min at 4°C.
5. Transfer 800 μL for immunoprecipitation and 150 μL for input (lysate).
6. Add 1–2 μg of mTOR antibodies (Santa Cruz Biotechnology, Santa Cruz, CA, USA, cat# SC1549) into 800 μL of extract and incubate for 2–3 h at 4°C with gentle rocking.
7. Add 20 μL of protein G sepharose beads (50% slurry in CHAPS lysis buffer, GE Healthcare, Little Chalfont, UK, cat# 17-068-01) and incubate for another hour.
8. Wash five times with CHAPS lysis buffer.
9. Remove the lysis buffer completely and denature immunoprecipitated proteins with 50 μL of 1 × SDS sample buffer for 5 min at 95°C.

10. After 10 min of incubation at room temperature, spin samples for 10 s and analyze those in 8% or lower SDS-PAGE, followed by Western blotting.
11. In Western blotting, co-IPed mTORC1 and mTORC2 components can be detected by using specific antibodies (Cell Signaling Technology, Danvers, MA, USA), such as mTOR (cat# 2983), Raptor (cat# 2280), mLST8 (cat# 3274), and Rictor (cat# 9476). All these antibodies are diluted 1/1000 in 5% BSA TBST and stored at −20°C. Frozen diluted antibodies are thawed at room temperature before use and incubated at 4°C for overnight.

3.8 In Vitro Kinase Assay

In order to measure the kinase activity of mTORC1 or mTORC2 directly, an in vitro kinase (IVK) assay can be performed using S6K1 or 4EBP1 as a substrate for mTORC1, and Akt for mTORC2 kinase assay.

3.8.1 Preparation of GST-S6K1 from Mammalian Cells

1. Grow HEK293T cells 60% confluent in 5 × 15-cm plates.
2. Transfect with 20 μg of mammalian expression GST-S6K1 using the calcium phosphate method.
3. Next, 48 h after transfection, starve cells with HBSS for 2 h or treat cells with 250 n*M* of torin1 or 100 n*M* of rapamycin for 1 h to completely dephosphorylate GST-S6K1 within the cells.
4. Rinse cells with ice-cold PBS one time.
5. Lyse cells with PBST buffer (PBS with 0.3% Tween-20, 1 mL per plate) with protease inhibitors.
6. Incubate on ice for 15 min.
7. Collect suspension into microcentrifuge tubes.
8. Spin at 4°C for 15 min at maximum rpm.
9. Transfer the supernatant into a new 15-mL tube.
10. Add 50 μL of PBST-washed 50% slurry of glutathione sepharose beads and rock it at 4°C for 4 h.
11. Wash three times with HNTG buffer (20 m*M* HEPES pH 7.5, 150 m*M* NaCl, 10% glycerol, 0.1% Triton-X100).
12. Wash twice with cold PBS.
13. Elute with 10 mg/mL of GSH solution (reduced glutathione in 100 m*M* Tris pH 8) for 4 h.
14. Dialyze the eluent with cold dialysis buffer (10 m*M* Tris pH 7.5, 50 m*M* NaCl, 0.5 m*M* EDTA, and 0.05% beta-mercaptoethanol).

15. Determine the concentration by SDS-PAGE and snap-freeze in liquid nitrogen and store at −80°C until use.

Note that GST-4EBP1 or GST-Akt can be prepared using the same method. For GST-Akt purification, cells transfected with GST-Akt should be treated with mTOR kinase inhibitors such as Torin1 before harvesting. Low-molecular-weight substrates such as GST-4EBP1 can also be prepared from bacteria using a bacterial GST expression vector.

3.8.2 mTOR in vitro Kinase Assay Using Phosphospecific Antibody

For kinase assays, mTORC1 can be immunoprecipitated by Raptor antibodies, while mTORC2 is immunoprecipitated by Rictor antibodies. For in vitro kinase assays using exogenous mTORC1 and mTORC2, each component of the complexes needs to be transfected.

1. Grow cells (e.g., HEK293T, MEF) at 80% confluency in 10-cm plates.
2. Remove media and briefly wash cells with cold PBS.
3. Lyse cells in CHAPS lysis buffer containing protease inhibitors and phosphatase inhibitors (1 mL per 10-cm plate) on ice for 15 min.
4. Collect lysates and spin at 4°C for 15 min at maximum rpm.
5. Transfer supernatant into a new tube.
6. Add 1 μg of Raptor or Rictor antibodies and rock it for 3 h at 4°C.
7. Add 20 μL of 50% slurry of protein G sepharose beads in CHAPS buffer.
8. Rock it for 1 h at 4°C.
9. Wash three times with CHAPS lysis buffer and wash once with HEPES washing buffer (25 m*M* HEPES–KOH, 20 m*M* KCl, pH 7.4).
10. Wash once with 1 × kinase reaction buffer without ATP (20 m*M* Tris–HCl pH 7.5, 10 m*M* $MgCl_2$).
11. Add 25 μL of kinase reaction mixture: 5 μL of 5 × kinase reaction buffer, 120 ng of GST-S6K1 for mTORC1 and GST-AKT for mTORC2, 200 μ*M* ATP.
12. Incubate at 37°C for 20 min with gentle rocking.
13. Terminate the reaction by adding 10 μL of 4 × SDS sample loading buffer and denature the samples at 95°C for 5 min.
14. Analyze the samples by Western blotting.

For mTORC1 kinase assays, levels of GST-S6K1 phosphorylation can be detected by using phospho-Thr389 S6K1 antibodies. Similarly, phospho-Ser473 Akt antibodies can be used for mTORC2-dependent GST-Akt phosphorylation.

3.9 mTOR Immunofluorescence Staining

Lysosomal localization of mTORC1 is necessary for its activation and stimulated in a manner dependent on Rag GTPases activity. Thus, monitoring cellular mTORC1 localization can be used for an indirect measurement of Rag and mTORC1 activity. To determine mTOR localization on lysosomes, cells need to be starved with amino acid-free media such as HBSS or DPBS (with calcium and magnesium) for 50 min to dissipate mTORC1 from lysosomal membranes. Replenishment of amino acids (DMEM without FBS) sufficiently induces lysosomal mTOR localization within 5 min in MEF cells.

1. Grow cells on a round-cover slide in a 12-well plate.
2. Wash with PBS once to remove media.
3. Fix cells with 1 mL of warmed PBS containing 4% paraformaldehyde at 37°C for 5 min.
4. Wash with PBS twice and permeabilize with 1 mL of permeabilizing buffer (PBS containing 0.05% Triton X-100) at room temperature for 5 min.
5. Incubate cells with PBS containing 0.25% BSA at room temperature for 1 h for blocking.
6. Add mTOR antibody (Cell Signaling Technology, Danvers, MA, USA, cat# 2983, 1:100 dilution) into blocking solution and incubate at room temperature for 1 h or at 4°C for 16 h.
7. Wash four times with PBS.
8. Incubate with fluorescence-conjugated secondary antibodies (1:1000 dilution) at room temperature for 30 min in the dark.
9. Wash four times with PBS and once with water.
10. Mount slides and keep in a slide box at room temperature until analysis using a microscope.

3.10 GTP Loading Assay of Small GTPase

To investigate events upstream of mTORC1, it is important to measure the activity of Rheb or Rag to dissect the molecular mechanism underlying mTORC1 activation (Fig. 4). The activity of these small GTPases can be assessed by determining amounts GTP and GDP that bind to Rheb or Rag in vivo. Generally, an increased ratio of GTP/GDP indicates the activation of small GTPases. This assay can be done for both endogenous and exogenous small GTPases. Next, we introduce the assay for measuring the

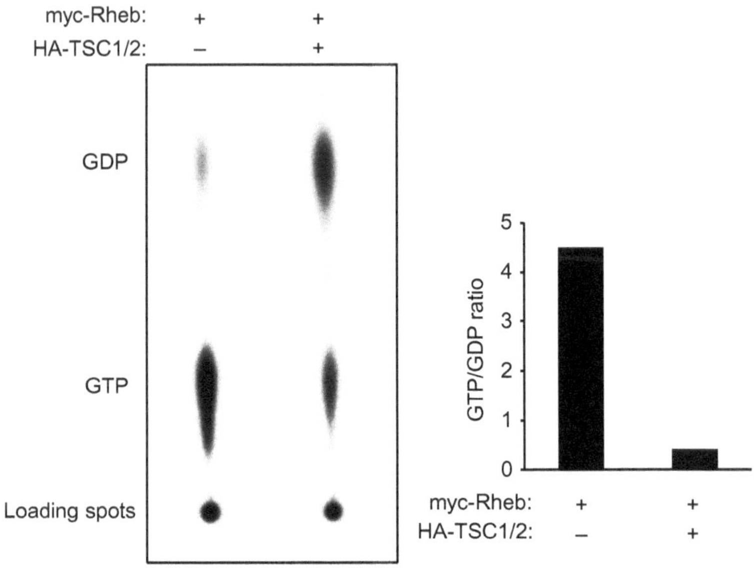

Fig. 4 GTP loading assay for Rheb GTPase. HEK293T cells were transfected with the indicated cDNA constructs. 48 h after transfection, cells were labeled with radioactive ^{32}P phosphate, and immunoprecipitated myc-Rheb was analyzed in the GTP loading assay. Rheb-bound radioactive GTP and GDP were visualized by a PhosphoImager and the volumes of Rheb-bound GTP and GDP were quantified by an ImageQuant. The data were expressed as a ratio (GTP moles/GDP moles, GTP moles = GTP volume/3, GDP moles = GDP volume/2) (*right panel*). Coexpression of HA-TSC2 with myc-Rheb largely stimulated GTP hydrolysis of Rheb.

activity of exogenous Rheb (myc-Rheb) and discuss appropriate approaches to measure the activity of Rag small GTPases.

3.10.1 Accessing Guanine Nucleotide Loading Status on Rheb In Vivo

1. On a six-well plate, transfect 50 ng of myc-Rheb construct using lipofectamine, as previously described.
2. Wash cells once with phosphate-free DMEM (ThermoFisher, Waltham, MA, USA, cat# 11971-025).
3. Incubate cells with 0.8 mL/well of phosphate-free DMEM at the humidified CO_2 incubator for 60 min.
4. During the incubation, prepare a labeling master mix: 245 μL of phosphate-free DMEM, 105 μL of ^{32}P orthophosphate (5 mCi/mL) per well of a six-well plate.
5. Add 50 μL of the mixture per well.
6. Incubate for 4 h.

7. Prepare antibody-protein G sepharose bead conjugates: Add 1–2 μg of myc antibodies into 10 μL of the beads (50% slurry in the lysis buffer) per well and rock it at 4°C for 2 h.
8. Remove labeling media and lyse the cells with 250 μL of lysis buffer (50 m*M* Tris pH 7.5, 0.5% NP-40, 100 m*M* NaCl, 10 m*M* $MgCl_2$, 1 m*M* DTT, and protease inhibitors) per well.
9. Gently rock the plate on ice for 30 s and transfer the lysate into a tube.
10. Centrifuge for 15 min with maximal rpm at 4°C.
11. Take 200 μL of supernatant and add 10 μL of the antibody-protein G bead conjugates and NaCl to a final concentration of 0.5 *M* to block any GAP activity in the immunoprecipitants.
12. Rock the tube for 2 h at 4°C.
13. Wash three times with washing buffer I (50 m*M* Tris–Cl pH 8.0, 500 m*M* $MgCl_2$, 500 m*M* NaCl, 1 m*M* DTT, 0.5% TX-100).
14. Wash three times with washing buffer II (50 m*M* Tris–Cl pH 8.0, 100 m*M* NaCl, 5 m*M* $MgCl_2$, 1 m*M* DTT, 0.1% TX-100).
15. Add 20 μL of elution buffer (2 m*M* EDTA, 0.2% SDS, 1 m*M* GDP, 1 m*M* GTP) and rock it at 68°C for 10 min.
16. Spin shortly and recover supernatants for analysis.
17. Apply 10 μL of each sample onto a PEI cellulose plate (approximately 3 cm above the bottom of the plate) and dry it completely with a regular hairdryer.
18. Soak the plate in MeOH and dry it.
19. Immerse the bottom portion of the plate (below where the samples are loaded) with MeOH.
20. Stand the plate in the TLC chamber that is filled to a depth of 1 cm with TLC running buffer (1 *M* LiCl, 1 *M* formic acid).
21. Close the chamber lid to keep humidified in the chamber and remove the plate from the chamber when the solvent ascends to the top of the plate.
22. Dry the plate with a hairdryer.
23. Expose the TLC plate on a PhosphorImager screen for 6 h and read the radioactive GTP and GDP in a PhosphorImager.
24. Determine the amount of radioactive GTP and GDP by the using Imagequant software and calculate GTP/GDP ratio using the formula. GTP moles (=GTP signal/3), GDP moles (=GDP signal/2). Note that overexpression of Rheb sufficiently activates mTORC1 in HEK293T cells, indicating that excess expression of Rheb overcomes the activity of endogenous TSC2, a specific GAP for Rheb. Therefore,

endogenous or lower levels of exogenous Rheb that do not affect basal mTORC1 activity need to be analyzed.

3.10.2 Accessing in vivo Guanine Nucleotide Loading Status on Rag

RagA or RagB forms an obligate heterodimer with RagC or RagD. In addition, in the most active state, RagA/B is GTP-charged, whereas RagC/D is the GDP-bound form. A Rag heterodimer is very stable; therefore, it is difficult to determine the amount of GTP and GDP that bind to one of the Rags within a heterodimer in vivo. Therefore, to analyze endogenous Rag activity in vivo, it may need to establish special cell lines. For instance, the RagC S75L mutant (Oshiro, Rapley, & Avruch, 2014), which is unable to bind guanyl nucleotides, can be stably expressing in cells lacking both RagC and RagD expression, and then endogenous RagA can be IPed and monitored levels of bound GTP and GDP.

4. CONCLUDING REMARKS

Over the past several decades, tremendous efforts to reveal molecular mechanisms underlying the regulation of mTOR signaling have been made, and hence numerous new members in the mTOR pathway have been discovered. Identification of new members in the mTORC1 signaling has shed light on the molecular mechanism by which mTORC1 is activated by amino acids. During the last decade, more than 20 new proteins have been identified in the regulation of amino acids-induced mTORC1 activation. Identifications of the molecular mechanisms by which mTORC1 receives signals from amino acids such as leucine and arginine are the most significant discoveries in the last few years. However, there are still unanswered questions. For example, the molecular mechanism by which glutamine induces mTORC1 activation remains elusive. It has been reported that glutamine is transported into cells through the SLC1A5 amino acid transporter; hence cellular glutamine in turn is used to import leucine via the antiporter SLC7A5-SLC3A2, thereby stimulating mTORC1 through Rag activation (Nicklin et al., 2009). In addition, α-ketoglutarate, a glutamine metabolite, can stimulate GTP loading of RagB. However, a recent study has shown that glutamine induces lysosomal mTORC1 localization and its activation in a manner independent of Rag small GTPases (Jewell et al., 2015). Secondarily, GEFs for Rheb and RagC/D have not been identified. It is possible that GEFs for Rheb or RagC/D may not be necessary for the regulation of these small GTPases. For instance, Rheb has little its own GTPase activity of its own, and high concentrations of cellular GTP may be spontaneously loaded to Rheb. Therefore, the regulation of the TSC complex may fulfill the sole mechanism of Rheb-induced mTORC1 activation. Finally, another interesting topic that has not been fully elucidated is where mTORC1 can physically interact with its distinct substrates, including S6K, 4EBPs, and ULK1. For instance, upon mTORC1 activation, the majority of these substrates can be sufficiently and fully phosphorylated. However, these substrates are not exclusively expressed at the surface or surrounding of the lysosomes where mTORC1 is activated. It remains unclear whether mTORC1 leaves lysosomes to find its substrates, and if it does, how the trafficking of mTORC1 from lysosomes is regulated.

In terms of the mechanism of mTORC2 activation, it has been shown that ribosomes play an important role in the PI3K-dependent mTORC2 activation (Zinzalla et al., 2011). However, it has not been fully understood how the association of mTORC2 with ribosomes stimulates mTORC2. In this chapter, we described the basic but critical techniques and methods to examine cellular mTOR activity, and hopefully, these methods can help to elucidate these questions that are still in mystery.

ACKNOWLEDGMENTS

We thank members of the Inoki lab for helpful discussions and this work was supported by National Institutes of Health (NIH) Grants (DK083491 and GM110019) and the Department of Defense CDMRP grant (DOD TS140055).

REFERENCES

Bar-Peled, L., Chantranupong, L., Cherniack, A. D., Chen, W. W., Ottina, K. A., Grabiner, B. C., et al. (2013). A tumor suppressor complex with GAP activity for the Rag GTPases that signal amino acid sufficiency to mTORC1. *Science*, *340*, 1100–1106.

Bar-Peled, L., Schweitzer, L. D., Zoncu, R., & Sabatini, D. M. (2012). Ragulator is a GEF for the rag GTPases that signal amino acid levels to mTORC1. *Cell*, *150*, 1196–1208.

Brown, E. J., Albers, M. W., Shin, T. B., Ichikawa, K., Keith, C. T., Lane, W. S., et al. (1994). A mammalian protein targeted by G1-arresting rapamycin-receptor complex. *Nature*, *369*, 756–758.

Cargnello, M., Tcherkezian, J., & Roux, P. P. (2015). The expanding role of mTOR in cancer cell growth and proliferation. *Mutagenesis*, *30*, 169–176.

Carroll, B., Maetzel, D., Maddocks, O. D., Otten, G., Ratcliff, M., Smith, G. R., et al. (2016). Control of TSC2-Rheb signaling axis by arginine regulates mTORC1 activity. *eLife*, *5*.

Chantranupong, L., Scaria, S. M., Saxton, R. A., Gygi, M. P., Shen, K., Wyant, G. A., et al. (2016). The CASTOR proteins are arginine sensors for the mTORC1 pathway. *Cell*, *165*, 153–164.

Chantranupong, L., Wolfson, R. L., Orozco, J. M., Saxton, R. A., Scaria, S. M., Bar-Peled,-L., et al. (2014). The Sestrins interact with GATOR2 to negatively regulate the amino-acid-sensing pathway upstream of mTORC1. *Cell Reports*, *9*, 1–8.

Demetriades, C., Doumpas, N., & Teleman, A. A. (2014). Regulation of TORC1 in response to amino acid starvation via lysosomal recruitment of TSC2. *Cell, 156*, 786–799.

Dibble, C. C., & Cantley, L. C. (2015). Regulation of mTORC1 by PI3K signaling. *Trends in Cell Biology, 25*, 545–555.

Dibble, C. C., Elis, W., Menon, S., Qin, W., Klekota, J., Asara, J. M., et al. (2012). TBC1D7 is a third subunit of the TSC1-TSC2 complex upstream of mTORC1. *Molecular Cell, 47*, 535–546.

Dibble, C. C., & Manning, B. D. (2013). Signal integration by mTORC1 coordinates nutrient input with biosynthetic output. *Nature Cell Biology, 15*, 555–564.

Feldman, M. E., Apsel, B., Uotila, A., Loewith, R., Knight, Z. A., Ruggero, D., et al. (2009). Active-site inhibitors of mTOR target rapamycin-resistant outputs of mTORC1 and mTORC2. *PLoS Biology, 7*, e38.

Garami, A., Zwartkruis, F. J., Nobukuni, T., Joaquin, M., Roccio, M., Stocker, H., et al. (2003). Insulin activation of Rheb, a mediator of mTOR/S6K/4E-BP signaling, is inhibited by TSC1 and 2. *Molecular Cell, 11*, 1457–1466.

Gaubitz, C., Oliveira, T. M., Prouteau, M., Leitner, A., Karuppasamy, M., Konstantinidou, G., et al. (2015). Molecular basis of the rapamycin insensitivity of target of rapamycin complex 2. *Molecular Cell, 58*, 977–988.

Geissler, E. K. (2015). Post-transplantation malignancies: Here today, gone tomorrow? *Nature Reviews. Clinical Oncology, 12*, 705–717.

Guertin, D. A., & Sabatini, D. M. (2007). Defining the role of mTOR in cancer. *Cancer Cell, 12*, 9–22.

Hara, K., Maruki, Y., Long, X., Yoshino, K., Oshiro, N., Hidayat, S., et al. (2002). Raptor, a binding partner of target of rapamycin (TOR), mediates TOR action. *Cell, 110*, 177–189.

Hsieh, A. C., Liu, Y., Edlind, M. P., Ingolia, N. T., Janes, M. R., Sher, A., et al. (2012). The translational landscape of mTOR signalling steers cancer initiation and metastasis. *Nature, 485*, 55–61.

Inoki, K., Li, Y., Xu, T., & Guan, K. L. (2003). Rheb GTPase is a direct target of TSC2 GAP activity and regulates mTOR signaling. *Genes & Development, 17*, 1829–1834.

Inoki, K., Li, Y., Zhu, T., Wu, J., & Guan, K. L. (2002). TSC2 is phosphorylated and inhibited by Akt and suppresses mTOR signalling. *Nature Cell Biology, 4*, 648–657.

Jacinto, E., Facchinetti, V., Liu, D., Soto, N., Wei, S., Jung, S. Y., et al. (2006). SIN1/MIP1 maintains rictor-mTOR complex integrity and regulates Akt phosphorylation and substrate specificity. *Cell, 127*, 125–137.

Jewell, J. L., Kim, Y. C., Russell, R. C., Yu, F. X., Park, H. W., Plouffe, S. W., et al. (2015). Metabolism. Differential regulation of mTORC1 by leucine and glutamine. *Science, 347*, 194–198.

Jewell, J. L., Russell, R. C., & Guan, K. L. (2013). Amino acid signalling upstream of mTOR. *Nature Reviews. Molecular Cell Biology, 14*, 133–139.

Kim, E., Goraksha-Hicks, P., Li, L., Neufeld, T. P., & Guan, K. L. (2008). Regulation of TORC1 by Rag GTPases in nutrient response. *Nature Cell Biology, 10*, 935–945.

Kim, J., Kundu, M., Viollet, B., & Guan, K. L. (2011). AMPK and mTOR regulate autophagy through direct phosphorylation of Ulk1. *Nature Cell Biology, 13*, 132–141.

Kim, J. S., Ro, S. H., Kim, M., Park, H. W., Semple, I. A., Park, H., et al. (2015). Sestrin2 inhibits mTORC1 through modulation of GATOR complexes. *Scientific Reports, 5*, 9502.

Kim, D. H., Sarbassov, D. D., Ali, S. M., King, J. E., Latek, R. R., Erdjument-Bromage, H., et al. (2002). mTOR interacts with raptor to form a nutrient-sensitive complex that signals to the cell growth machinery. *Cell, 110*, 163–175.

Laplante, M., & Sabatini, D. M. (2012). mTOR signaling in growth control and disease. *Cell, 149*, 274–293.

Lizcano, J. M., Goransson, O., Toth, R., Deak, M., Morrice, N. A., Boudeau, J., et al. (2004). LKB1 is a master kinase that activates 13 kinases of the AMPK subfamily, including MARK/PAR-1. *The EMBO Journal, 23*, 833–843.

Manning, B. D., Tee, A. R., Logsdon, M. N., Blenis, J., & Cantley, L. C. (2002). Identification of the tuberous sclerosis complex-2 tumor suppressor gene product tuberin as a target of the phosphoinositide 3-kinase/akt pathway. *Molecular Cell, 10*, 151–162.

Menon, S., Dibble, C. C., Talbott, G., Hoxhaj, G., Valvezan, A. J., Takahashi, H., et al. (2014). Spatial control of the TSC complex integrates insulin and nutrient regulation of mTORC1 at the lysosome. *Cell, 156*, 771–785.

Moschetta, M., Reale, A., Marasco, C., Vacca, A., & Carratu, M. R. (2014). Therapeutic targeting of the mTOR-signalling pathway in cancer: Benefits and limitations. *British Journal of Pharmacology, 171*, 3801–3813.

Nakashima, N., Noguchi, E., & Nishimoto, T. (1999). Saccharomyces cerevisiae putative G protein, Gtr1p, which forms complexes with itself and a novel protein designated as Gtr2p, negatively regulates the Ran/Gsp1p G protein cycle through Gtr2p. *Genetics, 152*, 853–867.

Nicklin, P., Bergman, P., Zhang, B., Triantafellow, E., Wang, H., Nyfeler, B., et al. (2009). Bidirectional transport of amino acids regulates mTOR and autophagy. *Cell, 136*, 521–534.

Oshiro, N., Rapley, J., & Avruch, J. (2014). Amino acids activate mammalian target of rapamycin (mTOR) complex 1 without changing Rag GTPase guanyl nucleotide charging. *The Journal of Biological Chemistry, 289*, 2658–2674.

Parmigiani, A., Nourbakhsh, A., Ding, B., Wang, W., Kim, Y. C., Akopiants, K., et al. (2014). Sestrins inhibit mTORC1 kinase activation through the GATOR complex. *Cell Reports, 9*, 1281–1291.

Peterson, T. R., Laplante, M., Thoreen, C. C., Sancak, Y., Kang, S. A., Kuehl, W. M., et al. (2009). DEPTOR is an mTOR inhibitor frequently overexpressed in multiple myeloma cells and required for their survival. *Cell, 137*, 873–886.

Potter, C. J., Pedraza, L. G., & Xu, T. (2002). Akt regulates growth by directly phosphorylating Tsc2. *Nature Cell Biology, 4*, 658–665.

Sancak, Y., Bar-Peled, L., Zoncu, R., Markhard, A. L., Nada, S., & Sabatini, D. M. (2010). Ragulator-Rag complex targets mTORC1 to the lysosomal surface and is necessary for its activation by amino acids. *Cell, 141*, 290–303.

Sancak, Y., Peterson, T. R., Shaul, Y. D., Lindquist, R. A., Thoreen, C. C., Bar-Peled, L., et al. (2008). The Rag GTPases bind raptor and mediate amino acid signaling to mTORC1. *Science, 320*, 1496–1501.

Sancak, Y., Thoreen, C. C., Peterson, T. R., Lindquist, R. A., Kang, S. A., Spooner, E., et al. (2007). PRAS40 is an insulin-regulated inhibitor of the mTORC1 protein kinase. *Molecular Cell, 25*, 903–915.

Sarbassov, D. D., Ali, S. M., Sengupta, S., Sheen, J. H., Hsu, P. P., Bagley, A. F., et al. (2006). Prolonged rapamycin treatment inhibits mTORC2 assembly and Akt/PKB. *Molecular Cell, 22*, 159–168.

Sarbassov, D. D., Guertin, D. A., Ali, S. M., & Sabatini, D. M. (2005). Phosphorylation and regulation of Akt/PKB by the rictor-mTOR complex. *Science, 307*, 1098–1101.

Sekiguchi, T., Hirose, E., Nakashima, N., Ii, M., & Nishimoto, T. (2001). Novel G proteins, Rag C and Rag D, interact with GTP-binding proteins, Rag A and Rag B. *The Journal of Biological Chemistry, 276*, 7246–7257.

Shaw, R. J., Kosmatka, M., Bardeesy, N., Hurley, R. L., Witters, L. A., DePinho, R. A., et al. (2004). The tumor suppressor LKB1 kinase directly activates AMP-activated kinase and regulates apoptosis in response to energy stress. *Proceedings of the National Academy of Sciences of the United States of America, 101*, 3329–3335.

Su, B., & Jacinto, E. (2011). Mammalian TOR signaling to the AGC kinases. *Critical Reviews in Biochemistry and Molecular Biology, 46*, 527–547.

Thoreen, C. C., Kang, S. A., Chang, J. W., Liu, Q., Zhang, J., Gao, Y., et al. (2009). An ATP-competitive mammalian target of rapamycin inhibitor reveals rapamycin-resistant functions of mTORC1. *The Journal of Biological Chemistry*, *284*, 8023–8032.

Vander Haar, E., Lee, S. I., Bandhakavi, S., Griffin, T. J., & Kim, D. H. (2007). Insulin signalling to mTOR mediated by the Akt/PKB substrate PRAS40. *Nature Cell Biology*, *9*, 316–323.

Wolfson, R. L., Chantranupong, L., Saxton, R. A., Shen, K., Scaria, S. M., Cantor, J. R., et al. (2016). Sestrin2 is a leucine sensor for the mTORC1 pathway. *Science*, *351*, 43–48.

Wullschleger, S., Loewith, R., & Hall, M. N. (2006). TOR signaling in growth and metabolism. *Cell*, *124*, 471–484.

Yang, Q., Inoki, K., Ikenoue, T., & Guan, K. L. (2006). Identification of Sin1 as an essential TORC2 component required for complex formation and kinase activity. *Genes & Development*, *20*, 2820–2832.

Zhang, Y., Gao, X., Saucedo, L. J., Ru, B., Edgar, B. A., & Pan, D. (2003). Rheb is a direct target of the tuberous sclerosis tumour suppressor proteins. *Nature Cell Biology*, *5*, 578–581.

Zinzalla, V., Stracka, D., Oppliger, W., & Hall, M. N. (2011). Activation of mTORC2 by association with the ribosome. *Cell*, *144*, 757–768.

CHAPTER TWENTY-FIVE

Methods to Study the BECN1 Interactome in the Course of Autophagic Responses

M. Antonioli*,†, F. Ciccosanti†, J. Dengjel*, G.M. Fimia†,‡,1
*Freiburg Institute for Advanced Studies (FRIAS), University of Freiburg, Freiburg, Germany
†National Institute for Infectious Diseases I.R.C.C.S. 'Lazzaro Spallanzani', Rome, Italy
‡University of Salento, Lecce, Italy
[1]Corresponding author: e-mail address: gianmaria.fimia@inmi.it

Contents

1. Introduction 430
2. Materials 431
   2.1 Cell Culture 431
   2.2 Cell Lysis and Complex Purification 432
   2.3 Gel Electrophoresis and Sample Preparation for LC-MS/MS Analysis 432
   2.4 MS Equipment 433
3. Methods 433
   3.1 Cell Culture and SILAC Labeling 435
   3.2 Cell Lysis 435
   3.3 IgG-Mouse, Anti-FLAG, and Anti-HA Agarose Beads Preparation 436
   3.4 Sample Preclearing 436
   3.5 Anti-FLAG IP and FLAG Elution 436
   3.6 Anti-HA IP and HA Elution 437
   3.7 Sample Preparation for In-Gel Digestion 437
   3.8 LC-MS/MS Analysis 439
   3.9 Data Analysis 439

Acknowledgments 443
References 443

Abstract

Autophagy is an extremely dynamic process that mediates the rapid degradation of intracellular components in response to different stress conditions. The autophagic response is executed by specific protein complexes, whose function is regulated by posttranslational modifications and interactions with positive and negative regulators. A comprehensive analysis of how autophagy complexes are temporally modified upon stress stimuli is therefore particularly relevant to understand how this pathway is regulated. Here, we describe a method to define the protein–protein interaction network

Methods in Enzymology, Volume 587
ISSN 0076-6879
http://dx.doi.org/10.1016/bs.mie.2016.09.069

of a central complex involved in autophagy induction, the Beclin 1 complex. This method is based on the quantitative comparison of protein complexes immunopurified at different time points using a stable isotope labeling by amino acids in cell culture approach. Understanding how the Beclin 1 complex dynamically changes in response to different stress stimuli may provide useful insights to disclose novel molecular mechanisms responsible for the dysregulation of autophagy in pathological conditions, such as cancer, neurodegeneration, and infections.

1. INTRODUCTION

The Beclin 1 complex represents the focal point of autophagy induction, and its activity is crucial in many physiological processes (Antonioli, Albiero, Fimia, & Piacentini, 2015; Fimia, Stoykova, & Romagnoli, 2007; Yue, Jin, & Yang, 2003). Its deregulation has been linked to pathological conditions, including cancer, neurodegeneration, and immune response failure (Cianfanelli et al., 2014; Liang et al., 1999; Morris, Yip, Shi, Chait, & Wang, 2015; Vescovo, Refolo, & Romagnoli, 2014). Moreover, Beclin 1 plays also an important role in the maturation of both autophagosomes and endosomes (Wirawan et al., 2012). Due to its peculiar ability to orchestrate this pathway at different steps, the Beclin 1 complex is central to study the dynamics of autophagic responses. The Beclin 1 core complex is composed of Beclin 1 itself, VPS34 and its regulatory cofactor VPS15 (Itakura, Kishi, Inoue, & Mizushima, 2008; Kihara, Kabeya, Ohsumi, & Yoshimori, 2001; Liang et al., 1999). On the one hand, the core associates with Atg14 to form the complex I, which allows the formation of the phagophore assembly site through the synthesis of the phosphatidylinositol 3-phosphate (PI3P) (Itakura & Mizushima, 2009). On the other hand, complex II is composed of UVRAG (UV-radiation resistance-associated gene) in place of ATG14, and is mostly involved in vesicle trafficking contributing to autophagosome maturation and fusion with lysosome, as well as to endosome formation and maturation (Matsunaga et al., 2009; Thoresen & Pedersen, 2010). Interestingly, Beclin 1 associates with different affinities to a plethora of proteins able to positively or negatively influence its activity (He & Levine, 2010). For instance, cellular and viral prosurvival Bcl-2 family proteins associate to the BH3 domain of Beclin 1 and inhibit autophagosome formation (Pattingre, Tassa, Qu, & Garuti, 2005). Conversely, positive cofactors, like AMBRA1, interact with Beclin 1 to enhance complex stability and allow autophagy induction

(Di Bartolomeo & Corazzari, 2010; Fimia, Corazzari, Antonioli, & Piacentini, 2012; Fimia et al., 2007; Nazio, Strappazzon, & Antonioli, 2013). In addition, Beclin 1 complex is tightly regulated through posttranslational modifications (PTMs), such as phosphorylation and ubiquitination, which ensure a rapid modulation of its activity (Abrahamsen, Stenmark, & Platta, 2012). In light of these evidences, the analyses of the Beclin 1 interactome (Behrends, Sowa, Gygi, & Harper, 2010; Sun et al., 2008) and its PTMs, as well as of the other components of the Beclin 1 complexes, represent an important tool to unveil novel mechanisms involved at different steps of autophagy induction.

Mass spectrometry (MS)-based proteomics combined to protein complexes purifications through immunoprecipitation (IP) is a powerful method to reveal protein–protein interactions on a large scale (Blagoev, Kratchmarova, Ong, & Nielsen, 2003; Zimmermann, Zarei, Eiselein, & Dengjel, 2010). Here, we describe a method to efficiently immunopurify autophagy complexes and analyze their composition by stable isotope labeling by amino acids in cell culture (SILAC)-based, quantitative MS-based proteomics. In particular, the proposed method is based on LC-MS/MS technology, which allows to efficiently identify new protein interactors and to define temporal changes in protein complex composition and modification during autophagic response.

2. MATERIALS

2.1 Cell Culture

1. Stable Flag–HA-tagged proteins expressing cells.
2. SILAC-DMEM lacking arginine and lysine.
3. Unlabeled L-arginine (Arg0), L-lysine (Lys0); stable isotope-labeled L-arginine-$^{13}C_6$ hydrochloride (Arg6), L-lysine-4,4,5,5-D4 hydrochloride (LysD4), and L-arginine-$^{13}C_6$ $^{15}N_4$ hydrochloride (Arg10), L-lysine-$^{13}C_6$ $^{15}N_2$ hydrochloride (Lys8), and L-proline.
4. Dialyzed fetal bovine serum (dFBS).
5. 200 mM L-glutamine (100 ×).
6. PBS for cell cultures.
7. Earle's Balanced Salt Solution (EBSS) or Hanks' Balanced Salt solution (HBSS) for cell cultures.
8. Penicillin 10,000 U/mL and streptomycin 10,000 μg/mL (P/S 100 ×).
9. Trypsin–EDTA solution (200 mg/L trypsin, 500 mg/L EDTA).

2.2 Cell Lysis and Complex Purification

1. Tandem affinity purification lysis buffer (10 m*M* Tris, pH 7.4, 150 m*M* NaCl, 10% glycerol, 0.5% NP-40), TAP-LB.
2. Protease inhibitor cocktail (P8340, Sigma-Aldrich, St. Louis, MO, USA), sodium fluoride, sodium orthovanadate, sodium molibdate, 2-chloroacetamide, 1,10-phenanthroline monohydrate, PMSF.
3. Cell scrapers.
4. Rotating wheel at 4°C.
5. Microcentrifuge 2-mL tube.
6. Glycine 100 m*M*, pH 2.4
7. Tris–HCl 1 *M*, pH 8.0.
8. BSA 5% in TAP lysis buffer.
9. Bradford reagent and spectrophotometer (analysis at 595 nm).
10. Anti-FLAG M2 affinity gel (A2220, Sigma-Aldrich), monoclonal anti-HA (clone HA-7) agarose beads produced in mouse (A2095, Sigma-Aldrich), mouse-IgG agarose (A0919, Sigma-Aldrich).
11. Flag peptide (F3290, Sigma-Aldrich) to elute FLAG fusion proteins from anti-FLAG M2 antibodies.

2.3 Gel Electrophoresis and Sample Preparation for LC-MS/MS Analysis

1. Dithiothreitol (DTT) (100 m*M* stock concentration, stored at −20°C).
2. Iodoacetamide (550 m*M* stock concentration, stored at −20°C).
3. SDS-loading buffer 4× (Life Technologies, Waltham, MA, USA).
4. 4–12% Bis-Tris gradient gels NuPAGE® Novex (Life Technologies).
5. MOPS running buffer 20× (Life Technologies).
6. Antioxidants (Life Technologies).
7. Colloidal Blue Stain (Invitrogen/Life Technologies).
8. ABC buffer: 100 m*M* ammonium bicarbonate, pH 7.5.
9. Ethanol (HPLC grade).
10. Trifluoroacetic acid (TFA), 2%.
11. Buffer A: 0.5% acetic acid in water.
12. Buffer A*: 3% acetonitrile (ACN) and 0.3% TFA in water.
13. Buffer B: 0.5% acetic acid and 80% ACN in water.
14. Empore Discs—C18 material for STAGE tips (3M, IVA Analysentechnik, Meerbusch, Germany).
15. Modified sequencing-grade trypsin, 12.5 ng/μL in ABC buffer.
16. Speedvac.

2.4 MS Equipment

1. LTQ Orbitrap XL mass spectrometers (Thermo Fisher Scientific, Bremen, Germany).
2. Agilent 1200 nanoflow-HPLC (Agilent Technologies GmbH, Waldbronn, Germany). HPLC-column tips (fused silica): 75 μm inner diameter (New Objective, Woburn, MA, USA) are self-packed with Reprosil-Pur 120 ODS–3 (Dr. Maisch, Ammerbuch, Germany) to a length of 20 cm.

3. METHODS

Analysis of protein complexes composition is one of the most important approaches to understand molecular functions of specific proteins of interest. In particular, quality of experimental data is enormously increased by combining IP approaches to quantitative proteomics (e.g., SILAC labeling), thus allowing the identification of complexes compositions in different cellular conditions, as well as highlighting protein abundances and PTM changes. Through the proposed method, we intend to give a guide for the purification and LC-MS/MS analysis of protein complexes during autophagy. This method is based on SILAC labeling and IP approaches, and it can be extended to different stimuli of interest. Most important aspects to succeed in an interactome study are to have an adequate amount of proteins of interest and high-affinity antibodies. To this aim, we strongly suggest to apply this protocol starting from stable cell lines expressing proteins of interest fused to tag sequences, such as HA and Flag. We recommend using stable cell lines obtained by retro/lentiviral expression, which guarantee homogeneous expression levels in the cell population, without the need of clone selection. Here, we choose a human fibrosarcoma cell line (i.e., 2FTGH) as example, but every cell type can be used after having monitored the responsiveness to autophagy stimuli.

As reported in Fig. 1, 2FTGH cells are labeled using isotopically labeled variants of arginine and lysine for seven cell doublings, and then cells are lysed and subjected to a double purification of the protein of interest by using anti-HA and anti-FLAG conjugated beads, respectively (see Notes 1 and 2). After the elution of immunopurified complexes, sample are combined 1:1:1 (see Note 3), reduced, alkylated, resolved by SDS-PAGE, in-gel digested with trypsin, and prepared for LC-MS/MS analysis (Dengjel, Kratchmarova, & Blagoev, 2010).

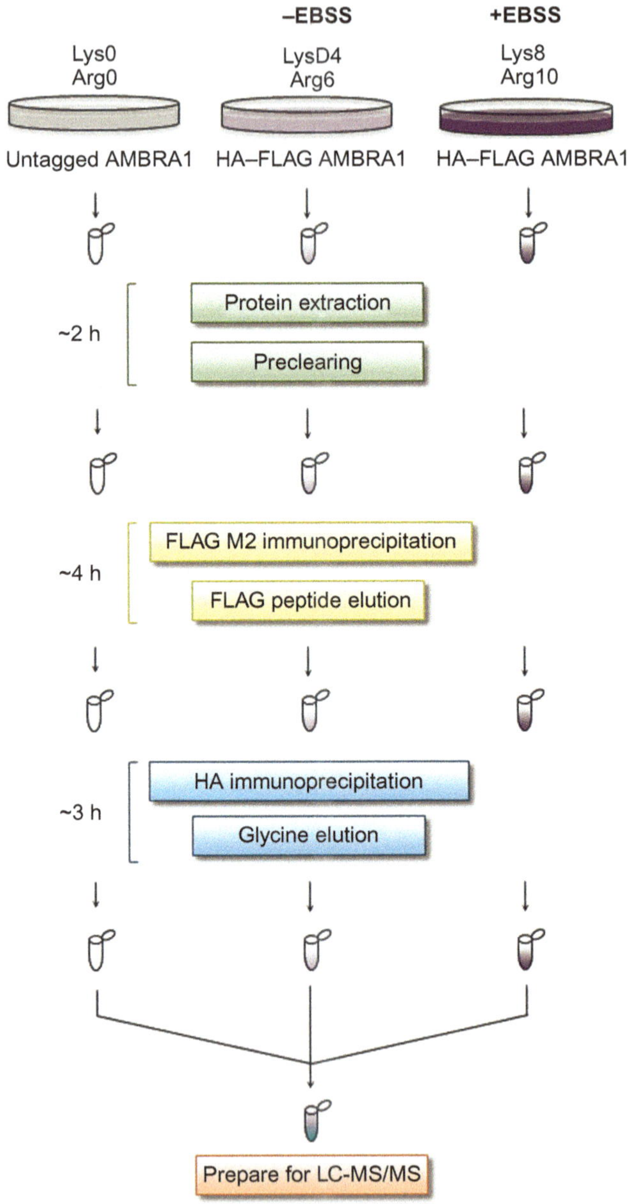

Fig. 1 Flowchart of the purification protocol of AMBRA1-interacting proteins, with an indication of the experimental timeline. See text for details.

3.1 Cell Culture and SILAC Labeling

1. 2FTGH cell lines stably expressing a Beclin 1 complex member (e.g., AMBRA1) (Antonioli et al., 2014) in fusion, or not, with HA and Flag tags are cultivated in 10% FCS DMEM with 5% of CO_2 at 37°C.
2. To ensure a proper control for the purification specificity, we recommend to produce also a stable cell line expressing your protein of interest without tags.
3. For SILAC labeling, 2FTGH cells are grown in SILAC-DMEM for at least 2 weeks to allow seven cell doublings, thus ensuring the proper incorporation of amino acids (Sprenger et al., 2010). Labeling time should be evaluated for individual cell lines and should be a short as possible, since the dialyzed serum could affect cell viability.
4. SILAC medium is supplemented with: (a) 0.2 m*M* Arg0 and 0.4 m*M* Lys0 (i.e., "light") in case of untagged protein-expressing cells, (b) 0.2 m*M* Arg6 and 0.4 m*M* LysD4 (i.e., "medium"), or (c) 0.2 m*M* Arg10 and 0.4 m*M* Lys8 (i.e., "heavy") in case of HA–Flag protein-expressing cells. To complete the medium, 0.68 m*M* L-proline, 10% of dFBS, L-glutamine, and P/S are added (see also Section 2.1).
5. During SILAC labeling, cells are first cultivated in small dishes and expanded to 15 cm dishes only shortly before the purification step, in order to reduce the consumption of SILAC medium.
6. An amount of 10 dishes per condition with a confluence of 75% (approx. 1×10^8 cells) is recommended to obtain an amount of approx. 10 mg of proteins extract.
7. Before starting the large-scale experiment for LC-MS/MS analysis, it is important to evaluate the conversion of arginine to proline and the labeling efficiency, as previously shown (Lößner, Warnken, Pscherer, & Schnölzer, 2011) (see Note 4).
8. To induce autophagy, EBSS or HBSS for cell culture can be used to starve cells (for the duration of treatment, see Note 5).

3.2 Cell Lysis

1. Harvest cells by scraping on ice, centrifuge them at 4°C, and collect the same condition in one tube (see Note 6).
2. Discard the supernatant (SN) and resuspend the pellet in at least 1.5 mL of TAP lysis buffer (TAP-LB) adding protease, phosphatase, and deubiquitinase inhibitors (Protease inhibitor cocktail 1:200, 0.5 m*M* PMSF; 5 m*M* sodium fluoride, 0.5 m*M* sodium orthovanadate,

1 m*M* sodium molibdate, 50 m*M* 2-chloroacetamide, 2 m*M* 1,10-phenanthroline monohydrate). All compounds are from Sigma-Aldrich.

3. Incubate in rotation at 4°C for at least 45 min.
4. Centrifuge the crude lysate at 17,000 × *g* for 10 min at 4°C.
5. Transfer the SN and measure protein concentrations of difference samples to load an equal amount of total protein for the purification. At least 10 mg of total proteins are recommended to successfully purify complexes for LC-MS/MS analysis (see Note 7).
6. Save 250 μg from each lysate before proceeding with the IP (see Note 4).

3.3 IgG-Mouse, Anti-FLAG, and Anti-HA Agarose Beads Preparation

1. An important step to optimize the proposed protocol is represented by the use of procedures aimed at cleaning beads from contaminants (such as loosely linked antibodies) and at preventing their nonspecific binding. To this aim, 150 μL per sample of dried IgG-mouse, anti-FLAG, and anti-HA agarose conjugated beads are incubated for 5 min at RT with 500 μL with Glycine 100 m*M*, pH 2.4 and centrifuged at 600 × *g* for 5 min at 4°C.
2. Discard the SN and wash twice the beads by centrifugation at 600 × *g* for 5 min at 4°C with 1 mL TAP-LB supplemented with 0.05% of BSA without protein inhibitors. Before the second centrifugation, incubate TAP-LB with BSA for 5 min at RT in order to saturate beads.
3. Discard the SN and wash twice with 1 mL of TAP-LB without neither protein inhibitors nor BSA centrifuging at 600 × *g*.
4. Discard the SN and resuspend beads in 150 μL of TAB-LB.

3.4 Sample Preclearing

1. In order to avoid contamination of proteins nonspecifically binding mouse-IgG beads, we strongly recommend including this step. 10 mg of total proteins derived from step 5 of Section 3.2 are incubated for 45–60 min in rotation at 4°C with 150 μL of IgG-mouse agarose conjugated beads prepared as reported in Section 3.3.
2. Centrifuge at 600 × *g* for 5 min at 4°C.

3.5 Anti-FLAG IP and FLAG Elution

1. Transfer the SN to tubes containing 150 μL of anti-FLAG M2 affinity gel prepared as reported in Section 3.3.

2. Incubate in rotation for 2–3 h at 4°C.
3. Pellet immune complexes by centrifugation at 600 × *g* for 5 min at 4°C.
4. Discard SN and wash at least four times with 1 mL of TAP-LB centrifuging at 600 × *g* 5 min at 4°C.
5. Following the last washing, dry the beads with a narrow-end pipette tip or a Hamilton syringe, paying attention not to transfer any resin.
6. Elute proteins under native condition by competition adding 150 μL of FLAG peptide (0.2 mg/mL) specific for FLAG M2 affinity gel binding resuspended in TAP-LB.
7. Incubate at RT for 30 min gently resuspending beads every 5–10 min and centrifuge at 600 × *g* for 5 min at 4°C.
8. Transfer the SN into new tube, dry the beads, and repeat step 7.
9. Combine immune complexes from two elutions and proceed with anti-HA IP (see Note 3).

3.6 Anti-HA IP and HA Elution

1. Transfer eluted complexes into tubes containing 150 μL of anti-HA agarose beads prepared as reported in Section 3.3.
2. Incubate in rotation for 2–3 h at 4°C.
3. Centrifuge at 600 × *g* for 5 min at 4°C.
4. Discard SN and wash at least four times with 1 mL of TAP-LB centrifuging at 600 × *g* 5 min at 4°C.
5. Combine differentially labeled purified complexes 1:1:1 derived from immunopurification of "light-":"medium-":"heavy-" labeled cells (see Note 2).
6. Dry beads and elute by adding 300 μL of Glycine 100 m*M*, pH 2.4, then incubate at RT for 5 min maintaining beads resuspended by gently swirling.
7. Centrifuge at 5000 × *g* for 5 min at 4°C and transfer the SN avoiding transfer of beads. If necessary, repeat this step to clean the samples from beads residues.
8. Add 15 μL of Tris–HCl 1 *M* pH 8.0 to increase the pH.
9. Samples at this point can be stored at −80°C until processing for LC-MS/MS.

3.7 Sample Preparation for In-Gel Digestion

Although at this step samples could be directly analyzed by LC-MS/MS, here we suggest performing an SDS-PAGE and in-gel digestion in order to increase the possibility to identify low-abundant proteins that could be masked by residual IP immunoglobulins if complexes are analyzed without further fractionation.

1. Laemmli buffer and 1 m*M* DTT are added to denature proteins and reduce disulfide bonds, respectively.
2. Incubate samples 15 min at 75°C shaking at 800 rpm.
3. Reach RT, and then alkylate thiols by adding 5.5 m*M* iodoacetamide and by incubating 20 min at RT in darkness at 800 rpm.
4. Resolve mixed precipitated complexes on 4–12% Bis-Tris gradient gels NuPAGE running in MOPS buffer.
5. Fix the gel incubating with 50% methanol/10% acid acetic in deionized water for 15 min on a rocking platform at RT.
6. Use Colloidal Blue to stain proteins in order to evaluate the SDS-PAGE quality.
7. Cut the lane into 10 equal slices, and then each slice into cubes of approx. 1 mm^3 and transfer the cubes from each slice into a new tube.
8. Remove the Colloidal Blue by covering cubes with ABC buffer (approx. 100–150 μL), incubate at RT 10 min shacking at 200 × *g*, and discard the ABC solution.
9. Add ethanol 100% to cover cubes, incubate at RT 10 min shacking at 200 × *g*, and discard ethanol.
10. Repeat steps 10 and 11 for at least two more times.
11. Let cubes swell in 50–80 μL of trypsin solution (12.5 μg/mL in ABC buffer) verifying that they are covered by the trypsin solution and then incubate overnight at 37°C.
12. Stop the activity of trypsin by adding 50 μL of 2% TFA. Shake gel cubes at 200 × *g* for 15 min at RT, transfer solution to new tube.
13. To extract peptides completely, repeat step 14 adding twice 100–150 μL of pure ethanol to gel pieces. Combine SNs of respective slices after each step.
14. Using a speedvac, concentrate the collected peptide solution to less than 50 μL in to remove ethanol and add 150 μL of buffer A.
15. In order to remove salts (see Note 8), we take advantage of C18 material affinity to peptides. Prepare STAGE tips (Rappsilber, Mann, & Ishihama, 2007) by punching out two 0.5 mm discs from a C18 material and pack them tightly in tips of a 200-μL pipette.
16. Prepare STAGE tips by allowing the different solutions to pass through the discs by centrifugation, as follow: add 50 μL of methanol to reconstitute the material, 50 μL of buffer B to remove impurities, and twice

50 μL of buffer A to remove buffer B. Then, load the sample and wash one time with 100 μL buffer A. To elute samples add 50 μL of buffer B into the STAGE tip and collect peptides into new tube.

17. Using a speedvac, concentrate eluates to less than 5 μL to remove ACN and add 15 μL of buffer A/A* (75/25). Store samples at −80°C until loading on HPLC for the MS analysis.

3.8 LC-MS/MS Analysis

LTQ Orbitrap XL mass spectrometer (Thermo Fisher Scientific) coupled to an Agilent 1200 nanoflow–HPLC is used to perform MS measurements (other high-resolution instruments can be used for the same purpose). Fused-silica HPLC-column, with an inner diameter of 75 μm, is self-packed with Reprosil–Pur 120 ODS–3 to a length of 20 cm. Samples are applied directly to the column (without precolumn), and an increased concentration of organic proportion (buffer B) is generated by mixing with buffer A, thus allowing peptide separation and elution. Sample loading is performed with 2% of buffer B at 500 nL/min flow rate, and then the separation ramp is executed with a gradient from 10 to 30% of buffer B within 80 min and a flow rate of 250 nL/min. As ESI source a Nanospray Flex Ion Source from Thermo Scientific is used. The mass spectrometer runs in the data-dependent mode and switches automatically between MS (max. of 1×10^6 ions) and MS/MS. Each MS scan is followed by a maximum of five MS/MS scans in the linear ion trap using normalized collision energy of 35% and a target value of 5000. Parent ions with a charge state from $z = 1$ and unassigned charge states are excluded for fragmentation. The mass range for MS is 370–2000 m/z. The resolution is set to 60,000. MS parameters are as follows: spray voltage 2.3 kV, no sheath and auxiliary gas flow, and ion transfer tube temperature at 125°C.

3.9 Data Analysis

The freely available program MaxQuant (Cox et al., 2009) is recommended for an efficient analysis of raw MS data for both peptide identification and protein quantification (Table 1). The quality of experiment can be addressed, and data analysis can be performed using Perseus Software and Graphical Proteomics Data Explorer (GProX) (Rigbolt, Vanselow, & Blagoev, 2011), these area freely available platforms for comprehensive and integrated analysis and visualization of large proteomics datasets.

Table 1 MaxQuant Settings

Parameter	Settings
Version	1.4.1.2 (or newer)
Fixed modifications	Carbamidomethyl (C)
Variable modifications	Oxidation (M)
	Acetylation (protein N-term)
	Gly–Gly (K, not C-term)
	Phosphorylation (STY)
Multiplicity	3
Enzyme	Trypsin/P
Maximum missed cleavages	3
Special amino acids	K
Include contaminants	True
First search p.p.m.	20
Main search p.p.m.	6
MS/MS tolerance (Fourier transform MS, p.p.m.)	20
Top MS/MS peaks per 100 Da	10
Peptide false discovery rate	0.01
Protein false discovery rate	0.01
Site false discovery rate	0.01
Minimum peptide length (residues)	7
Minimum unique peptide	1
Maximum peptide mass (Da)	4600
Minimum ratio count	2

1. Experimental reproducibility is evaluated by plotting two biological replicates in a scatter plot reporting the corresponding ratios transformed into logarithm to the basis of 2. As example, AMBRA1 interactors detected in high nutrients or after 2 h of starvation are reported in Fig. 2A and B. The Pearson correlation and *p*-values are used to evaluate the accuracy of independent experiments.

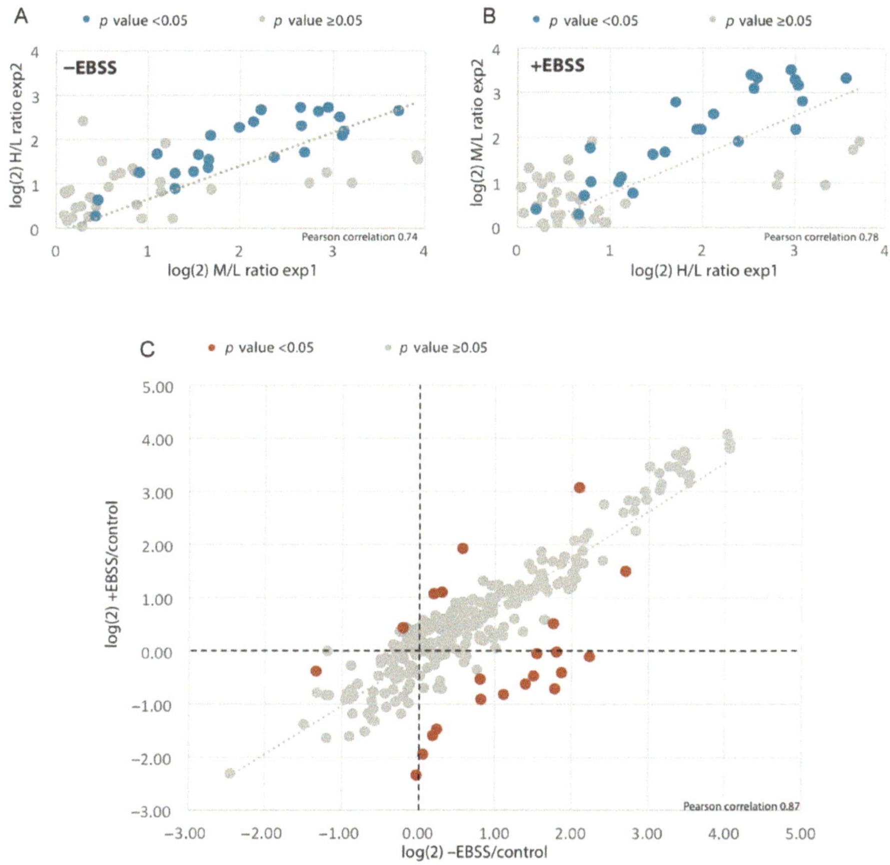

Fig. 2 Example of data analysis of AMBRA1 interactors. Data were obtained from experiments performed in 2FTGH expressing cells grown in nutrient-rich conditions (−EBSS) or starved (+EBSS) for 2 h. (A and B) Analysis of identified AMBRA1 interactors was performed by evaluating data derived from two independent experiments. To analyze reproducibility of experiments, log(2) of SILAC ratios was used to calculate the Pearson's correlation and the *p*-value was calculated based on a two side *t*-test and randomized FDR calculation. *Highlighted* are reproducibly identified proteins showing *p*-value < 0.05 and a log(2) > 0. True interactors are located in the *upper right* corner of the quadrant. (C) Analysis of changing interactors. The scatter plot reports log(2) of −EBSS/control vs log(2) +EBSS/control SILAC ratios of identified proteins among two experiments. *Highlighted* are changing proteins showing *p*-value < 0.05 calculated based on significance A.

2. The evaluation of identified interactors is performed by calculating the significance that proteins are outliers of a normal distributed dataset ($p < 0.05$; significance A in MaxQuant, data not shown). In the proposed experimental settings, the ratios between "medium/light" and

"heavy/light" represent the amount of purified protein compared to the control IP.

3. To evaluate changes in interactors between different experimental conditions, significance can be used to compare log(2) of SILAC ratios obtained from two independent experiments (Fig. 2C). Before proceeding with this evaluation, two important aspects should be evaluated. First, the amount of immunoprecipitated protein (e.g., AMBRA1) should be similar or normalized between different experiments. Second, starvation can dramatically reduce the amount of total proteins, thus influencing stoichiometry of identified interactors. In the light of these, it is important to consider that increased interactions are more plausible than reduced ones. Also, it is recommended to normalize IP data to respective whole cell lysate data to account for abundance changes of proteins. For these reasons, immunoblotting validation of changes in protein complex composition observed by IP/MS data is recommended.
4. The biological functions, cellular compartment, and metabolic pathways, in which identified binding partners take part, can be highlighted through Gene Ontology analysis comparing upregulated vs downregulated proteins. A heat map encoding statistical significance as color intensity can easily visualize Gene Ontology terms (Sprenger, Küttner, Bruckner-Tuderman, & Dengjel, 2013).

Notes

1. Depending on experimental approaches, a single round of IP is recommended if the interest is mainly focused on PTMs and/or on weak interactions. Although, by avoiding the second IP, contaminants will be more abundant, this will speed up the protocol reducing the loss of transient modifications and interactions.
2. The proposed method can be useful to compare data from more than two experimental conditions (e.g., nutrient starvation for 0, 2, 4, 8 h). This could be achieved by performing parallel MS analyses in which one condition (e.g., cells at time point 0 h) is chosen as a reference and present in each LC-MS/MS run. Data from the different time points could be indirectly compared after normalization with the reference sample.
3. The mixing of samples can be performed before or after the IP. To highlight not only stable interactions but also week/dynamic ones, we strongly suggest performing independent IPs and, then combine eluates before proceeding with samples preparation for in-gel digestion (Küttner, Mack, & Gretzmeier, 2014; Wang & Huang, 2008).

4. Unmixed lysate deriving from each SILAC sample can be used to evaluate the amount of labeled arginine and lysine incorporated at peptide level as well as the arginine to proline conversion. To minimize this conversion, unlabeled proline can be titrated to the medium. Moreover, we recommend swapping SILAC labeling in case of biological replicates to reduce artifacts coming from the label per se.
5. The proposed protocol is aimed at studying protein–protein interactions during autophagy induction. As previously reported for 2FTGH cells (Antonioli et al., 2014), 30 min of starvation with EBSS are suggested to monitor early steps of autophagy induction. On the other hands, 2 and 4 h represent the peak and the switching off the process, respectively.
6. We recommend to store cells pellets, but not protein lysates, at −80°C maximally for 1–3 months to avoid loss of weak interactions.
7. The amount of 10 mg of total proteins extract subjected to IP indicated in this protocol is required to ensure a proper amount of protein identification by MS, taking into consideration that proteins involved in autophagy regulation are normally not highly expressed, and the double-purification protocol using anti-HA and anti-FLAG beads reduces the quantity of unstable interactors.
8. This step can be avoided in case the LC-MS/MS system includes a precolumn.

ACKNOWLEDGMENTS

This work is supported by Ricerca Corrente and Ricerca Finalizzata (RF-2011-02349395) from the Italian Ministry of Health and AIRC IG-2015 17404, FRIAS COFUND Fellowship Programme (University of Freiburg, Germany), and People Programme (Marie Curie Actions) of the European Union's Seventh Framework Programme (FP/2007-2013) under REA Grant agreement no. [609305]; the Swiss National Science Foundation, Grant 31003A-166482/1.

REFERENCES

Abrahamsen, H., Stenmark, H., & Platta, H. W. (2012). Ubiquitination and phosphorylation of Beclin 1 and its binding partners: Tuning class III phosphatidylinositol 3-kinase activity and tumor suppression. *FEBS Letters*, *586*(11), 1584–1591. http://dx.doi.org/10.1016/j.febslet.2012.04.046.

Antonioli, M., Albiero, F., Fimia, G. M., & Piacentini, M. (2015). AMBRA1-regulated autophagy in vertebrate development. *The International Journal of Developmental Biology*, *59*(1–3), 109–117. http://dx.doi.org/10.1387/ijdb.150057mp.

Antonioli, M., Albiero, F., Nazio, F., Vescovo, T., Perdomo, A. B., Corazzari, M., … Fimia, G. M. (2014). AMBRA1 interplay with cullin E3 ubiquitin ligases regulates autophagy dynamics. *Developmental Cell*, *31*(6), 734–746. http://dx.doi.org/10.1016/j.devcel.2014.11.013.

Behrends, C., Sowa, M. E., Gygi, S. P., & Harper, J. W. (2010). Network organization of the human autophagy system. *Nature*, *466*(7302), 68–76. http://dx.doi.org/10.1038/nature09204.

Blagoev, B., Kratchmarova, I., Ong, S., & Nielsen, M. (2003). A proteomics strategy to elucidate functional protein-protein interactions applied to EGF signaling. *Nature Biotechnology*, *21*, 315–318. Retrieved from http://www.nature.com/nbt/journal/v21/n3/full/nbt790.html.

Cianfanelli, V., Fuoco, C., Lorente, M., Salazar, M., Quondamatteo, F., Gherardini, P. F., … Cecconi, F. (2014). AMBRA1 links autophagy to cell proliferation and tumorigenesis by promoting c-Myc dephosphorylation and degradation. *Nature Cell Biology*, *17*(1), 20–30. http://dx.doi.org/10.1038/ncb3072.

Cox, J., Matic, I., Hilger, M., Nagaraj, N., Selbach, M., Olsen, J. V., & Mann, M. (2009). A practical guide to the MaxQuant computational platform for SILAC-based quantitative proteomics. *Nature Protocols*, *4*(5), 698–705. http://dx.doi.org/10.1038/nprot.2009.36.

Dengjel, J., Kratchmarova, I., & Blagoev, B. (2010). Mapping protein-protein interactions by quantitative proteomics. *Methods in Molecular Biology (Clifton, NJ)*, *658*, 267–278. http://dx.doi.org/10.1007/978-1-60761-780-8_16.

Di Bartolomeo, S., & Corazzari, M. (2010). The dynamic interaction of AMBRA1 with the dynein motor complex regulates mammalian autophagy. *The Journal of Cell Biology*, *191*(1), 155–168. Retrieved from http://jcb.rupress.org/content/191/1/155.short.

Fimia, G., Corazzari, M., Antonioli, M., & Piacentini, M. (2012). Ambra1 at the crossroad between autophagy and cell death. *Oncogene*, *32*(28), 3311–3318. Retrieved from http://www.nature.com/onc/journal/v32/n28/abs/onc2012455a.html.

Fimia, G., Stoykova, A., & Romagnoli, A. (2007). Ambra1 regulates autophagy and development of the nervous system. *Nature*, *447*, 1121–1125. Retrieved from http://www.nature.com/nature/journal/v447/n7148/abs/nature05925.html.

He, C., & Levine, B. (2010). The Beclin 1 interactome. *Current Opinion in Cell Biology*, *22*(2), 140–149. http://dx.doi.org/10.1016/j.ceb.2010.01.001.

Itakura, E., Kishi, C., Inoue, K., & Mizushima, N. (2008). Beclin 1 forms two distinct phosphatidylinositol 3-kinase complexes with mammalian Atg14 and UVRAG. *Molecular Biology of the Cell*, *19*(12), 5360–5372. http://dx.doi.org/10.1091/mbc.E08-01-0080.

Itakura, E., & Mizushima, N. (2009). Atg14 and UVRAG: Mutually exclusive subunits of mammalian Beclin 1-PI3K complexes. *Autophagy*, *5*(4), 534–536. Retrieved from http://www.ncbi.nlm.nih.gov/pubmed/19223761.

Kihara, A., Kabeya, Y., Ohsumi, Y., & Yoshimori, T. (2001). Beclin-phosphatidylinositol 3-kinase complex functions at the trans-Golgi network. *EMBO Reports*, *2*(4), 330–335. http://dx.doi.org/10.1093/embo-reports/kve061.

Küttner, V., Mack, C., & Gretzmeier, C. (2014). Loss of collagen VII is associated with reduced transglutaminase 2 abundance and activity. *The Journal of Investigative Dermatology*, *134*, 2381–2389. Retrieved from http://www.nature.com/jid/journal/v134/n9/abs/jid2014185a.html.

Liang, X. H., Jackson, S., Seaman, M., Brown, K., Kempkes, B., Hibshoosh, H., & Levine, B. (1999). Induction of autophagy and inhibition of tumorigenesis by Beclin 1. *Nature*, *402*(6762), 672–676. http://dx.doi.org/10.1038/45257.

Lößner, C., Warnken, U., Pscherer, A., & Schnölzer, M. (2011). Preventing arginine-to-proline conversion in a cell-line-independent manner during cell cultivation under stable isotope labeling by amino acids in cell culture (SILAC). *Analytical Biochemistry*, *412*, 123–125. Retrieved from http://www.sciencedirect.com/science/article/pii/S000326971100025X.

Matsunaga, K., Saitoh, T., Tabata, K., Omori, H., Satoh, T., Kurotori, N., … Yoshimori, T. (2009). Two Beclin 1-binding proteins, Atg14L and Rubicon, reciprocally regulate autophagy at different stages. *Nature Cell Biology*, *11*(4), 385–396. http://dx.doi.org/10.1038/ncb1846.

Morris, D. H., Yip, C. K., Shi, Y., Chait, B. T., & Wang, Q. J. (2015). Beclin 1-VPS34 complex architecture: Understanding the nuts and bolts of therapeutic targets. *Frontiers in Biology*, *10*(5), 398–426. http://dx.doi.org/10.1007/s11515-015-1374-y.

Nazio, F., Strappazzon, F., & Antonioli, M. (2013). mTOR inhibits autophagy by controlling ULK1 ubiquitylation, self-association and function through AMBRA1 and TRAF6. *Nature Cell Biology*, *15*(4), 406–416. Retrieved from http://www.nature.com/ncb/journal/v15/n4/abs/ncb2708.html.

Pattingre, S., Tassa, A., Qu, X., & Garuti, R. (2005). Bcl-2 antiapoptotic proteins inhibit Beclin 1-dependent autophagy. *Cell*, *122*, 927–939. Retrieved from http://www.sciencedirect.com/science/article/pii/S0092867405006926.

Rappsilber, J., Mann, M., & Ishihama, Y. (2007). Protocol for micro-purification, enrichment, pre-fractionation and storage of peptides for proteomics using StageTips. *Nature Protocols*, *2*(8), 1896–1906. http://dx.doi.org/10.1038/nprot.2007.261.

Rigbolt, K. T. G., Vanselow, J. T., & Blagoev, B. (2011). GProX, a user-friendly platform for bioinformatics analysis and visualization of quantitative proteomics data. *Molecular & Cellular Proteomics: MCP*. *10*(8). http://dx.doi.org/10.1074/mcp.O110.007450O110.007450.

Sprenger, A., Küttner, V., Biniossek, M. L., Gretzmeier, C., Boerries, M., Mack, C., … Dengjel, J. (2010). Comparative quantitation of proteome alterations induced by aging or immortalization in primary human fibroblasts and keratinocytes for clinical applications. *Molecular BioSystems*, *6*(9), 1579–1582. http://dx.doi.org/10.1039/c003962d.

Sprenger, A., Küttner, V., Bruckner-Tuderman, L., & Dengjel, J. (2013). Global proteome analyses of SILAC-labeled skin cells. *Methods in Molecular Biology (Clifton, NJ)*, *961*, 179–191. http://dx.doi.org/10.1007/978-1-62703-227-8_10.

Sun, Q., Fan, W., Chen, K., Ding, X., Chen, S., & Zhong, Q. (2008). Identification of Barkor as a mammalian autophagy-specific factor for Beclin 1 and class III phosphatidylinositol 3-kinase. *Proceedings of the National Academy of Sciences of the United States of America*, *105*(49), 19211–19216. http://dx.doi.org/10.1073/pnas.0810452105.

Thoresen, S., & Pedersen, N. (2010). A phosphatidylinositol 3-kinase class III sub-complex containing VPS15, VPS34, Beclin 1, UVRAG and BIF-1 regulates cytokinesis and degradative endocytic traffic. *Experimental Cell Research*, *316*(20), 3368–3378. Retrieved from http://www.sciencedirect.com/science/article/pii/S0014482710003757.

Vescovo, T., Refolo, G., & Romagnoli, A. (2014). Autophagy in HCV infection: Keeping fat and inflammation at bay. *BioMed Research International*, *2014*, 265353. Retrieved from http://www.hindawi.com/journals/bmri/2014/265353/abs/.

Wang, X., & Huang, L. (2008). Identifying dynamic interactors of protein complexes by quantitative mass spectrometry. *Molecular & Cellular Proteomics: MCP*, 7(1), 46–57. http://dx.doi.org/10.1074/mcp.M700261-MCP200.

Wirawan, E., Lippens, S., Vanden Berghe, T., Romagnoli, A., Fimia, G. M., Piacentini, M., & Vandenabeele, P. (2012). Beclin1: A role in membrane dynamics and beyond. *Autophagy*, *8*(1), 6–17. http://dx.doi.org/10.4161/auto.8.1.16645.

Yue, Z., Jin, S., & Yang, C. (2003). Beclin 1, an autophagy gene essential for early embryonic development, is a haploinsufficient tumor suppressor. *Proceedings of the National Academy of Sciences of the United States of America*, *100*, 15077–15082. Retrieved from http://www.pnas.org/content/100/25/15077.short.

Zimmermann, A., Zarei, M., Eiselein, S., & Dengjel, J. (2010). Quantitative proteomics for the analysis of spatio-temporal protein dynamics during autophagy. *Autophagy*, *6*, 1009–1016. Retrieved from http://www.tandfonline.com/doi/abs/10.4161/auto.6.8.12786.

CHAPTER TWENTY-SIX

In Vitro Characterization of VPS34 Lipid Kinase Inhibition by Small Molecules

F. Fassy, C. Dureuil, A. Lamberton, M. Mathieu, N. Michot, B. Ronan, B. Pasquier[1]

Sanofi, Quai Jules Guesde, Vitry Sur Seine, Paris, France

[1]Corresponding author: e-mail address: benoit.pasquier@sanofi.com

Contents

1. Introduction 448
2. Purification of Recombinant VPS34 Proteins 450
   2.1 Protein for Catalytic and Binding Assays 450
   2.2 Protein for Crystallization 450
3. Catalytic Assay 451
   3.1 Calibration for the Catalytic Assay 451
   3.2 Assay Conditions 452
   3.3 K_m for ATP 453
   3.4 IC_{50} of Compounds 453
4. Binding Assay 454
   4.1 Assay Principle and Protocol 454
   4.2 Binding Kinetics of Compounds 456
5. Crystallization 457
6. Cell Assay 458
   6.1 Generation of HeLa GFP-2xFYVE Clone 459
   6.2 Cellular Assay 459
7. Conclusion 461

Acknowledgments 462

References 462

Abstract

VPS34 is a class III phosphoinositide 3-kinase that acts on vesicle trafficking. This kinase has recently attracted significant attention because of the function it plays in the machinery involved in the early steps of autophagy. Moreover, because significant progress had been made in the optimization of specific kinase inhibitors, its potential to be targeted by catalytic inhibitors has been investigated by different groups. The aim of this review is to present the key in vitro assays necessary for characterizing inhibitors of the catalytic activity of VPS34. The review covers catalytic (IC_{50} on purified recombinant

Methods in Enzymology, Volume 587
ISSN 0076-6879
http://dx.doi.org/10.1016/bs.mie.2016.09.070

protein) and binding assays (K_D, k_a, k_d on purified recombinant protein), and a cell-based assay (IC_{50} in GFP-FYVE expressing cell line). The methodology for crystallization of VPS34 protein is also presented as it can provide guidance for the design by medicinal chemistry of small molecular mass kinase inhibitor.

1. INTRODUCTION

In mammalian cells, phosphoinositide 3-kinases, PI3Ks, are divided into three classes based on their structural features and substrate specificities (Marat & Haucke, 2016; Sasaki et al., 2009; Vanhaesebroeck, Guillermet-Guibert, Graupera, & Bilanges, 2010; Yu, Long, & Shen, 2015). The class III enzyme (gene name PIK3C3) also called VPS34 (vacuolar protein sorting 34) was originally identified in *Saccharomyces cerevisiae* (Herman & Emr, 1990; Schu et al., 1993). VPS34 phosphorylates phosphatidylinositol (PtdIns) into phosphatidylinositol 3-phosphate (PtdIns3P) (Fig. 1).

As observed for the other lipids produced by the other members of the PI3K family, the pool of PtdIns3P is modulated by numerous proteins (Marat & Haucke, 2016). This can be done by phosphatases from the myotubularin family, such as MTM1, MTMR2, MTMR3, and MTMR14/JUMPY

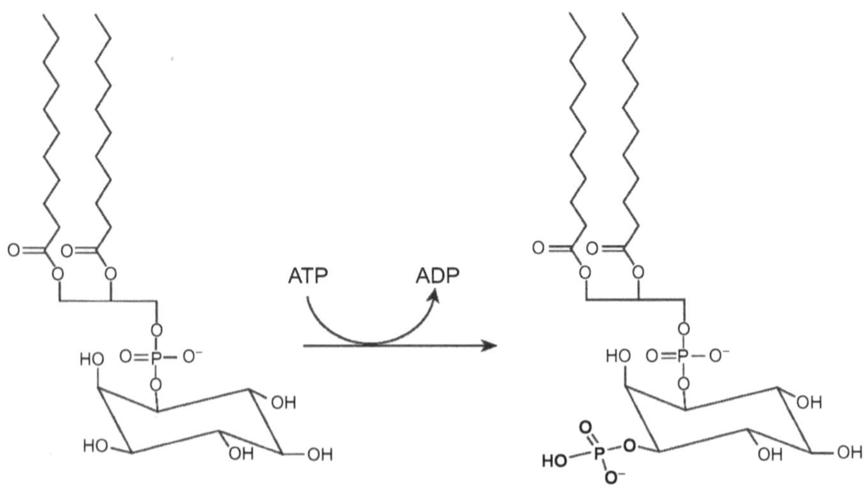

Fig. 1 The reaction catalyzed by VPS34, E.C. 2.7.1.137, is the phosphorylation by ATP of phosphatidylinositol at the three position of the inositol ring, into phosphatidylinositol 3-phosphate, with production of ADP.

and the dual-domain protein tyrosine phosphatase σ (PTPσ) (Ketel et al., 2016; Marat & Haucke, 2016; Martin et al., 2010; Vergne et al., 2009). PtdIns3P can also be converted to PI(3,5)P2 by the kinase PIKfyve (Marat & Haucke, 2016; Raiborg, Schink, & Stenmark, 2013). Once produced, PtdIns3P acts as a second messenger and recruits downstream proteins containing either a FYVE (Fab1, YOTB, Vac1, and EEA1) zinc-finger domain or a PX (Phox homology) domain, including phosphatases, kinases, and adaptor proteins (Marat & Haucke, 2016; O'Farrell, Rusten, & Stenmark, 2013; Raiborg et al., 2013; Yu et al., 2015). This downstream signaling is not only involved in a broad range of vesicular trafficking pathways including autophagy but also in the endosomal trafficking of receptors, for example, the epidermal growth factor receptor, the platelet-derived growth factor receptor, or the transferrin receptor (Jaber et al., 2012; Johnson, Overmeyer, Gunning, & Maltese, 2006; O'Farrell et al., 2013; Raiborg et al., 2013). These multiple functions of the reaction product of VPS34 can be explained at the molecular level. Indeed, VPS34 is active within complexes with several accessory proteins (Funderburk, Wang, & Yue, 2010). At least two distinct complexes are described, both encompass VPS34, Vps15, beclin-1 plus either Atg14L/Barkor or UVRAG. The Atg14L-containing VPS34 complex is localized in autophagosomal structures and participates in the generation of autophagosomes, while the UVRAG-containing VPS34 complex is found at the endosomes (Funderburk et al., 2010; Marat & Haucke, 2016; Yu et al., 2015). Allosteric regulation of these two distinct VPS34 complexes has also been proposed (Baskaran et al., 2014; Rostislavleva et al., 2015).

Although different molecules have been described as potential inhibitors of VPS34, most of them are poorly selective, limiting their application (Pasquier, 2016). However, more recently significant progress has been made with the discovery of specific catalytic inhibitors of VPS34 (Bago et al., 2014; Dowdle et al., 2014; Honda et al., 2015; Pasquier et al., 2015; Ronan et al., 2014). The aim of this review is to present suitable in vitro assays supporting drug discovery on VPS34 target. Protein-based assays, depending on catalysis (inhibition by the compound described by IC_{50}) or on binding properties (K_D, k_a, k_d of the compound—protein complex), and a cell-based assay (IC_{50} in a stable GFP-FYVE-expressing human cell line) will be described. In addition, crystallization of VPS34 protein with the catalytic inhibitors is also detailed as it provides structural data to guide rational design by medicinal chemistry.

2. PURIFICATION OF RECOMBINANT VPS34 PROTEINS

Two constructs have been prepared for the protein-based assays. The full-length GST-tagged protein (residues 2–887) was used for the catalytic and binding assays, and a truncated untagged version (residues 282–879) for crystallization.

2.1 Protein for Catalytic and Binding Assays

Human full-length VPS34 protein (accession number NM_002647.2), residues 2–887 with a N-terminal GST tag was expressed in insect Sf-21 cells (Invitrogen). Sf-21 cells were cultured in SF900II medium (Gibco) and infected with VPS34 baculovirus in an optimal virus/insect cells ratio. Cells were harvested 72 h after infection; pellets were kept at −80°C. All the purification steps were carried out at 4°C. The frozen cell paste was resuspended in a lysis buffer consisting of 50 m*M* Tris, pH 8.0, 150 m*M* NaCl, 1 m*M* tris (2-carboxyethyl)phosphine (TCEP), benzonase nuclease (Novagen), and complete protease inhibitors (Roche) and lysed by stirring during 1 h (100 g of Sf-21 cell pellet for 350 mL of lysis buffer). After centrifugation at 48,000 × *g* for 2 h, the lysate was loaded onto a 10 mL glutathione sepharose column (GE Healthcare), followed by wash with 20 column volumes of the lysis buffer and elution of bound material with a step gradient of lysis buffer in presence of 10 m*M* GSH extemporaneously in 10 column volumes. Fractions containing GST-VPS34 were pooled and further purified up to 98% of purity on Superdex 200 in the lysis buffer containing 10% glycerol.

2.2 Protein for Crystallization

The truncated human VPS34 protein (residues 282–879) with N-terminal 6xHis tag and TEV cleavage site was expressed in *Escherichia coli* BL21 (DE3). This construct was chosen based on the Structural Genomics Consortium study of the same protein (http://www.thesgc.org/). This construct was devoid of the N-terminal C2 domain and the last eight residues of the protein. Cells were grown at 37°C in LB medium to $OD_{600} = 0.6$, and protein expression was induced by addition of 0.5 m*M* IPTG. Cells were harvested after 3 h by centrifugation and pellets were kept at −20°C. The *E. coli* cell pellet (50 g) was mixed with 250 mL of lysis buffer, 50 m*M* HEPES, pH 7.5, 500 m*M* NaCl, 1 m*M* TCEP, 10% glycerol, 15 m*M* imidazole, pH 7.5, benzonase nuclease (Novagen), and complete EDTA-free protease inhibitors (Roche). It was then lysed mechanically on Emusiflex. After centrifugation at 76,000 × *g* for 1 h, the lysate was loaded onto a 2 × 5-mL HisTrap FF column (GE Healthcare), followed by wash with 40 m*M* imidazole and elution of bound material with a linear gradient to 500 m*M* imidazole in 20 column volumes. Fractions containing VPS34 were pooled and tag cleavage was performed by addition of 100 U of TEV protease per mg of VPS34 during 2 h at 4°C, followed by dialysis overnight at room temperature against 2 L of lysis buffer using 8–10 kDa MWCO Spectra/Por Dialysis Tubing (Spectrumlabs 131264). Then the protein was passed again over a HisTrap column. Cleaved material was eluted in the flow through and in the 15 m*M* imidazole wash. Cleaved material eluted was concentrated and loaded to Superdex 75 26/60 gel-filtration column equilibrated with 50 m*M* HEPES, 500 m*M* NaCl, 1 m*M* TCEP, 10% glycerol, pH 7.5 buffer. The peak corresponding to the estimated molecular mass was pooled and concentrated to 10 mg/mL. The protein purity was estimated to 98%.

3. CATALYTIC ASSAY

The characterization of inhibitors of an enzymatic reaction requires an assay with the highest possible sensitivity. Many techniques are available for the follow up of a kinase reaction. The quantification of the ADP product by a TR–FRET method (time resolved–fluorescence resonance energy transfer) was preferred because of its extreme sensitivity and excellent reproducibility. In practice, we used the Transcreener ADP kit from Cisbio, using the HTRF (homogeneous time resolved fluorescence) technology. The drawback of the assay is the nonlinearity of the signal as a function of the ADP concentration. It is due to the principle of the assay: a competition between the ADP produced by the kinase reaction and a labeled ADP which is detected by the HTRF reagents. This kind of assay is well adapted to determine IC_{50} (concentration of inhibitor that corresponds to 50% inhibition) but is not well suited for K_i determinations.

3.1 Calibration for the Catalytic Assay

An ADP standard curve was prepared for each experiment, using seven ADP concentrations ranging from 10 to 0.156 μ*M* in two-step serial dilution. The signals of fluorescence at 665 and 620 nm after incubation with the HTRF revelation reagents were recorded on a Rubystar (bMG) instrument. The results were expressed as fluorescence signal ratio (665/620 * 10,000)

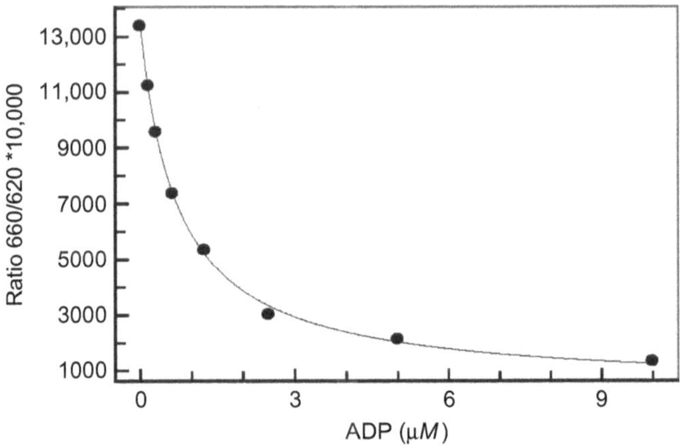

Fig. 2 Typical ADP calibration curve obtained with the Transcreener ADP kit from Cisbio.

(Fig. 2). The signal ratio (Y) was fit to the ADP standard concentrations (X) according to Eq. (1):

$$Y = C + A^*\left(1 - \left(\frac{X}{(B + X)}\right)\right). \tag{1}$$

Using this standard curve, the signal ratios of all the samples were then converted to ADP concentrations. We considered that the concentration of produced ADP had to remain between 1 and 4 μM to allow robust quantification.

3.2 Assay Conditions

It is important to find conditions where the enzyme is strongly active, and therefore it was used at low concentration. Indeed, the lowest measurable IC_{50} is equal to half of the active enzyme concentration [E] according to Eq. (2):

$$IC_{50} = K_{i,\mathrm{app}} + \frac{[E]}{2} \tag{2}$$

and $K_{i,\mathrm{app}} = K_i{}^*(1 + [S]/K_m)$ when the inhibitor competes with the substrate S.

The VPS34 protein, 2–887 construct with N-terminal GST tag was purified in-house at a large scale, allowing us to work on a single batch with constant specific activity.

In our assay conditions, the commercial substrate: L-α-phosphatidylinositol ammonium salt (natural lipid extracted from soybean, *Glycine max*) was selected (Sigma-Aldrich P5766). The lipid was handled in particular conditions: to 0.5 mL of 10 mg/mL lipid in chloroform, 4 mL of chloroform was added. The mixture was aliquoted in to 130 μL (144 μg) samples in polypropylene tubes and evaporated under argon. Tubes were tightly closed and stored at −80°C for a maximum of 6 months. Before use, the lipid was resuspended in the reaction buffer and sonicated for a few seconds. We worked at a final concentration in the reaction of 55 μg/mL.

The complete reaction buffer was 50 m*M* HEPES, pH 7.3, 2 m*M* TCEP, 0.1% CHAPS, and 5 m*M* $MnCl_2$. CHAPS was added to reduce nonspecific binding to surfaces and participated to the solubilization of the lipid. The reaction was sensitive to the presence of salts: the activity was higher with manganese than with magnesium; NaCl and KCl were inhibitory. Assays were performed in 96-well microtiter plates with 20 μL reaction volume and 20 μL HTRF revelation reagents. The assay is transposable to 384-well plate, using 5 μL reaction, and 5 μL revelation volumes.

3.3 K_m for ATP

The K_m for ATP was determined in the complete reaction buffer, using 55 μg/mL lipid, and 5 or 7.5 n*M* VPS34, at room temperature. For each ATP concentration, the concentration of ADP produced was measured at 4, 8, 12, 16, 20, and 30 min. The reaction rate was then plotted as a function of the ATP concentration and fitted to the Michaelis Menten model according to Eq. (3) (Fig. 3):

$$\text{rate} = \frac{V_{max}[\text{ATP}]}{K_m + [\text{ATP}]} \quad (3)$$

The K_m from eight independent experiments was 65 ± 31 μ*M*. The V_{max}, from five independent experiments at 7.5 n*M* VPS34, was 0.26 ± 0.14 μ*M*/min, corresponding to a turnover of 0.58 s^{-1}.

3.4 IC_{50} of Compounds

The IC_{50} determinations were performed in the presence of serial dilutions of inhibitor (10 μ*M* to 0.51 n*M*, 3% DMSO, final) and a mixture of 55 μg/mL lipid substrate, 5 n*M* VPS34, 10 μ*M* ATP, in complete reaction buffer. After 60-min incubation at room temperature, the reaction

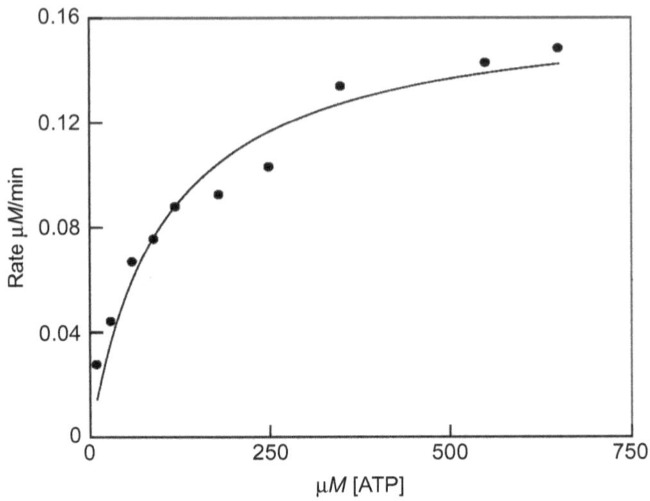

Fig. 3 K_m ATP determination, using 7.5 n*M* VPS34, 55 μg/mL L-α-phosphatidylinositol ammonium salt, Sigma-Aldrich P5766.

was stopped by addition of the HTRF revelation mixture. Because the ATP concentration was below the K_m, the inhibitors that competed with ATP were favored.

The percent inhibition as a function of each concentration of compound was calculated and the set of data was fitted using a four-parameter logistic model. The IC_{50} value was defined as the concentration of compound where percent inhibition is equal to 50 and was the mean from at least two independent experiments.

According to Eq. (2), the minimal IC_{50} should be 2.5 n*M* since the concentration of VPS34 was 5 n*M*. In practice, the lowest we could measure was 1 n*M*, which was in line with the hypothesis that some protein was lost by nonspecific binding to surfaces. We suspected that our best inhibitors had subnanomolar affinity. The determination of a *Ki* below the enzyme concentration is possible by using tight binding equations (Morrison, 1969) but was not envisaged because of practical difficulties. We measured the affinity of our best inhibitors by surface plasmon resonance.

4. BINDING ASSAY

4.1 Assay Principle and Protocol

The binding kinetics of compounds were evaluated by surface plasmon resonance (SPR) on the N-terminal GST tag full-length protein. In addition, a specific study was performed to compare the full-length and the truncated 282–879 protein used for crystallization, to ascertain that both forms had similar molecular interactions with the compounds.

The experiments were achieved on ProteOn XPR36 (Biorad). SPR sensors measure refractive index changes occurring at the sensor surface during a course of biomolecular interaction. SPR involves immobilization of one of the binding partner (usually named ligand, i.e., VPS34 protein) on the sensor surface, while the other binding partner (named analyte, i.e., small molecule) remains in solution. The interaction of an analyte with a ligand increases the refractive index at the surface and is directly proportional to the amount of bound analyte. This change in refractive index is expressed in terms of response units (RU). The interaction between an analyte and the ligand is monitored in real time and is represented by a plot, or sensorgram, of RU vs time. The ProteOn XPR36 has a unique approach to multiplexing; this system generates a 6 × 6 interaction array for the simultaneous analysis of up to six ligands with up to six analytes.

Each new sensor chips were preconditioned by three subsequent washes using 50 m*M* NaOH, 10 m*M* HCl, and 0.1% SDS. The protein immobilization was performed on GLM sensor chip using standard amine coupling procedure in PBS, 0.05% Tween 20, immobilization running buffer (Biorad). Amine coupling requires preparation of the protein in a low-salt buffer at least 0.5 pH unit below the protein isoelectric point. After an 8-min activation of carboxylic groups by a 1:1 mixture of 1 *M* 1-ethyl-3-(3-dimethylaminopropyl)-carbodiimide and 0.25 *M* *N*-hydroxysuccinimide sulfo to give reactive succinimide esters, the protein was injected at a concentration of 16 μg/mL (full length) or 20 μg/mL (truncated) in 10 m*M* MES buffer at pH 6 in the presence of an inhibitor of intermediate affinity to protect the exposed lysines of the active site. During this step, the esters reacted spontaneously with the amino or other nucleophilic groups to link the ligand to the surface of the chip. Free surface activated carboxyl groups remaining after covalent modification were quenched with 1 *M* ethanolamine pH 8.5 during 5 min. Typical immobilization levels ranged from 8500 to 11,500 RU and from 6500 to 7300 RU for the truncated and full-length protein, respectively. A nonderivatized surface (without ligand) was used as reference for bulk refractive index correction.

All interactions of small molecules (analytes) were studied at 25°C. The compounds were tested at six concentrations (typically from $10 \times K_D$ to $0.3 \times K_D$). The running buffer was made of 50 m*M* HEPES-NaOH, pH

7.1, 5 m*M* $MgCl_2$, 150 m*M* NaCl, 2 m*M* TCEP, and it was supplemented with 2% DMSO. Analyte injections were performed at a flow rate of 100 μL/min for a 1-min association phase, followed by a 5-min dissociation phase. DMSO has a high refractive index, e.g., a 0.1% variation in DMSO concentration between the sample and the running buffer translates to 100 RU of SPR signal. It was therefore necessary to add a DMSO calibration curve for bulk effect and "excluded volume" corrections. Typical DMSO calibration curve ranged from 1.5% to 3.5% DMSO. A buffer blank control was also included for baseline drift correction.

The determination of the kinetic parameters was performed according to the recommendations of the constructor (http://www.bioradiations.com/guide-to-spr-data-analysis-on-the-proteon-xpr36-system/). First, the reference surface and the buffer blank responses were subtracted from all the interaction data collected over the different reaction surfaces and the "excluded volume" corrections was applied. From the corrected sensorgram data, the rate constants were determined by the ProteOn software using a Langmuir 1:1 binding model, in which one analyte molecule (A) interacts reversibly with one ligand molecule (B), with the following assumptions: pseudo-first-order kinetics, equivalence and independence of binding reactions, no limitation by mass transport. The equilibrium dissociation constant (K_D) is the ratio between two rate constants, $K_D = k_d/k_a$, with k_a, the association rate constant, and k_d, the dissociation rate constant. The kinetic analysis of the association and dissociation phases of the sensorgram was based on Eqs. (4) and (5), respectively.

$$R_t = \frac{R_{max}[A]}{K_D + [A]}\left[1 - e^{-(k_a[A] + k_d)t}\right] \tag{4}$$

$$R_t = R_0 e^{-k_d t}, \tag{5}$$

where R_t, R_{max}, and R_0 are the responses as time t, at the maximum of the association phase for a saturating concentration of analyte, and at the beginning of the dissociation phase, respectively.

4.2 Binding Kinetics of Compounds

Using the full-length construct, we measured the kinetic parameters of a large number of compounds, in particular those that displayed IC_{50}s in the catalytic assay around 1 n*M*. These compounds, that were comparable in the catalytic assay, became differentiated in the binding assay: K_D as low as 0.1 n*M* were measured for the most potent inhibitors.

In addition, compounds were selected for a measure on the truncated VPS34 construct that was used for structural studies. The association rate constant and dissociation rate constant, as well as the equilibrium constant were similar for 15 inhibitors (data not shown). An example of the data obtained with one compound is depicted in Fig. 4. We concluded that the truncated form was similar in terms of binding properties to the full length protein and that the X-ray data obtained with the crystals were relevant.

Truncated VPS34			**Full-Length VPS34**		
k_a ($M^{-1}s^{-1}$)	k_d (s^{-1})	K_D (M)	k_a ($M^{-1}s^{-1}$)	k_d (s^{-1})	K_D (M)
6.61×10^5	2.85×10^{-4}	4.28×10^{-10}	7.40×10^5	3.88×10^{-4}	5.25×10^{-10}

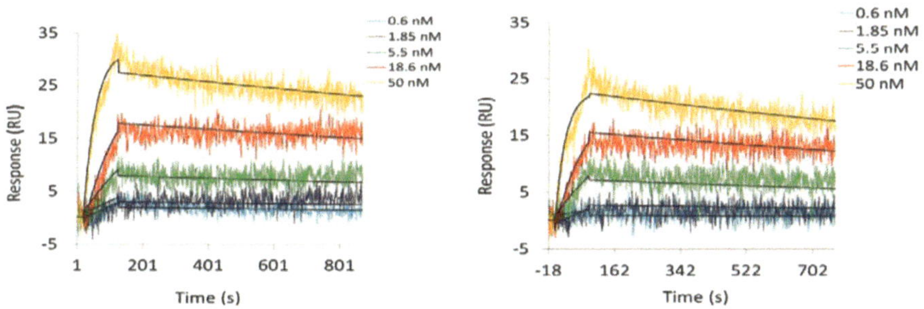

Fig. 4 SPR sensorgrams of the binding of one compound to (*left*) truncated or (*right*) full-length VPS34 protein, immobilized at the level of 8500 and 6500 RU, respectively. Increasing concentrations (0.6, 1.85, 5.5, 18.5, and 50 nM) of the compound were injected. The experimental data points and the fitted curves are superimposed. The binding parameters of the compound are given in the table above.

5. CRYSTALLIZATION

Crystallization assays were performed on the purified VPS34, containing residues 282–879. Compounds of interest (ligands) were solubilized at 100 mM in 100% DMSO. The protein was concentrated to 10 mg/mL in 20 mM Hepes, pH 7.5, 100 mM NaCl, 1 mM TCEP, and incubated overnight with 1 mM ligand (1% DMSO). Crystallization conditions were set-up by the hanging drop method. Briefly, 1 μL of protein was mixed with 1 μL of crystallization buffer on a coverslip, which was returned and sealed (either by manually applying Dow Corning vacuum grease or by using pregreased coverslips) over a well containing 1 mL of the crystallization buffer. Different crystallization conditions were systematically tested for each compound: 1.7–2 *M* ammonium sulfate, 100 m*M* Hepes, pH 7.5 or Tris, pH 8.5; 1 *M* NaCitrate, pH 8; 1.4 *M* NaMalonate Tris, pH 8.0. Multiple clustered needle-like crystals grew rapidly.

The size and volume of these crystals were ligand-dependent and this was loosely related to the affinity of the ligand for the protein. Crystal volume (and diffraction quality) could usually be improved by seeding. Briefly, a drop containing protein crystals was harvested and seeds were prepared using the seed bead method (Hampton Research, user guide HR2-320). Serial dilutions were done in the crystallization buffer, up to 1 in 30,000. Hanging drops were then prepared by adding 0.2 μL of seed solution to the crystallization drops.

Crystals were frozen using 22% glycerol as cryoprotectant included in the crystallization buffer. A crystal was separated from the crystal cluster and "fished out" of the crystal drop using a cryoloop, transferred for a few seconds in the cryoprotectant buffer before being frozen in liquid nitrogen for transportation and data collection. Usually, the size of the loop is adapted to the size of the crystal but VPS34 crystals were particularly fragile and prompt to break. Consequently, we preferred to use larger loops than usual. Cryoloops are loops of 25 μm to 1 mm diameter made of 20-μm diameter nylon and staked to hollow microtubes usually glued to a magnetic cap, used commonly to freeze and transport protein crystals.

Helicoidal data collection was the preferred method of data collection, enabling a full dataset to be collected to good resolution prior to crystal decay. In this approach, the crystal is translated perpendicular to the beam during data collection, to decrease the time a given volume of crystal remains in the beam. This approach might not be necessary on the newer synchrotron beamlines with collection strategies taking into account crystal decay.

All crystals diffracted in the same spacegroup ($P2_12_12$) with cell parameters varying only by a few Ås depending on the crystallization conditions.

Inhibitor optimization based on the crystal structure of VPS34 with compounds was described in Pasquier et al. (2015).

6. CELL ASSAY

With the aim to develop a dedicated cell assay to evaluate VPS34 inhibitors, we benefited from the catalytic activity of VPS34 to generate

PtdIns3P. Of note, some alternative proteins can also contribute to the production of PtdIns3P such as the class II PI3K (Devereaux et al., 2013; Falasca & Maffucci, 2007). As described in the introduction, PtdIns3P can bind to proteins containing a FYVE finger or PX domain with high affinity and specificity (Stenmark, Aasland, & Driscoll, 2002). FYVE finger is a cysteine-rich double-zinc binding domain that contains a module of approximately 70 residues that specifically recognizes PtdIns3P. When bound to PtdIns3P, FYVE domains are localized at early endosome membranes (Gillooly et al., 2000; Stenmark & Aasland, 1999; Stenmark et al., 2002). In response to VPS34 inhibition, FYVE-containing proteins which were localized at the endosomes are expected to diffuse into the cytoplasm.

A stable human cell line was engineered with tetracycline-inducible (TET-on) GFP-2xFYVE expression allowing to measure the redistribution of fluorescence with an imaging cytometer upon inhibition of VPS34 catalytic activity (Gillooly et al., 2000; Ronan et al., 2014).

6.1 Generation of HeLa GFP-2xFYVE Clone

The FYVE finger domain from human hepatocyte growth factor-regulated tyrosine kinase substrate (HGS, Pro158–Asn220) was duplicated in tandem, fused to the C-terminal of GFP protein and then cloned into the pcDNA4/TO (Life Technologies) to generate the GFP-2xFYVE plasmid. T-Rex-HeLa cells (Life Technologies), expressing the Tet-repressor, were stably transfected with the GFP-2xFYVE plasmid using Fugene 6 (Roche). Cells were cultured under antibiotics selection then sorted with the GFP intensity using a FACSDiva (Becton Dickinson). Single-cell-derived clonal populations with high GFP expression were generated. HeLa GFP-2xFYVE, clone C7, was selected on the basis of fluorescence intensity using a FACSCalibur (Becton Dickinson) and on the basis of GFP-2xFYVE spots observed on the iCyte® automated Imaging Cytometer (Compucyte) under doxycycline induction. The lack of P-glycoprotein expression was confirmed in this clone, avoiding the problem of multidrug resistance during compound evaluation.

6.2 Cellular Assay

The HeLa GFP-2xFYVE (C7 clone) stable cell line was maintained in MEM medium supplemented with 1% glutamine, 10% fetal bovine serum tetracycline-free (Clontech), 5 μg/mL blasticidin, and 200 μg/mL zeocin (Life Technologies).

For compounds evaluation, the cells were seeded in 96-well poly-D-lysine cell culture microplates, with clear bottom (Greiner) at 1×10^4 cells per well in 180 μL MEM medium supplemented with 1% glutamine, 10% fetal bovine serum tetracycline-free. Cells were incubated 48 h at 37°C, 5% CO_2 then induced by 20 μL doxycycline (Sigma) (1 μg/mL final). Doxycycline was preferred to tetracycline since it has a longer half-life (48 vs 24 h, respectively). Cells were incubated 48 h at 37°C, 5% CO_2. Compounds were diluted in 100% DMSO prior to dilution in complete medium, then added to the cells under 10 μL to generate final concentrations of 10 μ*M* to 0.3 n*M* (threefold serial dilution) and 0.1% DMSO. Wortmannin (Sigma) was used as positive control at a final concentration of 100 n*M*. Cells were incubated with compounds for 2 h at 37°C, 5% CO_2. The medium was carefully removed from each well and cells were fixed by adding 100 μL of 4% PFA (Sigma). After a 10 min incubation at room temperature in the dark, cells were carefully washed twice with 250 μL PBS. Nuclei were stained with 1 μg/mL Hoechst 33342 (Life Technologies) or with 7.5 μ*M* Draq5 (Biostatus). Cells were incubated at least 15 min at room temperature or 37°C in the dark before measuring fluorescence with the iCyte®.

The iCyte® Imaging Cytometer automatically segmented and quantified nuclei and subcellular events, i.e., endosomes labeled with GFP-2xFYVE probe. An analysis strategy was set-up as followed. First, a "threshold" contour was drawn based on the data computed from the photomultiplier measuring blue fluorescence of Hoechst or red fluorescence of Draq5; it defined the nuclei. Second, a "peripheral" contour was drawn by extending an area around the nuclei. Then detection of subcellular events within the peripheral contour, based on green fluorescence, was performed. For each well, the total cell count and the average number GFP-2xFYVE labeled endosomes per cell were measured.

Under doxycycline induction, the GFP-2xFYVE probe appeared as green dots indicating PtdIns3P localization at the endosomes (Fig. 5, left). Without doxycycline induction, none or very low level of green dots was observed (Fig. 5, right). As described treatment with active compound triggered the relocalization of GFP-2xFYVE probe into the cytoplasm as a diffuse green fluorescence (Ronan et al., 2014). This confirms that VPS34 is a key protein involved in the production of PtdIns3P. A minimum of 100 cells were counted. Cells were considered positive, e.g., cells containing PtdIns3P localized at the endosomes, when the number of green spots was greater than four per cell. The percent inhibition in reference to untreated cells was calculated for each concentration of compound. IC_{50}

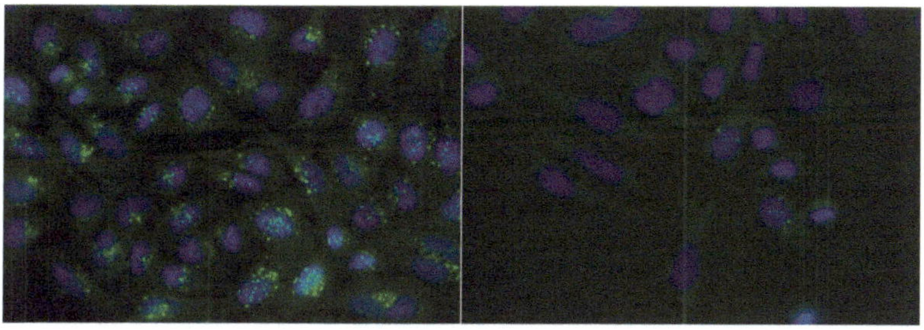

Fig. 5 Hela GFP-2xFYVE clone C7 was incubated with 1 μ*M* doxycycline for 48 h (*left*) or without doxycycline (*right*). Images were acquired with the iCyte® automated Imaging Cytometer. Nuclei are in *blue* and GFP-2xFYVE *dots* in *green*.

values were obtained from a dose–response curve with 10 concentrations tested in single replicate and fitted using a four-parameter logistic model.

7. CONCLUSION

In this review, we described the assays that we used to develop VPS34 inhibitors, within a rational design approach based on catalysis, binding, and crystallization results obtained with a recombinant purified protein. The catalytic assay could not characterize our most potent inhibitors because of its limit of sensitivity (i.e., 1 n*M*). The SPR-binding assay, with lower throughput, was better adapted for the compounds with high affinity that displayed K_D as low as 0.1 n*M*. A cellular assay also participated to the optimization of VPS34 inhibitors. The IC_{50} obtained with this assay was compared to the data obtained with the protein-based assays. Interestingly, the dissociation rate constant (k_d) was the parameter that best correlated with IC_{50} generated with the cellular assay (Fig. 6).

The assays presented here correspond to early stage of drug discovery. Additional properties must be optimized for compound that aimed to advance toward preclinical stage, such as the selectivity vs the kinases with similar active sites, physicochemical, and ADME properties. Selective VPS34 inhibitors were recently described and were used to demonstrate, in vitro, the key role of the catalytic activity of VPS34 in autophagy, induced either by starvation or by inhibition of mTOR (Dowdle et al., 2014; Ronan et al., 2014). Two scaffolds were further optimized into VPS34 inhibitors with excellent drug-like properties: compound 31 (Pasquier et al., 2015) and compound 19 (Honda et al., 2015). Interestingly, it was shown that the compound 19 prevented autophagy in vivo using LC3 readout

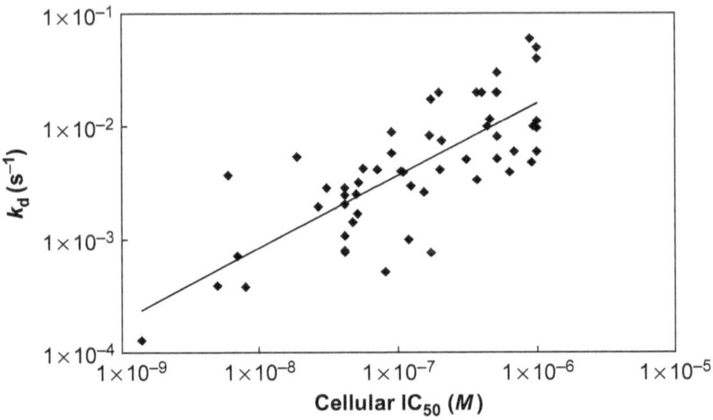

Fig. 6 Comparison between IC_{50} from GFP-2xFYVE cellular assay and the dissociation rate constant k_d. The correlation coefficient r^2 is 0.59 on 50 data points.

(Honda et al., 2015). Considering the synergistic antiproliferative activity of a specific VPS34 inhibitor with an mTOR inhibitor, measured in vitro in renal tumor cell lines (Ronan et al., 2014), there is now good evidence that inhibiting VPS34 has potential therapeutic perspectives.

ACKNOWLEDGMENTS

We thank M.-F. Bachelot, T. Bertrand, C. Castell, C. Delorme, L. Durand, O. Flamand, J. Houtmann, and J.-P. Marquette for their valuable input and discussion.

REFERENCES

Bago, R., Malik, N., Munson, M. J., Prescott, A. R., Davies, P., Sommer, E., et al. (2014). Characterization of VPS34-IN1, a selective inhibitor of Vps34, reveals that the phosphatidylinositol 3-phosphate-binding SGK3 protein kinase is a downstream target of class III phosphoinositide 3-kinase. *Biochemical Journal*, *463*, 413–427.

Baskaran, S., Carlson, L. A., Stjepanovic, G., Young, L. N., Kim do, J., Grob, P., et al. (2014). Architecture and dynamics of the autophagic phosphatidylinositol 3-kinase complex. *Elife*, *3*:e05115.

Devereaux, K., Dall'Armi, C., Alcazar-Roman, A., Ogasawara, Y., Zhou, X., Wang, F., et al. (2013). Regulation of mammalian autophagy by class II and III PI 3-kinases through PI3P synthesis. *PLoS One*, *8*:e76405.

Dowdle, W. E., Nyfeler, B., Nagel, J., Elling, R. A., Liu, S., Triantafellow, E., et al. (2014). Selective VPS34 inhibitor blocks autophagy and uncovers a role for NCOA4 in ferritin degradation and iron homeostasis in vivo. *Nature Cell Biology*, *16*, 1069–1079.

Falasca, M., & Maffucci, T. (2007). Role of class II phosphoinositide 3-kinase in cell signalling. *Biochemical Society Transactions*, *35*, 211–214.

Funderburk, S. F., Wang, Q. J., & Yue, Z. (2010). The beclin 1-Vps34 complex—At the crossroads of autophagy and beyond. *Trends in Cell Biology*, *20*, 355–362.

Gillooly, D. J., Morrow, I. C., Lindsay, M., Gould, R., Bryant, N. J., Gaullier, J. M., et al. (2000). Localization of phosphatidylinositol 3-phosphate in yeast and mammalian cells. *EMBO Journal, 19*, 4577–4588.

Herman, P. K., & Emr, S. D. (1990). Characterization of *Vps34*, a gene required for vacuolar protein sorting and vacuole segregation in *Saccharomyces cerevisiae*. *Molecular and Cellular Biology, 10*, 6742–6754.

Honda, A., Harrington, E., Cornella-Taracido, I., Furet, P., Knapp, M. S., Glick, M., et al. (2015). Potent, selective, and orally bioavailable inhibitors of VPS34 provide chemical tools to modulate autophagy in vivo. *ACS Medicinal Chemistry Letters, 7*, 72–76.

Jaber, N., Dou, Z., Chen, J. S., Catanzaro, J., Jiang, Y. P., Ballou, L. M., et al. (2012). Class III PI3K Vps34 plays an essential role in autophagy and in heart and liver function. *Proceedings of the National Academy of Sciences of the United States of America, 109*, 2003–2008.

Johnson, E. E., Overmeyer, J. H., Gunning, W. T., & Maltese, W. A. (2006). Gene silencing reveals a specific function of hVps34 phosphatidylinositol 3-kinase in late versus early endosomes. *Journal of Cell Science, 119*, 1219–1232.

Ketel, K., Krauss, M., Nicot, A. S., Puchkov, D., Wieffer, M., Müller, R., et al. (2016). A phosphoinositide conversion mechanism for exit from endosomes. *Nature, 529*, 408–412.

Marat, A. L., & Haucke, V. (2016). Phosphatidylinositol 3-phosphates at the interface between cell signalling and membrane traffic. *EMBO Journal, 35*, 561–579.

Martin, K. R., Xu, Y., Looyenga, B. D., Davis, R. J., Wu, C. L., Tremblay, M. L., et al. (2010). Identification of PTPσ as an autophagic phosphatase. *Journal of Cell Science, 124*, 812–819.

Morrison, J. F. (1969). Kinetics of the reversible inhibition of enzyme-catalysed reactions by tight-binding inhibitors. *Biochimica et Biophysica Acta, 185*, 269–286.

O'Farrell, F., Rusten, T. E., & Stenmark, H. (2013). Phosphoinositide 3-kinases as accelerators and brakes of autophagy. *FEBS Journal, 280*, 6322–6337.

Pasquier, B. (2016). Autophagy inhibitors. *Cellular and Molecular Life Sciences, 73*, 985–1001.

Pasquier, B., El-Ahmad, Y., Filoche-Rommé, B., Dureuil, C., Fassy, F., Abecassis, P.-Y., et al. (2015). Discovery of (2S)-8-[(3R)-3-methylmorpholin-4-yl]-1-(3-methyl-2-oxo-butyl)-2-(trifluoromethyl)-3,4-dihydro-2H-pyrimido[1,2-a]pyrimidin-6-one: A novel potent and selective inhibitor of Vps34 for the treatment of solid tumors. *Journal of Medicinal Chemistry, 58*, 376–400.

Raiborg, C., Schink, K. O., & Stenmark, H. (2013). Class III phosphatidylinositol 3-kinase and its catalytic product PtdIns3P in regulation of endocytic membrane traffic. *FEBS Journal, 280*, 2730–2742.

Ronan, B., Flamand, O., Vescovi, L., Dureuil, C., Durand, L., Fassy, F., et al. (2014). A highly potent and selective Vps34 inhibitor alters vesicle trafficking and autophagy. *Nature Chemical Biology, 10*, 1013–1019.

Rostislavleva, K., Soler, N., Ohashi, Y., Zhang, L., Pardon, E., Burke, J. E., et al. (2015). Structure and flexibility of the endosomal Vps34 complex reveals the basis of its function on membranes. *Science, 350*(6257). aac7365-1–aac7365-11.

Sasaki, T., Takasuga, S., Sasaki, J., Kofuji, S., Eguchi, S., Yamazaki, M., et al. (2009). Mammalian phosphoinositide kinases and phosphatases. *Progress in Lipid Research, 48*, 307–343.

Schu, P. V., Takegawa, K., Fry, M. J., Stack, J. H., Waterfield, M. D., & Emr, S. D. (1993). Phosphatidylinositol 3-kinase encoded by yeast VPS34 gene essential for protein sorting. *Science, 264*, 88–91.

Stenmark, H., & Aasland, R. (1999). FYVE-finger proteins—Effectors of an inositol lipid. *Journal of Cell Science, 112*, 4174–4183.

Stenmark, H., Aasland, R., & Driscoll, P. C. (2002). The phosphatidylinositol 3-phosphate-binding FYVE finger. *FEBS Letters, 513*, 77–84.

Vanhaesebroeck, B., Guillermet-Guibert, J., Graupera, M., & Bilanges, B. (2010). The emerging mechanisms of isoform-specific PI3K signaling. *Nature Reviews. Molecular Cell Biology*, *11*, 329–341.

Vergne, I., Roberts, E., Elmaoued, R. A., Tosch, V., Delgado, M. A., Proikas-Cezanne, T., et al. (2009). Control of autophagy initiation by phosphoinositide 3-phosphatase jumpy. *EMBO Journal*, *28*, 2244–2258.

Yu, X., Long, Y. C., & Shen, H. M. (2015). Differential regulatory functions of three classes of phosphatidylinositol and phosphoinositide 3-kinases in autophagy. *Autophagy*, *11*, 1711–1728.

CHAPTER TWENTY-SEVEN

Methods to Study Lysosomal AMPK Activation

C.-S. Zhang, M. Li, S.-C. Lin[1]
State Key Laboratory of Cellular Stress Biology, Innovation Center for Cell Signaling Network, School of Life Sciences, Xiamen University, Xiamen, China
[1]Corresponding author: e-mail address: linsc@xmu.edu.cn

Contents

1. Introduction 466
2. Analysis of Lysosomal Localization of AXIN/LKB1 467
   2.1 Starvation of Cultured Cells and Mice 467
   2.2 Preparation of Detergent-Resistant Membrane 469
   2.3 Preparation of Lysosomes 470
   2.4 Immunofluorescent Analysis of Lysosomal Localization of AXIN 471
   2.5 Coimmunoprecipitational Analysis of the Complex Formation Between v-ATPase–Ragulator and AXIN/LKB1–AMPK 472
3. In Vitro Reconstitution of Lysosomal AMPK Activation 473
   3.1 Purification of His-AXIN and GST-ACC1 for Reconstitution 473
   3.2 Purification of Light Organelles 475
   3.3 In Vitro Reconstitution Assay for Lysosomal Association of AXIN/LKB1 475
   3.4 Light Organelle-Based AMPK Phosphorylation Assays 476
   3.5 Light Organelle-Based ACC Phosphorylation Assays 477
Acknowledgment 477
References 477

Abstract

The AMP-activated protein kinase (AMPK) is a master regulator of metabolic homeostasis. It is activated by the upstream kinase LKB1 (liver kinase B1) when the AMP/ATP ratio is increased during starvation or heightened exercises. Based on reconstitution experiments using purified individual proteins, AMPK was demonstrated to be directly phosphorylated on its conserved residue Thr172 by LKB1, which was promoted by increased levels of AMP. However, recent studies have engendered a paradigm shift for how AMPK is activated inside the cell or animal tissues, unraveling that AXIN binds to LKB1 and tethers it to AMPK located on the surface of late endosome and lysosome (hereafter, only lysosome is discussed) in response to glucose starvation. Moreover, AXIN extends its interaction with the v-ATPase–Ragulator complex, which is paradoxically also required for activation of mTORC1 despite the fact that the two kinases AMPK and mTORC1 are inversely activated. Here, we summarize the experimental procedures

Methods in Enzymology, Volume 587
ISSN 0076-6879
http://dx.doi.org/10.1016/bs.mie.2016.09.071

of the assays for translocation of AXIN/LKB1 to the detergent-resistant lipid fractions of lysosomal membrane and the assembly of AMPK-activating complexes thereon. These methods will be useful for determining whether AMPK activation induced by various metabolic stresses or by pharmacological stimuli is mediated by the v-ATPase–Ragulator–AXIN/LKB1 axis.

1. INTRODUCTION

Levels of AMP and ADP, as a result of consuming ATP, will rise under nutrient-poor conditions or when exercise is heightened, manifesting energy stress (Hardie, Ross, & Hawley, 2012). AMP-activated protein kinase (AMPK) is not only a master sensor of, but also plays crucial roles in adaptive responses to, energy stress (Carling, Thornton, Woods, & Sanders, 2012; Hardie et al., 2012; Steinberg & Kemp, 2009). Upon activation, AMPK phosphorylates a series of substrates, increasing catabolic activities to generate more ATP, and meanwhile reducing energy consumption by switching off anabolic pathways (Hardie, 2014a, 2014b).

AMPK is comprised of a catalytic α subunit and regulatory β and γ subunits. During energy stress, the increased level of AMP binds to γ subunits and allosterically activates AMPK (Gowans, Hawley, Ross, & Hardie, 2013; Kemp, 2004; Moore, Weekes, & Hardie, 1991; Xiao et al., 2007). Importantly, AMP binding also leads to AMPK activation through promoting the upstream kinase LKB1 (liver kinase B1) to phosphorylate Thr172 directly in the α subunit of AMPK, inducing tens of fold increase of its kinase activity (Chen et al., 2009; Hawley et al., 1996, 2003; Oakhill et al., 2010; Sakamoto et al., 2005; Shaw et al., 2004, 2005; Woods et al., 2003). In addition, AMP/ADP binding to AMPK protects p-Thr172 from being dephosphorylated (Davies, Helps, Cohen, & Hardie, 1995; Gowans et al., 2013; Oakhill et al., 2011). These properties of AMPK have led to the traditional view that once AMPK is bound with AMP, AMPK is allosterically activated and becomes more easily phosphorylated by LKB1. However, it is now clear that activation of AMPK in vivo entails much more complex steps involving many more factors. Recently, through analyzing the physiological changes in the mouse liver depleted of AXIN, it was found that AXIN plays a necessary role in starvation-induced AMPK activation in that AXIN serves as a bridge linking LKB1 to AMPK in response to starvation (Zhang et al., 2013). It was subsequently found that LAMTOR1, along with other members of the Ragulator complex and the vacuolar ATPase, serve as docking sites for AXIN, allowing the formation of the AMPK-activating complex of v-ATPase–Ragulator–AXIN/LKB1–AMPK complex on the lysosome (Zhang et al., 2014). Through analysis of p-AMPK and total AMPK on the lysosomal membranes isolated from cells and mouse liver before and after starvation, we found a significant portion of AMPK is constitutively localized in lysosomes, consistent with a previous proteomic analysis of lysosomal proteins identifying the γ subunit of AMPK as a lysosomal residential protein (Chapel et al., 2013). Interestingly, the lysosomally localized AMPK is preferentially activated under glucose starvation. By applying a series of reconstruction methods, we have also demonstrated that under glucose starvation, the lysosomal v-ATPase–Ragulator complex becomes accessible to the scaffold protein AXIN in complex with LKB1, forming the v-ATPase–Ragulator–AXIN/LKB1–AMPK complex for AMPK phosphorylation by LKB1. Importantly, the starvation-induced conformational change in the lysosomal v-ATPase–Ragulator complex lowers the threshold concentration of AMP for stimulating Thr172 phosphorylation of AMPK, consistent with the increased localization of AXIN/LKB1 on lysosomes in starved cells (Zhang et al., 2014). These findings open up several new avenues and pose new challenges for studying the activation of AMPK under various physiological and pharmaceutical settings. Next, we describe detailed methodological features for analyzing lysosomal AMPK activation.

2. ANALYSIS OF LYSOSOMAL LOCALIZATION OF AXIN/LKB1

As described earlier, lysosomal translocation of AXIN in complex with LKB1 is essential for AMPK activation upon glucose starvation. In this section, the methods for starving cultured cells and mice, and the methods for directly detecting the dynamic association of AXIN/LKB1 with the lysosomal v-ATPase–Ragulator complex and measuring AMPK activation by determining p-AMPK and p-ACC are described.

2.1 Starvation of Cultured Cells and Mice

AMPK activation could be determined in both cultured cells and animal tissues. We describe them separately in the following sections. As AMPK could be artificially activated during collection of cells or dissection of animal tissues, caution must be taken to avoid elevation of basal AMPK activities. Cells should not reach 100% confluence and never be left with phosphate-buffered saline (PBS) during rinsing for too long. In addition, contamination of microplasma would also inflict an increase of basal AMPK activity. As for animal tissues, decapitation must be avoided to prevent draining of blood because AMPK will otherwise be phosphorylated by CaMKKβ in response to ischemia (McCullough et al., 2013; Rousset et al., 2015).

2.1.1 Glucose Starvation and Determination of p-AMPK in Mouse Embryonic Fibroblasts

1. Mouse embryonic fibroblasts (MEFs) are cultured to 70–80% confluence in six-well dishes containing 2 mL of complete Dulbecco's modified Eagle's medium (DMEM; Gibco, 11965) with 10% fetal bovine serum (FBS) in a 5% CO_2 incubator.
2. Aspirate the medium and rinse cells twice with 2 mL of PBS (at room temperature).
3. Add 2 mL of complete DMEM with 10% FBS and glucose-free DMEM (Gibco, 11966) with 10% FBS into the control cells and cells to be starved, respectively. Incubate cells for desired periods of time at 37°C. The FBS used is not dialyzed, the reason being that the glucose brought by 10% FBS is not sufficient to affect the activation of AMPK during glucose starvation.
4. Aspirate the medium, and quickly add 250 μL of ice-cold lysis buffer [20 m*M* Tris–HCl, pH 7.5, 150 m*M* NaCl, 1 m*M* EDTA, 1 m*M* EGTA, 1% Triton X-100, 2.5 m*M* sodium pyrophosphate, 1 m*M* β-glycerolphosphate, with protease inhibitor cocktail (Roche, 04693116001)] to each well, then place the dishes on ice.
5. Scrap and collect the lysed cells, and then sonicate the cells with a sonicator (SONICS, VCX130PB) for 3–5 s at 30% maximum power on ice. The lysates are then centrifuged at 20,000 × *g* for 10 min at 4°C.
6. Transfer 250 μL of each supernatant into a new tube containing 250 μL of 2 × SDS sample buffer, mix well, and then boil the mixture.
7. Perform immunoblotting with specific antibodies to analyze the levels of p-AMPKα-Thr172 or p-ACC-Ser79 in the variously treated cells.

2.1.2 Starvation of Mice and Determination of p-AMPK in Mouse Liver

1. Male mice (C57BL/6) at 4–6 weeks old are housed individually with free access to water and standard diet (65% carbohydrate, 11% fat, 24% protein). The light–dark cycle for mice is from 8 am to 8 pm
2. Withdraw the diet from the cage at 5 pm.
3. Sacrifice the mice at 9 am the next day by cervical dislocation or anesthesia. Liver tissues are excised and instantly frozen in liquid nitrogen.

4. To begin homogenization, thaw the frozen liver tissue on ice and add ice-cold lysis buffer (10 μL/mg liver weight) into the liver tissues, and homogenize and sonicate the lysates on ice.
5. Centrifuge the lysates at 20,000 × *g* for 10 min at 4°C and add an equal amount of 2 × SDS sample buffer to the supernatant, and then analyze p-AMPK and p-ACC as described in steps 6–7 of Section 2.1.1.

2.2 Preparation of Detergent-Resistant Membrane

It has been reported that Ragulator complex is localized on the detergent-resistant membrane (DRM) region (lipid raft) of lysosome (Bar-Peled, Schweitzer, Zoncu, & Sabatini, 2012; Nada et al., 2009; Sancak et al., 2010). The method for isolating DRM was described in Kawabuchi et al. (2000) and is depicted in Fig. 1. Through this method, we determined the dynamic association of the AXIN–LKB1 and mTORC1 with the lysosomal DRM, as well as the phosphorylation status of AMPK on the DRM before and after starvation. All procedures, unless indicated, should be carried out at 4°C or on ice.

1. Homogenize MEFs collected from five 10-cm dishes (60–80% confluence), or 0.3 g of fresh liver tissue (handled as described in steps 1–3 of Section 2.1.2) with 2 mL of Buffer A (50 m*M* Tris–HCl, pH 7.4, 150 m*M* NaCl, 1 m*M* EDTA, 5 m*M* β-mercaptoethanol, 0.25% Triton X-100 containing protease inhibitor cocktail).
2. Incubate the homogenates on rotator at 40 rpm for 1 h.

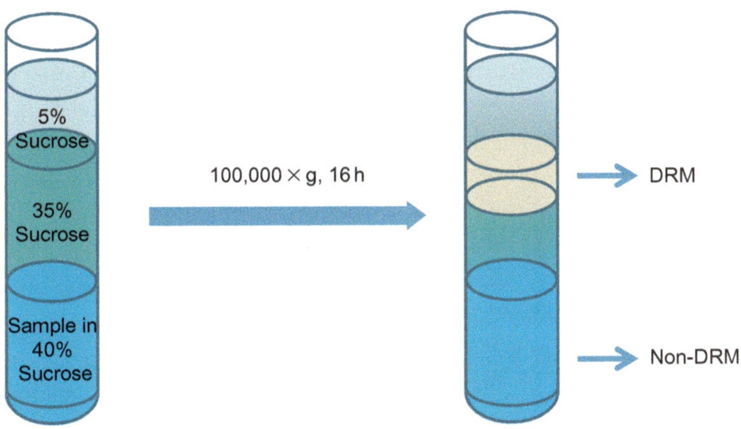

Fig. 1 Preparation of DRM by sucrose gradient centrifugation. Mouse livers are homogenized in Buffer A and mixed with an equal volume of 80% sucrose dissolved in Buffer A. After ultracentrifugation, DRM is approximately located at the top portion of the 35% sucrose cushions, and non-DRM is in the bottom fractions.

3. The homogenates are well mixed with 2 mL of 80% sucrose (dissolved in Buffer A) by gentle pipetting.
4. Load the mixture to the bottom of a 14 × 89-mm centrifuge tube (Beckman, ref. 344059) and overlay sequentially with 5 mL of 35% and 1 mL of 5% sucrose (dissolved in Buffer A).
5. Install each tube onto precooled SW41 rotor (Beckman), and then separate the mixture by ultracentrifugation at 100,000 × *g* for 16 h.
6. The gradient is collected into 10 fractions, 1 mL of each. Fractions 2 and 3 contain DRM, and fractions 7–10 (the bottom fractions) are cytosolic fractions (non-DRM).
7. The DRM proteins are dissolved with an equal volume of Buffer A supplemented with octyl β-D-glucopyranoside (ODG) (2%) and Nonidet P-40 (NP-40) (1%), and are then added with 5 × SDS sample buffer for immunoblotting.

2.3 Preparation of Lysosomes

As mentioned earlier, the lysosome is where AMPK phosphorylation occurs after AXIN in complex with LKB1 interacts with v-ATPase–Ragulator upon glucose starvation. Gorvel, Chavrier, Zerial, and Gruenberg (1991) developed a convenient method for isolating lysosomes from liver tissues. This method was further modified by Kobayashi et al. (2002) and is depicted in Fig. 2. All procedures, unless otherwise indicated, should be carried out at 4°C or on ice.

1. Some 0.15 g of each freshly excised liver tissue (handled as described in steps 1–3 of Section 2.1.2) is homogenized in 800 μL of the homogenization buffer (HB) containing a 250-m*M* sucrose, 3-m*M* imidazole, pH 7.4, and protease inhibitor cocktail.
2. Pass each homogenate through a 22-G needle attached to a 1-mL syringe six times, then spin down the homogenates at 2000 × *g* for 10 min, yielding postnuclear supernatants (PNSs).
3. Load an 11 × 60-mm centrifuge tube (Beckman, ref. 344062) sequentially with 1 mL of 40.6% sucrose (dissolved in HB), 1 mL of 35% sucrose, and 1 mL of 25% sucrose, and then the PNS.
4. The sample tubes are centrifuged for 1 h at 35,000 rpm using an SW60 rotor (Beckman).
5. The top fraction (200 μL) is collected as cytosolic fraction, and the fraction at the top of 25% sucrose cushion (100 μL) is collected as the lysosome fraction.
6. The individual fractions are added with an equal volume of 2 × SDS sample buffer for immunoblotting.

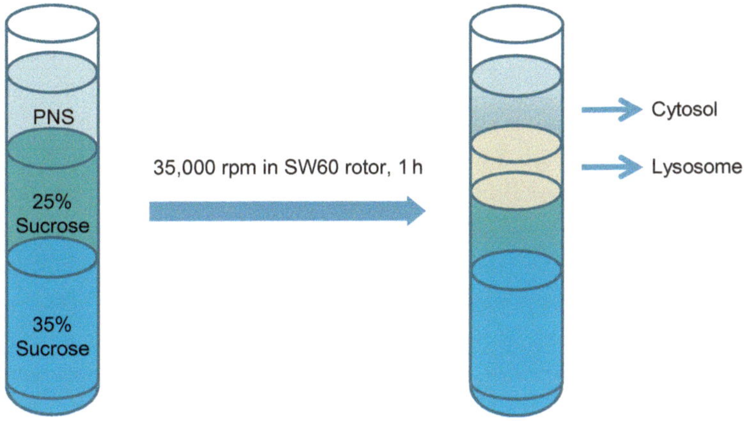

Fig. 2 Preparation of lysosomes by sucrose gradient centrifugation. Postnuclear supernatants (PNS) from mouse liver are loaded onto 25% sucrose cushion. After ultracentrifugation, lysosomes are positioned in the top portion of the 25% sucrose cushion, while the membrane-cleared cytosol remains on top.

2.4 Immunofluorescent Analysis of Lysosomal Localization of AXIN

For monitoring dynamic translocation of AXIN onto the surface of lysosome, immunofluorescence staining is performed, with LAMP2 as a lysosomal marker.

1. Cells in a different medium are placed on glass coverslips in six-well dishes and cultured to 60–80% confluence. Incubate the cells with complete DMEM or glucose-free DMEM.
2. Aspirate the medium and fix the cells with 1 mL of 4% formaldehyde (diluted in PBS) at room temperature for 20 min.
3. Rinse the cells twice with 1 mL of PBS (room temperature) and permeabilize cells with 1 mL of 0.1% Triton X-100 (diluted in PBS) for 5 min at 4°C.
4. Rinse the permeabilized cells twice with 1 mL of PBS, and incubate the cells with primary antibodies (for AXIN, Santa Cruz Biotechnology, sc-8567, diluted 1:60 in PBS; for LAMP2, Abcam, ab13524, diluted 1:120 in PBS) overnight at 4°C.
5. Rinse the cells three times with 1 mL of PBS, and then incubate the cells with Alexa-Fluor 488-conjugated antigoat secondary antibody (Molecular Probes, A11058; diluted 1:100 in PBS) and Alexa-Fluor 594-conjugated antirat secondary antibody (Molecular Probes, A21209; diluted 1:100 in PBS) for 8 h at room temperature in the dark.

6. Wash the cells four times with 1 mL of PBS, and then mount the coverslips on slides by ProLong Diamond Antifade Mountant (Molecular Probes, P36970).
7. Visualize the cells under a Zeiss Laser Scanning Microscope (LSM) 780. The samples are excited with Ar gas laser (Zeiss, laser module LGK 7812) using a 488-nm laser line for Alexa-Fluor 488 dye (green channel), and with HeNe gas laser (Zeiss, LGK 7512 PF) using a 594-nm laser line for Alexa-Fluor 594 dye (red channel).
8. Confocal microscopic pictures are taken with a 63 × oil objective using a confocal microscope (Zeiss, LSM 780). The parameters, including PMT voltage, Offset, Pinhole, and Gain are kept unchanged between each picture taken. The colocalization percentages are analyzed by using ZEN 2010 software (Zeiss).

2.5 Coimmunoprecipitational Analysis of the Complex Formation Between v-ATPase–Ragulator and AXIN/LKB1–AMPK

Coimmunoprecipitational (co-IP) analysis has proven to be a powerful approach to detect the complex formation of v-ATPase–Ragulator–AXIN/LKB1–AMPK during starvation. Cell lysis and immunoprecipitation are carried out as described in Rui et al. (2004), with some modifications. In particular, to ensure that the membrane-associated proteins such as Ragulator are adequately extracted in the lysis buffer, a series of detergents were screened and titrated. We have found ODG (Sigma, O8001) as the most adequate detergent to be included in the lysis buffer.

1. For immunoprecipitating AXIN or LAMTOR1, 4–10 dishes (15 cm) of MEFs grown to 80% confluence are collected.
2. Incubate cells with complete DMEM medium or glucose-free DMEM.
3. Aspirate the medium and quickly lyse the cells with 750-μL/dish of ice-cold ODG buffer (50 m*M* Tris–HCl, pH 8.0, 50 m*M* NaCl, 1 m*M* EDTA, 1 m*M* EGTA, 2% ODG, 5 m*M* β-mercaptoethanol, and protease inhibitor cocktail).
4. Sonicate the cells, and centrifuge the lysates at 20,000 × g for 15 min at 4°C.
5. Incubate the centrifuged supernatants with appropriate antibodies overnight on a rotator at 40 rpm inside a refrigerator. The antibody against AXIN was purchased from Santa Cruz Biotech., sc-8567 (dilution 1:50–100), and the LAMTOR1 antibody was raised with purified bacterially expressed GST-tagged fragment of aa 1–64 of human LAMTOR1 (1:100).

6. Overnight protein aggregates are precleared by centrifugation at 20,000 × g at 4°C for 10 min, and protein A/G beads (1:100 for immunoprecipitating AXIN and 1:250 for LAMTOR1) are added into the supernatant and mixed on a rotator at 40 rpm for 3 h inside a refrigerator.
7. Wash the beads with 100 times volume of ODG buffer three times at 4 C. Aspirate the ODG buffer and mix the beads with an equal volume of 2 × SDS sample buffer and boil the mixture for immunoblotting.

3. IN VITRO RECONSTITUTION OF LYSOSOMAL AMPK ACTIVATION

Although methods for the isolation of DRM and lysosomes, as described in Sections 2.2 and 2.3, allow to obtain relatively more specific membrane components, they are "less active" and are not suitable for reconstituting the binding of AXIN/LKB1 to v-ATPase–Ragulator in vitro. To obtain active membrane structures, Steinberg et al. (2010) and Zoncu et al. (2011) brilliantly developed a method to isolate crude cytosolic membranes, named *light organelles*, in relatively mild conditions and applied it to investigate mechanisms for regulation of v-ATPase and mTORC1. Here, the light organelles isolated from starved cells and cells cultured in complete medium represent "starved" and "unstarved" states, respectively, and preserve their activities for a series of in vitro reconstitution assays on AMPK regulation (Fig. 3). In this section, the methods for purification of light organelles, as well as the versatile applications of these membrane structures in monitoring the binding of AXIN/LKB1 to lysosome and activation of AMPK, are described.

3.1 Purification of His-AXIN and GST-ACC1 for Reconstitution

It has to be noted that relative to ACC1, full-length AXIN is difficult to be expressed in *E. coli* and is easily degraded during purification. Extra caution must be taken during the purification of AXIN, and each step should be performed as quickly as possible on ice. The bacteria for induction should be kept at a low optical density (0.4 at 600 nm); the induction period should not exceed 4 h; duration of sonication should be short; and the homogenates to be sonicated should not be too dense.

1. Full-length AXIN is cloned into the pET32a vector and the GST-tagged ACC1 (aa 34–100) is cloned into the pGEX4T-1 vector.

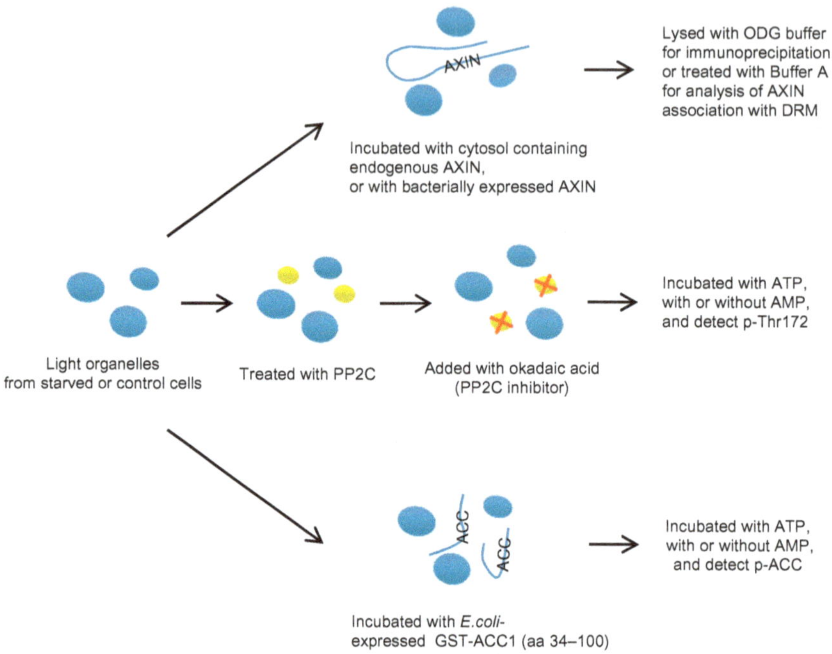

Fig. 3 Procedures for light organelle–based assays for AXIN association with the DRM and AMPK activation.

2. Transform plasmids separately into the *E. coli* strain BL21 (DE3). Single transformed colonies are picked for inoculation into culture flasks with the LB medium. Grow the bacteria at 37°C.
3. Induce the culture with 0.1 m*M* IPTG at an optical density of 0.4 at 600 nm for His-AXIN, and 1.0 at 600 nm for GST-ACC1. For His-AXIN, collect the cells after induction for 4 h at 16°C, and for GST-ACC1, collect the cells after induction for 16 h at 20°C.
4. For His-AXIN, collected cells are homogenized in an His binding buffer (50 m*M* sodium phosphate, pH 7.4, 150 m*M* NaCl, 1% Triton X-100, 5% glycerol, and 10 m*M* imidazole), and for GST-ACC1, with a GST binding buffer (PBS supplemented with 10 m*M* β-mercaptoethanol and 1% Triton X-100).
5. Sonicate the homogenates with a sonicator (SONICS, VCX750) for two rounds of 15 min each at 25% of maximum power on ice.
6. Subject the homogenates to ultracentrifugation at 150,000 × g for 30 min, followed by purification of His-AXIN with Nickel Affinity Gel (Sigma, P6611) or GST-ACC1 with Glutathione Sepharose 4 Fast Flow Gel (GE Healthcare, 17-5132).

7. Wash the Nickel Affinity Gel with 100 times the volume of His wash buffer (50 m*M* sodium phosphate, pH 7.4, 150 m*M* NaCl, and 20 m*M* imidazole); wash the Glutathione Sepharose 4B column with 100 times the volume of PBS.
8. His-AXIN is eluted from the resin by His elution buffer (50 m*M* sodium phosphate, pH 7.4, 150 m*M* NaCl, and 250 m*M* imidazole), and GST-ACC1 is eluted by GST elution buffer (50 m*M* Tris–HCl, pH 8.0, and 10 m*M* reduced glutathione).
9. Concentrate proteins to approximately 3 mg/mL by ultrafiltration (Millipore, UFC905096).
10. Further purify the proteins by using gel filtration columns (GE Healthcare, Superdex-200) in a buffer containing 50 m*M* Tris–HCl, pH 7.4, 150 m*M* NaCl, 5% glycerol, and 1 m*M* DTT.

3.2 Purification of Light Organelles

1. Incubate cells with a complete or glucose-free medium to 80% confluence.
2. Scrap and spin down cells at 200 × g at room temperature, and then resuspend them in 500 μL per 10-cm dish of fractionation buffer [50 m*M* KCl, 90 m*M* K-gluconate, 1 m*M* EGTA, 5 m*M* $MgCl_2$, 50 m*M* sucrose, 20 m*M* HEPES, pH 7.4, supplemented with 2.5 m*M* ATP, amino acids (Gibco, Cat. #11130-077) and protease inhibitor cocktail] at room temperature. The fractionation buffer is also suitable for AMPK/ACC phosphorylation assay (see Sections 3.4 and 3.5), as it contains magnesium ions.
3. Mechanically break the cells to obtain PNS, as described in step 2 of Section 2.3.
4. The PNS is then spun at maximal speed for 15 min in a tabletop centrifuge (Eppendorf, 5417R) at 4°C. The pellets are light organelles, and the supernatants are membrane-cleared cytosolic fractions.

3.3 In Vitro Reconstitution Assay for Lysosomal Association of AXIN/LKB1

To recapitulate starvation-induced translocation of AXIN onto the surface of lysosomes, we incubated light organelles purified from starved or unstarved cells with AXIN in the membrane-cleared cytosols or purified bacterially expressed AXIN (Fig. 3). We found that after glucose starvation or v-ATPase inhibition, the v-ATPase–Ragulator complex becomes more

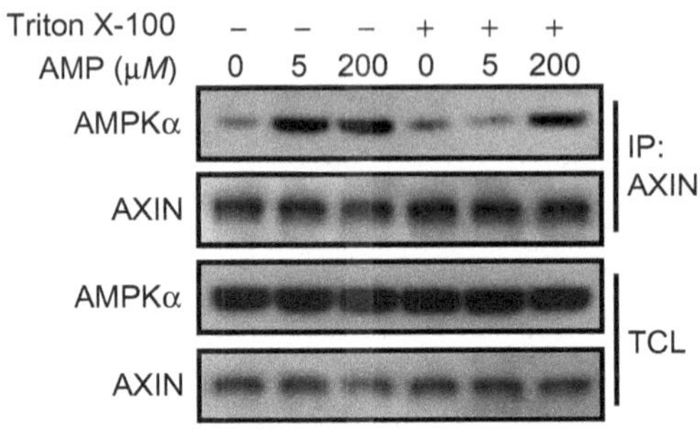

Fig. 4 Intact light organelles lower threshold concentrations of AMP to promoting AXIN to bind lysosomal AMPK.

accessible to AXIN–LKB1. Importantly, it was found that as little as 5 μ*M* AMP was sufficient to promote AXIN binding to AMPK using light organelles purified from starved cells, while in membrane-free systems using Triton X-100-treated light organelles (Fig. 4) or purified bacterially expressed AXIN and AMPK (Zhang et al., 2014), 200 μ*M* of AMP is required for this process.

1. Each reconstitution reaction requires light organelles from two 15-cm dishes of MEFs.
2. Resuspend the light organelles (isolated as described in Section 3.2) with 500 μL of cytosolic fractions purified from four 15-cm dishes of MEFs, or 500 μL of fractionation buffer containing 1 μg of bacterially expressed His-AXIN at 4°C.
3. Incubate the cytosolic fractions or His-AXIN with light organelles at 37°C, 600 rpm in a thermomixer for 25 min. The mixture can be used for:
   - **(a)** Preparation of DRM: Treat the mixture with 1.5 mL of ice-cold Buffer A and leave the tubes on a rotator at 40 rpm for 1 h at 4°C, and then prepare and analyze DRM fractions by following steps 3–7 of Section 2.2.
   - **(b)** Immunoprecipitation: Lyse the mixture with 500 μL of ODG buffer at 4°C, and then analyze the interaction between LAMTOR1 and AXIN as described in steps 4–7 of Section 2.5.

3.4 Light Organelle-Based AMPK Phosphorylation Assays

The light organelles from starved cells allow phosphorylation of AMPK by LKB1 bound to AXIN (Fig. 3).

1. For each reaction, one 10-cm dish of light organelle cells are resuspended in 100 μL of ice-cold fractionation buffer containing 5 μg of PP2C.
2. Incubate the mixtures in a thermomixer at 600 rpm for 30 min at 32°C to dephosphorylate AMPK.
3. Terminate the reaction by adding okadaic acid (10 ng/mL final concentration) to the mixture.
4. Add various concentrations of AMP (0–200 μ*M* final concentrations) to the mixture containing ATP (200 μ*M* final concentration), and incubate the mixture at 30°C for 20 min. The reaction is terminated by 5 × SDS sample buffer for immunoblotting. Of note, ATP is included to better reflect cellular situations of the ratio of AMP over ATP.

3.5 Light Organelle-Based ACC Phosphorylation Assays

ACC phosphorylation (S79 for ACC1 and S212 for ACC2) is primarily catalyzed by AMPK (Gowans et al., 2013). In vitro phosphorylation assays using ACC as the substrate for AMPK is thus referred to as the "gold standard" for measuring the activity of AMPK. Light organelles are applicable for determining AMPK activities toward ACC after incubation with variously treated cytosols that contain AXIN in complex with LKB1 (Fig. 3).

1. Light organelles with two 15-cm dishes worth of MEFs for each reaction are resuspended with 50 μL of fractionation buffer containing 1 μg of bacterially expressed GST-ACC1 (aa 34–100) as substrate at 4°C.
2. Add ATP (200 μ*M* final concentration), appropriate concentrations of AMP, or both and incubate the mixtures in a thermomixer at 800 rpm for 30 min at 30°C. The reactions are terminated by addition of 5 × SDS and are analyzed by immunoblotting.

ACKNOWLEDGMENT

This work was supported by Grants from National Natural Science Foundation of China (#31130016, #31430094, and #31370744).

REFERENCES

Bar-Peled, L., Schweitzer, L. D., Zoncu, R., & Sabatini, D. M. (2012). Ragulator is a GEF for the rag GTPases that signal amino acid levels to mTORC1. *Cell*, *150*(6), 1196–1208. http://dx.doi.org/10.1016/j.cell.2012.07.032.

Carling, D., Thornton, C., Woods, A., & Sanders, M. J. (2012). AMP-activated protein kinase: New regulation, new roles? *The Biochemical Journal*, *445*(1), 11–27. http://dx.doi.org/10.1042/BJ20120546.

Chapel, A., Kieffer-Jaquinod, S., Sagne, C., Verdon, Q., Ivaldi, C., Mellal, M., ... Journet, A. (2013). An extended proteome map of the lysosomal membrane reveals novel potential transporters. *Molecular & Cellular Proteomics*, *12*(6), 1572–1588. http://dx.doi.org/10.1074/mcp.M112.021980.

Chen, L., Jiao, Z. H., Zheng, L. S., Zhang, Y. Y., Xie, S. T., Wang, Z. X., & Wu, J. W. (2009). Structural insight into the autoinhibition mechanism of AMP-activated protein kinase. *Nature*, *459*(7250), 1146–1149. http://dx.doi.org/10.1038/nature08075.

Davies, S. P., Helps, N. R., Cohen, P. T., & Hardie, D. G. (1995). 5'-AMP inhibits dephosphorylation, as well as promoting phosphorylation, of the AMP-activated protein kinase. Studies using bacterially expressed human protein phosphatase-2C alpha and native bovine protein phosphatase-2AC. *FEBS Letters*, *377*(3), 421–425. http://dx.doi.org/10.1016/0014-5793(95)01368-7.

Gorvel, J. P., Chavrier, P., Zerial, M., & Gruenberg, J. (1991). rab5 controls early endosome fusion in vitro. *Cell*, *64*(5), 915–925.

Gowans, G. J., Hawley, S. A., Ross, F. A., & Hardie, D. G. (2013). AMP is a true physiological regulator of AMP-activated protein kinase by both allosteric activation and enhancing net phosphorylation. *Cell Metabolism*, *18*(4), 556–566. http://dx.doi.org/10.1016/j.cmet.2013.08.019.

Hardie, D. G. (2014a). AMP-activated protein kinase: Maintaining energy homeostasis at the cellular and whole-body levels. *Annual Review of Nutrition*, *34*, 31–55. http://dx.doi.org/10.1146/annurev-nutr-071812-161148.

Hardie, D. G. (2014b). AMPK—Sensing energy while talking to other signaling pathways. *Cell Metabolism*, *20*(6), 939–952. http://dx.doi.org/10.1016/j.cmet.2014.09.013.

Hardie, D. G., Ross, F. A., & Hawley, S. A. (2012). AMPK: A nutrient and energy sensor that maintains energy homeostasis. *Nature Reviews. Molecular Cell Biology*, *13*(4), 251–262. http://dx.doi.org/10.1038/nrm3311.

Hawley, S. A., Boudeau, J., Reid, J. L., Mustard, K. J., Udd, L., Makela, T. P., … Hardie, D. G. (2003). Complexes between the LKB1 tumor suppressor, STRAD alpha/beta and MO25 alpha/beta are upstream kinases in the AMP-activated protein kinase cascade. *Journal of Biology*, *2*(4), 28. http://dx.doi.org/10.1186/1475-4924-2-28.

Hawley, S. A., Davison, M., Woods, A., Davies, S. P., Beri, R. K., Carling, D., & Hardie, D. G. (1996). Characterization of the AMP-activated protein kinase kinase from rat liver and identification of threonine 172 as the major site at which it phosphorylates AMP-activated protein kinase. *The Journal of Biological Chemistry*, *271*(44), 27879–27887.

Kawabuchi, M., Satomi, Y., Takao, T., Shimonishi, Y., Nada, S., Nagai, K., … Okada, M. (2000). Transmembrane phosphoprotein Cbp regulates the activities of Src-family tyrosine kinases. *Nature*, *404*(6781), 999–1003. http://dx.doi.org/10.1038/35010121.

Kemp, B. E. (2004). Bateman domains and adenosine derivatives form a binding contract. *The Journal of Clinical Investigation*, *113*(2), 182–184. http://dx.doi.org/10.1172/JCI20846.

Kobayashi, T., Beuchat, M. H., Chevallier, J., Makino, A., Mayran, N., Escola, J. M., … Gruenberg, J. (2002). Separation and characterization of late endosomal membrane domains. *The Journal of Biological Chemistry*, *277*(35), 32157–32164. http://dx.doi.org/10.1074/jbc.M202838200.

McCullough, L. D., Tarabishy, S., Liu, L., Benashski, S., Xu, Y., Ribar, T., … Li, J. (2013). Inhibition of calcium/calmodulin-dependent protein kinase kinase beta and calcium/calmodulin-dependent protein kinase IV is detrimental in cerebral ischemia. *Stroke*, *44*(9), 2559–2566. http://dx.doi.org/10.1161/STROKEAHA.113.001030.

Moore, F., Weekes, J., & Hardie, D. G. (1991). Evidence that AMP triggers phosphorylation as well as direct allosteric activation of rat liver AMP-activated protein kinase. A sensitive mechanism to protect the cell against ATP depletion. *European Journal of Biochemistry*, *199*(3), 691–697.

Nada, S., Hondo, A., Kasai, A., Koike, M., Saito, K., Uchiyama, Y., & Okada, M. (2009). The novel lipid raft adaptor p18 controls endosome dynamics by anchoring the MEK-ERK pathway to late endosomes. *The EMBO Journal*, *28*(5), 477–489. http://dx.doi.org/10.1038/emboj.2008.308.

Oakhill, J. S., Chen, Z. P., Scott, J. W., Steel, R., Castelli, L. A., Ling, N., … Kemp, B. E. (2010). beta-Subunit myristoylation is the gatekeeper for initiating metabolic stress sensing by AMP-activated protein kinase (AMPK). *Proceedings of the National Academy of Sciences of the United States of America*, *107*(45), 19237–19241. http://dx.doi.org/10.1073/pnas.1009705107.

Oakhill, J. S., Steel, R., Chen, Z. P., Scott, J. W., Ling, N., Tam, S., & Kemp, B. E. (2011). AMPK is a direct adenylate charge-regulated protein kinase. *Science*, *332*(6036), 1433–1435. http://dx.doi.org/10.1126/science.1200094.

Rousset, C. I., Leiper, F. C., Kichev, A., Gressens, P., Carling, D., Hagberg, H., & Thornton, C. (2015). A dual role for AMP-activated protein kinase (AMPK) during neonatal hypoxic-ischaemic brain injury in mice. *Journal of Neurochemistry*, *133*(2), 242–252. http://dx.doi.org/10.1111/jnc.13034.

Rui, Y., Xu, Z., Lin, S., Li, Q., Rui, H., Luo, W., … Lin, S. C. (2004). Axin stimulates p53 functions by activation of HIPK2 kinase through multimeric complex formation. *The EMBO Journal*, *23*(23), 4583–4594. http://dx.doi.org/10.1038/sj.emboj.7600475.

Sakamoto, K., McCarthy, A., Smith, D., Green, K. A., Grahame Hardie, D., Ashworth, A., & Alessi, D. R. (2005). Deficiency of LKB1 in skeletal muscle prevents AMPK activation and glucose uptake during contraction. *The EMBO Journal*, *24*(10), 1810–1820. http://dx.doi.org/10.1038/sj.emboj.7600667.

Sancak, Y., Bar-Peled, L., Zoncu, R., Markhard, A. L., Nada, S., & Sabatini, D. M. (2010). Ragulator-Rag complex targets mTORC1 to the lysosomal surface and is necessary for its activation by amino acids. *Cell*, *141*(2), 290–303. http://dx.doi.org/10.1016/j.cell.2010.02.024.

Shaw, R. J., Kosmatka, M., Bardeesy, N., Hurley, R. L., Witters, L. A., DePinho, R. A., & Cantley, L. C. (2004). The tumor suppressor LKB1 kinase directly activates AMP-activated kinase and regulates apoptosis in response to energy stress. *Proceedings of the National Academy of Sciences of the United States of America*, *101*(10), 3329–3335. http://dx.doi.org/10.1073/pnas.0308061100.

Shaw, R. J., Lamia, K. A., Vasquez, D., Koo, S. H., Bardeesy, N., Depinho, R. A., … Cantley, L. C. (2005). The kinase LKB1 mediates glucose homeostasis in liver and therapeutic effects of metformin. *Science*, *310*(5754), 1642–1646. http://dx.doi.org/10.1126/science.1120781.

Steinberg, B. E., Huynh, K. K., Brodovitch, A., Jabs, S., Stauber, T., Jentsch, T. J., & Grinstein, S. (2010). A cation counterflux supports lysosomal acidification. *The Journal of Cell Biology*, *189*(7), 1171–1186. http://dx.doi.org/10.1083/jcb.200911083.

Steinberg, G. R., & Kemp, B. E. (2009). AMPK in health and disease. *Physiological Reviews*, *89*(3), 1025–1078. http://dx.doi.org/10.1152/physrev.00011.2008.

Woods, A., Johnstone, S. R., Dickerson, K., Leiper, F. C., Fryer, L. G. D., Neumann, D., … Carling, D. (2003). LKB1 is the upstream kinase in the AMP-activated protein kinase cascade. *Current Biology*, *13*(22), 2004–2008. http://dx.doi.org/10.1016/j.cub.2003.10.031.

Xiao, B., Heath, R., Saiu, P., Leiper, F. C., Leone, P., Jing, C., … Gamblin, S. J. (2007). Structural basis for AMP binding to mammalian AMP-activated protein kinase. *Nature*, *449*(7161), 496–500. http://dx.doi.org/10.1038/nature06161.

Zhang, Y. L., Guo, H., Zhang, C. S., Lin, S. Y., Yin, Z., Peng, Y., … Lin, S. C. (2013). AMP as a low-energy charge signal autonomously initiates assembly of AXIN-AMPK-LKB1 complex for AMPK activation. *Cell Metabolism*, *18*(4), 546–555. http://dx.doi.org/10.1016/j.cmet.2013.09.005.

Zhang, C. S., Jiang, B., Li, M., Zhu, M., Peng, Y., Zhang, Y. L., … Lin, S. C. (2014). The lysosomal v-ATPase-Ragulator complex is a common activator for AMPK and mTORC1, acting as a switch between catabolism and anabolism. *Cell Metabolism*, *20*(3), 526–540. http://dx.doi.org/10.1016/j.cmet.2014.06.014.

Zoncu, R., Bar-Peled, L., Efeyan, A., Wang, S., Sancak, Y., & Sabatini, D. M. (2011). mTORC1 senses lysosomal amino acids through an inside-out mechanism that requires the vacuolar H(+)-ATPase. *Science*, *334*(6056), 678–683. http://dx.doi.org/10.1126/science.1207056.

CHAPTER TWENTY-EIGHT

Allosteric Modulation of AMPK Enzymatic Activity: In Vitro Characterization

J. Ward*,[1], A.R. Reyes*, R.G. Kurumbail[†]
*Cardiovascular and Metabolic Diseases Research Unit, Pfizer Worldwide Research and Development, Cambridge, MA, United States
[†]Worldwide Medicinal Chemistry, Pfizer Worldwide Research and Development, Groton, CT, United States
[1]Corresponding author: e-mail address: jessica.ward2@pfizer.com

Contents

1. Introduction 482
2. Notes About Key Reagents 489
   2.1 Buffer/Solution Conditions 489
   2.2 Sources of AMPK Protein 489
   2.3 Peptide Substrates 490
   2.4 Protein Phosphatase 2a 491
   2.5 LKB1 491
3. Assays for Measuring Allosteric Activation of AMPK 491
   3.1 ^{33}P-ATP Filter Assay 492
   3.2 Nonradioactive Assay Formats 495
4. Steady-State Kinetic Analysis of AMPK Activators 496
5. Assays for Monitoring Phosphorylation at Thr172 of the α-Subunit 498
   5.1 Protection Assay 498
   5.2 Phosphorylation Modulation Assay 500
6. Activation–Protection Assay 501
7. Summary 503
References 503

Abstract

AMP-activated protein kinase (AMPK) is a heterotrimeric serine/threonine protein kinase found in nearly all eukaryotes that functions as a master energy sensor in cells. During times of cell stress and changes in the AMP/ATP ratio, AMPK becomes activated and phosphorylates a multitude of protein substrates involved in various cellular processes such as metabolism, cell growth and autophagy. The endogenous ligand AMP is known to bind to the γ-subunit and activates the enzyme via three distinct mechanisms (1) enhancing phosphorylation by upstream kinases of Thr172 in the activation loop (a site critical for AMPK activity), (2) protecting Thr172 from dephosphorylation by phosphatases, and (3) allosteric activation of the kinase activity. Given the important regulatory

Methods in Enzymology, Volume 587
ISSN 0076-6879
http://dx.doi.org/10.1016/bs.mie.2016.10.010

role for AMPK in various cellular processes and the multiple known modes of activation, there is great interest in identifying small-molecule activators of this kinase and a need for assays to identify and characterize compounds. Here we describe several assay formats that have been used for identifying and characterizing small-molecule AMPK activators.

1. INTRODUCTION

One of the fundamental tenets of life is the conservation of cellular and organismal energy. All species have evolved specialized molecular machineries to balance the consumption of ATP with its synthesis. This metabolic homeostasis is orchestrated through the coordinated action of a number of signaling pathways in cells. Of these, the one mediated by AMP-activated protein kinase (AMPK) has a pivotal role (Hardie, 2016). It is an evolutionarily conserved enzyme found in essentially all eukaryotes that functions as an extremely sensitive fuel gauge which can detect even small changes in cellular energy levels and rapidly respond to rebalance it. Nature has bestowed a molecular sensor on AMPK to detect the overall AMP/ATP or ADP/ATP ratios in cells. Whenever the adenine nucleotide ratios drop under conditions of metabolic stress, AMPK gets activated which in turn switches on ATP-generating catabolic processes while simultaneously shutting off ATP-consuming anabolic processes. This is accomplished by direct phosphorylation of a plethora of metabolic enzymes and other proteins as well as long-term adaptive changes through transcriptional regulation. The latter is achieved through direct or indirect phosphorylation of a number of transcriptional activators and coactivators such as SREBP1, HDACs, p300, PGC1α, and FOXO (Carling & Viollet, 2015; Hardie, 2015; Mounier, Theret, Lantier, Foretz, & Viollet, 2015; Neumann et al., 2007; Oakhill, Scott, & Kemp, 2012).

In addition to its critical role in metabolism, AMPK also regulates cell growth through its action on mTOR (mammalian target of rapamycin) (Mihaylova & Shaw, 2011). In response to growth-inducing stimuli and nutrient availability, mTOR coordinates a diverse range of anabolic processes such as protein synthesis, cell growth, and proliferation (Cargnello, Tcherkezian, & Roux, 2015). mTOR exists as part of two large complexes, mTORC1 and mTORC2. Of these, mTORC1 (~600 kDa) serves as a critical node that integrates signaling from insulin, growth factors, nutrients, normoxia, cytokines, and Wnt pathway to promote translation and protein synthesis (Tan & Miyamoto, 2016). Under conditions of nutrient deprivation or starvation, AMPK suppresses growth by direct phosphorylation of TSC2 (tuberous sclerosis complex 2) or raptor (regulatory-associated protein of target of rapamycin), two well-known regulators of mTOR (Alers, Loffler, Wesselborg, & Stork, 2012). AMPK also promotes cell migration and differentiation by control of cell polarity and cytoskeletal dynamics (Mihaylova & Shaw, 2011).

To conserve nutrients and energy under conditions of stress or starvation, eukaryotic cells undergo autophagy which allows them to recycle cytosolic components or organelles (e.g., mitochondria) (Apel, Zentgraf, Büchler, & Herr, 2009). This is an evolutionarily conserved process that exists from yeast to mammals. In mammals, autophagy is coordinated through the action of three principal components: mTOR, ULK1/2, and AMPK (Alers et al., 2012; Kim, Kundu, Viollet, & Guan, 2011; Shang & Wang, 2011). ULK-1 and -2 are the mammalian orthologues of Atg1 which have been well recognized as the initiators of autophagy in yeast along with Atg13 and Atg17 (Bach, Larance, James, & Ramm, 2011; Chan, Kir, & Tooze, 2007; Chen et al., 2014). Like many other signaling molecules, ULK-1 and -2 can be activated or inhibited through phosphorylation at distinct sites. Unlike in yeast, ULK-1 and -2 exist in a stable complex with Atg13, Atg101, and FIP200 under fed as well as starvation conditions (Behrends, Sowa, Gygi, & Wade Harper, 2010; Ganley et al., 2009; Hara & Mizushima, 2009; Hosokawa et al., 2009; Jung et al., 2009). When sufficient nutrients are available, mTORC1 phosphorylates ULK1/2 and Atg13 thereby suppressing the kinase activity of ULK1/2 and preventing autophagy. Upon starvation, mTORC1-dependent phosphorylation sites on ULK1/2 are rapidly dephosphorylated which promotes autophosphorylation of ULK1/2 and subsequent phosphorylation of Atg13 and FIP200. This triggers translocation of the entire complex to the preautophagosomal structure, the initial event in the autophagy process that ultimately results in recycling of critical nutrients and metabolites (Chan et al., 2007; Chang & Neufeld, 2009; Ganley et al., 2009; Hara & Mizushima, 2009; Hara et al., 2008; Hosokawa et al., 2009; Jung et al., 2009). Through inhibition of mTOR activity via phosphorylation of TSC2 and raptor, AMPK plays a critical regulatory role in induction of autophagy. In addition, AMPK also directly initiates autophagy via phosphorylation of activating sites on ULK-1 and -2 (Bach et al., 2011; Dorsey et al., 2009; Egan et al., 2011; Kim et al., 2011; Shang et al., 2011). In turn, it has been shown that ULK-1 is involved in additional feedback regulation of autophagy by direct phosphorylation of both

AMPK and mTORC1 (Chang & Neufeld, 2009; Ganley et al., 2009; Hosokawa et al., 2009; Jung et al., 2009; Kim et al., 2011; Löffler et al., 2011). Thus autophagy is a tightly regulated process with several checks and balances in place to ensure that this rescue mechanism is properly utilized to conserve nutrients.

AMPK is a heterotrimeric serine/threonine kinase that consists of a catalytic α subunit that is tightly associated with two regulatory subunits, β and γ. Multiple isoforms of each of the subunits are present in mammals—two α subunits (α1 and α2), two β subunits (β1 and β2), and three γ subunits (γ1, γ2, and γ3). These can be mixed and matched in different combinations to generate 12 possible AMPK isoforms which are differentially expressed in tissues and species. The protein kinase module resides on the N-terminal portion of the α subunit while its C-terminal portion forms a regulatory anchor along with β and γ subunits (Calabrese et al., 2014; Li et al., 2015; Xiao et al., 2013). These two ends of the α subunit are connected by a flexible linker that allows the catalytic activity of AMPK to respond to changes in adenine nucleotide levels (Chen et al., 2013; Xiao et al., 2011). The γ subunit of AMPK harbors three nucleotide-binding sites which are variably occupied by AMP, ADP, or ATP depending on their relative concentrations (Chen et al., 2012; Xiao et al., 2007). AMP is known to regulate AMPK by three distinct mechanisms: (1) it promotes phosphorylation of AMPK on a critical threonine residue (Thr174 on α1 and Thr172 on α2; usually referred to as Thr172 in the literature and here onward in the current document) on its activation loop by upstream kinases such as LKB1 or calmodulin-dependent protein kinase kinase 2 (CaMKK2); (2) protection of pThr172 from dephosphorylation by phosphatases; (3) allosteric activation (Dale, Wilson, Edelman, & Grahame Hardie, 1995; Gowans, Hawley, Ross, & Hardie, 2013; Hawley et al., 2003, 1996, 2005; Hurley et al., 2005; Oakhill et al., 2010, 2012, 2011; Shaw et al., 2004, 2005; Suter et al., 2006; Warden et al., 2001; Woods, Johnstone, et al., 2003; Woods, Vertommen, et al., 2003; Woods et al., 2005). Some of these functions can also be carried out by ADP although there has been conflicting reports regarding this (Gowans et al., 2013; Oakhill et al., 2012; Ross, Jensen, and Hardie, 2016; Xiao et al., 2011). Compared to ADP, AMP appears to be ~10× more potent in protecting AMPK from dephosphorylation (Gowans et al., 2013). In contrast, ATP antagonizes the effects of AMP and ADP through binding to some of the same sites on the γ subunits. Among the three nucleotides, AMP alone is capable of allosteric activation of AMPK which could be as much as 10 × even at physiological ATP concentrations and might be an important contributor for the fine tuning of AMPK activity in vivo (Gowans et al., 2013). Similarly, only AMP appears to promote phosphorylation of AMPK by the upstream kinases even though conflicting data have been reported by different labs on this subject (Gowans et al., 2013; Oakhill et al., 2011).

Activation of AMPK is triggered when cells undergo metabolic stress and suffer energy depletion such as during exercise, hypoxia conditions, and rapid proliferation (Hardie, 2015). This occurs via elevation of the cellular AMP/ATP or ADP/ATP ratio which allows AMP or ADP to bind to the γ subunit of AMPK and elicit appropriate conformational changes in the enzyme. AMPK that is unphosphorylated on its activation loop possesses minimal catalytic activity toward its substrates. Upon phosphorylation by LKB1 or CaMKK2 on Thr172 of the activation loop of AMPK, its specific activity increases by 500- to 1000-fold (Suter et al., 2006). Recently, it was reported that AMPK that is not phosphorylated on Thr172 of the α subunit can be synergistically activated by the synthetic activator, A769662 and AMP if Ser 108 of the β subunit is phosphorylated (Scott et al., 2014). It is not clear, however, whether the observed activity was due to the presence of small amounts of phosphorylated AMPK as a contaminant that was not detectable by pThr172 antibody or mass spec. In our hands, recombinant AMPK produced from *E. coli* has very little catalytic activity prior to activation by upstream kinases. This has prevented us from systematically exploring the effect of small-molecule activators on nonphosphorylated AMPK. Normally, phosphorylated AMPK is rapidly dephosphorylated by phosphatases to keep its activity in check. Upon binding of AMP and ADP, conformational changes occur via the intervening regulatory segments of the α subunit known as α-RIMs that leads to restricted access of pThr172 to phosphatases (Chen et al., 2013; Xiao et al., 2011). This protection from dephosphorylation allows AMPK to retain its high activity state which leads to phosphorylation of downstream substrates. One of the earliest reported substrates of AMPK is acetyl-CoA carboxylase (ACC) which converts acetyl-CoA to malonyl-CoA which is further utilized for the biosynthesis of fatty acids (Carling, Clarke, Zammit, & Hardie, 1989). There are two known isoforms of ACC, ACC1, and ACC2 which have distinct roles in fatty acid synthesis and oxidation, respectively. AMPK phosphorylates Ser79 of ACC1 and Ser212 (mice) or Ser222 (human) of ACC2 which leads to their downregulation (Fullerton et al., 2013). A 15 amino acid peptide known as the SAMS peptide corresponding to

His73 to Lys85 of ACC1 with mutation of Ser77 to Ala and incorporation of two additional arginine residues at the C-terminus was designed as a tool to investigate the function of AMPK (Dale et al., 1995). This is one of the most commonly used peptide substrates to monitor AMPK activity in vitro. Many studies rely on evaluation of pThr172 levels as a measure of AMPK activation in cells or tissues. However, this may not reflect a complete picture since allosteric effects of compounds can be significant and not captured by estimation of pThr172 levels. Monitoring of phosphorylation of ACC or some other AMPK substrates, in contrast, might be a better strategy.

While AMP and ADP are the endogenous small-molecule activators of AMPK, several other direct and indirect activators of AMPK are known (Fig. 1) (Hardie, 2016). AICAR (5-aminoimidazole-4-carboxamide-1-β-D-ribofuranoside) is a cell-permeable prodrug of ZMP (AICAR monophosphate) which functions as an AMP-mimetic and activates AMPK by binding to its γ subunit. The well-known diabetes medication metformin has been reported to be an indirect activator of AMPK (Foretz, Guigas, Bertrand, Pollak, & Viollet, 2014; Fryer, Parbu-Patel, & Carling, 2002; Fullerton et al., 2013; Hawley, Gadalla, Olsen, & Hardie, 2002;

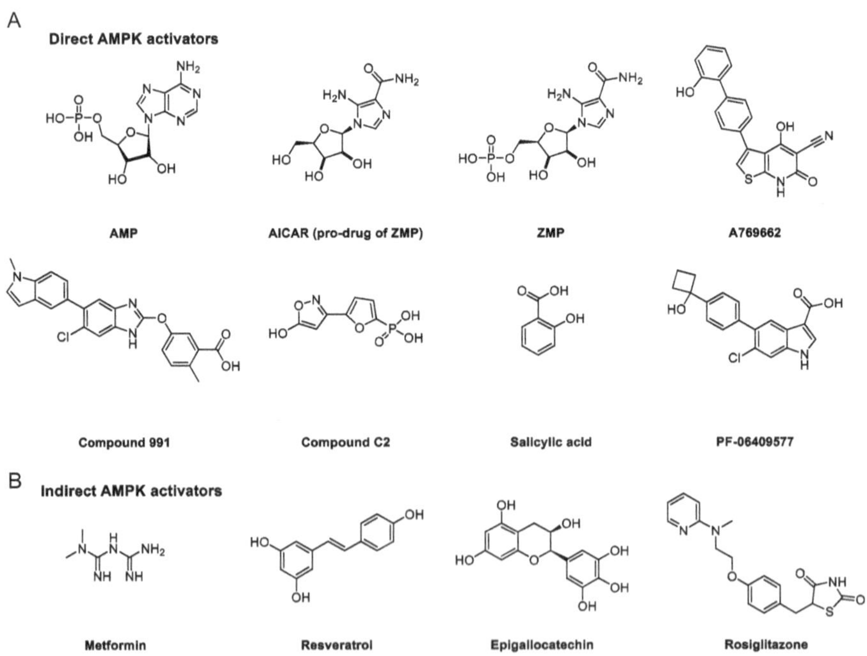

Fig. 1 Chemical structures of direct and indirect AMPK activators.

Sliwinska & Drzewoski, 2015; Zhou et al., 2001). It has been shown to be an inhibitor of Complex I of the mitochondrial respiratory chain which results in inhibition of ATP production and an increase in AMP/ATP ratio (Owen, Doran, & Halestrap, 2000). Metformin provides glucose-lowering benefits primarily by suppressing hepatic gluconeogenesis (Foretz et al., 2010; Foretz & Viollet, 2015; Fullerton et al., 2013; Sliwinska & Drzewoski, 2015; Stephenne et al., 2011; Viollet et al., 2012). While other AMPK-independent mechanisms of action have been reported for metformin, there are no doubts that it is a legitimate AMPK activator in cells at the pharmacological doses that are used clinically (Fullerton et al., 2013; Madiraju et al., 2014; Madsen, Bozickovic, Bjune, Mellgren, & Sagen, 2015; Miller et al., 2013). Several natural products have been reported to be activators of AMPK. The list includes resveratrol from red grapes, epigallocatechin from green tea, flavonoids such as quercetin and genistein from fruits and vegetables and berberine which has been used in traditional Chinese medicine for centuries (Sharma & Kumar, 2016). The vast majority of these function as indirect AMPK activators by elevating the cellular AMP/ATP ratio (Hawley et al., 2010). Over the last few years, several direct synthetic AMPK activators have been reported. Cool et al. from Abbott labs reported the discovery of A769662 that activates AMPK in a completely different manner compared to AMP (Figs. 1 and 2) (Cool et al., 2006; Landgraf et al., 2013). Since then, the pharmaceutical industry has been actively pursuing identification of novel direct AMPK activators as evidenced by the rich patent literature on this topic (Kim, Shin, & Ha, 2015; Kim, Yang, Kim, Kim, & Ha, 2016; Miglianico, Nicolaes, & Neumann, 2015; Rana, Blowers, & Natarajan, 2015). Small molecule synthetic activators based on benzimidazole (e.g., compound 991 or Ex 229), indole (PF-06409577), or indazole scaffold have been described that also activate AMPK by a mechanism similar to A769662 (Cameron et al., 2016; Xiao et al., 2013). Phosphonic derivatives of furanyl isoxazole-based AMPK activators have been reported that function as AMP mimetics and function via the known nucleotide-binding sites on the γ subunit (Gomez-Galeno et al., 2010; Langendorf et al., 2016). Given the heterotrimeric organization and the conformational flexibility, it is not surprising that AMPK can be activated through multiple binding sites some of which are yet to be discovered.

Recent crystallographic studies from multiple laboratories have revealed that A769662 or its chloro analog bind at a novel allosteric site called "ADaM" (allosteric drug and metabolite) site at the interface of the kinase module of the α subunit and the carbohydrate binding module of the β

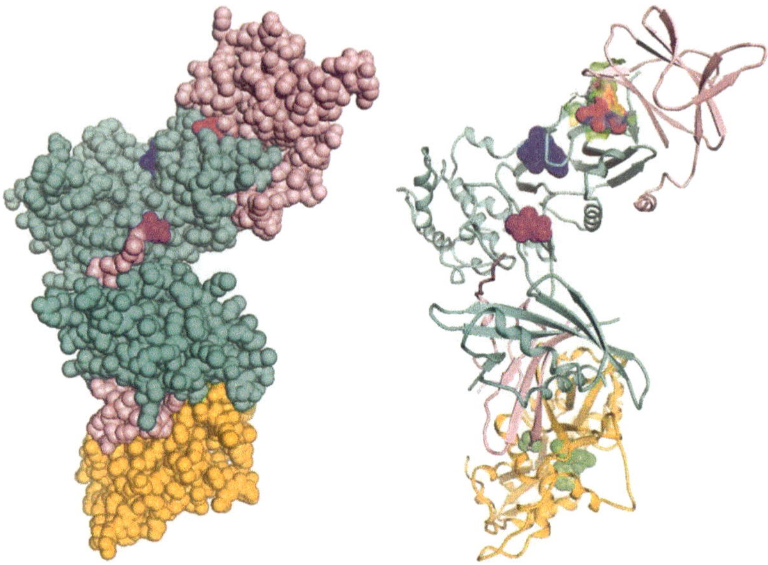

Fig. 2 Crystal structure of an analog of A769662 bound to AMPK α1β1γ1. The panel on the *left* is rendered in space-filling representation while it is shown in ribbon representation on the *right*. α Subunit is in *cyan* color, β subunit in *purple*, and the γ subunit in *orange*. Chloro-A769662 is shown in *pink* at the *top* at the intersection of the kinase module of the α subunit and the glycogen binding domain of the β subunit. Staurosporine bound at the ATP of the kinase module is shown in *blue*. The AMP sites on the γ subunit are shown in *green* at the *bottom* (invisible on the *left* panel). Also shown is pThr172 on the activation loop of the kinase module (*deep purple*).

subunit (Calabrese et al., 2014; Langendorf & Kemp, 2015; Xiao et al., 2013). The ADaM site is created by juxtaposition of two orthogonal β sheets, one each from the α and β subunit. It is physically separated from the nucleotide-binding sites of the γ subunit by ~70 Å. The thienopyridone core of A769662 is anchored at the ADaM site via electrostatic, hydrogen bonding, and cation–pi interactions formed between the ligand and the protein atoms. Compounds ex991 and PF-06409577 also bind at the ADaM site in a similar binding mode and activate by a similar mechanism as employed by A769662 (Cameron et al., 2016; Xiao et al., 2013). Enzymology and kinetic studies have shown that A769662 and PF-06409577 activate AMPK by lowering the K_m for the SAMS peptide with minimal effects on the K_m for ATP or the overall V_{max} (k_{cat}) of the reaction (Calabrese et al., 2014; Cameron et al., 2016). In contrast, AMP allosterically activates AMPK by increasing the V_{max} of the reaction without altering the K_m for ATP or SAMS peptide. Recent studies have suggested that there could be potential synergistic effects between AMP and synthetic activators that operate via the ADaM site. Scott et al. have recently shown that A769662 and AMP can bring about synergistic activation of even nonphosphorylated AMPK (Scott et al., 2014). Similarly, addition of salicylate on top of metformin seems to result in increased efficacy in lowering lipid accumulation in tumor models (O'Brien et al., 2015). Salicylate has been shown to be a weak AMPK activator that operates via the ADaM site while increased activation of AMPK by elevating cellular AMP levels is one of the many modalities proposed for the pharmacological action of metformin (Calabrese et al., 2014; Foretz et al., 2014; Hawley et al., 2012).

Given the critical regulatory role played by AMPK in glucose uptake, metabolism, lipid modulation, cell growth, and autophagy, it is no surprise to see the heightened interest in the discovery of AMPK activators for cardiovascular and metabolic diseases and cancer. The complexity of AMPK with three subunits, each existing as multiple isoforms presents unique opportunities and challenges for drug discovery. Moreover, there has been only limited experience in the discovery of enzyme activators as a therapeutic modality (Zorn & Wells, 2010). While hundreds of enzyme inhibitor drugs are known, there has been only one enzyme activator drug (riociguat) that has been discovered prospectively (Hambly & Granton, 2015). The availability of numerous conformational states and the ability to selectively modulate different functions presents multiple options for developing appropriate biochemical assays targeting AMPK. We describe below some of the common assay platforms that have been developed to identify and screen for AMPK activators.

2. NOTES ABOUT KEY REAGENTS

2.1 Buffer/Solution Conditions

We typically run all AMPK biochemical assays in 1 × kinase buffer (50 m*M* HEPES, 1 m*M* EGTA, 10 m*M* $MgCl_2$, 0.25 m*M* DTT, 0.01% Tween-20, and 0.01% BSA, pH 7.4) at room temperature. All purchased reagents used are of the highest purity available and all solutions are made using HPLC-grade water.

2.2 Sources of AMPK Protein

Several different sources of AMPK protein can be used in biochemical activity assays. We primarily use AMPK expressed and purified in *E. coli* and containing various tags such as His or BAP (biotin acceptor peptide) which is convenient for assays requiring the protein to be immobilized such as on a nickel or streptavidin-coated plate. Large amounts of highly purified, full-length, mammalian AMPK have been successfully produced using a tricistronic bacterial expression system (Neumann, Woods, Carling, Wallimann, & Schlattner, 2003; Rajamohan et al., 2010). Following purification, the protein is treated with an upstream kinase (LKB1 or CaMKK2) to generate active AMPK fully phosphorylated at Thr172 on the α-subunit (pAMPK). One drawback to this system is the lack of certain posttranslational modifications such as myristoylation on the β-subunit that has been shown to be important for AMP stimulation of α-Thr172 phosphorylation by upstream kinases (Oakhill et al., 2010; Sanders, Grondin, Hegarty, Snowden, & Carling, 2007; Warden et al., 2001). To circumvent this issue, frequently AMPK has been expressed and characterized in mammalian cells such as Cos7 or sf21 insect cells (Iseli et al., 2008; Warden et al., 2001).

Procedures are also available for purifying AMPK from rat liver (Carling et al., 1989; Davies et al., 1994; Hawley et al., 1996). Unlike the bacterially expressed material, AMPK purified from rat liver has all the necessary posttranslational modifications, however, it contains a mixture of both α1 and α2 AMPK heterotrimers and not a single AMPK isoform (Michell et al., 1996; Woods, Salt, Scott, Hardie, & Carling, 1996). Another potential advantage of native rat liver kinase could be its increased sensitivity to allosteric activation by AMP (Gowans et al., 2013).

2.3 Peptide Substrates

The most commonly used peptide substrate for monitoring AMPK activity is the 15 amino acid SAMS peptide (His-Met-Arg-Ser-Ala-Met-Ser-Gly-Leu-His-Leu-Val-Lys-Arg-Arg) which is based on the sequence surrounding Ser79 of the known AMPK substrate, rat ACC (Davies, Carling, & Hardie, 1989; Munday, Campbell, Carling, & Hardie, 1988). The SAMS peptide was designed to be relatively specific for AMPK by eliminating a known cyclic-AMP-dependent protein kinase phosphorylation site. The AMARA peptide (Ala-Met-Ala-Arg-Ala-Ala-Ser-Ala-Ala-Ala-Leu-Ala-Arg-Arg-Arg) is also occasionally used and it has been shown to be a better AMPK substrate compared to SAMS peptide (Dale et al., 1995). Substrate phosphorylation efficiency may vary depending on the substrate and heterotrimeric AMPK complex used in the assay. In addition to short peptide substrates, more extended domains of substrates such as the ACC catalytic domain have also been used to evaluate AMPK activity (Scott, Norman, Hawley, Kontogiannis, & Hardie, 2002).

2.4 Protein Phosphatase 2a

Protein phosphatase 2a (PP2a) is one of the phosphatases known to dephosphorylate Thr172 of the α-subunit (Wu, Song, Xu, Zhang, & Zou, 2007). When running assays used to evaluate the ability of a compound to protect AMPK from dephosphorylation at Thr172, we use recombinant PP2a. This reagent is commercially available or can be expressed and purified as described previously (Ikehara, Shinjo, Ikehara, Imamura, & Yasumoto, 2006). To quench the PP2a reaction, we use the potent PP2a inhibitor okadaic acid.

2.5 LKB1

Two upstream kinases have been reported to activate AMPK via phosphorylation at Thr172 in cell-free assays, the tumor suppressor LKB1 and CaMKK2 (Fogarty & Hardie, 2009; Hawley et al., 2003, 2005; Hurley et al., 2005; Shaw et al., 2004; Woods et al., 2005; Woods, Johnstone, et al., 2003). In vitro assays monitoring phosphorylation by LKB1 require a biologically active heterotrimeric complex containing LKB1 and the accessory subunits STRAD (STE-20-related kinase adapter protein) and MO25 (also known as Calcium-binding protein 39) (Boudeau et al., 2003). The recombinant complex is available commercially or can be expressed and purified as previously described (Hawley et al., 2003; Sanders et al., 2007). We describe the use of LKB1 here, but CaMKK2 can also be used with the advantage of not requiring a heterotrimeric protein complex for activity.

3. ASSAYS FOR MEASURING ALLOSTERIC ACTIVATION OF AMPK

Allosteric activation of AMPK by a small molecule is evaluated by monitoring an increase in pAMPK catalytic activity as measured by the phosphorylation of the SAMS or AMARA peptide substrate. Several different assay technologies can be used to measure this kinase activity including radioactive and nonradioactive methods. For initial compound identification, we usually perform the assays with peptide substrate and ATP fixed at K_m concentrations. The absolute concentrations will vary depending on the AMPK isoform being tested as the various heterotrimers have different substrate K_m values (Michell et al., 1996; Rajamohan et al., 2016). In order to ensure assay linearity and to prevent substrate depletion, one should be careful about the concentration of AMPK that is used in the assay. The concentration of pAMPK used in each of the assays also varies depending on the specific activity of the AMPK source and should be set for only about 5–10% substrate conversion over the course of the experiment to allow for the greatest assay window for activation. This should be determined ahead of time by running a time course and enzyme titration curve.

3.1 ^{33}P-ATP Filter Assay

Assay formats utilizing radioactivity are frequently used to measure AMPK kinase activity. The traditional filter format (described later) is the most common, but Abbott Laboratories developed an alternative microarrayed compound screening technology (μARCS) and used this method for the initial discovery of A769662 (Anderson et al., 2004). The ^{33}P-ATP filter assay protocol is a modification of a lower-throughput method that has been previously described and used to identify and characterize AMPK activators (Davies et al., 1989; Gomez-Galeno et al., 2010; Zhou et al., 2001). Aside from the obvious drawbacks of utilizing radioactivity, in our experience this is a robust and versatile assay that can be easily adapted for testing at a range of peptide or ATP concentrations (Calabrese et al., 2014; Rajamohan et al., 2016). In this format, the kinase reaction is quenched with phosphoric acid which not only inactivates AMPK but simultaneously protonates the basic residues on the peptide. Under these acidic conditions, the phosphorylated SAMS peptide product is retained on the phosphocellulose filter via charge interaction while the ^{33}P-ATP is removed by filtration. The protocol below is for a 384-well assay plate; however volumes can be adjusted for other plate sizes.

Equipment and stock solutions

- 1 × kinase assay buffer (50 m*M* HEPES, 1 m*M* EGTA, 10 m*M* $MgCl_2$, 0.25 m*M* DTT, 0.01% Tween-20, 0.01% BSA, pH 7.4)
- Wash buffer (50 m*M* HEPES, 1 m*M* EGTA, 10 m*M* $MgCl_2$, 0.1% Tween-20, 0.01% BSA, pH 7.4)
- 20 m*M* SAMS peptide
- 10 m*M* ATP
- Polypropylene assay plates
- Phosphocellulose filter plates

- 2% phosphoric acid
- 10 mCi/mL, 3000 Ci/mmol ^{33}P ATP
- Vacuum manifold
- Scintillation fluid
- Liquid scintillation counter such as PerkinElmer MicroBeta Trilux

Working solutions

1. Activator mixture. Dilute compound to 4 × the desired final compound concentration in 1 × kinase buffer.
2. AMPK enzyme mix. Prepare a 4 × AMPK enzyme solution by diluting AMPK in 1 × kinase buffer to desired concentration which gives 5–10% substrate conversion within the timeframe of the assay. A preliminary time course experiment at multiple enzyme concentrations is recommended to select the ideal final AMPK concentration.
3. SAMS substrate mix. Prepare SAMS peptide solution by diluting SAMS peptide in 1 × kinase buffer to a concentration equal to 4 × the reported K_m value for the specific isoform.
4. ATP mix. Prepare ATP solution by diluting ATP in 1 × kinase buffer to a final concentration equal to 4 × the reported K_m value for the specific isoform and to a final ^{33}P-ATP specific activity of 0.25 uCi/nmol (for example 20 μ*M* unlabeled ATP/0.2 uCi ^{33}P ATP in 40 μL reaction).

Method

1. Add 10 μL of activator mix and 10 μL of the AMPK enzyme mix to a 384-well polypropylene plate being sure to include vehicle/DMSO and AMP as negative and positive controls, respectively. Incubate plate at room temperature for 15 min to allow enzyme and compound to come to equilibrium.
2. Add 10 μL of the SAMS peptide mix.
3. Initiate kinase reaction by the addition of 10 μL of the ATP mix.
4. Incubate reaction at room temperature for 1 h.
5. Terminate reaction with the addition of 20 μL of 2% phosphoric acid.
6. Prewet phosphocellulose filter plate with 2% phosphoric acid and filter by applying vacuum using vacuum manifold. Transfer the quenched AMPK reaction mixture to the filter plate and wash four times with wash buffer.
7. Allow filter plate to dry. Add 25 μL of scintillation fluid and measure the amount of radioactivity in each well using liquid scintillation counting on an instrument such as a PerkinElmer MicroBeta Trilux.
8. Raw data (counts per minute) are converted to fold change from basal by subtracting the value of the DMSO control well from the compound

treated well and dividing by the value of the DMSO control well. When generating full compound dose responses, the fold change from basal value is plotted against the concentrations of activator and curves can be fit using standard nonlinear regression analysis.

The AC_{50} is defined as the concentration of compound required to reach 50% of the maximal activation and is a value routinely used to evaluate and compare activator compounds. The maximal activation is the fold change from basal activity at saturating concentration of compound. It is interesting to note that there can be quite a bit of variability in the maximal activation between different kinds of AMPK activators. As reported previously and shown in Fig. 3, the activator A769662 which binds in a cleft between the α and β subunits has been shown to have a larger maximal activation value than the native ligand AMP which binds in the γ-domain

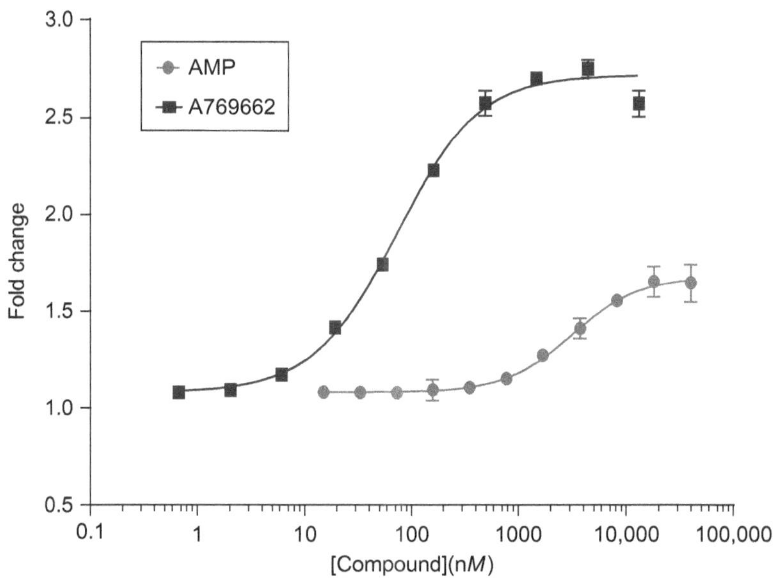

Fig. 3 Allosteric activation of AMPK by AMP and A769662. Effect of AMP (*gray*) and A769662 (*black*) on the α1β1γ1 AMPK isoform as monitored using the TR-FRET allosteric activation assay. Different AMPK activators display different levels of maximal activation. AMP, which binds to the γ subunit has a lower maximal activation than A769662 which binds between the α- and β-subunits. *Data are from Rajamohan, F., Reyes, A. R., Frisbie, R. K., Hoth, L. R., Sahasrabudhe, P., Magyar, R., et al. (2016). Probing the enzyme kinetics, allosteric modulation and activation of alpha1- and alpha2-subunit-containing AMP-activated protein kinase (AMPK) heterotrimeric complexes by pharmacological and physiological activators.* Biochemical Journal, 473, *581–592.*

(Rajamohan et al., 2016). This becomes even more prominent with the use of nonphosphorylated AMPK in the biochemical assay (Scott et al., 2014).

3.2 Nonradioactive Assay Formats

In addition to the radioactive format, there are several nonradioactive assay technologies that have been used to monitor AMPK kinase activity. These alternative formats include (but are not limited to) fluorescent assays such as Alphascreen, FRET (Förster resonance energy transfer), TR-FRET (time-resolved Förster resonance energy transfer), and CHEF (chelate-enhanced fluorescence) (Li, Cummings, Cunningham, Chen, & Zhou, 2003; Reichling et al., 2008). Many of the reagents for these various assay formats are commercially available. The method described later utilizes TR-FRET technology and is based on PerkinElmer's LANCE kinase assay.

In this format, the SAMS peptide is labeled with the Ulight™ acceptor moiety. Following the kinase reaction, an europium donor labeled anti-phospho-ACC (Ser79) antibody is added into the mixture. The newly phosphorylated SAMS peptide is recognized by the anti-phospho-ACC antibody, brings the Ulight™ acceptor and europium donor in close proximity and results in an increase in signal at 665 nm following excitation at 320 nm. The intensity of the signal is proportional to the level of AMPK-mediated SAMS peptide phosphorylation. The protocol below is for a 384-well assay plate; however, volumes can be adjusted for other plate sizes.

Equipment and stock solutions

- 1 × kinase assay buffer (50 m*M* HEPES, 1 m*M* EGTA, 10 m*M* $MgCl_2$, 0.25 m*M* DTT, 0.01% Tween-20, 0.01% BSA, pH 7.4)
- 10 m*M* ATP
- PerkinElmer Optiplate
- LANCE detection buffer
- **Ulight™-acetyl-CoA carboxylase (Ser79) peptide (SAMS peptide).
- Europium-anti-phospho-acetyl-CoA carboxylase (Ser79) antibody.
- Envision® Multiplate Reader.

Working solutions

- Activator mixture. Dilute compound to 4 × the desired final compound concentration in 1 × kinase buffer.
- AMPK enzyme mix. Prepare a 2 × AMPK enzyme solution by diluting AMPK in 1 × kinase buffer to desired concentration which gives 5–10% substrate conversion within the timeframe of the assay. A preliminary

time course experiment at multiple enzyme concentrations is recommended to select the ideal final AMPK concentration.

- Substrate mix. Prepare substrate mixture by diluting ATP in 1 × kinase buffer to a final concentration equal to 4 × the K_m value for the specific isoform and the Ulight SAMS peptide to 200 n*M* (50 n*M* final in the reaction).
- Stop/detection mix. Prepare 5 × detection mix by diluting Eu-anti-phospho-ACC antibody to 10 n*M* in 1 × LANCE detection buffer containing 50 m*M* EDTA.

Method

1. Add 2.5 μL of activator mixture.
2. Add 5 μL of the AMPK enzyme mix to an Optiplate. If measuring activity modulation by ligands, the complex should be in equilibrium prior to addition of substrates.
3. Initiate kinase reaction by the addition of 2.5 μL of substrate mix.
4. Incubate reaction at room temperature for 1 h.
5. Terminate reaction with the addition of 2.5 μL of stop/detection mixture.
6. Incubate plate for 1 h at room temperature.
7. Read Signal with EnVision® MultiReader in TR-FRET mode (excitation at 320 nm and emission at 665 nm).

4. STEADY-STATE KINETIC ANALYSIS OF AMPK ACTIVATORS

Allosteric enzyme activation usually occurs through effects on the V_{max} (the enzyme velocity at saturating substrate concentrations) and/or K_m (Michaelis constant, concentration of substrate required to reach ½ V_{max}) of the enzyme (Segel, 1993). The effects on the steady state kinetic parameters can be explored in detail for AMPK using the ^{33}P-ATP filter assay given that the concentrations of peptide substrate and ATP can be easily varied over a wide range of concentrations in that assay format.

Steady-state kinetic experiments to determine catalytic parameters can be performed by varying one substrate while keeping the other fixed at 10-fold the K_m value as described previously (Calabrese et al., 2014). The varied substrate is tested in a serial one-to-one dilution scheme while ensuring a substrate concentration at least 5- to 10-fold above and below the expected K_m for the varied substrate. The experiment should be performed in the presence and absence of a saturating concentration of the allosteric activator compound.

In the case of [^{33}P]-ATP, the specific activity was maintained at 0.25 μCi/nmoL. Assay linearity was ensured in separate experiments by determining that <30% substrate was depleted for all concentrations tested.

A similar analysis has been performed for activators of other enzymes (Bebernitz et al., 2009; Milne et al., 2007). For AMPK, it has been reported that activators with various binding sites can have differing effects on K_m and V_{max} (Calabrese et al., 2014). As shown in Fig. 4, the endogenous ligand AMP which binds in the γ-domain increases the V_{max} with very little effect on the K_m for peptide substrate. In contrast, A769662 which binds between the α- and β-subunits decreases the K_m for peptide substrate with little effect on the V_{max}.

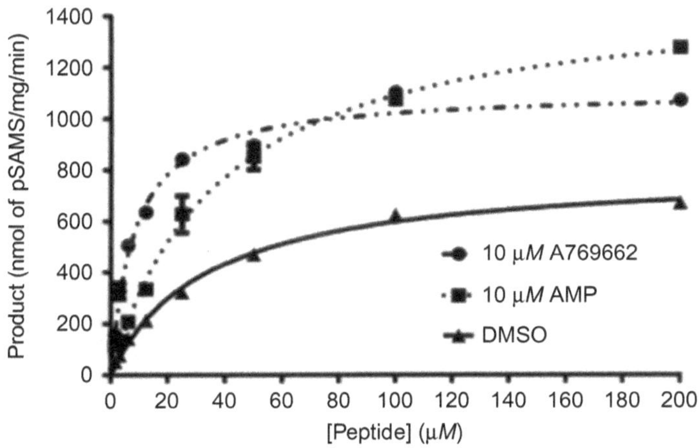

	K_m	V_{max}
	(μmol/L of peptide)	nmol of pSAMS/mg/min
DMSO	33.7 ± 2.4	799.9 ± 19.8
10 μM AMP	38.0 ± 3.6	1507 ± 50.5
10 μM A769662	8.24 ± 0.8	1105 ± 29.2

Fig. 4 Steady-state kinetic analysis of AMPK activators. AMPK activity was monitored using the ^{33}P-ATP filter assay with saturating concentrations of ATP and varying concentrations of SAMS peptide substrate to determine the SAMS peptide K_m and V_{max} in the presence and absence of various AMPK activators. Different AMPK activators display different mechanisms of activation. AMP activates by increasing the V_{max} of the kinase while A769662 decreases the K_m of the SAMS peptide substrate. *Data and figure are from Calabrese, M. F., Rajamohan, F., Harris, M. S., Caspers, N. L., Magyar, R., Withka, J. M., et al. (2014). Structural basis for AMPK activation: Natural and synthetic ligands regulate kinase activity from opposite poles by different molecular mechanisms.* Structure, *22, 1161–1172.*

5. ASSAYS FOR MONITORING PHOSPHORYLATION AT THR172 OF THE α-SUBUNIT

Given that the endogenous ligand AMP has been shown to increase phosphorylation of AMPK by the upstream kinase as well as protect it from dephosphorylation by phosphatase, it is desirable to have assays to measure the extent of Thr172 phosphorylation following treatment with a kinase or phosphatase. It is important to note, however that simply monitoring pThr172 does not necessarily reflect the entire activity of the compound on AMPK because it does not take into effect allosteric activation. Often Thr172 phosphorylation is monitored via standard western blotting techniques. The following protocols were established as an alternative to standard western blot procedures to assess the phosphorylation state of Thr172 of the α-subunit (Rajamohan et al., 2016). The protection assay measures residual pThr172 content following treatment of pAMPK with PP2a. The phosphorylation modulation assay measures pThr172 following treatment with an upstream kinase. Both assays utilize the dot-blot technique for immunodetection. Quenched reactions are transferred to a nitrocellulose or polyvinylidene fluoride (PVDF) membrane using a dot-blot apparatus and subjected to immunoblotting. Blots are initially probed with a rabbit α-pT172-specific antibody and signal is visualized and quantified using IRDye® infrared fluorescence secondary antibodies and the LiCOR Odyssey imaging system. The dot-blot method does not require separation by electrophoresis and is therefore a simplified detection technique that does not require time-consuming and complex gel blotting procedures.

5.1 Protection Assay

This assay measures the degree to which a small molecule binds to pAMPK and protects it from dephosphorylation. Fully phosphorylated AMPK is incubated with compound, treated with PP2a to dephosphorylate the enzyme by about 50% in the DMSO control reaction and the amount of pAMPK remaining is measured via immunodetection.

Equipment and stock solutions

- 1 × kinase assay buffer (50 m*M* HEPES, 1 m*M* EGTA, 10 m*M* $MgCl_2$, 0.25 m*M* DTT, 0.01% Tween-20, 0.01% BSA, pH 7.4).
- Polypropylene assay plates

- Rabbit anti-α subunit-pThr172 antibody
- Goat anti-rabbit 800CW (Rockland)
- Western blot blocking buffer reagent (Li-Cor)
- 3 × SDS loading buffer
- Nitrocellulose or PVDF membranes
- Dot-blot apparatus
- Wash buffer-PBS with 0.1% Tween-20

Working solutions

- Activator mixture. Dilute compound to 4 × the desired final compound concentration in 1 × kinase buffer.
- AMPK enzyme mix. Prepare AMPK enzyme solution by diluting AMPK in 1 × kinase buffer to desired concentration.
- PP2a mix. Dilute PP2a to 4 × desired final concentration in 1 × kinase buffer. The optimal PP2a concentration should dephosphorylate pAMPK by about 50% in 1 h.
- Primary anti-α subunit-pThr172 antibody mixture—dilute pThr172 antibody 1:500 in blocking buffer.
- IRDye secondary antibody mixture—dilute goat anti-rabbit 800CW 1:10,000 in blocking buffer.

Method

1. Add 10 μL of activator mixture.
2. Add 20 μL of the AMPK enzyme mix to a polypropylene assay plate.
3. Add 10 μL of the PP2a mix.
4. Incubate reaction at room temperature for 1 h.
5. Terminate reaction with the addition 20 μL of SDS loading buffer.
6. Transfer reaction to nitrocellulose/PVDF membrane using a dot-blot apparatus.
7. Incubate with Western blot blocking buffer reagent (Li-Cor) for 1 to 2 h at room temperature.
8. Wash three times using 42°C prewarmed wash buffer.
9. Incubate with primary antibody mixture for 2 h at room temperature.
10. Wash three times using 42°C prewarmed wash buffer.
11. Incubate with secondary antibody mixture for 1 h at room temperature.
12. Wash three times using 42°C prewarmed wash buffer.
13. Visualize blots using the Li-Cor Odyssey infrared imaging system and quantify signal using the Odyssey software.

5.2 Phosphorylation Modulation Assay

The phosphorylation modulation assay measures the degree to which a small-molecule modulator induces a conformational change in AMPK which enhances (or prevents) phosphorylation at Thr172 by the upstream kinase. In the phosphorylation modulation assay, nonphosphorylated AMPK is incubated with compound and the LKB1/MO25/STRAD complex for a set period of time followed by probing for pThr172 with an epitope-specific antibody. We routinely use the LKB1 complex as the upstream kinase, but CaMKK2 can also be used. Previous work has demonstrated that myristoylation on the β-subunit is required for AMP-stimulated phosphorylation on pThr172 (Oakhill et al., 2010). Therefore, it is important to use a source of AMPK with this specific posttranslational modification for the phosphorylation modulation assay.

Equipment and stock solutions

- 1 × kinase assay buffer (50 m*M* HEPES, 1 m*M* EGTA, 10 m*M* $MgCl_2$, 0.25 m*M* DTT, 0.01% Tween-20, 0.01% BSA, pH 7.4)
- Polypropylene assay plates
- Rabbit anti-α subunit-pThr172 antibody
- Goat anti-rabbit 800CW (Rockland)
- Western blot blocking buffer reagent (Li-Cor)
- 3 × SDS loading buffer nitrocellulose/PVDF membranes
- Dot-blot apparatus
- Wash buffer-1×PBS/0.1% Tween-20

Working solutions

- Activator mixture. Dilute compound to 4 × the desired final compound concentration in 1 × kinase buffer
- AMPK enzyme mix. Prepare AMPK enzyme solution by diluting AMPK in 1 × kinase buffer to desired concentration. Note that AMPK protein should be myristoylated on the β-subunit and not phosphorylated on Thr172 of the α-subunit.
- LKB1 mix. Prepare a 4 × LKB1/ATP solution by diluting LKB1 complex in 1 × kinase buffer containing 400 μ*M* ATP. The final LKB1 complex concentration should be predetermined from optimization assays to ensure the final phosphorylation percent allows accurate assessment of modulation. The reaction should be in the linear range and in the lower end of the final pAMPK content. In previous efforts, enough LKB1 for 30% phospho-AMPK has been used.

- Primary anti-α subunit-pThr172 antibody mix. Dilute pThr172 antibody 1:500 in blocking buffer.
- IRDye secondary antibody mix. Dilute goat anti-rabbit 800CW 1:10,000 in blocking buffer.

Method

1. Add 10 μL of activator mixture and 20 μL of AMPK enzyme mix to a polypropylene assay plate.
2. Add 10 μL of LKB1/ATP mixture.
3. Incubate reaction at room temperature for 1 h.
4. Terminate reaction with the addition 20 μL of SDS loading buffer. And proceed with detection protocol as noted in steps 6–13 of the protection assay.

6. ACTIVATION–PROTECTION ASSAY

Since many of the known AMPK activators are both protectors and activators (Cool et al., 2006; Gomez-Galeno et al., 2010; Goransson et al., 2007), it is desirable to have an assay capable of assessing both functions simultaneously. The activation–protection assay is a relatively novel assay that is able to simultaneously evaluate small molecules for allosteric activation of AMPK and/or protection from dephosphorylation by introducing a phosphatase treatment into the normal activity assay protocol. In the first part of this two-step assay, recombinant pAMPK fully phosphorylated at Thr172 on the α-subunit is incubated with a compound of interest and PP2a under conditions which will decrease the basal AMPK activity in the DMSO control reaction by about 50% due to dephosphorylation at Thr172. This condition was chosen since it reduces the population of pAMPK sufficiently to generate an acceptable protection assay window, while preserving enough pAMPK to measure activity and allosteric activation. PP2a is used because it has previously been shown to inactivate/dephosphorylate AMPK in cells (Wu et al., 2007). The phosphatase treatment is terminated with okadaic acid and the second step (kinase reaction) is simultaneously initiated by the addition of the ATP and SAMS peptide substrate mixtures. The activity of AMPK is monitored using the standard ^{33}P-ATP filter assay or some other assay format capable of monitoring AMPK kinase activity.

When setting up this assay there are several initial experiments that should be performed to ensure the optimal assay conditions. First, AMPK titrations should be performed to identify the concentration of AMPK that yields approximately 10–20% substrate conversion (5–10% conversion after PP2a treatment). Once the target AMPK concentration is identified, a phosphatase titration should be done to identify a condition where the PP2a treatment is stringent enough to reduce the population of pAMPK sufficiently to generate an acceptable protection assay window, while preserving enough pAMPK to measure activity and allosteric activation. Typically, 50% dephosphorylation is an acceptable final condition.

Equipment and stock solutions

- PP2a
- Okadaic acid—prepare 100 μ*M* stock in 100% EtOH
- All equipment and stock solutions from allosteric activation assay

Working solutions

1. Activator mixture. Dilute compound to 4 × the desired final compound concentration in 1 × kinase buffer.
2. AMPK enzyme mix. Prepare a 4 × AMPK enzyme solution by diluting AMPK in 1 × kinase buffer.
3. PP2a mix. Prepare a 4 × PP2a solution by diluting PP2a in 1 × kinase buffer.
4. Okadaic acid/substrate mix. Dilute okadaic acid to 10 × the phosphatase concentration in 1 × kinase buffer containing ATP and SAMS peptide at 4 × their respective K_m concentrations.
5. Stop solution. Dependent on assay format. Refer to the ^{33}P-ATP filter assay or TR-FRET assay.

Method

1. Add 10 μL of activator mixture.
2. Add 10 μL of the AMPK enzyme mix to a polypropylene assay plate. If measuring activity modulation by ligands, the complex should be in equilibrium prior to addition of substrates.
3. Add 10 μL of the PP2a mix to initiate the phosphatase reaction.
4. Incubate for 1 h.
5. Terminate phosphatase reaction and initiate kinase reaction by the addition of 10 μL of the okadaic acid/ATP/SAMS mix.
6. Incubate reaction at room temperature for 1 h.
7. Terminate kinase reaction with stop solution and proceed with detection.

7. SUMMARY

Given the importance of AMPK in regulating many cellular processes there is much interest in the identification of small-molecule AMPK activator compounds. Over the last several years, exciting data have been generated that suggest the potential therapeutic applications of AMPK activators for the treatment of a diverse array of cardiovascular and metabolic diseases. AMPK is a heterotrimeric protein with multiple isoforms that is activated via three distinct mechanisms by the endogenous ligand AMP. In order to drive a successful drug discovery program targeting AMPK, it is necessary to develop in vitro assays that can offer insights into the specific molecular mode of action of compounds. Here we have described in detail several assay formats that can be used to individually characterize these different modes of activation or protection. Direct allosteric activation is best evaluated by monitoring the direct phosphorylation of SAMS peptide substrate using a radioactive or nonradioactive method. We have also developed sensitive assays to quantitate levels of pThr172 following treatment of AMPK with the upstream kinase LKB1 or by the phosphatase PP2a. Taken together, the assay formats described here are valuable tools that can be used for the identification and characterization of AMPK activators.

REFERENCES

Alers, S., Loffler, A. S., Wesselborg, S., & Stork, B. (2012). Role of AMPK-mTOR-Ulk1/2 in the regulation of autophagy: Cross talk, shortcuts, and feedbacks. *Molecular and Cellular Biology*, *32*, 2–11.

Anderson, S. N., Cool, B. L., Kifle, L., Chiou, W., Egan, D. A., Barrett, L. W., et al. (2004). Microarrayed compound screening (microARCS) to identify activators and inhibitors of AMP-activated protein kinase. *Journal of Biomolecular Screening*, *9*, 112–121.

Apel, A., Zentgraf, H., Büchler, M. W., & Herr, I. (2009). Autophagy—A double-edged sword in oncology. *International Journal of Cancer*, *125*, 991–995.

Bach, M., Larance, M., James, D. E., & Ramm, G. (2011). The serine/threonine kinase ULK1 is a target of multiple phosphorylation events. *The Biochemical Journal*, *440*, 283–291.

Bebernitz, G. R., Beaulieu, V., Dale, B. A., Deacon, R., Duttaroy, A., Gao, J., et al. (2009). Investigation of functionally liver selective glucokinase activators for the treatment of type 2 diabetes. *Journal of Medicinal Chemistry*, *52*, 6142–6152.

Behrends, C., Sowa, M. E., Gygi, S. P., & Harper, J. W. (2010). Network organization of the human autophagy system. *Nature*, *466*, 68–76.

Boudeau, J., Baas, A. F., Deak, M., Morrice, N. A., Kieloch, A., Schutkowski, M., et al. (2003). MO25alpha/beta interact with STRADalpha/beta enhancing their ability to bind, activate and localize LKB1 in the cytoplasm. *The EMBO Journal*, *22*, 5102–5114.

Calabrese, M. F., Rajamohan, F., Harris, M. S., Caspers, N. L., Magyar, R., Withka, J. M., et al. (2014). Structural basis for AMPK activation: Natural and synthetic ligands regulate kinase activity from opposite poles by different molecular mechanisms. *Structure*, *22*, 1161–1172.

Cameron, K. O., Kung, D. W., Kalgutkar, A. S., Kurumbail, R. G., Miller, R., Salatto, C. T., et al. (2016). Discovery and preclinical characterization of 6-chloro-5-[4-(1-hydroxycyclobutyl)phenyl]-1h-indole-3-carboxylic acid (PF-06409577), a direct activator of adenosine monophosphate-activated protein kinase (AMPK), for the potential treatment of diabetic nephropathy. *Journal of Medicinal Chemistry*, *59*, 8068–8081. http://dx.doi.org/10.1021/acs.jmedchem.6b00866.

Cargnello, M., Tcherkezian, J., & Roux, P. P. (2015). The expanding role of mTOR in cancer cell growth and proliferation. *Mutagenesis*, *30*, 169–176.

Carling, D., Clarke, P. R., Zammit, V. A., & Hardie, D. G. (1989). Purification and characterization of the AMP-activated protein kinase. Copurification of acetyl-CoA carboxylase kinase and 3-hydroxy-3-methylglutaryl-CoA reductase kinase activities. *European Journal of Biochemistry*, *182*, 129–136.

Carling, D., & Viollet, B. (2015). Beyond energy homeostasis: The expanding role of AMP-activated protein kinase in regulating metabolism. *Cell Metabolism*, *21*, 799–804.

Chan, E. Y. W., Kir, S., & Tooze, S. A. (2007). siRNA screening of the kinome identifies ULK1 as a multidomain modulator of autophagy. *The Journal of Biological Chemistry*, *282*, 25464–25474.

Chang, Y. Y., & Neufeld, T. P. (2009). An Atg1/Atg13 complex with multiple roles in TOR-mediated autophagy regulation. *Molecular Biology of the Cell*, *20*, 2004–2014.

Chen, Y., He, J., Tian, M., Zhang, S. Y., Guo, M. R., Kasimu, R., et al. (2014). UNC51-like kinase 1, autophagic regulator and cancer therapeutic target. *Cell Proliferation*, *47*, 494–505.

Chen, L., Wang, J., Zhang, Y. Y., Yan, S. F., Neumann, D., Schlattner, U., et al. (2012). AMP-activated protein kinase undergoes nucleotide-dependent conformational changes. *Nature Structural & Molecular Biology*, *19*, 716–718.

Chen, L., Xin, F. J., Wang, J., Hu, J., Zhang, Y. Y., Wan, S., et al. (2013). Conserved regulatory elements in AMPK. *Nature*, *498*, E8–E10.

Cool, B., Zinker, B., Chiou, W., Kifle, L., Cao, N., Perham, M., et al. (2006). Identification and characterization of a small molecule AMPK activator that treats key components of type 2 diabetes and the metabolic syndrome. *Cell Metabolism*, *3*, 403–416.

Dale, S., Wilson, W. A., Edelman, A. M., & Grahame Hardie, D. (1995). Similar substrate recognition motifs for mammalian AMP-activated protein kinase, higher plant HMG-CoA reductase kinase-A, yeast SNF1, and mammalian calmodulin-dependent protein kinase I. *FEBS Letters*, *361*, 191–195.

Davies, S. P., Carling, D., & Hardie, D. G. (1989). Tissue distribution of the AMP-activated protein kinase, and lack of activation by cyclic-AMP-dependent protein kinase, studied using a specific and sensitive peptide assay. *European Journal of Biochemistry*, *186*, 123–128.

Davies, S. P., Hawley, S. A., Woods, A., Carling, D., Haystead, T. A., & Hardie, D. G. (1994). Purification of the AMP-activated protein kinase on ATP-gamma-sepharose and analysis of its subunit structure. *European Journal of Biochemistry*, *223*, 351–357.

Dorsey, F. C., Rose, K. L., Coenen, S., Prater, S. M., Cavett, V., Cleveland, J. L., et al. (2009). Mapping the phosphorylation sites of Ulk1. *Journal of Proteome Research*, *8*, 5253–5263.

Egan, D. F., Shackelford, D. B., Mihaylova, M. M., Gelino, S., Kohnz, R. A., Mair, W., et al. (2011). Phosphorylation of ULK1 (hATG1) by AMP-activated protein kinase connects energy sensing to mitophagy. *Science*, *331*, 456–461.

Fogarty, S., & Hardie, D. G. (2009). C-terminal phosphorylation of LKB1 is not required for regulation of AMP-activated protein kinase, BRSK1, BRSK2, or cell cycle arrest. *The Journal of Biological Chemistry*, *284*, 77–84.

Foretz, M., Guigas, B., Bertrand, L., Pollak, M., & Viollet, B. (2014). Metformin: From mechanisms of action to therapies. *Cell Metabolism*, *20*, 953–966.

Foretz, M., Hébrard, S., Leclerc, J., Zarrinpashneh, E., Soty, M., Mithieux, G., et al. (2010). Metformin inhibits hepatic gluconeogenesis in mice independently of the LKB1/AMPK pathway via a decrease in hepatic energy state. *The Journal of Clinical Investigation*, *120*, 2355–2369.

Foretz, M., & Viollet, B. (2015). Therapy: Metformin takes a new route to clinical efficacy. *Nature Reviews. Endocrinology*, *11*, 390–392.

Fryer, L. G., Parbu-Patel, A., & Carling, D. (2002). The anti-diabetic drugs rosiglitazone and metformin stimulate AMP-activated protein kinase through distinct signaling pathways. *The Journal of Biological Chemistry*, *277*, 25226–25232.

Fullerton, M. D., Galic, S., Marcinko, K., Sikkema, S., Pulinilkunnil, T., Chen, Z. P., et al. (2013). Single phosphorylation sites in Acc1 and Acc2 regulate lipid homeostasis and the insulin-sensitizing effects of metformin. *Nature Medicine*, *19*, 1649–1654.

Ganley, I. G., Lam, D. H., Wang, J., Ding, X., Chen, S., & Jiang, X. (2009). ULK1·ATG13·FIP200 complex mediates mTOR signaling and is essential for autophagy. *The Journal of Biological Chemistry*, *284*, 12297–12305.

Gomez-Galeno, J. E., Dang, Q., Nguyen, T. H., Boyer, S. H., Grote, M. P., Sun, Z., et al. (2010). A potent and selective AMPK activator that inhibits de novo lipogenesis. *ACS Medicinal Chemistry Letters*, *1*, 478–482.

Goransson, O., McBride, A., Hawley, S. A., Ross, F. A., Shpiro, N., Foretz, M., et al. (2007). Mechanism of action of A-769662, a valuable tool for activation of AMP-activated protein kinase. *The Journal of Biological Chemistry*, *282*, 32549–32560.

Gowans, G. J., Hawley, S. A., Ross, F. A., & Hardie, D. G. (2013). AMP is a true physiological regulator of AMP-activated protein kinase by both allosteric activation and enhancing net phosphorylation. *Cell Metabolism*, *18*, 556–566.

Hambly, N., & Granton, J. (2015). Riociguat for the treatment of pulmonary hypertension. *Expert Review of Respiratory Medicine*, *9*, 679–695.

Hara, T., & Mizushima, N. (2009). Role of ULK-FIP200 complex in mammalian autophagy: FIP200, a counterpart of yeast Atg17? *Autophagy*, *5*, 85–87.

Hara, T., Takamura, A., Kishi, C., Iemura, S.-I., Natsume, T., Guan, J.-L., et al. (2008). FIP200, a ULK-interacting protein, is required for autophagosome formation in mammalian cells. *The Journal of Cell Biology*, *181*, 497–510.

Hardie, D. G. (2015). AMPK: Positive and negative regulation, and its role in whole-body energy homeostasis. *Current Opinion in Cell Biology*, *33*, 1–7.

Hardie, D. G. (2016). Regulation of AMP-activated protein kinase by natural and synthetic activators. *Acta Pharmaceutica Sinica B*, *6*, 1–19.

Hawley, S. A., Boudeau, J., Reid, J. L., Mustard, K. J., Udd, L., Makela, T. P., et al. (2003). Complexes between the LKB1 tumor suppressor, STRAD alpha/beta and MO25 alpha/beta are upstream kinases in the AMP-activated protein kinase cascade. *Journal of Biology*, *2*, 28.

Hawley, S. A., Davison, M., Davies, S. P., Beri, R. K., Carling, D., & Hardie, D. G. (1996). Characterization of the AMP-activated protein kinase from rat liver and identification of threonine 172 as the major site at which it phosphorylates AMP-activated protein kinase. *The Journal of Biological Chemistry*, *271*, 27879–27887.

Hawley, S. A., Fullerton, M. D., Ross, F. A., Schertzer, J. D., Chevtzoff, C., Walker, K. J., et al. (2012). The ancient drug salicylate directly activates AMP-activated protein kinase. *Science*, *336*, 918–922.

Hawley, S. A., Gadalla, A. E., Olsen, G. S., & Hardie, D. G. (2002). The antidiabetic drug metformin activates the AMP-activated protein kinase cascade via an adenine nucleotide-independent mechanism. *Diabetes*, *51*, 2420–2425.

Hawley, S. A., Pan, D. A., Mustark, K. J., Ross, L., Bain, J., Edelman, A. M., et al. (2005). Calmodulin-dependent protein kinase kinase-beta is an alternative upstream kinase for AMP-activated protein kinase. *Cell Metabolism, 2*, 9–19.

Hawley, S. A., Ross, F. A., Chevtzoff, C., Green, K. A., Evans, A., Fogarty, S., et al. (2010). Use of cells expressing gamma subunit variants to identify diverse mechanisms of AMPK activation. *Cell Metabolism, 11*, 554–565.

Hosokawa, N., Hara, T., Kaizuka, T., Kishi, C., Takamura, A., Miura, Y., et al. (2009). Nutrient-dependent mTORC1 association with the ULK1-Atg13-FIP200 complex required for autophagy. *Molecular Biology of the Cell, 20*, 1981–1991.

Hurley, R. L., Anderson, K. A., Franzone, J. M., Kemp, B. E., Means, A. R., & Witters, L. A. (2005). The Ca2+/calmodulin-dependent protein kinase kinases are AMP-activated protein kinase kinases. *The Journal of Biological Chemistry, 280*, 29060–29066.

Ikehara, T., Shinjo, F., Ikehara, S., Imamura, S., & Yasumoto, T. (2006). Baculovirus expression, purification, and characterization of human protein phosphatase 2A catalytic subunits alpha and beta. *Protein Expression and Purification, 45*, 150–156.

Iseli, T. J., Oakhill, J. S., Bailey, M. F., Wee, S., Walter, M., van Denderen, B. J., et al. (2008). AMP-activated protein kinase subunit interactions: Beta1:gamma1 association requires beta1 Thr-263 and Tyr-267. *The Journal of Biological Chemistry, 283*, 4799–4807.

Jung, C. H., Jun, C. B., Ro, S.-H., Kim, Y.-M., Otto, N. M., Cao, J., et al. (2009). ULK-Atg13-FIP200 complexes mediate mTOR signaling to the autophagy machinery. *Molecular Biology of the Cell, 20*, 1992–2003.

Kim, J., Kundu, M., Viollet, B., & Guan, K. L. (2011). AMPK and mTOR regulate autophagy through direct phosphorylation of Ulk1. *Nature Cell Biology, 13*, 132–141.

Kim, J., Shin, J., & Ha, J. (2015). Screening methods for AMP-activated protein kinase modulators: A patent review. *Expert Opinion on Therapeutic Patents, 25*, 261–277.

Kim, J., Yang, G., Kim, Y., Kim, J., & Ha, J. (2016). AMPK activators: Mechanisms of action and physiological activities. *Experimental & Molecular Medicine, 48*. e224.

Landgraf, R. R., Goswami, D., Rajamohan, F., Harris, M. S., Calabrese, M. F., Hoth, L. R., et al. (2013). Activation of AMP-activated protein kinase revealed by hydrogen/deuterium exchange mass spectrometry. *Structure, 21*, 1942–1953.

Langendorf, C. G., & Kemp, B. E. (2015). Choreography of AMPK activation. *Cell Research, 25*, 5–6.

Langendorf, C. G., Ngoei, K. R., Scott, J. W., Ling, N. X., Issa, S. M., Gorman, M. A., et al. (2016). Structural basis of allosteric and synergistic activation of AMPK by furan-2-phosphonic derivative C2 binding. *Nature Communications, 7*, 10912. http://dx.doi.org/10.1038/ncomms10912.

Li, Y., Cummings, R. T., Cunningham, B. R., Chen, Y., & Zhou, G. (2003). Homogeneous assays for adenosine 5′-monophosphate-activated protein kinase. *Analytical Biochemistry, 321*, 151–156.

Li, X., Wang, L., Zhou, X. E., Ke, J., de Waal, P. W., Gu, X., et al. (2015). Structural basis of AMPK regulation by adenine nucleotides and glycogen. *Cell Research, 25*, 50–66.

Löffler, A. S., Alers, S., Dieterle, A. M., Keppeler, H., Franz-Wachtel, M., Kundu, M., et al. (2011). Ulk1-mediated phosphorylation of AMPK constitutes a negative regulatory feedback loop. *Autophagy, 7*, 696–706.

Madiraju, A. K., Erion, D. M., Rahimi, Y., Zhang, X. M., Braddock, D. T., Albright, R. A., et al. (2014). Metformin suppresses gluconeogenesis by inhibiting mitochondrial glycerophosphate dehydrogenase. *Nature, 510*, 542–546.

Madsen, A., Bozickovic, O., Bjune, J. I., Mellgren, G., & Sagen, J. V. (2015). Metformin inhibits hepatocellular glucose, lipid and cholesterol biosynthetic pathways by transcriptionally suppressing steroid receptor coactivator 2 (SRC-2). *Scientific Reports, 5*, 16430. http://dx.doi.org/10.1038/srep16430.

Michell, B. J., Stapleton, D., Mitchelhill, K. I., House, C. M., Katsis, F., Witters, L. A., et al. (1996). Isoform-specific purification and substrate specificity of the 5′-AMP-activated protein kinase. *The Journal of Biological Chemistry*, *271*, 28445–28450.
Miglianico, M., Nicolaes, G. A., & Neumann, D. (2015). Pharmacological targeting of AMP-activated protein kinase and opportunities for computer-aided drug design. *Journal of Medicinal Chemistry*, *59*, 2879–2893.
Mihaylova, M. M., & Shaw, R. J. (2011). The AMPK signalling pathway coordinates cell growth, autophagy and metabolism. *Nature Cell Biology*, *13*, 1016–1023.
Miller, R. A., Chu, Q., Xie, J., Foretz, M., Viollet, B., & Birnbaum, M. J. (2013). Biguanides suppress hepatic glucagon signalling by decreasing production of cyclic AMP. *Nature*, *494*, 256–260.
Milne, J. C., Lambert, P. D., Schenk, S., Carney, D. P., Smith, J. J., Gagne, D. J., et al. (2007). Small molecule activators of SIRT1 as therapeutics for the treatment of type 2 diabetes. *Nature*, *450*, 712–716.
Mounier, R., Theret, M., Lantier, L., Foretz, M., & Viollet, B. (2015). Expanding roles for AMPK in skeletal muscle plasticity. *Trends in Endocrinology and Metabolism*, *26*, 275–286.
Munday, M. R., Campbell, D. G., Carling, D., & Hardie, D. G. (1988). Identification by amino acid sequencing of three major regulatory phosphorylation sites on rat acetyl-CoA carboxylase. *European Journal of Biochemistry*, *175*, 331–338.
Neumann, D., Wallimann, T., Rider, M. H., Tokarska-Schlattner, M., Grahame Hardie, D., & Schlattner, U. (2007). Signaling by AMP-activated protein kinase. In V. Saks (Ed.), *Molecular system bioenergetics* (pp. 303–338). Weinheim: Wiley-VCH Verlag GmbH & Co. KGaA.
Neumann, D., Woods, A., Carling, D., Wallimann, T., & Schlattner, U. (2003). Mammalian AMP-activated protein kinase: Functional, heterotrimeric complexes by co-expression of subunits in *Escherichia coli*. *Protein Expression and Purification*, *30*, 230–237.
Oakhill, J. S., Chen, Z. P., Scott, J. W., Steel, R., Castelli, L. A., Ling, N., et al. (2010). beta-Subunit myristoylation is the gatekeeper for initiating metabolic stress sensing by AMP-activated protein kinase (AMPK). *Proceedings of the National Academy of Sciences of the United States of America*, *107*, 19237–19241.
Oakhill, J. S., Scott, J. W., & Kemp, B. E. (2012). AMPK functions as an adenylate charge-regulated protein kinase. *Trends in Endocrinology and Metabolism*, *23*, 125–132.
Oakhill, J. S., Steel, R., Chen, Z.-P., Scott, J. W., Ling, N., Tam, S., et al. (2011). AMPK is a direct adenylate charge-regulated protein kinase. *Science*, *332*, 1433–1435.
O'Brien, A. J., Villani, L. A., Broadfield, L. A., Houde, V. P., Galic, S., Blandino, G., et al. (2015). Salicylate activates AMPK and synergizes with metformin to reduce the survival of prostate and lung cancer cells ex vivo through inhibition of de novo lipogenesis. *The Biochemical Journal*, *469*, 177–187.
Owen, M. R., Doran, E., & Halestrap, A. P. (2000). Evidence that metformin exerts its anti-diabetic effects through inhibition of complex 1 of the mitochondrial respiratory chain. *The Biochemical Journal*, *348*(Pt. 3), 607–614.
Rajamohan, F., Harris, M. S., Frisbie, R. K., Hoth, L. R., Geoghegan, K. F., Valentine, J. J., et al. (2010). *Escherichia coli* expression, purification and characterization of functional full-length recombinant $\alpha 2\beta 2\gamma 3$ heterotrimeric complex of human AMP-activated protein kinase. *Protein Expression and Purification*, *73*, 189–197.
Rajamohan, F., Reyes, A. R., Frisbie, R. K., Hoth, L. R., Sahasrabudhe, P., Magyar, R., et al. (2016). Probing the enzyme kinetics, allosteric modulation and activation of alpha1- and alpha2-subunit-containing AMP-activated protein kinase (AMPK) heterotrimeric complexes by pharmacological and physiological activators. *The Biochemical Journal*, *473*, 581–592.
Rana, S., Blowers, E. C., & Natarajan, A. (2015). Small molecule adenosine 5′-monophosphate activated protein kinase (AMPK) modulators and human diseases. *Journal of Medicinal Chemistry*, *58*, 2–29.

Reichling, L. J., Riddle, S. M., Mei, B., Bruinsma, R., Goossens, T. A., Huwiler, K. G., et al. (2008). Homogenous fluorescent assays for characterizing small-molecule activators of AMP-activated protein kinase (AMPK). *Current Chemical Genomics*, *1*, 34–42.

Ross, F. A., Jensen, T. E., & Hardie, D. G. (2016). Differential regulation by AMP and ADP of AMPK complexes containing different γ subunit isoforms. *The Biochemical Journal*, *473*, 189–199.

Sanders, M. J., Grondin, P. O., Hegarty, B. D., Snowden, M. A., & Carling, D. (2007). Investigating the mechanism for AMP activation of the AMP-activated protein kinase cascade. *The Biochemical Journal*, *403*, 139–148.

Scott, J. W., Ling, N., Issa, S. M. A., Dite, T. A., O'Brien, M. T., Chen, Z. P., et al. (2014). Small molecule drug A-769662 and AMP synergistically activate naive AMPK independent of upstream kinase signaling. *Chemistry & Biology*, *21*, 1–9.

Scott, J. W., Norman, D. G., Hawley, S. A., Kontogiannis, L., & Hardie, D. G. (2002). Protein kinase substrate recognition studied using the recombinant catalytic domain of AMP-activated protein kinase and a model substrate. *Journal of Molecular Biology*, *317*, 309–323.

Segel, I. H. (1993). Enzyme activation. In *Enzyme kinetics: Behavior and analysis of rapid equilibrium and steady-state enzyme systems*. New York: Wiley-Interscience.

Shang, L., Chen, S., Du, F., Li, S., Zhao, L., & Wang, X. (2011). Nutrient starvation elicits an acute autophagic response mediated by Ulk1 dephosphorylation and its subsequent dissociation from AMPK. *Proceedings of the National Academy of Sciences of the United States of America*, *108*, 4788–4793.

Shang, L., & Wang, X. (2011). AMPK and mTOR coordinate the regulation of Ulk1 and mammalian autophagy initiation. *Autophagy*, *7*, 924–926.

Sharma, H., & Kumar, S. (2016). Natural AMPK activators: An alternate approach for treatment and management of metabolic syndrome. *Current Medicinal Chemistry*, *23*. http://dx.doi.org/10.2174/0929867323666160406120814.

Shaw, R. J., Kosmatka, M., Bardeesy, N., Hurley, R. L., Witters, L. A., DePinho, R. A., et al. (2004). The tumor suppressor LKB1 kinase directly activates AMP-activated kinase and regulates apoptosis in response to energy stress. *Proceedings of the National Academy of Sciences of the United States of America*, *101*, 3329–3335.

Shaw, R. J., Lamia, K. A., Vasquez, D., Koo, S. H., Bardeesy, N., Depinho, R. A., et al. (2005). The kinase LKB1 mediates glucose homeostasis in liver and therapeutic effects of metformin. *Science*, *310*, 1642–1646.

Sliwinska, A., & Drzewoski, J. (2015). Molecular action of metformin in hepatocytes: An updated insight. *Current Diabetes Reviews*, *11*, 175–181.

Stephenne, X., Foretz, M., Taleux, N., van der Zon, G. C., Sokal, E., Hue, L., et al. (2011). Metformin activates AMP-activated protein kinase in primary human hepatocytes by decreasing cellular energy status. *Diabetologia*, *54*, 3101–3110.

Suter, M., Riek, U., Tuerk, R., Schlattner, U., Wallimann, T., & Neumann, D. (2006). Dissecting the role of 5′-AMP for allosteric stimulation, activation, and deactivation of AMP-activated protein kinase. *The Journal of Biological Chemistry*, *281*, 32207–32216.

Tan, V. P., & Miyamoto, S. (2016). Nutrient-sensing mTORC1: Integration of metabolic and autophagic signals. *Journal of Molecular and Cellular Cardiology*, *95*, 31–41.

Viollet, B., Guigas, B., Sanz Garcia, N., Leclerc, J., Foretz, M., & Andreelli, F. (2012). Cellular and molecular mechanisms of metformin: An overview. *Clinical Science (London, England)*, *122*, 253–270.

Warden, S. M., Richardson, C., O'Donnell, J., Jr., Stapleton, D., Kemp, B. E., & Witters, L. A. (2001). Post-translational modifications of the β1 subunit of AMP-activated protein kinase affect enzyme activity and cellular localization. *The Biochemical Journal*, *354*, 275–283.

Woods, A., Dickerson, K., Heath, R., Hong, S. P., Momcilovic, M., Johnstone, S. R., et al. (2005). Ca^{2+}/calmodulin-dependent protein kinase kinase-β acts upstream of AMP-activated protein kinase in mammalian cells. *Cell Metabolism*, *2*, 21–33.

Woods, A., Johnstone, S. R., Dickerson, K., Leiper, F. C., Fryer, L. G., Neumann, D., et al. (2003). LKB1 is the upstream kinase in the AMP-activated protein kinase cascade. *Current Biology*, *13*, 2004–2008.

Woods, A., Salt, I., Scott, J., Hardie, D. G., & Carling, D. (1996). The alpha1 and alpha2 isoforms of the AMP-activated protein kinase have similar activities in rat liver but exhibit differences in substrate specificity in vitro. *FEBS Letters*, *397*, 347–351.

Woods, A., Vertommen, D., Neumann, D., Turk, R., Bayliss, J., Schlattner, U., et al. (2003). Identification of phosphorylation sites in AMP-activated protein kinase (AMPK) for upstream AMPK kinases and study of their roles by site-directed mutagenesis. *The Journal of Biological Chemistry*, *278*, 28434–28443.

Wu, Y., Song, P., Xu, J., Zhang, M., & Zou, M. H. (2007). Activation of protein phosphatase 2A by palmitate inhibits AMP-activated protein kinase. *The Journal of Biological Chemistry*, *282*, 9777–9788.

Xiao, B., Heath, R., Saiu, P., Leiper, F. C., Leone, P., Jing, C., et al. (2007). Structural basis for AMP binding to mammalian AMP-activated protein kinase. *Nature*, *449*, 496–500.

Xiao, B., Sanders, M. J., Carmena, D., Bright, N. J., Haire, L. F., Underwood, E., et al. (2013). Structural basis of AMPK regulation by small molecule activators. *Nature Communications*, *4*, 2017.

Xiao, B., Sanders, M. J., Underwood, E., Heath, R., Mayer, F. V., Carmena, D., et al. (2011). Structure of mammalian AMPK and its regulation by ADP. *Nature*, *472*, 230–233.

Zhou, G., Myers, R., Li, Y., Chen, Y., Shen, X., Fenyk-Melody, J., et al. (2001). Role of AMP-activated protein kinase in mechanism of metformin action. *The Journal of Clinical Investigation*, *108*, 1167–1174.

Zorn, J. A., & Wells, J. A. (2010). Turning enzymes ON with small molecules. *Nature Chemical Biology*, *6*, 179–188.

CHAPTER TWENTY-NINE

Assessing the Catalytic Activity of Transglutaminases in the Context of Autophagic Responses

M. D'Eletto*, M.G. Farrace*, M. Piacentini*,†,1, F. Rossin*
*University of Rome "Tor Vergata", Rome, Italy
†National Institute for Infectious Diseases, IRCCS "Lazzaro Spallanzani", Rome, Italy
[1]Corresponding author: e-mail address: mauro.piacentini@uniroma2.it

Contents

1. Introduction 512
2. TG Transamidating Activity Assay 513
   2.1 Assay Procedure 515
3. Analysis of TG2 Degradation During Autophagy 516
   3.1 Assay Procedure 516
4. Analysis of TG2 Interaction With p62 During Autophagy 517
   4.1 Assay Procedure 517
References 519

Abstract

The human transglutaminases (TGases) are a widely distributed and peculiar group of enzymes that catalyze the posttranslational modification of proteins by the formation of isopeptide bonds. Tissue or type 2 transglutaminase (TG2) represents the most ubiquitous isoform belonging to TGases family. The vast array of biochemical functions catalyzed by TG2 distinguishes it from the other members of the TGase family. In the presence of high calcium levels TG2 catalyzes a vast array of protein posttranslational modifications, including protein–protein cross-linking, incorporation of primary amines into proteins, as well as glutamine deamination.

In the last few years, it has become evident that TG2 is involved in the final maturation of autolysosomes. The TG2 regulation of autophagy occurs by its transamidating activity and its inhibition results in the intracellular increase of ubiquitinated protein aggregates. In this chapter, we describe the methods used in our laboratories to assess the catalytic activity of TG2 in the autophagic process.

Methods in Enzymology, Volume 587
ISSN 0076-6879
http://dx.doi.org/10.1016/bs.mie.2016.10.011

1. INTRODUCTION

Tissue or type 2 transglutaminase (TG2) (Grosso & Mouradian, 2012) is a multifunctional ubiquitously expressed member of TG (TGs) family that catalyzes posttranslational modifications of proteins through Ca^{2+}-dependent reactions (Folk & Cole, 1965; Schrode & Folk, 1978).

TG2 has also a protein disulphide isomerase as well as isopeptidase activity and can hydrolyze the isopeptide bond. Beyond its main transamidating activity, TG2 presents other several enzymatic functions (Nurminskaya & Belkin, 2012). Its GTPase activity has been shown to impact signal transduction as well as the ability to bind and hydrolyze GTP distinguishing TG2 from other TGs (Fesus & Piacentini, 2002). Furthermore, TG2 interacts with several proteins as an adhesion or scaffold protein (Nurminskaya & Belkin, 2012).

In the presence of calcium, TG2 adopts an "open" and active conformation that cross-links selected glutamine residues on the surfaces of target proteins to a variety of small molecules or macromolecular amines (Pinkas, Strop, Brunger, & Khosla, 2007) (Fig. 1). Conversely, in the absence of calcium TG2 assumes a "closed" and transamidating catalytically inactive conformation.

Accumulating evidences have proposed a role of TG2 in autophagy (Akar et al., 2007; D'Eletto et al., 2009; Rossin, D'Eletto, Macdonald, Farrace, & Piacentini, 2012). The first direct involvement of TG2 protein in the autophagic process has been revealed in highly metastatic pancreatic carcinoma cells, where the knockdown of TG2 protein leads to the accumulation of LC3II and autophagic vacuoles in the cytoplasm (Akar et al., 2007). Moreover, it has been demonstrated that the ablation of TG2 protein results in an evident accumulation of LC3II on preautophagic vesicles associated with a lack of acidification of the autophagosome–lysosomes suggesting an impairment in the autophagosomes maturation process (D'Eletto et al., 2009).

In particular, TG2 and its transamidating activity have shown to be important for the recruitment of cargos into the autophagolysosomes, indeed its ablation leads to intracellular accumulation of ubiquitinated protein aggregates and defective mitochondria, which are not cleared by autophagy (D'Eletto et al., 2012; Rossin et al., 2015). In keeping with this assumption, specific activation of the enzyme's transamidating activity upon mitochondrial damage by treatment with carbonyl cyanide *m*-chlorophenyl hydrazine (CCCP) has been demonstrated (Rossin et al., 2015). This TG2

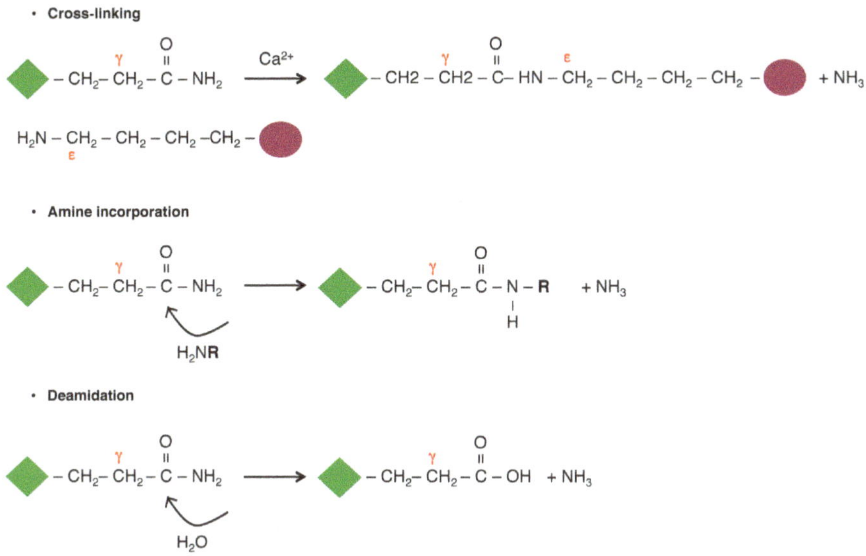

Fig. 1 Enzymatic reactions catalyzed by transglutaminases (TGs). TGs catalyze Ca^{2+}-dependent acyl-transfer reaction between γ-carboxamide group of a specific protein-bound glutamine and either the ε-amino group of a distinct protein-bound lysine residue (cross-linking) or primary amines such as polyamines and histamine (amine incorporation). Water can replace amine donor substrates, leading to deamidation of the recognized glutamines (deamidation).

action is mediated by its interaction with key molecules controlling the clearance of autophagic substrates such as: LC3II, p62, and NBR1 (D'Eletto et al., 2012).

A recent study has revealed a new mechanism by which TG2 cross-links p53 and transports it to the autophagosomes suggesting that TG2 acts as a chaperone of p53 with a cross-linking catalytic activity (Kang et al., 2016).

2. TG TRANSAMIDATING ACTIVITY ASSAY

TG2-dependent posttranslational modification of its substrates is increased during autophagy induction. In this regard, it is accepted that during the early stage of autophagy there is an accumulation of free calcium ions in the cytoplasm that may activate the TG2's transamidating activity (Høyer-Hansen et al., 2007). It has been shown that TG2 activity is stimulated either by starvation (Fig. 2) or rapamycin (Altuntas et al., 2015) but also by other stimuli known to induce the autophagic process, such as proteasome inhibition by MG132 (D'Eletto et al., 2012),

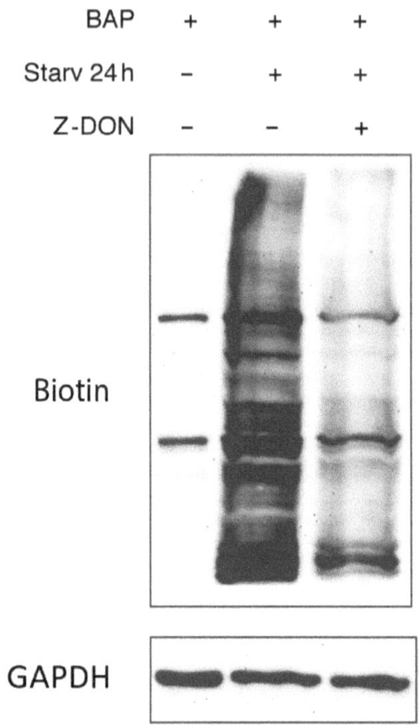

Fig. 2 Analysis of TG2 transamidating activity. Representative blot of the TG-catalyzed incorporation of 5-(biotinamido)pentylamine (BAP) into proteins in 2fTGH (human fibrosarcoma) as a measure of intracellular TG2 activity. The cells were labeled with BAP and subjected to starvation for 24 h in the absence or presence of Z-DON. After separation by SDS-PAGE, biotinylated proteins were revealed with an antibiotin primary antibody. GAPDH was used as a loading control.

mitochondrial damage by CCCP, and glycolysis inhibition by 2-deoxy-D-glucose (Rossin et al., 2015).

The transamidating activity of TGs has been shown to catalyze the incorporation of primary amines into protein substrates through a replacement reaction to yield a modified protein and ammonia (Mycek, Clarke, Neidle, & Waelsch, 1959). The methods used to measure the transamidating activity are usually based on the TGs dependent incorporation of radioactive, fluorescent, or colorimetric mono- and polyamines into substrates. Artificial polyamine substrates of TGs such as [14C]-labeled putrescine (Folk et al., 1980), monodansylcadaverine (Lorand, Murthy, Velasco, & Karush, 1986), biotin-cadaverine (Kunioka & Ando, 1996), and fluorescein-cadaverine (Gross, Balklava, & Griffin, 2003) were widely used to evaluate the TGs activity. In this regard, in vitro and in situ assay can be performed using TGs synthetic substrates such as *N,N′*-dimethylcasein.

A simple protocol to detect intracellular transamidating activity, by evaluating intracellular incorporation of amines into TGs substrates, can be performed using the 5-(biotinamido)pentylamine compound (BAP) associated to SDS-electrophoresis (Ikura, Kita, Fujita, Hashimoto, & Kawabata, 1998). BAP is a biotin molecule that has been conjugated to pentylenediamine to add a short, amine-terminated spacer to its valeric acid group, providing a simple amine-modified compound.

2.1 Assay Procedure

- Dissolve BAP (Soltec Ventures, Beverly, MA, USA, B110) in complete medium or EBSS (Earle's balanced salt solution), for starvation treatments, to prepare a 100 m*M* stock solution.
- Plate the cells and add the BAP in the culture medium to a final concentration of 2 m*M*. BAP can be added in the medium together with the drugs used to induce autophagy.
- Collect the cells in a lysis buffer containing 20 m*M* Tris–HCl, pH 7.4, 150 m*M* NaCl, and 1% Triton X-100 with protease inhibitors.
- No centrifugation of the cell lysate is suggested to avoid the loss of high-molecular-weight protein aggregates. Sonicate the samples to shear DNA. Adjust sonication time to your type of sample.
- After quantification, resolve 60 μg of proteins on SDS-polyacrylamide (10%) gel and transfer to a nitrocellulose membrane.
- Block the blots in 5% nonfat dry milk for 1 h at room temperature.
- To detect the biotinylated substrates of TGs, the blots can be probed with HRP-conjugated streptavidin (Amersham Biosciences) for 2 h at room temperature and then developed following the standard ECL protocol. Otherwise, the blots can be incubated overnight with an antibiotin primary antibody (Abcam, 1227) followed by incubation with HRP-conjugated secondary antibody for 1 h at room temperature.

To assess if the transamidating activity is due to TG2 specific activation and not to other transglutaminases, it is possible to use Z-DON, a highly specific TG2 inhibitor.

- Prepare a 40 m*M* stock solution in DMSO of the inhibitor Z-DON (Zedira, Darmstadt, Germany; Z006).
- Add Z-DON to the culture medium at the final concentration 40 μ*M* in combination with BAP.
- Proceed as described earlier.

3. ANALYSIS OF TG2 DEGRADATION DURING AUTOPHAGY

It is accepted that TG2 is degraded inside the autophagosomes during prolonged autophagy stimuli both in vivo and in vitro. Moreover, it has been shown that the inhibition of the autophagic flux blocks TG2 degradation (D'Eletto et al., 2012).

3.1 Assay Procedure

Preparation of protein extract from liver:

- Starve the mice for 24 or 48 h.
- Kill the animal by decapitation and immediately wash the tissue in PBS (phosphate-buffered saline).
- Cut the tissue into small pieces using scissors.
- Discard the bloody and wash once again in PBS.
- Transfer the tissue pieces to the 10-mL glass/Teflon Potter homogenizer. Add "embryo lysis buffer," containing 50 m*M* Tris–HCl, pH 7.4, 50 m*M* NaCl, 1% Triton X-100, 10% glycerol, 320 m*M* sucrose plus protease inhibitor, in the ratio 3 mL of buffer per gram of liver.
- Homogenization as well as the following steps must be carried out at 4°C (i.e., in cold room) to minimize the activation of proteases and phospholipases.
- Homogenize the liver pieces using a Teflon pestle by six strokes at 1300 rpm.
- Keep the samples on ice for 30 min.
- Transfer the homogenate in tubes and centrifuge at 10,000 × *g* for 30 min at 4°C to pellet cell debris.
- Collect and transfer the supernatant to a fresh tube without disturbing the pellet.
- Determine protein concentration of the lysate.

Preparation of protein extract from cells:

- Plate the cells in the appropriate medium and treat them with the drugs used to induce autophagy.
- Collect the cells in a lysis buffer containing 20 m*M* Tris–HCl, pH 7.4, 150 m*M* NaCl, and 1% Triton X-100 with protease inhibitors.
- No centrifugation of the cell lysate is suggested to avoid the loss of high-molecular-weight protein aggregates. Sonicate the samples to shear DNA. Adjust sonication time to your type of sample.

– Determine protein concentration of the lysate.

TG2 protein degradation can be assessed by western blotting analysis.

– After quantification, resolve the proteins (30–100 μg) on SDS-polyacrylamide (10%) gel and transfer to a nitrocellulose membrane.
– Block the blots in 5% nonfat dry milk for 1 h at room temperature.
– Probe the blots overnight with an anti-TG2 antibody (NeoMarkers, CUB7402 and Cell Signaling, 3557 to detect, respectively, the human and the mouse TG2).
– Incubate the blots with HRP-conjugated secondary antibody for 1 h at room temperature and then develop following the standard ECL protocol.

4. ANALYSIS OF TG2 INTERACTION WITH P62 DURING AUTOPHAGY

Several studies report the interaction of TG2 with the autophagy cargo protein p62. This cargo protein contains a UBA domain at its C-terminal region, allowing it to interact with the ubiquitinated proteins. Subsequently, by a LIR domain p62 interacts with the ATG8-family members inside the nascent preautophagic vesicles. It has been shown the interaction of the two proteins upon starvation and inhibition of the proteasome (D'Eletto et al., 2012), a condition known to favor the formation of ubiquitinated aggregates then removed by the autophagic process. A binding assay using a series of deletion mutants of p62 and TG2 revealed that the PB1 (Phox and Bem1p-1) domain of p62 (residues 85–110) directly interacts with the β-barrel domains of TG2 (residues 592–687) during starvation (Kang et al., 2016). It is possible to study autophagy induction using a coimmunoprecipitation assay to reveal the interaction between TG2 and p62 (Fig. 3).

4.1 Assay Procedure

– Plate the cells in the appropriate medium and treat them with the drugs used to induce autophagy.
– Collect the cells in a lysis buffer (IP buffer) containing 150 m*M* NaCl, 50 m*M* Tris–HCl, pH 7.5, 1% NP 40, and 2 m*M* EDTA with protease inhibitors.
– No centrifugation of the cell lysate is suggested to avoid the loss of high-molecular-weight protein aggregates. Sonicate the samples to shear DNA. Adjust sonication time to your type of sample.

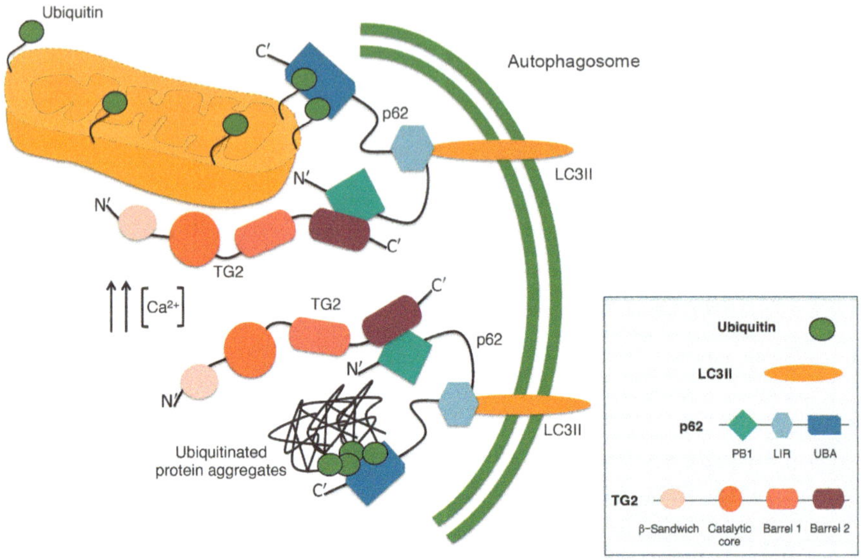

Fig. 3 Model of the interactions of TG2 and p62. The accumulation of ubiquitinated proteins and organelles leads to the activation of autophagy causing the release of free calcium ions in the cytoplasm. This event induces the activation of TG2 transamidating activity associated with its structural change from the inactive "closed conformation" to the active "opened conformation." TG2 catalyzes the misfolded protein cross-linking leading to the formation of ubiquitinated aggregates. p62, by interacting with TG2, recognizes these complexes via the UBA domain. Finally, p62 interacts with LC3II resulting in the uploading of aggregates inside the autophagosome.

- Determine protein concentration of the lysate.
- Preclearing the lysate (1 mg of proteins in 1 mL of IP buffer) adding 30 μL of Protein G PLUS-Agarose (Santa Cruz, sc-2002). Incubate for 1 h at 4°C under rotary agitation.
- Spin in centrifuge at 1000 × *g* at 4°C for 5 min.
- Discard bead pellet and keep supernatant for immunoprecipitation.
- Prepare an antibody conjugate mixture containing 80 μL of Protein G PLUS-Agarose and 2 μg of antibody in 500 μL of PBS with protease inhibitors and incubate for 4 h under rotary agitation.
- Wash the mixture two times with 1 mL of PBS with protease inhibitors.
- Mix the cleared lysate with the antibody conjugate mixture and incubate at 4°C under rotary agitation overnight.
- Wash the beads two times with 1 mL of IP buffer with protease inhibitors and perform the last wash in 1 mL of PBS with protease inhibitors.
- Carefully remove as much buffer as possible from the beads.

– Add 40 μL of 4 × SDS loading buffer with β-mercaptoethanol and boil for 10 min at 95°C.
– Pellet beads and transfer supernatant to a new tube.
– Analyze content of the sample by western blot.

REFERENCES

Akar, U., Ozpolat, B., Mehta, K., Fok, J., Kondo, Y., & Lopez-Berestein, G. (2007). Tissue transglutaminase inhibits autophagy in pancreatic cancer cells. *Molecular Cancer Research*, *5*, 241–249.

Altuntas, S., Rossin, F., Marsella, C., D'Eletto, M., Diaz-Hidalgo, L., Farrace, M. G., et al. (2015). The transglutaminase type 2 and pyruvate kinase isoenzyme M2 interplay in autophagy regulation. *Oncotarget*, *6*, 44941–44954.

D'Eletto, M., Farrace, M. G., Falasca, L., Reali, V., Oliverio, S., Melino, G., et al. (2009). Transglutaminase 2 is involved in autophagosome maturation. *Autophagy*, *5*, 1145–1154.

D'Eletto, M., Farrace, M. G., Rossin, F., Strappazzon, F., Giacomo, G. D., Cecconi, F., et al. (2012). Type 2 transglutaminase is involved in the autophagy-dependent clearance of ubiquitinated proteins. *Cell Death and Differentiation*, *19*, 1228–1238.

Fesus, L., & Piacentini, M. (2002). Transglutaminase 2: An enigmatic enzyme with diverse functions. *Trends in Biochemical Sciences*, *27*, 534–539.

Folk, J. E., & Cole, P. W. (1965). Structural requirements of specific substrates for guinea pig liver transglutaminase. *The Journal of Biological Chemistry*, *240*, 2951–2960.

Folk, J. E., Park, M. H., Chung, S. I., Schrode, J., Lester, E. P., & Cooper, H. L. (1980). Polyamines as physiological substrates for transglutaminases. *The Journal of Biological Chemistry*, *255*, 3695–3700.

Gross, S. R., Balklava, Z., & Griffin, M. (2003). Importance of tissue transglutaminase in repair of extracellular matrices and cell death of dermal fibroblasts after exposure to a solarium ultraviolet A source. *The Journal of Investigative Dermatology*, *121*, 412–423.

Grosso, H., & Mouradian, M. M. (2012). Transglutaminase 2: Biology, relevance to neurodegenerative diseases and therapeutic implications. *Pharmacology & Therapeutics*, *133*, 392–410.

Høyer-Hansen, M., Bastholm, L., Szyniarowski, P., Campanella, M., Szabadkai, G., Farkas, T., et al. (2007). Control of macroautophagy by calcium, calmodulin-dependent kinase kinase-beta, and Bcl-2. *Molecular Cell*, *25*, 193–205.

Ikura, K., Kita, K., Fujita, I., Hashimoto, H., & Kawabata, N. (1998). Identification of amine acceptor protein substrates of transglutaminase in liver extracts: Use of 5-(biotinamido) pentylamine as a probe. *Archives of Biochemistry and Biophysics*, *356*, 280–286.

Kang, J. H., Lee, J. S., Hong, D., Lee, S. H., Kim, N., Lee, W. K., et al. (2016). Renal cell carcinoma escapes death by p53 depletion through transglutaminase 2-chaperoned autophagy. *Cell Death & Disease*, *7*, e2163.

Kunioka, Y., & Ando, T. (1996). Innocuous labeling of the subfragment-2 region of skeletal muscle heavy meromyosin with a fluorescent polyacrylamide nanobead and visualization of individual heavy meromyosin molecules. *Journal of Biochemistry*, *119*, 1024–1032.

Lorand, L., Murthy, S. N., Velasco, P. T., & Karush, F. (1986). Identification of transglutaminase substrates in inside-out vesicles from human erythrocytes: Immunoblotting with anti-dansyl antibody. *Biochemical and Biophysical Research Communications*, *134*, 685–689.

Mycek, M. J., Clarke, D. D., Neidle, A., & Waelsch, H. (1959). Amine incorporation into insulin as catalyzed by transglutaminase. *Archives of Biochemistry and Biophysics*, *84*, 528–540.

Nurminskaya, M. V., & Belkin, A. M. (2012). Cellular functions of tissue transglutaminase. *International Review of Cell and Molecular Biology, 294*, 1–97.

Pinkas, D. M., Strop, P., Brunger, A. T., & Khosla, C. (2007). Transglutaminase 2 undergoes a large conformational change upon activation. *PLoS Biology, 5*, e327.

Rossin, F., D'Eletto, M., Falasca, L., Sepe, S., Cocco, S., Fimia, G. M., et al. (2015). Transglutaminase 2 ablation leads to mitophagy impairment associated with a metabolic shift towards aerobic glycolysis. *Cell Death and Differentiation, 22*, 408–418.

Rossin, F., D'Eletto, M., Macdonald, D., Farrace, M. G., & Piacentini, M. (2012). TG2 transamidating activity acts as a reostat controlling the interplay between apoptosis and autophagy. *Amino Acids, 42*, 1793–1802.

Schrode, J., & Folk, J. E. (1978). Transglutaminase-catalyzed cross-linking through diamines and polyamines. *The Journal of Biological Chemistry, 253*, 4837–4840.

AUTHOR INDEX

Note: Page numbers followed by "*f*" indicate figures, and "*t*" indicate tables.

A

Aasland, R., 259–260, 458–459
Abada, A., 378–379
Abdelmohsen, K., 23, 44, 56–58, 66–67, 73–74, 77–80, 92, 186–187, 190, 352, 362–363
Abe, A., 23, 44, 56–58, 66–67, 73–74, 77–80, 92, 186–187, 190, 352, 362–363
Abecassis, P.-Y., 449, 458, 461–462
Abedin, M.J., 23, 44, 56–58, 66–67, 73–74, 77–80, 92, 186–187, 190, 352, 362–363
Abeliovich, H., 23, 44, 56–58, 66–67, 73–74, 77–80, 92, 186–187, 190, 352, 362–363
Abert, C., 378–379
Abrahamsen, H., 172–173, 430–431
Abrams, J.M., 73
Abu-Asab, M., 332
Acevedo Arozena, A., 23, 44, 56–58, 66–67, 73–74, 77–80, 92, 186–187, 190, 352, 362–363
Adachi, W., 134, 144–145, 378
Adijanto, J., 47
Adjemian, S., 74
Adrian, M., 334
Agostinis, P., 74
Aguirre, A., 192–193
Ahlberg, J., 273–274
Ahlen, G., 200–203
Ahn, H.J., 212–213
Ai, H.W., 89
Akar, U., 512
Akin, D., 209–210
Akira, S., 72
Akopiants, K., 407–408
Akutsu, M., 144–146, 158–159, 160–161*f*
Albanesi, J., 272–274, 285
Albers, M.W., 406–407
Albiero, F., 430–431, 435, 443
Albright, R.A., 486–487
Alcazar-Roman, A., 259, 458–459
Alemu, E.A., 155–156, 259–260, 313
Alers, S., 392, 394–395, 398, 400–402, 482–484
Alessi, D.R., 466–467
Ali, S.M., 406–407, 409
Allart, S., 312
Almendinger, J., 44–45
Alnemri, E.S., 73
Altuntas, S., 513–514
Alvarez-Venegas, R., 312
Amaravadi, R.K., 72, 248–249
Amin, P., 312–313
An, Z., 172–173
Anderson, D.E., 122–123
Anderson, K.A., 484–485, 491
Anderson, K.E., 296
Anderson, M., 172–173, 259–260
Anderson, S.N., 492–494
Ando, T., 514
Andreelli, F., 486–487
Andrews, N.W., 51–52
Andrews, S.B., 333–334, 342–343
Antkowiak, A., 293–294
Antonioli, M., 430–443
Anwar, T., 332, 334, 336, 341–342
Apel, A., 483–484
Aplin, A., 344–345
Apsel, B., 406–407
Archange, C., 3
Artioli, G., 117–118
Asai, E., 99, 209–210
Asara, J.M., 393–394, 407
Ashkenazi, A., 272–274, 293–295, 313, 314–315*f*, 316, 318, 320, 322, 327*f*
Ashworth, A., 466–467
Astle, C.M., 83
Auer, M., 333–334
Augustyn, S., 200–203
Authier, F., 52
Avruch, J., 424
Axe, E., 3, 258
Axe, E.L., 22, 260–261, 273–274

B

Baalbaki, M., 312
Baas, A.F., 491
Baba, M., 144, 208–209
Baccino, F.M., 361
Bach, M., 172–173, 394, 483–484
Backer, J.M., 260–261
Bae, J.Y., 99
Bae, W.R., 374
Baehrecke, E.H., 72–73, 248–249
Bagley, A.F., 406–407
Bago, R., 449
Bailey, M.F., 489–490
Bailly, Y., 132–133
Bain, J., 484–485, 491
Bakula, D., 352–353, 355, 356*t*, 358, 361–363
Balasubramani, M., 172–173
Balklava, Z., 514
Ball, G., 13–14
Balla, T., 258–259, 273–274, 312, 315
Ballou, L.M., 448–449
Baluska, F., 312
Bampton, E.T.W., 209–210
Bandhakavi, S., 406
Bank, J.A., 126
Bara, M.A., 230
Bardeesy, N., 410–411, 466–467, 484–485, 491
Barfeld, S.J., 352–353, 355, 356*t*, 358, 361–363
Barnes, C.A., 400–401
Bar-Peled, L., 406–408, 469–470, 473
Barr, F.A., 23
Barret, C., 277–278
Barrett, J.C., 228–230
Barrett, L.W., 492–494
Barrnett, R.J., 333–334, 342–343
Bashford, D., 127
Baskaran, S., 313, 378, 448–449
Bastholm, L., 513–514
Bates, M., 14–19
Baumgart, T., 378–379
Bavro, V.N., 130–132, 233–234*t*
Bax, A., 123
Bayer, P., 147
Bayliss, J., 484–485
Bayly, C.I., 127
Beau, I., 258–259
Beauchamp, K.A., 127
Beaulieu, V., 497
Beberg, A.L., 126
Bebernitz, G.R., 497
Becken, U., 276, 289
Becker, A.C., 397–400
Becker, K., 125–126, 132–133
Becker, N., 172–173
Bedford, F.K., 132–133
Behl, C., 144
Behrends, C., 36–37, 158–159, 165–166, 430–431, 483–484
Beier, V., 145, 378–379
Bekes, M., 211–213, 220
Belkin, A.M., 512
Bellott, M., 127
Benashski, S., 467–468
Bennett, E.J., 172–173
Bento, C.F., 23, 293–294
Berendsen, H.J.C., 128
Berg, T.O., 273–274, 344–345
Berger, I., 99–100
Bergeron, J.J., 52
Bergman, P., 424–425
Beri, R.K., 466–467, 484–485, 490
Berissi, H., 172–173
Berleth, N., 394–395
Berne, B.J., 129–130
Bernreuther, D., 208
Beron, W., 260
Bertrand, L., 486–489
Bessereau, J.L., 334–335
Bestebroer, J., 352
Betin, V.M.S., 211
Betz, H., 130–132, 233–234*t*
Betzel, C., 233–234*t*
Beuchat, M.H., 470–471
Bhagat, G., 172–173
Biagini Myers, J.M., 228–230
Biazik, J.M., 332–346
Bilanges, B., 448
Billiar, T.R., 195
Binda, O., 315–316
Biniossek, M.L., 435
Birgisdottir, Å., 22–23
Birgisdottir, Å.B., 132–133, 144–166
Birkeland, H.C., 259–260
Birnbaum, M.J., 486–487
Bjorkoy, G., 144–146, 165–166

Bjune, J.I., 486–487
Blagoev, B., 431, 433, 439–442
Blagosklonny, M.V., 73
Blandino, G., 487–489
Blankson, H., 344–345
Blenis, J., 407
Bligh, E.G., 283
Blomgren, K., 74
Bloor, S., 144–145
Blowers, E.C., 486–487
Boal, F., 294–295, 312, 315, 318–319
Bockus, A., 66
Bodenhausen, G., 123
Boeckers, T.M., 333
Boerries, M., 435
Boesze-Battaglia, K., 44–53
Bogdanov, A.M., 4
Bohdanowicz, M., 260–261
Bohler, P., 394–395
Bohley, P., 352–353, 355, 358, 361–363
Bolte, S., 93
Bordallo, J., 208, 210–211
Bos, E., 277–278
Bosc, D., 211–213
Boscheron, C., 296, 306
Botbol, Y., 56
Boudeau, J., 410–411, 466–467, 484–485, 491
Boudier, T., 333
Bouley, R., 172–173
Boulianne, G.L., 272
Boura, E., 313
Bowen, M., 312–328
Boxer, S.G., 374
Boya, P., 248–249, 272
Boyer, S.H., 486–487, 492–494, 501
Boyer-Guittaut, M., 172–173, 208
Bozickovic, O., 486–487
Bracher, A., 130–132, 233–234*t*
Braddock, D.T., 486–487
Brady, N.R., 144, 164, 279
Bragin, A., 44–47, 50–52
Brancato, J., 44–47, 50–52
Brandon, N.J., 132–133
Bravo-San Pedro, J.M., 57–58, 66–67, 72–84, 248–249
Breakefield, X.O., 212
Brech, A., 23, 58, 144–145, 165–166, 259–260, 313
Bredschneider, M., 208
Brewster, A.S., 138
Bridges, D., 318
Bright, N.J., 484–489
Broadfield, L.A., 487–489
Brodovitch, A., 473
Brown, E.J., 406–407
Brown, H.M., 334–335
Brown, K., 430–431
Brown, N.G., 101
Bruckner-Tuderman, L., 442
Bruinsma, R., 495
Brünger, A.T., 119, 366, 373–374, 512
Bruun, J.A., 58, 144–146, 259–260, 313
Bryant, N.J., 260–261, 274–275, 458–459
Bua, D.J., 315–316
Bucci, C., 260, 332
Büchler, M.W., 483–484
Bulyk, M.L., 312
Buque, A., 73
Burd, C.G., 259–260
Burke, J.E., 448–449
Burman, C., 258, 285–287, 293–294
Buser, C., 334
Butsch Kovacic, M., 228–230
Buys, A.V., 333
Byfield, M.P., 260–261

C

Cabanillas, R., 192–194, 209–210
Cabrera, S., 192–194, 209–210
Cacheux, W., 230
Cadwell, K., 98–99
Cai, W., 260–261
Calabrese, M.F., 484–489, 492–494, 496–497
Caldwell-Busby, J., 393–394
Cameron, K.O., 486–489
Campanella, M., 513–514
Campbell, D.G., 490–491
Campbell, R.E., 89
Cande, W.Z., 322, 325–326
Cantaloube, I., 344–345
Cantley, L.C., 312, 406–407, 466–467
Cantor, J.R., 407–408
Cao, J., 44, 394–395, 398, 400–401, 483–484
Cao, N., 486–487, 501
Cao, Y., 144, 208–209

Cargnello, M., 406–407, 482–483
Carling, D., 466–468, 482, 484–487, 489–494
Carlson, L.A., 448–449
Carlsson, S.R., 249
Carlton, J.G., 273–274
Carlton, P.M., 322, 325–326
Carmena, D., 484–489
Carney, D.P., 497
Carratu, M.R., 409
Carroll, B., 408
Caspers, N.L., 484–485, 487–489, 492–494, 496–497
Castelli, L.A., 466–467, 484–485, 489–490, 500–501
Catanzaro, J., 448–449
Cavanagh, J., 122
Cavett, V., 172–173, 393–394, 483–484
Cebollero, E., 293–294
Ceccato, L., 293–294, 312
Cecconi, F., 72, 248–249, 430–431, 512–514, 516–517
Ceizar, M., 98–99
Cha, J., 315–316
Chait, B.T., 430–431
Chakrama, F.Z., 192–193, 208
Chan, C.S., 56
Chan, E.Y., 23, 312–313, 344–345, 392, 394–395, 400–401
Chan, E.Y.W., 483–484
Chan, F.L., 333
Chan, Y.H.M., 374
Chandak, P.G., 52
Chandra, A., 99
Chandra, P., 3, 22, 258, 260–261, 273–274
Chang, J.J., 334
Chang, J.W., 92, 406–407
Chang, Y.Y., 172–173, 394–395, 400–401, 483–484
Chantranupong, L., 407–408
Chapel, A., 466–467
Chatham, J., 172–173
Chaumorcel, M., 44
Chavrier, P., 374, 470–471
Chen, B.C., 323–324
Chen, C.X., 312
Chen, G., 144
Chen, J., 98–99, 273, 320
Chen, J.S., 448–449
Chen, K., 430–431
Chen, L., 466–467, 484–486
Chen, M., 144
Chen, Q., 144
Chen, S., 172–173, 393–395, 400–401, 430–431, 483–484
Chen, W.W., 407–408
Chen, X., 193, 208, 212–214, 216–220, 222
Chen, Y., 483–484, 486–487, 492–495
Chen, Z., 195
Chen, Z.-P., 466–467, 484–490, 494–495, 500–501
Cheng, Y., 99, 211
Cheong, H., 392
Cherniack, A.D., 407–408
Chernomordik, L.V., 366
Cherra, S.J., 172–173
Chevallier, J., 470–471
Chevtzoff, C., 486–489
Chicanne, G., 293–308, 312, 315–316
Chiou, W., 486–487, 492–494, 501
Chiquet, M., 333–334
Chiu, S.M., 98–99
Cho, E.H., 93
Cho, J.H., 228–230
Choi, A.M., 72, 190
Choi, K.M., 212–213
Chowdhury, S., 127
Chu, K., 172–173
Chu, Q., 486–487
Chu, S., 373
Chuang, R.Y., 101
Chun, M.G., 394–395, 400–402
Chung, S.I., 514
Cianfanelli, V., 73–74, 430–431
Ciccosanti, F., 430–443
Cieplak, P., 127
Cipriano, D.J., 374
Ciruela, A., 296, 308
Clague, M.J., 22, 313
Clark, J., 296
Clarke, D.D., 514
Clarke, J.H., 272–274, 294–295, 313, 314–315*f*, 316, 318–320, 322, 327*f*
Clarke, N.G., 283–284
Clarke, P.A., 332
Clarke, P.R., 485–486, 490
Clarke, R., 332
Clausen, T.H., 58, 144–145, 259–260

Cleveland, J.L., 172–173, 393–394, 483–484
Coats, J.E., 99
Cocco, S., 512–514
Codogno, P., 44, 72, 248–249, 258–259, 272, 312–313, 358
Coenen, S., 172–173, 393–394, 483–484
Cohen, A., 44–47, 50–52
Cohen, G.M., 209–210
Cohen, P.T., 466–467
Cole, P.W., 512
Colecchia, D., 172–173
Collinson, L.M., 23, 31, 36–37
Colombo, M.I., 260
Comb, W.C., 83
Connell, S., 44–45
Conway, O.J., 401–402
Cook, K.L., 332
Cool, B.L., 486–487, 492–494, 501
Cooper, H.L., 514
Corazzari, M., 430–431, 435, 443
Cordelieres, F.P., 93
Cornell, W.D., 127
Cornella-Taracido, I., 449, 461–462
Corrotte, M., 51–52
Cosio, G., 260–261
Costelli, P., 361
Costes, S.V., 93
Coutts, A.S., 56, 66
Cowtan, K., 119
Cox, J., 439–442
Coyle, J.E., 130–132, 134
Crawshaw, S.G., 23
Cregg, J.M., 191–192
Crespo, J.L., 194–195, 209–210
Cristo, A., 265
Cuervo, A.M., 56, 72, 116, 190, 194, 258
Cullen, P.J., 273–274
Cummings, R.T., 495
Cunha, L.D., 44–45
Cunningham, B.R., 495
Curak, J., 146
Cypionka, A., 374

D

Daelemans, D., 93
Dahl, R., 342–343
Dahlquist, F.W., 122–123
Dai, X.F., 212–213
Dale, B.A., 497
Dale, S., 484–486, 490–491
Dall'Armi, C., 172–173, 258–259, 312–313, 458–459
Dancourt, J., 99, 378–379
Dang, Q., 486–487, 492–494, 501
Darden, T., 128
Darley-Usmar, V., 172–173
Das, R., 127
Davidson, M.W., 89, 323–324
Davies, P., 449
Davies, S.P., 466–467, 484–485, 490–494
Davis, I., 13–14
Davis, R.J., 448–449
Davison, M., 466–467, 484–485, 490
Dawson, R.M., 283–284
Day, B.W., 172–173
De Angelis, D.A., 89–90
De Camilli, P., 258–260, 282
de Duve, C., 293–294
de Lartigue, J., 22
De Matteis, M.A., 272
de Waal, P.W., 484–485
Deacon, R., 497
Deak, M., 410–411, 491
Debnath, J., 72
Defacque, H., 277–278
Degani, I., 344–345
Delage-Mourroux, R., 192–193, 208
D'Eletto, M., 512–519
Delgado, M.A., 293–294, 448–449
Delvecchio, K., 318, 320
Demetriades, C., 407–408
Demmerle, J., 13–14
Deneroff, M.M., 126
Deng, A., 99, 232, 233–234*t*
Dengjel, J., 392–402, 430–443
DePinho, R.A., 410–411, 466–467, 484–485, 491
Deretic, V., 72, 313
Devereaux, K.A., 258–259, 312–313, 458–459
DeWald, D.B., 273, 320
Dezelak, M., 98–99
Dhaliwal, J.S., 98–99
Dhingra, A., 44–53
Di Bartolomeo, S., 430–431
Di Lello, P., 312
Di Paolo, G., 258–260, 312–313

Diao, J.J., 366–374
Diaz-Hidalgo, L., 513–514
Dibble, C.C., 406–407
Dice, J.F., 293–294
Dickerson, K., 466–467, 484–485, 491
Diedrich, B., 397–400
Dieterle, A.M., 392, 394–395, 398, 400–402, 483–484
Dikic, I., 144, 146–147, 160–161*f*, 163–164
Dillon, C.P., 44–45
Diment, S., 276
Dimova, R., 378–379
Ding, B., 407–408
Ding, H., 217–218, 222
Ding, W., 212–213
Ding, W.-X., 66, 273–274, 285–287
Ding, X., 394–395, 400–401, 430–431, 483–484
Dinglasan, R.R., 233–234*t*
Dingley, A.J., 123
Dite, T.A., 485–489, 494–495
Divecha, N., 302
Djavaheri-Mergny, M., 191
Dobbie, I.M., 13–14
Dobbin, Z., 93
Dobson, R.C., 121, 233–234*t*
Dodson, M., 172–173
Doggett, T.A., 44–45
Donnes, P., 259, 312–313
Dooley, H.C., 22–23, 31, 36–37, 73, 272
Doran, E., 486–487
Dorsey, F., 44–45
Dorsey, F.C., 172–173, 393–395, 483–484
Dotsch, V., 144–166, 160–161*f*
Dou, Z., 448–449
Doumpas, N., 407–408
Dowdle, W.E., 144, 449, 461–462
Drag, M., 211–213, 220
Drakew, A., 334–335
Driessen, S., 394–395
Driscoll, P.C., 259–260, 458–459
Dror, R.O., 126–127
Drzewoski, J., 486–487
Du, F., 172–173, 393–394, 483–484
Du, J., 47
Du, Y., 172–173
Duan, Y., 127
Dubochet, J., 334, 342–343
Duckworth, B.C., 312
Duda, D.M., 233–234*t*
Duerr, R.H., 228–230
Dull, T., 91
Dunbrack, R.L., 127
Dunham, W.H., 146
Dunn, W.A., 191–192, 209–210, 344–345
Dupuis-Coronas, S., 312, 315–316
Durand, C.M., 333
Durand, L., 449, 459–462
Dureuil, C., 448–462
Durrant, T.N., 296
Duszenko, M., 233–234*t*
Duttaroy, A., 497
Dyer, W.J., 283

E

Eastman, P., 126
Eastwood, M.P., 127
Edelman, A.M., 484–486, 490–491
Edelmann, L., 342–343
Edgar, B.A., 407
Edlind, M.P., 406–407
Efeyan, A., 83, 258–259, 473
Egan, D.A., 492–494
Egan, D.F., 393–395, 400–402, 483–484
Egner, R., 272
Eguchi, S., 272, 448
Eidelman, D., 228–230
Eils, R., 279
Eiselein, S., 431
Eisenstein, M., 172–173
El-Ahmad, Y., 449, 458, 461–462
Elazar, Z., 116–117, 130–132, 135, 145, 193–195, 208–210, 216–217, 258, 344–345, 378–379
Eleanore Seibert, T.S.T., 219
Eliceiri, K.W., 93
Elis, W., 407
Elling, R.A., 144, 449, 461–462
Elmaoued, R.A., 293–294, 448–449
Emoto, K., 384–385
Emr, S.D., 98–99, 191–192, 259–260, 448
Engedal, N., 352–363, 356*t*
Entzeroth, R., 334–335
Erdjument-Bromage, H., 406–407
Erion, D.M., 486–487
Ernst, A., 144

Esclatine, A., 44, 258–259
Escola, J.M., 470–471
Eskelinen, E.L., 260, 332–346
Espanol, Y., 192–193
Evans, A., 486–487
Evanseck, J.D., 127
Ezaki, J., 145, 208, 222

F

Facchinetti, V., 406
Fairbrother, W.J., 122
Falasca, L., 512–514
Falasca, M., 458–459
Faller, J.W., 333–334, 342–343
Fan, W., 172–173, 430–431
Fang, C., 172–173
Fang, Z., 89
Farhan, H., 144, 164
Farkas, T., 513–514
Farrace, M.G., 512–519
Farre, J.C., 144
Fass, D., 130–132, 135, 216–217
Fass, E., 194–195, 208–210, 344–345
Fasshauer, D., 366
Fassy, F., 448–462
Favre, C., 312, 315–316
Feenstra, K.A., 128
Feldman, M.E., 406–407
Fellinger, E., 344–345
Feng, D., 144
Feng, Y., 56, 172, 191–192, 272
Fengsrud, M., 273–274, 344–345
Fenyk-Melody, J., 486–487, 492–494
Ferguson, D.M., 127
Ferguson, T.A., 44–45
Ferkolj, I., 98–99
Fernandes, M.C., 51–52
Fernandez, A.F., 192–194, 209–210
Fernandez, J.L., 212
Ferraris, G., 117–118
Fesus, L., 512
Fetalvero, K.M., 293–294
Field, M.J., 127
Fields, S., 146
Filimonenko, M., 259–260
Filios, C., 318, 320
Filoche-Rommé, B., 449, 458, 461–462
Fimia, G.M., 430–443, 512–514
Finley, K.D., 259–260
Firaguay, G., 312, 315–316
Fiskin, E., 233–234*t*
Fitzgerald, D.J., 99–100
Fitzgerald, P., 44–45
Flamand, O., 449, 459–462
Fleming, A., 72
Fletcher, D.A., 387–388
Florey, O., 44–45
Flurkey, K., 83
Foeglein, A., 145
Fogarty, S., 486–487, 491
Fok, J., 512
Folk, J.E., 512, 514
Fong, W.M., 99, 378–379
Foretz, M., 482, 486–489, 501
Fosse, M., 352, 358
Fracchiolla, D., 378–388
Fraichard, A., 192–193, 208
Frank, R., 155–156
Franke, A., 228–230
Franquelim, H.G., 145, 378–379
Franzone, J.M., 484–485, 491
Franz-Wachtel, M., 398, 400–402, 483–484
Fraser, J.S., 138
Fraudeau, M., 230
Freire, E., 162–165
Frelin, L., 200–203
Freyer, M.W., 162–165
Friedrichs, M.S., 126
Friesner, R.A., 129–130
Frisbie, R.K., 489–495, 498
Frith, D., 36–37
Frost, L.S., 44–53
Frotscher, M., 334–335
Fry, M.J., 259, 448
Fryer, L.G., 484–487, 491
Fryer, L.G.D., 466–467
Fu, Y., 208–222
Fuchs, Y., 73
Fuentes, J.M., 56–69, 258–267
Fueyo, A., 192–193, 209–210
Fugier, C., 312
Fujioka, Y., 98–99, 101, 135–137, 210–212, 233–234*t*, 378–379
Fujita, I., 515
Fujita, N., 332, 335, 337–341
Fullerton, M.D., 485–489

Funderburk, S.F., 44, 448–449
Fuoco, C., 430–431
Furet, P., 449, 461–462

G

Gadalla, A.E., 486–487
Gagne, D.J., 497
Gaits, F., 312
Gaits-Iacovoni, F., 293–308, 312, 318–319
Galic, S., 485–489
Galluzzi, L., 72–84, 248–249
Gamblin, S.J., 466–467
Gambotto, A.A., 273–274, 285–287
Ganley, I.G., 392–395, 400–402, 483–484
Gao, J., 497
Gao, W., 195, 273–274, 285–287
Gao, X., 407
Gao, Y., 92, 406–407
Garami, A., 407
Garuti, R., 430–431
Garvalov, B., 277–278
Gatica, D., 98–99, 228–230, 244
Gaubitz, C., 406–407
Gaullier, J.M., 260–261, 274–275, 458–459
Gayral, M., 312, 315
Ge, L., 378
Geddes, S., 130–132, 233–234*t*
Geeraert, C., 344–345
Geerts, W.J., 334–335
Geissler, E.K., 406–407
Gelato, K.A., 312
Gelin, B.R., 126
Gelino, S., 393–394, 483–484
Geng, J., 228–230
Geoghegan, K.F., 489–490
George, M.D., 172–173, 378
Gerard, A., 312, 315–316
Gerke, V., 52
Gerner, L., 352–363
Geuze, H.J., 333, 343–344
Gherardini, P.F., 430–431
Ghislat, G., 293–294
Giacomo, G.D., 512–514, 516–517
Giacovazzo, C., 117–118
Giardoglou, P., 312
Gibson, D.G., 101
Gil, L., 194–195, 208–210
Gillooly, D.J., 259–261, 274–275, 458–459
Gingras, A.C., 146
Girardin, S.E., 22, 272
Giuriato, S., 312
Glaumann, H., 273–274
Glick, M., 449, 461–462
Godi, A., 272
Goebeler, V., 52
Goeritzer, M., 52
Goetschkes, M., 293–294
Golubovskaya, I.N., 322, 325–326
Gomez-Galeno, J.E., 486–487, 492–494, 501
Gomez-Sanchez, R., 57–58, 66–67, 265
Gong, H.Y., 212–213
Gong, Y., 212–213
Gonzalez, A., 138
Gonzalez-Polo, R.A., 56–69, 258–267
Goossens, T.A., 495
Goraksha-Hicks, P., 407–408
Goransson, O., 410–411, 501
Gordon, P.B., 352–353, 355, 358, 361–363
Gorman, M.A., 486–487
Gorvel, J.-P., 374, 470–471
Goswami, D., 486–487
Gottlieb, R.A., 48
Gottschalk, B., 52
Gould, I.R., 127
Gould, R., 260–261, 274–275, 458–459
Gould, T., 293–294
Gowans, G.J., 466–467, 477, 484–485, 490
Goyette, P., 228–230
Gozani, O., 315–316
Graber, W., 334–335
Grabiner, B.C., 407–408
Gragera-Hidalgo, A., 265
Graham, D.R., 233–234*t*
Grahame Hardie, D., 466–467, 482, 484–486, 490–491
Granton, J., 489
Grasso, D., 3
Gratacap, M.P., 296, 306
Gratian, M.J., 312–328
Graupera, M., 448
Gray, N.S., 92
Green, D.R., 44–45, 72, 82
Green, K.A., 466–467, 486–487
Gressens, P., 467–468
Gretzmeier, C., 435, 442

Griffin, M., 514
Griffin, T.J., 406
Griffith, J., 23
Griffith, J.M., 334–335
Grinde, B., 355
Grindheim, A.K., 312
Grinstein, S., 260–261, 473
Grob, P., 448–449
Grondin, P.O., 489–491
Gross, S.R., 514
Grossman, J.P., 126
Grosso, H., 512
Grote, M.P., 486–487, 492–494, 501
Gruenberg, J., 374, 470–471
Grumati, P., 144, 146
Grunwald, D., 394–395
Gu, X., 484–485
Guan, J., 144
Guan, J.-L., 483–484
Guan, K.L., 44, 172–173, 393–395, 400–401, 406–409, 483–484
Gubas, A., 273
Guertin, D.A., 406, 409
Guigas, B., 486–489
Guillermet-Guibert, J., 448
Guittard, G., 312, 315–316
Guldvik, I.J., 352–353, 355, 356*t*, 358, 361–363
Gulley, R.L., 333
Gunderson, J.E.C., 98–111
Gunning, W.T., 448–449
Guntert, P., 160–161*f*, 163
Guo, H., 466–467
Guo, M.R., 483–484
Gupta, J., 228–230
Gurkar, A.U., 172–173
Gust, A.A., 98–99
Gustafsson, M.G., 322, 325–326
Gutierrez, M.G., 260
Gygi, M.P., 407–408
Gygi, S.P., 144, 165–166, 172–173, 273–274, 430–431, 483–484

H

Ha, J., 486–487
Ha, S.D., 46, 52
Ha, T., 374
Haas, A., 272–289
Håberg, K., 23
Habermann, A., 22, 260–261, 273–274
Habibzadegah-Tari, P., 209–210
Habisov, S., 145, 158–159, 160–161*f*
Hagberg, H., 467–468
Hagen, L.K., 352–353, 356*t*, 358, 361–363
Haglund, K., 147
Hain, A.U., 233–234*t*
Haire, L.F., 484–489
Hait, W.N., 99
Halestrap, A.P., 486–487
Hall, M.N., 406, 409, 425
Ham, B., 46, 52
Hamacher-Brady, A., 279
Hamasaki, M., 249, 293–294
Hambly, N., 489
Hamburger, M., 282, 284
Hamid, Q., 228–230
Hammarback, J.A., 344–345
Hammer, C., 312
Hammond, G.R., 273–274, 315
Hammond, H., 233–234*t*
Hampe, J., 228–230
Han, B.X., 312–313
Han, L., 217–218, 222
Han, S., 244
Hanada, K., 90–91
Hanada, T., 98–99, 101, 233–234*t*, 378–379
Haneburger, I., 276–278, 285
Hansoul, S., 228–230
Hao, Y., 260–261
Hara, K., 406–407
Hara, T., 394–395, 400–401, 483–484
Hardie, D.G., 466–467, 477, 482, 484–487, 490–494
Harper, J.W., 144, 165–166, 172–173, 228–230, 273–274, 430–431, 483–484
Harrington, E., 449, 461–462
Harris, M.S., 484–490, 492–494, 496–497
Harris, R., 130–132, 233–234*t*
Harris, V.M., 175
Harrison, D.E., 83
Hartmann, R., 120–121, 125–126, 129–130, 132–137, 233–234*t*
Hasegawa, H., 312–313
Hashimoto, H., 515
Hatano, M., 73
Haucke, V., 448–449

Hawley, S.A., 466–467, 477, 484–491, 501
Hayashi-Nishino, M., 332, 335, 337–341
Haynes, C.M., 44–45
Haystead, T.A., 490
He, C., 293–294, 430–431
He, D., 56, 172, 191–192, 272
He, F.T., 212–213
He, J., 483–484
Heath, R., 466–467, 484–486, 491
Hébrard, S., 486–487
Hecker, C.M., 147
Hegarty, B.D., 489–491
Heilemann, M., 14–19
Heinonen, T.Y., 312
Heinrich, T., 144, 146
Helms, J.B., 293–294
Helps, N.R., 466–467
Hemelaar, J., 208
Hennessy-Walters, K.M., 394–395
Herman, P.K., 448
Herr, I., 483–484
Hess, B., 126, 128
Hettema, E.H., 144
Hibshoosh, H., 430–431
Hidayat, S., 406–407
Hieke, N., 394–395
High, S., 23
Hilger, M., 439–442
Hill, J.A., 48
Hinchliffe, K.A., 296, 308
Hintze, U., 334–335
Hiragami-Hamada, K., 312
Hirata, M., 130–132
Hirose, E., 407–408
Hirschberg, K., 344–345
Hlavacka, A., 312
Hnia, K., 293–308
Hoekstra, D., 366–367
Hoffmann, Y., 125–126, 132–133
Hofmann, K., 99
Hohenberg, H., 334–335
Holen, I., 344–345, 352, 358
Hollas, H., 312
Holmstrom, F., 200–203
Holthuis, J.C., 334–335
Homo, J.C., 334
Honda, A., 449, 461–462
Hondo, A., 469–470
Hong, B., 208–210
Hong, D., 513, 517
Hong, S., 406–425
Hong, S.B., 233–234*t*, 244
Hong, S.P., 484–485, 491
Honson, N.S., 211–213
Horazdovsky, B., 259
Hosokawa, N., 394–395, 400–401, 483–484
Hoth, L.R., 486–487, 489–495, 498
Hou, Y., 193, 208, 213, 219–220, 222
Houde, V.P., 487–489
House, C.M., 490–492
Houston, M., 126
Hoxhaj, G., 407
Høyer-Hansen, M., 513–514
Hsieh, A.C., 406–407
Hsu, P.P., 406–407
Hu, C., 233–234*t*
Hu, H.X., 233–234*t*
Hu, J., 484–486
Hu, P., 260–261
Hu, X.W., 23, 344–345
Huang, C., 48
Huang, C.M., 217–218, 222
Huang, J., 172–173
Huang, L., 200, 442
Huang, X., 397–400
Huber, J., 144–166, 160–161*f*
Hue, L., 486–487
Huebner, A.K., 144, 146
Huett, A., 228–230
Huis In 't Veld, P.J., 101
Humbel, B.M., 333–335
Hundeshagen, P., 279
Hurle, J.B., 47
Hurley, J.H., 228–230, 313, 378
Hurley, R.L., 410–411, 466–467, 484–485, 491
Huse, K., 228–230
Hutchins, M.U., 144
Hutchison, C.A., 101
Huwiler, K.G., 495
Huynh, K.K., 473

I

Iavarone, C., 172–173
Ibiricu, I., 378–379, 384–385, 387–388
Ichikawa, K., 406–407

Ichimura, Y., 98–99, 101, 116, 145, 158–159, 160–161*f*, 172–173, 208, 211–212, 222, 233–234*t*, 378–379, 384–385
Idelchuk, Y., 172–173
Idevall-Hagren, O., 282
Iemura, S.-I., 483–484
Ii, M., 407–408
Ikehara, S., 491
Ikehara, T., 491
Ikenoue, T., 406
Ikonomov, O.C., 318, 320
Ikura, K., 515
Imamura, S., 491
Imamura, Y., 384–385
Inagaki, F., 72, 98–99, 132–133, 144–145, 210–212, 233–234*t*, 378
Ingolia, N.T., 406–407
Inoki, K., 406–425
Inoue, K., 430–431
Inoue, S., 333
Irvine, R.F., 296, 308, 318–319
Isakson, P., 259–260
Iseli, T.J., 489–490
Ishihama, Y., 400, 438
Ishihara, N., 98–99, 116, 172–173, 378
Ishii, J., 99, 144–145, 209–210, 378
Ishii, T., 172–173, 378
Ishitsuka, Y., 374
Ishiura, S., 90–91
Issa, S.M.A., 485–489, 494–495
Itakura, E., 58, 66–67, 366, 430–431
Ivaldi, C., 466–467
Ivanov, D., 315–316
Ivarsson, Y., 312
Iwai-Kanai, E., 48
Iwanari, H., 293–294
Izzo, V., 72–84

J

Jaber, N., 448–449
Jabs, S., 473
Jacinto, E., 406
Jackson, S., 430–431
Jager, S., 260, 332
Jahn, R., 366
Jahreiss, L., 44, 317, 344–345
Jain, A., 155–156
James, D.E., 172–173, 394, 483–484
Janes, M.R., 406–407
Jang, S., 129–130
Janis, J., 312
Jaravine, V., 159–162
Jasionowski, T., 344–345
Jayabalasingham, B., 233–234*t*
Jefferies, H.B., 73
Jensen, E.C., 173
Jensen, T.E., 484–485
Jentsch, S., 144
Jentsch, T.J., 473
Jeon, H., 233–234*t*
Jeong, H., 259–260
Jeschke, A., 272–289
Jewell, J.L., 407, 424–425
Jia, C.C., 98–99
Jiang, B., 466–467, 475–476
Jiang, F.J., 208–210
Jiang, P., 56–58, 66–67, 230
Jiang, X., 392–395, 400–401, 483–484
Jiang, Y., 312
Jiang, Y.P., 448–449
Jiao, Z.H., 466–467
Jin, J.P., 318
Jin, S., 430–431
Jing, C., 466–467, 484–485
Jo, Y.H., 98–99
Joachim, J., 22–41, 73
Joaquin, M., 407
Johansen, T., 22–23, 58, 132–133, 144–166, 259–260
Johnson, E.E., 448–449
Johnstone, S.R., 466–467, 484–485, 491
Jokitalo, E., 332–346
Jones, D.R., 302, 312, 315–316
Jones, K., 66
Jones, M., 315–316, 319
Jones, T.N., 312
Joo, J.H., 394–395
Jorgensen, E.M., 334–335
Jorgensen, W.L., 127
Joshi, A., 394–395
Journet, A., 466–467
Jouvenot, M., 192–193, 208
Judith, D., 36–37, 73
Jun, C.B., 44, 394–395, 398, 400–401, 483–484

Jung, C.H., 44, 394–395, 398, 400–401, 483–484
Jung, S.Y., 406
Juvin, V., 296
Jyothula, S.S., 228–230

K

Kabeya, Y., 22, 73, 89, 208, 211–212, 285–287, 378, 430–431
Kaboord, B., 181–182
Kain, S.R., 89
Kaiser, S.E., 99, 228–244, 233–234*t*
Kaizuka, T., 394–395, 400–401, 483–484
Kalgutkar, A.S., 486–489
Kalvari, I., 145–146
Kamada, Y., 116, 228–230
Kamal, A., 144, 401–402
Kaminskyy, V., 191
Kaminskyy, V.O., 209–210
Kaneko, S., 312–313
Kanematsu, T., 130–133, 216–217, 344–345
Kang, J.H., 273–274, 285–287, 513, 517
Kang, R., 172–187
Kang, S.A., 92, 406–407
Kanki, T., 144
Karanasios, E., 2–19, 293–294
Karpe, F., 296
Karplus, M., 126
Karuman, P., 315–316
Karuppasamy, M., 406–407
Karush, F., 514
Kasai, A., 469–470
Kasimu, R., 483–484
Kasper, R., 14–19
Katakam, A.K., 191
Katayama, H., 89
Kato, R., 121, 233–234*t*
Katsis, F., 490–492
Kaufmann, A., 145, 378–379
Kaufmann, R., 13–14
Kaur, J., 72
Kaushik, S., 194
Kawabata, N., 515
Kawabuchi, M., 469–470
Kawamura, T., 293–294
Kawano, K., 130–132, 216–217
Kawasaki, M., 121, 233–234*t*
Kay, L.E., 138
Ke, J., 484–485
Kehrl, J.H., 172–173
Keith, C.T., 406–407
Kellenberger, E., 334
Kelly, M., 91
Kemp, A.J., 144
Kemp, B.E., 466–467, 482, 484–485, 487–491
Kempkes, B., 430–431
Kenworthy, A.K., 165–166
Kepp, O., 72–84
Keppeler, H., 392, 394–395, 398, 400–402, 483–484
Kessler, B.M., 208
Ketel, K., 448–449
Ketteler, R., 212
Khaminets, A., 144, 146
Kharaziha, P., 248–255
Khosla, C., 512
Kichev, A., 467–468
Kiefer, S., 282, 284
Kieffer-Jaquinod, S., 466–467
Kielkowska, A., 296
Kieloch, A., 491
Kifle, L., 486–487, 492–494, 501
Kihara, A., 430–431
Kikkert, M., 334, 336
Kim do, J., 448–449
Kim, B., 129–130
Kim, B.W., 233–234*t*
Kim, D.E., 212
Kim, D.H., 44, 98–99, 394–395, 398, 400–401, 406–407
Kim, E., 407–408
Kim, I., 98–99
Kim, J., 44, 144, 172–173, 233–234*t*, 393–395, 400–401, 409, 483–484, 486–487
Kim, J.H., 172–173, 233–234*t*, 244
Kim, J.S., 407–408
Kim, J.Y., 44–45
Kim, K., 212–213
Kim, K.H., 394–395
Kim, L., 48
Kim, M., 98–99, 172–173, 228–230, 244, 407–408
Kim, N., 513, 517

Kim, S.E., 44–45
Kim, S.O., 46, 52
Kim, S.W., 212–213, 233–234*t*
Kim, S.Y., 212–213
Kim, Y., 486–487
Kim, Y.B., 172–173
Kim, Y.C., 172–173, 407–408, 424–425
Kim, Y.-M., 44, 394–395, 398, 400–401, 483–484
Kimmelman, A.C., 144, 273–274
Kimura, S., 89, 273, 344–345
King, J.E., 406–407
Kir, S., 483–484
Kirisako, H., 99
Kirisako, T., 73, 89, 98–99, 116, 172–173, 208, 211–212, 285–287, 378
Kirkin, V., 144–166, 160–161*f*
Kisen, G.Ø., 352–353, 355, 358, 361–363
Kishi, C., 394–395, 400–401, 430–431, 483–484
Kishi-Itakura, C., 366
Kisseleva, M.V., 293–294
Kita, K., 515
Kittanakom, S., 146
Kleijmeer, M.J., 333, 343–344
Klein, A.F., 312
Klekota, J., 407
Klenin, K., 129–130
Klepeis, J.L., 127
Klionsky, D.J., 2, 23, 44, 56–58, 66–67, 73–74, 77–80, 92, 98–99, 116, 144, 146, 172–173, 186–187, 190–192, 208–209, 228–230, 258, 272, 293–294, 344–345, 352, 362–363, 366, 378
Knævelsrud, H., 23
Knapp, M.S., 449, 461–462
Knecht, E., 192–193, 209–210
Kneussel, M., 130–132, 233–234*t*
Knight, D., 130–132, 233–234*t*
Knight, Z.A., 406–407
Kniss, A., 144–166, 233–234*t*
Knobel, S.M., 89
Knoops, K., 334, 336
Knop, M., 11
Knorr, R.L., 378–379
Kobayashi, S., 98
Kobayashi, T., 470–471
Kobayashi, Y., 73
Kobor, M.S., 312
Koch, B., 272
Köchl, R., 23, 344–345
Kodama, T., 293–294
Koder, S., 98–99
Koenig, B.W., 125–126, 132–133
Kofuji, S., 272, 448
Kohlbacher, O., 259, 312–313
Kohnz, R.A., 393–394, 483–484
Koike, M., 469–470
Kolesnikov, A.V., 44–45
Komatsu, M., 88, 98–99, 145, 191–193, 233–234*t*, 384–385
Kominami, E., 89–94, 98–100, 193–194, 208, 222, 260, 332, 384–385
Kondo, C., 144
Kondo, Y., 512
Kondo-Okamoto, N., 144
Konstantinidou, G., 406–407
Kontogiannis, L., 490–491
Koo, S.H., 466–467, 484–485
Koopmann, R., 233–234*t*
Kopitz, J., 352–353, 355, 358, 361–363
Korbelius, M., 52
Koren, I., 172–173
Kornblau, S., 333
Korolchuk, V.I., 194, 272–274, 294–295, 313, 314–315*f*, 316, 318, 320, 322, 327*f*
Koryakina, A., 282, 284
Kosmatka, M., 410–411, 466–467, 484–485, 491
Koster, A.J., 333–334, 336, 343–344
Kouno, T., 130–133, 216–217, 344–345
Kovacic, S., 211–213
Kovacs, A.L., 343–345
Kovacs, J., 343–345
Kovacsics, D., 200–203
Kozlov, M.M., 366
Kraft, C., 228–230, 378–379, 384–385, 387–388, 400–401
Kraft, L.J., 165–166
Kratchmarova, I., 431, 433
Kratky, D., 52
Krauss, M., 448–449
Kremer, J.R., 336
Krishnappa, L., 23
Kroemer, G., 72–84, 191, 248–249
Krupp, J., 276–278, 285

Ktistakis, N.T., 2–19, 22, 258, 260–261, 285–287, 293–294
Kudryavtseva, E.I., 4
Kuehl, W.M., 406
Kulich, S.M., 172–173
Kuma, A., 194, 259–260
Kumanomidou, T., 145, 233–234*t*
Kumar, K.L., 98–99
Kumar, N.S., 211–213
Kumar, S., 486–487
Kumeta, H., 134–137, 144–145, 210–211, 233–234*t*, 378
Kundu, M., 44, 172–173, 393–395, 398, 400–402, 409, 483–484
Kung, D.W., 486–489
Kunioka, Y., 514
Kurotori, N., 430–431
Kurth, T., 334
Kurumbail, R.G., 482–503
Kuskin, J.S., 126
Küttner, V., 435, 442
Kutzner, C., 126
Kyoung, M.J., 373

L

L'Hermitte, A., 230
La Thangue, N.B., 56, 66
Lagace, D.C., 98–99
Lagarrigue, F., 312, 315–316
Lai, Y., 366–374
Lam du, H., 394–395, 400–401
Lam, D.H., 483–484
Lamark, T., 22–23, 58, 132–133, 144–146, 155–156, 165–166, 259–260, 313
Lamb, C.A., 22–41, 72–73
Lambert, P.D., 497
Lamberton, A., 448–462
Lamia, K.A., 466–467, 484–485
Lampert, F., 101
Landgraf, R.R., 486–487
Lane, J.D., 211
Lane, W.S., 406–407
Lang, T., 208, 366
Langenberg, D.M., 334–335
Langendorf, C.G., 486–489
Lantier, L., 482
Laplante, M., 406, 409
Laprise, C., 228–230
Larance, M., 172–173, 394, 483–484
Larsen, K.B., 155–156, 259–260
Larson, R.H., 126
Laszlo, L., 344–345
Latek, R.R., 406–407
Lattime, E.C., 98–99
Lauber, K., 392, 394–395
Law, B., 209–210
Lazarus, M.B., 401–402
Le Grand, J.N., 192–193, 208
Leavitt, S., 162–165
Lebbink, M.N., 334–335
Lechaire, J.P., 333
Lecher, J., 121, 129–130, 135–137, 233–234*t*
Leclerc, J., 486–487
Lee, J.A., 116–117
Lee, J.K., 244
Lee, J.S., 513, 517
Lee, K.E., 233–234*t*, 244
Lee, M.C., 127
Lee, S.H., 172–173, 513, 517
Lee, S.I., 406
Lee, W.K., 513, 517
Lee, W.W., 318
Lee, Y.K., 116–117
Leech, M.S., 276
Legant, W.R., 323–324
Legault, P., 312
Legrand, S., 126
Leiper, F.C., 466–468, 484–485, 491
Leitner, A., 406–407
Lelyveld, V.S., 208
Lemaire, S.D., 194–195, 209–210
Lemmon, M.A., 259–260
Lenk, S.E., 344–345
Lenz, H.D., 98–99
Leone, P., 466–467, 484–485
Lepage, P., 230
Lepault, J., 334
Lester, E.P., 514
Letcher, A.J., 296, 308
Levine, B., 72, 83, 116, 172, 190, 228–230, 248–249, 258, 272–274, 279, 285, 312–313, 366, 430–431
Levy, J., 230
Lewis, E.A., 162–165
Li, H.-S., 276, 289

Li, J., 211, 467–468
Li, L., 48, 128, 366–374, 407–408
Li, M., 193, 208–222, 466–477
Li, Q., 472–473
Li, S., 172–173, 393–394, 483–484
Li, W.F., 233–234*t*
Li, X., 98–99, 484–485
Li, Y., 172–173, 191, 407, 486–487, 492–495
Liang, G., 293–294
Liang, X.H., 430–431
Liao, Y., 273–274, 285–287
Lin, F., 48
Lin, J., 89
Lin, S., 46, 52, 472–473
Lin, S.C., 466–477
Lin, S.Y., 466–467
Lin, Y.-S., 127
Lin, Z.X., 98–99
Lindahl, E., 126
Lindner, B., 276–278, 285
Lindorff-Larsen, K., 127
Lindquist, R.A., 406–408
Lindsay, M., 260–261, 274–275, 458–459
Lindsten, T., 392
Ling, N.X., 466–467, 484–490, 494–495, 500–501
Lipari, G., 124, 135
Lipowsky, R., 378–379
Lippai, M., 58
Lippens, S., 430–431
Liu, B., 211
Liu, C., 217–218, 222
Liu, D., 406
Liu, H., 312
Liu, J., 172–173
Liu, L., 138, 144, 467–468
Liu, P., 129–130
Liu, P.F., 217–218, 222
Liu, Q., 92, 406–407
Liu, S., 144, 449, 461–462
Liu, X., 98–99, 228–230, 244
Liu, Y., 406–407
Liu, Y.J., 312
Liu, Z.Q., 312
Lizcano, J.M., 410–411
Lo Re, A., 3
Lockett, S., 93
Loehr, F., 145, 158–159, 160–161*f*
Loewith, R., 406–407
Löffler, A.S., 392, 394–395, 398, 400–402, 482–484
Logsdon, M.N., 407
Lohr, F., 159–163, 160–161*f*
Long, X., 406–407
Long, Y.C., 448–449
Longatti, A., 23, 394–395, 400–401
Looyenga, B.D., 448–449
Lopes, V.S., 44–47, 50–52
Lopez-Berestein, G., 512
Lopez-Otin, C., 190–204, 208–211
Lorand, L., 514
Lorente, M., 430–431
Lorenzo, O., 313
Lößner, C., 435
Low, P., 58
Lowe, M., 302
Lu, C., 392
Lu, G., 312
Lu, K., 144
Lu, Q., 44–45
Luge, C., 123
Lugovskoy, A.A., 315–316
Luhr, M., 352–363
Lukacs, N.W., 98–99
Lukyanov, K.A., 4
Lundberg, Y.W., 192–194, 209–210
Lundström, P., 138
Luo, W., 472–473
Lux, R., 334–335

M

Ma, C., 374
Ma, P., 120–121, 129–130, 133–137, 233–234*t*
Macdonald, D., 512
Macian, F., 56
Mack, C., 435, 442
Mack, H.I., 393–394
Mackenzie, K.M., 126
MacKerell, A.D., 127
Maddocks, O.D., 408
Madiraju, A.K., 486–487
Madsen, A., 486–487
Maetzel, D., 408
Maffucci, T., 458–459

Magill, S., 312
Magyar, R., 484–485, 487–489, 491–498
Mahmood, T., 173
Maiolica, A., 400–401
Mair, W., 393–394, 483–484
Maiuri, M.C., 72–84, 191
Majerus, P.W., 293–294
Makela, T.P., 466–467, 484–485, 491
Makino, A., 470–471
Malhotra, V., 352
Malik, N., 449
Malik, S.A., 73–74
Maltese, W.A., 448–449
Mancias, J.D., 144, 273–274
Mandel, R.J., 91
Mangeat, P., 277–278
Manifava, M., 2–3, 22, 258, 260–261, 273–274
Manjithaya, R., 144
Mann, M., 400, 438–442
Mann, S.S., 344–345
Manning, B.D., 407
Mannweiler, K., 334–335
Mansour, R., 293–294, 312, 315
Mao, K., 99, 233–234*t*
Maragakis, P., 127
Marasco, C., 409
Marat, A.L., 448–449
Marchand, C.H., 194–195, 209–210
Marcinko, K., 485–487
Mardones, G.A., 344–345
Mari, M., 23, 144, 146, 208–209
Marino, G., 73–74, 190–204, 208–211
Marjanovic, J., 293–294
Markhard, A.L., 407–408, 469–470
Marsella, C., 513–514
Martens, S., 228–230, 378–388
Martin, G.M., 315–316
Martin, K.R., 448–449
Martin, L.J., 228–230
Martinez, J., 44–45
Martinez-Chacon, G., 58, 66–67, 258–267
Maruki, Y., 406–407
Marzella, L., 273–274
Mastronarde, D.N., 336
Mataruga, A., 125–126, 132–133
Mathewson, R.D., 144
Mathieu, M., 448–462
Matic, I., 439–442
Matoba, K., 99, 233–234*t*
Matsui, M., 73, 194
Matsunaga, K., 258, 293–294, 430–431
Matthies, H.J., 334–335
Mauthe, M., 352
Maxwell, D.S., 127
Mayer, F.V., 484–486
Mayinger, P., 258
Mayr, L.M., 123, 130–132, 134
Mayran, N., 470–471
Mazzone, R.W., 333
McAlister, M.S., 130–132, 233–234*t*
McAlpine, F., 312–313
McBride, A., 501
McCammon, J.A., 126
McCarthy, A., 466–467
McCartney, A.J., 313
McCastlain, K., 394–395
McCullough, L.D., 467–468
McDonald, K
McDonald, K.L., 333–335, 342–343
McDowall, A.W., 334
McEwan, D.G., 144, 146, 160–161*f*, 164, 273
McIntosh, J.R., 336
McIntosh, L.P., 122–123
McIver, E.G., 144
McKnight, N.C., 394–395, 400–401
McPhee, C.K., 344–345
Means, A.R., 484–485, 491
Mehnert, J.M., 56
Mehrpour, M., 72, 258–259, 312–313
Mehta, K., 512
Mei, B., 495
Melia, T.J., 259–260, 378–379
Melino, G., 512
Mellal, M., 466–467
Mellgren, G., 486–487
Melone, A., 282, 284
Menon, S., 407
Menzies, F.M., 44, 72, 272–274, 294–295, 313, 314–315*f*, 316, 318, 320, 322, 327*f*, 344–345
Mertz, A.C., 282, 284
Merz, K.M., 127
Metlagel, Z., 99, 101, 233–234*t*, 249
Michel, M., 342–343

Michell, B.J., 490–492
Michot, N., 448–462
Miesenbock, G., 89–90
Miglianico, M., 486–487
Mihaylova, M.M., 393–394, 482–484
Milaneso, M., 117–118
Milkie, D.E., 323–324
Millarte, V., 158–159
Miller, C.J., 394–395, 400–402
Miller, R., 486–489
Miller, R.A., 486–487
Mills, I.G., 352–353, 355, 356*t*, 357–358, 361–363
Milne, J.C., 497
Minematsu-Ikeguchi, N., 89, 92–94, 193–194, 208, 222
Minoux, H., 74
Mitchelhill, K.I., 490–492
Mitchell, C.H., 44–45
Mithieux, G., 486–487
Miura, K., 130–132
Miura, Y., 394–395, 400–401, 483–484
Miyamoto, S., 482–483
Miyawaki, A., 89
Mizrachy, L., 172–173
Mizuguchi, M., 132–133, 216–217, 344–345
Mizushima, N., 2, 22, 56–58, 66–67, 72–73, 88–89, 98, 116, 135–137, 144–145, 172–173, 191–194, 208–212, 228–230, 233–234*t*, 249, 258–260, 272–273, 279, 285–287, 366, 378, 392, 394–395, 400–401, 430–431, 483–484
Mizushima, T., 145, 233–234*t*
Mochizuki, Y., 293–294
Moffat, C., 47
Mogridge, J., 46, 52
Mohrlüder, J., 120–121, 125–126, 132–134, 144–145
Moloughney, J.G., 98–99
Momcilovic, M., 484–485, 491
Monaco, H.L., 117–118
Monastyrska, I., 344–345
Mongis, C., 11
Monken, C.E., 98–99
Moore, F., 466–467
Moravcevic, K., 259–260
Moreau, K., 23, 272, 317
Mori, Y., 132–133, 216–217, 344–345
Morimoto, M., 144
Morita, E., 121, 233–234*t*
Morphew, M.K., 334–335
Morrice, N.A., 410–411, 491
Morris, D.H., 430–431
Morris, J.B., 296, 308
Morris, S.H., 98–99
Morrison, J.F., 454
Morrow, I.C., 260–261, 274–275, 458–459
Mortier, E., 312, 315–316
Moschetta, M., 409
Moshiach, S., 44–45
Moss, S.J., 130–133, 233–234*t*
Motley, A.M., 144
Mounier, J., 315–316
Mounier, R., 482
Mountzouris, J., 172–173
Mouradian, M.M., 512
Mpamhanga, C., 401–402
Muchmore, D.C., 122–123
Muhammad, K., 233–234*t*
Mujalli, A., 293–294
Mukherjee, A., 14–19
Mulakkal, N.C., 145–146
Muller, M., 333–335, 342–343
Müller, R., 448–449
Mullin, M., 146
Munafo, D.B., 260
Munday, M.R., 490–491
Munnik, T., 296
Muno, D., 92
Munson, M.J., 449
Murk, J.L., 333–335, 343–344
Murphy, L.O., 144
Murray, J.T., 260–261
Murthy, A., 191
Murthy, S.N., 514
Mustard, K.J., 466–467, 484–485, 491
Mustark, K.J., 484–485, 491
Muzammil, S., 162–165
Mycek, M.J., 514
Myers, R., 486–487, 492–494

N

Na, J.H., 212–213
Nada, S., 407–408, 469–470
Nag, S., 99, 378–379

Nagai, K., 469–470
Nagaraj, N., 439–442
Nagel, J., 144, 449, 461–462
Nahoum, V., 293–294
Nair, U., 172–173, 208–209, 366, 374, 378
Naito, M., 293–294
Nakashima, N., 407–408
Nakatogawa, H., 98, 116, 134, 144–145, 209–210, 222, 228–230, 233–234*t*, 378–379
Naldini, L., 91
Nam, H.Y., 212–213
Natarajan, A., 486–487
Nath, S., 99, 378–379
Natsume, T., 483–484
Nazio, F., 430–431, 435, 443
Neidle, A., 514
Nelson, J.F., 83
Ness, R., 44–45
Neudecker, P., 138
Neufeld, T.P., 394–395, 407–408, 483–484
Neumann, D., 466–467, 482, 484–487, 489–491
Newman, A.C., 144
Newman, M., 144
Nezami, A., 162–165
Nezis, I.P., 145–146
Ngoei, K.R., 486–487
Nguyen, B.D., 312
Nguyen, M., 91
Nguyen, T.G., 211–213
Nguyen, T.H., 486–487, 492–494, 501
Ni, H.M., 66
Ni, T., 208–210
Ni, Z.H., 212–213
Nickel, W., 374
Nicklin, P., 424–425
Nicolae, D.L., 228–230
Nicolaes, G.A., 486–487
Nicot, A.S., 448–449
Niebuhr, K., 312
Nielsen, M., 431
Niewczas, I., 296
Niimi, K., 293–294
Nikolov, D.B., 130–132, 134
Nishimoto, T., 407–408
Niso-Santano, M., 56–69, 73–74, 258–267
Nobukuni, T., 407
Noda, N.N., 72, 98–99, 101, 132–133, 135–137, 144–145, 210–211, 233–234*t*, 378–379
Noda, T., 73, 89, 172–173, 211–212, 258, 273, 285–287, 293–294, 332, 335, 337–341, 344–345, 378, 384–385
Noguchi, E., 407–408
Noh, M.Y., 98–99
Norman, D.G., 490–491
Norman, J.M., 209–210
Noubade, R., 191
Nourbakhsh, A., 407–408
Nourse, A., 99, 232, 233–234*t*
Novak, I., 144, 146, 160–161*f*, 164
Novotny, C.J., 401–402
Nühlen, S., 36–37
Nunn, J.L., 144, 146
Nurminskaya, M.V., 512
Nurnberger, T., 98–99
Nuttall, J.M., 144
Nyfeler, B., 144, 424–425, 449, 461–462
Nymann-Andersen, J., 120–121

O

O'Farrell, F., 448–449
Oakhill, J.S., 466–467, 482, 484–485, 489–490, 500–501
Obara, K., 293–294
Oberst, A., 44–45
O'Brien, A.J., 487–489
O'Brien, M.T., 485–489, 494–495
O'Donnell, J., 484–485, 489–490
Ogasawara, Y., 259, 458–459
Ogris, E., 378–379, 384–385, 387–388
Ogura, K., 134, 233–234*t*
Ohashi, K., 99
Ohashi, R., 293–294
Ohashi, Y., 448–449
Ohsumi, Y., 2, 22, 73, 98–99, 101, 116, 132–133, 135–137, 144–145, 208–212, 222, 228–230, 233–234*t*, 249, 272, 293–294, 378–379, 384–385, 430–431
Ohtaka, H., 162–165
Okada, H., 208, 211–212, 378
Okada, M., 469–470
Okamoto, K., 144
Okamoto, Y., 129–130
Oleinick, N.L., 98–99

Olenych, S.G., 89
Oliveira, T.M., 406–407
Oliverio, S., 512
Olsen, G.S., 486–487
Olsen, J.V., 439–442
Olsen, R.W., 120–121, 132–133
Olszewski, J.L., 233–234*t*
Omori, H., 293–294, 430–431
Ong, M.S., 312
Ong, S., 431
Oppliger, W., 409, 425
Orchard, R., 44–45
Orozco, J.M., 407–408
Orsi, A., 23, 31, 36–37
Osborne, S.A., 401–402
Osgood, R., 145–146
Oshiro, N., 406–407, 424
Oshitani-Okamoto, S., 22, 73, 208
Ostrov, D., 209–210
Otomo, C., 99, 101, 233–234*t*, 249
Otomo, T., 99, 101, 233–234*t*, 249
O'Toole, E., 334–335
Otten, G., 408
Ottina, K.A., 407–408
Otto, N.M., 44, 394–395, 398, 400–401, 483–484
Outzen, H., 58, 144–145
Ouyang, L., 211
Øverbye, A., 352–353, 357–358, 361–363
Overholtzer, M., 44–45
Overmeyer, J.H., 448–449
Overvatn, A., 145, 165–166, 259–260, 313
Owczarczyk, A.B., 98–99
Owen, M.R., 486–487
Oxley, C.L., 259–260
Ozaki, S., 273, 320
Ozpolat, B., 512

P

Paasch, F., 392, 394–395
Pacheco, V., 120–121
Pagano, R.E., 366–367
Pak, Y., 129–130
Palfia, Z., 343–344
Palmer, A.G., 122
Palmo, K., 127
Palsdottir, H., 334–335
Pan, D., 407
Pan, D.A., 484–485, 491
Pan, H., 260–261
Panaretakis, T., 248–255
Pande, V.S., 127
Pankiv, S., 58, 144–145, 165–166, 259–260, 313
Papinski, D., 400–401
Parbu-Patel, A., 486–487
Pardo, R., 3
Pardon, E., 448–449
Park, H., 407–408
Park, H.H., 99
Park, H.W., 172–173, 394–395, 400–401, 407–408, 424–425
Park, J.M., 394–395
Park, M.H., 514
Park, S., 44–45
Parmigiani, A., 407–408
Pasquier, B., 448–462
Patel, K.K., 98–99
Patnaik, B.B., 98–99
Patterson, G.H., 89
Patterson, T.L., 228–230
Pattingre, S., 358, 430–431
Pavlakis, G., 93
Payrastre, B., 293–308, 312, 318–319
Paz, Y., 130–132, 135, 216–217
Pedersen, L., 128
Pedersen, N., 430–431
Pedraza, L.G., 407
Pedron, T., 312
Peng, I., 191
Peng, J., 344–345
Peng, Y., 466–467, 475–476
Perbandt, M., 233–234*t*
Perdiz, D., 344–345
Perdomo, A.B., 435, 443
Perera, L., 128
Perez-Perez, M.E., 194–195, 209–210
Perham, M., 486–487, 501
Perozzo, R., 191
Perr, M., 181–182
Perry-Garza, C.N., 48
Peterson, T.R., 406–408
Petherick, K.J., 401–402
Petiot, A., 358
Petzold, G., 101
Pfeiffer, S., 334–335

Pfisterer, S.G., 344–345
Phelan, J.P., 130–132, 233–234*t*
Philp, N.J., 47
Philpott, D.J., 312
Piacentini, M., 430–431, 512–519
Piana, S., 127
Pietrocola, F., 72–84, 248–249
Pietrokovski, S., 145
Pinkas, D.M., 512
Piston, D.W., 89
Pizarro-Estrella, E., 57–58, 66–67, 265
Platta, H.W., 172–173, 430–431
Ploegh, H.L., 208
Plouffe, S.W., 424–425
Poeter, M., 52
Pollak, M., 486–489
Polson, H.E., 22, 272
Pons, V., 293–294, 315–316
Poon, A., 228–230
Popelka, H., 146, 172–173
Popovic, D., 273
Posner, B.I., 52
Posthuma, G., 333, 343–344
Potempa, K., 312
Potocnik, U., 98–99
Potter, C.J., 407
Prater, S.M., 172–173, 393–394, 483–484
Prescott, A.R., 449
Prestwich, G.D., 273, 320
Pretorius, E., 333
Proikas-Cezanne, T., 72, 259, 293–294, 312–313, 448–449
Promponas, V.J., 145–146
Prouteau, M., 406–407
Psakhye, I., 144
Pscherer, A., 435
Puchkov, D., 448–449
Puente, C., 392–394
Puente, G., 99, 378–379
Puente, X.S., 208, 210–211
Pulinilkunnil, T., 485–487
Puri, C., 23, 272–274, 293–295, 313, 314–315*f*, 316–318, 320, 322, 327*f*
Puri, P., 99

Q

Qamar, S., 130–132, 134
Qin, W., 407
Qiu, Y., 98–111, 228–244
Qiu, Z., 217–218, 222
Qu, X., 430–431
Quesada, V., 208, 210–211
Quondamatteo, F., 430–431

R

Rabiller, M., 147
Ragusa, M.J., 313
Rahimi, Y., 486–487
Raiborg, C., 23, 258–259, 448–449
Raj, L., 172–173
Rajamohan, F., 484–498
Rajashankar, K.R., 130–132, 134
Rameh, L.E., 296, 312, 315–316, 319
Ramel, D., 312, 315–316
Ramirez, I.B., 302
Ramirez, J., 312
Ramm, G., 172–173, 394, 483–484
Rana, S., 486–487
Randow, F., 144–145
Raper, J., 200–203
Rapley, J., 424
Rappolee, D., 318
Rappsilber, J., 400, 438
Rasband, W.S., 93
Rasmuson, F., 23
Ratcliff, M., 408
Ratier, A., 344–345
Ravenhill, B.J., 145
Ravikumar, B., 194, 317
Razi, M., 22–23, 31, 36–37, 272
Real, E., 333
Reale, A., 409
Reali, V., 512
Reed, J.C., 211–213, 220
Reed, M., 98–99
Reedijk, M., 22
Reese, T.S., 333
Refolo, G., 430–431
Reggiori, F., 23, 190, 248–249, 272, 344–345, 352
Reichelt, M., 191
Reichling, L.J., 495
Reid, J.L., 466–467, 484–485, 491
Reiter, W., 400–401
Remis, J.P., 334–335
Ren, X., 99

Renna, M., 23, 272, 293–294
Repnik, K., 98–99
Repnik, U., 191
Rescher, U., 52
Reyes, A.R., 482–503
Reyes-Reveles, J., 44–53
Rez, G., 343–345
Ribar, T., 467–468
Richardson, C., 484–485, 489–490
Richmond, D.L., 387–388
Richmond, T.J., 99–100
Riddle, S.M., 495
Rider, M.H., 482
Riehle, E., 318
Riek, U., 484–486
Rieter, E., 23, 293–294, 344–345
Rigbolt, K.T., 397–400
Rigbolt, K.T.G., 439–442
Rigden, D.J., 22
Rioux, J.D., 228–230
Rizo, J., 374
Ro, S.-H., 44, 394–395, 398, 400–401, 407–408, 483–484
Roberts, D.D., 332
Roberts, E., 293–294, 448–449
Robinson, D., 23, 31, 36–37
Roccio, M., 407
Roderick, H.L., 22, 260–261, 273–274
Rodriguez, F., 192–194, 209–210
Rodriguez-Arribas, M., 56–69, 265
Rogger, J., 282, 284
Rogov, V., 144–145, 158–159
Rogov, V.V., 144–166, 160–161*f*, 233–234*t*
Rogova, N.Y., 145, 158–163, 160–161*f*
Roh, K.H., 244
Romagnoli, A., 430–431
Romanov, J., 378–379, 384–385, 387–388
Romero, G., 276, 289
Ronan, B., 448–462
Rong, J., 394–395, 400–402
Rong, Y., 344–345
Ropolo, A., 3
Rose, K.L., 172–173, 393–395, 483–484
Rosenstiel, P., 228–230
Ross, F.A., 466–467, 477, 484–490, 501
Ross, L., 484–485, 491
Rossi, M., 172–173
Rossin, F., 512–519
Rostaing, P., 333–335
Rostislavleva, K., 448–449
Rothman, J.E., 89–90
Rotnes, J.S., 352, 358
Rouault, A., 344–345
Rousset, C.I., 467–468
Roux, P.P., 406–407, 482–483
Roy, C., 277–278
Rozenknop, A., 144, 146, 159–164, 160–161*f*, 233–234*t*
Ru, B., 407
Ruan, D.Y., 98–99
Ruben, D.J., 123
Rubinsztein, D.C., 23, 44, 72, 194, 258, 272, 312–328, 344–345
Ruggero, D., 406–407
Rui, H., 472–473
Rui, Y., 472–473
Rupp, B., 117–118
Russell, C.B., 122–123
Russell, R.C., 172–173, 394–395, 400–401, 407, 424–425
Rust, M.J., 14–19
Rusten, T.E., 273–274, 282, 448–449
Rutherford, T.J., 145
Ryter, S.W., 72, 190
Ryzhakov, G., 144

S

Sabatini, D.M., 83, 258–259, 406–409, 469–470, 473
Sable, J., 312
Saccheti, M.L., 3
Sadder, M., 312
Sætre, F., 352–363, 356*t*
Saftig, P., 46, 52, 260, 332
Sagen, J.V., 486–487
Sagiv, Y., 193, 208
Sagne, C., 466–467
Sahani, M.H., 58, 66–67
Sahasrabudhe, P., 491–495, 498
Saito, K., 469–470
Saitoh, T., 72, 430–431
Saiu, P., 466–467, 484–485
Sakai, Y., 191–192
Sakamoto, K., 466–467
Sakoh-Nakatogawa, M., 99
Saland, E., 312, 315

Salatto, C.T., 486–489
Salazar, M., 430–431
Sallberg, M., 200–203
Salmon, J.K., 126
Salt, I., 490
Salvador-Montoliu, N., 192–193, 209–210
Salvesen, G.S., 211–213, 220
Sancak, Y., 406–408, 469–470, 473
Sanders, M.J., 466, 484–491
Sandford, E., 98–99, 228–230, 244
Sandoval, C.P., 44–45
Sandoval, I.V., 191–192
Sanjuan, M.A., 44–45
Sanz Garcia, N., 486–487
Sanzone, S., 172–173
Saraste, J., 312
Sarbassov, D.D., 406–407, 409
Sarkar, S., 194
Sarkes, D., 296, 315–316, 319
Sasaki, J., 272, 448
Sasaki, T., 272, 448
Satoh, T., 430–431
Satomi, Y., 98–99, 101, 116, 172–173, 378–379, 469–470
Satoo, K., 135–137, 144–145, 210–211, 233–234*t*, 378
Satterlee, A., 200
Saucedo, L.J., 407
Sauter, N.K., 138
Sauvat, A., 72–84
Sawada, R., 233–234*t*
Sawa-Makarska, J., 378–379
Saxton, R.A., 407–408
Saxty, B., 401–402
Sayen, M.R., 48
Sbrissa, D., 318, 320
Scapozza, L., 191
Scaria, S.M., 407–408
Schaeffeler, E., 208
Schaffner, T., 191
Schaudinn, C., 334–335
Schekman, R., 366, 378
Schenk, S., 497
Schermelleh, L., 13–14
Schertzer, J.D., 487–489
Scherz-Shouval, R., 193–195, 208–210
Schillinger, O., 121, 129–130, 135–137, 233–234*t*
Schink, K.O., 258–259, 448–449
Schlager, S., 52
Schlattner, U., 482, 484–486, 489–490
Schlumpberger, M., 272
Schmid, E.M., 387–388
Schmid, I., 191
Schneider, C.A., 93
Schnölzer, M., 435
Scholefield, C.L., 144
Schrode, J., 512, 514
Schu, P.V., 259, 448
Schuchner, S., 378–379, 384–385, 387–388
Schulman, B.A., 98–111, 228–244
Schuschnig, M., 400–401
Schutkowski, M., 491
Schüttpelz, M., 14–19
Schwarten, M., 116–139, 233–234*t*
Schwarze, P.E., 361
Schweitzer, L.D., 407–408, 469–470
Schwudke, D., 272–289
Scott, J.W., 466–467, 482, 484–491, 494–495, 500–501
Scott, S.V., 144
Seaman, M., 430–431
Seed, B., 212
Seefeldt, B., 14–19
Seemann, E., 334
Segel, I.H., 496
Seglen, P.O., 273–274, 344–345, 352–363, 356*t*
Seguin-Py, S., 192–193, 208
Seifert, E.L., 47
Sekiguchi, T., 407–408
Sekito, T., 293–294
Selbach, M., 439–442
Semple, I.A., 407–408
Sengupta, S., 406–407
Senovilla, L., 74
Seo, M., 394–395
Sepe, S., 512–514
Seven, A.B., 374
Severin, S., 296, 306
Shackelford, D.B., 393–394, 483–484
Shang, L., 172–173, 393–394, 483–484
Shao, L., 322–326
Shao, Z.M., 193, 208, 213, 219–220, 222
Sharif, W.D., 89
Sharma, H., 486–487

Sharp, D.J., 334–335
Sharp, Z.D., 83
Shaul, Y.D., 406–408
Shaw, D.E., 126–127
Shaw, R.J., 393–395, 400–402, 410–411, 466–467, 482–485, 491
Sheen, J.H., 406–407
Shen, H.M., 448–449
Shen, K., 407–408
Shen, S., 74
Shen, X., 486–487, 492–494
Shen, Y., 123
Sheng, F., 89
Sher, A., 406–407
Shi, C.S., 172–173
Shi, L., 172–173
Shi, W., 334–335
Shi, Y., 430–431
Shibutani, S.T., 249, 258
Shimonishi, Y., 98–99, 116, 172–173, 378, 469–470
Shimron, F., 378–379
Shin, J., 486–487
Shin, S., 129–130
Shin, T.B., 406–407
Shin, Y.K., 374
Shinjo, F., 491
Shirakawa, M., 130–132
Shisheva, A., 318–320
Shokat, K.M., 401–402
Shope, J.C., 273, 320
Shorer, H., 193–195, 208–210
Shpilka, T., 145, 258, 378–379
Shpiro, N., 501
Shteyn, V., 99, 378–379
Shu, C.W., 211–213, 217–218, 220, 222
Shvets, E., 194–195, 208–210, 344–345, 378–379
Sica, V., 72–84
Siggers, T., 312
Sikkema, S., 485–487
Siksou, L., 333
Silverberg, M.S., 228–230
Simonsen, A., 23, 249, 258–260
Sindlinger, J., 312
Singh, R., 190
Singh, S.K., 133–134
Skelton, N.J., 122
Skwarek, L.C., 272
Sliwinska, A., 486–487
Slobodkin, M.R., 116–117
Smetham, G., 200–203
Smirnoff, N., 315–316, 319
Smith, D., 466–467
Smith, G.R., 408
Smith, H.O., 101
Smith, J.J., 497
Smith, T.S., 296
Snider, J., 146
Snijders, A.P., 36–37
Snowden, M.A., 489–491
Sokal, E., 486–487
Sola, M., 130–132, 233–234*t*
Soler, N., 448–449
Solheim, A.E., 355
Solinhac, R., 293–308
Sommer, E., 449
Sommer, L., 312
Song, H.K., 233–234*t*
Song, O., 146
Song, P., 144, 491, 501
Søreng, K., 23
Soto, N., 406
Soto-Pantoja, D.R., 332
Soty, M., 486–487
Sou, Y.S., 98–99, 144–146, 193–194, 208, 222, 233–234*t*, 384–385
Sowa, M.E., 165–166, 172–173, 430–431, 483–484
Sponheim, M., 352–353, 358, 361–363
Spooner, E., 406
Sprenger, A., 397–400, 435, 442
Sridhar, S., 56
Stack, J.H., 259, 448
Stadel, D., 158–159, 273
Staehelin, L.A., 342–343
Stagljar, I., 146
Stahl, P.D., 276
Stang, M.T., 195
Stangler, T., 123, 125–126, 130–134
Stapleton, D., 484–485, 489–492
Stapleton, E., 2–3
Stappenbeck, T.S., 98–99
Stauber, T., 473
Steel, R., 466–467, 484–485, 489–490, 500–501

Steinberg, B.E., 473
Steinberg, G.R., 466
Steinbrecht, R.A., 334
Steller, H., 73
Stenmark, H., 172–173, 258–261, 273–274, 282, 430–431, 448–449, 458–459
Stephenne, X., 486–487
Stephens, L.R., 296
Stijf-Bultsma, Y., 312
Stjepanovic, G., 448–449
Stocker, H., 407
Stoka, V., 191
Stoldt, M., 121, 129–130, 133–137, 233–234*t*
Stolz, A., 144
Stolz, D.B., 276, 289
Stork, B., 56, 392–402, 482–484
Stoykova, A., 430–431
Stracka, D., 409, 425
Strambi, A., 172–173
Strappazzon, F., 430–431, 512–514, 516–517
Straub, M., 272
Strodel, B., 116–139
Stromhaug, P.E., 273–274, 344–345
Strong, R., 83
Strop, P., 512
Struck, D.K., 366–367
Studer, D., 333–335, 342–343
Su, B., 406
Su, L., 374
Su, Z.L., 374
Suarez, M.F., 192–193
Subramani, S., 144, 222, 352
Sudhof, T.C., 366
Sugawara, K., 211–212, 233–234*t*
Sugita, Y., 129–130
Sumpter, R., 172–173
Sun, H.-Q., 272–274, 285
Sun, L., 172–173
Sun, Q., 430–431
Sun, X., 172–173
Sun, Z., 486–487, 492–494, 501
Suter, M., 484–486
Sutton, R.B., 366
Suzuki, A., 233–234*t*
Suzuki, H., 121, 233–234*t*, 273
Suzuki, K., 73, 116, 144, 228–230
Suzuki, N.N., 211–212, 233–234*t*
Szabadkai, G., 513–514
Szabo, A., 124, 135
Szalai, P., 352–363
Szczesny, P.J., 333
Szyniarowski, P., 513–514

T

Tabata, K., 121, 233–234*t*, 430–431
Taguchi-Atarashi, N., 258, 293–294
Taherbhoy, A.M., 99, 228–244, 233–234*t*
Tait, S.W., 44–45, 99, 232, 233–234*t*
Takacs, Z., 259, 312–313
Takaesu, G., 99, 101, 233–234*t*, 249
Takahashi, H., 407
Takamura, A., 394–395, 400–401, 483–484
Takao, T., 98–99, 101, 116, 172–173, 378–379, 469–470
Takasuga, S., 272, 448
Takegawa, K., 259, 448
Takehara, T., 74
Talbott, G., 407
Taleux, N., 486–487
Tam, C., 51–52
Tam, S., 466–467, 484–485
Tamin-Yecheskel, B.C., 158–159
Tan, V.P., 482–483
Tanaka, Y., 172–173, 378
Tang, D., 172–187
Tang, P.H., 44–45
Tanida, I., 88–95, 98–100, 132–133, 193–194, 208, 216–217, 222, 233–234*t*, 260, 332, 344–345, 384–385
Tannous, B.A., 212
Tao, H., 98–99
Tarabishy, S., 467–468
Tassa, A., 430–431
Tauber, M., 312
Taylor, K.D., 228–230
Tcherkezian, J., 406–407, 482–483
Tee, A.R., 407
Teleman, A.A., 407–408
Teng, Y.B., 233–234*t*
Terawaki, S., 273
Terrisse, A.D., 296, 306
Tessitore, L., 361
Teuber, M., 228–230
Theer, P., 11

Theret, M., 482
Thielmann, Y., 120–121, 125–126, 132–133, 144–145
Thomas, J.D., 98–99
Thomas, S.M., 393–394
Thompson, C.B., 392
Thoreen, C.C., 92, 406–408
Thoresen, S., 430–431
Thoresen, S.B., 172–173
Thornton, C., 466–468
Thumm, M., 208, 272
Thurston, T.L., 144
Tian, M., 483–484
Tian, Y., 172–173, 394–395, 400–401
Tikole, S., 159–162
Till, A., 222, 228–230
Tillmann, K.D., 158–159
Tindwa, H., 98–99
Tirado-Rives, J., 127
Tokarska-Schlattner, M., 482
Tolias, K.F., 312
Tomova, C., 334–335
Tooze, S.A., 22–41, 72–73, 258, 272, 312–313, 344–345, 392, 394–395, 400–401, 483–484
Torgersen, K.M., 155–156
Torgersen, M.L., 352–353, 355, 356*t*, 358, 361–363
Tosch, V., 293–294, 448–449
Toth, R., 410–411
Toussaint, A., 312
Trave, G., 312
Tremblay, M.L., 448–449
Triantafellow, E., 144, 424–425, 449, 461–462
Triller, A., 334–335
Tronchere, H., 294–295, 312, 315–316, 318–319
Trono, D., 91
Tsompanis, S., 145–146
Tsukada, M., 272
Tuerk, R., 484–486
Turco, E., 378–379
Turk, B., 191
Turk, R., 484–485
Turk, V., 191
Tuttle, D.L., 344–345

U

Uchiyama, Y., 88–95, 469–470
Udd, L., 466–467, 484–485, 491
Uechi, G., 172–173
Ueno, T., 73, 88–95, 98–100, 132–133, 145, 193–194, 208, 216–217, 222, 260, 285–287, 332, 344–345, 384–385
Uero, T., 211–212
Umeda, M., 384–385
Underwood, E., 484–489
Uotila, A., 406–407
Urbé, S., 22, 313
Uria, J.A., 208, 210–211
Uribe-Carretero, E., 58, 66–67

V

Vacca, A., 409
Vacchelli, E., 74
Vaculik, M., 98–99
Vaidyanathan, V., 126
Valdez, R.A., 293–294
Vale, R.D., 334–335
Valentine, J.J., 489–490
Valet, C., 293–294
Valjakka, J., 312
Valvezan, A.J., 407
Vamos, M., 394–395, 400–402
Van Benschoten, A.H., 138
van de Linde, S., 14–19
van Denderen, B.J., 489–490
van der Meer, Y., 334, 336
van der Spoel, D., 126
van der Vaart, A., 293–294
van der Zon, G.C., 486–487
van Lengerich, B., 374
Vanden Berghe, T., 430–431
Vandenabeele, P., 73, 430–431
Vander Haar, E., 406
VanderLinden, R., 101
Vanhaesebroeck, B., 448
Vanselow, J.T., 439–442
Vasquez, D., 466–467, 484–485
Vassilopoulos, S., 312
Vedeler, A., 312
Veenhuis, M., 272
Velasco, P.T., 514
Velazquez-Campoy, A., 162–165

Vellai, T., 343–344
Veltman, K., 276, 289
Venter, J.C., 101
Verdon, Q., 466–467
Vergne, I., 293–294, 313, 448–449
Verkleij, A.J., 333–335, 343–344
Verlet, L., 128
Verma, R., 318
Vertommen, D., 484–485
Vescovi, L., 449, 459–462
Vescovo, T., 430–431, 435, 443
Vezenkov, L., 211–213
Viaud, J., 293–308, 312, 318–319
Vicinanza, M., 272–274, 293–295, 312–328, 314–315*f*, 327*f*
Vielhaber, G., 334–335
Vierstra, R.D., 98–99
Vietzke, J.P., 334–335
Vihinen, H., 332–346
Viiri, K.M., 312
Villani, L.A., 487–489
Viollet, B., 44, 172–173, 393–394, 409, 482–484, 486–489
Virgin, H.W., 72
Virgin, H.W.T., 98–99
Vitale, I., 73–74
Viterbo, D., 117–118
V'Kovski, P., 352
Vogt, A., 212–214, 216–219, 222
von Muhlinen, N., 144–145
Vujic, N., 52

W

Waelsch, H., 514
Wagner, S., 144, 164
Waguri, S., 92
Wakelam, M.J., 296
Walczak, C., 334–335
Walczak, M., 378–379, 384–385, 387–388
Wales, D.J., 129–130
Walker, K.J., 487–489
Walker, S.A., 2–19, 22, 258, 260–261, 273–274
Wallimann, T., 482, 484–486, 489–490
Walter, M., 489–490
Walther, P., 333–334
Wan, S., 484–486
Wang, B., 212–213
Wang, C.J., 322, 325–326
Wang, F., 259, 458–459
Wang, H., 120–121, 132–133, 272–274, 285, 424–425
Wang, H.Y., 208–210
Wang, J., 193, 208, 213, 219–220, 222, 394–395, 400–401, 483–486
Wang, K., 144, 323–324
Wang, L., 312–313, 484–485
Wang, M., 318–319
Wang, Q.J., 44, 430–431, 448–449
Wang, R.C., 172–173
Wang, S., 172–173, 473
Wang, S.K., 209–210
Wang, W., 195, 407–408
Wang, X., 144, 172–173, 273–274, 393–394, 442, 483–484
Wang, Z.V., 48
Wang, Z.X., 466–467
Wani, W.Y., 172–173
Ward, J., 482–503
Warden, S.M., 484–485, 489–490
Warnken, U., 435
Watanabe, M., 134
Waterfield, M.D., 259, 448
Watkins, S.C., 273–274, 285–287
Weber, T., 366–367, 374
Wee, S., 489–490
Weekes, J., 466–467
Wei, S., 406
Wei, Y., 89, 172–173
Weidberg, H., 145, 378–379
Weiergräber, O.H., 116–139, 144–145
Weimer, R.M., 334–335
Weinman, S., 66
Weisman, L.S., 313
Weissenhorn, W., 130–132, 233–234*t*
Weissleder, R., 212
Weissmann, F., 101
Wells, J.A., 489
Weltzer, R.R., 233–234*t*
Wenzel, W., 129–130
Wepf, R., 334–335
Wesselborg, S., 56, 392–393, 482–484
Weston, A.E., 23, 31, 36–37
White, D.L., 333–334, 342–343
White, E., 56
Wickner, W., 366

Wieffer, M., 448–449
Wiesehan, K., 125–126, 132–133
Wild, P., 144, 146, 160–161*f*, 164, 233–234*t*
Wileman, T., 98
Wilhelm, L., 400–401
Wilkinson, S., 144
Willbold, D., 116–139, 144–145
Williams, A.H., 99, 232, 233–234*t*
Williamson, L.E., 312–313
Wilson, M.I., 22, 272
Wilson, M.P., 293–294
Wilson, W.A., 484–486, 490–491
Winter, S., 312
Wirawan, E., 430–431
Withka, J.M., 484–485, 487–489, 492–494, 496–497
Wittern, K.P., 334–335
Witters, L.A., 410–411, 466–467, 484–485, 489–492
Wolf, D.H., 208
Wolfson, R.L., 407–408
Wollert, T., 145, 378–379
Wong, E., 56, 72
Wong, P., 89
Wong, P.M., 392–394
Woods, A., 466–467, 484–485, 489–491
Worm, S.H., 334, 336
Wozniak, A.L., 66
Wu, C., 127
Wu, C.L., 448–449
Wu, D.H., 98–99
Wu, J., 392, 407
Wu, J.W., 466–467
Wu, Y., 491, 501
Wullschleger, S., 406
Wüthrich, K., 122
Wyant, G.A., 407–408

X

Xavier, R.J., 228–230
Xi, Y., 98–99
Xia, H., 172–173
Xia, P., 172–173
Xia, Y., 312
Xiao, B., 466–467, 484–489
Xiao, G., 172–173
Xie, J., 486–487
Xie, S.T., 466–467
Xie, Y., 172–187, 318
Xie, Z., 172–173, 208–210, 378
Xin, F.J., 484–486
Xiong, G.M., 127
Xu, D., 217–218, 222
Xu, J., 491, 501
Xu, L., 172–173
Xu, T., 407
Xu, Y., 374, 448–449, 467–468
Xu, Z., 217–218, 222, 472–473
Xue, L.Y., 98–99
Xuereb, J.M., 312, 315

Y

Yakhine-Diop, S.M., 56–69, 258–267
Yamada, Y., 233–234*t*
Yamaguchi, A., 332, 335, 337–341
Yamaguchi, M., 134, 233–234*t*
Yamaji, T., 90–91
Yamamoto, A., 22, 73, 89, 208, 211–212, 259–260, 285–287, 332, 335, 337–341
Yamamoto, H., 233–234*t*
Yamazaki, M., 272, 448
Yan, S.F., 484–485
Yang, C., 430–431
Yang, G., 486–487
Yang, J.M., 99
Yang, M., 44–45
Yang, P.C., 173
Yang, Q., 406
Yang, Z., 2, 172, 190, 208–222
Yao, Z., 56, 172, 191–192, 272
Yasumoto, T., 491
Ye, Q.Z., 212–214, 216–219, 222
Yen, W.L., 208–209
Yin, X.M., 66, 193, 208–222
Yin, Z., 466–467
Yip, C.K., 430–431
Yla-Anttila, P., 332, 335–344
Yoon, T.Y., 374
Yoshii, S.R., 56–58
Yoshimori, T., 2, 22, 57–58, 67, 72–73, 89, 144–145, 172–173, 208, 211–212, 228–230, 249, 258–260, 272–273, 279, 293–294, 332, 335, 337–341, 344–345, 378, 430–431
Yoshimoto, K., 233–234*t*
Yoshimura, S., 23

Yoshino, K., 406–407
Young, A.R., 23
Young, L., 101
Young, L.N., 448–449
Young, R.N., 211–213
Yousefi, S., 191
Yu, F.X., 424–425
Yu, J., 260–261
Yu, L., 344–345
Yu, S., 233–234*t*
Yu, X., 448–449
Yu, Y., 293–294
Yu, Z.Q., 208–210
Yuan, H., 48, 172–173, 394–395, 400–401
Yuan, J., 260–261
Yue, Z., 44, 430–431, 448–449

Z

Zaffagnini, G., 378
Zaffagnini, M., 194–195, 209–210
Zalckvar, E., 172–173
Zammit, V.A., 485–486, 490
Zarei, M., 397–400, 431
Zarrinpashneh, E., 486–487
Zehethofer, N., 272–289
Zens, B., 378–388
Zentgraf, H., 483–484
Zerial, M., 374, 470–471
Zeuschner, D., 52
Zevenhoven-Dobbe, J.C., 334, 336
Zhai, D., 211–213, 220
Zhang, B., 424–425
Zhang, C., 293–294
Zhang, C.S., 466–477
Zhang, H., 98–99
Zhang, J., 92, 144, 146, 160–161*f*, 164, 172–173, 406–407
Zhang, L., 172–173, 211, 260–261, 272–274, 285, 448–449
Zhang, M., 491, 501
Zhang, S.Y., 483–484
Zhang, W., 127
Zhang, X., 233–234*t*
Zhang, X.M., 486–487
Zhang, Y., 200, 313, 318, 407
Zhang, Y.L., 466–467, 475–476
Zhang, Y.X., 373
Zhang, Y.Y., 466–467, 484–486
Zhao, H., 44–45
Zhao, L., 172–173, 393–394, 483–484
Zhao, M., 293–294, 374
Zhao, S., 334–335
Zhao, Z., 172–173
Zheng, B., 393–394
Zheng, L., 344–345
Zheng, L.S., 466–467
Zheng, Q., 144
Zheng, Y., 98–111, 228–244
Zhivotovsky, B., 191, 209–210
Zhong, M., 172–173
Zhong, Q., 366–374, 430–431
Zhong, W., 89
Zhou, C., 89
Zhou, G., 486–487, 492–495
Zhou, J., 89
Zhou, X., 259, 312–313, 458–459
Zhou, X.E., 484–485
Zhou, Z., 217–218, 222
Zhu, M., 466–467, 475–476
Zhu, T., 407
Zhu, X., 272–274, 285
Zhuang, X., 14–19
Ziemiecki, A., 191
Zierold, K., 334
Zimmermann, A., 431
Zimmermann, P., 312, 315–316
Zinker, B., 486–487, 501
Zinzalla, V., 409, 425
Zitvogel, L., 73
Zolov, S.N., 318
Zoncu, R., 258–259, 407–408, 469–470, 473
Zorn, J.A., 489
Zou, H., 394–395, 400–402
Zou, J., 293–294
Zou, M.H., 491, 501
Zou, S., 208–210
Zou, Z., 172–173
Zufferey, R., 91
Zughaier, S.M., 23, 44, 56–58, 66–67, 92, 352, 362–363
Zwartkruis, F.J., 407

SUBJECT INDEX

Note: Page numbers followed by "*f*" indicate figures, and "*t*" indicate tables.

A

Accessory proteins, 372
ACC phosphorylation assays, light organelle, 477
Accumax, 356–357
Acetyl-CoA carboxylase (ACC), 485–486
Acetyl-Gly-Thr-Phe-Gly-AFC (Ac-GTFG-AFC), 211–212
Adherent cells, 335
  cell pellet preparation from, 303
  lipid extraction, 301–302
Affinity purification, ULK1, 397–400, 399*f*
Allosteric drug and metabolite (ADaM), 487–489
Amicon Ultra centrifugal filter, 380–384
Amino acids, 407–408
5-Aminoimidazole-4-carboxamide-1-β-D-ribofuranoside (AICAR), 486–487
AMP-activated protein kinase (AMPK), 466
  lysosomes activation, 466–467
    co-IP analysis, 472–473
    DRM region, 469–470, 469*f*
    immunofluorescent analysis, 471–472
    light organelle–based assays, 474*f*, 476–477
    localization analysis, 467–473
    p-AMPK, 468–469
    in vitro reconstitution, 473–477
  properties, 466–467
  ULK1 phosphorylation, 393–394, 400–401
  in vitro assays, 482–486
    activation–protection assay, 501–502
    allosteric activation, 491–496, 494*f*
    buffer/solution conditions, 489
    crystal structure, 488*f*
    direct and indirect activators, 486*f*
    enzyme activation, 489, 496
    LKB1, 491
    peptide substrates, 490–491
    phosphorylation modulation assay, 500–501
    protection assay, 498–499
    protein phosphatase 2a, 491
    protein sources, 489–490
    reagents, 489–491
    steady-state kinetic analysis, 496–497, 497*f*
    Thr172 phosphorylation, α-subunit, 498–501
Anti-flag and anti-HA agarose beads, 436
Antigen staining, autophagy induction, 25–26
Artifact, 333, 342–344
ATG3
  protein expression
    in *E. coli*, 103
    expression constructs, 100–102
    in insect cells, 103–104
    and purification, 99–100
  protein purification
    cell lysate preparation, 105
    gel filtration chromatography for, 110–111
    glutathione-affinity chromatography for, 105–106
    glutathione-mediated, 106–108
    ion exchange chromatography, 109–110
    nickel-affinity chromatography, 108–109
    protease-mediated elution from beads for, 106
  storing purified proteins, 111
Atg4, 208
  activity in vitro and ex vivo, 211–213, 215–219
  construct design for, 213–214
  deconjugation reaction with, 387
  enzymatic kinetics analysis, 219–221
  in live animals, 200–202
  protein expression and purification, 214

Atg4 (*Continued*)
recombinant, 213–214
S. cerevisiae, 384
structure and regulatory machinery, 210–211
substrates, 209*f*
Atg5, 66–67
ATG7
protein expression
in *E. coli*, 103
expression constructs, 100–102
in insect cells, 103–104
and purification, 99–100
protein purification
cell lysate preparation, 105
gel filtration chromatography for, 110–111
glutathione-affinity chromatography for, 105–106
glutathione-mediated, 106–108
ion exchange chromatography, 109–110
nickel-affinity chromatography, 108–109
protease-mediated elution from beads for, 106
storing purified proteins, 111
ATG9, 23
quantification
cell cultures preparation, 37–38
in live cells, 36–41
proximity analysis, 38–41
redistribution
cell culture, 32
confocal imaging, 33
HEK293A cells, 32*f*
image analysis, 33–36
indirect immunofluorescence labeling, 32
induction of autophagy, 32
in peripheral regions, 31–32
Atg7–Atg10
Saccharomyces cerevisiae
cross-linking, 238
crystallization, 238–239
Atg5–Atg12–Atg16L1, 248–249
immunoprecipitation, 251–252
model systems, 250
protein expression
in *E. coli*, 103
expression constructs, 100–102
in insect cells, 103–104
and purification, 99–100
protein purification
cell lysate preparation, 105
gel filtration chromatography for, 110–111
glutathione-affinity chromatography for, 105–106
glutathione-mediated, 106–108
ion exchange chromatography, 109–110
nickel-affinity chromatography, 108–109
protease-mediated elution from beads for, 106
proximity ligation assays, 253–254
stable cell lines, generation of, 250–251
storing purified proteins, 111
ATG5–ATG16L1 (1–69) E122D disease-associated mutant
bacmid preparation, 240–241
baculovirus/insect cell expression system, 239–240
crystallization, 244
insect cells transfection, 241–242
P2/P3 amplification, 242
purification, 243
suspension Hi5 insect cells, 242–243
transposition, 240
Atg7–Atg3, *Saccharomyces cerevisiae*
cross-linking, 237–238
crystallization, 238
protein preparation, 237
Atg4–Atg8 system, 191–194, 192*f*
Atg8-family interacting motif (AIM), 144–145
Atg8/LC3/GABARAP, 151–155
LIR-containing peptides, 158–165
Y2H methods, 147–150
$Atg7^{NTD}/Atg7^{CTD}$, *Saccharomyces cerevisiae*
crystallization, 235–236
expression, 232–233
purification, 234–235
$Atg7^{NTD}$–$Atg3FR^{pep}$, 236

Atg8 proteins, 208, 352
applications
assessing dynamic properties, 134–137
investigating lipidated Atg8 proteins, 133–134
nonlipidated Atg8 proteins, 130–133
autophagic flux analysis, 194–195
biological functions, 132–133
cDNA sequence analysis, 195–196
cellular functions, 116–117
conjugation, 379*f*, 386*f*
eGFP-Atg8 to GUVs, 387–388
expression and purification, 379–384
SUVs, 384–387
ubiquitin-like protein, 378–380
deconjugation reactions, 379*f*
hydrodynamic delivery-based expression, 200–202
mammalian expression vectors, 195–197
MD simulations
AMBER force field, 127
force fields, 127
Lennard-Jones potential, 127
linking theory and experiments, 126
Newton's second law of motion, 128–129
sampling techniques in, 129–130
NMR spectroscopy, 122
investigating dynamics, 124
mapping interactions, 125–126
structure determination, 122–124
phosphatidylethanolamine and, 208–209, 249, 378–379, 379*f*, 386*f*, 388
protein expression and purification, 214
recombinant, 213–214
S. cerevisiae, 378, 380
in SDS-PAGE analyses, 195
sequence and structure data, 132*t*
structures, 210*f*
subcloning, 197
three-dimensional structures, 131*f*
ubiquitin-like proteins in, 98
X-ray crystallography
asymmetric unit, 117–118
Bragg's law, 117–118
experimental phase determination, 118–119
practical notes, 120–121
structure determination, 118–120
structure factor, 118
ATG14 proteins, 373
Autolysosome formation
evaluation of, 48–50
monitoring, 47–50
rapamycin stimulation, 48
serum starvation, 48
Autophagic flux, 56–58, 194–195
Autophagic–lysosomal pathway, 352–353, 355–356
Autophagic puncta
confocal immunofluorescence microscopy, coverslips preparation, 24–26
cell culture and autophagy induction, 24–25
confocal imaging, 26
direct fluorescence, 26
permeablization and antigen staining, 25–26
detection and quantification, 24–31
Imaris software package
cell segmentation, 28–31, 30*f*
cytosol using, 28–31
nuclei using, 28–31
"Spots" function, 27–31, 28*f*
Autophagosome–lysosome fusion, 272–273, 285
Autophagosomes, 2–3, 56, 72, 344–345, 378
biogenesis, 208
content of, 190
double-membraned, 312–313
dSTORM image data, 14–19, 15*f*
formation, 249
HeLa cell enrichment, 279–280
in HPF–FS samples, 337
ImageJ analysis, 281–282
incubation with lipid-binding probes, 280
labeling of samples, 10
markers, 317
microscope stage
fixation of samples, 7–10
for structured illumination imaging, 10–12
NRK-52E cells, 338*f*
PIPs compositions, 273

Autophagosomes (*Continued*)
quantification
cell cultures preparation, 37–38
in live cells, 36–41
proximity analysis, 38–41
relocate cell(s) of interest, 12–13
staining for bound lipid probes, 280–281
super resolution imaging of, 6–19
temporal evolution, 8*f*
3D structured illumination
images and assess for artifacts, 14
raw image data, 13–14
in vitro reconstitution, 366–367
fluorescent measurement, 372–373
protein reconstitution, 370–372
single-vesicle assay, 373
SNARE protein purification, 367–370
wide-field fluorescence microscopy, 9*f*
Autophagy, 44–45, 172, 208, 248–249, 272, 332, 366, 378
cell culture, 62–63
of cytosolic cargo, 352
DQ-BSA
autolysosome formation evaluation, 48–50
confocal imaging, 51
conjugates, 47
experimental design, 51–52
labeling endogenous LC3, 51
LC3-associated phagolysosome formation, 50
materials, 45–46
polarized epithelial cell cultures, 47
rapamycin stimulation, 48
serum starvation, 48
electrophoresis, 64
eukaryotic cells, 98
execution, 190, 249
induction
ATG9 redistribution, 32
autophagic puncta, 24–25
interpretation, 66–67
in mammals, 483–484
materials
antibodies, 60
cell lines, 58
equipment, 61–62
labware, 62
reagents, 59–60
solutions, 60–61
nonselective sequestration, cytosol, 353–362, 356*t*
phagophores, 88, 88*f*
pHlurorin-mKate2-HUMAN LC3, 90–95
phosphatidylinositol 5-phosphate, 312–315, 314–315*f*
autophagosomal markers, 317
cell culture conditions, 316
confocal microscopy, 317–318
fluorescent analogs, 320–321
genetic perturbation, 320
manipulation of intracellular, 318–321
microscopy-based detection, 315–318
pharmacological manipulations, 320
SR-SIM, 321–327, 323*f*
phosphoinositides in, 295*f*
sample preparation, 63–64
selective pathway, 144
sequestration assays, 352–353
TG2
degradation, 516–517
with P62, 517–519
treatments, 63
types, 56
types of, 72
UBL conjugation pathway in, 231, 233–234*t*
Western blotting, 64–65
Autophagy-related (Atg) proteins, 272
ATG13, 392, 394–397
ATG101, 392, 394–395
PTMs
immunofluorescence assays, 183–186
immunoprecipitation assays, 181–183
Simple Western™ assays, 175–181
Western blotting, 173–175
S. cerevisiae
Atg3, 381–382
Atg7, 381
Atg12-Atg5, 382
Atg12–Atg5–Atg16, 383–384
Atg16, 383
Atg16-mCherry, 383
in UBL conjugation pathway, 230–231, 232*f*

B

Bafilomycin A1, 57–58
Beclin 1 complex, 430–431
  anti-flag IP and FLAG elution, 436–437
  anti-HA IP and HA elution, 437
  cell culture, 431, 435
  cell lysis, 432, 435–436
  data analysis, 439–443, 441*f*
  gel electrophoresis, 432
  IgG-mouse, anti-flag and anti-HA agarose beads, 436
  in-gel digestion, sample preparation, 437–439
  LC-MS/MS analysis, 432, 434*f*, 439
  materials, 431–433
  methods, 433–443
  MS equipment, 433
  posttranslational modifications, 430–431, 433
  SILAC labeling, 435
Binding assays, VPS34, 450
  binding kinetics of compounds, 456–457
  principle and protocol, 454–456
Biotin acceptor peptide (BAP), 489–490, 515

C

Calcium phosphate transfection, 412–414
Carbonyl cyanide *m*-chlorophenyl hydrazine (CCCP), 512–513
Catalytic assays, VPS34, 450–454
  calibration, 451–452, 452*f*
  conditions, 452–453
  IC_{50}, 453–454
  K_m for ATP, 453, 454*f*
cDNA, 149, 195–196, 200–203
Cell assay, VPS34, 458–461
Cell culture
  ATG9 redistribution, 32
  autophagic puncta, 24–25
  conditions, 316
  cytosol in, 353–362
  GFP-LC3, 75
  materials, 431
  SILAC labeling, 435
  starvation of, 467–469
  transfection for Atg4-like activity, 197–200
  treatments, 410–412
Cell lysis, 105, 432, 435–436, 472–473
Cell pellet
  from adherent cells, 303
  from suspension cells, 303
CellProfiler, 265
Cell transfection, 197
Cellular assay, 459–461
Chemical fixation, 333–335
  vs. cryoimmobilization, 341–342
Chemical shift perturbation (CSP) experiments, 125–126
Chloroquine, 57–58
Co-immunoprecipitation (co-IP), 181–182, 472–473
  mTOR pathway complex 1 and 2, 418–419
Confocal microscopy
  ATG9 redistribution, 33
  autophagic puncta, 24–26
  autophagy induction, 26
  cell cultures preparation, 37–38
  phosphatidylinositol 5-phosphate, 317–318
Content-mixing vesicles, 367*f*, 371
Coomassie Brilliant Blue (CBB) staining, 214
Cryoimmobilization, 334–335, 342–343
  chemical fixation *vs.*, 341–342
  electron tomography on, 337–341
  and freeze substitution, 335–336
Crystallization assay, VPS34, 450–451, 457–458

D

Density modification, 119
DEP domain-containing mTOR-interacting protein (Deptor), 406
Detergent-resistant membrane (DRM), 469–470, 469*f*, 472–473
Direct fluorescence, autophagy induction, 26
Dodecylmaltoside (DDM), 367–368
DQ-BSA, autophagy
  autolysosome formation evaluation, 48–50
  confocal imaging, 51
  conjugates, 47
  experimental design, 51–52

DQ-BSA, autophagy (*Continued*)
labeling endogenous LC3, 51
LC3-associated phagolysosome formation, 50
materials, 45–46
polarized epithelial cell cultures, 47
rapamycin stimulation, 48
serum starvation, 48
dSTORM image data, 14–19, 15*f*

E

Earle's balanced salt solution (EBSS), 83, 264
E. coli, protein expression in, 103
eGFP-Atg8 conjugation, 387–388
Electron microscopy (EM), 332
PI(5)P, 321–322
Electron tomography (ET), 332, 336
on cryoimmobilization, 337–341
with HPF–FS method, 342–343, 345–346
Electroporation, 357–358, 357*f*, 360
Endoplasmic reticulum (ER), 332, 343–344
Endosomes, 448–449, 458–461
Enzymatic kinetics analysis, Atg4, 219–221
Enzyme activation, 489, 496

F

Fluorescence microscopy
PI(5)P, 320–321
PtdIns3P, 264
Fluorescence resonance energy transfer (FRET), 366–367
Fluorescent measurement, lysosome, 372–373
Force fields, 127
Free energy surface (FES), 135–137
Freeze substitution (FS), 334–336, 343–344
FRET-based assay
design, expression, and purification, 214–215
ex vivo, 212–213
HTS assay, 217–218
kinetics measurement with, 221, 222*f*
under in vitro conditions, 216–217
FYVE domain, PtdIns3P, 274–276
in autophagosome biogenesis, 258
autophagy assay conditions, 264
cell culture conditions, 261–262
fluorescence microscopy, detection by, 264
quantification, 263*f*, 265
signaling pathway of, 258–261, 259*f*
synthesis, 259
transient transfection of plasmid, 262–263

G

$GABA_A$ receptor-associated protein (GABARAP), 208
β-Galactosidase Assay, 149–150
Gaussia princeps luciferase (GLUC), 212
Gel-based assay
kinetics measurement with, 220–221
under in vitro conditions, 215–216
Gel electrophoresis, 155, 432
Gel filtration chromatography, 110–111
GFP-LC3, 74
aggregation, 74–75
cell culture, 75
high-throughput screening approaches, 74, 79*f*
image analysis, 77, 78*f*
stable GFP-LC3-expressing cells, 75–76
treatments, 76–77
GFP-2xFYVE, 459–461, 461*f*
Giant unilamellar vesicles (GUVs)
eGFP-Atg8 conjugation, 387–388
mGFP-Atg8 conjugation, 388
preparation, 387–388
Glutaraldehyde, 333–334
Glutathione-affinity chromatography, 105–106
Glutathione-sepharose beads, 153
GST-ACC1, 473–475
GST-Atg8/LC3/GABARAP
immobilization testing on beads, 153
lysates from bacteria expressing, 151–152
production and purification, 156–158
purification and immobilization, 153
GST-PIP4KIIα recombinant proteins, 299–301
GST pulldown assay, 151–155, 152*f*
GST-S6K1, from mammalian cells, 419–420
GUVs. *See* Giant unilamellar vesicles (GUVs)

H

Hanks' balanced salt solution (HBSS), 76, 83
HEK-293A cells, 219
HeLa cells, 279–280, 317, 321, 325, 327*f*, 410–411
HeLa GFP-2xFYVE clone, 459, 461*f*
Heteronuclear ^{-15}N nuclear Overhauser effects (hetNOEs), 124
High-pressure freezing (HPF), 334
High-throughput screening (HTS), FRET, 217–218
His-AXIN, purification for reconstitution, 473–475
HPF–FS method, 341–343
  autophagosomes in, 337
  electron tomography with, 342–343, 345–346
  phagophore, 344
  tomographic data, 344–345

I

iCyte® Imaging Cytometer, 460
Ifdotmeter® software, 265
IgG-mouse, 436
Image analysis, GFP-LC3, 77, 78*f*
ImageJ, 265, 281–282
ImageXpress Micro XLS Widefield High-Content Analysis System, 77, 78–79*f*
Imaris software package
  autophagic puncta identification using, 27–31
  manually creating cell surfaces in, 38–39
  mRFP-ATG9 and GFP-LC3B structures using, 38–41
Immunofluorescence analysis
  DQ-BSA
    confocal imaging, 51
    experimental design, 51–52
  lysosomal AMPK activation, 471–472
  PTMs of ATG proteins using, 183–186
Immunoprecipitation (IP), 181–182
  anti-flag, 436–437
  anti-HA IP, 437
  ATG antibodies for, 184–185*t*
  Atg5–Atg12–Atg16L1 system, 251–252
  lysosomal AMPK activation, 472–473, 476
  PTMs of ATG proteins using, 181–183
Indirect immunofluorescence, 32
In-gel digestion, sample preparation, 437–439
Insect cells, protein expression in, 103–104
Interactome, Beclin 1 complex. *See* Beclin 1 complex
In vitro assays, AMPK, 482–486
  activation–protection assay, 501–502
  allosteric activation, 491–496, 494*f*
  buffer/solution conditions, 489
  crystal structure, 488*f*
  direct and indirect activators, 486*f*
  enzyme activation, 489, 496
  LKB1, 491
  peptide substrates, 490–491
  phosphorylation modulation assay, 500–501
  protection assay, 498–499
  protein phosphatase 2a, 491
  protein sources, 489–490
  reagents, 489–491
  steady-state kinetic analysis, 496–497, 497*f*
  Thr172 phosphorylation, α-subunit, 498–501
In vitro kinase (IVK) assay
  mTOR pathway, 419–420
  ULK1 inhibitors, 400–402, 401*f*
In vitro reconstitution
  Atg8 conjugation, 379*f*, 386*f*
    eGFP-Atg8 to GUVs, 387–388
    expression and purification, 379–384
    phosphatidylethanolamine and, 378–379, 379*f*, 386*f*, 388
    S. cerevisiae, 378, 380
    SUVs, 384–387
    ubiquitin-like protein, 378–380
  Atg8 deconjugation reactions, 379*f*
  autophagosome, 366–367
    fluorescent measurement, 372–373
    protein reconstitution, 370–372
    single-vesicle assay, 373
    SNARE protein purification, 367–370
  lysosomal AMPK activation, 475–476
Ion exchange chromatography, 109–110
IP. *See* Immunoprecipitation (IP)
Isolation membranes, 378
Isothermal titration calorimetry (ITC), 162–165

J

J-coupling, 122–123
J774E macrophage-like cells, 276–277

L

Lactate dehydrogenase (LDH), 354–355*f*
advantage, 352–353
measurement, 359*f*
in sediment, 356, 361
sequestration rate, 361
LC3-associated phagolysosome formation, 50
LC3B
protein expression
in *E. coli*, 103
expression constructs, 100–102
in insect cells, 103–104
and purification, 99–100
protein purification
cell lysate preparation, 105
gel filtration chromatography for, 110–111
glutathione-affinity chromatography for, 105–106
glutathione-mediated, 106–108
ion exchange chromatography, 109–110
nickel-affinity chromatography, 108–109
protease-mediated elution from beads for, 106
storing purified proteins, 111
LC3/GABARAP proteins, 22–23, 98
LC3-interacting region (LIR), 144–158
peptide arrays, 155–158, 157*f*
Light chain 3 (LC3)
cell culture, 62–63
electrophoresis, 64
interpretation, 66–67
LC3-I, 56–58, 57*f*, 89
LC3-II, 56–58, 57*f*, 89
materials
antibodies, 60
cell lines, 58
equipment, 61–62
labware, 62
reagents, 59–60
solutions, 60–61
sample preparation, 63–64
treatments, 63
types, 56
Western blotting, 64–65
Light organelle, 473
ACC phosphorylation assays, 477
AMPK phosphorylation assays, 476–477
purification, 475, 476*f*
Lipari-Szabo model-free analysis, 124
Lipidation of Atg8, 384–387
Lipid-binding domain
autophagosomes
ImageJ, 281–282
incubation with, 280
staining, 280–281
lysosomes
ImageJ, 281–282
incubation with, 277–278
staining, 280–281
Lipid extraction
from cell extracts, 301–303
from tissues, 303–304
Lipid-mixing vesicles, 367*f*, 370–371
Lipofectamine, 412, 414
Liposome, transfection using, 412, 414
Liquid chromatography–mass spectrometry (LC-MS) analysis, 432, 434*f*, 439
deacylated PIPs, 284
LIR-containing peptides
Atg8/LC3/GABARAP and, 158–165
expression and purification, 159–162
LIR-containing proteins, 146–158
Live cell imaging, 3–6, 4*f*
cell cultures preparation, 37–38
imaging medium, 4
imaging parameters, 6
microscopy, 5
proximity analysis, 38–41
transfection, 3
Liver kinase B1 (LKB1), 466–467, 472–473, 475–476, 491
LY294002, 264
Lysosomal AMPK activation, 466–467
co-IP analysis, 472–473
DRM region, 469–470, 469*f*
immunofluorescent analysis, 471–472
light organelle-based assays
ACC phosphorylation assays, 477

AMPK phosphorylation assays, 476–477
light organelle–based assays, 474*f*
localization analysis, 467–473
p-AMPK
mouse embryonic fibroblasts, 468
mouse liver, 468–469
in vitro reconstitution, 473–477
Lysosomes, 366–367
fluorescent measurement, 372–373
fluorescent proteolytic products in, 50–51
inhibitor, 74
from J774E macrophage-like cells, 276–277
LC3-containing autophagosomes, 44
LC3-OS-lysosome, 52*f*
lipid-binding probes
ImageJ analysis, 281–282
incubation, 277–278
staining for, 280–281
PIPs extraction from, 282–283
preparation of, 470, 471*f*
protein reconstitution, 370–372
RP-HPLC-MS, 282–284
single-vesicle assay, 373
SNARE protein purification, 367–370

M

Macroautophagy, 44, 72, 116, 190, 272. *See also* Autophagy
Mammalian target of rapamycin (mTOR), 406, 482–483
calcium phosphate transfection, 412–414
cell culture and treatments, 410–412
complex 1 and 2, 406–409
coimmunoprecipitation, 418–419
diagnostic posttranslational modifications, 411*f*
inhibitors, 409–410
Western blotting, 417
functional readouts, 409–410
GTP loading assay of small GTPase, 421–424, 422*f*
immunofluorescence staining, 421
inhibitors, 409–410, 410*f*
membrane stripping, 418
multiprotein kinase complexes, 406–407
pharmacological/physiological inhibition, 413*f*
protein extraction for Western blotting, 415
reprobing membrane, 418
signaling, 410–411
transfection, 412–414
ULK1 phosphorylation, 393–394
using phosphospecific antibodies, 415–416
in vitro kinase assay, 419–420
Mass assay
PtdIns3P
quantification, 297*f*, 306–307, 307*f*
recombinant PIKfyve preparation, 298–299
PtdIns5P, 297*f*, 308
Mass spectrometry (MS), 146, 432, 434*f*, 439
equipment, 433
ULK1 phosphorylation, 397–400, 399*f*
Mass spectrometry (MS)-based proteomics, 431
MaxQuant, 439–442, 440*t*
MD simulations, Atg8 protein
AMBER force field, 127
force fields, 127
Lennard-Jones potential, 127
linking theory and experiments, 126
Newton's second law of motion, 128–129
sampling techniques in, 129–130
Membrane contact sites (MCSs), 332, 342
Membrane fusion, 366–367
Metformin, 486–487
3-Methyladenine (3-MA) treatment, 264
Microscopy-based detection, phosphatidylinositol 5-phosphate, 315–318
Microtubule-associated protein (MAP) 1 light chain 3 (MAP1-LC3), 344–345
Microtubules, 345–346
array of, 337–341
HFP–FS samples, 341–342
trajectory, 344–345
Molecular replacement technique, 119
Monomeric eGFP (mGFP)-Atg8, 380
Monomeric ras homolog enriched in the brain (Rheb), 407
guanine nucleotide loading status on, 422–424

Mouse embryonic fibroblasts (MEFs), 199*f*, 395–397, 468
mTOR. *See* Mammalian target of rapamycin (mTOR)
Multicloning site (MCS), 195–196

N

Newton's second law of motion, 128–129
Nickel-affinity chromatography, 108–109
Nonradioactive assay formats, 495–496
Nuclear magnetic resonance (NMR) spectroscopy, Atg8 protein, 122
  investigating dynamics, 124
  mapping interactions, 125–126
  structure determination, 122–124
Nuclear Overhauser effect (NOE), 123

O

Oligonucleotide design, 197
Osmium tetroxide postfixation, 333, 343–344

P

p-AMPK
  mouse embryonic fibroblasts, 468
  mouse liver, 468–469
Particle-mesh Ewald summation method, 128
^{33}P-ATP filter assay, 492–496
PE. *See* Phosphatidylethanolamine (PE)
Peptide arrays
  LIR, 155–158, 157*f*
  synthesis of, 156
Peptide overlay assay, 158
Peptide substrates, AMPK, 490–491
Permeablization, autophagy induction, 25–26
Phagophores, 72, 172, 190, 312–313, 337–341, 344, 378
  machinery involved in, 352
  morphology, 333
pHlurorin-mKate2-hLC3
  autophagic responses monitoring, 92–93
  fluorescence, 88*f*
  interpretation of results, 93–95
  pH sensitivity, 91*f*
  plasmids for expression, 90–91
  puncta formation, 94*f*
Phosphatase and tensin homolog (PTEN), 410–411
Phosphatidylethanolamine (PE), 73, 98–99, 208–209, 249, 378–379, 379*f*, 386*f*, 388
Phosphatidylinositol monophosphates, 304–306, 305*f*
Phosphatidylinositol 3-phosphate (PI3P and PtdIns3P), 22, 430–431, 448–449, 458–459
  ectopic expression and purification, 274–276
  FYVE domain
    in autophagosome biogenesis, 258
    autophagy assay conditions, 264
    cell culture conditions, 261–262
    fluorescence microscopy, detection by, 264
    quantification, 263*f*, 265
    signaling pathway of, 258–261, 259*f*
    synthesis, 259
    transient transfection of plasmid, 262–263
Phosphatidylinositol 4-phosphate (PI4P), 278, 280–282, 285–289
Phosphatidylinositol 5-phosphate (PI(5)P), autophagy, 312–315, 314–315*f*
  autophagosomal markers, 317
  cell culture conditions, 316
  confocal microscopy, 317–318
  fluorescent analogs, 320–321
  genetic perturbation, 320
  manipulation of intracellular, 318–321
  microscopy-based detection, 315–318
  pharmacological manipulations, 320
  SR-SIM, 321–327, 323*f*
Phosphoinositide 3-kinases (PI3Ks), 448–449
Phosphoinositide lipids (PIPs), 272–274
  autophagosome compositions, 273
  coincidence detection, 273–274
  deacylation of extracted lipids, 283–284
  extraction from purified lysosomes, 282–283
  LC-MS analysis of deacylated, 284
  lysosome preparations by RP-HPLC-MS, 282–284
  results and discussion, 285–289

Phosphoinositides (PIs), 293–294
in autophagy, 295*f*
metabolism, 312
Phospholipids, 293–294
Phosphorylation of ULK1
mass spectrometry, 397–400, 399*f*
on Ser757, 409
Ser556, Ser638 and Ser758, 393–397, 395*f*
Phosphospecific antibodies, mTOR, 415–416, 420
PIK3C3. *See* Vacuolar protein sorting 34 (VPS34)
PIKfyve, 298–299, 318–319
PIs. *See* Phosphoinositides (PIs)
Plasma membrane, electrodisruption, 353, 358–361
Polyethylenamine (PEI), transfection using, 414
Postfixation in osmium tetroxide, 333, 343–344
Posttranslational modifications (PTMs), 172–173, 430–431, 433
ATG proteins
immunofluorescence assays, 183–186
immunoprecipitation assays, 181–183
Simple Western™ assays, 175–181
Western blotting, 173–175
PP2a. *See* Protein phosphatase 2a (PP2a)
p62 protein, 57*f*, 58, 66*f*
Prey interaction, 149–150
Proline-rich Akt substrate of 40 kDa (PRAS40), 406
Proteases, 191
Protein extraction, for Western blotting, 415
Protein phosphatase 2a (PP2a), 491
Proteolysis, 191, 194–195, 203
ProteOn XPR36 (Biorad), 455–456
Proximity ligation assays, 253–254
PtdIns3P mass assay
quantification, 297*f*, 306–307, 307*f*
recombinant PIKfyve preparation, 298–299
PtdIns5P mass assay, 297*f*, 308
P62, TG2 with, 517–519
PTMs. *See* Posttranslational modifications (PTMs)
Purification
cell lysis and, 432
GST-PIP4KIIα recombinant proteins, 300–301
His-AXIN/GST-ACC1 for reconstitution, 473–475
light organelles, 475, 476*f*
phosphatidylinositol monophosphates, 304–306, 305*f*
recombinant GST-PIKfyve, 298–299
S. cerevisiae
Atg3, 381–382
Atg4, 384
Atg7, 381
Atg8, 380
Atg12–Atg5, 382
Atg12–Atg5–Atg16, 383–384
Atg16 and Atg16-mCherry, 383
SNARE protein, 367–370
VPS34, 450–451

R

Rag
GTPases activity, 421
heterodimer, 424
in vivo guanine nucleotide loading status on, 424
Ragulator, v-ATPase, 472–473
Rapamycin, 79*f*, 83, 264
bacteria-produced, 406–407
with FKBP12, 406–407
RB1-inducible coiled-coil protein 1 (RB1CC1), 392
Recombinant proteins
Atg4/Atg8, 213–214
Atg8 conjugation, 379–384
GST-PIP4KIIα, 299–301
PIKfyve, 298–299
preparation, 298–301
VPS34 purification, 450–451
Reconstitution
SNAREs, 371–372
in vitro (*see* In vitro reconstitution)
Replica exchange molecular dynamics (REMD) simulations, 129–130

Reverse-phase high-performance liquid chromatography – mass spectrometry (RP-HPLC-MS), 274
  lysosomes preparations by, 282–284
  PIPs extracted from purified lysosomes, 282–284

S

Saccharomyces cerevisiae
  Atg3, 381–382
  Atg4, 384
  Atg7, 381
  Atg8, 378, 380
  Atg12–Atg5, 382
  Atg12–Atg5–Atg16, 383–384
  Atg16 and Atg16-mCherry, 383
  crystallization
    Atg7–Atg3, 237–238
    Atg7–Atg10, 238–239
    $Atg7^{NTD}$ and $Atg7^{CTD}$, 232–236
    $Atg7^{NTD}$–$Atg3FR^{pep}$, 236
  genetic studies in, 272
SAMS peptide, 485–491, 495–496
Saturation transfer difference (STD) NMR spectroscopy, 125
Scalar coupling, 122–123
SDS-PAGE, 211–212, 214–215, 217
  Atg4 enzymes kinetics analysis, 219
  Atg8 proteins in, 195
  FRET-based assay, 217
Sf9 cells, 298
Simple Western™ assays
  ATG antibodies for, 179–180*t*
  PTMs of ATG proteins using, 175–181
Single-vesicle assay, 373
siRNA transfection, 319
^{35}S-labeled proteins, 154–155
Small unilamellar vesicles (SUVs)
  Atg8 conjugation/lipidation, 384–387
  conjugation reaction, 385–387
  preparation of, 385
Soluble *N*-ethylmaleimide-sensitive factor attachment protein receptors (SNAREs)
  purification, 367–370
  reconstitution, 371–372
Spin–spin coupling, 122–123
SPR. *See* Surface plasmon resonance (SPR)
Stable cell lines, generation of, 250–251
Stable isotope labeling by amino acids in cell culture (SILAC), 393, 398
Starvation, 513–514, 517
Starvation-induced autophagy, 260
Steady-state kinetic analysis, AMPK, 496–497, 497*f*
Super-resolution structured illumination microscopy (SR-SIM), PI(5)P, 321–327, 323*f*
  disadvantage, 323–324
  image processing, 326–327
  instrument preparation, 324–325
  sample imaging, 325–326
  sample preparation, 325
Surface plasmon resonance (SPR), 454–455, 457*f*
Suspension cells, 302–303
SUVs. *See* Small unilamellar vesicles (SUVs)
Synaptosome-associated protein 29 kDa (SNAP-29), 369–370
Syntaxin 17, 368–369

T

TG2. *See* Transglutaminase type 2 (TG2)
Thr172 phosphorylation, 498–501
Tissues, lipid extraction from, 303–304
Transfection, 412–414
  cultured cells, 197–200
  siRNA, 319
  using liposome, 414
  using PEI, 414
Transglutaminase type 2 (TG2), 512–513
  degradation during autophagy, 516–517
  enzymatic reactions, 513*f*
  with P62 during autophagy, 517–519
  transamidating activity assay, 513–515, 514*f*
Trypan blue, 357–360, 357*f*
TSC complex, 407–408

U

Ubiquitin (Ub)-like proteins (UBLs), 144–145, 164
  conjugation pathway
    ATG proteins in, 230–231, 232*f*
    in autophagy, 231, 233–234*t*
Ulight™ acceptor, 495–496

Unc51-like kinase 1 (ULK1), 392
affinity purification, 397–400, 399*f*
AMPK, 393–394, 400–401
ATG13 and, 392, 394–397
ATG101 and, 392, 394–395
inhibitors, in vitro kinase assays, 400–402, 401*f*
phosphorylation
mass spectrometry, 397–400, 399*f*
mouse embryonic fibroblasts, 395–397
Ser556, Ser638 and Ser758, 393–397, 395*f*
Ser757, 409

V

Vacuolar protein sorting 34 (VPS34), 259, 312–313, 448
binding assays, 450, 454–457
catalytic assays, 450–454
cell assay, 458–461
crystallization, 450–451, 457–458
HeLa GFP-2xFYVE clone, 459, 461*f*
inhibitors, 449
purification, 450–451
reaction catalyzed by, 448*f*
v-ATPase–Ragulator–AXIN/LKB1–AMPK, 472–473
Vesicle-associated membrane protein 8 (VAMP8), 368–369
Vesicular trafficking pathways, 448–449

W

Western blotting (WB)
ATG phosphorylation antibodies for, 176–178*t*
autophagy, 64–65
mTOR complex 1 and 2, 417
protein extraction for, 415
PTMs of ATG proteins using, 173–175
WIPI2b, 23

X

X-ray crystallography, Atg8 protein
asymmetric unit, 117–118
Bragg's law, 117–118
experimental phase determination, 118–119
practical notes, 120–121
structure determination, 118–120
structure factor, 118

Y

Yeast-two-hybrid (Y2H) methods, 146–150, 150*f*

FSC
www.fsc.org
MIX
Papier | Fördert gute Waldnutzung
FSC® C083411

Kolibri 360 GmbH
Ferdinand-Jühlke-Straße 7
99095 Erfurt, Deutschland
produktsicherheit@kolibri360.de